T0320927

Many-body Quantum Theory in
Condensed Matter Physics

Many-body Quantum Theory in Condensed Matter Physics

an introduction

HENRIK BRUUS

Department of Physics
Technical University of Denmark

and

KARSTEN FLENSBERG

Niels Bohr Institute,
University of Copenhagen

Copenhagen, 14 July 2004
Corrected version: 14 January 2016

OXFORD
UNIVERSITY PRESS

OXFORD

UNIVERSITY PRESS

Great Clarendon Street, Oxford OX2 6DP

Oxford University Press is a department of the University of Oxford.
It furthers the University's objective of excellence in research, scholarship,
and education by publishing worldwide in

Oxford New York

Auckland Cape Town Dar es Salaam Hong Kong Karachi
Kuala Lumpur Madrid Melbourne Mexico City Nairobi
New Delhi Shanghai Taipei Toronto

With offices in

Argentina Austria Brazil Chile Czech Republic France Greece
Guatemala Hungary Italy Japan South Korea Poland Portugal
Singapore Switzerland Thailand Turkey Ukraine Vietnam

Oxford is a registered trade mark of Oxford University Press
in the UK and in certain other countries

Published in the United States
by Oxford University Press Inc., New York

© Oxford University Press 2004

The moral rights of the author have been asserted

Database right Oxford University Press (maker)

First published 2004

All rights reserved. No part of this publication may be reproduced,
stored in a retrieval system, or transmitted, in any form or by any means,
without the prior permission in writing of Oxford University Press,
or as expressly permitted by law, or under terms agreed with the appropriate
reprographics rights organization. Enquiries concerning reproduction
outside the scope of the above should be sent to the Rights Department,
Oxford University Press, at the address above.

You must not circulate this book in any other binding or cover
and you must impose this same condition on any acquirer.

A catalogue record for this title is available from the British Library

Library of Congress Cataloging in Publication Data
(Data available)

ISBN 978-0-19-856633-5 (Hbk)

Printed in Great Britain
by CPI Group (UK) Ltd, Croydon, CR0 4YY

PREFACE

This introduction to many-body quantum theory in condensed matter physics has emerged from a set of lecture notes used in our courses *Many-particle Physics I* and *II* for graduate and advanced undergraduate students at the Niels Bohr Institute, University of Copenhagen, held six times between 1999 and 2004. The notes have also been used twice in the course *Transport in Nanostructures* taught at the Technical University of Denmark. The courses have been followed by students of both theoretical and experimental physics and it is our experience that both groups have benefited from the notes. The theory students gained a good background for further studies, while the experimental students obtained a familiarity with theoretical concepts they encounter in research papers.

We have gone through the trouble of writing this textbook, because we felt the pedagogical need for putting an emphasis on the physical contents and applications of the machinery of quantum field theory without loosing mathematical rigor. We hope we have succeeded, at least to some extent, in reaching this goal.

Since our main purpose is to provide a pedagogical introduction, and *not* to present a review of the physical examples presented, we do not give comprehensive references to these topics. Instead, we refer the reader to the review papers and topical books mentioned in the text and in the bibliography.

We would like to thank our ever enthusiastic students for their valuable help throughout the years improving the notes preceding this book.

Copenhagen, July 2004.

Karsten Flensberg   Henrik Bruus
Ørsted Laboratory   *MIC – Department of*
Niels Bohr Institute   *Micro and Nanotechnology*
University of Copenhagen   *Technical University of Denmark*

Preface to corrected edition January 2016.
    The book has been corrected for an, unfortunately, rather large number of misprints. We would like to thank all the colleagues and readers who have sent corrections to us and in particular the many students and the teachers of the courses *Condensed Matter Theory I* at University of Copenhagen and *Transport in nanostructures* at the Technical University of Denmark for helping in locating the misprints. We have not made major changes to the book other than Section 10.5 has been rewritten somewhat.

Karsten Flensberg   Henrik Bruus
Niels Bohr Institute   *Department of Physics*
University of Copenhagen   *Technical University of Denmark*

CONTENTS

List of symbols   xiv

1 First and second quantization   1
   1.1 First quantization, single-particle systems   2
   1.2 First quantization, many-particle systems   4
     1.2.1 Permutation symmetry and indistinguishability   5
     1.2.2 The single-particle states as basis states   6
     1.2.3 Operators in first quantization   8
   1.3 Second quantization, basic concepts   10
     1.3.1 The occupation number representation   10
     1.3.2 The boson creation and annihilation operators   10
     1.3.3 The fermion creation and annihilation operators   13
     1.3.4 The general form for second quantization operators   14
     1.3.5 Change of basis in second quantization   16
     1.3.6 Quantum field operators and their Fourier transforms   17
   1.4 Second quantization, specific operators   18
     1.4.1 The harmonic oscillator in second quantization   18
     1.4.2 The electromagnetic field in second quantization   19
     1.4.3 Operators for kinetic energy, spin, density and current   21
     1.4.4 The Coulomb interaction in second quantization   23
     1.4.5 Basis states for systems with different particles   25
   1.5 Second quantization and statistical mechanics   26
     1.5.1 The distribution function for non-interacting fermions   29
     1.5.2 The distribution function for non-interacting bosons   29
   1.6 Summary and outlook   30

2 The electron gas   32
   2.1 The non-interacting electron gas   33
     2.1.1 Bloch theory of electrons in a static ion lattice   33
     2.1.2 Non-interacting electrons in the jellium model   36
     2.1.3 Non-interacting electrons at finite temperature   39
   2.2 Electron interactions in perturbation theory   40
     2.2.1 Electron interactions in 1st-order perturbation theory   42
     2.2.2 Electron interactions in 2nd-order perturbation theory   44
   2.3 Electron gases in 3, 2, 1 and 0 dimensions   45
     2.3.1 3D electron gases: metals and semiconductors   45
     2.3.2 2D electron gases: GaAs/GaAlAs heterostructures   47
     2.3.3 1D electron gases: carbon nanotubes   49
     2.3.4 0D electron gases: quantum dots   50
   2.4 Summary and outlook   51

3   Phonons; coupling to electrons   52
  3.1   Jellium oscillations and Einstein phonons   52
  3.2   Electron–phonon interaction and the sound velocity   53
  3.3   Lattice vibrations and phonons in 1D   54
  3.4   Acoustical and optical phonons in 3D   57
  3.5   The specific heat of solids in the Debye model   59
  3.6   Electron–phonon interaction in the lattice model   61
  3.7   Electron–phonon interaction in the jellium model   64
  3.8   Summary and outlook   65

4   Mean-field theory   66
  4.1   Basic concepts of mean-field theory   66
  4.2   The art of mean-field theory   69
  4.3   Hartree–Fock approximation   70
        4.3.1   H–F approximation for the homogenous electron gas   71
  4.4   Broken symmetry   72
  4.5   Ferromagnetism   74
        4.5.1   The Heisenberg model of ionic ferromagnets   74
        4.5.2   The Stoner model of metallic ferromagnets   76
  4.6   Summary and outlook   78

5   Time dependence in quantum theory   80
  5.1   The Schrödinger picture   80
  5.2   The Heisenberg picture   81
  5.3   The interaction picture   81
  5.4   Time-evolution in linear response   84
  5.5   Time-dependent creation and annihilation operators   84
  5.6   Fermi's golden rule   86
  5.7   The T-matrix and the generalized Fermi's golden rule   87
  5.8   Fourier transforms of advanced and retarded functions   88
  5.9   Summary and outlook   90

6   Linear response theory   92
  6.1   The general Kubo formula   92
        6.1.1   Kubo formula in the frequency domain   94
  6.2   Kubo formula for conductivity   95
  6.3   Kubo formula for conductance   97
  6.4   Kubo formula for the dielectric function   98
        6.4.1   Dielectric function for translation-invariant system   100
        6.4.2   Relation between dielectric function and conductivity   100
  6.5   Summary and outlook   101

7	**Transport in mesoscopic systems**	102
	7.1 The S-matrix and scattering states	103
	7.1.1 Definition of the S-matrix	103
	7.1.2 Definition of the scattering states	106
	7.1.3 Unitarity of the S-matrix	106
	7.1.4 Time-reversal symmetry	107
	7.2 Conductance and transmission coefficients	108
	7.2.1 The Landauer formula, heuristic derivation	109
	7.2.2 The Landauer formula, linear response derivation	111
	7.2.3 Landauer–Büttiker formalism for multiprobe systems	112
	7.3 Electron wave guides	113
	7.3.1 Quantum point contact and conductance quantization	113
	7.3.2 The Aharonov–Bohm effect	117
	7.4 Summary and outlook	118
8	**Green's functions**	120
	8.1 "Classical" Green's functions	120
	8.2 Green's function for the one-particle Schrödinger equation	120
	8.2.1 Example: from the S-matrix to the Green's function	123
	8.3 Single-particle Green's functions of many-body systems	124
	8.3.1 Green's function of translation-invariant systems	125
	8.3.2 Green's function of free electrons	125
	8.3.3 The Lehmann representation	127
	8.3.4 The spectral function	129
	8.3.5 Broadening of the spectral function	130
	8.4 Measuring the single-particle spectral function	131
	8.4.1 Tunneling spectroscopy	132
	8.5 Two-particle correlation functions of many-body systems	135
	8.6 Summary and outlook	138
9	**Equation of motion theory**	139
	9.1 The single-particle Green's function	139
	9.1.1 Non-interacting particles	141
	9.2 Single level coupled to a continuum	141
	9.3 Anderson's model for magnetic impurities	142
	9.3.1 The equation of motion for the Anderson model	144
	9.3.2 Mean-field approximation for the Anderson model	145
	9.4 The two-particle correlation function	148
	9.4.1 The random phase approximation	148
	9.5 Summary and outlook	150
10	**Transport in interacting mesoscopic systems**	151
	10.1 Model Hamiltonians	151
	10.2 Sequential tunneling: the Coulomb blockade regime	153
	10.2.1 Coulomb blockade for a metallic dot	154
	10.2.2 Coulomb blockade for a quantum dot	157

10.3 Coherent many-body transport phenomena   158
    10.3.1 Cotunneling   158
    10.3.2 Inelastic cotunneling for a metallic dot   159
    10.3.3 Elastic cotunneling for a quantum dot   160
10.4 The conductance for Anderson-type models   161
    10.4.1 The conductance in linear response   162
    10.4.2 Calculation of Coulomb blockade peaks   165
10.5 The Kondo effect in quantum dots   168
    10.5.1 From the Anderson model to the Kondo model   168
    10.5.2 Comparing Kondo effect in metals and quantum dots   173
    10.5.3 Kondo-model conductance to second order in $H_S^{(2)}$ 173
    10.5.4 Kondo-model conductance to third order in $H_S^{(2)}$ 174
    10.5.5 Origin of the logarithmic divergence   179
    10.5.6 The Kondo problem beyond perturbation theory   181
10.6 Summary and outlook   182

11 Imaginary-time Green's functions   184
11.1 Definitions of Matsubara Green's functions   187
    11.1.1 Fourier transform of Matsubara Green's functions   188
11.2 Connection between Matsubara and retarded functions   189
    11.2.1 Advanced functions   191
11.3 Single-particle Matsubara Green's function   192
    11.3.1 Matsubara Green's function, non-interacting particles   192
11.4 Evaluation of Matsubara sums   193
    11.4.1 Summations over functions with simple poles   194
    11.4.2 Summations over functions with known branch cuts   196
11.5 Equation of motion   197
11.6 Wick's theorem   198
11.7 Example: polarizability of free electrons   201
11.8 Summary and outlook   202

12 Feynman diagrams and external potentials   204
12.1 Non-interacting particles in external potentials   204
12.2 Elastic scattering and Matsubara frequencies   206
12.3 Random impurities in disordered metals   208
    12.3.1 Feynman diagrams for the impurity scattering   209
12.4 Impurity self-average   211
12.5 Self-energy for impurity scattered electrons   216
    12.5.1 Lowest-order approximation   217
    12.5.2 First-order Born approximation   217
    12.5.3 The full Born approximation   220
    12.5.4 Self-consistent full Born approximation and beyond   222
12.6 Summary and outlook   224

13 Feynman diagrams and pair interactions   226
   13.1 The perturbation series for $\mathcal{G}$ 227
   13.2 The Feynman rules for pair interactions   228
      13.2.1 Feynman rules for the denominator of $\mathcal{G}(b,a)$ 229
      13.2.2 Feynman rules for the numerator of $\mathcal{G}(b,a)$ 230
      13.2.3 The cancellation of disconnected Feynman diagrams   231
   13.3 Self-energy and Dyson's equation   233
   13.4 The Feynman rules in Fourier space   233
   13.5 Examples of how to evaluate Feynman diagrams   236
      13.5.1 The Hartree self-energy diagram   236
      13.5.2 The Fock self-energy diagram   237
      13.5.3 The pair-bubble self-energy diagram   238
   13.6 Cancellation of disconnected diagrams, general case   239
   13.7 Feynman diagrams for the Kondo model   241
      13.7.1 Kondo model self-energy, second order in J 243
      13.7.2 Kondo model self-energy, third order in J 244
   13.8 Summary and outlook   245

14 The interacting electron gas   246
   14.1 The self-energy in the random phase approximation   246
      14.1.1 The density dependence of self-energy diagrams   247
      14.1.2 The divergence number of self-energy diagrams   248
      14.1.3 RPA resummation of the self-energy   248
   14.2 The renormalized Coulomb interaction in RPA   250
      14.2.1 Calculation of the pair-bubble   251
      14.2.2 The electron-hole pair interpretation of RPA   253
   14.3 The groundstate energy of the electron gas   253
   14.4 The dielectric function and screening   256
   14.5 Plasma oscillations and Landau damping   260
      14.5.1 Plasma oscillations and plasmons   262
      14.5.2 Landau damping   263
   14.6 Summary and outlook   264

15 Fermi liquid theory   266
   15.1 Adiabatic continuity   266
      15.1.1 Example: one-dimensional well   267
      15.1.2 The quasiparticle concept and conserved quantities   268
   15.2 Semi-classical treatment of screening and plasmons   269
      15.2.1 Static screening   270
      15.2.2 Dynamical screening   271
   15.3 Semi-classical transport equation   272
      15.3.1 Finite lifetime of the quasiparticles   276
   15.4 Microscopic basis of the Fermi liquid theory   278
      15.4.1 Renormalization of the single-particle Green's function   278
      15.4.2 Imaginary part of the single-particle Green's function   280
      15.4.3 Mass renormalization?   283
   15.5 Summary and outlook   283

16 Impurity scattering and conductivity   285
16.1 Vertex corrections and dressed Green's functions   286
16.2 The conductivity in terms of a general vertex function   291
16.3 The conductivity in the first Born approximation   293
16.4 Conductivity from Born scattering with interactions   296
16.5 The weak localization correction to the conductivity   298
16.6 Disordered mesoscopic systems   308
     16.6.1 Statistics of quantum conductance,
                random matrix theory   308
     16.6.2 Weak localization in mesoscopic systems   309
     16.6.3 Universal conductance fluctuations   310
16.7 Summary and outlook   312

17 Green's functions and phonons   313
17.1 The Green's function for free phonons   313
17.2 Electron–phonon interaction and Feynman diagrams   314
17.3 Combining Coulomb and electron–phonon interactions   316
     17.3.1 Migdal's theorem   317
     17.3.2 Jellium phonons and the effective
                electron–electron interaction   318
17.4 Phonon renormalization by electron screening in RPA   319
17.5 The Cooper instability and Feynman diagrams   322
17.6 Summary and outlook   324

18 Superconductivity   325
18.1 The Cooper instability   325
18.2 The BCS groundstate   327
18.3 Microscopic BCS theory   329
18.4 BCS theory with Matsubara Green's functions   331
     18.4.1 Self-consistent determination of
                the BCS order parameter $\Delta_\mathbf{k}$ 332
     18.4.2 Determination of the critical temperature T_c 333
     18.4.3 Determination of BCS quasiparticle density of states   334
18.5 The Nambu formalism of the BCS theory   335
     18.5.1 Spinors and Green's functions in Nambu formalism   335
     18.5.2 The Meissner effect and the London equation   336
     18.5.3 Zero paramagnetic current response in BCS theory   337
18.6 Gauge symmetry breaking and zero resistivity   341
     18.6.1 Gauge transformations   341
     18.6.2 Broken gauge symmetry and dissipationless current   342
18.7 The Josephson effect   343
18.8 Summary and outlook   346

19 1D electron gases and Luttinger liquids   347
   19.1 What is a Luttinger liquid?   347
   19.2 Experimental realizations of Luttinger liquid physics   348
      19.2.1 Example: Carbon Nanotubes   348
      19.2.2 Example: semiconductor wires   348
      19.2.3 Example: quasi 1D materials   348
      19.2.4 Example: Edge states in fractional quantum Hall effect   348
   19.3 A first look at the theory of interacting electrons in 1D   348
      19.3.1 The "quasiparticles" in 1D   350
      19.3.2 The lifetime of the "quasiparticles" in 1D   351
   19.4 The spinless Luttinger–Tomonaga model   352
      19.4.1 The Luttinger–Tomonaga model Hamiltonian   352
      19.4.2 Inter-branch interaction   354
      19.4.3 Intra-branch interaction and charge conservation   355
      19.4.4 Umklapp processes in the half-filled band case   356
   19.5 Bosonization of the Tomonaga model Hamiltonian   357
      19.5.1 Derivation of the bosonized Hamiltonian   357
      19.5.2 Diagonalization of the bosonized Hamiltonian   360
      19.5.3 Real space representation   360
   19.6 Electron operators in bosonized form   363
   19.7 Green's functions   368
   19.8 Measuring local density of states by tunneling   369
   19.9 Luttinger liquid with spin   373
   19.10 Summary and outlook   374

A Fourier transformations   376
   A.1 Continuous functions in a finite region   376
   A.2 Continuous functions in an infinite region   377
   A.3 Time and frequency Fourier transforms   377
   A.4 Some useful rules   377
   A.5 Translation-invariant systems   378

Exercises   380

Bibliography   423

Index   426

LIST OF SYMBOLS

Symbol	Meaning	Definition
$\hat{\heartsuit}$	operator $\heartsuit$ in the interaction picture	Section 5.3
$\dot{\heartsuit}$	time derivative of $\heartsuit$	
$|\nu\rangle$	Dirac ket notation for a quantum state ν	Section 1.1
$\langle\nu|$	Dirac bra notation for an adjoint quantum state ν	Section 1.1
$|0\rangle$	vacuum state	Section 1.3
a	annihilation operator for particle (fermion or boson)	Section 1.3
$a^\dagger$	creation operator for particle (fermion or boson)	Section 1.3
$a_\nu, a_\nu^\dagger$	annihilation/creation operators (state ν)	Section 1.3
$a_n^\pm$	amplitudes of wavefunctions to the left	Section 7.1
a_0	Bohr radius	Eq. (2.36)
$\mathbf{A}(\mathbf{r}, t)$	electromagnetic vector potential	Section 1.4.2
$A(\nu, \omega)$	spectral function in frequency domain (state ν)	Section 8.3.4
$A(\mathbf{r}, \omega), A(\mathbf{k}, \omega)$	spectral function (real space, Fourier space)	Section 8.3.4
$A_0(\mathbf{r}, \omega), A_0(\mathbf{k}, \omega)$	spectral function for free particles	Section 8.3.4
$A, A^\dagger$	phonon annihilation and creation operator	Section 17.1
b	annihilation operator for particle (boson, phonon)	Section 1.3
β	inverse temperature	Eq. (1.113)
$b^\dagger$	creation operator for particle (boson, phonon)	Section 1.3
$b_n^\pm$	amplitudes of wavefunctions to the right	Section 7.1
$\mathbf{B}$	magnetic field	
c	annihilation operator for particle (fermion, electron)	Section 1.3
$c^\dagger$	creation operator for particle (fermion, electron)	Section 1.3
$c_\nu, c_\nu^\dagger$	annihilation/creation operators (state ν)	Section 1.3
$C_{AB}^R(t, t')$	retarded correlation function between A and B (time)	Section 6.1
$C_{AB}^A(t, t')$	advanced correlation function between A and B (time)	Section 11.2.1
$C_{II}^R(\omega)$	retarded current–current correlation function (frequency)	Section 6.3
$\mathcal{C}_{AB}$	Matsubara correlation function	Section 11.1
$\mathcal{C}(\mathbf{Q}, ik_n, ik_n + iq_n)$	Cooperon in the Matsubara domain	Section 16.5
$C^R(\mathbf{Q}, \varepsilon, \varepsilon)$	Cooperon in the real time domain	Section 16.5
C_V^{ion}	specific heat for ions (constant volume)	Section 3.5

$d(\epsilon)$	density of states (including spin degeneracy)	Eq. (2.31)
D	band width	
$D^R(\mathbf{r}t, \mathbf{r}t')$	retarded phonon propagator	Chapter 17
$D^R(\mathbf{q}, \omega)$	retarded phonon propagator (Fourier space)	Chapter 17
$\mathcal{D}(\mathbf{r}\tau, \mathbf{r}\tau')$	Matsubara phonon propagator	Chapter 17
$\mathcal{D}(\mathbf{q}, iq_n)$	Matsubara phonon propagator (Fourier space)	Chapter 17
$D^R(\nu t, \nu't')$	retarded many particle Green's function	Eq. (9.9b)
$D_{\alpha\beta}(\mathbf{r})$	phonon dynamical matrix	Section 3.4)
$\delta(\mathbf{r})$	Dirac delta function	Eq. (1.12)
δ_{ij}	Kronecker's delta function	Eq. (1.10)
$\Delta_{\mathbf{k}}$	superconducting orderparameter	Eq. (18.11)
e	elementary charge	
e_0^2	electron interaction strength	Eq. (1.100a)
$\mathbf{E}(\mathbf{r}, t)$	electric field	
E	total energy of the electron gas	Chapter 2
$E^{(1)}$	interaction energy, first-order perturbation	Section 2.2.1
$E^{(2)}$	interaction energy, second-order perturbation	Section 2.2.2
E_0	Rydberg energy	Eq. (2.36)
$E_{\mathbf{k}}$	dispersion relation for BCS quasiparticles	Eq. (18.14)
ε	energy variable	
ϵ_0	the dielectric constant of vacuum	
$\varepsilon_{\mathbf{k}}$	dispersion relation	
ε_ν	energy of quantum state ν	
ε_F	Fermi energy	Chapter 2
$\boldsymbol{\epsilon}_{\mathbf{k}\lambda}$	phonon polarization vector	Eq. (3.20)
$\varepsilon(\mathbf{r}t, \mathbf{r}t')$	dielectric function in real space	Section 6.4
$\varepsilon(\mathbf{k}, \omega)$	dielectric function in Fourier space	Section 6.4
$\varepsilon(\mathbf{k}, \omega)$	dielectric function in Fourier space	Section 6.4
ϵ_{ijk}	Levi–Civita symbol	Eq. (1.11)
F	free energy	Section 1.5
$\mathcal{F}$	Anomalous Green's function	Eq. (18.18)
$|\text{FS}\rangle$	the filled Fermi sea N–particle quantum state	Eq. (2.22)
$\phi(x)$	displacement field operator	Eq. (19.49)
$\phi(\mathbf{r}, t)$	electric potential	Section 6.4
ϕ_{ext}	external electric potential	Section 6.4
ϕ_{ind}	induced electric potential	Section 6.4
$\phi, \tilde{\phi}$	wavefunctions with different normalizations	Eq. (7.4)
$\phi^{\pm}_{LnE}, \phi^{\pm}_{RnE}$	wavefunctions in the left and right leads	Section 7.1
$g_{\mathbf{q}\lambda}$	electron–phonon coupling constant (lattice model)	Eq. (3.38)
$g_{\mathbf{q}}$	electron–phonon coupling constant (jellium model)	Eq. (3.42)
G	conductance	Section 6.3

$G_0^<(\mathbf{r}t,\mathbf{r}'t')$	free lesser Green's function	Section 8.3.1
$G_0^>(\mathbf{r}t,\mathbf{r}'t')$	free grater Green's function	Section 8.3.1
$G_0^A(\mathbf{r}t,\mathbf{r}'t')$	free advanced Green's function	Section 8.3.1
$G_0^R(\mathbf{r}t,\mathbf{r}'t')$	free retarded Green's function	Section 8.3.1
$G_0^R(\mathbf{k},\omega)$	free retarded Green's function (Fourier space)	Section 8.3.1
$G^<(\mathbf{r}t,\mathbf{r}'t')$	lesser Green's function	Section 8.3
$G^>(\mathbf{r}t,\mathbf{r}'t')$	greater Green's function	Section 8.3
$G^A(\mathbf{r}t,\mathbf{r}'t')$	advanced Green's function	Section 8.3
$G^R(\mathbf{r}t,\mathbf{r}'t')$	retarded Green's function (real space)	Section 8.3
$G^R(\mathbf{k},\omega)$	retarded Green's function in Fourier space	Section 8.3
$G^R(\mathbf{k},\omega)$	retarded Green's function (Fourier space)	Section 8.3.1
$G^R(\nu t,\nu't')$	retarded single–particle Green's function ($\{\nu\}$ basis)	Eq. (8.34)
$\bar{\bar{\mathcal{G}}}(\mathbf{k},\tau)$	Nambu Green's function	Eq. (18.44)
$\mathcal{G}(\mathbf{r}\sigma\tau,\mathbf{r}'\sigma'\tau')$	Matsubara Green's function (real space)	Section 11.3
$\mathcal{G}(\nu\tau,\nu'\tau')$	Matsubara Green's function ($\{\nu\}$ basis)	Section 11.3
$\mathcal{G}(1,1')$	Matsubara Green's function (real space four–vectors)	Section 12.1
$\mathcal{G}(\tilde{k},\tilde{k}')$	Matsubara Green's function (four–momentum notation)	Section 13.4
$\mathcal{G}_0(\mathbf{r}\sigma\tau,\mathbf{r}'\sigma'\tau')$	Matsubara Green's function (real space, free particles)	Section 11.3.1
$\mathcal{G}_0(\nu\tau,\nu'\tau')$	Matsubara Green's function ($\{\nu\}$ basis, free particles)	Section 11.3.1
$\mathcal{G}_0(\mathbf{k},ik_n)$	Matsubara Green's function (Fourier space, free particles)	Section 11.3
$\mathcal{G}_0(\nu,ik_n)$	Matsubara Green's function (free particles)	Section 11.3
$\mathcal{G}_0^{(n)}$	n–particle Green's function (free particles)	Section 11.6
$\mathcal{G}(\mathbf{k},ik_n)$	Matsubara Green's function (Fourier space)	Section 11.3
$\mathcal{G}(\nu,ik_n)$	Matsubara Green's function ($\{\nu\}$ basis, frequency domain)	Section 11.3
γ,γ^{RA}	scalar vertex function	Section 16.3
Γ	imaginary part of self–energy	
$\Gamma_x(\tilde{k},\tilde{k}+\tilde{q})$	vertex function (x–component, four vector notation)	Eq. (16.21b)
$\Gamma_{0,x}$	free (undressed) vertex function	Eq. (16.20)
Γ_{fi}	transition rate	Eq. (5.34)
$\hbar$	Planck's constant ($h/2\pi$), $\hbar \to 1$ in Chap. 5 and onwards	
H	a general Hamiltonian	
H_0	unperturbed part of an Hamiltonian	
H'	perturbative part of an Hamiltonian	
H_{ext}	external potential part of an Hamiltonian	
H_{int}	interaction part of an Hamiltonian	
H_{ph}	phonon part of an Hamiltonian	
H_T	tunneling Hamiltonian	Eq. (8.65)
η	positive infinitisimal	Section 5.8
I	current operator (particle current)	Section 6.3
I_e	electrical current (charge current)	Section 6.3

$J_\sigma(\mathbf{r})$	current density operator	Eq. (1.98a)
$J_\sigma^\Delta(\mathbf{r})$	current density operator, paramagnetic term	Eq. (1.98a)
$J_\sigma^A(\mathbf{r})$	current density operator, diamagnetic term	Eq. (1.98a)
$J_\sigma(\mathbf{q})$	current density operator (momentum space)	Eq. (1.98a)
$J_e(\mathbf{r}, t)$	electric current density operator	Section 6.2
J_{ij}	interaction strength in the Heisenberg model	Section 4.5.1
$J_{\alpha\beta}$	interaction strength in the Kondo model	Eq. (10.91a)
k_B	Boltzmann's constant	
k_n	Matsubara frequency (fermions)	Eq. (11.42)
k_F	Fermi wave number	Chapter 2
$\mathbf{k}$	general momentum or wave vector variable	
$\ell, \ell_\mathbf{k}$	mean free path or scattering length	Chapter 12
ℓ_0	$v_\mathbf{k}\tau_0$ mean free path (first Born approximation)	
ℓ_ϕ	phase breaking mean free path	Chapter 7
$\mathcal{L}$	normalization length or system size in 1D	
λ_F	Fermi wave length	Eq. (2.23)
Λ^{irr}	irreducible four–point function	Eq. (16.18)
m	mass (electrons and general particles)	
m^*	effective interaction renormalized mass	Section 15.4.1
μ	chemical potential	Eq. (1.120)
μ	general quantum number label	
n	particle density	
$n_F(\varepsilon)$	Fermi–Dirac distribution function	Section 1.5.1
$n_B(\varepsilon)$	Bose–Einstein distribution function	Section 1.5.2
n_{imp}	impurity density	
N	number of particles	
N_{imp}	number of impurities	
ν	general quantum number label	
ω	frequency variable	
$\omega_\mathbf{q}$	phonon dispersion relation	
ω_n	Matsubara frequency (boson)	Eq. (11.28)
Ω	thermodynamic potential	Section 1.5
$\mathbf{p}$	general momentum or wave number variable	
p_n	Matsubara frequency (fermion)	Eq. (11.28)
$P(x)$	momentum field operator	Eq. (19.50)
$\mathcal{P}$	principle part	
$\Pi_{\alpha\beta}^R(\mathbf{r}t, \mathbf{r}'t')$	retarded current–current correlation function	Eq. (6.25)
$\Pi_{\alpha\beta}^R(\mathbf{q}, \omega)$	retarded current–current correlation function	
$\Pi_{\alpha\beta}(\mathbf{q}, i\omega_n)$	Matsubara current–current correlation function	Chapter 16
$\Pi^0(\mathbf{q}, iq_n)$	free pair–bubble diagram	Eq. (13.37)

$\mathbf{q}$	general momentum variable	
q_n	Matsubara frequency (bosons)	Eq. (11.28)
$\mathbf{r}$	general space variable	
$\mathbf{r}$	reflection matrix coming from left	Section 7.1
$\mathbf{r}'$	reflection matrix coming from right	Section 7.1
r_s	electron gas density parameter	Eq. (2.37)
ρ	density matrix	Section 1.5
ρ_0	unperturbed density matrix	Eq. (6.3b)
$\rho_\sigma(\mathbf{r})$	particle density operator (real space)	Eq. (1.94)
$\rho_\sigma(\mathbf{q})$	particle density operator (momentum space)	Eq. (1.94)
ρ_L, ρ_R	left and right mover density operators	Eq. (19.20)
$\mathbf{S}$	scattering matrix	Section 7.1
S^x	spin operator	Eq. (1.92b)
σ	general spin index	
$\sigma_{\alpha\beta}(\mathbf{r}t, \mathbf{r}'t')$	conductivity tensor	Section 6.2
$\Sigma^R(\mathbf{q}, \omega)$	retarded self–energy (Fourier space)	
$\Sigma(\mathbf{q}, ik_n)$	Matsubara self–energy	
$\Sigma_{\mathbf{k}}$	impurity scattering self–energy	Section 12.5
$\Sigma_{\mathbf{k}}^{1BA}$	first Born approximation	Section 12.5.1
$\Sigma_{\mathbf{k}}^{FBA}$	full Born approximation	Section 12.5.3
$\Sigma_{\mathbf{k}}^{SCBA}$	self–consistent Born approximation	Section 12.5.4
$\Sigma(l, j)$	general electron self–energy	
$\Sigma_\sigma(\mathbf{k}, ik_n)$	general electron self–energy	
$\Sigma_\sigma^F(\mathbf{k}, ik_n)$	Fock self–energy	Section 13.5
$\Sigma_\sigma^H(\mathbf{k}, ik_n)$	Hartree self–energy	Section 13.5
$\Sigma_\sigma^P(\mathbf{k}, ik_n)$	pair–bubble self–energy	Section 13.5
$\Sigma_\sigma^{RPA}(\mathbf{k}, ik_n)$	RPA electron self–energy	Eq. (14.11)
t	general time variable	
$\mathbf{t}$	tranmission matrix coming from left	Section 7.1
$\mathbf{t}'$	transmission matrix coming from right	Section 7.1
T	temperature	
T	kinetic energy	
T	T–matrix	Section 5.7
τ	general imaginary time variable	Chapter 11
$\tau_{\sigma\sigma'}^i$	Pauli's spin matrixes	Eq. (1.91)
τ^{tr}	transport scattering time	Eq. (15.38)
$\tau_0, \tau_{\mathbf{k}}$	life–time in the first Born approximation	Section 12.5.2
u_j	ion displacement (1D)	Eq. (3.8)
$\mathbf{u}(\mathbf{R}_0)$	ion displacement (3D)	Section 3.4
$u_{\mathbf{k}}$	BCS coherence factor	Section 18.3
U	general unitary matrix	Section 16.6
$\hat{U}(t, t')$	real time–evolution operator, interaction picture	Section 5.3
$\hat{U}(\tau, \tau')$	imaginary time–evolution operator, interaction picture	Eq. (11.12)

$v_{\mathbf{k}}$	BCS coherence factor	Section 18.3
$V(\mathbf{r}), V(\mathbf{q})$	general single impurity potential	Eq. (12.1)
$V(\mathbf{r}), V(\mathbf{q})$	Coulomb interaction	Eq. (1.100a)
V_{eff}	combined Coulomb and phonon–mediated interaction	Section 14.2
$\mathcal{V}$	normalization volume	
W	pair interaction Hamiltonian	Chapter 13
$W(\mathbf{r}), W(\mathbf{q})$	pair interaction	Chapter 13
W^{RPA}	RPA–screened Coulomb interaction	Section 14.2
$\xi_{\mathbf{k}}$	$\varepsilon_{\mathbf{k}} - \mu$	
ξ_{ν}	$\varepsilon_{\nu} - \mu$	
$\chi(\mathbf{q}, iq_n)$	Matsubara charge–charge correlation function	Section 14.4
$\chi^{\mathrm{RPA}}(\mathbf{q}, iq_n)$	RPA Matsubara charge–charge correlation function	Section 14.4
$\chi^{\mathrm{irr}}(\mathbf{q}, iq_n)$	irreducible Matsubara charge–charge correlation function	Section 14.4
$\chi_0(\mathbf{q}, iq_n)$	free Matsubara charge–charge correlation function	Section 14.4
$\chi^R(\mathbf{r}t, \mathbf{r}'t')$	retarded charge–charge correlation function	Eq. (6.37b)
$\chi^R(\mathbf{q}, \omega)$	retarded charge–charge correlation function (Fourier)	Eq. (8.81)
$\chi_n(y)$	transverse wavefunction	Section 7.1
$\psi_{\nu}(\mathbf{r})$	single–particle wave function, quantum number ν	Section 1.1
$\psi_{nE}^{\pm}$	single–particle scattering states	Section 7.1
$\psi(\mathbf{r}_1, \mathbf{r}_2, \dots, \mathbf{r}_n)$	n–particle wave function (first quantization)	Section 1.1
$\Psi_{\sigma}(\mathbf{r})$	quantum field annihilation operator	Section 1.3.6
$\Psi_{\sigma}^{\dagger}(\mathbf{r})$	quantum field creation operator	Section 1.3.6
$\theta(x)$	Heaviside's step function	Eq. (1.13)

1

FIRST AND SECOND QUANTIZATION

Quantum theory is the most complete microscopic theory we have today describing the physics of energy and matter. It has successfully been applied to explain phenomena ranging over many orders of magnitude, from the study of elementary particles on the sub-nucleonic scale to the study of neutron stars and other astrophysical objects on the cosmological scale. Only the inclusion of gravitation stands out as an unsolved problem in fundamental quantum theory.

Historically, quantum physics first dealt only with the quantization of the motion of particles, leaving the electromagnetic field classical, hence the name quantum mechanics. Later also the electromagnetic field was quantized, and even the particles themselves became represented by quantized fields, resulting in the development of quantum electrodynamics (QED) and quantum field theory (QFT) in general. By convention, the original form of quantum mechanics is denoted first quantization, while quantum field theory is formulated in the language of second quantization.

Regardless of the representation, be it first or second quantization, certain basic concepts are always present in the formulation of quantum theory. The starting point is the notion of quantum states and the observables of the system under consideration. Quantum theory postulates that all quantum states are represented by state vectors in a Hilbert space, and that all observables are represented by Hermitian operators acting on that space. Parallel state vectors represent the same physical state, and therefore one mostly deals with normalized state vectors. Any given Hermitian operator A has a number of eigenstates $|\psi_\alpha\rangle$ that are left invariant by the action of the operator up to a real scale factor α, i.e., $A|\psi_\alpha\rangle = \alpha|\psi_\alpha\rangle$. The scale factors are denoted the eigenvalues of the operator. It is a fundamental theorem of Hilbert space theory that the set of all eigenvectors of any given Hermitian operator forms a complete basis set of the Hilbert space. In general, the eigenstates $|\psi_\alpha\rangle$ and $|\phi_\beta\rangle$ of two different Hermitian operators A and B are not the same. By measurement of the type B the quantum state can be prepared to be in an eigenstate $|\phi_\beta\rangle$ of the operator B. This state can also be expressed as a superposition of eigenstates $|\psi_\alpha\rangle$ of the operator A as $|\phi_\beta\rangle = \sum_\alpha |\psi_\alpha\rangle C_{\alpha\beta}$. If one measures the dynamical variable associated with the operator A in this state, one cannot in general predict the outcome with certainty. It is only described in probabilistic terms. The probability of having any given $|\psi_\alpha\rangle$ as the outcome is given as the absolute square $|C_{\alpha\beta}|^2$ of the associated expansion coefficient. This non-causal element of quantum theory is also known as the collapse of the wavefunction. However, between collapse events the time evolution of quantum states is perfectly deterministic. The time evolution of a state vector $|\psi(t)\rangle$ is governed by the central operator in quantum mechanics, the Hamiltonian H (the operator associated with the total energy of the system), through Schrödinger's equation

$$i\hbar\,\partial_t\big|\psi(t)\big\rangle = H\big|\psi(t)\big\rangle. \tag{1.1}$$

Each state vector $|\psi\rangle$ is associated with an adjoint state vector $(|\psi\rangle)^\dagger \equiv \langle\psi|$. One can form inner products, "bra(c)kets", $\langle\psi|\phi\rangle$ between adjoint "bra" states $\langle\psi|$ and "ket" states $|\phi\rangle$, and use standard geometrical terminology; e.g., the norm squared of $|\psi\rangle$ is given by $\langle\psi|\psi\rangle$, and $|\psi\rangle$ and $|\phi\rangle$ are said to be orthogonal if $\langle\psi|\phi\rangle = 0$. If $\{|\psi_\alpha\rangle\}$ is an orthonormal basis of the Hilbert space, then the above-mentioned expansion coefficient $C_{\alpha\beta}$ is found by forming inner products: $C_{\alpha\beta} = \langle\psi_\alpha|\phi_\beta\rangle$. A further connection between the direct and the adjoint Hilbert space is given by the relation $\langle\psi|\phi\rangle = \langle\phi|\psi\rangle^*$, which also leads to the definition of adjoint operators. For a given operator A the adjoint operator $A^\dagger$ is defined by demanding $\langle\psi|A^\dagger|\phi\rangle = \langle\phi|A|\psi\rangle^*$ for any $|\psi\rangle$ and $|\phi\rangle$.

In this chapter, we will briefly review standard first quantization for one- and many-particle systems. For more complete reviews the reader is referred to standard textbooks by, for instance, Dirac (1989), Landau and Lifshitz (1977), and Merzbacher (1970). Based on this we will introduce second quantization. This introduction, however, is not complete in all details, and we refer the interested reader to the textbooks by Mahan (1990), Fetter and Walecka (1971), and Abrikosov $et\ al.$ (1975).

1.1   First quantization, single-particle systems

For simplicity consider a non-relativistic particle, say an electron with charge $-e$, moving in an external electromagnetic field described by the potentials $\varphi(\mathbf{r}, t)$ and $\mathbf{A}(\mathbf{r}, t)$. The corresponding Hamiltonian is

$$H = \frac{1}{2m}\left(\frac{\hbar}{i}\nabla_{\mathbf{r}} + e\mathbf{A}(\mathbf{r}, t)\right)^2 - e\,\varphi(\mathbf{r}, t). \tag{1.2}$$

An eigenstate describing a free spin-up electron traveling inside a box of volume $\mathcal{V}$ can be written as a product of a propagating plane wave and a spin-up spinor. Using the Dirac notation the state ket can be written as $|\psi_{\mathbf{k},\uparrow}\rangle = |\mathbf{k}, \uparrow\rangle$, where one simply lists the relevant quantum numbers in the ket. The state function (also denoted the wave function) and the ket are related by

$$\psi_{\mathbf{k},\sigma}(\mathbf{r}) = \langle\mathbf{r}|\mathbf{k}, \sigma\rangle = \tfrac{1}{\sqrt{\mathcal{V}}}\,e^{i\mathbf{k}\cdot\mathbf{r}}\chi_\sigma \quad \text{(free particle orbital)}, \tag{1.3}$$

i.e., by the inner product of the position bra $\langle\mathbf{r}|$ with the state ket.

The plane wave representation $|\mathbf{k}, \sigma\rangle$ is not always a useful starting point for calculations. For example in atomic physics, where electrons orbiting a point-like positively charged nucleus are considered, the hydrogenic eigenstates $|n, l, m, \sigma\rangle$ are much more useful. Recall that

$$\langle\mathbf{r}|n, l, m, \sigma\rangle = R_{nl}(r)Y_{l,m}(\theta, \phi)\chi_\sigma \quad \text{(hydrogen orbital)}, \tag{1.4}$$

where $R_{nl}(r)$ is a radial Coulomb function with $n-l$ nodes, while $Y_{l,m}(\theta, \phi)$ is a spherical harmonic representing angular momentum l with a z component m.

A third example is an electron moving in a constant magnetic field $\mathbf{B} = B\,\mathbf{e}_z$, which in the Landau gauge $\mathbf{A} = xB\,\mathbf{e}_y$ leads to the Landau eigenstates $|n, k_y, k_z, \sigma\rangle$,

(a) 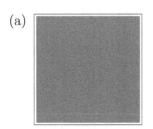 (b) 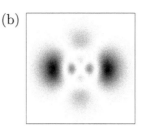 (c)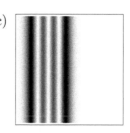

FIG. 1.1. The probability density $|\langle \mathbf{r}|\psi_\nu\rangle|^2$ in the xy plane for (a) any plane wave $\nu = (k_x, k_y, k_z, \sigma)$, (b) the hydrogen orbital $\nu = (4, 2, 0, \sigma)$, and (c) the Landau orbital $\nu = (3, k_y, 0, \sigma)$.

where n is an integer, k_y (k_z) is the y (z) component of $\mathbf{k}$, and σ the spin variable. Recall that

$$\langle \mathbf{r}|n, k_y, k_z, \sigma\rangle = H_n(x/\ell - k_y\ell)e^{-\frac{1}{2}(x/\ell - k_y\ell)^2}\frac{1}{\sqrt{L_yL_z}}e^{i(k_yy+k_zz)}\chi_\sigma \quad \text{(Landau orbital)}$$
(1.5)

where $\ell = \sqrt{\hbar/eB}$ is the magnetic length and H_n is the normalized Hermite polynomial of order n associated with the harmonic oscillator potential induced by the magnetic field. Examples of each of these three types of electron orbitals are shown in Fig. 1.1.

In general a complete set of quantum numbers is denoted ν. The three examples given above correspond to $\nu = (k_x, k_y, k_z, \sigma)$, $\nu = (n, l, m, \sigma)$, and $\nu = (n, k_y, k_z, \sigma)$, each yielding a state function of the form $\psi_\nu(\mathbf{r}) - \langle \mathbf{r}|\nu\rangle$. The completeness of a basis state as well as the normalization of the state vectors plays a central role in quantum theory. Loosely speaking, the normalization condition means that with probability unity a particle in a given quantum state $\psi_\nu(\mathbf{r})$ must be somewhere in space: $\int d\mathbf{r}\, |\psi_\nu(\mathbf{r})|^2 = 1$, or in the Dirac notation: $1 = \int d\mathbf{r}\, \langle\nu|\mathbf{r}\rangle\langle\mathbf{r}|\nu\rangle = \langle\nu|\left(\int d\mathbf{r}\, |\mathbf{r}\rangle\langle\mathbf{r}|\right)|\nu\rangle$. From this we conclude

$$\int d\mathbf{r}\, |\mathbf{r}\rangle\langle\mathbf{r}| = 1.$$
(1.6)

Similarly, the completeness of a set of basis states $\psi_\nu(\mathbf{r})$ means that if a particle is in some state $\psi(\mathbf{r})$ it must be found with probability unity within the orbitals of the basis set: $\sum_\nu |\langle\nu|\psi\rangle|^2 = 1$. Again using the Dirac notation we find $1 = \sum_\nu \langle\psi|\nu\rangle\langle\nu|\psi\rangle = \langle\psi|\left(\sum_\nu |\nu\rangle\langle\nu|\right)|\psi\rangle$, and we conclude

$$\sum_\nu |\nu\rangle\langle\nu| = 1.$$
(1.7)

We shall often use the completeness relation (1.7). A simple example is the expansion of a state function in a given basis: $\psi(\mathbf{r}) = \langle\mathbf{r}|\psi\rangle = \langle\mathbf{r}|1|\psi\rangle = \langle\mathbf{r}|\left(\sum_\nu |\nu\rangle\langle\nu|\right)|\psi\rangle = \sum_\nu\langle\mathbf{r}|\nu\rangle\langle\nu|\psi\rangle$, which can be expressed as

$$\psi(\mathbf{r}) = \sum_\nu \psi_\nu(\mathbf{r})\left(\int d\mathbf{r}'\, \psi_\nu^*(\mathbf{r}')\psi(\mathbf{r}')\right) \quad \text{or} \quad \langle\mathbf{r}|\psi\rangle = \sum_\nu \langle\mathbf{r}|\nu\rangle\langle\nu|\psi\rangle.$$
(1.8)

It should be noted that the quantum label ν can contain both discrete and continuous quantum numbers. In that case the symbol $\sum_\nu$ is to be interpreted as a combination of both summations and integrations. For example, in the case in Eq. (1.5) with Landau orbitals in a box with side lengths L_x, L_y and L_z, we have

$$\sum_\nu = \sum_{\sigma=\uparrow,\downarrow} \sum_{n=0}^{\infty} \int_{-\infty}^{\infty} \frac{L_y}{2\pi} dk_y \int_{-\infty}^{\infty} \frac{L_z}{2\pi} dk_z. \tag{1.9}$$

In the mathematical formulation of quantum theory we shall often encounter the following special functions:
- Kronecker's delta-function δ_{ij} for discrete variables,

$$\delta_{ij} = \begin{cases} 1, & \text{for } i = j, \\ 0, & \text{for } i \neq j. \end{cases} \tag{1.10}$$

- The Levi–Civita symbol ϵ_{ijk} for discrete variables,

$$\epsilon_{ijk} = \begin{cases} +1, & \text{if } (ijk) \text{ is an even permutation of } (123) \text{ or } (xyz), \\ -1, & \text{if } (ijk) \text{ is an odd permutation of } (123) \text{ or } (xyz), \\ 0, & \text{otherwise.} \end{cases} \tag{1.11}$$

- Dirac's delta-function $\delta(\mathbf{r})$ for continuous variables,

$$\delta(\mathbf{r}) = 0, \text{ for } \mathbf{r} \neq 0, \quad \text{while } \int d\mathbf{r}\, \delta(\mathbf{r}) = 1, \tag{1.12}$$

- and, finally, Heaviside's step-function $\theta(x)$ for continuous variables,

$$\theta(x) = \begin{cases} 0, & \text{for } x < 0, \\ 1, & \text{for } x > 0. \end{cases} \tag{1.13}$$

1.2 First quantization, many-particle systems

When turning to N-particle systems, i.e., systems containing N identical particles, say, electrons, three more assumptions are added to the basic assumptions defining quantum theory. The first assumption is the natural extension of the single-particle state function $\psi(\mathbf{r})$, which (neglecting the spin degree of freedom for the time being) is a complex wave function in 3-dimensional space, to the N-particle state function $\psi(\mathbf{r}_1, \mathbf{r}_2, \ldots, \mathbf{r}_N)$, which is a complex function in the $3N$-dimensional configuration space. As for one particle, this N-particle state function is interpreted as a probability amplitude such that its absolute square is related to a probability:

$$|\psi(\mathbf{r}_1, \mathbf{r}_2, \ldots, \mathbf{r}_N)|^2 \prod_{j=1}^{N} d\mathbf{r}_j = \left\{ \begin{array}{l} \text{The probability for finding the } N \text{ particles} \\ \text{in the } 3N-\text{dimensional volume } \prod_{j=1}^{N} d\mathbf{r}_j \\ \text{surrounding the point } (\mathbf{r}_1, \mathbf{r}_2, \ldots, \mathbf{r}_N) \text{ in} \\ \text{the } 3N-\text{dimensional configuration space} \end{array} \right\}. \tag{1.14}$$

1.2.1 *Permutation symmetry and indistinguishability*

A fundamental difference between classical and quantum mechanics concerns the concept of indistinguishability of identical particles. In classical mechanics each particle can be equipped with an identifying marker (e.g. a colored spot on a billiard ball) without influencing its behavior, and moreover it follows its own continuous path in phase space. Thus, in principle, each particle in a group of identical particles can be identified. This is not so in quantum mechanics. Not even in principle is it possible to mark a particle without influencing its physical state, and worse, if a number of identical particles are brought to the same region in space, their wavefunctions will rapidly spread out and overlap with one another, thereby soon rendering it impossible to say which particle is where.

The second fundamental assumption for N-particle systems is therefore that identical particles, i.e., particles characterized by the same quantum numbers such as mass, charge and spin, are in principle indistinguishable.

From the indistinguishability of particles it follows that if two coordinates in an N-particle state function are interchanged the same physical state results, and the corresponding state function can at most differ from the original one by a simple prefactor λ. If the same two coordinates then are interchanged a second time, we end up with the exact same state function,

$$\psi(\mathbf{r}_1, .., \mathbf{r}_j, .., \mathbf{r}_k, .., \mathbf{r}_N) = \lambda\ \psi(\mathbf{r}_1, .., \mathbf{r}_k, .., \mathbf{r}_j, .., \mathbf{r}_N)$$
$$= \lambda^2 \psi(\mathbf{r}_1, .., \mathbf{r}_j, .., \mathbf{r}_k, .., \mathbf{r}_N), \qquad (1.15)$$

and we conclude that $\lambda^2 = 1$, or $\lambda = \pm 1$. Only two species of particles are thus possible in quantum physics, the so-called bosons and fermions:[1]

$$\psi(\mathbf{r}_1, \ldots, \mathbf{r}_j, \ldots, \mathbf{r}_k, \ldots, \mathbf{r}_N) = +\psi(\mathbf{r}_1, \ldots, \mathbf{r}_k, \ldots, \mathbf{r}_j, \ldots, \mathbf{r}_N) \quad \text{(bosons)}, \qquad (1.16a)$$
$$\psi(\mathbf{r}_1, \ldots, \mathbf{r}_j, \ldots, \mathbf{r}_k, \ldots, \mathbf{r}_N) = -\psi(\mathbf{r}_1, \ldots, \mathbf{r}_k, \ldots, \mathbf{r}_j, \ldots, \mathbf{r}_N) \quad \text{(fermions)}. \qquad (1.16b)$$

The importance of the assumption of indistinguishability of particles in quantum physics cannot be exaggerated, and it has been introduced due to overwhelming experimental evidence. For fermions it immediately leads to the Pauli exclusion principle, stating that two fermions cannot occupy the same state, because if in Eq. (1.16b) we let $\mathbf{r}_j = \mathbf{r}_k$ then $\psi = 0$ follows. It thus explains the periodic table of the elements, and consequently the starting point in our understanding of atomic physics, condensed matter physics, and chemistry. It furthermore plays a fundamental role in the studies of the nature of stars and of scattering processes in high energy physics. For bosons, the assumption is necessary to understand Planck's radiation law for the electromagnetic field, and spectacular phenomena such as Bose–Einstein condensation, superfluidity, and laser light.

---

[1] This discrete permutation symmetry is always obeyed. However, some quasiparticles in 2D exhibit *any* phase $e^{i\phi}$, a so-called Berry phase, upon adiabatic interchange; they are therefore called *anyons*.

1.2.2 *The single-particle states as basis states*

We now show that the basis states for the N-particle system can be built from any complete orthonormal single-particle basis $\{\psi_\nu(\mathbf{r})\}$,

$$\sum_\nu \psi_\nu^*(\mathbf{r}')\psi_\nu(\mathbf{r}) = \delta(\mathbf{r}-\mathbf{r}'), \qquad \int d\mathbf{r}\, \psi_\nu^*(\mathbf{r})\psi_{\nu'}(\mathbf{r}) = \delta_{\nu,\nu'}. \qquad (1.17)$$

Starting from an arbitrary N-particle state $\psi(\mathbf{r}_1,\ldots,\mathbf{r}_N)$ we form the $(N-1)$-particle function $A_{\nu_1}(\mathbf{r}_2,\ldots,\mathbf{r}_N)$ by projecting on to the basis state $\psi_{\nu_1}(\mathbf{r}_1)$:

$$A_{\nu_1}(\mathbf{r}_2,\ldots,\mathbf{r}_N) \equiv \int d\mathbf{r}_1\, \psi_{\nu_1}^*(\mathbf{r}_1)\psi(\mathbf{r}_1,\ldots,\mathbf{r}_N). \qquad (1.18)$$

This can be inverted by multiplying with $\psi_{\nu_1}(\tilde{\mathbf{r}}_1)$ and summing over ν_1,

$$\psi(\tilde{\mathbf{r}}_1,\mathbf{r}_2,\ldots,\mathbf{r}_N) = \sum_{\nu_1} \psi_{\nu_1}(\tilde{\mathbf{r}}_1)A_{\nu_1}(\mathbf{r}_2,\ldots,\mathbf{r}_N). \qquad (1.19)$$

Now define, analogously, $A_{\nu_1,\nu_2}(\mathbf{r}_3,\ldots,\mathbf{r}_N)$ from $A_{\nu_1}(\mathbf{r}_2,\ldots,\mathbf{r}_N)$:

$$A_{\nu_1,\nu_2}(\mathbf{r}_3,\ldots,\mathbf{r}_N) \equiv \int d\mathbf{r}_2\, \psi_{\nu_2}^*(\mathbf{r}_2)A_{\nu_1}(\mathbf{r}_2,\ldots,\mathbf{r}_N). \qquad (1.20)$$

As before, we can invert this expression to give A_{ν_1} in terms of A_{ν_1,ν_2}, which upon insertion into Eq. (1.19) leads to

$$\psi(\tilde{\mathbf{r}}_1,\tilde{\mathbf{r}}_2,\mathbf{r}_3\ldots,\mathbf{r}_N) = \sum_{\nu_1,\nu_2} \psi_{\nu_1}(\tilde{\mathbf{r}}_1)\psi_{\nu_2}(\tilde{\mathbf{r}}_2)A_{\nu_1,\nu_2}(\mathbf{r}_3,\ldots,\mathbf{r}_N). \qquad (1.21)$$

Continuing all the way through $\tilde{\mathbf{r}}_N$ (and then writing $\mathbf{r}$ instead of $\tilde{\mathbf{r}}$) we end up with

$$\psi(\mathbf{r}_1,\mathbf{r}_2,\ldots,\mathbf{r}_N) = \sum_{\nu_1,\ldots,\nu_N} A_{\nu_1,\nu_2,\ldots,\nu_N}\psi_{\nu_1}(\mathbf{r}_1)\psi_{\nu_2}(\mathbf{r}_2)\ldots\psi_{\nu_N}(\mathbf{r}_N), \qquad (1.22)$$

where $A_{\nu_1,\nu_2,\ldots,\nu_N}$ are just complex numbers. Thus any N-particle state function can be written as a (rather complicated) linear superposition of product states containing N factors of single-particle basis states.

Even though the product states $\prod_{j=1}^N \psi_{\nu_j}(\mathbf{r}_j)$, in a mathematical sense, form a perfectly valid basis for the N-particle Hilbert space, we know from the discussion on indistinguishability that physically it is not a useful basis since the coordinates have to appear in a symmetric way. No physical perturbation can ever break the fundamental fermion or boson symmetry, which therefore ought to be explicitly incorporated in the basis states. The symmetry requirements from Eqs. (1.16a) and (1.16b) are in Eq. (1.22) hidden in the coefficients $A_{\nu_1,\ldots,\nu_N}$. A physical meaningful basis bringing the N coordinates on equal footing in the products $\psi_{\nu_1}(\mathbf{r}_1)\psi_{\nu_2}(\mathbf{r}_2)\ldots\psi_{\nu_N}(\mathbf{r}_N)$ of single-particle state functions is obtained by applying the bosonic symmetrization operator $\hat{S}_+$ or the fermionic anti-symmetrization operator $\hat{S}_-$ defined by the following determinants and permanent:[2]

$$\hat{S}_\pm \prod_{j=1}^{N} \psi_{\nu_j}(\mathbf{r}_j) = \begin{vmatrix} \psi_{\nu_1}(\mathbf{r}_1) & \psi_{\nu_1}(\mathbf{r}_2) & \cdots & \psi_{\nu_1}(\mathbf{r}_N) \\ \psi_{\nu_2}(\mathbf{r}_1) & \psi_{\nu_2}(\mathbf{r}_2) & \cdots & \psi_{\nu_2}(\mathbf{r}_N) \\ \vdots & \vdots & \ddots & \vdots \\ \psi_{\nu_N}(\mathbf{r}_1) & \psi_{\nu_N}(\mathbf{r}_2) & \cdots & \psi_{\nu_N}(\mathbf{r}_N) \end{vmatrix}_\pm . \tag{1.23}$$

The fermion case involves ordinary determinants, which in physics are denoted Slater determinants,

$$\begin{vmatrix} \psi_{\nu_1}(\mathbf{r}_1) & \psi_{\nu_1}(\mathbf{r}_2) & \cdots & \psi_{\nu_1}(\mathbf{r}_N) \\ \psi_{\nu_2}(\mathbf{r}_1) & \psi_{\nu_2}(\mathbf{r}_2) & \cdots & \psi_{\nu_2}(\mathbf{r}_N) \\ \vdots & \vdots & \ddots & \vdots \\ \psi_{\nu_N}(\mathbf{r}_1) & \psi_{\nu_N}(\mathbf{r}_2) & \cdots & \psi_{\nu_N}(\mathbf{r}_N) \end{vmatrix}_- = \sum_{p \in S_N} \left(\prod_{j=1}^{N} \psi_{\nu_j}(\mathbf{r}_{p(j)}) \right) \text{sign}(p), \tag{1.24}$$

while the boson case involves a sign-less determinant, a so-called permanent,

$$\begin{vmatrix} \psi_{\nu_1}(\mathbf{r}_1) & \psi_{\nu_1}(\mathbf{r}_2) & \cdots & \psi_{\nu_1}(\mathbf{r}_N) \\ \psi_{\nu_2}(\mathbf{r}_1) & \psi_{\nu_2}(\mathbf{r}_2) & \cdots & \psi_{\nu_2}(\mathbf{r}_N) \\ \vdots & \vdots & \ddots & \vdots \\ \psi_{\nu_N}(\mathbf{r}_1) & \psi_{\nu_N}(\mathbf{r}_2) & \cdots & \psi_{\nu_N}(\mathbf{r}_N) \end{vmatrix}_+ = \sum_{p \in S_N} \left(\prod_{j=1}^{N} \psi_{\nu_j}(\mathbf{r}_{p(j)}) \right). \tag{1.25}$$

Here S_N is the group of the $N!$ permutations p on the set of N coordinates,[3] and $\text{sign}(p)$, used in the Slater determinant, is the sign of the permutation p. Note how in the fermion case $\nu_j = \nu_k$ leads to $\psi = 0$, i.e., the Pauli principle. Using the symmetrized basis states the expansion in Eq. (1.22) gets replaced by the following, where the new expansion coefficients $B_{\nu_1,\nu_2,\ldots,\nu_N}$ are completely symmetric in their ν-indices,

$$\psi(\mathbf{r}_1, \mathbf{r}_2, \ldots, \mathbf{r}_N) = \sum_{\nu_1,\ldots,\nu_N} B_{\nu_1,\nu_2,\ldots,\nu_N} \, \hat{S}_\pm \psi_{\nu_1}(\mathbf{r}_1) \psi_{\nu_2}(\mathbf{r}_2) \ldots \psi_{\nu_N}(\mathbf{r}_N). \tag{1.26}$$

We need not worry about the precise relation between the two sets of coefficients A and B since we are not going to use it.

---

[2] Note that to obtain a normalized state on the right-hand side in Eq. (1.23) it should be multiplied by a prefactor: $\frac{1}{\prod_{\nu'} \sqrt{n_{\nu'}!}} \frac{1}{\sqrt{N!}}$, where $n_{\nu'}$ is the number of times the state $|\nu'\rangle$ appears in the set $\{|\nu_1\rangle, |\nu_2\rangle, \ldots |\nu_N\rangle\}$, i.e. 0 or 1 for fermions (which means $n_{\nu'}! = 1$) and between 0 and N for bosons. For fermions the prefactor thus reduces to $\frac{1}{\sqrt{N!}}$.

[3] For $N = 3$ we have, with the signs of the permutations as subscripts,

$$S_3 = \left\{ \begin{pmatrix} 1 \\ 2 \\ 3 \end{pmatrix}_+ , \begin{pmatrix} 1 \\ 3 \\ 2 \end{pmatrix}_- , \begin{pmatrix} 2 \\ 1 \\ 3 \end{pmatrix}_- , \begin{pmatrix} 2 \\ 3 \\ 1 \end{pmatrix}_+ , \begin{pmatrix} 3 \\ 1 \\ 2 \end{pmatrix}_+ , \begin{pmatrix} 3 \\ 2 \\ 1 \end{pmatrix}_- \right\}$$

1.2.3 *Operators in first quantization*

We now turn to the third assumption needed to complete the quantum theory of
N-particle systems. It states that single- and few-particle operators defined for single-
and few-particle states remain unchanged when acting on N-particle states. In this
book we will only work with one- and two-particle operators.

Let us begin with one-particle operators. A given local one-particle operator $T =
T(\mathbf{r}, \nabla_{\mathbf{r}})$, say the kinetic energy operator or an external potential, takes the following
form in the $|\nu\rangle$-representation for a particle number j:

$$T_j = \sum_{\nu_a,\nu_b} T_{\nu_b\nu_a} \left|\nu_b\right\rangle_j \left\langle\nu_a\right|_j, \tag{1.27a}$$

where the subscripts on $\left|\nu_b\right\rangle_j$ and $\left\langle\nu_a\right|_j$ refer to the states of particle number j, and
where

$$T_{\nu_b\nu_a} = \int d\mathbf{r}\ \psi_{\nu_b}^*(\mathbf{r})\, T(\mathbf{r}, \nabla_{\mathbf{r}})\, \psi_{\nu_a}(\mathbf{r}). \tag{1.27b}$$

In a system with N identical particles all coordinates must appear in a symmetric way,
hence the proper kinetic energy operator in this case must be the total (symmetric)
kinetic energy operator T_{tot} associated with all the coordinates,

$$T_{\text{tot}} = \sum_{j=1}^N T_j. \tag{1.28}$$

The action of T_{tot} on any simple product of N single-particle states, ψ_{ν_1} to ψ_{ν_N}, is

$$T_{\text{tot}}\left|\nu_1\right\rangle_1 \left|\nu_2\right\rangle_2 \cdots \left|\nu_N\right\rangle_N \tag{1.29}$$

$$= \sum_{j=1}^N \sum_{\nu_a\nu_b} T_{\nu_b\nu_a} \delta_{\nu_a\nu_j}\left|\nu_1\right\rangle_1 \cdots \left|\nu_b\right\rangle_j \cdots \left|\nu_N\right\rangle_N.$$

Here the Kronecker delta comes from $\langle\nu_a|\nu_j\rangle = \delta_{\nu_a\nu_j}$. It is straight forward to extend
this result to the proper symmetrized basis states.

We move on to discuss symmetric two-particle operators V_{jk}, such as the Coulomb
interaction $V(\mathbf{r}_j - \mathbf{r}_k) = \frac{e^2}{4\pi\epsilon_0} \frac{1}{|\mathbf{r}_j - \mathbf{r}_k|}$ between a pair of electrons. For a two-particle
system described by the coordinates $\mathbf{r}_j$ and $\mathbf{r}_k$ in the $|\nu\rangle$-representation with basis
states $|\nu_a\rangle_j|\nu_b\rangle_k$ we have the definition of V_{jk}:

$$V_{jk} = \sum_{\substack{\nu_a\nu_b \\ \nu_c\nu_d}} V_{\nu_c\nu_d,\nu_a\nu_b} \left|\nu_c\right\rangle_j \left|\nu_d\right\rangle_k \left\langle\nu_a\right|_j \left\langle\nu_b\right|_k, \tag{1.30a}$$

where

$$V_{\nu_c\nu_d,\nu_a\nu_b} = \int d\mathbf{r}_j d\mathbf{r}_k\ \psi_{\nu_c}^*(\mathbf{r}_j)\psi_{\nu_d}^*(\mathbf{r}_k)V(\mathbf{r}_j - \mathbf{r}_k)\psi_{\nu_a}(\mathbf{r}_j)\psi_{\nu_b}(\mathbf{r}_k). \tag{1.30b}$$

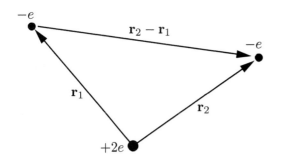
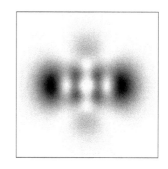

FIG. 1.2. The position vectors ($\mathbf{r}_1$ and $\mathbf{r}_2$) of the two electrons ($-e$) orbiting the helium nucleus ($+2e$) and the single-particle probability density (grayscale plot) $P(\mathbf{r}_1) = \int d\mathbf{r}_2 \frac{1}{2}|\psi_{\nu_1}(\mathbf{r}_1)\psi_{\nu_2}(\mathbf{r}_2) + \psi_{\nu_2}(\mathbf{r}_1)\psi_{\nu_1}(\mathbf{r}_2)|^2$ for the symmetric two-particle state based on the single-particle orbitals $|\nu_1\rangle = |(3,2,1,\uparrow)\rangle$ and $|\nu_2\rangle = |(4,2,0,\downarrow)\rangle$. Compare with the single orbital $|(4,2,0,\downarrow)\rangle$ depicted in Fig. 1.1(b).

In the N-particle system we must again take the symmetric combination of the coordinates, i.e., introduce the operator of the total interaction energy V_{tot},

$$V_{\text{tot}} = \sum_{j>k}^{N} V_{jk} = \frac{1}{2}\sum_{j\neq k}^{N} V_{jk}, \tag{1.31}$$

which acts as follows:

$$V_{\text{tot}}|\nu_1\rangle_1|\nu_2\rangle_2 \ldots |\nu_N\rangle_N = \frac{1}{2}\sum_{j\neq k}^{N}\sum_{\substack{\nu_a\nu_b \\ \nu_c\nu_d}} V_{\nu_c\nu_d,\nu_a\nu_b}\delta_{\nu_a\nu_j}\delta_{\nu_b\nu_k}$$
$$\times |\nu_1\rangle_1 \ldots |\nu_c\rangle_j \ldots |\nu_d\rangle_k \ldots |\nu_N\rangle_N. \tag{1.32}$$

A typical Hamiltonian for an N-particle system thus takes the form

$$H = T_{\text{tot}} + V_{\text{tot}} = \sum_{j=1}^{N} T_j + \frac{1}{2}\sum_{j\neq k}^{N} V_{jk}. \tag{1.33}$$

A specific example is the Hamiltonian for the helium atom, which, in a simple form neglecting spin interactions, can be thought of as two electrons with coordinates $\mathbf{r} = \mathbf{r}_1$ and $\mathbf{r} = \mathbf{r}_2$ orbiting around a nucleus with charge $Z = +2$ at $\mathbf{r} = 0$,

$$H_{\text{He}} = \left(-\frac{\hbar^2}{2m}\nabla_1^2 - \frac{Ze^2}{4\pi\epsilon_0}\frac{1}{r_1}\right) + \left(-\frac{\hbar^2}{2m}\nabla_2^2 - \frac{Ze^2}{4\pi\epsilon_0}\frac{1}{r_2}\right) + \frac{e^2}{4\pi\epsilon_0}\frac{1}{|\mathbf{r}_1 - \mathbf{r}_2|}. \tag{1.34}$$

This Hamiltonian consists of four one-particle operators and one two-particle operator; see also Fig. 1.2.

1.3 Second quantization, basic concepts

Many-particle physics is formulated in terms of the so-called second quantization representation also known by the more descriptive name occupation number representation. The starting point of this formalism is the notion of indistinguishability of particles discussed in Section 1.2.1 combined with the observation in Section 1.2.2 that determinants or permanents of single-particle states form a basis for the Hilbert space of N-particle states. As we shall see, quantum theory can be formulated in terms of occupation numbers of these single-particle states.

1.3.1 *The occupation number representation*

The first step in defining the occupation number representation is to choose any ordered and complete single-particle basis $\{|\nu_1\rangle, |\nu_2\rangle, |\nu_3\rangle, \ldots\}$, the ordering being of paramount importance for fermions. From the form $\hat{S}_\pm \psi_{\nu_{n_1}}(\mathbf{r}_1)\psi_{\nu_{n_2}}(\mathbf{r}_2)\ldots\psi_{\nu_{n_N}}(\mathbf{r}_N)$ of the basis states in Eq. (1.26), it is clear that in each term only the occupied single-particle states $|\nu_{n_j}\rangle$ play a role. It must somehow be simpler to formulate a representation where one just counts how many particles there are in each orbital $|\nu\rangle$. This simplification is achieved with the occupation number representation.

The basis states for an N-particle system in the occupation number representation are obtained simply by listing the occupation numbers of each basis state,

$$N\text{-particle basis states:}\quad |n_{\nu_1}, n_{\nu_2}, n_{\nu_3}, \ldots\rangle, \quad \sum_j n_{\nu_j} = N. \tag{1.35}$$

It is therefore natural to define occupation number operators $\hat{n}_{\nu_j}$ which as eigenstates have the basis states $|n_{\nu_j}\rangle$, and as eigenvalues have the number n_{ν_j} of particles occupying the state ν_j,

$$\hat{n}_{\nu_j}|n_{\nu_j}\rangle = n_{\nu_j}|n_{\nu_j}\rangle. \tag{1.36}$$

We shall show later that for fermions n_{ν_j} can be 0 or 1, while for bosons it can be any non-negative number,

$$n_{\nu_j} = \begin{cases} 0, 1 & \text{(fermions)}, \\ 0, 1, 2, \ldots & \text{(bosons)}. \end{cases} \tag{1.37}$$

Naturally, the question arises how to connect the occupation number basis Eq. (1.35) with the first quantization basis Eq. (1.24). This will be answered in the next section.

The space spanned by the occupation number basis, denoted the Fock space $\mathcal{F}$, is defined as $\mathcal{F} = \mathcal{F}_0 \oplus \mathcal{F}_1 \oplus \mathcal{F}_2 \oplus \cdots$, where $\mathcal{F}_N = \text{span}\{|n_{\nu_1}, n_{\nu_2}, \ldots\rangle \mid \sum_j n_{\nu_j} = N\}$. In Table 1.1 some of the fermionic and bosonic basis states in the occupation number representation are shown. Note how by virtue of the direct sum, states containing a different number of particles are defined to be orthogonal.

1.3.2 *The boson creation and annihilation operators*

To connect first and second quantization we first treat bosons. Given the occupation number operator it is natural to introduce the creation operator $b_{\nu_j}^\dagger$ that raises the occupation number in the state $|\nu_j\rangle$ by 1,

Table 1.1 *Some occupation number basis states for N-particle systems.*

N	Fermion basis states $|n_{\nu_1}, n_{\nu_2}, n_{\nu_3}, \ldots\rangle$
0	$|0,0,0,0,..\rangle$
1	$|1,0,0,0,..\rangle$, $|0,1,0,0,..\rangle$, $|0,0,1,0,..\rangle$, ..
2	$|1,1,0,0,..\rangle$, $|0,1,1,0,..\rangle$, $|1,0,1,0,..\rangle$, $|0,0,1,1,..\rangle$, $|0,1,0,1,..\rangle$, $|1,0,0,1,..\rangle$, ..
$\vdots$	$\qquad\vdots\qquad\qquad\vdots\qquad\qquad\vdots\qquad\qquad\vdots\qquad\qquad\vdots$

N	Boson basis states $|n_{\nu_1}, n_{\nu_2}, n_{\nu_3}, \ldots\rangle$
0	$|0,0,0,0,..\rangle$
1	$|1,0,0,0,..\rangle$, $|0,1,0,0,..\rangle$, $|0,0,1,0,..\rangle$, ..
2	$|2,0,0,0,..\rangle$, $|0,2,0,0,..\rangle$, $|1,1,0,0,..\rangle$, $|0,0,2,0,..\rangle$, $|0,1,1,0,..\rangle$, $|1,0,1,0,..\rangle$, ..
$\vdots$	$\qquad\vdots\qquad\qquad\vdots\qquad\qquad\vdots\qquad\qquad\vdots\qquad\qquad\vdots$

$$b_{\nu_j}^\dagger |\ldots, n_{\nu_{j-1}}, n_{\nu_j}, n_{\nu_{j+1}}, \ldots\rangle = B_+(n_{\nu_j}) |\ldots, n_{\nu_{j-1}}, n_{\nu_j} + 1, n_{\nu_{j+1}}, \ldots\rangle, \qquad (1.38)$$

where $B_+(n_{\nu_j})$ is a normalization constant to be determined. The only non-zero matrix elements of $b_{\nu_j}^\dagger$ are $\langle n_{\nu_j} + 1|b_{\nu_j}^\dagger|n_{\nu_j}\rangle$, where for brevity we only explicitly write the occupation number for ν_j. The adjoint of $b_{\nu_j}^\dagger$ is found by complex conjugation as $\langle n_{\nu_j} + 1|b_{\nu_j}^\dagger|n_{\nu_j}\rangle^* = \langle n_{\nu_j}|(b_{\nu_j}^\dagger)^\dagger|n_{\nu_j} + 1\rangle$. Consequently, one defines the annihilation operator $b_{\nu_j} \equiv (b_{\nu_j}^\dagger)^\dagger$, which lowers the occupation number of state $|\nu_j\rangle$ by 1,

$$b_{\nu_j} |\ldots, n_{\nu_{j-1}}, n_{\nu_j}, n_{\nu_{j+1}}, \ldots\rangle = B_-(n_{\nu_j}) |\ldots, n_{\nu_{j-1}}, n_{\nu_j} - 1, n_{\nu_{j+1}}, \ldots\rangle. \qquad (1.39)$$

The creation and annihilation operators $b_{\nu_j}^\dagger$ and b_{ν_j} are the fundamental operators in the occupation number formalism. As we will demonstrate later any operator can be expressed in terms of them.

Let us proceed by investigating the properties of $b_{\nu_j}^\dagger$ and b_{ν_j} further. Since bosons are symmetric in the single-particle state index ν_j we of course demand that $b_{\nu_j}^\dagger$ and $b_{\nu_k}^\dagger$ must commute, and hence by Hermitian conjugation that also b_{ν_j} and b_{ν_k} commute. The commutator $[A, B]$ for two operators A and B is defined as

$$[A, B] \equiv AB - BA, \quad \text{so that} \quad [A, B] = 0 \quad \Rightarrow \quad BA = AB. \qquad (1.40)$$

We demand further that if $j \neq k$ then b_{ν_j} and $b_{\nu_k}^\dagger$ commute. However, if $j = k$ we must be careful. It is evident that since an unoccupied state can not be emptied further we must demand $b_{\nu_j}|\ldots, 0, \ldots\rangle = 0$, i.e., $B_-(0) = 0$. We also have the freedom to normalize the operators by demanding $b_{\nu_j}^\dagger|\ldots, 0, \ldots\rangle = |\ldots, 1, \ldots\rangle$, i.e., $B_+(0) = 1$. But since $\langle 1|b_{\nu_j}^\dagger|0\rangle^* = \langle 0|b_{\nu_j}|1\rangle$, it also follows that $b_{\nu_j}|\ldots, 1, \ldots\rangle = |\ldots, 0, \ldots\rangle$, i.e., $B_-(1) = 1$.

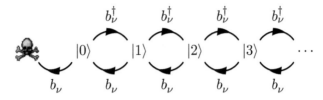

FIG. 1.3. The action of the bosonic creation operator $b^\dagger_\nu$ and adjoint annihilation operator b_ν in the occupation number space. Note that $b^\dagger_\nu$ can act indefinitely, while b_ν eventually hits $|0\rangle$ and annihilates it yielding 0.

It is clear that b_{ν_j} and $b^\dagger_{\nu_j}$ do not commute: $b_{\nu_j} b^\dagger_{\nu_j} |0\rangle = |0\rangle$ while $b^\dagger_{\nu_j} b_{\nu_j} |0\rangle = 0$, i.e., we have $[b_{\nu_j}, b^\dagger_{\nu_j}] |0\rangle = |0\rangle$. We now define this commutation relation to be valid as an operator identity in general and below derive the consequences of this construction. In summary, we define the operator algebra for the bosonic creation and annihilation operators by the following three commutation relations:

$$[b^\dagger_{\nu_j}, b^\dagger_{\nu_k}] = 0, \quad [b_{\nu_j}, b_{\nu_k}] = 0, \quad [b_{\nu_j}, b^\dagger_{\nu_k}] = \delta_{\nu_j \nu_k}. \tag{1.41}$$

By definition $b^\dagger_\nu$ and b_ν are not Hermitian. However, the product $b^\dagger_\nu b_\nu$ is, and by using the operator algebra Eq. (1.41) we show below that this operator in fact is the occupation number operator $\hat{n}_\nu$. First, Eq. (1.41) leads immediately to the following two very important commutation relations:

$$[b^\dagger_\nu b_\nu, b_\nu] = -b_\nu, \quad [b^\dagger_\nu b_\nu, b^\dagger_\nu] = b^\dagger_\nu. \tag{1.42}$$

Second, for any state $|\phi\rangle$ we note that $\langle \phi | b^\dagger_\nu b_\nu | \phi \rangle$ is the norm of the state $b_\nu |\phi\rangle$ and hence a positive real number (unless $|\phi\rangle = |0\rangle$, for which $b_\nu |0\rangle = 0$). Let $|\phi_\lambda\rangle$ be any eigenstate of $b^\dagger_\nu b_\nu$, i.e., $b^\dagger_\nu b_\nu |\phi_\lambda\rangle = \lambda |\phi_\lambda\rangle$ with $\lambda > 0$. Now choose a particular λ_0 and study $b_\nu |\phi_{\lambda_0}\rangle$. We find that

$$(b^\dagger_\nu b_\nu) b_\nu |\phi_{\lambda_0}\rangle = (b_\nu b^\dagger_\nu - 1) b_\nu |\phi_{\lambda_0}\rangle = b_\nu (b^\dagger_\nu b_\nu - 1) |\phi_{\lambda_0}\rangle = b_\nu (\lambda_0 - 1) |\phi_{\lambda_0}\rangle, \tag{1.43}$$

i.e., $b_\nu |\phi_{\lambda_0}\rangle$ is also an eigenstate of $b^\dagger_\nu b_\nu$, but with the eigenvalue reduced by 1 to $(\lambda_0 - 1)$. As illustrated in Fig. 1.3, if λ_0 is not a non-negative integer this lowering process can continue until a negative eigenvalue is encountered, but this violates the condition $\lambda_0 > 0$, and we conclude that $\lambda = n = 0, 1, 2, \ldots$. Writing $|\phi_\lambda\rangle = |n_\nu\rangle$ we have shown that $b^\dagger_\nu b_\nu |n_\nu\rangle = n_\nu |n_\nu\rangle$ and $b_\nu |n_\nu\rangle \propto |n_\nu - 1\rangle$. Analogously, we find that

$$(b^\dagger_\nu b_\nu) b^\dagger_\nu |n_\nu\rangle = (n + 1) b^\dagger_\nu |n_\nu\rangle, \tag{1.44}$$

i.e., $b^\dagger_\nu |n_\nu\rangle \propto |n_\nu + 1\rangle$. The normalization factors for $b^\dagger_\nu$ and b_ν are found from

$$\|b_\nu |n_\nu\rangle\|^2 = (b_\nu |n_\nu\rangle)^\dagger (b_\nu |n_\nu\rangle) = \langle n_\nu | b^\dagger_\nu b_\nu | n_\nu \rangle = n_\nu, \tag{1.45a}$$

$$\|b^\dagger_\nu |n_\nu\rangle\|^2 = (b^\dagger_\nu |n_\nu\rangle)^\dagger (b^\dagger_\nu |n_\nu\rangle) = \langle n_\nu | b_\nu b^\dagger_\nu | n_\nu \rangle = n_\nu + 1. \tag{1.45b}$$

Hence we arrive at

$$b_\nu^\dagger b_\nu = \hat{n}_\nu, \quad b_\nu^\dagger b_\nu |n_\nu\rangle = n_\nu |n_\nu\rangle, \quad n_\nu = 0, 1, 2, \dots \tag{1.46a}$$

$$b_\nu |n_\nu\rangle = \sqrt{n_\nu} |n_\nu - 1\rangle, \quad b_\nu^\dagger |n_\nu\rangle = \sqrt{n_\nu + 1} |n_\nu + 1\rangle, \quad (b_\nu^\dagger)^{n_\nu} |0\rangle = \sqrt{n_\nu!} |n_\nu\rangle, \tag{1.46b}$$

and we can therefore identify the equivalence between first and second quantized basis states,

$$\hat{S}_+ |\nu_{n_1}\rangle_1 |\nu_{n_2}\rangle_2 \cdots |\nu_{n_N}\rangle_N \quad \leftrightarrow \quad b_{\nu_{n_1}}^\dagger b_{\nu_{n_2}}^\dagger \cdots b_{\nu_{n_N}}^\dagger |0\rangle, \tag{1.47}$$

where both sides contain N-particle kets symmetric in the single-particle state index ν_{n_j}.

1.3.3 The fermion creation and annihilation operators

Also for fermions it is natural to introduce creation and annihilation operators, now denoted $c_{\nu_j}^\dagger$ and c_{ν_j}, which are the Hermitian adjoints of each other:

$$c_{\nu_j}^\dagger |\dots, n_{\nu_{j-1}}, n_{\nu_j}, n_{\nu_{j+1}}, \dots\rangle = C_+(n_{\nu_j}) |\dots, n_{\nu_{j-1}}, n_{\nu_j} + 1, n_{\nu_{j+1}}, \dots\rangle, \tag{1.48}$$

$$c_{\nu_j} |\dots, n_{\nu_{j-1}}, n_{\nu_j}, n_{\nu_{j+1}}, \dots\rangle = C_-(n_{\nu_j}) |\dots, n_{\nu_{j-1}}, n_{\nu_j} - 1, n_{\nu_{j+1}}, \dots\rangle. \tag{1.49}$$

But to maintain the fundamental fermionic antisymmetry upon exchange of orbitals, apparent in Eq. (1.24), it is not sufficient in the fermionic case just to list the occupation numbers of the states, also the order of the occupied states has a meaning. We must therefore demand

$$|\dots, n_{\nu_j} = 1, \dots, n_{\nu_k} = 1, \dots\rangle = -|\dots, n_{\nu_k} = 1, \dots, n_{\nu_j} = 1, \dots\rangle. \tag{1.50}$$

and consequently we must have that $c_{\nu_j}^\dagger$ and $c_{\nu_k}^\dagger$ anti-commute, and hence, by Hermitian conjugation, that also c_{ν_j} and c_{ν_k} anti-commute. The anti-commutator $\{A, B\}$ for two operators A and B is defined as

$$\{A, B\} \equiv AB + BA, \quad \text{so that} \quad \{A, B\} = 0 \quad \Rightarrow \quad BA = -AB. \tag{1.51}$$

For $j \neq k$ we also demand that c_{ν_j} and $c_{\nu_k}^\dagger$ anti-commute. However, if $j = k$ we again must be careful. It is evident that since an unoccupied state can not be emptied further we must demand $c_{\nu_j} |\dots, 0, \dots\rangle = 0$, i.e., $C_-(0) = 0$. We also have the freedom to normalize the operators by demanding $c_{\nu_j}^\dagger |\dots, 0, \dots\rangle = |\dots, 1, \dots\rangle$, i.e., $C_+(0) = 1$. But since $\langle 1 | c_{\nu_j}^\dagger | 0\rangle^* = \langle 0 | c_{\nu_j} | 1\rangle$ it follows that $c_{\nu_j} |\dots, 1, \dots\rangle = |\dots, 0, \dots\rangle$, i.e., $C_-(1) = 1$.

It is clear that c_{ν_j} and $c_{\nu_j}^\dagger$ do not anti-commute: $c_{\nu_j} c_{\nu_j}^\dagger |0\rangle = |0\rangle$ while $c_{\nu_j}^\dagger c_{\nu_j} |0\rangle = 0$, i.e., we have $\{c_{\nu_j}, c_{\nu_j}^\dagger\} |0\rangle = |0\rangle$. We now define this commutation relation to be valid as an operator identity in general and below derive the consequences of this construction. In summary, we define the operator algebra for the fermionic creation and annihilation operators by the following three anti-commutation relations:

$$\{c_{\nu_j}^\dagger, c_{\nu_k}^\dagger\} = 0, \quad \{c_{\nu_j}, c_{\nu_k}\} = 0, \quad \{c_{\nu_j}, c_{\nu_k}^\dagger\} = \delta_{\nu_j \nu_k}. \tag{1.52}$$

An immediate consequence of the anti-commutation relations (1.52) is

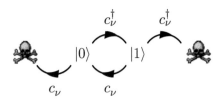

FIG. 1.4.   The action of the fermionic creation operator $c_\nu^\dagger$ and the adjoint annihilation operator c_ν in the occupation number space. Note that both $c_\nu^\dagger$ and c_ν can act at most twice before annihilating a state completely.

$$(c_{\nu_j}^\dagger)^2 = 0, \quad (c_{\nu_j})^2 = 0. \tag{1.53}$$

Now, as for bosons, we introduce the Hermitian operator $c_\nu^\dagger c_\nu$, and by using the operator algebra Eq. (1.52) we show below that this operator, in fact, is the occupation number operator $\hat{n}_\nu$. In analogy with Eq. (1.42) we find

$$[c_\nu^\dagger c_\nu, c_\nu] = -c_\nu, \quad [c_\nu^\dagger c_\nu, c_\nu^\dagger] = c_\nu^\dagger, \tag{1.54}$$

so that $c_\nu^\dagger$ and c_ν step the eigenvalues of $c_\nu^\dagger c_\nu$ up and down by one, respectively. From Eqs. (1.52) and (1.53) we have $(c_\nu^\dagger c_\nu)^2 = c_\nu^\dagger (c_\nu c_\nu^\dagger) c_\nu = c_\nu^\dagger (1 - c_\nu^\dagger c_\nu) c_\nu = c_\nu^\dagger c_\nu$, so that $c_\nu^\dagger c_\nu (c_\nu^\dagger c_\nu - 1) = 0$, and $c_\nu^\dagger c_\nu$ thus only has 0 and 1 as eigenvalues; leading to a simple normalization for $c_\nu^\dagger$ and c_ν. In summary, as illustrated in Fig. 1.4, we have

$$c_\nu^\dagger c_\nu = \hat{n}_\nu, \quad c_\nu^\dagger c_\nu |n_\nu\rangle = n_\nu |n_\nu\rangle, \quad n_\nu = 0, 1 \tag{1.55}$$

$$c_\nu |0\rangle = 0, \quad c_\nu^\dagger |0\rangle = |1\rangle, \quad c_\nu |1\rangle = |0\rangle, \quad c_\nu^\dagger |1\rangle = 0, \tag{1.56}$$

and we can readily identify the first and second quantized basis states,

$$\hat{S}_- |\nu_{n_1}\rangle_1 |\nu_{n_2}\rangle_2 \cdots |\nu_{n_N}\rangle_N \quad \leftrightarrow \quad c_{\nu_{n_1}}^\dagger c_{\nu_{n_2}}^\dagger \cdots c_{\nu_{n_N}}^\dagger |0\rangle, \tag{1.57}$$

where both sides contain normalized N-particle kets anti-symmetric in the single-particle state index ν_{n_j} in accordance with the Pauli exclusion principle.

1.3.4   *The general form for second quantization operators*

In second quantization all operators can be expressed in terms of the fundamental creation and annihilation operators defined in the previous two sections. This rewriting of the first quantized operators in Eqs. (1.29) and (1.32) into their second quantized form is achieved by using the basis state identities (1.47) and (1.57) linking the two representations.

For simplicity, let us first consider the single-particle operator T_{tot} from Eq. (1.29) acting on a bosonic N-particle system. In this equation we then act with the bosonic symmetrization operator S_+ on both sides. Utilizing that T_{tot} and S_+ commute and invoking the basis state identity (1.47) we obtain

$$T_{\text{tot}} b_{\nu_{n_1}}^\dagger \cdots b_{\nu_{n_N}}^\dagger |0\rangle = \sum_{\nu_a \nu_b} T_{\nu_b \nu_a} \sum_{j=1}^{N} \delta_{\nu_a \nu_{n_j}} b_{\nu_{n_1}}^\dagger \cdots \overbrace{b_{\nu_b}^\dagger}^{\text{site } n_j} \cdots b_{\nu_{n_N}}^\dagger |0\rangle, \tag{1.58}$$

where, on the right-hand side of the equation, the operator $b^\dagger_{\nu_b}$ stands on the site n_j. To make the kets on the two sides of the equation look alike, we would like to reinsert the operator $b^\dagger_{\nu_{n_j}}$ at the site n_j on the right. To do this we focus on the state $\nu \equiv \nu_{n_j}$. Originally, i.e., on the left-hand side, the state ν may appear, say, p times leading to a contribution $(b^\dagger_\nu)^p|0\rangle$. We have $p > 0$, since otherwise both sides would yield zero. On the right-hand side the corresponding contribution has changed into $b^\dagger_{\nu_b}(b^\dagger_\nu)^{p-1}|0\rangle$. This is then rewritten by use of Eqs. (1.41), (1.46a) and (1.46b) as

$$b^\dagger_{\nu_b}(b^\dagger_\nu)^{p-1}|0\rangle = b^\dagger_{\nu_b}\left(\frac{1}{p}b_\nu b^\dagger_\nu\right)(b^\dagger_\nu)^{p-1}|0\rangle = \left(\frac{1}{p}b^\dagger_{\nu_b}b_\nu\right)(b^\dagger_\nu)^p|0\rangle. \qquad (1.59)$$

Now, the p operators $b^\dagger_\nu$ can be redistributed to their original places as they appear on the left-hand side of Eq. (1.58). The sum over j together with $\delta_{\nu_a\nu_{n_j}}$ yields p identical contributions cancelling the factor $1/p$ in Eq. (1.59), and we arrive at the simple result

$$T_{\text{tot}}\left[b^\dagger_{\nu_{n_1}}\cdots b^\dagger_{\nu_{n_N}}|0\rangle\right] = \sum_{a,b} T_{\nu_b\nu_a}b^\dagger_{\nu_b}b_{\nu_a}\left[b^\dagger_{\nu_{n_1}}\cdots b^\dagger_{\nu_{n_N}}|0\rangle\right]. \qquad (1.60)$$

Since this result is valid for any basis state $b^\dagger_{\nu_{n_1}}\cdots b^\dagger_{\nu_{n_N}}|0\rangle$, it is actually an operator identity stating $T_{\text{tot}} = \sum_{ij} T_{\nu_i\nu_j}b^\dagger_{\nu_i}b_{\nu_j}$.

It is straightforward to generalize this result to two-particle (or any-number-of-particle) operators acting on boson states, and a similar reasoning can be made for the fermion case (see Exercise 1.1) when the necessary care is taken regarding the sign appearing from the anti-commutators in this case. If we let $a^\dagger$ denote either a boson operator $b^\dagger$ or a fermion operator $c^\dagger$ we can state the general form for one- and two-particle operators in second quantization:

$$T_{\text{tot}} = \sum_{\nu_i,\nu_j} T_{\nu_i\nu_j}\, a^\dagger_{\nu_i}a_{\nu_j}, \qquad (1.61)$$

$$V_{\text{tot}} = \frac{1}{2}\sum_{\substack{\nu_i\nu_j \\ \nu_k\nu_l}} V_{\nu_i\nu_j,\nu_k\nu_l}\, a^\dagger_{\nu_i}a^\dagger_{\nu_j}a_{\nu_l}a_{\nu_k}. \qquad (1.62)$$

In Fig. 1.5 a graphical representation of these fundamental operator expressions is shown.

Operators in second quantization are thus composed of linear combinations of products of creation and annihilation operators, weighted by the appropriate matrix elements of the operator calculated in first quantization. Note the order of the indices, which is extremely important in the case of two-particle fermion operators. The first quantization matrix element can be read as a transition induced from the initial state $|\nu_k\nu_l\rangle$ to the final state $|\nu_i\nu_j\rangle$. In second quantization, the initial state is annihilated by first annihilating the state $|\nu_k\rangle$ and then the state $|\nu_l\rangle$, while the final state is created by first creating the state $|\nu_j\rangle$ and then state $|\nu_i\rangle$:

$$|0\rangle = a_{\nu_l}a_{\nu_k}|\nu_k\nu_l\rangle, \quad |\nu_i\nu_j\rangle = a^\dagger_{\nu_i}a^\dagger_{\nu_j}|0\rangle. \qquad (1.63)$$

FIG. 1.5. A graphical representation of the one- and two-particle operators in second
quantization. The incoming and outgoing arrows represent initial and final states,
respectively. The dashed and wiggled lines represent the transition amplitudes for
the one- and two-particle processes contained in the operators.

Note how all the permutation symmetry properties are taken care of by the operator algebra of $a_\nu^\dagger$ and a_ν. The matrix elements are all in the simple non-symmetrized
form of Eq. (1.30b).

1.3.5 *Change of basis in second quantization*

Different quantum operators are most naturally expressed in different representations
making basis changes a central issue in quantum physics. In this section we give the
general transformation rules, which are to be exploited throughout this book.

Let $\{|\nu_1\rangle, |\nu_2\rangle, \ldots\}$ and $\{|\mu_1\rangle, |\mu_2\rangle, \ldots\}$ be two different complete and ordered
single-particle basis sets. From the completeness condition (1.7) we have the basic
transformation law for single-particle states:

$$|\mu\rangle = \sum_\nu |\nu\rangle\langle\nu|\mu\rangle = \sum_\nu \langle\mu|\nu\rangle^* |\nu\rangle. \qquad (1.64)$$

In the case of single-particle systems we define, quite naturally, creation operators
$a_\mu^\dagger$ and $a_\nu^\dagger$ corresponding to the two basis sets, and find directly from Eq. (1.64) that
$a_\mu^\dagger|0\rangle = |\mu\rangle = \sum_\nu\langle\mu|\nu\rangle^* a_\nu^\dagger|0\rangle$, which guides us to the transformation rules for creation
and annihilation operators (see also Fig. 1.6):

$$a_\mu^\dagger = \sum_\nu \langle\mu|\nu\rangle^* a_\nu^\dagger, \quad a_\mu = \sum_\nu \langle\mu|\nu\rangle\, a_\nu. \qquad (1.65)$$

The general validity of Eq. (1.65) follows when applying the first quantization result
Eq. (1.64) to the N-particle first quantized basis states $\hat{S}_\pm|\nu_{n_1}\rangle_1 \ldots |\nu_{n_N}\rangle_N$ leading to

$$a_{\mu_{n_1}}^\dagger a_{\mu_{n_2}}^\dagger \ldots a_{\mu_{n_N}}^\dagger |0\rangle = \left(\sum_{\nu_{n_1}}\langle\mu_{n_1}|\nu_{n_1}\rangle^* a_{\nu_{n_1}}^\dagger\right) \ldots \left(\sum_{\nu_{n_N}}\langle\mu_{n_N}|\nu_{n_N}\rangle^* a_{\nu_{n_N}}^\dagger\right)|0\rangle. \quad (1.66)$$

The transformation rules Eq. (1.65) lead to two very desirable results. First, that the
basis transformation preserves the bosonic or fermionic particle statistics,

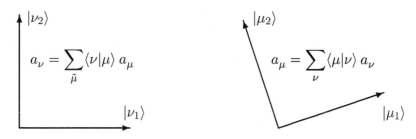

FIG. 1.6. The transformation rules for annihilation operators a_ν and a_μ upon change of basis between $\{|\nu\rangle\}$ and $\{|\mu\rangle\}$.

$$[a_{\mu_1}, a_{\mu_2}^\dagger]_\pm = \sum_{\nu_j \nu_k} \langle \mu_1|\nu_j\rangle\langle\mu_2|\nu_k\rangle^* [a_{\nu_j}, a_{\nu_k}^\dagger]_\pm$$

$$= \sum_{\nu_j \nu_k} \langle\mu_1|\nu_j\rangle\langle\nu_k|\mu_2\rangle\delta_{\nu_j,\nu_k}$$

$$= \sum_{\nu_j} \langle\mu_1|\nu_j\rangle\langle\nu_j|\mu_2\rangle = \delta_{\mu_1,\mu_2}, \tag{1.67}$$

and second, that it leaves the total number of particles unchanged,

$$\sum_\mu a_\mu^\dagger a_\mu = \sum_\mu \sum_{\nu_j \nu_k} \langle\nu_j|\mu\rangle\langle\mu|\nu_k\rangle a_{\nu_j}^\dagger a_{\nu_k} = \sum_{\nu_j \nu_k}\langle\nu_j|\nu_k\rangle a_{\nu_j}^\dagger a_{\nu_k} = \sum_{\nu_j} a_{\nu_j}^\dagger a_{\nu_j}. \tag{1.68}$$

1.3.6 *Quantum field operators and their Fourier transforms*

In particular, one second quantization representation requires special attention, namely, the real space representation leading to the definition of quantum field operators. If we in Section 1.3.5 let the transformed basis set $\{|\tilde{\psi}_\mu\rangle\}$ be the continuous set of position kets $\{|\mathbf{r}\rangle\}$ and, suppressing the spin index, denote $\tilde{a}_\mu^\dagger$ by $\Psi^\dagger(\mathbf{r})$ we obtain from Eq. (1.65)

$$\Psi^\dagger(\mathbf{r}) \equiv \sum_\nu \langle\mathbf{r}|\nu\rangle^* a_\nu^\dagger = \sum_\nu \psi_\nu^*(\mathbf{r}) a_\nu^\dagger, \quad \Psi(\mathbf{r}) \equiv \sum_\nu \langle\mathbf{r}|\nu\rangle a_\nu = \sum_\nu \psi_\nu(\mathbf{r}) a_\nu. \tag{1.69}$$

Note that $\Psi^\dagger(\mathbf{r})$ and $\Psi(\mathbf{r})$ are second quantization operators, while the coefficients $\psi_\nu^*(\mathbf{r})$ and $\psi_\nu(\mathbf{r})$ are ordinary first quantization wavefunctions. Loosely speaking, $\Psi^\dagger(\mathbf{r})$ is the sum of all possible ways to add a particle to the system at position $\mathbf{r}$ through any of the basis states $\psi_\nu(\mathbf{r})$. Since $\Psi^\dagger(\mathbf{r})$ and $\Psi(\mathbf{r})$ are second quantization operators defined in every point in space they are called quantum field operators. From Eq. (1.67) it is straight forward to calculate the following fundamental commutator and anti-commutator,

$$[\Psi(\mathbf{r}_1), \Psi^\dagger(\mathbf{r}_2)] = \delta(\mathbf{r}_1 - \mathbf{r}_2), \quad \text{boson fields}, \tag{1.70a}$$

$$\{\Psi(\mathbf{r}_1), \Psi^\dagger(\mathbf{r}_2)\} = \delta(\mathbf{r}_1 - \mathbf{r}_2), \quad \text{fermion fields}. \tag{1.70b}$$

In some sense the quantum field operators express the essence of the wave/particle duality in quantum physics. On the one hand they are defined as fields, like waves, but on the other hand they exhibit the commutator properties associated with particles.

The introduction of quantum field operators makes it easy to write down operators in the real space representation. By applying the definition (1.69) to the second quantized single-particle operator of Eq. (1.61) one obtains

$$T = \sum_{\nu_i \nu_j} \left(\int d\mathbf{r} \, \psi^*_{\nu_i}(\mathbf{r}) T_{\mathbf{r}} \psi_{\nu_j}(\mathbf{r}) \right) a^\dagger_{\nu_i} a_{\nu_j}$$

$$= \int d\mathbf{r} \left(\sum_{\nu_i} \psi^*_{\nu_i}(\mathbf{r}) a^\dagger_{\nu_i} \right) T_{\mathbf{r}} \left(\sum_{\nu_j} \psi_{\nu_j}(\mathbf{r}) a_{\nu_j} \right) = \int d\mathbf{r} \, \Psi^\dagger(\mathbf{r}) T_{\mathbf{r}} \Psi(\mathbf{r}). \quad (1.71)$$

So, in the real space representation, i.e., using quantum field operators, operators in second quantization have a form analogous to matrix elements in first quantization.

Finally, when working with homogeneous systems it is often desirable to transform between the real space and the momentum representations, i.e., to perform a Fourier transformation. Substituting in Eq. (1.69) the $|\psi_\nu\rangle$ basis with the momentum basis $|\mathbf{k}\rangle$ yields

$$\Psi^\dagger(\mathbf{r}) = \frac{1}{\sqrt{\mathcal{V}}} \sum_{\mathbf{k}} e^{-i\mathbf{k}\cdot\mathbf{r}} a^\dagger_{\mathbf{k}}, \quad \Psi(\mathbf{r}) = \frac{1}{\sqrt{\mathcal{V}}} \sum_{\mathbf{k}} e^{i\mathbf{k}\cdot\mathbf{r}} a_{\mathbf{k}}. \quad (1.72)$$

The inverse expressions are obtained by multiplying by $e^{\pm i\mathbf{q}\cdot\mathbf{r}}$ and integrating over $\mathbf{r}$,

$$a^\dagger_{\mathbf{q}} = \frac{1}{\sqrt{\mathcal{V}}} \int d\mathbf{r} \, e^{i\mathbf{q}\cdot\mathbf{r}} \Psi^\dagger(\mathbf{r}), \quad a_{\mathbf{q}} = \frac{1}{\sqrt{\mathcal{V}}} \int d\mathbf{r} \, e^{-i\mathbf{q}\cdot\mathbf{r}} \Psi(\mathbf{r}). \quad (1.73)$$

1.4 Second quantization, specific operators

In this section we will use the general second quantization formalism to derive some expressions for specific second quantization operators that we are going to use repeatedly in this book.

1.4.1 The harmonic oscillator in second quantization

The 1D harmonic oscillator in first quantization is characterized by two conjugate variables appearing in the Hamiltonian: the position x and the momentum p,

$$H = \frac{1}{2m} p^2 + \frac{1}{2} m\omega^2 x^2, \quad [p, x] = \frac{\hbar}{i}. \quad (1.74)$$

This can be rewritten in second quantization by identifying two operators $a^\dagger$ and a satisfying the basic boson commutation relations (1.41). By inspection it can be verified that the following operators do the job,

$$\left. \begin{array}{l} a \equiv \dfrac{1}{\sqrt{2}} \left(\dfrac{x}{\ell} + i\dfrac{p}{\hbar/\ell} \right) \\[2ex] a^\dagger \equiv \dfrac{1}{\sqrt{2}} \left(\dfrac{x}{\ell} - i\dfrac{p}{\hbar/\ell} \right) \end{array} \right\} \Rightarrow \left\{ \begin{array}{l} x \equiv \ell \dfrac{1}{\sqrt{2}} (a^\dagger + a), \\[2ex] p \equiv \dfrac{\hbar}{\ell} \dfrac{i}{\sqrt{2}} (a^\dagger - a), \end{array} \right. \quad (1.75)$$

where x is given in units of the harmonic oscillator length $\ell = \sqrt{\hbar/m\omega}$ and p in units of the harmonic oscillator momentum $\hbar/\ell$. Mnemotechnically, one can think of a as

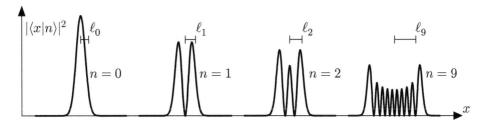

FIG. 1.7.   The probability density $|\langle \mathbf{r}|n\rangle|^2$ for $n = 0, 1, 2$, and 9 quanta in the oscillator state. Note that the width of the wave function is $\sqrt{\langle n|x^2|n\rangle} = \sqrt{n + 1/2}\,\ell$.

being the ($1/\sqrt{2}$-normalized) complex number formed by the real part x/ℓ and the imaginary part $p/(\hbar/\ell)$, while $a^\dagger$ is found as the adjoint operator to a. From Eq. (1.75) we obtain the Hamiltonian H and the eigenstates $|n\rangle$:

$$H = \hbar\omega\left(a^\dagger a + \frac{1}{2}\right) \quad \text{and} \quad |n\rangle = \frac{(a^\dagger)^n}{\sqrt{n!}}\,|0\rangle, \text{with } H|n\rangle = \hbar\omega\left(n + \frac{1}{2}\right)|n\rangle. \quad (1.76)$$

The excitation of the harmonic oscillator can thus be interpreted as filling the oscillator with bosonic quanta created by the operator $a^\dagger$. This picture is particularly useful in the studies of the photon and phonon fields, as we shall see throughout the book. If we as a measure of the amplitude of the oscillator in the state $|n\rangle$ with n quanta use the square-root of the expectation value of $x^2 = \ell^2(a^\dagger a^\dagger + a^\dagger a + aa^\dagger + aa)/2$, we find $\sqrt{\langle n|x^2|n\rangle} = \sqrt{n + 1/2}\,\ell$. Thus the width of the oscillator wavefunction scales roughly with the square-root of the number of quanta in the oscillator, as sketched in Fig. 1.7.

The creation operator can also be used to generate the specific form of the eigenfunctions $\psi_n(x)$ of the oscillator starting from the groundstate wavefunction $\psi_0(x)$:

$$\psi_n(x) = \langle x|n\rangle = \langle x|\frac{(a^\dagger)^n}{\sqrt{n!}}\,|0\rangle = \frac{1}{\sqrt{n!}}\langle x|\left(\frac{x}{\sqrt{2}\ell} - i\frac{p}{\frac{\hbar}{\ell}\sqrt{2}}\right)^n|0\rangle = \frac{1}{\sqrt{2^n n!}}\left(\frac{x}{\ell} - \ell\frac{d}{dx}\right)^n\psi_0(x).$$
$$(1.77)$$

1.4.2   *The electromagnetic field in second quantization*

Historically, the electromagnetic field was the first example of second quantization. The quantum nature of the radiation field, and the associated concept of photons play a crucial role in the theory of interactions between matter and light. In the applications in this book we shall, however, treat the electromagnetic field classically.

The quantization of the electromagnetic field is based on the observation that the eigenmodes of the classical field can be thought of as a collection of harmonic oscillators. These are then quantized. In the free field case the electromagnetic field is completely determined by the vector potential $\mathbf{A}(\mathbf{r}, t)$ in a specific gauge. Traditionally, the transversality condition $\nabla\cdot\mathbf{A} = 0$ is chosen, in which case $\mathbf{A}$ is denoted the radiation field, and we have

$$\mathbf{B} = \nabla \times \mathbf{A} \qquad\qquad \nabla\cdot\mathbf{A} = 0$$
$$\mathbf{E} = -\partial_t\mathbf{A} \qquad \nabla^2\mathbf{A} - \frac{1}{c^2}\,\partial_t^2\mathbf{A} = 0. \qquad (1.78)$$

We assume periodic boundary conditions for $\mathbf{A}$ enclosed in a huge box taken to be a cube of volume $\mathcal{V}$ and hence side length $L = \sqrt[3]{\mathcal{V}}$. The dispersion law is $\omega_{\mathbf{k}} = kc$ and the two-fold polarization of the field is described by polarization vectors $\boldsymbol{\epsilon}_\lambda$, $\lambda = 1, 2$. The normalized eigenmodes $\mathbf{u}_{\mathbf{k},\lambda}(\mathbf{r}, t)$ of the wave equation (1.78) are seen to be

$$\mathbf{u}_{\mathbf{k},\lambda}(\mathbf{r}, t) = \tfrac{1}{\sqrt{\mathcal{V}}}\boldsymbol{\epsilon}_\lambda e^{i(\mathbf{k}\cdot\mathbf{r}-\omega_{\mathbf{k}}t)}, \quad \lambda = 1, 2, \quad \omega_{\mathbf{k}} = ck$$
$$k_x = \tfrac{2\pi}{L}n_x, \quad n_x = 0, \pm 1, \pm 2, \dots \text{(same for } y \text{ and } z\text{)}. \tag{1.79}$$

The set $\{\boldsymbol{\epsilon}_1, \boldsymbol{\epsilon}_2, \mathbf{k}/k\}$ forms a right-handed orthonormal basis set.

Bearing in mind that $\mathbf{A}$ is real, a general solution can be written as an expansion in the eigenmodes:

$$\mathbf{A}(\mathbf{r}, t) = \frac{1}{\sqrt{\mathcal{V}}} \sum_{\mathbf{k}} \sum_{\lambda=1,2} \left(A_{k,\lambda} e^{i(\mathbf{k}\cdot\mathbf{r}-\omega_{\mathbf{k}}t)} + A_{k,\lambda}^* e^{-i(\mathbf{k}\cdot\mathbf{r}-\omega_{\mathbf{k}}t)} \right) \boldsymbol{\epsilon}_\lambda, \tag{1.80}$$

where $A_{k,\lambda}$ are the complex expansion coefficients. We now turn to the Hamiltonian H of the system, which is simply the field energy known from electromagnetism. Using Eq. (1.78) we can express H in terms of the radiation field $\mathbf{A}$,

$$H = \frac{1}{2} \int d\mathbf{r} \left(\epsilon_0 |\mathbf{E}|^2 + \frac{1}{\mu_0} |\mathbf{B}|^2 \right). \tag{1.81}$$

Inserting the expansion Eq. (1.80) and using Parceval's theorem the Hamiltonian becomes

$$H = 2\epsilon_0 \sum_{\mathbf{k},\lambda} \omega_{\mathbf{k}}^2 |A_{\mathbf{k},\lambda}|^2 = 2\epsilon_0 \sum_{\mathbf{k},\lambda} \omega_{\mathbf{k}}^2 \left(|A_{\mathbf{k},\lambda}^R|^2 + |A_{\mathbf{k},\lambda}^I|^2 \right), \tag{1.82}$$

where we have introduced the real and the imaginary parts, $A_{\mathbf{k},\lambda} = A_{\mathbf{k},\lambda}^R + iA_{\mathbf{k},\lambda}^I$. Merging the time dependence and the coefficients as $A_{\mathbf{k},\lambda}(t) = A_{\mathbf{k},\lambda}e^{-i\omega_{\mathbf{k}}t}$, the time dependence for the real and imaginary parts are seen to be

$$\dot{A}_{\mathbf{k},\lambda}^R = +\omega_{\mathbf{k}} A_{\mathbf{k},\lambda}^I, \quad \dot{A}_{\mathbf{k},\lambda}^I = -\omega_{\mathbf{k}} A_{\mathbf{k},\lambda}^R. \tag{1.83}$$

From Eqs. (1.82) and (1.83) it thus follows that, up to some normalization constants, $A_{\mathbf{k},\lambda}^R$ and $A_{\mathbf{k},\lambda}^I$ are conjugate variables: $\frac{\partial H}{\partial A_{\mathbf{k},\lambda}^R} = -4\epsilon_0\omega_{\mathbf{k}}\dot{A}_{\mathbf{k},\lambda}^I$ and $\frac{\partial H}{\partial A_{\mathbf{k},\lambda}^I} = +4\epsilon_0\omega_{\mathbf{k}}\dot{A}_{\mathbf{k},\lambda}^R$. Proper normalized conjugate variables $Q_{\mathbf{k},\lambda}$ and $P_{\mathbf{k},\lambda}$ are therefore introduced:

$$\left.\begin{matrix} Q_{\mathbf{k},\lambda} \equiv 2\sqrt{\epsilon_0}A_{\mathbf{k},\lambda}^R \\[2mm] P_{\mathbf{k},\lambda} \equiv 2\omega_{\mathbf{k}}\sqrt{\epsilon_0}A_{\mathbf{k},\lambda}^I \end{matrix}\right\} \Rightarrow \begin{cases} H = \sum_{\mathbf{k},\lambda} \frac{1}{2}\left(P_{\mathbf{k},\lambda}^2 + \omega_{\mathbf{k}}^2 Q_{\mathbf{k},\lambda}^2 \right), \\[2mm] \dot{Q}_{\mathbf{k},\lambda} = P_{\mathbf{k},\lambda}, \quad \dot{P}_{\mathbf{k},\lambda} = -\omega_{\mathbf{k}}^2 Q_{\mathbf{k},\lambda}, \\[2mm] \frac{\partial H}{\partial Q_{\mathbf{k},\lambda}} = -\dot{P}_{\mathbf{k},\lambda}, \quad \frac{\partial H}{\partial P_{\mathbf{k},\lambda}} = \dot{Q}_{\mathbf{k},\lambda}. \end{cases} \tag{1.84}$$

This ends the proof that the radiation field $\mathbf{A}$ can be thought of as a collection of classical harmonic oscillator, where each oscillator is characterized by the conjugate variables $Q_{\mathbf{k},\lambda}$ and $P_{\mathbf{k},\lambda}$. Quantization is now introduced by imposing the condition for the commutator of the conjugate variables and subsequently introducing the second quantized Bose operators $a_{\mathbf{k},\lambda}^{\dagger}$ for each quantized oscillator:

$$[P_{\mathbf{k},\lambda}, Q_{\mathbf{k},\lambda}] = \frac{\hbar}{i} \Rightarrow \begin{cases} H = \sum_{\mathbf{k},\lambda} \hbar\omega_{\mathbf{k}}(a_{\mathbf{k},\lambda}^{\dagger} a_{\mathbf{k},\lambda} + \frac{1}{2}), \quad [a_{\mathbf{k},\lambda}, a_{\mathbf{k},\lambda}^{\dagger}] = 1, \\[2ex] Q_{\mathbf{k},\lambda} = \sqrt{\frac{\hbar}{2\omega_{\mathbf{k}}}}\,(a_{\mathbf{k},\lambda}^{\dagger} + a_{\mathbf{k},\lambda}), \quad P_{\mathbf{k},\lambda} = \sqrt{\frac{\hbar\omega_{\mathbf{k}}}{2}}\,i(a_{\mathbf{k},\lambda}^{\dagger} - a_{\mathbf{k},\lambda}). \end{cases}$$

(1.85)

To obtain the final expression for $\mathbf{A}$ in second quantization we simply express $\mathbf{A}_{\mathbf{k},\lambda}$ in terms of $P_{\mathbf{k},\lambda}$ and $Q_{\mathbf{k},\lambda}$, which in turn is expressed in terms of $a_{\mathbf{k},\lambda}^{\dagger}$ and $a_{\mathbf{k},\lambda}$:

$$A_{\mathbf{k},\lambda} = A_{\mathbf{k},\lambda}^{R} + iA_{\mathbf{k},\lambda}^{I} \rightarrow \frac{Q_{\mathbf{k},\lambda}}{2\sqrt{\epsilon_0}} + i\frac{P_{\mathbf{k},\lambda}}{2\omega_{\mathbf{k}}\sqrt{\epsilon_0}} = \sqrt{\frac{\hbar}{2\epsilon_0\omega_{\mathbf{k}}}}\,a_{\mathbf{k},\lambda}, \quad \text{and} \quad A_{\mathbf{k},\lambda}^{*} \rightarrow \sqrt{\frac{\hbar}{2\epsilon_0\omega_{\mathbf{k}}}}\,a_{\mathbf{k},\lambda}^{\dagger}.$$

(1.86)

Substituting this into the expansion (1.80) our final result is:

$$\mathbf{A}(\mathbf{r},t) = \frac{1}{\sqrt{\mathcal{V}}} \sum_{\mathbf{k}} \sum_{\lambda=1,2} \sqrt{\frac{\hbar}{2\epsilon_0\omega_{\mathbf{k}}}} \left(a_{\mathbf{k},\lambda} e^{i(\mathbf{k}\cdot\mathbf{r} - \omega_{\mathbf{k}}t)} + a_{\mathbf{k},\lambda}^{\dagger} e^{-i(\mathbf{k}\cdot\mathbf{r} - \omega_{\mathbf{k}}t)} \right) \boldsymbol{\epsilon}_{\lambda}.$$

(1.87)

1.4.3 Operators for kinetic energy, spin, density and current

In the following we establish the second quantization representation of the four important single-particle operators associated with kinetic energy, spin, particle density, and particle current density.

First, we study the kinetic energy operator T, which is independent of spin and hence diagonal in the spin indices. In first quantization it has the representations

$$T_{\mathbf{r},\sigma'\sigma} = -\frac{\hbar^2}{2m}\nabla_{\mathbf{r}}^2\,\delta_{\sigma',\sigma}, \quad \text{real space representation,} \tag{1.88a}$$

$$\langle \mathbf{k}'\sigma'|T|\mathbf{k}\sigma\rangle = \frac{\hbar^2 k^2}{2m}\delta_{\mathbf{k}',\mathbf{k}}\,\delta_{\sigma',\sigma}, \quad \text{momentum representation.} \tag{1.88b}$$

Its second quantized forms with spin indices follow directly from Eqs. (1.61) and (1.71)

$$T = \sum_{\mathbf{k},\sigma} \frac{\hbar^2 k^2}{2m} a_{\mathbf{k},\sigma}^{\dagger} a_{\mathbf{k},\sigma} = -\frac{\hbar^2}{2m} \sum_{\sigma} \int d\mathbf{r}\, \Psi_{\sigma}^{\dagger}(\mathbf{r}) \left(\nabla_{\mathbf{r}}^2 \Psi_{\sigma}(\mathbf{r}) \right).$$

(1.89)

The second equality can also be proven directly by inserting $\Psi^{\dagger}(\mathbf{r})$ and $\Psi(\mathbf{r})$ from Eq. (1.72). For particles with charge q, a magnetic field can be included in the expression for the kinetic energy by substituting the canonical momentum $\mathbf{p}$ with the kinetic momentum[4] $\mathbf{p} - q\mathbf{A}$,

---

[4]In analytical mechanics $\mathbf{A}$ enters through the Lagrangian: $L = \frac{1}{2}mv^2 - V + q\mathbf{v}\cdot\mathbf{A}$, since this by the Euler–Lagrange equations yields the Lorentz force. But then $\mathbf{p} = \partial L/\partial\mathbf{v} = m\mathbf{v} + q\mathbf{A}$, and by means of a Legendre transform we get $H(\mathbf{r},\mathbf{p}) = \mathbf{p}\cdot\mathbf{v} - L(\mathbf{r},\mathbf{v}) = \frac{1}{2}mv^2 + V = \frac{1}{2m}(\mathbf{p} - q\mathbf{A})^2 + V$. Considering infinitesimal variations $\delta\mathbf{A}$ we get $\delta H = H(\mathbf{A} + \delta\mathbf{A}) - H(\mathbf{A}) = -q\mathbf{v}\cdot\delta\mathbf{A} = -q\int d\mathbf{r}\,\mathbf{J}\cdot\delta\mathbf{A}$, an expression used to find $\mathbf{J}$.

$$T_{\mathbf{A}} = \frac{1}{2m} \sum_{\sigma} \int d\mathbf{r} \, \Psi_{\sigma}^{\dagger}(\mathbf{r}) \left(\frac{\hbar}{i} \nabla_{\mathbf{r}} - q\mathbf{A} \right)^2 \Psi_{\sigma}(\mathbf{r}). \qquad (1.90)$$

Next, we treat the spin operator $\mathbf{s}$ for electrons. In first quantization it is given by the Pauli matrices

$$\mathbf{s} = \frac{\hbar}{2}\boldsymbol{\tau}, \quad \text{with } \boldsymbol{\tau} = \left\{ \begin{pmatrix} 0 & 1 \\ 1 & 0 \end{pmatrix}, \begin{pmatrix} 0 & -i \\ i & 0 \end{pmatrix}, \begin{pmatrix} 1 & 0 \\ 0 & -1 \end{pmatrix} \right\}. \qquad (1.91)$$

To obtain the second quantized operator we pull out the spin index explicitly in the basis kets, $|\nu\rangle = |\mu\rangle|\sigma\rangle$, and obtain with fermion operators the following vector expression,

$$\mathbf{S} = \sum_{\mu\sigma\mu'\sigma'} \langle \mu'|\langle \sigma'|\mathbf{s}|\sigma\rangle|\mu\rangle \, c_{\mu'\sigma'}^{\dagger} c_{\mu\sigma} = \frac{\hbar}{2} \sum_{\mu} \sum_{\sigma'\sigma} \langle \sigma'|(\tau^x, \tau^y, \tau^z)|\sigma\rangle \, c_{\mu\sigma'}^{\dagger} c_{\mu\sigma}, \qquad (1.92a)$$

with components

$$\left(S^x, \, S^y, \, S^z \right) = \frac{\hbar}{2} \sum_{\mu} \left([c_{\mu\downarrow}^{\dagger} c_{\mu\uparrow} + c_{\mu\uparrow}^{\dagger} c_{\mu\downarrow}], \, i[c_{\mu\downarrow}^{\dagger} c_{\mu\uparrow} - c_{\mu\uparrow}^{\dagger} c_{\mu\downarrow}], \, [c_{\mu\uparrow}^{\dagger} c_{\mu\uparrow} - c_{\mu\downarrow}^{\dagger} c_{\mu\downarrow}] \right). \qquad (1.92b)$$

We then turn to the particle density operator $\rho(\mathbf{r})$. In first quantization the fundamental interpretation of the wave function $\psi_{\mu,\sigma}(\mathbf{r})$ gives us $\rho_{\mu,\sigma}(\mathbf{r}) = |\psi_{\mu,\sigma}(\mathbf{r})|^2$ which can also be written as $\rho_{\mu,\sigma}(\mathbf{r}) = \int d\mathbf{r}' \, \psi_{\mu,\sigma}^*(\mathbf{r}')\delta(\mathbf{r}' - \mathbf{r})\psi_{\mu,\sigma}(\mathbf{r}')$, and thus the density operator for spin σ is given by $\rho_{\sigma}(\mathbf{r}) = \delta(\mathbf{r}' - \mathbf{r})$. In second quantization this combined with Eq. (1.61) yields

$$\rho_{\sigma}(\mathbf{r}) = \int d\mathbf{r}' \, \Psi_{\sigma}^{\dagger}(\mathbf{r}')\delta(\mathbf{r}' - \mathbf{r})\Psi_{\sigma}(\mathbf{r}') = \Psi_{\sigma}^{\dagger}(\mathbf{r})\Psi_{\sigma}(\mathbf{r}). \qquad (1.93)$$

From Eq. (1.73) the momentum representation of this is found to be

$$\rho_{\sigma}(\mathbf{r}) = \frac{1}{\mathcal{V}} \sum_{kk'} e^{i(\mathbf{k}-\mathbf{k}')\cdot\mathbf{r}} a_{k'\sigma}^{\dagger} a_{k\sigma} = \frac{1}{\mathcal{V}} \sum_{kq} e^{-i\mathbf{q}\cdot\mathbf{r}} a_{k+q\sigma}^{\dagger} a_{k\sigma} = \frac{1}{\mathcal{V}} \sum_{q} \left[\sum_{k} a_{k\sigma}^{\dagger} a_{k+q\sigma} \right] e^{i\mathbf{q}\cdot\mathbf{r}}, \qquad (1.94)$$

where the momentum transfer $\mathbf{q} = \mathbf{k}' - \mathbf{k}$ has been introduced. The explicit expression for the Fourier transform $\rho_{\sigma}(\mathbf{q})$ of $\rho_{\sigma}(\mathbf{r})$ is

$$\rho_{\sigma}(\mathbf{q}) = \sum_{k} a_{k\sigma}^{\dagger} a_{k+q\sigma}. \qquad (1.95)$$

The fourth and last operator to be treated is the particle current density operator $\mathbf{J}(\mathbf{r})$. It is related to the particle density operator $\rho(\mathbf{r})$ through the continuity equation $\partial_t \rho + \nabla \cdot \mathbf{J} = 0$. This relationship can be used to actually define $\mathbf{J}$. However, we shall take a more general approach based on analytical mechanics, see Eq. (1.90) and the associated footnote. This allows us in a simple way to take the magnetic field, given by the vector potential $\mathbf{A}$, into account. By analytical mechanics it is found that variations δH in the Hamiltonian function due to variations $\delta \mathbf{A}$ in the vector potential is given by

$$\delta H = -q \int d\mathbf{r}\, \mathbf{J} \cdot \delta \mathbf{A}. \qquad (1.96)$$

We use this expression with H given by the kinetic energy Eq. (1.90). Variations due to a varying parameter are calculated as derivatives if the parameter appears as a simple factor. Expanding the square in Eq. (1.90) and writing only the $\mathbf{A}$-dependent terms of the integrand, $-\Psi_\sigma^\dagger(\mathbf{r}) \frac{q\hbar}{2mi}[\nabla \cdot \mathbf{A} + \mathbf{A} \cdot \nabla]\Psi_\sigma(\mathbf{r}) + \frac{q^2}{2m}\mathbf{A}^2 \Psi_\sigma^\dagger(\mathbf{r})\Psi_\sigma(\mathbf{r})$, reveals one term where ∇ is acting on $\mathbf{A}$. By partial integration this ∇ is shifted to $\Psi^\dagger(\mathbf{r})$, and we obtain

$$H = T + \sum_\sigma \int d\mathbf{r} \left\{ \frac{q\hbar}{2mi} \mathbf{A} \cdot \left[\left(\nabla \Psi_\sigma^\dagger(\mathbf{r}) \right) \Psi_\sigma(\mathbf{r}) - \Psi_\sigma^\dagger(\mathbf{r}) \left(\nabla \Psi_\sigma(\mathbf{r}) \right) \right] + \frac{q^2}{2m} \mathbf{A}^2 \Psi_\sigma^\dagger(\mathbf{r})\Psi_\sigma(\mathbf{r}) \right\}.$$
$$(1.97)$$

The variations of Eq. (1.96) can in Eq. (1.97) be performed as derivatives and $\mathbf{J}$ is immediately read off as the prefactor to $\delta \mathbf{A}$. The two terms in the current density operator are denoted the paramagnetic and the diamagnetic term, $\mathbf{J}^\nabla$ and $\mathbf{J}^A$, respectively:

$$\mathbf{J}_\sigma(\mathbf{r}) = \mathbf{J}_\sigma^\nabla(\mathbf{r}) + \mathbf{J}_\sigma^A(\mathbf{r}), \qquad (1.98a)$$

$$\text{paramagnetic}: \quad \mathbf{J}_\sigma^\nabla(\mathbf{r}) = \frac{\hbar}{2mi} \left[\Psi_\sigma^\dagger(\mathbf{r}) \left(\nabla \Psi_\sigma(\mathbf{r}) \right) - \left(\nabla \Psi_\sigma^\dagger(\mathbf{r}) \right) \Psi_\sigma(\mathbf{r}) \right], \quad (1.98b)$$

$$\text{diamagnetic}: \quad \mathbf{J}_\sigma^A(\mathbf{r}) = -\frac{q}{m} \mathbf{A}(\mathbf{r})\Psi_\sigma^\dagger(\mathbf{r})\Psi_\sigma(\mathbf{r}). \qquad (1.98c)$$

The momentum representation of $\mathbf{J}$ is found in complete analogy with that of ρ

$$\mathbf{J}_\sigma^\nabla(\mathbf{r}) = \frac{\hbar}{m\mathcal{V}} \sum_{\mathbf{kq}} (\mathbf{k} + \tfrac{1}{2}\mathbf{q}) e^{i\mathbf{q}\cdot\mathbf{r}} a_{\mathbf{k}\sigma}^\dagger a_{\mathbf{k}+\mathbf{q},\sigma}, \qquad \mathbf{J}_\sigma^A(\mathbf{r}) = \frac{-q}{m\mathcal{V}} \mathbf{A}(\mathbf{r}) \sum_{\mathbf{kq}} e^{i\mathbf{q}\cdot\mathbf{r}} a_{\mathbf{k}\sigma}^\dagger a_{\mathbf{k}+\mathbf{q},\sigma}.$$
$$(1.99)$$

The expression for $\mathbf{J}$ in an arbitrary basis is treated in Exercise 1.2.

1.4.4   The Coulomb interaction in second quantization

The Coulomb interaction operator V is a two-particle operator not involving spin, and is thus diagonal in the spin indices of the particles. Using the same reasoning that led us from Eq. (1.61) to Eq. (1.71) we can go directly from Eq. (1.62) to the following quantum field operator form of V:

$$V = \frac{1}{2} \sum_{\sigma_1 \sigma_2} \int d\mathbf{r}_1 d\mathbf{r}_2\, \frac{e_0^2}{|\mathbf{r}_2 - \mathbf{r}_1|}\, \Psi_{\sigma_1}^\dagger(\mathbf{r}_1)\Psi_{\sigma_2}^\dagger(\mathbf{r}_2)\Psi_{\sigma_2}(\mathbf{r}_2)\Psi_{\sigma_1}(\mathbf{r}_1), \qquad (1.100a)$$

where we have introduced the abbreviation

$$e_0^2 = \frac{e^2}{4\pi\epsilon_0}. \qquad (1.100b)$$

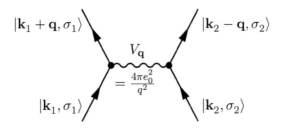

FIG. 1.8. A graphical representation of the Coulomb interaction in second quanti-
zation. Under momentum and spin conservation the incoming states $|\mathbf{k}_1, \sigma_1\rangle$ and
$|\mathbf{k}_2, \sigma_2\rangle$ are, with probability amplitude $V_{\mathbf{q}}$, scattered into the outgoing states
$|\mathbf{k}_1 + \mathbf{q}, \sigma_1\rangle$ and $|\mathbf{k}_2 - \mathbf{q}, \sigma_2\rangle$.

We can also write the Coulomb interaction directly in the momentum basis by us-
ing Eqs. (1.30b) and (1.62) with $|\nu\rangle = |\mathbf{k}, \sigma\rangle$ and $\psi_{\mathbf{k},\sigma}(\mathbf{r}) = \frac{1}{\sqrt{\mathcal{V}}} e^{i\mathbf{k}\cdot\mathbf{r}} \chi_\sigma$. We can
interpret the Coulomb matrix element as describing a transition from an initial state
$|\mathbf{k}_1\sigma_1, \mathbf{k}_2\sigma_2\rangle$ to a final state $|\mathbf{k}_3\sigma_1, \mathbf{k}_4\sigma_2\rangle$ without flipping any spin, and we obtain

$$V = \frac{1}{2} \sum_{\sigma_1\sigma_2} \sum_{\substack{\mathbf{k}_1\mathbf{k}_2 \\ \mathbf{k}_3\mathbf{k}_4}} \langle \mathbf{k}_3\sigma_1, \mathbf{k}_4\sigma_2 | V | \mathbf{k}_1\sigma_1, \mathbf{k}_2\sigma_2 \rangle \, a^\dagger_{\mathbf{k}_3\sigma_1} a^\dagger_{\mathbf{k}_4\sigma_2} a_{\mathbf{k}_2\sigma_2} a_{\mathbf{k}_1\sigma_1} \tag{1.101}$$

$$= \frac{1}{2} \sum_{\sigma_1\sigma_2} \sum_{\substack{\mathbf{k}_1\mathbf{k}_2 \\ \mathbf{k}_3\mathbf{k}_4}} \left(\frac{e_0^2}{\mathcal{V}^2} \int d\mathbf{r}_1 d\mathbf{r}_2 \, \frac{e^{i(\mathbf{k}_1\cdot\mathbf{r}_1 + \mathbf{k}_2\cdot\mathbf{r}_2 - \mathbf{k}_3\cdot\mathbf{r}_1 - \mathbf{k}_4\cdot\mathbf{r}_2)}}{|\mathbf{r}_2 - \mathbf{r}_1|} \right) a^\dagger_{\mathbf{k}_3\sigma_1} a^\dagger_{\mathbf{k}_4\sigma_2} a_{\mathbf{k}_2\sigma_2} a_{\mathbf{k}_1\sigma_1}.$$

Since $\mathbf{r}_2 - \mathbf{r}_1$ is the relevant variable for the interaction, the exponential is rewritten
as $e^{i[(\mathbf{k}_1 - \mathbf{k}_3)\cdot\mathbf{r}_1 + (\mathbf{k}_2 - \mathbf{k}_4)\cdot\mathbf{r}_2]} = e^{i(\mathbf{k}_1 - \mathbf{k}_3 + \mathbf{k}_2 - \mathbf{k}_4)\cdot\mathbf{r}_1} e^{i(\mathbf{k}_2 - \mathbf{k}_4)\cdot(\mathbf{r}_2 - \mathbf{r}_1)}$ leaving us with two
integrals, which with the definitions $\mathbf{q} \equiv \mathbf{k}_2 - \mathbf{k}_4$ and $\mathbf{r} \equiv \mathbf{r}_2 - \mathbf{r}_1$ become

$$\int d\mathbf{r}_1 \, e^{i(\mathbf{k}_1 - \mathbf{k}_3 + \mathbf{q})\cdot\mathbf{r}_1} = \mathcal{V} \, \delta_{\mathbf{k}_3, \mathbf{k}_1 + \mathbf{q}}, \qquad V_{\mathbf{q}} \equiv \int d\mathbf{r} \, \frac{e_0^2}{r} e^{i\mathbf{q}\cdot\mathbf{r}} = \frac{4\pi e_0^2}{q^2}. \tag{1.102}$$

These integrals express the Fourier transform of the Coulomb interaction[5] and the
explicit momentum conservation obeyed by the interaction. The momenta $\mathbf{k}_3$ and $\mathbf{k}_4$
of the final states can now be written as $\mathbf{k}_3 = \mathbf{k}_1 + \mathbf{q}$ and $\mathbf{k}_4 = \mathbf{k}_2 - \mathbf{q}$. The final
second quantized form of the Coulomb interaction in momentum space is

$$V = \frac{1}{2\mathcal{V}} \sum_{\sigma_1\sigma_2} \sum_{\mathbf{k}_1\mathbf{k}_2\mathbf{q}} V_{\mathbf{q}} \, a^\dagger_{\mathbf{k}_1 + \mathbf{q}\sigma_1} a^\dagger_{\mathbf{k}_2 - \mathbf{q}\sigma_2} a_{\mathbf{k}_2\sigma_2} a_{\mathbf{k}_1\sigma_1}. \tag{1.103}$$

We shall study this operator thoroughly in Section 2.2 in connection with the interacting electron gas. Here, in Fig. 1.8, we just show a graphical representation of the
operator.

[5] We show in Exercise 1.6 how to calculate the Fourier transform $V_{\mathbf{q}}^{k_s}$ of the Yukawa potential
$V^{k_s}(\mathbf{r}) = (e_0^2/r) e^{-k_s r}$. The result is $V_{\mathbf{q}}^{k_s} = 4\pi e_0^2/(q^2 + k_s^2)$ from which Eq. (1.102) follows by setting
$k_s = 0$.

In Eqs. (1.100a) and (1.103) we have seen the explicit form of the second quantized Coulomb interaction in real space and in **k**-space. These equations are easily generalized to any basis $|\nu, \sigma\rangle$, where the spin appears explicitly. A straightforward generalization of Eq. (1.101) yields

$$V = \frac{1}{2} \sum_{\substack{\sigma_1 \sigma_2 \\ \nu_3 \nu_4}} \sum_{\nu_1 \nu_2} \langle \nu_3 \sigma_1, \nu_4 \sigma_2| \frac{e_0^2}{|\mathbf{r}_2 - \mathbf{r}_1|} |\nu_1 \sigma_1, \nu_2 \sigma_2\rangle \, a_{\nu_3 \sigma_1}^\dagger a_{\nu_4 \sigma_2}^\dagger a_{\nu_2 \sigma_2} a_{\nu_1 \sigma_1}, \qquad (1.104)$$

where the matrix element is found by use of Eqs. (1.30b) and (1.62).

1.4.5 Basis states for systems with different kinds of particles

In the previous sections we have derived different fermion and boson operators. But so far we have not treated systems where different kinds of particles are coupled. In this book one important example of such a system is the fermionic electrons in a metal interacting with the bosonic lattice vibrations (phonons). We study this system in Chapter 3. Another example is electrons interacting with the photon field. Here we will briefly clarify how to construct the basis set for such composed systems in general.

Let us for simplicity just study two different kinds of particles. The arguments are easily generalized to include more complicated systems. The starting point is the case where the two kinds of particles do not interact with each other. Let the first kind of particles be described by the Hamiltonian H_1 and a complete set of basis states $\{|\nu\rangle\}$. Likewise we have H_2 and $\{|\mu\rangle\}$ for the second kind of particles. For the two decoupled systems an example of separate occupation number basis sets is

$$|\psi^{(1)}\rangle = |n_{\nu_1}, n_{\nu_2}, \ldots, n_{\nu_j}, \ldots\rangle, \qquad (1.105a)$$

$$|\psi^{(2)}\rangle = |n_{\mu_1}, n_{\mu_2}, \ldots, n_{\mu_j}, \ldots\rangle. \qquad (1.105b)$$

When a coupling H_{12} between the two systems is introduced, we need to enlarge the Hilbert space. The natural definition of basis states is the outer product states written as

$$\begin{aligned} |\psi\rangle &= |\psi^{(1)}\rangle |\psi^{(2)}\rangle \\ &= |n_{\nu_1}, n_{\nu_2}, \ldots, n_{\nu_j}, \ldots\rangle |n_{\mu_1}, n_{\mu_2}, \ldots, n_{\mu_l}, \ldots\rangle \\ &= |n_{\nu_1}, n_{\nu_2}, \ldots, n_{\nu_j}, \ldots; \, n_{\mu_1}, n_{\mu_2}, \ldots, n_{\mu_l}, \ldots\rangle. \end{aligned} \qquad (1.106)$$

In the last line all the occupation numbers are simply listed within the same ket but the two groups are separated by a semicolon. A general state $|\Phi\rangle$ can, of course, be any superposition of the basis states:

$$|\Phi\rangle = \sum_{\{\nu_j\}\{\mu_l\}} C_{\{\nu_j\},\{\mu_l\}} |n_{\nu_1}, n_{\nu_2}, \ldots, n_{\nu_j}, \ldots; \, n_{\mu_1}, n_{\mu_2}, \ldots, n_{\mu_l}, \ldots\rangle. \qquad (1.107)$$

As a concrete example we can write down the basis states for interacting electrons and photons in the momentum representation. The electronic basis states are the plane wave orbitals $|\mathbf{k}\sigma\rangle$ of Eq. (1.3), and the photon states are $|\mathbf{q}\lambda\rangle$ given in Eq. (1.79). We let $n_{\mathbf{k}\sigma}$ and $N_{\mathbf{q}\lambda}$ denote the occupation numbers for electrons and photons, respectively. A basis state $|\psi\rangle$ in this representation has the form:

$$|\psi\rangle = |n_{\mathbf{k}_1\sigma_1}, n_{\mathbf{k}_2\sigma_2}, \ldots, n_{\mathbf{k}_j\sigma_j}, \ldots; N_{\mathbf{q}_1\lambda_1}, N_{\mathbf{q}_2\lambda_2}, \ldots, N_{\mathbf{q}_l\lambda_l}, \ldots\rangle. \qquad (1.108)$$

1.5 Second quantization and statistical mechanics

The basic conjecture of statistical mechanics is that our ignorance about a system is best described by the following assumption: all states that comply with a given set of external constraints, e.g., have a definite total energy, are equally likely.[6] Furthermore, to give a meaningful definition to the term "all states", ergodicity is assumed. The ergodic assumption states that, as time evolves, a system visits all allowed states complying with the constraints. The time it takes for the system to visit all of the allowed phase space is the ergodicity time, which is assumed to be smaller than typical time scales of the observation.

Suppose we are interested in some small system connected to the outside world, the so-called reservoir, and assume that, taken as a whole, they constitute a closed system with total energy E_T. Let us call the energy of the small system E_s and that of the reservoir E_r, i.e., $E_T = E_s + E_r$. Based on the ergodicity assumption it is natural to conjecture that the probability for a subsystem to have a definite energy E_s is proportional to the number of ways that the subsystem can have that energy. The density of states is defined as $d(E) = dN(E)/dE$, where $N(E)$ is the number of states with energy less than E. We denote the density of states of the total system at a given total energy $d(E_T)$, while the small system and the reservoir have the densities of states $d_s(E_s)$ and $d_r(E_r)$, respectively. Since for a given small energy interval ΔE the number of states in the reservoir is much larger than the number of states in the smaller subsystem, the total density of states is dominated by that of the reservoir and hence $d(E_T) \approx d_r(E_r)$. The probability $P(E_s)$ for the subsystem to have energy E_s then follows from our fundamental assumption, namely that it is proportional to the number of reservoir states that have $E_T - E_s$, thus

$$P(E_s) \propto d_r(E_T - E_s)\,\Delta E. \qquad (1.109)$$

Now, we do not expect this probability to be dependent on the size of the reservoir, i.e., if we make it smaller by cutting it in two by some wall, nothing should happen to the state of the small system, provided of course that it is still much smaller than the new reservoir. This means that if we consider the ratio of two probabilities

$$\frac{P(E_s)}{P(E_s')} = \frac{d_r(E_T - E_s)}{d_r(E_T - E_s')}, \qquad (1.110)$$

it must only depend on the energies E_s and E_s' and neither on the total energy E_T nor on d_r. Furthermore, because the energy is only defined up to an additive constant, it

---

[6]This is consistent with a kinetic model where transitions between the different states are included. If a particular state is seldom visited, it also takes a long time to leave the state. Likewise, if there is large transition rate for going to particular state, there is also a large rate for leaving it. The net result is thus that the average time spend in a state is the same for all states.

can thus only depend on the difference $E_s - E'_s$. The only function $P(E)$ that satisfies the condition

$$\frac{P(E_s)}{P(E'_s)} = \frac{d_r(E_T - E_s)}{d_r(E_T - E'_s)} = f(E_s - E'_s),$$

(1.111)

is

$$P(E) \propto e^{-\beta E}.$$

(1.112)

We have thus arrived at the famous Boltzmann or Gibbs distribution, which of course should be normalized. In conclusion: from statistical mechanics we know that for both classical and quantum mechanical systems that are connected to a heat bath the probability for a given state s with energy E_s to be occupied is determined by Boltzmann distribution

$$P(E_s) = \frac{1}{Z} \exp(-\beta E_s),$$

(1.113)

where $\beta = 1/(k_{\mathrm{B}}T)$ is the inverse temperature and where the normalization factor Z is the partition function

$$Z = \sum_s \exp(-\beta E_s).$$

(1.114)

When we sum over states, we must sum over a set of states which cover the entire space of possible states, i.e. the basis set that we use to compute the energy must be a complete set. For a quantum system with many particles, the states s are, as we have seen, in general quite complicated to write down, and it is therefore an advantage to have a form which is independent of the choice of basis states. Also for a quantum system it is not clear what is meant by the energy of a given state, unless of course it is an eigenstate of the Hamiltonian. Therefore, the only meaningful interpretation of Eq. (1.114) is that the sum of states runs over eigenstates of the Hamiltonian. Using the basis states $|\nu\rangle$ defined by

$$H|\nu\rangle = E_\nu|\nu\rangle,$$

(1.115)

it is now quite natural to introduce the so-called density matrix operator ρ corresponding to the classical Boltzmann factor $e^{-\beta E}$,

$$\rho \equiv e^{-\beta H} = \sum_\nu |\nu\rangle e^{-\beta E_\nu} \langle \nu|.$$

(1.116)

We can thus write the expression Eq. (1.114) for the partition function as

$$Z = \sum_\nu \langle \nu|\rho|\nu\rangle = \mathrm{Tr}[\rho].$$

(1.117)

Likewise, the thermal average of any quantum operator A is easily expressed using the density matrix ρ. Following the elementary definition, we have

$$\langle A \rangle = \frac{1}{Z} \sum_\nu \langle \nu|A|\nu\rangle e^{-\beta E_\nu} = \frac{1}{Z} \mathrm{Tr}[\rho A] = \frac{\mathrm{Tr}[\rho A]}{\mathrm{Tr}[\rho]}.$$

(1.118)

Eqs. (1.117) and (1.118) are basis-independent expressions, since the sum over states is identified with the trace operation.[7] This is of course true whatever formalism we use to evaluate the trace. In first quantization the trace runs over for example the determinant basis, which in second quantization translates to the Fock space of the corresponding quantum numbers. For the canonical ensemble the trace is however restricted to run over states with a given number of particles.

For the grand canonical ensemble the number of particles is not conserved. The small system is allowed to exchange particles with the reservoir while keeping its average particle number constant, and we introduce a chemical potential μ of the reservoir to accommodate this constraint. Essentially, the result obtained from the canonical ensemble is carried over to the grand canonical ensemble by the substitution $H \to H - \mu N$, where N is the particle number operator. The corresponding density matrix ρ_G and partition function Z_G are defined as:

$$\rho_G \equiv e^{-\beta(H-\mu N)}, \quad Z_G = \mathrm{Tr}[\rho_G], \tag{1.119}$$

where the trace now includes states with any number of particles. Likewise, it is useful to introduce the Hamiltonian H_G corresponding to the grand canonical ensemble,

$$H_G \equiv H - \mu N. \tag{1.120}$$

Unfortunately, the symbol H is often used instead of H_G for the grand canonical Hamiltonian, so the reader must always carefully check whether H refers to the canonical or to the grand canonical ensemble. In this book, we shall for brevity write H in both cases. This ought not cause any problems, since most of the times we are working in the grand canonical ensemble, i.e., we include the term μN in the Hamiltonian.

The partition functions Z and Z_G are more than mere normalization factors. Their role as fundamental quantities in statistical mechanics is reflected by the following observation: Z and Z_G are directly related to the free energy $F \equiv U - TS$ in the canonical ensemble and to the thermodynamic potential $\Omega \equiv U - TS - \mu N$ in the grand canonical ensemble, respectively,

$$Z = e^{-\beta F}, \quad \text{with} \quad F = U - TS, \tag{1.121a}$$

$$Z_G = e^{-\beta\Omega}, \quad \text{with} \quad \Omega = U - TS - \mu N. \tag{1.121b}$$

Let us now study the free energy, which is minimal when the entropy is maximal. Recall that

$$F = U - TS = \langle H \rangle - TS. \tag{1.122}$$

In various approximation schemes, for example the mean field approximation in Chapter 4, we shall use the principle of minimizing the free energy. This is based on the following inequality

$$F \leq \langle H \rangle_0 - TS_0, \tag{1.123}$$

---

[7]Remember that if $t_\nu = \mathrm{Tr}\,[A]$ is the trace of A in the basis $|\nu\rangle$, then in the transformed basis $U|\nu\rangle$ we have $t_{U\nu} = \mathrm{Tr}\left[UAU^{-1}\right] = \mathrm{Tr}\left[AU^{-1}U\right] = \mathrm{Tr}\,[A] = t_\nu$. Here we have used that the trace is invariant under cyclic permutation, i.e., $\mathrm{Tr}\,[ABC] = \mathrm{Tr}\,[BCA]$.

where both $\langle H \rangle_0$ and S_0 are calculated in the approximation $\rho \approx \rho_0 = \exp(-\beta H_0)$, for example

$$\langle H \rangle_0 = \frac{\mathrm{Tr}[\rho_0 H]}{\mathrm{Tr}[\rho_0]}. \tag{1.124}$$

This inequality insures that by minimizing the free energy calculated from the approximate Hamiltonian, we are guaranteed to make the best possible approximation based on the trial Hamiltonian, H_0.

1.5.1 The distribution function for non-interacting fermions

As the temperature is raised from zero in a system of non-interacting fermions, the occupation number for the individual energy eigenstates begins to fluctuate rather than being constantly 0 or constantly 1. Using the grand canonical ensemble, we can derive the famous Fermi–Dirac distribution $n_{\mathrm{F}}(\varepsilon)$.

Consider the electron state $|\mathbf{k}\sigma\rangle$ with energy $\varepsilon_{\mathbf{k}}$. The state can contain either 0 or 1 electron. The average occupation $n_{\mathrm{F}}(\varepsilon_{\mathbf{k}})$ is therefore

$$n_{\mathrm{F}}(\varepsilon_{\mathbf{k}}) = \frac{\mathrm{Tr}[\rho_G n_{\mathbf{k}}]}{\mathrm{Tr}[\rho_G]} = \frac{\displaystyle\sum_{n_{\mathbf{k}}=0,1} n_{\mathbf{k}} e^{-\beta(n_{\mathbf{k}}\varepsilon_{\mathbf{k}} - \mu n_{\mathbf{k}})}}{\displaystyle\sum_{n_{\mathbf{k}}=0,1} e^{-\beta(n_{\mathbf{k}}\varepsilon_{\mathbf{k}} - \mu n_{\mathbf{k}})}} = \frac{0 + e^{-\beta(\varepsilon_{\mathbf{k}}-\mu)}}{1 + e^{-\beta(\varepsilon_{\mathbf{k}}-\mu)}} = \frac{1}{e^{\beta(\varepsilon_{\mathbf{k}}-\mu)} + 1}.$$

$$\tag{1.125}$$

We shall study the properties of the Fermi–Dirac distribution in Section 2.1.3. Note that the Fermi–Dirac distribution is defined in the grand canonical ensemble. The proper Hamiltonian is therefore $H_G = H - \mu N$. This is reflected in the single-particle energy variable. From Eq. (1.125) we see that the natural single-particle energy variable is not $\varepsilon_{\mathbf{k}}$ but rather $\xi_{\mathbf{k}}$ given by

$$\xi_{\mathbf{k}} \equiv \varepsilon_{\mathbf{k}} - \mu \tag{1.126}$$

For small excitation energies $\varepsilon_{\mathbf{k}}$ varies around μ whereas $\xi_{\mathbf{k}}$ varies around 0.

1.5.2 The distribution function for non-interacting bosons

Next we find the distribution function for non-interacting bosons. Again using the grand canonical ensemble we derive the equally famous Bose–Einstein distribution $n_{\mathrm{B}}(\varepsilon)$. It is derived like its fermionic counterpart, the Fermi–Dirac distribution $n_{\mathrm{F}}(\varepsilon)$.

Consider a bosonic state characterized by its fundamental energy $\varepsilon_{\mathbf{k}}$. The occupation number of the state can be any non-negative integer $n_{\mathbf{k}} = 0, 1, 2, \ldots$. In the grand canonical ensemble the average occupation number $n_{\mathrm{B}}(\varepsilon_{\mathbf{k}})$ is found by writing $\lambda_{\mathbf{k}} = e^{-\beta(\varepsilon_{\mathbf{k}}-\mu)}$ and using the formulas $\sum_{n=0}^{\infty} n\lambda^n = \lambda\frac{d}{d\lambda}\sum_{n=0}^{\infty} \lambda^n$ and $\sum_{n=0}^{\infty} \lambda^n = \frac{1}{1-\lambda}$:

$$n_{\mathrm{B}}(\varepsilon_{\mathbf{k}}) = \frac{\displaystyle\sum_{n_{\mathbf{k}}=0}^{\infty} n_{\mathbf{k}}\, e^{-\beta(n_{\mathbf{k}}\varepsilon_{\mathbf{k}}-\mu n_{\mathbf{k}})}}{\displaystyle\sum_{n_{\mathbf{k}}=0}^{\infty} e^{-\beta(n_{\mathbf{k}}\varepsilon_{\mathbf{k}}-\mu n_{\mathbf{k}})}} = \frac{\lambda_{\mathbf{k}}\dfrac{d}{d\lambda_{\mathbf{k}}}\displaystyle\sum_{n_{\mathbf{k}}=0}^{\infty}\lambda_{\mathbf{k}}^{n_{\mathbf{k}}}}{\displaystyle\sum_{n_{\mathbf{k}}=0}^{\infty}\lambda_{\mathbf{k}}^{n_{\mathbf{k}}}} = \frac{\dfrac{\lambda_{\mathbf{k}}}{(1-\lambda_{\mathbf{k}})^{2}}}{\dfrac{1}{1-\lambda_{\mathbf{k}}}} = \frac{1}{e^{\beta(\varepsilon_{\mathbf{k}}-\mu)}-1}.$$

$$(1.127)$$

The Bose–Einstein distribution differs from the Fermi–Dirac distribution in having -1 in the denominator instead of $+1$. Both distributions converge towards the classical Maxwell–Boltzmann distribution, $n_{\mathbf{k}} = e^{-\beta(\varepsilon_{\mathbf{k}}-\mu)}$, for very small occupation numbers, e.g. $n_{\mathbf{k}} \ll 1$, where the particular particle statistics is not felt very strongly.

1.6 Summary and outlook

In this chapter we have introduced second quantization, the representation of quantum mechanics we are going to use throughout this book. The basic concepts are the occupation number basis states and the fundamental creation and annihilation operators, $b_\nu^\dagger$ and b_ν in the bosonic case (see Eq. (1.41)), and $c_\nu^\dagger$ and c_ν in the fermionic case (see Eq. (1.52)). The intricate permutation symmetries are manifestly insured by the basic (anti-)commutator relations of these fundamental operators,

$$[b_{\nu_j}^\dagger, b_{\nu_k}^\dagger] = 0, \quad [b_{\nu_j}, b_{\nu_k}] = 0, \quad [b_{\nu_j}, b_{\nu_k}^\dagger] = \delta_{\nu_j \nu_k} \quad \text{bosons,}$$

$$\{c_{\nu_j}^\dagger, c_{\nu_k}^\dagger\} = 0, \quad \{c_{\nu_j}, c_{\nu_k}\} = 0, \quad \{c_{\nu_j}, c_{\nu_k}^\dagger\} = \delta_{\nu_j \nu_k} \quad \text{fermions.}$$

The main result of the chapter is the derivation of the general form of one- and two-particle operators, Eqs. (1.61) and (1.62) and Fig. 1.5,

$$T_{\mathrm{tot}} = \sum_{\nu_i, \nu_j} T_{\nu_i \nu_j}\, a_{\nu_i}^\dagger a_{\nu_j} \quad \text{and} \quad V_{\mathrm{tot}} = \frac{1}{2} \sum_{\substack{\nu_i \nu_j \\ \nu_k \nu_l}} V_{\nu_i \nu_j, \nu_k \nu_l}\, a_{\nu_i}^\dagger a_{\nu_j}^\dagger a_{\nu_l} a_{\nu_k}.$$

In fact, perhaps after some measure of acquaintance, this main result appears so simple and intuitively clear that one could choose to define quantum theory directly in second quantization rather than going the cumbersome way from first to second quantization. However, students usually learn basic quantum theory in first quantization, so for pedagogical reasons we have chosen to start from the usual first quantization representation.

In Section 1.4 we presented a number of specific examples of second quantization operators, and we got a first glimpse of how second quantization leads to a formulation of quantum physics in terms of creation and annihilation of particles and field quanta. Note in particular the trace-formular Eq. (1.118) for thermal averages of quantum operators,

$$\langle A \rangle = \frac{1}{Z} \sum_\nu \langle \nu|A|\nu\rangle e^{-\beta E_\nu} = \frac{1}{Z}\, \mathrm{Tr}[\rho A] = \frac{\mathrm{Tr}[\rho A]}{\mathrm{Tr}[\rho]}.$$

For textbooks on statistical mechanics, see Feynman (1972), Landau and Lifshitz (1982), or Kittel and Kroemer (2000).

In the following three chapters we shall get more acquainted with second quantization through studies of simplified stationary problems for non-interacting systems or systems where a given particle only interacts with the mean field of the other particles. First in Chapter 5 will the question be raised of how to treat time evolution in second quantization. With an answer to that question we can proceed with the very interesting but also rather difficult studies of the full time dependent dynamics of many-particle quantum systems.

Some useful textbooks on first quantization are Dirac (1989), Landau and Lifshitz (1977), or Merzbacher (1970). Some classic textbooks on second quantization are Mahan (1990), Fetter and Walecka (1971), and Abrikosov, Gorkov, and Dzyaloshinskii (1975). For good introductions to statistical physics the reader can consult Landau and Lifshitz (1982), Kittel and Kroemer (2000), and Feynman (1972).

2

THE ELECTRON GAS

The study of the interacting electron gas moving in a charge-compensating background of positively charged ions is central in this book. Not only is this system a model of the solids that surround us, such as metals, semiconductors, and insulators, but historically this system played a major role as testing ground for the development of quantum field theory. In this chapter we shall study the basic properties of this system using the formalism of time-independent second quantization as developed in Chap. 1. The main emphasis will be on the non-interacting electron gas, since it will be clear that we need to develop our theoretical tools further to deal with the electron-electron interactions in full.

Any atom in a metal consists of three parts: the positively charged heavy nucleus at the center, the light cloud of the many negatively charged core electrons tightly bound to the nucleus, and finally, the outermost few valence electrons. The nucleus with its core electrons is denoted an ion. The ion mass is denoted M, and if the atom has Z valence electrons the charge of the ion is $+Ze$. To a large extent the inner degrees of freedom of the ions do not play a significant role, leaving the center of mass coordinates R_j and total spin S_j of the ions as the only dynamical variables. In contrast to the core electrons, the Z valence electrons, each with mass m and charge $-e$, are often free to move away from their respective host atoms forming a gas of electrons swirling around among the ions. This is true for the alkali metals. The formation of a metal from N independent atoms is sketched in Fig. 2.1.

The Hamiltonian H of the system is written as the sum of kinetic and potential energy of the ionic system and the electronic system treated independently, and the Coulomb interaction between the two systems,

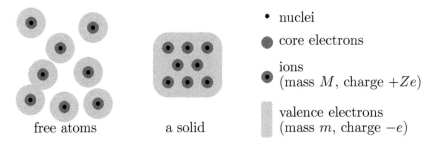

- nuclei
- core electrons

ions
(mass M, charge $+Ze$)

valence electrons
(mass m, charge $-e$)

free atoms   a solid

FIG. 2.1.   A sketch showing N free atoms merging into a metal. The ions are unchanged during the process where they end up by forming a periodic lattice. The valence electrons are freed from their host atoms and form an electron gas holding the ionic lattice together.

$$H = (T_{\text{ion}} + V_{\text{ion}-\text{ion}}) + (T_{\text{el}} + V_{\text{el}-\text{el}}) + V_{\text{el}-\text{ion}}. \tag{2.1}$$

The individual terms are easily written down in second quantization:

$$T_{\text{ion}} + V_{\text{ion}-\text{ion}} = \int d\mathbf{R}\ \Psi_{\text{ion}}^\dagger(\mathbf{R})\left(-\frac{\hbar^2}{2M}\nabla_\mathbf{R}^2\right)\Psi_{\text{ion}}(\mathbf{R}) \tag{2.2}$$

$$+\frac{1}{2}\int d\mathbf{R}_1 d\mathbf{R}_2\ \Psi_{\text{ion}}^\dagger(\mathbf{R}_1)\Psi_{\text{ion}}^\dagger(\mathbf{R}_2)\frac{Z^2 e_0^2}{|\mathbf{R}_1 - \mathbf{R}_2|}\Psi_{\text{ion}}(\mathbf{R}_2)\Psi_{\text{ion}}(\mathbf{R}_1),$$

$$T_{\text{el}} + V_{\text{el}-\text{el}} = \sum_\sigma \int d\mathbf{r}\ \Psi_\sigma^\dagger(\mathbf{r})\left(-\frac{\hbar^2}{2m}\nabla_\mathbf{r}^2\right)\Psi_\sigma(\mathbf{r}) \tag{2.3}$$

$$+\frac{1}{2}\sum_{\sigma_1\sigma_2}\int d\mathbf{r}_1 d\mathbf{r}_2\ \Psi_{\sigma_1}^\dagger(\mathbf{r}_1)\Psi_{\sigma_2}^\dagger(\mathbf{r}_2)\frac{e_0^2}{|\mathbf{r}_1 - \mathbf{r}_2|}\Psi_{\sigma_2}(\mathbf{r}_2)\Psi_{\sigma_1}(\mathbf{r}_1),$$

$$V_{\text{el}-\text{ion}} = \sum_\sigma \int d\mathbf{r}d\mathbf{R}\ \Psi_\sigma^\dagger(\mathbf{r})\Psi_{\text{ion}}^\dagger(\mathbf{R})\frac{(-Ze_0^2)}{|\mathbf{R} - \mathbf{r}|}\Psi_{\text{ion}}(\mathbf{R})\Psi_\sigma(\mathbf{r}). \tag{2.4}$$

Note that no double counting is involved in $V_{\text{el}-\text{ion}}$ since two different types of fields, $\Psi_\sigma^\dagger(\mathbf{r})$ and $\Psi_{\text{ion}}^\dagger(\mathbf{R})$ are involved, hence no factor $\frac{1}{2}$.

At zero temperature the ground state of the system is a periodic ion lattice held together by the cohesive forces of the surrounding electron gas. In principle it is possible in *ab initio* calculations to minimize the energy of the system and find the crystal structure and lattice parameters, i.e., the equilibrium positions $\mathbf{R}_j$ of the ions in the lattice. From the obtained ground state one can then study the various excitations of the system: phonons (ion vibrations), electron-hole excitations (single-particle excitations), plasmons (collective electronic charge density waves), magnons (spin waves), etc. In this book we will not plunge into such full fledged *ab initio* calculations. Two approximation schemes will be used instead. One is the phenomenological lattice approach. We take the experimental determination of the crystal structure, lattice parameters and elasticity constants as input to the theory, and from there calculate the electronic and phononic properties. The other approximation scheme, the so-called jellium model, is in fact an *ab initio* calculation where, however, the discrete nature of the ionic system is approximated by a positively charged, continuous and homogeneous fluid, the ion "jellium." Fortunately, most electronic and phononic properties of the system can be derived with good accuracy from the Hamiltonian describing the ion jellium combined with the electron gas.

2.1 The non-interacting electron gas

We first study the lattice model and the jellium model in the case of no electron-electron interaction. Later in Section 2.2 we attempt to include this interaction.

2.1.1 *Bloch theory of electrons in a static ion lattice*

Let us first consider the phenomenological lattice model. X-ray experiments show that the equilibrium positions of the ions form a periodic lattice. This lattice has an energy E_{latt} and an electrical potential $V_{\text{el-lat}}$ associated with it, both originating from a combination of T_{ion}, $V_{\text{ion}-\text{ion}}$, and $V_{\text{el}-\text{ion}}$ in the original Hamiltonian Eq. (2.1). At finite temperature the ions can vibrate about their equilibrium positions with the total electric field acting as the restoring force. As will be demonstrated in Chapter 3, these vibrations can be described in terms of quantized harmonic oscillators (much like the photon field of Section 1.4.2) giving rise to the concept of phonons. The non-interacting part of the phonon field is described by a Hamiltonian H_{ph}. Finally, the electrons are described by their kinetic energy T_{el}, their mutual interaction $V_{\text{el}-\text{el}}$, and their interaction with both the static part of the lattice, $V_{\text{el-lat}}$, and the vibrating part, i.e., the phonons, $V_{\text{el}-\text{ph}}$. The latter term must be there, since a vibrating ion is giving rise to a vibrating electrical potential influencing the electrons. Thus the Hamiltonian for the phenomenological lattice model changes H of Eq. (2.1) into

$$H = (E_{\text{latt}} + H_{\text{ph}}) + (T_{\text{el}} + V_{\text{el}-\text{el}}) + (V_{\text{el-lat}} + V_{\text{el}-\text{ph}}). \tag{2.5}$$

At zero temperature the ions are not vibrating except for their quantum mechanical zero point motion. Thus we can drop all the phonon-related terms of the Hamiltonian. If one furthermore neglects the electron-electron interaction (in Section 2.2 we study when this is reasonable) one arrives at the Hamiltonian H_{Bloch} used in Bloch's theory of non-interacting electrons moving in a static, periodic ion lattice:

$$H_{\text{Bloch}} = T_{\text{el}} + V_{\text{el-lat}}(\mathbf{r}), \quad \begin{cases} V_{\text{el-lat}}(\mathbf{r} + \mathbf{R}) = V_{\text{el-lat}}(\mathbf{r}) \\ \text{for any lattice vector } \mathbf{R}. \end{cases} \tag{2.6}$$

To solve the corresponding Schrödinger equation, and later the phonon problem, we have to understand the Fourier transform of periodic functions.

Let the static ion lattice be described by the ionic equilibrium positions $\mathbf{R}$ in terms of integer linear combinations of the lattice basis vectors $\{\mathbf{a}_1, \mathbf{a}_2, \mathbf{a}_3\}$:

$$\mathbf{R} = n_1\mathbf{a}_1 + n_2\mathbf{a}_2 + n_3\mathbf{a}_3, \quad n_1, n_2, n_3 \in \mathbb{Z}. \tag{2.7}$$

When working with periodic lattices, it is often convenient to Fourier transform from the direct space to $\mathbf{k}$-space, also known as the reciprocal space, RS. It is useful to introduce the reciprocal lattice, RL, in RS defined by

$$\text{RL} = \left\{ \mathbf{G} \in \text{RS} \,\middle|\, e^{i\mathbf{G}\cdot\mathbf{R}} = 1 \right\} \Rightarrow \mathbf{G} = m_1\mathbf{b}_1 + m_2\mathbf{b}_2 + m_3\mathbf{b}_3, \quad m_1, m_2, m_3 \in \mathbb{Z}, \tag{2.8}$$

where the basis vectors $\{\mathbf{b}_1, \mathbf{b}_2, \mathbf{b}_3\}$ in RL are defined as

$$\mathbf{b}_1 = 2\pi \frac{\mathbf{a}_2 \times \mathbf{a}_3}{\mathbf{a}_1 \cdot \mathbf{a}_2 \times \mathbf{a}_3}, \quad \mathbf{b}_2 = 2\pi \frac{\mathbf{a}_3 \times \mathbf{a}_1}{\mathbf{a}_2 \cdot \mathbf{a}_3 \times \mathbf{a}_1}, \quad \mathbf{b}_3 = 2\pi \frac{\mathbf{a}_1 \times \mathbf{a}_2}{\mathbf{a}_3 \cdot \mathbf{a}_1 \times \mathbf{a}_2}. \tag{2.9}$$

An important concept is the first Brillouin zone, FBZ, defined as all $\mathbf{k}$ in RS lying closer to $\mathbf{G} = 0$ than to any other reciprocal lattice vector $\mathbf{G} \neq 0$. Using vectors $\mathbf{k} \in$ FBZ, any wavevector $\mathbf{q} \in$ RS can be decomposed (the figure shows the FBZ for a 2D square lattice):

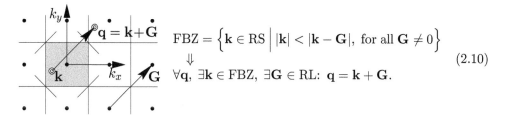

$$\text{FBZ} = \Big\{ \mathbf{k} \in \text{RS} \ \Big| \ |\mathbf{k}| < |\mathbf{k} - \mathbf{G}|, \text{ for all } \mathbf{G} \neq 0 \Big\}$$

$$\Downarrow \tag{2.10}$$

$$\forall \mathbf{q}, \ \exists \mathbf{k} \in \text{FBZ}, \ \exists \mathbf{G} \in \text{RL}: \ \mathbf{q} = \mathbf{k} + \mathbf{G}.$$

The Fourier transform of any function periodic in the lattice is as follows:

$$V(\mathbf{r} + \mathbf{R}) = V(\mathbf{r}), \text{ for all } \mathbf{R} \quad \Leftrightarrow \quad V(\mathbf{r}) = \sum_{\mathbf{G} \in \text{RL}} V_{\mathbf{G}} e^{i\mathbf{G} \cdot \mathbf{r}}. \tag{2.11}$$

The solution of the Schrödinger equation $H_{\text{Bloch}} \psi = E \psi$ can be found in the plane wave basis $|\mathbf{k}\sigma\rangle$, which separates in spatial part $e^{i\mathbf{k} \cdot \mathbf{r}}$ and a spin part χ_σ, e.g., $\chi_\uparrow = \binom{1}{0}$:

$$\psi_\sigma(\mathbf{r}) \equiv \frac{1}{\mathcal{V}} \sum_{\mathbf{k}'} c_{\mathbf{k}'} \, e^{i\mathbf{k}' \cdot \mathbf{r}} \chi_\sigma \quad \Rightarrow \quad \langle \mathbf{k}\sigma | H_{\text{Bloch}} | \psi_\sigma \rangle = \sum_{\mathbf{k}'} \Big(\varepsilon_{\mathbf{k}} \delta_{\mathbf{k}\mathbf{k}'} + \sum_{\mathbf{G}} V_{\mathbf{G}} \delta_{\mathbf{k}\mathbf{k}'+\mathbf{G}} \Big) c_{\mathbf{k}'},$$

$$\tag{2.12}$$

so the Schrödinger equation for a given $\mathbf{k}$ is

$$c_{\mathbf{k}} \varepsilon_{\mathbf{k}} + \sum_{\mathbf{G}} V_{\mathbf{G}} c_{\mathbf{k}-\mathbf{G}} = E \, c_{\mathbf{k}}. \tag{2.13}$$

We see that any given coefficient $c_{\mathbf{k}}$ only couples to other coefficients of the form $c_{\mathbf{k}+\mathbf{G}}$, i.e., each Schrödinger equation of the form Eq. (2.13) for $c_{\mathbf{k}}$ couples to an infinite, but countable, number of similar equations for $c_{\mathbf{k}-\mathbf{G}}$. Each such infinite family of equations has exactly one representative $\mathbf{k} \in \text{FBZ}$, while any $\mathbf{k}$ outside FBZ does not give rise to a new set of equations. The infinite family of equations generated by a given $\mathbf{k} \in \text{FBZ}$ gives rise to a discrete spectrum of eigenenergies $\varepsilon_{n\mathbf{k}}$, where $n \in \mathbb{N}$, as sketched in Fig. 2.2. The corresponding eigenfunctions $\psi_{n\mathbf{k}\sigma}$ are given by:

$$\psi_{n\mathbf{k}\sigma}(\mathbf{r}) = \frac{1}{\mathcal{V}} \sum_{\mathbf{G}} c_{\mathbf{k}+\mathbf{G}}^{(n)} \, e^{i(\mathbf{G}+\mathbf{k}) \cdot \mathbf{r}} \chi_\sigma = \Big(\frac{1}{\mathcal{V}} \sum_{\mathbf{G}} c_{\mathbf{G}}^{(n\mathbf{k})} \Big) e^{i\mathbf{k} \cdot \mathbf{r}} \chi_\sigma \equiv u_{n\mathbf{k}}(\mathbf{r}) \, e^{i\mathbf{k} \cdot \mathbf{r}} \chi_\sigma. \tag{2.14}$$

According to Eq. (2.11) the function $u_{n\mathbf{k}}(\mathbf{r})$ is periodic in the lattice, and thus we end up with Bloch's theorem:[8]

$$H_{\text{Bloch}} \psi_{n\mathbf{k}\sigma} = \varepsilon_{n\mathbf{k}} \psi_{n\mathbf{k}\sigma}, \quad \psi_{n\mathbf{k}\sigma}(\mathbf{r}) = u_{n\mathbf{k}}(\mathbf{r}) e^{i\mathbf{k} \cdot \mathbf{r}} \chi_\sigma, \quad \begin{cases} \mathbf{k} \in \text{FBZ}, \\ n \text{ is the band index}, \\ u_{n\mathbf{k}}(\mathbf{r}+\mathbf{R}) = u_{n\mathbf{k}}(\mathbf{r}). \end{cases}$$

$$\tag{2.15}$$

The eigenfunctions are seen to be plane waves modulated by a periodic function $u_{n\mathbf{k}}(\mathbf{r})$ having the same periodicity as the lattice. For many applications it turns out that the

---

[8] An alternative derivation of Bloch's theorem with emphasis on the group theoretic aspects builds on the translation operator $T_{\mathbf{R}}$, with $T_{\mathbf{R}} f(\mathbf{r}) \equiv f(\mathbf{r}+\mathbf{R})$. We get $[H, T_{\mathbf{R}}] = 0 \Rightarrow T_{\mathbf{R}} \psi = \lambda_{\mathbf{R}} \psi$ for an eigenstate ψ. Applying $T_{\mathbf{P}}$ after $T_{\mathbf{R}}$ leads to $\lambda_{\mathbf{P}} \lambda_{\mathbf{R}} = \lambda_{\mathbf{P}+\mathbf{R}} \Rightarrow \lambda_{\mathbf{R}} = e^{i\mathbf{k} \cdot \mathbf{R}} \Rightarrow \psi_{n\mathbf{k}}(\mathbf{r}) = u_{n\mathbf{k}}(\mathbf{r}) e^{i\mathbf{k} \cdot \mathbf{r}}$.

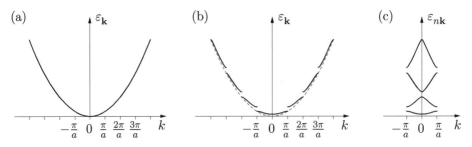

FIG. 2.2.   Bloch's theorem illustrated for a 1D lattice with lattice constant a. (a) The parabolic energy band for free electrons. (b) The Bloch bands viewed as a break-up of the parabolic free electron band in Brillouin zones (the extended zone scheme, $\mathbf{k} \in$ RS). (c) All wavevectors are equivalent to those in the FBZ, so it is most natural to displace all the energy branches into the FBZ (the reduced zone scheme, $\mathbf{k} \in$ FBZ).

Bloch electrons described by $\psi_{n\mathbf{k}\sigma}(\mathbf{r})$ can be approximated by plane waves if at the same time the electronic mass m is changed into a material-dependent effective mass m^*. We shall use this so-called effective mass approximation throughout the book[9],

$$\text{The effective mass approximation:} \quad \begin{cases} \psi_{n\mathbf{k}\sigma} \to \frac{1}{\sqrt{\mathcal{V}}} e^{i\mathbf{k}\cdot\mathbf{r}} \chi_\sigma, \\ m \quad \to m^*, \\ \mathbf{k} \text{ unrestricted.} \end{cases} \tag{2.16}$$

In the following, when no confusion is possible, m^* is often simply written as m.

2.1.2   Non-interacting electrons in the jellium model

In the effective mass approximation of the lattice model the electron eigenstates are plane waves. Also the jellium model results in plane wave solutions, which are therefore of major interest to study.

In the jellium model the ion charges are imagined to be smeared out to form a homogeneous and, to begin with, static positive charge density, $+Z\rho_{\text{jel}}$, the ion jellium. The periodic potential, $V_{\text{el-lat}}$, present in a real lattice becomes the constant potential $V_{\text{el-jel}}$ as sketched in Fig. 2.3. If we concentrate on the homogeneous part of the electron gas, i.e., discard the part of $V_{\text{el-el}}$ that leads to inhomogeneities, we notice that this part together with the ion jellium forms a completely charge neutral system. In other words, in H of Eq. (2.1) we have $V_{\text{ion-ion}} + V_{\text{el-el}} + V_{\text{el-ion}} = 0$, and we simply end up with

$$H_{\text{jel}} = T_{\text{el}}. \tag{2.17}$$

For a box with side lengths L_x, L_y, and L_z and volume $\mathcal{V} = L_x L_y L_z$ the single-particle basis states are the simple plane wave solutions to the free particle Schrödinger equation with periodic boundary conditions, $\psi(L, y, z) = \psi(0, y, z)$ and $\psi'(L, y, z) = \psi'(0, y, z)$ (likewise for the y and z directions). We prefer the periodic boundary

---

[9]For a derivation of the effective mass approximation, see e.g., Kittel (1995) or Ashcroft and Mermin (1981).

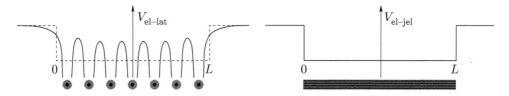

FIG. 2.3. A sketch showing the periodic potential, $V_{\text{el-lat}}$, present in a real lattice, and the imagined smeared out potential $V_{\text{el-jel}}$ of the jellium model.

conditions to the Dirichlet boundary conditions $\psi(0, y, z) = 0$ and $\psi(L, y, z) = 0$ (likewise for the y and z directions), since the former give current carrying eigenstates well suited for the description of transport phenomena, while the latter yield standing waves carrying no current. The single-particle basis states are thus

$$H_{\text{jel}}\psi_{\mathbf{k}\sigma} = \frac{\hbar^2 k^2}{2m}\psi_{\mathbf{k}\sigma}, \quad \psi_{\mathbf{k}\sigma}(\mathbf{r}) = \frac{1}{\sqrt{\mathcal{V}}} e^{i\mathbf{k}\cdot\mathbf{r}}\chi_\sigma, \quad \begin{cases} k_x = \frac{2\pi}{L_x}n_x \text{ (same for } y \text{ and } z), \\ n_x = 0, \pm 1, \pm 2, \ldots, \\ \mathcal{V} = L_x L_y L_z, \end{cases}$$

(2.18)

and with this basis we obtain H_{jel} in second quantization:

$$H_{\text{jel}} = \sum_\sigma \int d\mathbf{r}\, \Psi_\sigma^\dagger(\mathbf{r})\left(-\frac{\hbar^2}{2m}\nabla^2\right)\Psi_\sigma(\mathbf{r}) = \sum_{\mathbf{k}\sigma} \frac{\hbar^2 k^2}{2m} c_{\mathbf{k}\sigma}^\dagger c_{\mathbf{k}\sigma}.$$

(2.19)

Note how the quantization of $\mathbf{k}$ means that one state fills a volume $\frac{2\pi}{L_x}\frac{2\pi}{L_y}\frac{2\pi}{L_z} = \frac{(2\pi)^3}{\mathcal{V}}$ in $\mathbf{k}$-space, from which we obtain the following important rule of great practical value:

$$\sum_{\mathbf{k}} \to \frac{\mathcal{V}}{(2\pi)^3}\int d\mathbf{k}\,.$$

(2.20)

For further analysis on second quantization it is natural to order the single-particle states $\psi_{\mathbf{k}\sigma}(\mathbf{r}) = |\mathbf{k}\sigma\rangle$ according to their energies $\varepsilon_{\mathbf{k}} = \frac{\hbar^2 k^2}{2m}$ in ascending order,

$$|\mathbf{k}_1, \uparrow\rangle, |\mathbf{k}_1, \downarrow\rangle |\mathbf{k}_2, \uparrow\rangle, |\mathbf{k}_2, \downarrow\rangle, \ldots, \quad \text{where } \varepsilon_{\mathbf{k}_1} \leq \varepsilon_{\mathbf{k}_2} \leq \varepsilon_{\mathbf{k}_3} \leq \cdots$$

(2.21)

The ground state for N electrons at zero temperature is denoted the Fermi sea or the Fermi sphere $|\text{FS}\rangle$. It is obtained by filling up N states with the lowest possible energy,

$$|\text{FS}\rangle \equiv c_{\mathbf{k}_{N/2}\uparrow}^\dagger c_{\mathbf{k}_{N/2}\downarrow}^\dagger \cdots c_{\mathbf{k}_2\uparrow}^\dagger c_{\mathbf{k}_2\downarrow}^\dagger c_{\mathbf{k}_1\uparrow}^\dagger c_{\mathbf{k}_1\downarrow}^\dagger |0\rangle.$$

(2.22)

The energy of the topmost occupied state is denoted the Fermi energy, ε_{F}. Associated with ε_{F} is the Fermi wavenumber k_{F}, the Fermi wave length λ_{F}, and the Fermi velocity v_{F}:

$$k_{\text{F}} = \frac{1}{\hbar}\sqrt{2m\varepsilon_{\text{F}}}, \quad \lambda_{\text{F}} = \frac{2\pi}{k_{\text{F}}}, \quad v_{\text{F}} = \frac{\hbar k_{\text{F}}}{m}.$$

(2.23)

Thus in $|\text{FS}\rangle$ all states with $\varepsilon_{\mathbf{k}} < \varepsilon_{\text{F}}$ or $|\mathbf{k}| < k_{\text{F}}$ are occupied and the rest are unoccupied. A sketch of $|\text{FS}\rangle$ in energy- and $\mathbf{k}$-space is shown in Fig. 2.4.

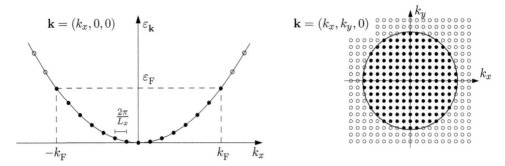

FIG. 2.4. Two aspects of $|\text{FS}\rangle$ in $\mathbf{k}$-space. To the left the dispersion relation $\varepsilon_{\mathbf{k}}$ is plotted along the line $\mathbf{k} = (k_x, 0, 0)$, and ε_F and k_F are indicated. To the right the occupation of the states is shown in the plane $\mathbf{k} = (k_x, k_y, 0)$. The Fermi sphere is shown as a circle with radius k_F. Filled and empty circles represent occupied and unoccupied states, respectively.

As a first exercise we calculate the relation between the macroscopic quantity $n = N/\mathcal{V}$, the density, and the microscopic quantity k_F, the Fermi wavenumber.

$$N = \langle \text{FS}|\hat{N}|\text{FS}\rangle = \langle \text{FS}|\sum_{\mathbf{k}\sigma} n_{\mathbf{k}\sigma}|\text{FS}\rangle = \sum_{\sigma} \frac{\mathcal{V}}{(2\pi)^3}\int d\mathbf{k}\,\langle \text{FS}|n_{\mathbf{k}\sigma}|\text{FS}\rangle. \qquad (2.24)$$

The matrix element is easily evaluated, since $n_{\mathbf{k}\sigma}|\text{FS}\rangle = |\text{FS}\rangle$ for $|\mathbf{k}| < k_F$ and 0 otherwise. This is written in terms of the theta function, see Eq. (1.13),

$$N = \sum_{\sigma}\frac{\mathcal{V}}{(2\pi)^3}\int d\mathbf{k}\,\theta(k_F - |\mathbf{k}|)\langle \text{FS}|\text{FS}\rangle = \frac{2\mathcal{V}}{(2\pi)^3}\int_0^{k_F}dk\,k^2\int_{-1}^{1}d(\cos\theta)\int_0^{2\pi}d\phi\,1 = \frac{\mathcal{V}}{3\pi^2}k_F^3,$$
$$(2.25)$$

and we arrive at the very important formula,

$$k_F^3 = 3\pi^2 n. \qquad (2.26)$$

This formula allows us to obtain the values of the microscopic parameters k_F, ε_F, and v_F. Hall measurements yield the electron density of copper[10], $n = 8.47\times10^{28}$ m^{-3}, and from Eqs. (2.23) and (2.26) it thus follows that for copper

$$k_F = 13.6 \text{ nm}^{-1}, \quad \varepsilon_F = 7.03 \text{ eV} = 81600 \text{ K},$$
$$\lambda_F = 0.46 \text{ nm}, \quad v_F = 1.57\times10^6 \text{ m/s} = 0.005\,c. \qquad (2.27)$$

Note that the Fermi energy corresponds to an extremely high temperature, which we shall return to shortly, and even though the Fermi velocity is large it is still less than one percent of the velocity of light, and we need not invoke relativistic considerations.

---

[10]The density can also be estimated as follows. The inter-atomic distances are typically $\simeq 2$ Å. In monovalent Cu one electron thus occupies a volume $\simeq (2\times10^{-10}$ m$)^3$, and $n \approx 10^{29}$ m^{-3} follows.

We move on to calculate the ground state energy $E^{(0)}$:

$$E^{(0)} = \langle \mathrm{FS}|H_{\mathrm{jel}}|\mathrm{FS}\rangle = \sum_{\mathbf{k}\sigma} \frac{\hbar^2 k^2}{2m} \langle \mathrm{FS}|n_{\mathbf{k}\sigma}|\mathrm{FS}\rangle = 2\frac{\mathcal{V}}{(2\pi)^3} \frac{\hbar^2}{2m} \int d\mathbf{k}\, k^2 \theta(k_F - |\mathbf{k}|)$$

$$= \frac{2\mathcal{V}}{(2\pi)^3} \frac{\hbar^2}{2m} \int_0^{k_F} dk\, k^4 \int_{-1}^{1} d(\cos\theta) \int_0^{2\pi} d\phi\, 1 = \frac{\mathcal{V}}{5\pi^2} \frac{\hbar^2}{2m} k_F^5 = \frac{3}{5} N\varepsilon_F. \quad (2.28)$$

In the last equation we again used Eq. (2.26). The result is reasonable, since the system consists of N electrons each with an energy $0 < \varepsilon_{\mathbf{k}} < \varepsilon_F$. The kinetic energy per particle becomes an important quantity when we, in the next section, begin to study the Coulomb interaction. By Eqs. (2.26) and (2.28) it can be expressed in terms of n:

$$\frac{E^{(0)}}{N} = \frac{3}{5} \frac{\hbar^2}{2m} k_F^2 = \frac{3}{5} \frac{\hbar^2}{2m} (3\pi^2)^{\frac{2}{3}} n^{\frac{2}{3}}. \quad (2.29)$$

The next concept to be introduced for the non-interacting electron gas is the density of states $D(\varepsilon) = \frac{dN}{d\varepsilon}$, counting the number ΔN of states in the energy interval $\Delta\varepsilon$ around the energy ε, $\Delta N = D(\varepsilon)\Delta\varepsilon$, and the density of states per volume $d(\varepsilon) = D(\varepsilon)/\mathcal{V} = \frac{dn}{d\varepsilon}$. Again using Eq. (2.26) we find

$$\varepsilon_F = \frac{\hbar^2}{2m} k_F^2 = \frac{\hbar^2}{2m} (3\pi^2)^{\frac{2}{3}} n^{\frac{2}{3}} \quad \Rightarrow \quad n(\varepsilon) = \frac{1}{3\pi^2} \left(\frac{2m}{\hbar^2}\right)^{\frac{3}{2}} \varepsilon^{\frac{3}{2}}, \text{ for } \varepsilon > 0, \quad (2.30)$$

and from this

$$d(\varepsilon) = \frac{dn}{d\varepsilon} = \frac{1}{2\pi^2} \left(\frac{2m}{\hbar^2}\right)^{\frac{3}{2}} \varepsilon^{\frac{1}{2}} \theta(\varepsilon), \quad D(\varepsilon) = \frac{dN}{d\varepsilon} = \frac{\mathcal{V}}{2\pi^2} \left(\frac{2m}{\hbar^2}\right)^{\frac{3}{2}} \varepsilon^{\frac{1}{2}} \theta(\varepsilon). \quad (2.31)$$

The density of states is a very useful function. In the following we shall, for instance demonstrate how, in terms of $D(\varepsilon)$, to calculate the particle number, $N = \int d\varepsilon\, D(\varepsilon) n_F(\varepsilon)$, and the total energy, $E^{(0)} = \int d\varepsilon\, \varepsilon\, D(\varepsilon) n_F(\varepsilon)$.

2.1.3 Non-interacting electrons at finite temperature

Finally, before turning to the problem of the Coulomb interaction, we study some basic temperature dependencies. As the temperature is raised from zero, the occupation number is given by the Fermi–Dirac distribution $n_F(\varepsilon_{\mathbf{k}})$, see Eq. (1.125). The main characteristics of this function are shown in Fig. 2.5. Note that, to be able to see any effects of the temperature in Fig. 2.5, $k_B T$ is set to $0.03\,\varepsilon_F$ corresponding to $T \approx 2400$ K. At room temperature $k_B T/\varepsilon_F \approx 0.003$, thus the low temperature limit of $n_F(\varepsilon_{\mathbf{k}})$ is of importance:

$$n_F(\varepsilon_{\mathbf{k}}) = \frac{1}{e^{\beta(\varepsilon_{\mathbf{k}}-\mu)}+1} \xrightarrow[T\to 0]{} \theta(\mu - \varepsilon_{\mathbf{k}}), \quad -\frac{\partial n_F}{\partial \varepsilon_{\mathbf{k}}} = \frac{\frac{\beta}{4}}{\cosh^2\left[\frac{\beta}{2}(\varepsilon_{\mathbf{k}}-\mu)\right]} \xrightarrow[T\to 0]{} \delta(\mu - \varepsilon_{\mathbf{k}}).$$

$$(2.32)$$

Note that, as mentioned in Section 1.5.1, the natural single-particle energy variable in these fundamental expressions actually is $\xi_{\mathbf{k}} = \varepsilon_{\mathbf{k}} - \mu$ and not $\varepsilon_{\mathbf{k}}$ itself.

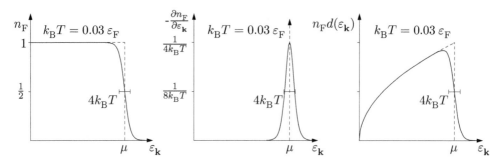

FIG. 2.5. The Fermi–Dirac distribution $n_F(\varepsilon_{\mathbf{k}})$, its derivative $-\frac{\partial n_F}{\partial \varepsilon_{\mathbf{k}}}$, and its product with the density of states, $n_F(\varepsilon_{\mathbf{k}})d(\varepsilon_{\mathbf{k}})$, shown at the temperature $k_B T = 0.03\,\varepsilon_F$, corresponding to $T = 2400$ K in metals. This rather high value is chosen in order to have a clearly observable deviation from the $T = 0$ K case, which is indicated by the dashed lines.

At $T = 0$ the chemical potential μ is identical to ε_F. But in fact μ varies slightly with temperature. A careful analysis based on the so-called Sommerfeld expansion combined with the fact that the number of electrons does not change with temperature yields

$$n(T = 0) = n(T) = \int_0^{\infty} d\varepsilon\, d(\varepsilon) f(\varepsilon) \;\Rightarrow\; \mu(T) = \varepsilon_F \left[1 - \frac{\pi^2}{12} \left(\frac{k_B T}{\varepsilon_F} \right)^2 + \cdots \right]. \quad (2.33)$$

Because ε_F according to Eq. (2.27) is around 80 000 K for metals, we find that, even at the melting temperature of metals, only a very limited number ΔN of electrons are affected by thermal fluctuations. Indeed, only the states within $2k_B T$ of ε_F are actually affected, and more precisely we have $\Delta N/N = 6 k_B T / \varepsilon_F$ ($\approx 10^{-3}$ at room temperature). The Fermi sphere is not destroyed by heating, it is only slightly smeared. Now we have at hand an explanation of the old paradox in thermodynamics, as to why only the ionic vibrational degrees of freedom contribute significantly to the specific heat of solids. The electronic degrees of freedom are simply "frozen" in. Only at temperatures comparable to ε_F they begin to play a major role. As we shall see in Section 2.3.1 this picture is not true for semiconductors, where the electron density is much smaller than in metals.

2.2 Electron interactions in perturbation theory

We now apply standard perturbation theory to take the inhomogeneous part of the electron-electron interaction $V_{\text{el}-\text{el}}$ of Eq. (2.3) into account. The homogeneous part, which in $\mathbf{k}$-space (see Eqs. (1.102) and (1.103)) corresponds to a vanishing wavevector $\mathbf{q} = 0$, has already been taken into account in the jellium model to cancel the homogeneous positive background. We thus exclude the $\mathbf{q} = 0$ term in the following sums, which is indicated by a prime:

$$V'_{\text{el}-\text{el}} = \frac{1}{2V} \sum_{\mathbf{k}_1 \mathbf{k}_2 \mathbf{q}}' \sum_{\sigma_1 \sigma_2} \frac{4\pi e_0^2}{q^2} c^{\dagger}_{\mathbf{k}_1+\mathbf{q}\sigma_1} c^{\dagger}_{\mathbf{k}_2-\mathbf{q}\sigma_2} c_{\mathbf{k}_2\sigma_2} c_{\mathbf{k}_1\sigma_1}. \quad (2.34)$$

However, as we shall see, the direct use of this interaction with the tools developed so far becomes the story of the rise and fall of simple-minded perturbation theory. The first-order calculation works well and good physical conclusions can be drawn, but already in second order the calculation collapses due to divergent integrals. It turns out that to get rid of these divergences the more powerful tools of quantum field theory must be invoked. But let us see how we arrive at these conclusions.

A natural question arises: under which circumstances can the non-interacting electron gas actually serve as a starting point for a perturbation expansion in the interaction potential? The key to the answer lies in the density dependence of the kinetic energy $E_{\text{kin}} = E^{(0)}/N \propto n^{\frac{2}{3}}$ displayed in Eq. (2.29). This is to be compared to the typical potential energy of particles with a mean distance $\bar{d}$, $E_{\text{pot}} \simeq e_0^2/\bar{d} \propto n^{\frac{1}{3}}$. So we find that

$$\frac{E_{\text{pot}}}{E_{\text{kin}}} \propto \frac{n^{\frac{1}{3}}}{n^{\frac{2}{3}}} = n^{-\frac{1}{3}} \xrightarrow[n\to\infty]{} 0, \tag{2.35}$$

revealing the following, perhaps somewhat counter-intuitive, fact: the importance of the electron-electron interaction diminishes as the density of the electron gas increases. Due to the Pauli exclusion principle the kinetic energy simply becomes the dominant energy scale in the interacting electron gas at high densities. Consequently, we approach the problem from this limit in the following analysis.

We begin the perturbation treatment by establishing the relevant length scale and energy scale for the problem of interacting charges. The prototypical example is of course the hydrogen atom, where a single electron orbits a proton. The ground state is a spherically symmetric s-wave with a radius denoted the Bohr radius a_0 and an energy E_0. The following considerations may be helpful mnemotechnically. The typical length scale a_0 yields a typical momentum $p = \hbar/a_0$. Writing E_0 as the sum of kinetic energy $p^2/2m$ and potential energy $-e_0^2/a_0$, we arrive at $E_0 = \frac{\hbar^2}{2ma_0^2} - \frac{e_0^2}{a_0}$. The values of a_0 and E_0 are found either by minimization, $\frac{\partial E_0}{\partial a_0} = 0$, or by using the virial theorem $E_{\text{kin}} = -\frac{1}{2}E_{\text{pot}}$:

$$a_0 = \frac{\hbar^2}{me_0^2} = 0.053 \text{ nm}, \quad E_0 = -\frac{e_0^2}{2a_0} = -13.6 \text{ eV}, \quad 1 \text{ Ry} = \frac{e_0^2}{2a_0} = 13.6 \text{ eV}. \tag{2.36}$$

Here we have also introduced the energy unit 1 Ry, often encountered in atomic physics as defining a natural energy scale. Lengths are naturally measured in units a_0, and the dimensionless measure r_s of the average inter-electronic distance in the electron gas is introduced as the radius in a sphere containing exactly one electron:

$$\frac{4\pi}{3}(r_s a_0)^3 = \frac{1}{n} = \frac{3\pi^2}{k_{\text{F}}^3} \Rightarrow a_0 k_{\text{F}} = \left(\frac{9\pi}{4}\right)^{\frac{1}{3}} r_s^{-1} \Rightarrow r_s = \left(\frac{9\pi}{4}\right)^{\frac{1}{3}} \frac{1}{a_0 k_{\text{F}}}. \tag{2.37}$$

Rewriting the energy $E^{(0)}$ of the non-interacting electron gas to these units we obtain:

$$\frac{E^{(0)}}{N} = \frac{3}{5}\frac{1}{2}\frac{\hbar^2}{m}k_{\text{F}}^2 = \frac{3}{5}\frac{1}{2}(a_0 e_0^2)\frac{(a_0 k_{\text{F}})^2}{a_0^2} = \frac{3}{5}\left(\frac{9\pi}{4}\right)^{\frac{2}{3}}\frac{e_0^2}{2a_0}r_s^{-2} \approx \frac{2.21}{r_s^2} \text{ Ry}. \tag{2.38}$$

This constitutes the zeroth order energy in our perturbation calculation.

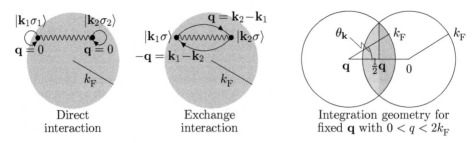

Direct interaction   Exchange interaction   Integration geometry for fixed $\mathbf{q}$ with $0 < q < 2k_F$

FIG. 2.6. The two possible processes in first-order perturbation theory for two states $|\mathbf{k}_1\sigma_1\rangle$ and $|\mathbf{k}_2\sigma_2\rangle$ in the Fermi sea. The direct process having $\mathbf{q} = 0$ is already taken into account in the homogeneous part, hence only the exchange process contributes to $V'_{\text{el}-\text{el}}$. Also the geometry for the $\mathbf{k}$-integration is shown for an arbitrary but fixed value of $\mathbf{q}$.

2.2.1 Electron interactions in first-order perturbation theory

The first-order energy $E^{(1)}$ is found by the standard perturbation theory procedure:

$$\frac{E^{(1)}}{N} = \frac{\langle\text{FS}|V'_{\text{el}-\text{el}}|\text{FS}\rangle}{N} = \frac{1}{2\mathcal{V}N}\sum_{\mathbf{q}}{}' \sum_{\mathbf{k}_1,\mathbf{k}_2}\sum_{\sigma_1,\sigma_2}\frac{4\pi e_0^2}{q^2}\langle\text{FS}|c^\dagger_{\mathbf{k}_1+\mathbf{q}\sigma_1}c^\dagger_{\mathbf{k}_2-\mathbf{q}\sigma_2}c_{\mathbf{k}_2\sigma_2}c_{\mathbf{k}_1\sigma_1}|\text{FS}\rangle.$$

(2.39)

The matrix element is evaluated as follows. First, the two annihilation operators can only give a non-zero result if both $|\mathbf{k}_1| < k_F$ and $|\mathbf{k}_2| < k_F$. Second, the factor $\langle\text{FS}|$ demands that the two creation operators bring us back to $|\text{FS}\rangle$, thus either $\mathbf{q} = 0$ (but that is excluded from $V'_{\text{el}-\text{el}}$) or $\mathbf{k}_2 = \mathbf{k}_1 + \mathbf{q}$ and $\sigma_2 = \sigma_1$. These possibilities are sketched in Fig. 2.6. For $\mathbf{q} \neq 0$ we therefore end with

$$\langle\text{FS}|c^\dagger_{\mathbf{k}_1+\mathbf{q}\sigma_1}c^\dagger_{\mathbf{k}_2-\mathbf{q}\sigma_2}c_{\mathbf{k}_2\sigma_2}c_{\mathbf{k}_1\sigma_1}|\text{FS}\rangle$$
$$= \delta_{\mathbf{k}_2,\mathbf{k}_1+\mathbf{q}}\delta_{\sigma_1,\sigma_2}\langle\text{FS}|c^\dagger_{\mathbf{k}_1+\mathbf{q}\sigma_1}c^\dagger_{\mathbf{k}_1\sigma_1}c_{\mathbf{k}_1+\mathbf{q}\sigma_1}c_{\mathbf{k}_1\sigma_1}|\text{FS}\rangle$$
$$= -\delta_{\mathbf{k}_2,\mathbf{k}_1+\mathbf{q}}\delta_{\sigma_1,\sigma_2}\langle\text{FS}|n_{\mathbf{k}_1+\mathbf{q}\sigma_1}n_{\mathbf{k}_1\sigma_1}|\text{FS}\rangle$$
$$= -\delta_{\mathbf{k}_2,\mathbf{k}_1+\mathbf{q}}\delta_{\sigma_1,\sigma_2}\theta(k_F-|\mathbf{k}_1+\mathbf{q}|)\theta(k_F-|\mathbf{k}_1|),$$
(2.40)

where $\mathbf{q} \neq 0$ leads to $\mathbf{k}_1 + \mathbf{q} \neq \mathbf{k}_1$, which results in a simple anticommutator yielding the occupation number operators with a minus in front. Since only one $\mathbf{k}$-vector appears we now drop the index 1.

The $\mathbf{k}$- and $\mathbf{q}$-sum are converted into integrals, and polar coordinates (q, θ_q, ϕ_q) and (k, θ_k, ϕ_k) are employed. First, note that the integral is independent of the direction of $\mathbf{q}$ so that $\int_{-1}^{1} d(\cos\theta_q)\int_0^{2\pi} d\phi_q = 4\pi$. Second, only for $0 < q < 2k_F$ does the theta function product give a non-zero result. For a given fixed value of q the rest of the integral is just the overlap volume between two spheres of radius k_F displaced by q. The geometry of this volume is sketched in Fig. 2.6, and is calculated by noting that $q/2k_F < \cos\theta_k < 1$, and that for a given $\cos\theta_k$ we have $q/(2\cos\theta_k) < k < k_F$. The last variable is free: $0 < \phi_k < 2\pi$. We thus obatin

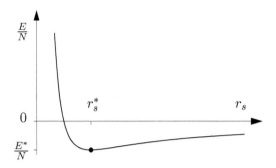

FIG. 2.7.   The energy per particle E/N of the 3D electron gas in first-order perturbation theory Eq. (2.43) as a function of the dimensionless inter-particle distance
r_s. Due to the exchange interaction the electron gas is stable at $r_s = r_s^* = 4.83$
with an ionization energy $E/N = E^*/N = -1.29$ eV.

$$\frac{E^{(1)}}{N} = -\frac{4\pi e_0^2}{2\mathcal{V}N} 2(4\pi) \frac{\mathcal{V}}{(2\pi)^3} \int_0^{2k_F} dq \frac{q^2}{q^2} \, 2(2\pi) \int_{\frac{q}{2k_F}}^1 d(\cos\theta_k) \frac{\mathcal{V}}{(2\pi)^3} \int_{\frac{q}{2\cos\theta_k}}^{k_F} dk\, k^2, \quad (2.41)$$

where the prefactors are a factor 2 for spin, 2 for symmetry, 4π for **q**-angles, 2π for
ϕ_k, and twice $\mathcal{V}/(2\pi)^3$ for the conversions of **k**- and **q**-sums to integrals. The integral
is elementary and results in

$$\frac{E^{(1)}}{N} - \frac{e_0^2}{2} \frac{\mathcal{V}}{N} \frac{k_F^4}{2\pi^3} - -\frac{e_0^2}{2a_0}(a_0 k_F)\frac{k_F^3}{2\pi^3 n} = -\frac{e_0^2}{2a_0}\left(\left(\frac{9\pi}{4}\right)^{\frac{1}{3}}\frac{1}{r_s}\right)\frac{3}{2\pi} \approx -\frac{0.916}{r_s}\ \text{Ry}.$$
$$(2.42)$$

The final result for the first-order perturbation theory is thus the simple expression

$$\frac{E}{N} \xrightarrow[r_s \to 0]{} \frac{E^{(0)} + E^{(1)}}{N} = \left(\frac{2.211}{r_s^2} - \frac{0.916}{r_s}\right)\ \text{Ry}. \qquad (2.43)$$

This result shows that the electron gas is stable when the repulsive Coulomb interaction is turned on; see Fig. 2.7. No external confinement potential is needed to hold
the electron gas in the ion jellium together. There exists an optimal density n^*, or
inter-particle distance r_s^*, which minimizes the energy and furthermore yields an energy $E^* < 0$. The negative exchange energy overcomes the positive kinetic energy.
The equilibrium situation is obtained from $\frac{\partial}{\partial r_s}(E^{(0)} + E^{(1)}) = 0$, and we can compare
the result with experiment:

$$r_s^* = 4.83, \quad \frac{E^*}{N} = -0.095\ \text{Ry} = -1.29\ \text{eV} \quad \text{(first-order perturbation theory)}$$
$$r_s = 3.96, \quad \frac{E}{N} = -0.083\ \text{Ry} = -1.13\ \text{eV} \quad \text{(experiment on Na)}$$
$$(2.44)$$

We note that the negative binding energy is due to the exchange energy of the
Coulomb interaction. Physically this can be interpreted as an effect of the Pauli
exclusion principle: the electrons are forced to avoid each other, since only one electron
at a time can be at a given point in space. The direct "classical" Coulomb interaction

does not take this into account and is therefore over-estimating the energy, and the exchange part corrects for this by being negative.

One may wonder what happens to the Fermi sphere as the interaction is turned on. We found before that thermal smearing occurs but is rather insignificant compared to the huge Fermi energy, $\varepsilon_F \approx 7$ eV. However, now we have learned that the interaction energy per particle is ≈ 1.3 eV, i.e., smaller than but certainly comparable to ε_F. One of the great results of quantum field theory, which we are going to study later in the book, is the explanation of why the Fermi surface is not destroyed by the strong Coulomb interaction between the electrons.

2.2.2 *Electron interactions in second-order perturbation theory*

One may try to improve on the first-order result by going to second-order perturbation theory. However, the result is disastrous. The matrix elements diverge without giving hope for a simple cure.

Here, we can only reveal what goes wrong, and then later learn how to deal correctly with the infinities occurring in the calculations. According to second-order perturbation theory, $E^{(2)}$ is given by

$$\frac{E^{(2)}}{N} = \frac{1}{N} \sum_{|\nu\rangle \neq |\mathrm{FS}\rangle} \frac{\langle \mathrm{FS}|V'_{\mathrm{el-el}}|\nu\rangle \langle \nu|V'_{\mathrm{el-el}}|\mathrm{FS}\rangle}{E^{(0)} - E_\nu}, \tag{2.45}$$

where all the intermediate states $|\nu\rangle$ must be different from $|\mathrm{FS}\rangle$. As sketched in Fig. 2.8, this combined with the momentum-conserving Coulomb interaction, yields intermediate states where two particles are injected out of the Fermi sphere. From such an intermediate state, $|\mathrm{FS}\rangle$ is restored by putting the excited electrons back into the holes they left behind. Only two types of processes are possible: the direct and the exchange process.

We now proceed to show that the direct interaction process gives a divergent contribution $E^{(2)}_{\mathrm{dir}}$ to $E^{(2)}$ due to the singular behavior of the Coulomb interaction at small momentum transfers $\mathbf{q}$. For the direct process the constraint $|\nu\rangle \neq |\mathrm{FS}\rangle$ leads to

$$|\nu\rangle = \theta(|\mathbf{k}_1+\mathbf{q}|-k_F)\theta(|\mathbf{k}_2-\mathbf{q}|-k_F)\theta(k_F-|\mathbf{k}_1|)\theta(k_F-|\mathbf{k}_2|)c^\dagger_{\mathbf{k}_1+\mathbf{q}\sigma_1}c^\dagger_{\mathbf{k}_2-\mathbf{q}\sigma_2}c_{\mathbf{k}_2\sigma_2}c_{\mathbf{k}_1\sigma_1}|\mathrm{FS}\rangle. \tag{2.46}$$

To restore $|\mathrm{FS}\rangle$, the same momentum transfer $\mathbf{q}$ must be involved in both $\langle \nu|V'_{\mathrm{el-el}}|\mathrm{FS}\rangle$ and $\langle \mathrm{FS}|V'_{\mathrm{el-el}}|\nu\rangle$, and writing $V_{\mathbf{q}} = \frac{4\pi e_0^2}{q^2}$ we find

$$E^{(2)}_{\mathrm{dir}} = \frac{1}{\mathcal{V}^2}\sum_{\mathbf{q}}\sum_{\substack{\mathbf{k}_1\sigma_1 \\ \mathbf{k}_2\sigma_2}}\frac{(\frac{1}{2}V_{\mathbf{q}})^2}{E^{(0)} - E_\nu}\theta(|\mathbf{k}_1+\mathbf{q}|-k_F)\theta(|\mathbf{k}_2-\mathbf{q}|-k_F)\theta(k_F-|\mathbf{k}_1|)\theta(k_F-|\mathbf{k}_2|). \tag{2.47}$$

The contribution from small values of $\mathbf{q}$ to $E^{(2)}_{\mathrm{dir}}$ is found by noting that

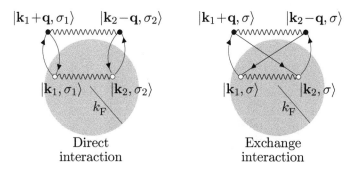

Direct interaction | Exchange interaction

FIG. 2.8. The two possible processes in second-order perturbation theory for two states $|\mathbf{k}_1\sigma_1\rangle$ and $|\mathbf{k}_2\sigma_2\rangle$ in the Fermi sea. The direct process gives a divergent contribution to E/N while the exchange process gives a finite contribution.

$$V_{\mathbf{q}}^2 \propto \frac{1}{q^4}, \tag{2.48a}$$

$$E_0 - E_\nu \propto \mathbf{k}_1^2 + \mathbf{k}_2^2 - (\mathbf{k}_1 + \mathbf{q})^2 - (\mathbf{k}_2 - \mathbf{q})^2 \underset{q\to 0}{\propto} q, \tag{2.48b}$$

$$\sum_{\mathbf{k}_1} \cdots \theta(|\mathbf{k}_1 + \mathbf{q}| - k_{\mathrm{F}})\theta(k_{\mathrm{F}} - |\mathbf{k}_1|) \underset{q\to 0}{\propto} q, \tag{2.48c}$$

from which we obtain

$$E_{\mathrm{dir}}^{(2)} \propto \int_0 dq\, q^2 \frac{1}{q^4} \frac{1}{q}\, q\, q = \int_0 dq\, \frac{1}{q} = \ln(q)\Big|_0 \propto \infty. \tag{2.49}$$

The exchange process does not lead to a divergence, since in this case the momentum transfer in the excitation part is $\mathbf{q}$, but in the relaxation part it is $\mathbf{k}_2 - \mathbf{k}_1 - \mathbf{q}$. Thus $V_{\mathbf{q}}^2$ is replaced by $V_{\mathbf{q}}V_{\mathbf{k}_2-\mathbf{k}_1-\mathbf{q}} \propto q^{-2}$ for $q \to 0$, which is less singular than $V_{\mathbf{q}}^2 \propto q^{-4}$.

This divergent behavior of second-order perturbation theory is a nasty surprise. We know that physically, the energy of the electron gas must be finite. The only hope for rescue lies in regularization of the divergent behavior by taking higher order perturbation terms into account. In fact, as we shall see in Chapter 14, it turns out that one has to consider perturbation theory to infinite order, which is possible using the full machinery of quantum field theory to be developed in the coming chapters.

2.3 Electron gases in 3, 2, 1 and 0 dimensions

We end this chapter on the electron gas by mentioning a few experimental realizations of electron gases in 3D, 2D and 1D. To work in various dimensions is a good opportunity to test one's understanding of the basic principles of the physics of electron gases. But as will become clear, this is not just an academic exercise. Experiments on electron gases at reduced dimensionality is of increasing importance.

2.3.1 *3D electron gases: metals and semiconductors*

Bloch's theory of non-interacting electrons moving in a periodic lattice provides an explanation for the existence of metals, semiconductors, and band insulators. The

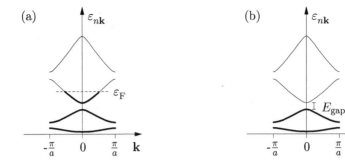

FIG. 2.9. (a) A generic band structure for a metal. The Fermi level ε_{F} lies in the middle of a band resulting in arbitrarily small possible excitations energies. (b) A generic band structure for an insulator or a semiconductor. The Fermi level ε_{F} lies at the top of the valence band resulting in possible excitations energies of at least E_{gap}, the distance up to the unoccupied conduction band.

important parameter is the position of the Fermi energy ε_{F} relative to the bands as sketched in Fig. 2.9. In the metallic case, ε_{F} lies in the middle of a band. Consequently, there is no energy gap between the last occupied level and the first unoccupied level, and any, however small, external field can excite the system and give rise to a significant response. In an insulator, ε_{F} is at the top of a band, the so-called filled valence band, and filled bands do not carry any electrical or thermal current.[11] The system can only be excited by providing sufficient energy for the electrons to overcome the energy band gap E_{gap} between the top of the valence band and the bottom of the next empty band, the so-called conduction band. This is not possible for small external fields, and hence the inability of insulators to conduct electronic thermal and electrical currents. Semiconductors are insulators at $T = 0$, but their band gap E_{gap} is relatively small, typically less than 2 eV, such that at room temperature a sufficient number of electrons are excited thermally up into the conduction band to yield a significant conductivity.

We emphasize that at room temperature the electron gas in a metal is a degenerate Fermi gas since $k_{\mathrm{B}}T \ll \varepsilon_{\mathrm{F}}$. A semiconductor, on the other hand, is normally described as a classical gas since for energies $\varepsilon_{\mathbf{k}}$ in the conduction band we have $\varepsilon_{\mathbf{k}} - \mu > E_{\mathrm{gap}}/2 \gg k_{\mathrm{B}}T$, and consequently $n_{\mathrm{F}}(\varepsilon_{\mathbf{k}}) \to e^{-(\varepsilon_{\mathbf{k}}-\mu)/k_{\mathrm{B}}T}$, i.e., the Maxwell–Boltzmann distribution.

Finally, we note that in a typical metal most of the electron states at the Fermi surface are far away from the regions in $\mathbf{k}$-space where the free electron dispersion relation is strongly distorted by the periodic lattice. Therefore, one finds effective masses m^*, see Eq. (2.16), close to the vacuum mass m. In contrast, all the electron states contributing to the transport properties in a semiconductor are close to these

---

[11]Transport properties are tightly connected to the electron velocity $v_{\mathbf{k}} = \frac{1}{\hbar}\frac{\partial \varepsilon_{\mathbf{k}}}{\partial \mathbf{k}}$. The current density is $\mathbf{J} = 2\sum_{\mathbf{k}\in\mathrm{FBZ}} \frac{1}{\mathcal{V}} n_{\mathbf{k}} v_{\mathbf{k}} = 2\int_{\mathrm{FBZ}} \frac{d\mathbf{k}}{(2\pi)^3} \frac{n_{\mathbf{k}}}{\hbar} \frac{\partial \varepsilon_{\mathbf{k}}}{\partial \mathbf{k}}$. Likewise, for the thermal current $\mathbf{J}_{\mathrm{th}} = 2\sum_{\mathbf{k}\in\mathrm{FBZ}} \frac{1}{\mathcal{V}} n_{\mathbf{k}} \varepsilon_{\mathbf{k}} v_{\mathbf{k}} = \int_{\mathrm{FBZ}} \frac{d\mathbf{k}}{(2\pi)^3} \frac{n_{\mathbf{k}}}{\hbar} \frac{\partial(\varepsilon_{\mathbf{k}}^2)}{\partial \mathbf{k}}$. For a filled band $n_{\mathbf{k}} = 1$. In that case both currents are integrals over FBZ of gradients of periodic functions, and they are therefore zero.

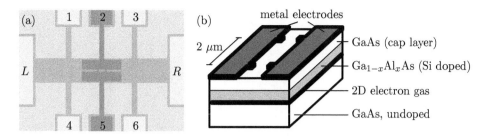

FIG. 2.10.   (a) A top-view of a GaAs-based semiconductor heterostructure containing a 2D electron gas (medium gray) situated on a substrate (light gray). Ohmic
contacts (white) are used to supply current (L and R) and to measure voltages
(1, 3, 4 and 6). Further control of the geometry is obtained by biasing the gate–
electrodes (dark gray) on the top at contacts 2 and 5. (b) A sketch showing the
different layers in the heterostructure and the metallic gate-electrodes at the top.

regions in **k**-space, and one finds strongly modified effective masses, typically $m^* \approx$
$0.1\ m$.

2.3.2   2D electron gases: GaAs/GaAlAs heterostructures

For the past four decades it has been possible to fabricate 2D electron gases at semiconductor interfaces. The first realization was inversion layers in the celebrated sil-
icon MOSFETs, the key component in integrated electronic circuits. More recent
realizations are in gallium-arsenide/gallium-aluminum-arsenide (GaAs/$Ga_{1-x}Al_xAs$)
heterostructures. In the latter system one can obtain extremely long mean free paths
(more than 10 μm), which is technologically important for high-speed electronics, and
which is essential for the basic research of many quantum effects in condensed matter
physics.

The interface between the GaAs and the $Ga_{1-x}Al_xAs$ semiconductor crystals in
the GaAs/$Ga_{1-x}Al_xAs$ heterostructure can be grown with mono-atomic-layer preci-
sion in molecular beam epitaxy (MBE) machines.   This is because the two semiconductor crystals have nearly the same crystal structure leading to a stress-free interface.
In Fig. 2.10, a top-view of a typical device is shown as well as a sketch of the various
layers in a GaAs heterostructure. The main difference between the two semiconductor
crystals is the values of the bottom of the conduction band. For $x = 0.3$, the conduction band in $Ga_{1-x}Al_xAs$ is 300 meV higher than the one in GaAs. Hence the
electrons in the former conduction band can gain energy by moving to the latter. At
$T = 0$ there are, of course, no free carriers in any of the conduction bands for pure
semiconductor systems, but by doping the $Ga_{1-x}Al_xAs$ with Si, conduction electrons
are provided, which then accumulate on the GaAs side of the interface due to the energy gain. However, not all donor electrons will be transferred. The ionized Si donors
left in the $Ga_{1-x}Al_xAs$ provide an electrostatic energy that grows with an increasing number of transferred electrons. At some point the energy gained by transferring
electrons to the GaAs layer is balanced by the growth in electrostatic energy. This is
sketched in Fig. 2.11 where the resulting conduction band in equilibrium is shown as
function of the position z perpendicular to the interface. The conduction band is not

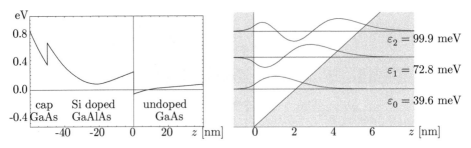

FIG. 2.11. The conduction band in a GaAs/GaAlAs heterostructure. Note the triangular well forming at the interface. The wavefunctions $\zeta_n(z)$ and eigenvalues of
the lowest three electron eigenstates in the triangular well.

flat due to the curvature induced by the charge densities, as calculated from Poisson's
equation: $\nabla^2 V = -en^{3D}/\epsilon^*$.

The key point to notice is the formation of the almost triangular quantum well at
the GaAs side of the interface. The well is so narrow that a significant size-quantization
is obtained. Without performing the full calculation, we can get a grasp of the order
of magnitude by the following estimate. We consider the positively charged layer of
the ionized Si donors as one plate of a plate capacitor, while the conduction electrons
at the GaAs/GaAlAs interface form the other plate. The charge density outside this
capacitor is zero. The electrical field $\mathbf{E}$ at the interface is then found simply by forming
a cylindrical Gauss box with its axis along the z direction and one circular "bottom
lid" at the interface and the other "top lid" deep into the GaAs. All the contributions
stems from the "bottom lid," since for symmetry reasons $\mathbf{E}$ must be perpendicular
to the z axis, yielding zero from the side of the cylindrical box, and since for the
reason of charge neutrality, $\mathbf{E} = 0$ at the "top lid." Thus at the interface $E = en/\epsilon^*$,
n being the 2D electron density at the interface. The typical length scale l for the
width of the triangular well is found by balancing the potential energy and the kinetic
quantum energy, $eEl = \hbar^2/(m^*l^2)$, which gives $l^3 = \frac{1}{4\pi}\frac{\epsilon^*/\epsilon_0}{m^*/m}\frac{a_0}{n}$, where we have used
the Bohr radius a_0 of Eq. (2.36) to bring in atomic units. The experimental input
for GaAs is $\epsilon^* = 13\epsilon_0$, $m^* = 0.067m$, and typically $n = 3 \times 10^{15}$ m^{-2}, which yields
$l \approx 5$ nm. From this we get the typical quantization energy ΔE due to the triangular
well: $\Delta E = (m/m^*)(a_0{}^2/l^2)$ 13.6 eV ≈ 20 meV.

The significance of this quantization energy is the following. Due to the triangular
well the 3D free electron wavefunction is modified,

$$\psi_{\mathbf{k}\sigma}(\mathbf{r}) = \frac{1}{\sqrt{\mathcal{V}}}e^{ik_x x}e^{ik_y y}e^{ik_z z}\chi_\sigma \quad \to \quad \psi_{k_x k_y n\sigma}(\mathbf{r}) = \frac{1}{\sqrt{A}}e^{ik_x x}e^{ik_y y}\zeta_n(z)\chi_\sigma, \quad (2.50)$$

where $\zeta_n(z)$ is the nth eigenfunction of the triangular well having the eigenenergy ε_n,
see Fig. 2.11. Only the z direction is quantized leaving the x and y direction unaltered,
and the total energy for all three spatial directions is

$$\varepsilon_{k_x,k_y,n} = \frac{\hbar^2}{2m^*}\left(k_x^2 + k_y^2\right) + \varepsilon_n, \quad k_F^2 = 2\pi n \quad \Rightarrow \quad \varepsilon_F \approx 10 \text{ meV}, \quad (2.51)$$

where we have given the 2D version of the fundamental relation between k_F and n (see Exercise 2.4 and compare to Eq. (2.26) for the 3D case). The highest occupied state has the energy $E_0 + \varepsilon_F$ while the lowest unoccupied state has the energy E_1. The difference is $E_1 - (E_0 + \varepsilon_F) = \Delta E - \varepsilon_F \approx 10$ meV ≈ 100 K, and we arrive at our conclusion: At temperatures $T \ll 100$ K all occupied electron states have the same orbital in the z direction, $\zeta_0(z)$. Any changes of this orbital requires an excitation energy of at least 10 meV. If this is not provided the system has effectively lost one spatial degree of freedom and is dynamically a 2D system. This means that theoretical studies of 2D electron gases is far from an academic exercise; 2D systems do indeed exist in reality.

2.3.3 1D electron gases: carbon nanotubes

Since the mid-1990s a new research field has developed involving studies of the cylindrically shaped carbon-based molecule, the so-called carbon nanotube. The carbon nanotube can be viewed as a normal graphite sheet rolled up into a cylinder with a radius $R_0 \approx 2$ nm and a length more than a thousand times R_0, see Fig. 2.12. These long and thin carbon molecules have some extraordinary material characteristics. They are believed to be the strongest material in the world, and depending on the specific way the cylinder is rolled up the nanotubes are either metallic, semiconducting or insulating. In the same dynamical sense as the GaAs heterostructure is a 2D metal sheet, a metallic nanotube is a nearly ideal 1D wire, i.e., two of the three spatial degrees of freedom are frozen in. We briefly sketch how this comes about.

The cylindrical symmetry of the nanotube makes it natural to change the basis functions from the 3D (x, y, z) plane waves to cylindrical (x, r, ϕ) wavefunctions:

$$\psi_{\mathbf{k}\sigma}(\mathbf{r}) = \frac{1}{\sqrt{\mathcal{V}}} e^{ik_x x} e^{ik_y y} e^{ik_z z} \chi_\sigma \quad \rightarrow \quad \psi_{k_x, n, l, \sigma}(\mathbf{r}) = \frac{1}{\sqrt{L}} e^{ik_x x} R_{nl}(r) Y_l(\phi) \chi_\sigma. \quad (2.52)$$

This is of course more than just a mathematical transformation. The electrons are strongly bound to the surface of the cylinder in quantum states arising from the original π-bonds of the graphite system. This means that the extension ΔR of the radial wave function $R_{nl}(r)$ around the mean value R_0 is of atomic scale, i.e., $\Delta R \approx 0.1$ nm, resulting in a radial confinement energy $E_0^R \sim \hbar^2/(2m\Delta R^2) \sim 10$ eV. Likewise, in the azimuthal angle coordinate ϕ, there is a strong confinement, since the perimeter must contain an integer number of electron wavelengths λ_n, i.e., $\lambda_n = 2\pi R_0/n < 2$ nm. The corresponding confinement energy is $E_n^\phi \sim \hbar^2/(2m\lambda_n^2) \sim 1$ eV n. There are no severe constraints along the cylinder axis, i.e., in the x direction. We therefore end up with a total energy

$$\varepsilon_{k_x, n, l} = E_0^R + E_n^\phi + \frac{\hbar^2}{2m} k_x^2, \quad (2.53)$$

with a considerable gap ΔE from the center of the $(n, l) = (0, 0)$ band (the position of ε_F for the metallic case) to the bottom of the $(n, l) = (0, 1)$ band:

$$\Delta E = \frac{1}{2}(E_1^\phi - E_0^\phi) \approx 1 \text{ eV} \sim 12\,000 \text{ K.} \quad (2.54)$$

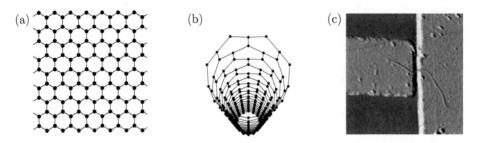

FIG. 2.12.   (a) Carbon atoms forming a sheet of graphite with a characteristic hexa-
gonal lattice. (b) A carbon nanotube molecule is formed by rolling up a graphite
sheet into a cylindrical geometry. (c) An atomic force micrograph recorded at the
Ørsted Laboratory, Niels Bohr Institute, showing a bundle of carbon nanotubes
placed across a gap between two metal electrodes, thereby connecting them and
allowing for electrical measurements on single molecules.

Thus, at room temperature the only available degree of freedom is the axial one
described by the continuous quantum number k_x and the associated plane waves
$e^{ik_x x}$.

Not only are nanotubes very interesting from an experimental point of view, they
also play an important role in purely theoretical studies. The nanotubes is one of only
a couple of systems exhibiting a nearly ideal 1D behavior. In particular that makes the
nanotubes a key testing ground for the diagonizable so-called Luttinger liquid model,
a central quantum model for describing interacting electrons in 1D, see Chapter 19.

2.3.4   0D electron gases: quantum dots

Naturally, one can think of confining the electrons in all three spatial dimensions. Any
isolated atom or molecule is an example of such a case. But also so-called artificial
atoms or quantum dots have been realized experimentally. One example of this is the
semiconductor device shown in Fig. 2.10(b), and likewise small metal clusters have
been studied. By careful preparation electrodes can be attached near the quantum
dot, so that small electrical probe currents can be sent through the quantum dot
without disturbing it too much.

Lately, it has been demonstrated that also large molecules like C_{60} or carbon
nanotubes can be placed in the gap between two electrodes, and thus play the role as
quantum dots.

Simplified models of quantum dots are studied in Chapter 10. But here we would
like to emphasize that the quantum dots usually do not have any nice spatial symmetries. Hence, the single-particle eigenstates must be characterized by a general index
ν for the spatial degrees of freedom and σ for the spin. The single particle energy
spectrum $\varepsilon_{\nu\sigma}$ is discrete with some typical non-zero level spacing $\Delta\varepsilon$.

Using the general theory of second quantized operators presented in Sections 1.3
and 1.4 we arrive at the following Hamiltonian H_D for an isolated quantum dot with
(possibly magnetic) electron-electron interaction:

$$H_D = \sum_{\nu_1\sigma_1} \varepsilon_{\nu_1\sigma_1}\, c^\dagger_{\nu_1\sigma_1} c_{\nu_1\sigma_1} \;+\; \frac{1}{2} \sum_{\substack{\sigma_1\sigma_2 \\ \sigma_3\sigma_4}} \sum_{\substack{\nu_1\nu_2 \\ \nu_3\nu_4}} V_{\nu_3\sigma_3,\nu_4\sigma_4;\nu_1\sigma_1,\nu_2\sigma_2}\, c^\dagger_{\nu_3\sigma_3} c^\dagger_{\nu_4\sigma_4} c_{\nu_2\sigma_2} c_{\nu_1\sigma_1}.$$

$$(2.55)$$

This Hamiltonian is the starting point for the study of many aspects of 0D systems. Especially, as we shall see when we return to it in Chapter 10, it is a central element in the theory of electron transport in systems containing quantum dots. The Hamiltonian H_D gives rise to a number of intricate quantum transport effects, such as the Coulomb blockade and the Kondo effect, once it is coupled to 3D electron reservoirs by some tunneling Hamiltonian describing the transfer of electrons between the quantum dot and the reservoirs. However, to be able to handle such complex problems we need to develop our theoretical tools further.

2.4   Summary and outlook

In this chapter we have introduced one of the main examples in this book, the electron gas in metals. In the rest of the book we shall study many different aspects of the electron gas.

An important concept is the ground state of the non-interacting electron gas, the so-called Fermi sea or Fermi sphere $|\mathrm{FS}\rangle$. This many-body state is defined as

$$|\mathrm{FS}\rangle \equiv c^\dagger_{\mathbf{k}_{N/2}\uparrow} c^\dagger_{\mathbf{k}_{N/2}\downarrow} \cdots c^\dagger_{\mathbf{k}_2\uparrow} c^\dagger_{\mathbf{k}_2\downarrow} c^\dagger_{\mathbf{k}_1\uparrow} c^\dagger_{\mathbf{k}_1\downarrow}|0\rangle,$$

i.e., the filling of all electron states up to a certain energy, the Fermi energy ε_{F} defined in terms of the Fermi wavenumber n_{F},

$$\varepsilon_{\mathrm{F}} = \frac{\hbar^2 k_{\mathrm{F}}^2}{2m}, \text{where} \quad k_{\mathrm{F}}^3 = 3\pi^2 n,$$

n being the density of valence electrons in the metal.

The Coulomb interaction was studied in first and second-order perturbation theory with $|\mathrm{FS}\rangle$ acting as the unperturbed state. It was shown that the first-order calculation could explain the stability of metals by the minimum of the energy

$$\frac{E}{N} \xrightarrow[r_s \to 0]{} \frac{E^{(0)} + E^{(1)}}{N} = \left(\frac{2.211}{r_s^2} - \frac{0.916}{r_s}\right) \mathrm{Ry},$$

where r_s is the dimensionless average distance between the electrons. However, a fundamental problem was revealed when going to second order in the perturbation. The supposedly small second-order correction to the energy diverged. The solution of this problem is one of the main achievements of many-body theory, and it is treated in Chapter 14.

Some good introductory textbooks on electron gases and metals are Kittel (1995) or Ashcroft and Mermin (1981).

3

PHONONS; COUPLING TO ELECTRONS

In this chapter we study the basic properties of ionic vibrations. These vibrations are well described by harmonic oscillators and, therefore, we can employ the results from Section 1.4.1 to achieve the second-quantized form of the corresponding Hamiltonian. The quantized vibrations are denoted phonons, a name pointing to the connection between sound waves and lattice vibrations. Phonons play a fundamental role in our understanding of sound, specific heat, elasticity, and electrical resistivity of solids. More surprising may be the fact that the electron-phonon coupling is the cause of conventional superconductivity. In the following sections we shall study the three types of matter oscillation sketched in Fig. 3.1. The ions will be treated using two models: the jellium model, where the ions are represented by a smeared-out continuous positive background, and the lattice model, where the ions oscillate around their equilibrium positions forming a regular crystal lattice.

Since phonons basically are harmonic oscillators, they are bosons according to the results of Section 1.4.1. Moreover, they naturally occur at finite temperature, so we will, therefore, often need the thermal distribution function for bosons, the Bose–Einstein distribution $n_B(\varepsilon)$ given in Eq. (1.127).

3.1 Jellium oscillations and Einstein phonons

Our first encounter with phonons will be that arising from a semiclassical treatment of the charge neutral jellium system. Let ρ_{ion}^0 be the particle density of the ion jellium, and $\rho_{el} = Z\rho_{ion}^0$ that of the homogeneous electron gas. We begin as depicted in Fig. 3.1(a) by studying oscillations in the smeared-out ion density while neglecting the electron dynamics, i.e., we keep ρ_{el} fixed. If we study the limit of small harmonic

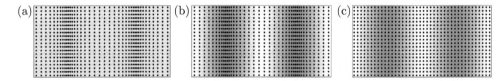

FIG. 3.1. Three types of oscillation in metals. The grayscale represents the electron density and the dots the ions. (a) Slow ionic density oscillations in a static electron gas (ion plasma oscillations). The restoring force is the long range Coulomb inter-action. (b) slow ion oscillations followed by the electron gas (sound waves, acoustic phonons). The restoring force is the compressibility of the disturbed electron gas. (c) Fast electronic plasma oscillations in a static ionic lattice (electronic plasma oscillations). The restoring force is the long range Coulomb interaction.

52

deviations from equilibrium $\delta\rho_{\text{ion}}(\mathbf{r},t) = \delta\rho_{\text{ion}}(\mathbf{r})\,e^{-i\Omega t}$, we obtain linear equations of motion with solutions of the form

$$\rho_{\text{ion}}(\mathbf{r},t) = \rho_{\text{ion}}^0 + \delta\rho_{\text{ion}}(\mathbf{r})\,e^{-i\Omega t}. \tag{3.1}$$

A non-zero $\delta\rho_{\text{ion}}$ corresponds to a charge density $Ze\,\delta\rho_{\text{ion}}$ and is hence associated with an electric field $\mathbf{E}$ obeying

$$\nabla\cdot\mathbf{E} = \frac{Ze}{\epsilon_0}\,\delta\rho_{\text{ion}} \quad\Rightarrow\quad \nabla\cdot\mathbf{f} = \frac{Z^2e^2\rho_{\text{ion}}^0}{\epsilon_0}\,\delta\rho_{\text{ion}}. \tag{3.2}$$

In the second equation we have introduced the force density $\mathbf{f}$, which to first order in $\delta\rho_{\text{ion}}$ becomes $\mathbf{f} = Ze\rho_{\text{ion}}\mathbf{E} \approx Ze\rho_{\text{ion}}^0\mathbf{E}$. This force equation is supplemented by the continuity equation, $\partial_t\rho_{\text{ion}} + \nabla\cdot(\rho_{\text{ion}}\mathbf{v}) = 0$, which to first order in $\delta\rho_{\text{ion}}$ becomes $\partial_t\delta\rho_{\text{ion}} + \rho_{\text{ion}}^0\nabla\cdot\mathbf{v} = 0$, since the velocity $\mathbf{v}$ is already a small quantity. Differentiating this with respect to time and using Newton's second law $\mathbf{f} = M\rho_{\text{ion}}\partial_t\mathbf{v}$, we obtain

$$\partial_t^2\delta\rho_{\text{ion}} + \frac{\nabla\cdot\mathbf{f}}{M} = 0 \Rightarrow \Omega^2\delta\rho_{\text{ion}} = \frac{Z^2e^2\rho_{\text{ion}}^0}{\epsilon_0 M}\delta\rho_{\text{ion}} \Rightarrow \Omega = \sqrt{\frac{Z^2e^2\rho_{\text{ion}}^0}{\epsilon_0 M}} = \sqrt{\frac{Ze^2\rho_{\text{el}}}{\epsilon_0 M}}. \tag{3.3}$$

Here Ω is the ionic plasma frequency. The ionic oscillations in the continuous jellium model are thus described by harmonic oscillators, which all have the same frequency Ω. Hence, the second quantization formalism leads to the following phonon Hamiltonian,

$$H_{\text{ph}} = \sum_{\mathbf{q}} \hbar\Omega\left(b_{\mathbf{q}}^\dagger b_{\mathbf{q}} + \frac{1}{2}\right). \tag{3.4}$$

These quantized ion oscillations are denoted phonons, and a model like this was proposed by Einstein in 1906 as the first attempt to explain the decrease of heat capacity C_V^{ion} of solids as a function of decreasing temperature (see Section 3.5). Note that the origin of the ionic plasma frequency is the long-range Coulomb interaction, which entered the analysis through the Maxwell equation $\nabla\cdot\mathbf{E} = Ze\delta\rho_{\text{ion}}/\epsilon_0$.

However, the Einstein phonons (also denoted optical phonons, see Section 3.3) are not a very good description of solids. Although it is correct that C_V^{ion} decreases at low temperature, the exact behavior is described by the Debye model, incorporating phonons with a photon-like dispersion $\omega_{\mathbf{q}} = v_s q$, where v_s is the sound velocity, instead of the Einstein dispersion $\omega_{\mathbf{q}} = \Omega$. These Debye phonons are also denoted acoustical phonons due to their relation to sound propagation. This is explained in details in Sections 3.3 and 3.5. To fully understand how the optical Einstein phonons get renormalized to become the acoustical Debye phonons requires the full machinery of quantum field theory, but we hint at the solution of the problem in Fig. 3.1(b) and in Section 3.2.

3.2   Electron-phonon interaction and the sound velocity

Compared to the light and very mobile valence electrons, the ions are much heavier, by a factor of more than 10^4, and much slower. Consequently, one would expect the electrons to follow the motion of the ions adiabatically and thereby always maintaining local charge neutrality and thus lowering the high ionic plasma frequency Ω, which is due to long-range charge Coulomb forces from the charge imbalance. This situation is depicted in Fig. 3.1(b), and to illustrate its correctness we now use it to estimate the sound velocity in metals. The kinetic energy density associated with a sound wave is of the order $\frac{1}{2}Mv_s^2 \, \rho_{\text{ion}}$, while the potential energy density associated with the restoring force must be related to the density dependent energy content of the compressed electron gas, i.e., of the order $\frac{3}{5}\rho_{\text{el}}\varepsilon_F$. In a stationary state these two energy densities must be of the same order of magnitude. This gives an estimate for v_s, which in a more detailed treatment (see Exercise 3.4) is expressed by the Bohm–Staver formula,

$$v_s = \sqrt{\frac{Zm}{3M}} \, v_F, \tag{3.5}$$

which for typical numbers yields $v_s \simeq 3000$ m/s as found experimentally. Note, how this estimate builds on classical considerations of the ionic motion while using the quantum result for the energy content of a degenerate electron gas. Surprisingly, an ordinary macroscopic phenomenon such as sound propagation is deeply rooted in quantum physics.

3.3   Lattice vibrations and phonons in 1D

Even though we are not yet able to demonstrate how to turn the optical ion plasma oscillations into acoustical phonons, we can, nevertheless, learn a lot by simply postulating the existence of a periodic ion lattice (as observed in nature), in which the ions can execute small oscillatory motion around their equilibrium positions. The surroundings somehow provide the restoring force.

We begin by a simple 1D quantum mechanical model consisting of a 1D box of length L containing N ions of mass M each interacting with its two neighbors through a linear force field (a spring) with the force constant K. The equilibrium position of the jth ion is denoted R_j^0, while its displacement away from this position is denoted u_j. The lattice spacing is denoted $a = R_j^0 - R_{j-1}^0$, so we have $L = Na$. This setup is shown in Fig. 3.2. The Hamiltonian is simply the sum of the kinetic energy of the ions and the potential energy of the springs, while the ion momentum p_j and the displacement u_j are canonical variables:

$$H_{\text{ph}} = \sum_{j=1}^{N}\left[\frac{1}{2M}p_j^2 + \frac{1}{2}K(u_j - u_{j-1})^2\right], \quad [p_{j_1}, u_{j_2}] = \frac{\hbar}{i}\delta_{j_1,j_2}. \tag{3.6}$$

As for the photon and jellium models, we impose periodic boundary conditions, $u_{N+1} = u_1$. Since the equilibrium system is periodic with the lattice spacing a, it is natural to solve the problem in k-space by performing a discrete Fourier transform. In analogy with electrons moving in a periodic lattice, also the present system of N ions forming a periodic lattice leads to a first Brillouin zone, FBZ, in reciprocal space. By Fourier transformation the N ion coordinates become N wave vectors in the FBZ:

$$\text{FBZ} = \left\{-\frac{\pi}{a} + \Delta k, \, -\frac{\pi}{a} + 2\Delta k, \ldots, -\frac{\pi}{a} + N\Delta k\right\}, \quad \Delta k = \frac{2\pi}{L} = \frac{2\pi}{a}\frac{1}{N}. \tag{3.7}$$

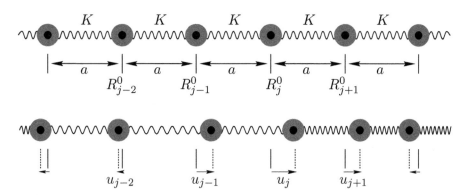

FIG. 3.2. A 1D lattice of ions with mass M, lattice constant a, and a nearest neighbor linear force coupling of strength K. The equilibrium positions R_j^0 and the displacements u_j are shown in top and bottom row, respectively.

The Fourier transforms of the conjugate variables are:

$$p_j \equiv \frac{1}{\sqrt{N}} \sum_{k \in \mathrm{FBZ}} p_k e^{ikR_j^0}, \quad u_j \equiv \frac{1}{\sqrt{N}} \sum_{k \in \mathrm{FBZ}} u_k e^{ikR_j^0}, \quad \delta_{R_j^0,0} = \frac{1}{N} \sum_{k \in \mathrm{FBZ}} e^{ikR_j^0},$$

$$p_k \equiv \frac{1}{\sqrt{N}} \sum_{j=1}^{N} p_j e^{-ikR_j^0}, \quad u_k \equiv \frac{1}{\sqrt{N}} \sum_{j=1}^{N} u_j e^{-ikR_j^0}, \quad \delta_{k,0} = \frac{1}{N} \sum_{j=1}^{N} e^{-ikR_j^0}.$$

$$(3.8)$$

By straightforward insertion of Eq. (3.8) into Eq. (3.6) we find

$$H = \sum_k \left[\frac{1}{2M} p_k p_{-k} + \frac{1}{2} M \omega_k^2 u_k u_{-k} \right], \quad \omega_k = \sqrt{\frac{K}{M}} \, 2 \left| \sin \frac{ka}{2} \right|, \quad [p_{k_1}, u_{k_2}] = \frac{\hbar}{i} \delta_{k_1, -k_2}.$$

$$(3.9)$$

This looks almost like the Hamiltonian for a set of harmonic oscillators except for some annoying details concerning k and $-k$. Note that while p_j in real space is a nice Hermitian operator, p_k in k-space is not self-adjoint. In fact, the hermiticity of p_j and the definition of the Fourier transform lead to $p_k^\dagger = p_{-k}$. Although the commutator in Eq. (3.9) tells us that u_k and p_{-k} form a pair of conjugate variables, we will not use this pair to form creation and annihilation operators in analogy with x and p of Eq. (1.75). The reason is that the Hamiltonian in the present case contains products like $p_k p_{-k}$ and not p_k^2 as in the original case. Instead we combine u_k and p_k in the definition of the annihilation and creation operators b_k and $b_{-k}^\dagger$:

$$b_k \equiv \frac{1}{\sqrt{2}} \left(\frac{u_k}{\ell_k} + i \frac{p_k}{\hbar/\ell_k} \right), \quad u_k \equiv \ell_k \frac{1}{\sqrt{2}} (b_{-k}^\dagger + b_k), \quad \ell_k = \sqrt{\frac{\hbar}{M\omega_k}},$$

$$b_{-k}^\dagger \equiv \frac{1}{\sqrt{2}} \left(\frac{u_k}{\ell_k} - i \frac{p_k}{\hbar/\ell_k} \right), \quad p_k \equiv \frac{\hbar}{\ell_k} \frac{i}{\sqrt{2}} (b_{-k}^\dagger - b_k).$$

$$(3.10)$$

Note how both the oscillator frequency, $\omega_k = \omega_{-k}$, and the oscillator length, $\ell_k = \ell_{-k}$, depend on the wavenumber k. Again, by direct insertion, it is readily verified that

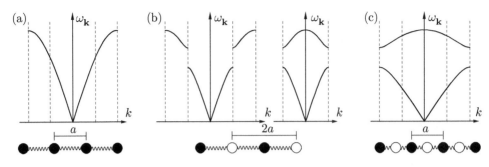

FIG. 3.3. The phonon dispersion relation for three different 1D lattices. (a) A system with lattice constant a and one ion of mass M_1 (black disks) per unit cell. (b) As in (a) but now substituting every second ion of mass M_1 with one of mass M_2 (white disks) resulting in two ions per unit cell and a doubling of the lattice constant. To the left is shown the extended zone scheme, and to the right the reduced zone scheme. (c) As in (a) but now with the addition of mass M_2 ions in between the mass M_1 ions resulting in two ions per unit cell, but the same lattice constant as in (a).

$$H_{ph} = \sum_k \hbar\omega_k \left(b_k^\dagger b_k + \frac{1}{2} \right), \quad [b_{k_1}, b_{k_2}^\dagger] = \delta_{k_1, k_2}. \tag{3.11}$$

This is finally the canonical form of a Hamiltonian describing a set of independent harmonic oscillators in second quantization. The quantized oscillations are denoted phonons. Their dispersion relation is shown in Fig. 3.3(a). It is seen from Eq. (3.9) that in the limit $k \to 0$ we have $\omega_k = \sqrt{K/M} \, ak$, so our solution Eq. (3.11) does in fact bring about the acoustical phonons. The sound velocity is found to be $v_s = \sqrt{K/M} \, a$, so upon measuring the value of it, one can determine the value of the free parameter K, the force constant in the model.

If, as shown in Fig. 3.3(b), the unit cell is doubled to hold two ions, the concept of phonon branches must be introduced. It is analogous to the Bloch bands for electrons. These came about as a consequence of breaking the translational invariance of the system by introducing a periodic lattice. Now we break the discrete translational invariance given by the lattice constant a. Instead the new lattice constant is $2a$. Hence the original BZ is halved in size and the original dispersion curve Fig. 3.3(a) is broken into sections. In the reduced zone scheme in Fig. 3.3(b) we of course find two branches, since no states can be lost. The lower branch resembles the original dispersion so it corresponds to acoustic phonons. The upper band never approaches zero energy, so to excite these phonons high energies are required. In fact they can be excited by light, so they are known as optical phonons. The origin of the energy difference between an acoustical and an optical phonon at the *same* wave length is sketched in Fig. 3.4 for the case of a two-ion unit cell. For acoustical phonons the size of the displacement of neighboring ions differs only slightly and its sign is the same, whereas for optical phonons the sign of the displacement alternates between the two types of ions.

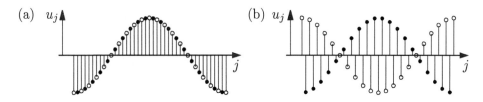

FIG. 3.4. (a) An acoustical and (b) an optical phonon having the *same* wave length for a 1D system with two ions, • and ○, per unit cell. In the acoustical case the two types of ions oscillate in phase, while in the optical case they oscillate π radians out of phase.

The generalization to p ions per unit cell is straightforward, and one finds the appearance of 1 acoustic branch and $(p-1)$ optical branches. The N appearing above, e.g. in Eq. (3.8), should be interpreted as the number of unit cells rather than the number of ions, so we have $N_{\text{ion}} = pN$. A branch index λ, analogous to the band index n for Bloch electrons is introduced to label the different branches, and in the general case the Hamiltonian Eq. (3.11) is changed into

$$H_{\text{ph}} = \sum_{k\lambda} \hbar\omega_{k\lambda} \left(b_{k\lambda}^{\dagger} b_{k\lambda} + \frac{1}{2} \right), \quad [b_{k_1\lambda_1}, b_{k_2\lambda_2}^{\dagger}] = \delta_{k_1,k_2}\,\delta_{\lambda_1,\lambda_2}. \qquad (3.12)$$

3.4   Acoustical and optical phonons in 3D

The fundamental principles for constructing the second quantized phonon fields established for the 1D case carries over to the 3D case, almost unchanged. The most notable difference is the appearance in 3D of polarization, in analogy to what we have already seen for the photon field. We treat the general case of any monatomic Bravais lattice. The ionic equilibrium positions are denoted $\mathbf{R}_j^0$ and the displacements by $\mathbf{u}(\mathbf{R}_j^0)$ with components $u_\alpha(\mathbf{R}_j^0)$, $\alpha = x, y, z$. The starting point of the analysis is a second order Taylor expansion in $u_\alpha(\mathbf{R}_j^0)$ of the potential energy $U[\mathbf{u}(\mathbf{R}_1^0), \ldots, \mathbf{u}(\mathbf{R}_N^0)]$,

$$U \approx U_0 + \frac{1}{2} \sum_{\mathbf{R}_1^0 \mathbf{R}_2^0} \sum_{\alpha\beta} u_\alpha(\mathbf{R}_1^0) \left. \frac{\partial^2 U}{\partial u_\alpha(\mathbf{R}_1^0)\,\partial u_\beta(\mathbf{R}_2^0)} \right|_{\mathbf{u}=0} u_\beta(\mathbf{R}_2^0). \qquad (3.13)$$

Note that nothing has been assumed about the range of the potential. It may very well go much beyond the nearest neighbor case studied in the 1D case. The central object in the theory is the force strength matrix $\frac{\partial^2 U}{\partial u_\alpha \partial u_\beta}$ (generalizing K from the 1D case) and its Fourier transform, the so-called dynamical matrix $\mathbf{D}(\mathbf{k})$ with components $D_{\alpha\beta}(\mathbf{k})$:

$$D_{\alpha\beta}(\mathbf{R}_1^0 - \mathbf{R}_2^0) = \left. \frac{\partial^2 U}{\partial u_\alpha(\mathbf{R}_1^0)\,\partial u_\beta(\mathbf{R}_2^0)} \right|_{\mathbf{u}=0}, \quad D_{\alpha\beta}(\mathbf{k}) = \sum_{\mathbf{R}} D_{\alpha\beta}(\mathbf{R})\, e^{-i\mathbf{k}\cdot\mathbf{R}}. \qquad (3.14)$$

The discrete Fourier transform in 3D is a straightforward generalization of the one in 1D, and for an arbitrary function $f(\mathbf{R}_j^0)$ we have

$$f(\mathbf{R}_j^0) \equiv \frac{1}{\sqrt{N}} \sum_{\mathbf{k} \in \mathrm{FBZ}} f(\mathbf{k})\, e^{i\mathbf{k}\cdot\mathbf{R}_j^0}, \qquad \delta_{\mathbf{R}_j^0,0} = \frac{1}{N} \sum_{\mathbf{k} \in \mathrm{FBZ}} e^{i\mathbf{k}\cdot\mathbf{R}_j^0},$$

$$f(\mathbf{k}) = \frac{1}{\sqrt{N}} \sum_{j=1}^{N} f(\mathbf{R}_j^0)\, e^{-i\mathbf{k}\cdot\mathbf{R}_j^0}, \qquad \delta_{\mathbf{k},0} = \frac{1}{N} \sum_{j=1}^{N} e^{-i\mathbf{k}\cdot\mathbf{R}_j^0}. \tag{3.15}$$

Due to the lattice periodicity $D_{\alpha\beta}(\mathbf{R}_1^0 - \mathbf{R}_2^0)$ depends only on the difference between any two ion positions. The D-matrix has the following three symmetry properties[12]

$$\boldsymbol{D}^{\mathrm{T}}(\mathbf{R}^0) = \boldsymbol{D}(\mathbf{R}^0), \qquad \sum_{\mathbf{R}^0} \boldsymbol{D}(\mathbf{R}^0) = 0, \qquad \boldsymbol{D}(-\mathbf{R}^0) = \boldsymbol{D}(\mathbf{R}^0). \tag{3.16}$$

Using these symmetries in connection with $\boldsymbol{D}(\mathbf{k})$ we obtain

$$\boldsymbol{D}(\mathbf{k}) = \sum_{\mathbf{R}^0} \boldsymbol{D}(\mathbf{R}^0)\, e^{-i\mathbf{k}\cdot\mathbf{R}^0} = \frac{1}{2}\left[\sum_{\mathbf{R}^0} \boldsymbol{D}(\mathbf{R}^0)\, e^{-i\mathbf{k}\cdot\mathbf{R}^0} + \sum_{\mathbf{R}^0} \boldsymbol{D}(-\mathbf{R}^0)\, e^{i\mathbf{k}\cdot\mathbf{R}^0} \right]$$

$$= \frac{1}{2} \sum_{\mathbf{R}^0} \boldsymbol{D}(\mathbf{R}^0) \left[e^{i\mathbf{k}\cdot\mathbf{R}^0} + e^{-i\mathbf{k}\cdot\mathbf{R}^0} - 2 \right] = -2 \sum_{\mathbf{R}^0} \boldsymbol{D}(\mathbf{R}^0) \sin^2\left[\tfrac{1}{2}\mathbf{k}\cdot\mathbf{R}^0\right]. \tag{3.17}$$

Thus $\boldsymbol{D}(\mathbf{k})$ is real and symmetric, hence diagonalizable in an orthonormal basis.

The classical equation of motions for the ions are simply

$$M\ddot{u}_\alpha(\mathbf{R}_1^0) = -\frac{\partial U}{\partial u_\alpha(\mathbf{R}_1^0)} \quad \Rightarrow \quad -M\ddot{\mathbf{u}}(\mathbf{R}_1^0) = \sum_{\mathbf{R}_2^0} \boldsymbol{D}(\mathbf{R}_2^0 - \mathbf{R}_1^0)\, \mathbf{u}(\mathbf{R}_2^0). \tag{3.18}$$

We seek simple harmonic solutions to the problem and find

$$\mathbf{u}(\mathbf{R}^0, t) \propto \boldsymbol{\epsilon}\, e^{i(\mathbf{k}\cdot\mathbf{R}^0 - \omega t)} \quad \Rightarrow \quad M\omega^2 \boldsymbol{\epsilon} = \boldsymbol{D}(\mathbf{k})\, \boldsymbol{\epsilon}. \tag{3.19}$$

Since $\boldsymbol{D}(\mathbf{k})$ is a real symmetric matrix, there exists for any value of $\mathbf{k}$ an orthonormal basis set of vectors $\{\boldsymbol{\epsilon}_{\mathbf{k},1}, \boldsymbol{\epsilon}_{\mathbf{k},2}, \boldsymbol{\epsilon}_{\mathbf{k},3}\}$, the so-called polarization vectors, that diagonalizes $\boldsymbol{D}(\mathbf{k})$, i.e., they are eigenvectors:

$$\boldsymbol{D}(\mathbf{k})\, \boldsymbol{\epsilon}_{\mathbf{k}\lambda} = K_{\mathbf{k}\lambda}\, \boldsymbol{\epsilon}_{\mathbf{k}\lambda}, \qquad \boldsymbol{\epsilon}_{\mathbf{k}\lambda}\cdot\boldsymbol{\epsilon}_{\mathbf{k}\lambda'} = \delta_{\lambda,\lambda'}, \qquad \lambda, \lambda' = 1, 2, 3. \tag{3.20}$$

We have now found the classical eigenmodes $\mathbf{u}_{\mathbf{k}\lambda}$ of the 3D lattice vibrations characterized by the wavevector $\mathbf{k}$ and the polarization vector $\boldsymbol{\epsilon}_{\mathbf{k}\lambda}$:

$$M\omega^2 \boldsymbol{\epsilon}_{\mathbf{k}\lambda} = K_{\mathbf{k}\lambda} \boldsymbol{\epsilon}_{\mathbf{k}\lambda} \quad \Rightarrow \quad \mathbf{u}_{\mathbf{k}\lambda}(\mathbf{R}^0, t) = \boldsymbol{\epsilon}_{\mathbf{k}\lambda}\, e^{i(\mathbf{k}\cdot\mathbf{R}^0 - \omega_{\mathbf{k}\lambda} t)}, \quad \omega_{\mathbf{k}\lambda} \equiv \sqrt{\frac{K_{\mathbf{k}\lambda}}{M}}. \tag{3.21}$$

Using as in Eq. (3.10) the now familiar second quantization procedure of harmonic oscillators we obtain

---

[12]The first follows from the interchangeability of the order of the differentiation in Eq. (3.14). The second follows from the fact that $U - U_0 = 0$ if *all* the displacements are the same, given by an arbitrary value, say $\mathbf{d}$, because then $0 = \sum_{\mathbf{R}_1^0 \mathbf{R}_2^0} \mathbf{d}\cdot\boldsymbol{D}(\mathbf{R}_1^0 - \mathbf{R}_2^0)\cdot\mathbf{d} = N\mathbf{d}\cdot[\sum_{\mathbf{R}^0} \boldsymbol{D}(\mathbf{R}^0)]\cdot\mathbf{d}$. The third follows from inversion symmetry always present in monatomic Bravais lattices.

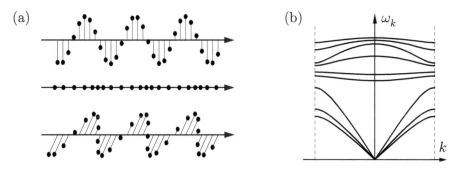

FIG. 3.5. (a) Three examples of polarization in phonon modes: transverse, longitudinal and general. (b) A generic phonon spectrum for a system with 3 ions in the unit cell. The 9 modes divides into 3 acoustical and 6 optical modes.

$$\mathbf{u}_{\mathbf{k}\lambda} \equiv \ell_{\mathbf{k}\lambda} \frac{1}{\sqrt{2}} \left(b^\dagger_{-\mathbf{k},\lambda} + b_{\mathbf{k},\lambda} \right) \boldsymbol{\epsilon}_{\mathbf{k}\lambda}, \quad \ell_{\mathbf{k}\lambda} \equiv \sqrt{\frac{\hbar}{M\omega_{\mathbf{k}\lambda}}}, \tag{3.22}$$

$$H_{\mathrm{ph}} = \sum_{\mathbf{k}\lambda} \hbar\omega_{\mathbf{k}\lambda} \left(b^\dagger_{\mathbf{k}\lambda} b_{\mathbf{k}\lambda} + \frac{1}{2} \right), \quad [b_{\mathbf{k}\lambda}, b^\dagger_{\mathbf{k}'\lambda'}] = \delta_{\mathbf{k},\mathbf{k}'} \, \delta_{\lambda,\lambda'}. \tag{3.23}$$

Now, what about acoustical and optical phonons in 3D? It is clear from Eq. (3.17) that $\mathbf{D}(\mathbf{k}) \propto k^2$ for $k \to 0$, so the same holds true for its eigenvalues $K_{\mathbf{k}\lambda}$. The dispersion relation in Eq. (3.19) therefore becomes $\omega_{\mathbf{k}\lambda} = v_\lambda(\theta_k, \phi_k) \, k$, which describes an acoustical phonon with a sound velocity $v_\lambda(\theta_k, \phi_k)$ in general depending on both the direction of $\mathbf{k}$ and the polarization λ. As in 1D the number of ions in the unit cell can be augmented from 1 to p. In that case it can be shown that of the resulting $3p$ modes 3 are acoustical and $3(p-1)$ are optical modes. The acoustical modes are appearing because it is always possible to construct modes where all the ions have been given nearly the same displacement resulting in an arbitrarily low energy cost associated with such a deformation of the lattice. The phonon modes for a unit cell with three ions are shown in Fig. 3.5.

A 3D lattice with N unit cells each containing p ions, each of which can oscillate in 3 directions, is described by $3pN$ modes. In terms of phonon modes we end up with $3p$ so-called phonon branches $\omega_{\mathbf{k}\lambda}$, which for each branch index λ are defined at N discrete points in $\mathbf{k}$-space. Thus, in 3D the index λ contains information on both which polarization and which of the acoustical or optical modes we are dealing with.

3.5 The specific heat of solids in the Debye model

Debye's phonon model is a simple model, which describes the temperature dependence of the heat capacitance $C_V = \frac{\partial E}{\partial T}$ of solids exceedingly well, although it contains just one material dependent free parameter. The phonon spectrum Fig. 3.5(b) in the reduced zone scheme has $3p$ branches. The acoustical and optical phonon branches in the extended zone scheme for a 1D chain with two ions per unit cell are shown in Fig. 3.6(a). Note, how the optical branch appears as an extension of the acoustical branch. In d dimensions a reasonable average of the spectrum can be obtained by representing all the phonon branches in the reduced zone scheme with d acoustical branches in the extended zone scheme, each with a linear dispersion relation $\omega_{\mathbf{k}\lambda} = v_\lambda k$. Furthermore, since we will use the model to calculate the specific heat by averaging over all modes, we can even employ a suitable average v_D over the polarization dependent velocities v_λ and use the *same* linear dispersion relation for *all* acoustical branches,

$$\omega_{\mathbf{k}\lambda} \equiv v_D k \Rightarrow \varepsilon = \hbar v_D\, k. \tag{3.24}$$

Even though we have deformed the phonon spectrum we may not change the number of phonon modes. In the 3D Debye model we have $3N_{\text{ion}}$ modes, in the form of 3 acoustic branches each with N_{ion} allowed wavevectors, where N_{ion} is the number of ions in the lattice. Since we are using periodic boundary conditions the counting of the allowed phonon wavevectors is equivalent to that of Section 2.1.2 for plane wave electron states, i.e., $N_{\text{ion}} = [\mathcal{V}/(2\pi)^3] \times [\text{volume in } \mathbf{k}\text{-space}]$. Since the Debye spectrum Eq. (3.24) is isotropic in $\mathbf{k}$-space, the Debye phonon modes must occupy a sphere in this space, i.e., all modes with $|\mathbf{k}| < k_D$, where k_D is denoted the Debye wave number determined by

$$N_{\text{ion}} = \frac{\mathcal{V}}{(2\pi)^3}\, \frac{4}{3}\pi\, k_D^3. \tag{3.25}$$

Inserting k_D into Eq. (3.24) yields the characteristic Debye energy, $\hbar\omega_D$ and hence the characteristic Debye temperature T_D:

$$\hbar\omega_D \equiv k_B T_D \equiv \hbar v_D\, k_D \Rightarrow 6\pi^2\, N_{\text{ion}}\, (\hbar v_D)^3 = \mathcal{V}\, (k_B T_D)^3. \tag{3.26}$$

Continuing the analogy with the electron case the density of phonon states $D_{\text{ph}}(\varepsilon)$ is found by combining Eqs. (3.24) and (3.25) and multiplying by 3 for the number of acoustic branches,

$$N_{\text{ion}}(\varepsilon) = \frac{\mathcal{V}}{6\pi^2}\, \frac{1}{(\hbar v_D)^3}\, \varepsilon^3 \Rightarrow D_{\text{ph}}(\varepsilon) = 3\frac{dN_{\text{ion}}(\varepsilon)}{d\varepsilon} = \frac{\mathcal{V}}{2\pi^2}\, \frac{1}{(\hbar v_D)^3}\, \varepsilon^2, \quad 0 < \varepsilon < k_B T_D. \tag{3.27}$$

The energy $E_{\text{ion}}(T)$ of the vibrating lattice is now easily computed using the Bose–Einstein distribution function $n_B(\varepsilon)$ Eq. (1.127) for the bosonic phonons:

$$E_{\text{ion}}(T) = \int_0^{k_B T_D} d\varepsilon\, \varepsilon D_{\text{ph}}(\varepsilon) n_B(\varepsilon) = \frac{\mathcal{V}}{2\pi^2}\, \frac{3}{(\hbar v_D)^3} \int_0^{k_B T_D} d\varepsilon\, \frac{\varepsilon^3}{e^{\beta\varepsilon} - 1}. \tag{3.28}$$

It is now straightforward to obtain C_V^{ion} from Eq. (3.28) by differentiation:

$$C_V^{\text{ion}}(T) = \frac{\partial E_{\text{ion}}}{\partial T} = 9N_{\text{ion}} k_B \left(\frac{T}{T_D}\right)^3 \int_0^{T_D/T} dx\, \frac{x^4\, e^x}{(e^x - 1)^2}, \tag{3.29}$$

where the integrand is rendered dimensionless by introducing T_D from Eq. (3.26). Note that T_D is the only free parameter in the Debye model of heat capacitance; v_D dropped out of the calculation. Note also how the model reproduces the classical Dulong–Petit value in the high temperature limit, where all oscillators are thermally excited. In the

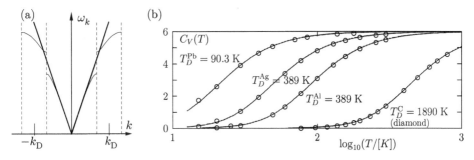

FIG. 3.6.   (a) The linear Debye approximation to the phonon spectrum with the Debye wave vector k_D shown. (b) Comparison between experiment and the Debye model of heat capacitance for lead, silver, aluminum, and diamond.

low temperature limit the oscillators "freeze out" and the heat capacitance drops as T^3,

$$C_V^{\mathrm{ion}}(T) \xrightarrow[T \ll T_\mathrm{D}]{} \frac{12\pi^4}{5} N_{\mathrm{ion}} k_\mathrm{B} \left(\frac{T}{T_\mathrm{D}}\right)^3, \qquad C_V^{\mathrm{ion}}(T) \xrightarrow[T \gg T_\mathrm{D}]{} 3 N_{\mathrm{ion}} k_\mathrm{B}. \qquad (3.30)$$

In Fig. 3.6(b) the Debye model is compared to experiment. A remarkable agreement is obtained over the wide temperature range from 10 to 1000 K after fitting T_D for each of the widely different materials lead, aluminum, silver and diamond.

We end this section by a historical remark. The very first published application of quantum theory to a condensed matter problem was in fact Einstein's work from 1906, reproduced in Fig. 3.7(a), explaining the main features of Weber's 1875 measurements on diamond. In analogy with Planck's quantization of the oscillators related to black body radiation, Einstein quantized the oscillators corresponding to the lattice vibrations, assuming that all oscillators had the same frequency ω_E. So instead of Eq. (3.27), Einstein employed the much simpler $D_{\mathrm{ion}}^E(\varepsilon) = \delta(\varepsilon - \hbar\omega_E)$, which immediately leads to

$$C_V^{\mathrm{ion},E}(T) = 3 N_{\mathrm{ion}} k_\mathrm{B} \left(\frac{T_E}{T}\right)^2 \frac{e^{T_E/T}}{(e^{T_E/T} - 1)^2}, \qquad T_E \equiv \hbar\omega_E / k_\mathrm{B}. \qquad (3.31)$$

While this theory also gives the classical result $3 N_{\mathrm{ion}} k_\mathrm{B}$ in the high temperature limit, it exaggerates the decrease of C_V^{ion} at low temperatures by predicting an exponential suppression. In Fig. 3.7(b) is shown a comparison of Debye's and Einstein's models. Nowadays, Einstein's formula is still in use, since it provides a fairly accurate description of the optical phonons which in many cases have a reasonably flat dispersion relation.

3.6   Electron-phonon interaction in the lattice model

In Chapter 2 we mentioned that in the lattice model the electron-ion interaction splits in two terms, one arising from the static lattice and the other from the ionic vibrations, $V_{\mathrm{el-ion}} = V_{\mathrm{el-latt}} + V_{\mathrm{el-ph}}$. The former has already been dealt with in

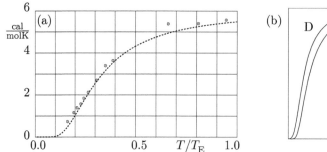
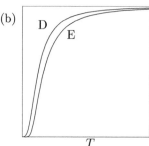

FIG. 3.7. (a) The first application of quantum theory to condensed matter physics. Einstein's 1906 theory of heat capacitance of solids. The theory is compared to Weber's 1875 measurements on diamond. (b) A comparison between Debye's and Einstein's model.

H_{Bloch}, so in this section the task is to derive the explicit second quantized form of the latter. Regarding the basis states for the combined electron and phonon system we are now in the situation discussed in Section 1.4.5. We will simply use the product states given in Eq. (1.108).

Our point of departure is the simple expression for the Coulomb energy of an electron density in the electric potential $V_{\text{ion}}(\mathbf{r} - \mathbf{R}_j)$ of an ion placed at the position $\mathbf{R}_j$,

$$V_{\text{el}-\text{ion}} = \int d\mathbf{r}\,(-e)\rho_{\text{el}}(\mathbf{r}) \sum_{j=1}^{N} V_{\text{ion}}(\mathbf{r} - \mathbf{R}_j). \qquad (3.32)$$

As before the actual ion coordinates are given by $\mathbf{R}_j = \mathbf{R}_j^0 + \mathbf{u}_j$, where $\mathbf{R}_j^0$ are the ionic equilibrium positions, i.e., the static periodic lattice, and where $\mathbf{u}_j$ denotes the lattice vibrations. The respective contributions from these two sets of coordinates are separated by a Taylor expansion, $V_{\text{ion}}(\mathbf{r} - \mathbf{R}_j) \approx V_{\text{ion}}(\mathbf{r} - \mathbf{R}_j^0) - \nabla_{\mathbf{r}} V_{\text{ion}}(\mathbf{r} - \mathbf{R}_j^0) \cdot \mathbf{u}_j$, note the sign of the second term, and we obtain

$$V_{\text{el}-\text{ion}} = \int d\mathbf{r}\,(-e)\rho_{\text{el}}(\mathbf{r}) \sum_{j=1}^{N} V_{\text{ion}}(\mathbf{r} - \mathbf{R}_j^0) - \int d\mathbf{r}\,(-e)\rho_{\text{el}}(\mathbf{r}) \sum_{j=1}^{N} \nabla_{\mathbf{r}} V_{\text{ion}}(\mathbf{r} - \mathbf{R}_j^0) \cdot \mathbf{u}_j. \qquad (3.33)$$

The first term is the one entering H_{Bloch} in Eq. (2.6), while the second is the electron-phonon interaction, also sketched in Fig. 3.8,

$$V_{\text{el}-\text{ph}} = \int d\mathbf{r}\, \rho_{\text{el}}(\mathbf{r}) \left\{ \sum_j e\, \mathbf{u}_j \cdot \nabla_{\mathbf{r}} V_{\text{ion}}(\mathbf{r} - \mathbf{R}_j^0) \right\}. \qquad (3.34)$$

$V_{\text{el}-\text{ph}}$ is readily defined in real space, but a lot easier to use in $\mathbf{k}$-space, so we will proceed by Fourier transforming it. Let us begin with the ionic part, the $\mathbf{u} \cdot \nabla V$-term. The Fourier transform of $\mathbf{u}_j$ is given in Eq. (3.22), where we note that the phonon wavevector $\mathbf{k}$ is restricted to the Brillouin zone $\mathbf{k} \in \text{FBZ}$. Defining $V_{\text{ion}}(\mathbf{r}) = \frac{1}{\mathcal{V}} \sum_{\mathbf{p}} V_{\mathbf{p}} e^{i\mathbf{p}\cdot\mathbf{r}}$, we see that $\nabla_{\mathbf{r}}$ simply brings down a factor $i\mathbf{p}$. To facilitate comparison

FIG. 3.8.   (a) Being in an eigenstate a Bloch electron moves through a perfect lattice without being scattered. (b) A displaced ion results in an electric dipole relative to the perfect background, and this can scatter Bloch electrons from $|\mathbf{k}, \sigma\rangle$ to $|\mathbf{k}', \sigma\rangle$.

to the phonon wavevector $\mathbf{k}$, we decompose $\mathbf{p}$ as in Eq. (2.10): $\mathbf{p} = \mathbf{q} + \mathbf{G}$, where $\mathbf{q} \in$ FBZ and $\mathbf{G} \in$ RL. All in all we have

$$\nabla_{\mathbf{r}} V_{\text{ion}}(\mathbf{r} - \mathbf{R}_j^0) = \frac{1}{\mathcal{V}} \sum_{\mathbf{q} \in \text{FBZ}} \sum_{\mathbf{G} \in \text{RL}} i(\mathbf{q} + \mathbf{G}) V_{\mathbf{q}+\mathbf{G}} e^{i(\mathbf{q}+\mathbf{G})\cdot(\mathbf{r}-\mathbf{R}_j^0)}, \tag{3.35}$$

$$\mathbf{u}_j = \frac{1}{\sqrt{N}} \sum_{\mathbf{k} \in \text{FBZ}} \sum_{\lambda} \frac{\ell_{\mathbf{k}\lambda}}{\sqrt{2}} \left(b_{\mathbf{k},\lambda} + b_{-\mathbf{k},\lambda}^{\dagger}\right) \boldsymbol{\epsilon}_{\mathbf{k}\lambda} \, e^{i\mathbf{k}\cdot\mathbf{R}_j^0}. \tag{3.36}$$

These expressions, together with $\sum_j e^{i\mathbf{k}\cdot\mathbf{R}_j} = N\delta_{\mathbf{k},0}$, and multiplying by e, lead to

$$\sum_j e\,\mathbf{u}_j \cdot \nabla_{\mathbf{r}} V_{\text{el}}(\mathbf{r} - \mathbf{R}_j^0) = \frac{1}{\mathcal{V}} \sum_{\substack{\mathbf{q} \in \text{FBZ} \\ \mathbf{G} \in \text{RL}, \lambda}} g_{\mathbf{q},\mathbf{G},\lambda} \left(b_{\mathbf{q},\lambda} + b_{-\mathbf{q},\lambda}^{\dagger}\right) e^{i(\mathbf{q}+\mathbf{G})\cdot\mathbf{r}}, \tag{3.37}$$

where we have introduced the phonon coupling strength $g_{\mathbf{q},\mathbf{G},\lambda}$ given by

$$g_{\mathbf{q},\mathbf{G},\lambda} = ie\sqrt{\frac{N\hbar}{2M\omega_{\mathbf{q}\lambda}}}(\mathbf{q} + \mathbf{G})\cdot\boldsymbol{\epsilon}_{\mathbf{q}\lambda}\, V_{\mathbf{q}+\mathbf{G}}. \tag{3.38}$$

The final result, $V_{\text{el-ph}}$, is now obtained by inserting the Fourier representation of the electron density, $\rho_{\text{el}}(\mathbf{r}) = \frac{1}{\mathcal{V}} \sum_{\mathbf{kp}\sigma} e^{-i\mathbf{p}\cdot\mathbf{r}} c_{\mathbf{k}+\mathbf{p}\sigma}^{\dagger} c_{\mathbf{k}\sigma}$, derived in Eq. (1.94), together with Eq. (3.37) into Eq. (3.34), and utilizing $\int d\mathbf{r}\, e^{i\mathbf{k}\cdot\mathbf{r}} = \mathcal{V}\delta_{\mathbf{k},0}$:

$$V_{\text{el-ph}} = \frac{1}{\mathcal{V}} \sum_{\mathbf{k}\sigma} \sum_{\mathbf{q}\lambda} \sum_{\mathbf{G}} g_{\mathbf{q},\mathbf{G},\lambda}\, c_{\mathbf{k}+\mathbf{q}+\mathbf{G},\sigma}^{\dagger} c_{\mathbf{k}\sigma} \left(b_{\mathbf{q},\lambda} + b_{-\mathbf{q},\lambda}^{\dagger}\right). \tag{3.39}$$

The interpretation of this formula is quite simple. Under momentum conservation (but only up to an undetermined reciprocal lattice vector due to the periodicity of the lattice) and spin conservation, the electrons can be scattered from any initial state $|\mathbf{k}, \sigma\rangle_{\text{el}}$ to the final state $|\mathbf{k} + \mathbf{q} + \mathbf{G}, \sigma\rangle_{\text{el}}$ either by absorbing a phonon from the state $|\mathbf{q}\lambda\rangle_{\text{ph}}$ or by emitting a phonon into the state $|-\mathbf{q}\lambda\rangle_{\text{ph}}$. A graphical representation of this fundamental process is shown in Fig. 3.9.

The normal processes, i.e., processes where by definition $\mathbf{G} = 0$, often tend to dominate over the so-called umklapp processes, where $\mathbf{G} \neq 0$, so in the following we shall completely neglect the latter.[13] Moreover, we shall treat only isotropic media,

---

[13]There are mainly two reasons why the umklapp processes often can be neglected: (1) $V_{\mathbf{q}+\mathbf{G}}$ is small due to the $1/(\mathbf{q}+\mathbf{G})^2$ dependence, and (2) At low temperatures the phase space available for umklapp processes is small.

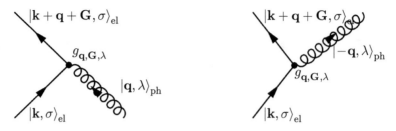

FIG. 3.9. A graphical representation of the fundamental electron-phonon coupling. The electron states are represented by the straight lines, the phonon states by curly spring-like lines, and the coupling strength by a dot. To the left the electron is scattered by absorbing a phonon, to the right by emitting a phonon.

where $\epsilon_{\mathbf{q}\lambda}$ is either parallel with or perpendicular to $\mathbf{q}$, i.e., $\mathbf{q}\cdot\epsilon_{\mathbf{q}\lambda}$ in $g_{\mathbf{q},\mathbf{G}=0,\lambda}$ is only non-zero for longitudinally polarized phonons. So in the isotropic case for normal (IN) phonon processes we have

$$V_{\mathrm{el-ph}}^{\mathrm{IN}} = \frac{1}{\mathcal{V}}\sum_{\mathbf{k}\sigma}\sum_{\mathbf{q}\lambda_l} g_{\mathbf{q},\lambda_l}\, c_{\mathbf{k}+\mathbf{q},\sigma}^{\dagger}c_{\mathbf{k}\sigma}\left(b_{\mathbf{q},\lambda_l}+b_{-\mathbf{q},\lambda_l}^{\dagger}\right). \tag{3.40}$$

Finally, the most significant physics of the electron-phonon coupling can often be extracted from considering just the acoustical modes. Due to their low energies they are excited significantly more than the high energy optical phonons at temperatures lower than the Debye temperature. Thus in the isotropic case for normal acoustical (INA) phonon processes only the longitudinal acoustical branch enters and we have

$$V_{\mathrm{el-ph}}^{\mathrm{INA}} = \frac{1}{\mathcal{V}}\sum_{\mathbf{k}\sigma}\sum_{\mathbf{q}} g_{\mathbf{q}}\, c_{\mathbf{k}+\mathbf{q},\sigma}^{\dagger}c_{\mathbf{k}\sigma}\left(b_{\mathbf{q}}+b_{-\mathbf{q}}^{\dagger}\right). \tag{3.41}$$

If we, for ions with charge $+Ze$, approximate $V_{\mathbf{q}}$ by a Yukawa potential, $V_{\mathbf{q}} = \frac{Ze}{\epsilon_0}\frac{1}{q^2+k_s^2}$ (see Exercise 1.6), the explicit form of the coupling constant $g_{\mathbf{q}}$ is particularly simple:

$$g_{\mathbf{q}} = i\frac{Ze^2}{\epsilon_0}\frac{q}{q^2+k_s^2}\sqrt{\frac{N\hbar}{2M\omega_{\mathbf{q}}}}. \tag{3.42}$$

3.7 Electron-phonon interaction in the jellium model

Finally, we return to the case of Einstein phonons in the jellium model treated in Section 3.1. The electron-phonon interaction in this case is derived in analogy with the that of normal lattice phonons in the isotropic case, Eq. (3.41). If we as in Section 3.1 neglect the weak dispersion of the Einstein phonons and simply assume that they all vibrate with the ion plasma frequency Ω of Eq. (3.3), the result for N vibrating ions in the volume $\mathcal{V}$ is

$$V_{\mathrm{el-ph}}^{\mathrm{jel}} = \frac{1}{\mathcal{V}}\sum_{\mathbf{k}\sigma}\sum_{\mathbf{q}} g_{\mathbf{q}}^{\mathrm{jel}}\, c_{\mathbf{k}+\mathbf{q},\sigma}^{\dagger}c_{\mathbf{k}\sigma}\left(b_{\mathbf{q}}+b_{-\mathbf{q}}^{\dagger}\right), \tag{3.43}$$

with

$$g_{\mathbf{q}}^{\text{jel}} = i\frac{Ze^2}{\epsilon_0}\frac{1}{q}\sqrt{\frac{N\hbar}{2M\Omega}}. \tag{3.44}$$

3.8   Summary and outlook

In this chapter we have derived the second quantized form of the Hamiltonian of the isolated phonon system and the electron-phonon coupling. The phonon problem is a real interacting many-particle system, since each ion is coupled to its neighbors,

$$H_{\text{ph}} = \sum_{j=1}^{N}\left[\frac{1}{2M}p_j^2 + \frac{1}{2}K(u_j - u_{j-1})^2\right], \quad [p_{j_1}, u_{j_2}] = \frac{\hbar}{i}\delta_{j_1, j_2}.$$

The solution of this problem (here for acoustical phonons) in second quantization,

$$H_{\text{ph}} = \sum_k \hbar\omega_k\left(b_k^\dagger b_k + \frac{1}{2}\right), \quad [b_{k_1}, b_{k_2}^\dagger] = \delta_{k_1, k_2}, \quad \omega_k = \sqrt{\frac{K}{M}}\, 2\left|\sin\frac{ka}{2}\right|,$$

thus constitutes our first solution of a real many-particle system with interactions.

Also the treatment of the electron-phonon coupling marks an important step forward: here we dealt for the first time with the coupling between two different kinds of particles, fermionic electrons and bosonic phonons. In the case of isotropic normal acoustical (INA) phonon proceses the electron-phonon coupling is

$$V_{\text{el-ph}}^{\text{INA}} = \frac{1}{\mathcal{V}}\sum_{\mathbf{k}\sigma}\sum_{\mathbf{q}} g_{\mathbf{q}}\, c_{\mathbf{k}+\mathbf{q},\sigma}^\dagger c_{\mathbf{k}\sigma}\left(b_{\mathbf{q}} + b_{-\mathbf{q}}^\dagger\right), \quad g_{\mathbf{q}} = i\frac{Ze^2}{\epsilon_0}\frac{q}{q^2 + k_s^2}\sqrt{\frac{N\hbar}{2M\omega_{\mathbf{q}}}}.$$

The electron-phonon coupling is a very important mechanism in condensed matter systems. It is the cause of a large part of electrical resistivity in metals and semiconductors, and it also plays a major role in studies of heat transport. Exercise 3.1 and Exercise 3.2 give a first idea of how the electron-phonon coupling leads to a scattering or relaxation time for electrons.

We shall return to the electron-phonon coupling in Chapter 17, and there study the remarkable instability of the electron gas induced by the intricate interplay between electrons and phonons. This instability is the starting point of the very successful microscopic theory of superconductivity, the so-called BCS theory. This theory will be treated in Chapter 18, and we shall see how indeed the electron-phonon scattering is the origin of conventional superconductivity.

For more details on the electron-phonon interaction see Mahan (1990).

4

MEAN-FIELD THEORY

The physics of interacting particles is often very complicated because the motions of the individual particles depend on the position of all the others, or in other words the particles motions become correlated. This is clearly the case for a system of charged particles interacting by Coulomb forces, such as e.g., the electron gas. There we expect the probability to find two electrons in close proximity to be small due to the strong repulsive interaction. Consequently, due to these correlation effects there is a suppressed density in the neighborhood of every electron, and one talks about a. "correlation hole"

Nevertheless, in spite of this complicated problem there are a number of cases where a more crude treatment, not fully including the correlations, gives a good physical model. In these cases it suffices to include correlations "on the average," which means that the effect of the other particles is included as a mean density (or mean-field), leaving an effective single-particle problem, which is soluble. This idea is illustrated in Fig. 4.1. The mean-fields are chosen as those which minimize the free energy, which in turn ensure that the method is consistent, as we shall see shortly. This approximation scheme is called "mean-field theory." Upon performing the mean-field approximation we can neglect the detailed dynamics and the time-independent second-quantization method described in Chapter 1 suffices.

4.1 Basic concepts of mean-field theory

There exist numerous examples of the success of the mean-field method and its ability to explain various physical phenomena. In this chapter, we shall discuss a few examples from condensed matter physics, but before going to specific examples let us discuss the mathematical structure of the mean-field theory.

First we consider a system with two kinds of particles, described by operators a_ν and b_μ, respectively. Let us assume that only interactions between different kind of particles are important. The Hamiltonian is

$$H = H_0 + V_{\text{int}}, \tag{4.1a}$$

$$H_0 = \sum_\nu \xi_\nu^a a_\nu^\dagger a_\nu + \sum_\mu \xi_\mu^b b_\mu^\dagger b_\mu, \tag{4.1b}$$

$$V_{\text{int}} = \sum_{\nu\nu',\mu\mu'} V_{\nu\mu,\nu'\mu'} a_\nu^\dagger b_\mu^\dagger b_{\mu'} a_{\nu'}. \tag{4.1c}$$

Now suppose that we expect, based on physical arguments, that the density operators $a_\nu^\dagger a_{\nu'}$ and $b_\mu^\dagger b_{\mu'}$ deviate only little from their average values, $\langle a_\nu^\dagger a_{\nu'}\rangle$ and $\langle b_\mu^\dagger b_{\mu'}\rangle$.

(a) 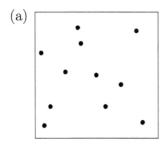   (b)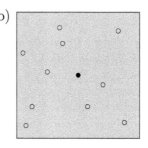

FIG. 4.1. Illustration of the mean-field idea. (a) In a real physical system interactions lead to correlated motion of the particles (black points). (b) In mean-field theory any individual particle (black point) interacts only with the average field (gray background) of the other particles (open circles).

It is then natural to use this deviation as a small parameter and perform an expansion. In order to do so we define the deviation operators

$$d_{\nu\nu'} = a_\nu^\dagger a_{\nu'} - \langle a_\nu^\dagger a_{\nu'} \rangle, \tag{4.2a}$$

$$e_{\mu\mu'} = b_\mu^\dagger b_{\mu'} - \langle b_\mu^\dagger b_{\mu'} \rangle, \tag{4.2b}$$

and insert them into Eq. (4.1a), which gives

$$H = H_0 + V_{\mathrm{MF}} + \overbrace{\sum_{\nu\nu',\mu\mu'} V_{\nu\mu,\nu'\mu'} d_{\nu\nu'} e_{\mu\mu'}}^{\text{neglected in mean-field}}, \tag{4.3}$$

where

$$V_{\mathrm{MF}} = \sum_{\nu\nu',\mu\mu'} V_{\nu\mu,\nu'\mu'} \left(a_\nu^\dagger a_{\nu'} \langle b_\mu^\dagger b_{\mu'} \rangle + b_\mu^\dagger b_{\mu'} \langle a_\nu^\dagger a_{\nu'} \rangle \right) - \sum_{\nu\nu',\mu\mu'} V_{\nu\mu,\nu'\mu'} \langle a_\nu^\dagger a_{\nu'} \rangle \langle b_\mu^\dagger b_{\mu'} \rangle, \tag{4.4}$$

Because $d_{\nu\nu'}$ and $e_{\mu\mu'}$ are assumed to be small the second term in Eq. (4.3) is neglected, and the interaction V_{int} is approximated by the mean-field interaction V_{MF} resulting in the so-called mean-field Hamiltonian H_{MF} given by

$$H_{\mathrm{MF}} = H_0 + V_{\mathrm{MF}}. \tag{4.5}$$

The mean-field Hamiltonian H_{MF} contains only single-particle operators, and thus the original many-body problem has been reduced to a single-particle problem, which in principle is always soluble (see Exercise 1.3).

Looking at Eq. (4.4) we can formulate the mean-field procedure in a different way: For an interaction term being the product of two operators A and B,

$$H_{AB} = AB, \tag{4.6}$$

the mean-field approximation is given by A coupled to the mean-field of B plus B coupled to the mean-field of A,

$$H_{AB}^{\text{MF}} = A\langle B\rangle + \langle A\rangle B - \langle A\rangle\langle B\rangle. \tag{4.7}$$

Here, the product of the mean-fields is subtracted on the right hand side to ensure the correct average: $\langle H_{AB}^{\text{MF}}\rangle = \langle A\rangle\langle B\rangle$.

The question is, however, how to find the averages $\langle a_\nu^\dagger a_{\nu'}\rangle$ and $\langle b_\mu^\dagger b_{\mu'}\rangle$. There are two possible routes which in fact are equivalent. Method 1: The average is to be determined self-consistently, i.e., when calculating the averages

$$\bar{n}_{\nu\nu'}^a \equiv \langle a_\nu^\dagger a_{\nu'}\rangle, \tag{4.8a}$$

$$\bar{n}_{\mu\mu'}^b \equiv \langle b_\mu^\dagger b_{\mu'}\rangle, \tag{4.8b}$$

using the new mean-field Hamiltonian, the same answer should come out. This means for $\bar{n}^a$ (and similarly for $\bar{n}^b$) that

$$\bar{n}_{\nu\nu'}^a \equiv \langle a_\nu^\dagger a_{\nu'}\rangle_{\text{MF}} = \frac{1}{Z_{\text{MF}}}\text{Tr}\left[e^{-\beta H_{\text{MF}}} a_\nu^\dagger a_{\nu'}\right], \tag{4.9}$$

where Z_{MF} is the mean-field partition function given by

$$Z_{\text{MF}} = \text{Tr}\left[e^{-\beta H_{\text{MF}}}\right]. \tag{4.10}$$

Eq. (4.9) and the similar one for $\bar{n}^b$ are called the self-consistency equations, because $\bar{n}^a$ and $\bar{n}^b$ are given by H_{MF} and Z_{MF}, which themselves depend on $\bar{n}^a$ and $\bar{n}^b$.

Next we turn to the alternative route. Method 2: Use the $n_{\nu\nu'}$ that minimizes the free energy F_{MF} of the mean-field Hamiltonian. Using the expression for the free energy given in Sec. 1.5, we obtain

$$\begin{aligned}
0 &= \frac{d}{d\bar{n}_{\nu\nu'}^a}F_{\text{MF}} = \frac{d}{d\bar{n}_{\nu\nu'}^a}\left(-\frac{1}{\beta}\ln Z_{\text{MF}}\right) \\
&= \frac{1}{Z_{\text{MF}}}\text{Tr}\left[e^{-\beta H_{\text{MF}}}\frac{d}{d\bar{n}_{\nu\nu'}^a}H_{\text{MF}}\right] \\
&= \frac{1}{Z_{\text{MF}}}\text{Tr}\left[e^{-\beta H_{\text{MF}}}\left\{\sum_{\mu\mu'}V_{\nu\mu,\nu'\mu'}\left(b_\mu^\dagger b_{\mu'} - \bar{n}_{\mu\mu'}^b\right)\right\}\right] \\
&= \sum_{\mu\mu'}V_{\nu\mu,\nu'\mu'}\left(\langle b_\mu^\dagger b_{\mu'}\rangle_{\text{MF}} - \bar{n}_{\mu\mu'}^b\right).
\end{aligned} \tag{4.11}$$

This should hold for any pair (ν,ν') and hence the last parenthesis has to vanish and we arrive at the self-consistency equation for $\bar{n}^b$. Similarly, by minimizing with respect to $\bar{n}^b$ we again obtain Eq. (4.9). Thus the two methods are equivalent.

We can gain some more understanding of the physical content of the mean-field approximation if we look at average interaction energy $\langle V_{\text{int}}\rangle$. A natural approximation would be to evaluate the expectation values of a and b operators separately,

$$\langle V_{\text{int}}\rangle \approx \sum_{\nu\nu',\mu\mu'}V_{\nu\mu,\mu'\nu'}\langle a_\nu^\dagger a_{\nu'}\rangle\langle b_\mu^\dagger b_{\mu'}\rangle, \tag{4.12}$$

which is equivalent to assuming that the a and b particles are uncorrelated.[14] This is in essence the approximation done in the mean-field approach. To see this let us evaluate $\langle V_{\text{int}} \rangle$ using the mean-field Hamiltonian

$$\langle V_{\text{int}} \rangle_{\text{MF}} = \frac{1}{Z_{\text{MF}}} \text{Tr} \left[e^{-\beta H_{\text{MF}}} V_{\text{int}} \right]. \tag{4.13}$$

Because the mean-field Hamiltonian can be separated into a part containing only a operators and a part containing only b operators, $H_{\text{MF}} = H_{\text{MF}}^a + H_{\text{MF}}^b$, the average factorizes exactly as in Eq. (4.12), and we get

$$\langle V_{\text{int}} \rangle_{\text{MF}} = \sum_{\nu\nu',\mu\mu'} V_{\nu\mu,\nu'\mu'} \langle a_\nu^\dagger a_{\nu'} \rangle_{\text{MF}} \langle b_\mu^\dagger b_{\mu'} \rangle_{\text{MF}}. \tag{4.14}$$

The mean-field approach hence provides a consistent and physically sensible method to study interacting systems where correlations are less important. Here "less important" should be quantified by checking the validity of the mean-field approximation. That is, one should check that d indeed is small by calculating $\langle d \rangle$, using the neglected term in Eq. (4.3) as a perturbation, and then compare the result to $\langle a_\nu^\dagger a_{\nu'} \rangle$. If it is not small, one has either chosen the wrong mean-field parameter, or the method simply fails and other tools more adequate to deal with the problem at hand must be applied.

4.2 The art of mean-field theory

In practice, one has to assume something about the averages $\langle a_\nu^\dagger a_{\nu'} \rangle$ and $\langle b_\mu^\dagger b_{\mu'} \rangle$ because even though Eq. (4.9) gives a recipe on how to find which averages are important, there are simply too many possible combinations. Suppose we have N different quantum numbers, then there are in principle N^2 different combinations, which gives N^2 coupled nonlinear equations, which of course is only tractable for small systems. With modern computers one can treat hundreds of particles in this way, but for a condensed matter system, it is out of the question. Therefore, one must provide some physical insight to reduce the number of mean-field parameters.

Often symmetry arguments can help in reducing the number of parameters. Suppose for example that the Hamiltonian that we are interested in has translational symmetry, such that momentum space is a natural choice. For a system of particles described by operators c and $c^\dagger$, we then have

$$\langle c_{\mathbf{k}}^\dagger c_{\mathbf{k'}} \rangle = \frac{1}{V} \int d\mathbf{r} \int d\mathbf{r'} e^{-i\mathbf{k'} \cdot \mathbf{r'}} e^{i\mathbf{k} \cdot \mathbf{r}} \langle \Psi^\dagger(\mathbf{r}) \Psi(\mathbf{r'}) \rangle. \tag{4.15}$$

It is natural to assume that the system is homogeneous, which means

$$\langle \Psi^\dagger(\mathbf{r}) \Psi(\mathbf{r'}) \rangle = f(\mathbf{r} - \mathbf{r'}) \quad \Rightarrow \quad \langle c_{\mathbf{k}}^\dagger c_{\mathbf{k'}} \rangle = \langle n_{\mathbf{k}} \rangle \delta_{\mathbf{k},\mathbf{k'}}. \tag{4.16}$$

This assumption about homogeneity is, however, not always true, because in some cases the symmetry of the system is lower than that of the Hamiltonian. For example,

[14]Remember from usual statistics that the correlation function between two stochastic quantities X and Y is defined by $\langle XY \rangle - \langle X \rangle \langle Y \rangle$.

if the system spontaneously orders into a state with a spatial density variation, like a wave formation then the average $\langle \Psi^\dagger(\mathbf{r})\Psi(\mathbf{r}') \rangle$ is not function of the $\mathbf{r} - \mathbf{r}'$ only. Instead it has a lower and more restricted symmetry, namely that

$$\langle \Psi^\dagger(\mathbf{r})\Psi(\mathbf{r}') \rangle = h(\mathbf{r}, \mathbf{r}'), \quad h(\mathbf{r}, \mathbf{r}') = h(\mathbf{r} + \mathbf{R}, \mathbf{r}' + \mathbf{R}), \quad (4.17)$$

with $\mathbf{R}$ being a lattice vector. The kind of crystal structure of course exists in Nature and when it happens, we talk about phenomena with broken symmetry. It is important to realize that this solution cannot be found if we assumed Eq. (4.16) as the starting point. Instead, we should have begun by assuming Eq. (4.17) leading to the possibility of $\langle c_{\mathbf{k}}^\dagger c_{\mathbf{k}+\mathbf{Q}} \rangle$ being finite, where $\mathbf{R} \cdot \mathbf{Q} = 2\pi$. Thus, the choice of the proper mean-field parameters requires physical motivation about which possible states one expects.

4.3   Hartree–Fock approximation

Above, we discussed the mean-field theory for interactions between different particles. Here, we go on to formulate the method for like particles. For the interaction term in Eq. (4.1a) we used the approximation to replace $a_\nu^\dagger a_{\nu'}$ and $b_\mu^\dagger b_{\mu'}$ by their average values plus small corrections. For interactions between identical particles this, however, does not exhaust the possibilities, since it only includes the so-called Hartree term. We now move on to discuss the more general approximation scheme, called the Hartree–Fock approximation.

Suppose we have a system of interacting particles described by the Hamiltonian

$$H = H_0 + V_{\text{int}}, \quad (4.18a)$$

$$H_0 = \sum_\nu \xi_\nu c_\nu^\dagger c_\nu, \quad (4.18b)$$

$$V_{\text{int}} = \frac{1}{2} \sum_{\nu\nu',\mu\mu'} V_{\nu\mu,\nu'\mu'} c_\nu^\dagger c_\mu^\dagger c_{\mu'} c_{\nu'}. \quad (4.18c)$$

In Eq. (4.14) we saw that the mean-field assumption amounted to separating the expectation values with respect to the a and b particles. In other words, the particles were treated as being independent. Performing the same approximation for the interaction term in Eq. (4.18c), we use a theorem called Wick's theorem, which is proved in Section 11.6. It states that if the particles can be treated as being independent (which is precisely the mean-field assumption) then the four-term average $\langle c_\nu^\dagger c_\mu^\dagger c_{\mu'} c_{\nu'} \rangle$ factorizes into two-term averages,

$$\langle c_\nu^\dagger c_\mu^\dagger c_{\mu'} c_{\nu'} \rangle_{\text{MF}} = \langle c_\nu^\dagger c_{\nu'} \rangle_{\text{MF}} \langle c_\mu^\dagger c_{\mu'} \rangle_{\text{MF}} \pm \langle c_\nu^\dagger c_{\mu'} \rangle_{\text{MF}} \langle c_\mu^\dagger c_{\nu'} \rangle_{\text{MF}}, \quad (4.19)$$

where the minus sign is for fermions while the plus sign is for bosons. Wick's theorem thus says that one should make all possible pairings while keeping track of sign changes if two fermions are interchanged.[15] The second term is new when compared to Eq. (4.14). It is denoted the exchange term. Note that when $\nu = \mu$ or $\nu' = \mu'$

---

[15]Another way to derive Eq. (4.19) is to assume that the wavefunction is a product wavefunction.

the product $c_\nu^\dagger c_\mu^\dagger c_{\mu'} c_{\nu'}$ is identically zero for fermions, which is consistent with the Pauli exclusion principle, because fermions are never in same state. This principle is fulfilled in the mean-field approximation in Eq. (4.19).

In order to find the self-consistency conditions, we must perform the approximation of independent particles on the Hamiltonian as well. This is done in the same spirit as we did for the case with interactions between different particles, except we must include the exchange terms as well and, of course, remember to subtract the average to avoid double counting. We thus arrive at the mean-field approximation for identical particles

$$
\begin{aligned}
c_\nu^\dagger c_\mu^\dagger c_{\mu'} c_{\nu'} \approx \quad & c_\nu^\dagger c_{\nu'} \langle c_\mu^\dagger c_{\mu'} \rangle_{\mathrm{MF}} + \langle c_\nu^\dagger c_{\nu'} \rangle_{\mathrm{MF}} c_\mu^\dagger c_{\mu'} \\
& \pm c_\nu^\dagger c_{\mu'} \langle c_\mu^\dagger c_{\nu'} \rangle_{\mathrm{MF}} \pm \langle c_\nu^\dagger c_{\mu'} \rangle_{\mathrm{MF}} c_\mu^\dagger c_{\nu'} \\
& - \langle c_\nu^\dagger c_{\nu'} \rangle_{\mathrm{MF}} \langle c_\mu^\dagger c_{\mu'} \rangle_{\mathrm{MF}} \mp \langle c_\nu^\dagger c_{\mu'} \rangle_{\mathrm{MF}} \langle c_\mu^\dagger c_{\nu'} \rangle_{\mathrm{MF}}.
\end{aligned}
\tag{4.20}
$$

It is easy to verify that the average of Eq. (4.20) is Eq. (4.19). The two first terms in Eqs. (4.19) and (4.20) represent the direct interaction, because they give the classical expectation value of the interaction between two densities. For example in the case of Coulomb interactions in Eq. (1.100a), we see that the first term in Eq. (4.19) corresponds to $\int d\mathbf{r}_1 d\mathbf{r}_2 V(\mathbf{r}_1 - \mathbf{r}_2) \langle \rho(\mathbf{r}_1) \rangle \langle \rho(\mathbf{r}_2) \rangle$. The exchange terms in Eqs. (4.19) and (4.20) represent a quantum correction to this. For example, we saw in Section 2.2 that the exchange energy explained the binding energy of metals.

The terms corresponding to the direct interactions constitute the Hartree approximation. Applying this to the interaction term in Eq. (4.18c), we obtain

$$
V_{\mathrm{int}}^{\mathrm{Hartree}} = \frac{1}{2} \sum V_{\nu\mu,\nu'\mu'} \bar{n}_{\mu\mu'} c_\nu^\dagger c_{\nu'} + \frac{1}{2} \sum V_{\nu\mu,\nu'\mu'} \bar{n}_{\nu\nu'} c_\mu^\dagger c_{\mu'} - \frac{1}{2} \sum V_{\nu\mu,\nu'\mu'} \bar{n}_{\nu\nu'} \bar{n}_{\mu\mu'}.
\tag{4.21}
$$

The exchange term is known as the Fock term, and again applying the factorization on Eq. (4.18c) we have

$$
V_{\mathrm{int}}^{\mathrm{Fock}} = -\frac{1}{2} \sum V_{\nu\mu,\nu'\mu'} \bar{n}_{\nu\mu'} c_\mu^\dagger c_{\nu'} - \frac{1}{2} \sum V_{\nu\mu,\nu'\mu'} \bar{n}_{\mu\nu'} c_\nu^\dagger c_{\mu'} + \frac{1}{2} \sum V_{\nu\mu,\nu'\mu'} \bar{n}_{\nu\mu'} \bar{n}_{\mu\nu'}.
\tag{4.22}
$$

The final mean-field Hamiltonian within the Hartree–Fock approximation is

$$
H^{\mathrm{HF}} = H_0 + V_{\mathrm{int}}^{\mathrm{Fock}} + V_{\mathrm{int}}^{\mathrm{Hartree}}.
\tag{4.23}
$$

4.3.1 Hartree-Fock approximation for the homogeneous electron gas

Consider now the example of a homogeneous electron gas which is translation invariant, which means that the expectation value $\langle c_\mathbf{k}^\dagger c_{\mathbf{k}'} \rangle$ is diagonal. We can now read off the corresponding Hartree–Fock Hamiltonian (see Exercise 4.1) from Eq. (1.103). The result is

$$H^{\mathrm{HF}} = \sum_{\mathbf{k}\sigma} \xi_{\mathbf{k}\sigma}^{\mathrm{HF}} c_{\mathbf{k}\sigma}^{\dagger} c_{\mathbf{k}\sigma}, \tag{4.24a}$$

$$\xi_{\mathbf{k}\sigma}^{\mathrm{HF}} = \xi_{\mathbf{k}} + \sum_{\mathbf{k}'\sigma'} \left[V(0) - \delta_{\sigma\sigma'} V(\mathbf{k} - \mathbf{k}') \right] \bar{n}_{\mathbf{k}'\sigma'},$$

$$= \xi_{\mathbf{k}} + V(0)N - \sum_{\mathbf{k}'} V(\mathbf{k} - \mathbf{k}') \bar{n}_{\mathbf{k}'\sigma}. \tag{4.24b}$$

The second term is the interaction with the average electron charge. As explained in Chapter 2, in condensed matter systems it is normally cancelled out by an equally large term due to the positively charged ionic background. The third term is the exchange correction.

Again we emphasize that the Hartree–Fock approximation depends crucially on what averages we assume to be finite, and these assumptions must be based on physical knowledge or clever guesswork. In deriving Eq. (4.24b) we assumed, e.g., that the spin symmetry is also maintained, which implies that $\langle c_{\mathbf{k}\downarrow}^{\dagger} c_{\mathbf{k}\downarrow} \rangle = \langle c_{\mathbf{k}\uparrow}^{\dagger} c_{\mathbf{k}\uparrow} \rangle$. If we allow them to be different we have the possibility of obtaining a ferromagnetic solution, which indeed happens in some cases. This is discussed in Sec. 4.5.2.

4.4   Broken symmetry

Mean-field theory is often used to study phase transitions and thus changes of symmetry. For a given Hamiltonian with some symmetry (e.g., translational symmetry in real space or rotational symmetry in real space or in spin space) there exists an operator which reflects this symmetry and therefore commutes with the Hamiltonian (e.g., the translation operator or the rotation operator in real space or spin space). Since the operator and the Hamiltonian commute, we know from the theory of Hermitian operators that a common set of eigenstates exists. Consider, for example, the case of a liquid of particles where the Hamiltonian of course has translation symmetry, which means that the translation operator $T(\mathbf{R})$ commutes with the Hamiltonian, $[H, T(\mathbf{R})] = 0$ for all $\mathbf{R}$. Here, $T(\mathbf{R})$ is an operator which displaces all particle coordinates by the amount $\mathbf{R}$. It can be written as $T(\mathbf{R}) = \exp\left(-\frac{i}{\hbar}\mathbf{R}\cdot\mathbf{P}\right)$, where $\mathbf{P}$ is the total momentum operator. The total momentum operator is thus a conserved quantity and it is given by

$$\mathbf{P} = \hbar \sum_{\mathbf{k}\sigma} \mathbf{k}\, c_{\mathbf{k}\sigma}^{\dagger} c_{\mathbf{k}\sigma}, \tag{4.25}$$

We can now choose an orthogonal basis of states with definite total momentum, $|\mathbf{P}\rangle$. This fact we can use to "prove" the unphysical result that a density wave can never exist. A density wave, with wave vector $\mathbf{Q}$, means that the Fourier transform of the density operator

$$\rho(\mathbf{Q}) = \sum_{\mathbf{k}\sigma} c_{\mathbf{k}\sigma}^{\dagger} c_{\mathbf{k}+\mathbf{Q}\sigma}, \tag{4.26}$$

has a finite expectation value, but

$$\langle c_{\mathbf{k}\sigma}^{\dagger} c_{\mathbf{k}+\mathbf{Q}\sigma} \rangle = \frac{1}{Z} \sum_{\mathbf{P}} e^{-\beta E_{\mathbf{P}}} \left\langle \mathbf{P} \left| c_{\mathbf{k}\sigma}^{\dagger} c_{\mathbf{k}+\mathbf{Q}\sigma} \right| \mathbf{P} \right\rangle = 0, \tag{4.27}$$

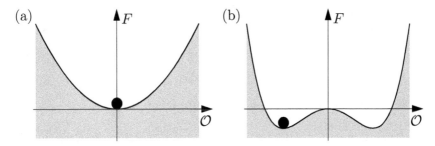

FIG. 4.2. The energetics of a phase transition in terms of the free energy F as a function of the order parameter $\mathcal{O}$ characterizing the phase. (a) Above the critical point the free energy has a well-defined minimum at the symmetry point, and the system is in a state of large symmetry. (b) Below the critical point a two-well potential develops and the system minimizes F by choosing one of the two possible values of $\mathcal{O}$. Even though the total potential is still symmetric the system will reside only in one well due to the macroscopically large energy barrier and thus the state of the system has "lower symmetry" than the potential.

because $c_{\mathbf{k}}^{\dagger} c_{\mathbf{k}+\mathbf{Q}} |\mathbf{P}\rangle$ has momentum $\mathbf{P} - \mathbf{Q}$ and is thus orthogonal to $|\mathbf{P}\rangle$. We have therefore reached the senseless result that crystals do not exist. In the same way, we could "prove" that magnetism, superconductivity, and other well-known physical phenomena cannot happen. What is wrong?

The proof above breaks down if the sum of states in the thermodynamical average is restricted. Even though crystals with different spatial reference points (or ferromagnets with magnetization in different directions) have formally the same energy, they are effectively decoupled due to the large energy barrier it takes to melt and then recrystallize into a new state with a shifted reference (or direction of magnetization). In those cases where many states of the system are degenerate, but separated by large energy barriers, it does not make sense to include them on equal footing in the statistical average as in Eq. (4.27) because they correspond to macroscopically totally different configurations. We are therefore forced to refrain from the fundamental ergodicity postulate of statistical mechanics, also discussed in Sec. 1.5, and build into the description that the phase space of the system falls into physically separated sections. This is often illustrated by the double barrier model of phase transitions shown in Fig. 4.2.

When at some critical temperature the thermodynamical state of the system develops a non-zero expectation value of some macroscopic quantity which has a symmetry lower than the original Hamiltonian it is called *spontaneous breaking of symmetry*. The quantity which signals that a phase transition has occurred is called the order parameter. Typical examples are listed in Table 4.1.

In order to arrive at the new phase in a calculation and to avoid the paradox in Eq. (4.27), one has to build in the possibility of the new phase into the theory. In the mean-field approach the trick is to include the order parameter in the choice of finite mean-fields and, of course, show that the resulting mean-field Hamiltonian leads to a self-consistent finite result. Next we study examples of symmetry breaking

Phenomena	Order parameter physical	Order parameter mathematical
Crystal	Density wave	$\sum_{\mathbf{k}} \langle c_{\mathbf{k}}^{\dagger} c_{\mathbf{k}+\mathbf{Q}} \rangle$
Ferromagnet	Magnetization	$\sum_{\mathbf{k}} \langle c_{\mathbf{k}\uparrow}^{\dagger} c_{\mathbf{k}\uparrow} - c_{\mathbf{k}\downarrow}^{\dagger} c_{\mathbf{k}\downarrow} \rangle$
Bose–Einstein condensate	Population of $\mathbf{k} = \mathbf{0}$ state	$\langle a_{\mathbf{k}=\mathbf{0}} \rangle$
Superconductor	Pair condensate	$\langle c_{\mathbf{k}\uparrow} c_{-\mathbf{k}\downarrow} \rangle$

Table 4.1 *Typical examples of spontaneous symmetry breakings and their corresponding order parameters. The Bose–Einstein condensate and the superconducting order parameter are not discussed in this Chapter. For the superconducting order parameter see Chapter 18*

phenomena and their corresponding order parameters.

4.5 Ferromagnetism

4.5.1 *The Heisenberg model of ionic ferromagnets*

In ionic magnetic crystals the interaction between the magnetic ions is due to the exchange interactions originating from the Coulomb interactions. Here we will not go into the details of this interaction but simply give the effective Hamiltonian,[16] known as the Heisenberg model for interaction between spins in a crystal. It reads

$$H = -\sum_{ij} J_{ij}\, \mathbf{S}_i \cdot \mathbf{S}_j, \tag{4.28}$$

where $\mathbf{S}_i$ is the spin operator for the ion on site i and J_{ij} is the strength of the interaction, between the magnetic moment of the ions on sites i and j. It depends only on the distance between the ions. The interaction is generally short ranged and we truncate it so that it, as in Fig. 4.3, is only non-zero for nearest neighbors

$$J_{ij} = \begin{cases} J_0, & \text{if } i \text{ and } j \text{ are neighbors}, \\ 0, & \text{otherwise}. \end{cases} \tag{4.29}$$

We immediately see that if $J < 0$, the spins tend to become antiparallel whereas for $J > 0$, it is energetically favorable to for the spins to be parallel. The first case corresponds to the antiferromagnetic case, while the latter to the ferromagnetic case. Here we consider only the ferromagnetic case, $J > 0$.

As the model Hamiltonian stands, although simple looking, it is immensely complicated and cannot be solved in general, the spins of the individual ions being strongly correlated. However, it is a good example where a mean-field solution gives an easy and also physically correct answer. The mean-field decomposition then gives

---

[16]The term "effective Hamiltonian" has a well-defined meaning. It means the Hamiltonian describing the important degrees of freedom on the relevant low-energy scale.

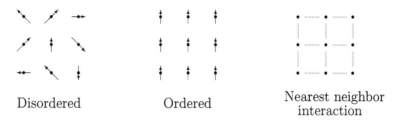

Disordered   Ordered   Nearest neighbor
                                       interaction

FIG. 4.3. The Heisenberg model in the disordered state (left panel), where there
is no preferred direction for spins, and in the ferromagnetic state (middle panel),
where the spins form a collective state with a finite macroscopic moment along one
direction. The model discussed here only includes interactions between adjacent
spins, as shown in the right panel.

$$H \approx H_{\mathrm{MF}} = -\sum_{ij} J_{ij} \langle \mathbf{S}_i \rangle \cdot \mathbf{S}_j - \sum_{ij} J_{ij} \mathbf{S}_i \cdot \langle \mathbf{S}_j \rangle + \sum_{ij} J_{ij} \langle \mathbf{S}_i \rangle \cdot \langle \mathbf{S}_j \rangle. \qquad (4.30)$$

Here $\langle \mathbf{S}_i \rangle$ is the average spin at site i. From symmetry arguments, we would expect
that the expectation value of this is zero, because all directions are equivalent. But
since this is not the right answer, we must assume that the symmetry is broken, i.e.,
allow for $\langle \mathbf{S}_i \rangle$ to be non-zero. Furthermore, because of the translation symmetry we
expect it to be independent of position coordinate i.[17] So we assume a finite but
spatially independent average spin polarization.

If we choose the z axis along the direction of the magnetization our mean-field
assumption is

$$\langle \mathbf{S}_i \rangle = \langle S_z \rangle \, \mathbf{e}_z, \qquad (4.31)$$

and the magnetic moment $\mathbf{m}$ (which by assumption is equal for all sites) felt by each
spin thus becomes

$$\mathbf{m} = \sum_j J_{ij} \langle S_z \rangle \, \mathbf{e}_z = n J_0 \langle S_z \rangle \, \mathbf{e}_z, \qquad (4.32)$$

where n is the number of neighbors. For a square lattice it is $n = 2d$, where d is the
dimension. The mean-field Hamiltonian

$$H_{\mathrm{MF}} = -2 \sum_i \mathbf{m} \cdot \mathbf{S}_i + mN\langle S_z \rangle, \qquad (4.33)$$

is now diagonal in the site index and hence easily solved. Here N is the number of
sites and $m = |\mathbf{m}|$. Suppose for simplicity that the ions have spin $S = \frac{1}{2}$. With this
simplification, the mean-field partition function is

$$Z_{\mathrm{MF}} = \left(e^{\beta m} + e^{-\beta m} \right)^N e^{-\beta Nm\langle S_z \rangle} = \left[\left(e^{\beta m} + e^{-\beta m} \right) e^{-\beta m^2/2nJ_0} \right]^N, \qquad (4.34)$$

with one term for each possible spin projection, $S_z = \pm\frac{1}{2}$.

---

[17]But had we reasons to believe that an antiferromagnetic solution (where the spins point in
opposite direction on even and odd sites) was relevant (if $J_{ij} < 0$), we would have to assume that
also this symmetry was broken.

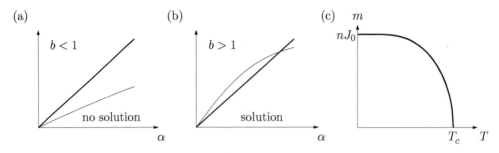

FIG. 4.4. (a) Graph showing the lack of non-zero solutions for the ferromagnetic moment $\alpha = m/n_0 J$ to the mean-field equation, $\alpha = \tanh(b\alpha)$, for the Heisenberg model in the high temperature limit $n_0 J/k_B T = b < 1$. (b) Graphical solution yielding a finite magnetic moment for the same model in the low temperature limit $T_c/T = b > 1$. (c) The resulting temperature dependence of the magnetization.

The self-consistency equation is found by minimizing the free energy

$$\frac{\partial F_{\mathrm{MF}}}{\partial m} = -\frac{1}{\beta}\frac{\partial}{\partial m}\ln Z_{\mathrm{MF}} = -N\left(\frac{e^{\beta m} - e^{-\beta m}}{e^{\beta m} + e^{-\beta m}}\right) + N\frac{m}{n J_0} = 0,$$

which has a solution given by the transcendental equation

$$\alpha = \tanh(b\alpha), \quad \text{with} \quad \alpha = \frac{m}{n J_0} \quad \text{and} \quad b = n J_0 \beta. \tag{4.35}$$

It is evident from an expansion for small α,

$$\alpha \approx b\alpha - \frac{1}{3}(b\alpha)^3, \tag{4.36}$$

that there is no solution for $b < 1$, and thus we can determine the critical temperature T_c where the magnetism disappears, by the condition $b_c = 1$ and hence $k_B T_c = n J_0$. Furthermore, for small α we find the solution for the magnetization, $\alpha \approx \frac{1}{b}\sqrt{3(b-1)/b} \Rightarrow m \approx n J_0 \sqrt{3(1 - T/T_c)}$, valid close to T_c. At $T = 0$, where $b = \infty$, the solution to Eq. (4.35) is $\alpha = 1$ and hence $m = n J_0$. For the functional form of the magnetization in the entire range of temperature one must solve Eq. (4.35) numerically, which of course is a simple task. The solution is shown in Fig. 4.4.

4.5.2 The Stoner model of metallic ferromagnets

In magnets where the electrons both generate the magnetic moments and also form conduction bands, the Heisenberg model cannot explain the magnetism. This is simply because the spins are not localized. Metallic magnetism happens, e.g., in transition metals where the conduction bands are formed by the narrower d or f orbitals. The interaction between two particles in those orbitals is stronger than that between electrons occupying the more spread-out s or p orbitals, and hence give a larger correlation between electrons. Typical metals where correlations between conduction band electrons are important are Fe and Ni.

Since the short range of the interaction is important it is relevant to study a model, the so-called Hubbard model, where this physical fact is reflected in a simple but extreme manner: the Coulomb interaction between electrons is taken to be point-like in real space and hence constant in momentum space

$$H_{\text{hub}} = \sum_{\mathbf{k}\sigma} \xi_{\mathbf{k}} c_{\mathbf{k}\sigma}^\dagger c_{\mathbf{k}\sigma} + \frac{U}{2\mathcal{V}} \sum_{\mathbf{k}'\mathbf{k}\mathbf{q},\sigma\sigma'} c_{\mathbf{k}+\mathbf{q}\sigma}^\dagger c_{\mathbf{k}'-\mathbf{q}\sigma'}^\dagger c_{\mathbf{k}'\sigma'} c_{\mathbf{k}\sigma}. \tag{4.37}$$

We now use the Hartree–Fock approximation scheme on this model, but search for a ferromagnetic solution by allowing for the expectation values to depend on the direction of the spin. The mean-field parameters are

$$\langle c_{\mathbf{k}\uparrow}^\dagger c_{\mathbf{k}'\uparrow} \rangle = \delta_{\mathbf{k}\mathbf{k}'} \bar{n}_{\mathbf{k}\uparrow}, \quad \langle c_{\mathbf{k}\downarrow}^\dagger c_{\mathbf{k}'\downarrow} \rangle = \delta_{\mathbf{k}\mathbf{k}'} \bar{n}_{\mathbf{k}\downarrow}, \tag{4.38}$$

and according to Eq. (4.23) the mean-field interaction Hamiltonian becomes

$$V_{\text{int}}^{\text{MF}} = \frac{U}{\mathcal{V}} \sum_{\mathbf{k}'\mathbf{k}\mathbf{q}\sigma\sigma'} c_{\mathbf{k}+\mathbf{q}\sigma}^\dagger \langle c_{\mathbf{k}'-\mathbf{q}\sigma'}^\dagger c_{\mathbf{k}'\sigma'} \rangle c_{\mathbf{k}\sigma} - \frac{U}{\mathcal{V}} \sum_{\mathbf{k}'\mathbf{k}\mathbf{q}\sigma\sigma'} \langle c_{\mathbf{k}+\mathbf{q}\sigma}^\dagger c_{\mathbf{k}'\sigma'} \rangle c_{\mathbf{k}'-\mathbf{q}\sigma'}^\dagger c_{\mathbf{k}\sigma}$$

$$- \frac{U}{2\mathcal{V}} \sum_{\mathbf{k}'\mathbf{k}\mathbf{q}\sigma\sigma'} \left[\langle c_{\mathbf{k}+\mathbf{q}\sigma}^\dagger c_{\mathbf{k}\sigma} \rangle \langle c_{\mathbf{k}'-\mathbf{q}\sigma'}^\dagger c_{\mathbf{k}'\sigma'} \rangle - \langle c_{\mathbf{k}+\mathbf{q}\sigma}^\dagger c_{\mathbf{k}'\sigma'} \rangle \langle c_{\mathbf{k}'-\mathbf{q}\sigma'}^\dagger c_{\mathbf{k}\sigma} \rangle \right]. \tag{4.39}$$

The factor $\frac{1}{2}$ disappeared because there are two identical terms in Eqs. (4.21) and (4.22). Using our mean-field assumptions Eq. (4.38), we obtain

$$V_{\text{int}}^{\text{MF}} = U \sum_{\mathbf{k}\sigma\sigma'} c_{\mathbf{k}\sigma}^\dagger c_{\mathbf{k}\sigma} \left[\bar{n}_{\sigma'} - \bar{n}_\sigma \delta_{\sigma\sigma'} \right] - \frac{U\mathcal{V}}{2} \sum_{\sigma\sigma'} \bar{n}_\sigma \bar{n}_{\sigma'} + \frac{U\mathcal{V}}{2} \sum_\sigma \bar{n}_\sigma^2, \tag{4.40}$$

where the spin-dependent densities have been defined as

$$\bar{n}_\sigma = \frac{1}{\mathcal{V}} \sum_{\mathbf{k}} \langle c_{\mathbf{k}\sigma}^\dagger c_{\mathbf{k}\sigma} \rangle. \tag{4.41}$$

The full mean-field Hamiltonian is now given by

$$H_{\text{MF}} = \sum_{\mathbf{k}\sigma} \xi_{\mathbf{k}\sigma}^{\text{MF}} c_{\mathbf{k}\sigma}^\dagger c_{\mathbf{k}\sigma} - \frac{U\mathcal{V}}{2} \sum_{\sigma\sigma'} \bar{n}_\sigma \bar{n}_{\sigma'} + \frac{U\mathcal{V}}{2} \sum_\sigma \bar{n}_\sigma^2, \tag{4.42a}$$

$$\xi_{\mathbf{k}\sigma}^{MF} = \xi_{\mathbf{k}} + U(\bar{n}_\uparrow + \bar{n}_\downarrow - \bar{n}_\sigma) = \xi_{\mathbf{k}} + U\bar{n}_{\bar{\sigma}}. \tag{4.42b}$$

The mean-field solution is found by minimization or by using the self-consistency condition

$$\bar{n}_\sigma = \frac{1}{\mathcal{V}} \sum_{\mathbf{k}} \langle c_{\mathbf{k}\sigma}^\dagger c_{\mathbf{k}\sigma} \rangle_{MF} = \frac{1}{\mathcal{V}} \sum_{\mathbf{k}} n_F(\xi_{\mathbf{k}\sigma}^{\text{MF}}). \tag{4.43}$$

We obtain at zero temperature

$$\bar{n}_\uparrow = \int \frac{d\mathbf{k}}{(2\pi)^3} \, \theta\left(\mu - \frac{\hbar^2 k^2}{2m} - U\bar{n}_\downarrow \right) = \frac{1}{6\pi^2} k_{F\uparrow}^3, \tag{4.44}$$

where $\frac{\hbar^2}{2m}k_{F\uparrow}^2 + U\bar{n}_\downarrow = \mu$, and of course a similar equation for spin down. The two equations are

$$\frac{\hbar^2}{2m}(6\pi)^{2/3}\bar{n}_\uparrow^{2/3} + U\bar{n}_\downarrow = \mu, \qquad \frac{\hbar^2}{2m}(6\pi)^{2/3}\bar{n}_\downarrow^{2/3} + U\bar{n}_\uparrow = \mu. \qquad (4.45)$$

Define the variables,

$$\zeta = \frac{\bar{n}_\uparrow - \bar{n}_\downarrow}{\bar{n}}, \qquad \gamma = \frac{2mUn^{1/3}}{(3\pi^2)^{2/3}\hbar^2}, \qquad \bar{n} = \bar{n}_\uparrow + \bar{n}_\downarrow. \qquad (4.46)$$

Then by subtracting the self-consistency conditions Eq. (4.45), we obtain

$$\bar{n}_\uparrow^{2/3} - \bar{n}_\downarrow^{2/3} = \frac{2mU}{\hbar^2}(6\pi)^{-2/3}\left(\bar{n}_\uparrow - \bar{n}_\downarrow\right), \qquad (4.47a)$$

or

$$\gamma\zeta = (1+\zeta)^{2/3} - (1-\zeta)^{2/3}. \qquad (4.47b)$$

The problem has three types of solutions:

$$\gamma < \frac{4}{3} : \quad \text{Isotropic solution (normal state)}, \qquad \qquad \zeta = 0, \qquad (4.48a)$$

$$\frac{4}{3} < \gamma < 2^{2/3} : \text{Partial polarization (weak ferromagnet)}, \ 0 < \zeta < 1, \qquad (4.48b)$$

$$\gamma > 2^{2/3} : \text{Full polarization (strong ferromagnet)}, \qquad \zeta = 1. \qquad (4.48c)$$

The different solutions are sketched in Fig. 4.5.

The possibility for a magnetic solution can be traced back to the spin-dependent energies Eq. (4.42a), where it is clear that the mean-field energy of a given spin direction depends on the occupation of the opposite spin direction, whereas the energy does not depend on the occupation of the same spin direction. This resulted from two things: the short range interaction and the exchange term. One can understand this simply from the Pauli principle which ensures that electrons with the same spin never occupy the same spatial orbital and therefore, if the interaction is short-range, they cannot interact. This leaves interactions between opposite spins as the only possibility. Thus the interaction energy is lowered by having a polarized ground state, which on the other hand for a fixed density costs kinetic energy. The competition between the potential and the kinetic energy contributions is what gives rise to the phase transition.

The Stoner model gives a reasonable account of metallic magnets. It is also capable of qualitatively explaining the properties of excitations in the spin polarized states, but this outside the scope of this section.

4.6 Summary and outlook

Mean-field theories are widely used to study phase transitions in condensed matter systems and also in, e.g., atomic physics to compute the energetics of finite-size systems. The mean-field approximation for two-particle interactions is in many cases

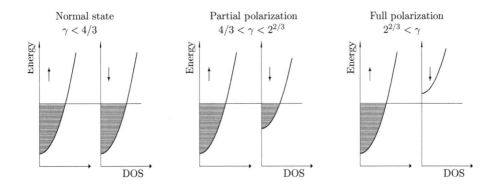

FIG. 4.5. The three possible solutions of the Stoner model. The polarization is thus
a function of the interaction strength; the stronger the interaction the larger the
polarization. The Stoner model provides a clear physical picture for how the exchange interactions induce a ferromagnetic phase transition in a metal with strong
on-site interactions.

sufficient to understand the important physical features, at least those that have to
do with static properties. In this chapter, we have seen how magnetism can be described by mean-field theory. Another example is superconductivity, to be treated in
Section 18.3. We have also introduced the important concept of symmetry breaking,
which means that the system can lower its free energy by being in a state of lower
symmetry than that of the original Hamiltonian.

This chapter is the last treating only stationary quantum states. In the remaining
part of the book we shall deal with the dynamical properties of many-particle systems. It turns out that also for the time-dependent systems different Hartree–Fock
type approximations can be evoked. One example is the so-called random phase approximation (RPA) treatment of the dielectric function in Sec. 8.5. In Chapter 14
the RPA result will be derived based on a more rigorous quantum field theoretical
approach.

Mean-field theory is treated in most books on many-body systems, also in those
quoted at the end of Chapter 1. An illuminating discussion on symmetry breaking in
condensed matter systems is presented by Anderson (1984), while a reader interested
in magnetism could consult Yosida (1996) or Jensen and Mackintosh (1991).

5

TIME DEPENDENCE IN QUANTUM THEORY

Using the second quantization procedure, we have so far only treated energy eigenstates with a trivial time dependence $e^{-i\omega t}$, instant processes at a single time, and systems where interactions are approximated by time-independent mean-field theory. But how does one then treat the general case of time dependence in second quantization? That question will be addressed in this chapter, where the formal theory of time evolution is discussed using three representations, or "pictures:" the Schrödinger picture, the Heisenberg picture, and the interaction picture.

We also discuss some applications towards the end of the chapter, namely the theory of scattering, which is needed in later chapters and, furthermore, we explain Fourier transformations of retarded correlation functions, which are used extensively in this book.

5.1 The Schrödinger picture

The Schrödinger picture is useful when dealing with a time-independent Hamiltonian H, i.e., $\partial_t H = 0$. Any other operator A may or may not depend on time. The state vectors $|\psi(t)\rangle$ do depend on time, and their time evolution is governed by Schrödinger's equation. The time-independence of H leads to a simple formal solution:

$$i\hbar\partial_t |\psi(t)\rangle = H |\psi(t)\rangle \quad \Rightarrow \quad |\psi(t)\rangle = e^{-\frac{i}{\hbar}Ht} |\psi_0\rangle. \qquad (5.1)$$

In the following we will measure the energy in units of frequency, such that $\hbar$ drops out of the time-evolution equations: $\varepsilon/\hbar \to \varepsilon$ and $H/\hbar \to H$. At the end of the calculations one can easily convert frequencies back to energies. With this notation we can summarize the Schrödinger picture with its states $|\psi(t)\rangle$ and operators A as:

$$\text{The Schrödinger picture} \begin{cases} \text{States:} \quad |\psi(t)\rangle = e^{-iHt} |\psi_0\rangle. \\ \text{Operators: } A, \text{ may or may not depend on time.} \\ \qquad\qquad\quad H, \text{ does not depend on time.} \end{cases} \quad (5.2)$$

To interpret the operator e^{-iHt} we recall that a function $f(B)$ of any operator B is defined by the Taylor expansion of f,

$$f(B) = \sum_{n=0}^{\infty} \frac{f^{(n)}(0)}{n!} B^n. \qquad (5.3)$$

While the Schrödinger picture is quite useful for time-independent operators A, it may sometimes be preferable to collect all time dependencies in the operators and work with time-independent state vectors. We can do that using the Heisenberg picture.

80

5.2   The Heisenberg picture

The central idea behind the Heisenberg picture is to obtain a representation where all the time dependence is transferred to the operators, $A(t)$, leaving the state vectors $|\psi_0\rangle$ time independent. The Hamiltonian H remains time independent in the Heisenberg picture. If the matrix elements of any operator between any two states are identical in the two representations, then the two representations are fully equivalent. By using Eq. (5.2) we obtain the identity

$$\langle\psi'(t)|A|\psi(t)\rangle = \langle\psi_0'|e^{iHt}Ae^{-iHt}|\psi_0\rangle \equiv \langle\psi_0'|A(t)|\psi_0\rangle. \qquad (5.4)$$

Thus we see that the correspondence between the Heisenberg picture with time-independent state vectors $|\psi_0\rangle$, but time-dependent operators $A(t)$, and the Schrödinger picture is given by the unitary transformation operator $\exp(iHt)$,

$$\text{The Heisenberg picture} \begin{cases} \text{States:} & |\psi_0\rangle \equiv e^{iHt}\,|\psi(t)\rangle. \\ \text{Operators:} & A(t) \equiv e^{iHt}A\,e^{-iHt}. \\ \quad H & \text{does not depend on time.} \end{cases} \qquad (5.5)$$

As before, the original operator A may be time dependent. The important equation of motion governing the time evolution of $A(t)$ is easily established. Since H is time independent, the total time derivative of A in the Heisenberg picture is denoted by a dot, as in $\dot{A}$, while the explicit time derivative of the original Schrödinger operator is denoted as $\partial_t A$:

$$\dot{A}(t) = e^{iHt}\Big(iHA - iAH + \partial_t A\Big)e^{-iHt} \Rightarrow \dot{A}(t) = i\Big[H, A(t)\Big] + (\partial_t A)(t), \qquad (5.6)$$

where $X(t)$ always means $e^{iHt}Xe^{-iHt}$ for any symbol X, in particular for $X = \partial_t A$. In this way, an explicit time dependence of A is taken into account. Note how carefully the order of the operators is kept during the calculation.

Both the Schrödinger and the Heisenberg picture require a time-independent Hamiltonian. In the general case of time-dependent Hamiltonians, we have to switch to the interaction picture.

5.3   The interaction picture

The third and last representation, the interaction picture, is introduced to deal with the situation where a system described by a time-independent Hamiltonian H_0, with known energy eigenstates $|n_0\rangle$, is perturbed by some, possibly time-dependent, interaction $V(t)$,

$$H = H_0 + V(t), \quad \text{with } H_0|n_0\rangle = \varepsilon_{n_0}|n_0\rangle. \qquad (5.7)$$

The key idea behind the interaction picture is to separate the trivial time evolution due to H_0 from the intricate one due to $V(t)$. This is obtained by using only H_0, not the full H, in the unitary transformation Eq. (5.5). As a result, in the interaction picture both the state vectors $|\hat{\psi}(t)\rangle$ and the operators $\hat{A}(t)$ depend on time. The defining equations for the interaction picture are

The interaction picture $\begin{cases} \text{States:} & |\hat{\psi}(t)\rangle \equiv e^{iH_0t}\,|\psi(t)\rangle. \\ \text{Operators:} \ \hat{A}(t) & \equiv e^{iH_0t}\,A\,e^{-iH_0t}. \\ H_0 & \text{does not depend on time.} \end{cases}$ (5.8)

The interaction picture and the Heisenberg picture coincide when $V = 0$, i.e., in the non-perturbed case. If $V(t)$ is a weak perturbation, then one can think of Eq. (5.8) as a way to pull out the fast, but trivial, time dependence due to H_0, leaving states that vary only slowly in time due to $V(t)$.

The first hint of the usefulness of the interaction picture comes from calculating the time derivative of $|\hat{\psi}(t)\rangle$ using the definition Eq. (5.8):

$$i\partial_t|\hat{\psi}(t)\rangle = \left(i\partial_t e^{iH_0t}\right)|\psi(t)\rangle + e^{iH_0t}\left(i\partial_t|\psi(t)\rangle\right) = e^{iH_0t}(-H_0 + H)|\psi(t)\rangle, \quad (5.9)$$

which by Eq. (5.8) is reduced to

$$i\partial_t|\hat{\psi}(t)\rangle = \hat{V}(t)\,|\hat{\psi}(t)\rangle. \quad (5.10)$$

The resulting Schrödinger equation for $|\hat{\psi}(t)\rangle$ thus contains explicit reference only to the interaction part $\hat{V}(t)$ of the full Hamiltonian H. This means that in the interaction picture the time evolution of a state $|\hat{\psi}(t_0)\rangle$ from time t_0 to t must be given in terms of a unitary operator $\hat{U}(t,t_0)$ which also only depends on $\hat{V}(t)$. $\hat{U}(t,t_0)$ is completely determined by

$$|\hat{\psi}(t)\rangle = \hat{U}(t,t_0)\,|\hat{\psi}(t_0)\rangle. \quad (5.11)$$

When V and thus H are time independent, an explicit form for $\hat{U}(t,t_0)$ is obtained by inserting $|\hat{\psi}(t)\rangle = e^{iH_0t}\,|\psi(t)\rangle = e^{iH_0t}\,e^{-iHt}\,|\psi_0\rangle$ and $|\hat{\psi}(t_0)\rangle = e^{iH_0t_0}\,e^{-iHt_0}\,|\psi_0\rangle$ into Eq. (5.11),

$$e^{iH_0t}e^{-iHt}\,|\psi_0\rangle = \hat{U}(t,t_0)\,e^{iH_0t_0}\,e^{-iHt_0}\,|\psi_0\rangle \quad \Rightarrow \quad \hat{U}(t,t_0) = e^{iH_0t}\,e^{-iH(t-t_0)}\,e^{-iH_0t_0}. \quad (5.12)$$

From this we observe that $\hat{U}^{-1} = \hat{U}^\dagger$, i.e., $\hat{U}$ is indeed a unitary operator.

In the general case with a time-dependent $\hat{V}(t)$, we must rely on the differential equation appearing when Eq. (5.11) is inserted in Eq. (5.10). We remark that Eq. (5.11) naturally implies the boundary condition $\hat{U}(t_0,t_0) = 1$, and we obtain:

$$i\partial_t\,\hat{U}(t,t_0) = \hat{V}(t)\,\hat{U}(t,t_0), \quad \hat{U}(t_0,t_0) = 1. \quad (5.13)$$

By integration of this differential equation we obtain the integral equation

$$\hat{U}(t,t_0) = 1 + \frac{1}{i}\int_{t_0}^{t} dt'\,\hat{V}(t')\,\hat{U}(t',t_0), \quad (5.14)$$

which we can solve iteratively for $\hat{U}(t,t_0)$ starting from $\hat{U}(t',t_0) = 1$. The solution is

$$\hat{U}(t, t_0) = 1 + \frac{1}{i} \int_{t_0}^{t} dt_1 \, \hat{V}(t_1) + \frac{1}{i^2} \int_{t_0}^{t} dt_1 \, \hat{V}(t_1) \int_{t_0}^{t_1} dt_2 \, \hat{V}(t_2) + \cdots \qquad (5.15)$$

Note that in the iteration the ordering of all operators is carefully kept. A more compact form is obtained by the following rewriting. Consider for example the second-order term, paying special attention to the dummy variables t_1 and t_2:

$$\int_{t_0}^{t} dt_1 \, \hat{V}(t_1) \int_{t_0}^{t_1} dt_2 \, \hat{V}(t_2)$$

$$= \frac{1}{2} \int_{t_0}^{t} dt_1 \, \hat{V}(t_1) \int_{t_0}^{t_1} dt_2 \, \hat{V}(t_2) + \frac{1}{2} \int_{t_0}^{t} dt_2 \, \hat{V}(t_2) \int_{t_0}^{t_2} dt_1 \, \hat{V}(t_1)$$

$$= \frac{1}{2} \int_{t_0}^{t} dt_1 \int_{t_0}^{t} dt_2 \, \hat{V}(t_1)\hat{V}(t_2)\theta(t_1 - t_2) + \frac{1}{2} \int_{t_0}^{t} dt_2 \int_{t_0}^{t} dt_1 \, \hat{V}(t_2)\hat{V}(t_1)\theta(t_2 - t_1)$$

$$= \frac{1}{2} \int_{t_0}^{t} dt_1 \int_{t_0}^{t} dt_2 \left[\hat{V}(t_1)\hat{V}(t_2)\theta(t_1 - t_2) + \hat{V}(t_2)\hat{V}(t_1)\theta(t_2 - t_1) \right]$$

$$\equiv \frac{1}{2} \int_{t_0}^{t} dt_1 \int_{t_0}^{t} dt_2 \, T_t[\hat{V}(t_1)\hat{V}(t_2)], \qquad (5.16)$$

where we have introduced the time ordering operator T_t. Time ordering is easily generalized to higher-order terms. In nth order, where n factors $\hat{V}(t_j)$ appear, all $n!$ permutations $p \in S_n$ of the n times t_j are involved, and we define[18]

$$T_t[\hat{V}(t_1)\hat{V}(t_2)\cdots\hat{V}(t_n)] \equiv \sum_{p \in S_n} \hat{V}(t_{p(1)})\hat{V}(t_{p(2)})\cdots\hat{V}(t_{p(n)}) \times \qquad (5.17)$$

$$\theta(t_{p(1)} - t_{p(2)}) \, \theta(t_{p(2)} - t_{p(3)}) \ldots \theta(t_{p(n-1)} - t_{p(n)}).$$

Using the time ordering operator, we obtain the final compact form (see also Exercise 5.2):

$$\hat{U}(t, t_0) = \sum_{n=0}^{\infty} \frac{1}{n!} \left(\frac{1}{i}\right)^n \int_{t_0}^{t} dt_1 \cdots \int_{t_0}^{t} dt_n \, T_t\left(\hat{V}(t_1)\cdots\hat{V}(t_n)\right) = T_t\left(e^{-i \int_{t_0}^{t} dt' \, \hat{V}(t')}\right).$$

$$(5.18)$$

Note the similarity with a usual time-evolution factor $e^{-i\varepsilon t}$. This expression for $\hat{U}(t, t_0)$ is the starting point for infinite-order perturbation theory and for introducing the concept of Feynman diagrams; it is therefore one of the central equations in quantum field theory. A graphical sketch of the contents of the formula is given in Fig. 5.1.

---

[18]For $n = 3$ we have $T_t[\hat{V}(t_1)\hat{V}(t_2)\hat{V}(t_3)] =$
$\hat{V}(t_1)\hat{V}(t_2)\hat{V}(t_3)\theta(t_1 - t_2)\theta(t_2 - t_3) + \hat{V}(t_1)\hat{V}(t_3)\hat{V}(t_2)\theta(t_1 - t_3)\theta(t_3 - t_2) +$
$\hat{V}(t_2)\hat{V}(t_3)\hat{V}(t_1)\theta(t_2 - t_3)\theta(t_3 - t_1) + \hat{V}(t_2)\hat{V}(t_1)\hat{V}(t_3)\theta(t_2 - t_1)\theta(t_1 - t_3) +$
$\hat{V}(t_3)\hat{V}(t_1)\hat{V}(t_2)\theta(t_3 - t_1)\theta(t_1 - t_2) + \hat{V}(t_3)\hat{V}(t_2)\hat{V}(t_1)\theta(t_3 - t_2)\theta(t_2 - t_1).$

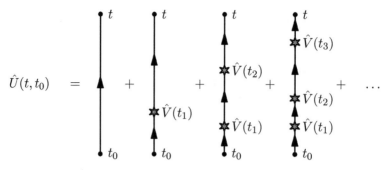

FIG. 5.1. The time-evolution operator $\hat{U}(t, t_0)$ in the form of Eq. (5.15) can be viewed as the sum of additional phase factors due to $\hat{V}$ on top of the trivial phase factors arising from H_0. The sum contains contributions from processes with $0, 1, 2, 3, \ldots$ scattering events $\hat{V}$, which happen during the evolution from time t_0 to time t.

5.4   Time-evolution in linear response

In many applications the perturbation $\hat{V}(t)$ is weak compared to H_0. It can therefore be justified to approximate $\hat{U}(t, t_0)$ by the first-order approximation

$$\hat{U}(t, t_0) \approx 1 + \frac{1}{i} \int_{t_0}^{t} dt' \, \hat{V}(t'). \tag{5.19}$$

This simple time-evolution operator forms the basis for the Kubo formula in linear response theory, which, as we shall see in the following chapters, is applicable to a wide range of physical problems.

5.5   Time-dependent creation and annihilation operators

It is of fundamental interest to study how the basic creation and annihilation operators $a_\nu^\dagger$ and a_ν evolve in time given some set of basis states $\{|\nu\rangle\}$ for a time-independent Hamiltonian H. As in Section 1.3.4 these operators can be taken to be either bosonic or fermionic. Let us first apply the definition of the Heisenberg picture, Eq. (5.5):

$$a_\nu^\dagger(t) \equiv e^{iHt} \, a_\nu^\dagger \, e^{-iHt}, \tag{5.20a}$$

$$a_\nu(t) \equiv e^{iHt} \, a_\nu \, e^{-iHt}. \tag{5.20b}$$

In the case of a general time-independent Hamiltonian with complicated interaction terms, the commutators $[H, a_\nu^\dagger]$ and $[H, a_\nu]$ are not simple, and consequently the fundamental (anti-)commutator $[a_\nu(t_1), a_\nu^\dagger(t_2)]_{F,B}$ involving two different times t_1 and t_2 cannot be given in a simple closed form:

$$[a_{\nu_1}(t_1), a_{\nu_2}^\dagger(t_2)]_{F,B} =$$

$$e^{iHt_1} a_{\nu_1} e^{-iH(t_1-t_2)} a_{\nu_2}^\dagger e^{-iHt_2} \pm e^{iHt_2} a_{\nu_2}^\dagger e^{-iH(t_2-t_1)} a_{\nu_1} e^{-iHt_1} \quad = \quad ?? \tag{5.21}$$

No further reduction is possible in the general case. In fact, as we shall see in the following chapters, calculating (anti-)commutators like Eq. (5.21) is *the* problem in many-particle physics.

But let us investigate some simple cases to get a grasp of the time-evolution pictures. Consider first a time-independent Hamiltonian H which is diagonal in the $|\nu\rangle$-basis,

$$H = \sum_{\nu} \varepsilon_{\nu} a_{\nu}^{\dagger} a_{\nu}. \tag{5.22}$$

The equation of motion, Eq. (5.6), is straightforward:[19]

$$\begin{aligned}
\dot{a}_{\nu}(t) = i[H, a_{\nu}(t)] \quad &= \quad ie^{iHt}[H, a_{\nu}]e^{-iHt} \\
= ie^{iHt} \sum_{\nu'} \varepsilon_{\nu'} \left[a_{\nu'}^{\dagger} a_{\nu'}, a_{\nu} \right] e^{-iHt} \quad &= \quad ie^{iHt} \sum_{\nu'} \varepsilon_{\nu'} \left(-\delta_{\nu\nu'} \right) a_{\nu'} e^{-iHt} \\
= -i\varepsilon_{\nu} e^{iHt} a_{\nu} e^{-iHt} \quad &= \quad -i\varepsilon_{\nu} a_{\nu}(t).
\end{aligned} \tag{5.23}$$

By integration we obtain

$$a_{\nu}(t) = e^{-i\varepsilon_{\nu}t} a_{\nu}, \tag{5.24}$$

which by Hermitian conjugation leads to

$$a_{\nu}^{\dagger}(t) = e^{+i\varepsilon_{\nu}t} a_{\nu}^{\dagger}. \tag{5.25}$$

In this very simple case the basic (anti-)commutator Eq. (5.21) can be evaluated directly:

$$[a_{\nu_1}(t_1), a_{\nu_2}^{\dagger}(t_2)]_{F,B} = e^{-i\varepsilon_{\nu_1}(t_1-t_2)} \delta_{\nu_1\nu_2}. \tag{5.26}$$

For the diagonal Hamiltonian the time evolution is thus seen to be given by trivial phase factors $e^{\pm i\varepsilon t}$.

We can also gain some insight into the interaction picture by a trivial extension of the simple model. Assume that

$$H = H_0 + \gamma H_0, \quad \gamma \ll 1, \tag{5.27}$$

where H_0 is diagonalized in the basis $\{|\nu\rangle\}$ with the eigenenergies ε_{ν}. Obviously, the full Hamiltonian H is also diagonalized in the same basis, but with the eigenenergies $(1+\gamma)\varepsilon$. Let us however try to treat γH_0 as a perturbation V to H_0, and then use the interaction picture of Section 5.3. From Eq. (5.8) we then obtain

$$|\hat{\nu}(t)\rangle = e^{i\varepsilon_{\nu}t} |\nu(t)\rangle. \tag{5.28}$$

But we actually know the time evolution of the Schrödinger state on the right-hand side of the equation, so

$$|\hat{\nu}(t)\rangle = e^{i\varepsilon_{\nu}t} e^{-i(1+\gamma)\varepsilon_{\nu}t} |\nu\rangle = e^{-i\gamma\varepsilon_{\nu}t} |\nu\rangle. \tag{5.29}$$

Here we clearly see that the fast Schrödinger time dependence given by the phase factor $e^{i\varepsilon_{\nu}t}$, is replaced in the interaction picture by the slow phase factor $e^{i\gamma\varepsilon_{\nu}t}$. The reader can try to obtain Eq. (5.29) directly from Eq. (5.18).

---

[19]We are using the identities $[AB, C] = A[B, C] + [A, C]B$ and $[AB, C] = A\{B, C\} - \{A, C\}B$, which are valid for any set of operators. Note that the first identity is particularly useful for bosonic operators and the second for fermionic operators (see Exercise 5.4).

Finally, we briefly point to the complications that arise when the interaction is given by a time-independent operator V not diagonal in the same basis as H_0. Consider for example the Coulomb-like interaction written symbolically as

$$H = H_0 + V = \sum_{\nu'} \varepsilon_{\nu'} a_{\nu'}^\dagger a_{\nu'} + \frac{1}{2} \sum_{\nu_1 \nu_2 q} V_q \, a_{\nu_1+q}^\dagger a_{\nu_2-q}^\dagger a_{\nu_2} a_{\nu_1}. \tag{5.30}$$

The equation of motion for fermionic operators $a_\nu(t)$ is (see also Exercise 5.5):

$$\dot{a}_\nu(t) = i[H, a_\nu(t)]$$

$$= -i\varepsilon_\nu \, a_\nu(t) + \frac{i}{2} \sum_{\nu_1 \nu_2 q} V_q \left[a_{\nu_1+q}^\dagger(t) \, a_{\nu_2-q}^\dagger(t), \, a_\nu(t) \right] a_{\nu_2}(t) \, a_{\nu_1}(t)$$

$$= -i\varepsilon_\nu \, a_\nu(t) + \frac{i}{2} \sum_{\nu_1 \nu_2} \left(V_{\nu_2-\nu} - V_{\nu-\nu_1} \right) a_{\nu_1+\nu_2-\nu}^\dagger(t) \, a_{\nu_2}(t) \, a_{\nu_1}(t). \tag{5.31}$$

The problem in this more general case is evident. The equation of motion for the single operator $a_\nu(t)$ contains terms with both one and three operators, and we do not know the time evolution of the three-operator product $a_{\nu_1+\nu_2-\nu}^\dagger(t) \, a_{\nu_2}(t) \, a_{\nu_1}(t)$. If we write down the equation of motion for this three-operator product we discover that terms are generated involving five-operator products. This feature is then repeated over and over again generating a never-ending sequence of products containing seven, nine, eleven, etc. operators. In the following chapters we will learn various approximate methods to deal with this problem.

5.6 Fermi's golden rule

Fermi's golden rule is a very useful expression in lowest-order perturbation theory for the transition rate Γ_{fi} associated with a transition from an initial state $|i\rangle$ to a different (and orthogonal) final state $|f\rangle$. The transition is due to some weak perturbation $V(t)$ that couples $|i\rangle$ and $|f\rangle$. The two states are eigenstates of the unperturbed system described by a Hamiltonian H_0, so $H_0|\nu\rangle = E_\nu|\nu\rangle$ with $\nu = i$ or $\nu = f$.

The total Hamiltonian is $H = H_0 + V(t)$, and the perturbation is thought to be adiabatically turned on at the distant past, so $V(t) = Ve^{\eta t}$, where V is a time-independent operator and η is a small positive rate. Suppose the system in question is in state $|i\rangle$ at time t_0. At some later time $t > t_0$ this initial state has evolved into $|i(t)\rangle$ according to Eqs. (5.8) and (5.11),

$$|i(t)\rangle = e^{-iH_0 t} \hat{U}(t, t_0) e^{iH_0 t_0} |i\rangle. \tag{5.32}$$

It is straightforward to determine the overlap between this state and a final state $|f\rangle$. To first order in V, after letting $t_0 \to -\infty$, we obtain (see Exercise 5.9)

$$\langle f|i(t)\rangle = -\langle f|V|i\rangle \frac{e^{-iE_i(t-t_0)} e^{\eta t}}{E_f - E_i - i\eta}. \tag{5.33}$$

The probability $P_f(t)$ to find the system in state $|f\rangle$ at time t is given by $|\langle f|i(t)\rangle|^2$. The time-derivative $dP_f(t)/dt$ is therefore the change in probability per unit time, which we interpret as the transition rate Γ_{fi} between initial and final states. As shown in Exercise 5.9, we arrive at Fermi's golden rule

$$\Gamma_{fi} = 2\pi \left| \langle f|V|i \rangle \right|^2 \delta(E_f - E_i). \tag{5.34}$$

If, furthermore, there is a continuum of equivalent final states one should sum over these and thus the density of states appears in the expression because $\rho(E_i) = \sum_f \delta(E_f - E_i)$. This version of Fermi's golden rule is often used in scattering theory. Here we give the more general expression and we emphasize that Eq. (5.34) is not limited to describing transitions between single-particle states, but is also valid for transitions between many-body states. This is exploited further in Chapter 10.

5.7   The T-matrix and the generalized Fermi's golden rule

In this section, Fermi's golden rule is generalized to include also higher-order processes where the initial and final states are coupled by multiple scatterings on the perturbation. However, the formal theory of scattering theory is rather involved, so we restrict ourselves to deriving the so-called T-matrix using a physically motivated formulation. It is our first encounter with higher-order perturbation theory in this book.

In order to derive a generalized Fermi's golden rule, we wish to include all higher orders in the perturbation $V(t)$. Thus, instead of using Eq. (5.19) in Eq. (5.32), we apply the expansion from Eq. (5.15). The overlap in Eq. (5.33) then becomes

$$\langle f|i(t)\rangle = \left\langle f \left| \frac{1}{i} \int_{t_0}^{t} dt_1\, \hat{V}(t_1) + \frac{1}{i^2} \int_{t_0}^{t} dt_1\, \hat{V}(t_1) \int_{t_0}^{t_1} dt_2\, \hat{V}(t_2) + \ldots \right| i \right\rangle e^{-iE_f t} e^{iE_i t_0},$$

$$\tag{5.35}$$

where we omitted the zeroth-order term in Eq. (5.15) because the initial and final states are considered to be orthogonal.

At this point, we have to be careful with the formulation of how the perturbation is turned on in the distant past. As above, we assume that the interaction is turned on slowly to avoid complication due to transients that are not relevant for the discussion. However, because many scattering events can take place between the time t_0 at which the perturbation is turned on and the time t of the measurement, we have to make sure that the "turning on" time η^{-1} is well separated from the duration $t - t_0$ of the interaction, i.e., $t - t_0 \gg \eta^{-1}$ and we thus take the limit $t_0 \to -\infty$ *before* the limit $\eta \to 0$, while keeping t finite. One possible way to incorporate these conditions is to use a functional form of $V(t)$ different from the exponential that we used in the derivation of the lowest-order Fermi's golden rule. We can for example use $V(t) = f(t)$, where $f(t) = \{1 + \exp[-\eta(t-t_0)]\}^{-1}$. This Fermi-function turns on the perturbation around time t_0 during a characteristic time interval η^{-1}.

We now incorporate the turning-on function $f(t)$ explicitly in the expansion of the overlap, Eq. (5.35). For a specific term $V^{(n)}$ in this expansion,

$$V^{(n)} = \frac{1}{i^n} \int_{-\infty}^{t} dt_1\, \hat{V}(t_1) f(t_1) \cdots \int_{-\infty}^{t_{i-1}} dt_i\, \hat{V}(t_i) f(t_i) \cdots \int_{-\infty}^{t_{n-1}} dt_n\, \hat{V}(t_n) f(t_n), \tag{5.36}$$

we note that if the last factor $f(t_n)$ is turned on, most likely so are all the other factors $f(t_i)$ because they have time arguments larger than t_n and because in the appropriate limits their range is between t_0 and t, which is much larger than η^{-1}. Using this observation, we can replace $f(t_i)$ by unity when $i \neq n$. Furthermore, now that we have only a single turning-on time, we can for simplicity go back to the exponential function, and let $f(t) \simeq \exp(\eta t)$ and $t_0 = -\infty$. With these steps Eq. (5.35) becomes

$$|\langle f|i(t)\rangle| = \left| \sum_{n=1}^{\infty} \frac{1}{i^n} \left\langle f \left| \int_{-\infty}^{t} dt_1 \int_{-\infty}^{t_1} dt_2 \cdots \int_{-\infty}^{t_{n-1}} dt_n\, \hat{V}(t_1)\hat{V}(t_2)\cdots \hat{V}(t_n)e^{\eta t_n} \right| i \right\rangle \right|.$$
(5.37)

Inserting the definition of the time-dependence in the interaction picture $\hat{V}(t) = e^{iH_0 t} V e^{-iH_0 t}$ and performing the integrations, we obtain

$$|\langle f|i(t)\rangle| = \left| \frac{e^{\eta t}}{E_i - E_f + i\eta} \langle f|T|i\rangle \right|,$$
(5.38)

where

$$T = V + V\frac{1}{E_i - H_0 + i\eta}V + V\frac{1}{E_i - H_0 + i\eta}V\frac{1}{E_i - H_0 + i\eta}V + \ldots,$$
(5.39)

is called the T-matrix . The T-matrix can also be written as

$$T = V + V\frac{1}{E_i - H_0 + i\eta}T,$$
(5.40)

because by iteration this expression generates the same series as in Eq. (5.39).

The transition rate is now evaluated in the same fashion as in the previous section, i.e., we identify Γ_{fi} with time derivative of $|\langle f|i(t)\rangle|^2$. One finds

$$\Gamma_{fi} = 2\pi |\langle f|T|i\rangle|^2 \delta(E_f - E_i).$$
(5.41)

This expression is the generalized Fermi's golden rule; clearly, to lowest order in V we recover Eq. (5.34). The T-matrix and Eq. (5.41) is used in Chapter 10, where we discuss higher-order tunneling effects.

5.8 Fourier transforms of advanced and retarded functions

In this book, we encounter different kinds of correlation functions and so-called Green's functions; the first appearing in the Kubo formula in the following chapter. Often we will perform Fourier transforms of these functions and therefore a discussion of some general aspects of this is useful.

We consider first a so-called retarded function. The name "retarded" means that, in accordance with normal causality, it is a measure of a physical observable due to the action of some force or interaction at times prior to the measurement. The general form of such functions is

$$C_{AB}^{R}(t, t') = -i\theta(t - t') \left\langle [A(t), B(t')] \right\rangle.$$
(5.42)

Here we have used an example, where the operators are bosonic operators, while for fermionic operators the retarded functions are defined with an anti-commutators instead of the usual commutators, see also Eq. (8.28). In Eq. (5.42), $A(t)$ and $B(t)$ are operators in the Heisenberg picture and the average $\langle \cdot \rangle$ means the thermal average defined in Eq. (1.118). The Hamiltonian thus enters in two places, namely through the time dependence of the operators and through the thermal average. In most cases, we will consider situation where the Hamiltonian is time independent and therefore the function $C_{AB}^R(t, t')$ only depends on the time difference $t - t'$. This is easily proved by writing out the definition of the average including the Heisenberg time-dependence,

$$\langle A(t)B(t') \rangle = \frac{1}{Z} \text{Tr} \left[e^{-\beta H} e^{iHt} A e^{-iHt} e^{iHt'} B e^{-iHt'} \right]$$
$$= \frac{1}{Z} \text{Tr} \left[e^{-\beta H} e^{iH(t-t')} A e^{-iH(t-t')} B \right], \tag{5.43}$$

where we have used the cyclic properties of the trace and that the exponentials $e^{-\beta H}$ and $e^{-iHt'}$ of course commute.

Since $C_{AB}^R(t, t')$ thus depends on one time variable, we write it as $C^R(t - t')$ and define the Fourier transform as

$$C_{AB}^R(\omega) = \int_{-\infty}^{\infty} dt \, e^{i\omega t} C_{AB}^R(t). \tag{5.44}$$

In order for the Fourier transform to be well-defined, the integrand must decay for both plus and minus infinity. Of course, for the retarded functions which are zero for negative times, only plus infinity could pose a problem. In the limit of infinite time difference, $t - t' \to \infty$, we expect on physical grounds that there can be no correlation between $A(t)$ and $B(t')$ and thus we expect that

$$C_{AB}^R(t, t') \to -i\theta(t - t') \left[\langle A(t) \rangle \langle B(t') \rangle - \langle B(t') \rangle \langle A(t) \rangle \right] = 0. \tag{5.45}$$

If this is true, the Fourier transform in Eq. (5.44) is well defined. Whether this asymptotic comes out correctly, depends on whether the theory or model that one uses has built in some relaxation mechanisms that destroy long time correlations. If not, correlations can exist at infinite ranges and the function will not decay and typically it will have some oscillatory dependence like $C_{AB}^R(t - t') \propto \exp[i\varepsilon(t - t')]$. Nevertheless, even if the function is not decaying one can perform a Fourier transform if the frequency is allowed to be complex. By replacing the real frequency ω in Eq. (5.44) with the complex frequency $\omega + i\eta$, where η is a *positive* real number, the Fourier transform then reads

$$C_{AB}^R(\omega) = \int_{-\infty}^{\infty} dt \, e^{i\omega t} e^{-\eta t} C_{AB}^R(t). \tag{5.46}$$

If the function is bounded in the sense that $|C_{AB}^R(t)| < M$ for all t and a positive M then the integral in Eq. (5.46) is convergent.[20] The inverse Fourier transform should, of course, be modified correspondingly. However, if we let η be a positive infinitesimal

---

[20]The transform can be defined even for an exponentially bounded function $|C^R(t)| < \exp(\alpha t)$ as long as $\eta > \alpha$.

and take the limit $\eta \to 0^+$ at the end of the calculation, we can use the usual inverse Fourier transformation, because

$$
\begin{aligned}
C_{AB}^R(t) &= \int_{-\infty}^{\infty} \frac{d\omega}{2\pi} e^{-i\omega t} C_{AB}^R(\omega) = \int_{-\infty}^{\infty} \frac{d\omega}{2\pi} e^{-i\omega t} \int_{-\infty}^{\infty} dt'\, e^{i\omega t'} e^{-\eta t'} C_{AB}^R(t') \\
&= e^{-\eta t} C_{AB}^R(t) \xrightarrow[\eta \to 0^+]{} C_{AB}^R(t).
\end{aligned}
\tag{5.47}
$$

Let us recapitulate: for retarded functions, which are not decaying at large times, we should define the Fourier transform with a complex frequency $\omega + i\eta$, where η is to be understood as a positive infinitesimal. The inverse Fourier transform is the same as usual, but in the end we must take the limit $\eta \to 0^+$.

One example of functions where the procedure is necessary is the unperturbed Green's functions that later in book turns out to be important building blocks in the formulation of many-body theory. Exercise 5.8 shows a typical function that we will encounter.

Another class of functions is the so-called advanced functions, which are defined similarly to Eq. (5.42)

$$
C_{AB}^A(t, t') = i\theta(t' - t) \left\langle \left[A(t), B(t') \right] \right\rangle.
\tag{5.48}
$$

In this case the Fourier transformation is carried out in the same way, except now the frequency should be have a small negative imaginary part, i.e., $\omega \to \omega - i\eta$.

5.9 Summary and outlook

In this chapter, we have introduced the fundamental representations used in the description of time evolution in many-particle systems: the Schrödinger picture, Eq. (5.2), the Heisenberg picture, Eq. (5.5), and the interaction picture, Eq. (5.8). Which picture to use depends on the problem at hand. The first two pictures rely on a time-independent Hamiltonian H, while the interaction picture involves a time-dependent Hamiltonian H of the form $H = H_0 + V(t)$, where H_0 is a time-independent Hamiltonian. For the state vectors and the operators in the interaction picture we found that

$$
i\partial_t |\hat{\psi}(t)\rangle = \hat{V}(t) \, |\hat{\psi}(t)\rangle, \quad \hat{A}(t) \equiv e^{iH_0 t} A \, e^{-iH_0 t}.
$$

The time-evolution operator $\hat{U}(t, t_0)$ describing the evolution of an interaction picture state $|\hat{\psi}(t_0)\rangle$, at time t_0, to $|\hat{\psi}(t)\rangle$, at time t, was introduced,

$$
|\hat{\psi}(t)\rangle = \hat{U}(t, t_0) \, |\hat{\psi}(t_0)\rangle, \quad i\partial_t \, \hat{U}(t, t_0) = \hat{V}(t) \, \hat{U}(t, t_0), \quad \text{with } \hat{U}(t_0, t_0) = 1.
$$

We shall see in the following chapters how the operator $\hat{U}(t, t_0)$, due to its form

$$
\hat{U}(t, t_0) = \sum_{n=0}^{\infty} \frac{1}{n!} \left(\frac{1}{i} \right)^n \int_{t_0}^{t} dt_1 \cdots \int_{t_0}^{t} dt_n \, T_t \left(\hat{V}(t_1) \cdots \hat{V}(t_n) \right) = T_t \left(e^{-i \int_{t_0}^{t} dt' \, \hat{V}(t')} \right).
$$

plays an important role in the formulation of infinite-order perturbation theory and the introduction of Feynman diagrams, and how its first-order approximation,

$$\hat{U}(t, t_0) \approx 1 + \frac{1}{i} \int_{t_0}^{t} dt' \, \hat{V}(t'),$$

forms the basis of the widely used linear response theory and the associated Kubo formalism to be treated in Chapter 6.

Finally, we studied a few examples where the full understanding of the time dependence was important, namely the transition rates between quantum states. We saw both the usual Fermi's golden rule, Eq. (5.34) which is a perturbative result and its generalization to all orders in the interaction through the T-matrix, Eq. (5.41).

6

LINEAR RESPONSE THEORY

Linear response theory is an extremely widely used concept in all branches of physics. It simply states that the response to a weak external perturbation is proportional to the perturbation, and therefore all one needs to understand is the proportionality constant. Below we derive the general formula for the linear response of a quantum system exerted by a perturbation. The physical question we ask is thus: given some external perturbation H', what is the measured consequence for an observable quantity, A. In other words, what is $\langle A \rangle$ to linear order in H'?

Among the numerous applications of the linear response formula, one can mention charge and spin susceptibilities of, for instance, electron systems due to external electric or magnetic fields. Responses to external mechanical forces or vibrations can also be calculated using the very same formula. Here we utilize the formalism to derive a general expression for the electrical conductivity and briefly mention other applications.

6.1 The general Kubo formula

Consider a quantum system described by the (time independent) Hamiltonian H_0 in thermodynamic equilibrium. According to Sec. 1.5 this means that an expectation value of a physical quantity, described by the operator A, can be evaluated as

$$\langle A \rangle = \frac{1}{Z_0} \text{Tr} \left[\rho_0 A \right] = \frac{1}{Z_0} \sum_n \langle n|A|n \rangle e^{-\beta E_n}, \tag{6.1a}$$

$$\rho_0 = e^{-\beta H_0} = \sum_n |n\rangle\langle n| e^{-\beta E_n}, \tag{6.1b}$$

where ρ_0 is the density operator and $Z_0 = \text{Tr} \left[\rho_0 \right]$ is the partition function. Here as in Sec. 1.5, we write the density operator in terms of a complete set of eigenstates $\{|n\rangle\}$ of the Hamiltonian H_0 with eigenenergies $\{E_n\}$.

Suppose now that at some time, $t = t_0$, an external perturbation is applied to the system, driving it out of equilibrium. The perturbation is described by an additional time-dependent term in the Hamiltonian

$$H(t) = H_0 + H'(t)\theta(t - t_0). \tag{6.2}$$

We emphasize that H_0 is the Hamiltonian describing the system before the perturbation was applied (see Fig. 6.1 for an illustration). Now, we wish to find the expectation value of the operator A at times t greater than t_0. In order to do so, we must find the time evolution of the density matrix or equivalently the time evolution of the

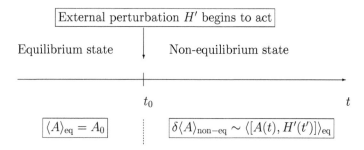

FIG. 6.1. Illustration of the linear response theory. At times prior to t_0 the system is in equilibrium, at times larger than t_0 the perturbation is turned on. The system is now evolving according to the new Hamiltonian and is in a non-equilibrium state. The Kubo formula relates the expectation value $\delta\langle A\rangle_{\text{non-eq}}$ in the non-equilibrium state to a equilibrium expectation value $\langle\cdots\rangle_{\text{eq}}$ of the more complicated time-dependent commutator $[\hat{A}(t), \hat{H}'(t')]$.

eigenstates of the unperturbed Hamiltonian. Once we know the $|n(t)\rangle$, we can obtain the time-dependent expectation value of the operator A as

$$\langle A\rangle(t) = \frac{1}{Z_0}\sum_n \langle n(t)|A|n(t)\rangle e^{-\beta E_n} = \frac{1}{Z_0}\text{Tr}\left[\rho(t)A\right], \tag{6.3a}$$

$$\rho(t) = \sum_n |n(t)\rangle\langle n(t)|e^{-\beta E_n}. \tag{6.3b}$$

The philosophy behind this expression is as follows. The initial states of the system are distributed according to the usual Boltzmann distribution $e^{-\beta E_{0n}}/Z_0$. At later times the system is described by the same distribution of states but the states are now time-dependent and they have evolved according to the new Hamiltonian. The time dependence of the states $|n(t)\rangle$ is of course governed by the Schrödinger equation

$$i\partial_t|n(t)\rangle = H(t)|n(t)\rangle. \tag{6.4}$$

Since H' is to be regarded as a small perturbation, it is convenient to utilize the interaction picture representation $|\hat{n}(t)\rangle$ introduced in Section 5.3. The time dependence in this representation is given by

$$|n(t)\rangle = e^{-iH_0 t}|\hat{n}(t)\rangle = e^{-iH_0 t}\hat{U}(t, t_0)|\hat{n}(t_0)\rangle, \tag{6.5}$$

where by definition $|\hat{n}(t_0)\rangle = e^{iH_0 t_0}|n(t_0)\rangle = |n\rangle$.

To linear order in H', Eq. (5.19) states that $\hat{U}(t, t_0) = 1 - i\int_{t_0}^t dt'\ \hat{H}'(t')$. Inserting this into (6.3a), one obtains the expectation value of A up to linear order in the perturbation

$$\langle A\rangle(t) = \langle A\rangle_0 - i\int_{t_0}^t dt' \frac{1}{Z_0}\sum_n e^{-\beta E_n}\langle n|\hat{A}(t)\hat{H}'(t') - \hat{H}'(t')\hat{A}(t)|n\rangle$$

$$= \langle A\rangle_0 - i\int_{t_0}^t dt'\langle[\hat{A}(t), \hat{H}'(t')]\rangle_0. \tag{6.6}$$

The brackets $\langle\ \rangle_0$ mean an equilibrium average with respect to the Hamiltonian H_0. This is in fact a remarkable and very useful result, because the inherently non-equilibrium quantity $\langle A(t)\rangle$ has been expressed as a retarded correlation function of the system in equilibrium.

The correlation function that appears in Eq. (6.6), is called a retarded correlation function or a retarded response function, and for later reference we rewrite the linear response result as

$$\delta\langle A\rangle(t) \equiv \langle A\rangle(t) - \langle A\rangle_0 = \int_{t_0}^{\infty} dt' C_{AH'}^R(t, t'), \tag{6.7}$$

where

$$C_{AH'}^R(t, t') = -i\theta(t - t')\left\langle\left[\hat{A}(t), \hat{H}'(t')\right]\right\rangle_0. \tag{6.8}$$

This is the famous Kubo formula which expresses the linear response to a perturbation, H'.

6.1.1 Kubo formula in the frequency domain

It is often convenient to express the response to an external disturbance in the frequency domain. This is particulary useful when the external perturbation has the form

$$H_B'(t) = Bf(t), \tag{6.9}$$

where B is a time-independent *operator* and where the time-dependent function $f(t)$ is *not an operator*. The response function $C_{AH_B'}^R(t, t')$, then becomes

$$C_{AH_B'}^R(t, t') = C_{AB}^R(t - t')f(t'), \tag{6.10}$$

where we anticipated that $C_{AB}^R(t, t')$ only depends on the time difference, see Eq. (5.43). When inserting Eq. (6.10) into the formula for the expectation value (6.7), and furthermore letting $t_0 = -\infty$ (which is reasonable if we are not interested in the transient behavior), we have

$$\delta\langle A\rangle(t) = \int_{-\infty}^{\infty} dt' C_{AB}^R(t - t')f(t'), \tag{6.11}$$

which is a convolution of the response function C_{AB}^R and the function f and hence the Fourier transform is

$$\delta\langle A\rangle(\omega) = C_{AB}^R(\omega)f(\omega). \tag{6.12}$$

We can specialize this result to cases where the external force depends on position and direction. For example, instead of Eq. (6.9), we consider an external perturbation of the form

$$H_B'(t) = \sum_\alpha \int d\mathbf{r}\, B^\alpha(\mathbf{r})f^\alpha(\mathbf{r}, t). \tag{6.13}$$

In this case, we can use the linear response formula (6.12) for each term in the sum and for each value of $\mathbf{r}$ such that

$$\delta\langle A\rangle(\omega) = \sum_\alpha \int d\mathbf{r}\, C^R_{AB^\alpha(\mathbf{r})}(\omega) f^\alpha(\mathbf{r},\omega). \tag{6.14}$$

This result is used in the following sections, where we discuss response to an external electromagnetic field.

6.2 Kubo formula for conductivity

Consider a system of charged particles, electrons say, which is subjected to an external electromagnetic field. The electromagnetic field induces a current, and the conductivity is the linear response coefficient. In the general case the conductivity may be non-local in both time and space, such that the electric current $\mathbf{J}_e$ at some point $\mathbf{r}$ at time t depends on the electric field at points $\mathbf{r}'$ at times t'. The linear relation between electrical current and electric field is

$$J_e^\alpha(\mathbf{r},t) = \int dt' \int d\mathbf{r}' \sum_\beta \sigma^{\alpha\beta}(\mathbf{r},\mathbf{r}',t-t')\, E^\beta_{\mathrm{ext}}(\mathbf{r}',t'), \tag{6.15}$$

where $\sigma^{\alpha\beta}(\mathbf{r},\mathbf{r}',t-t')$ is the conductivity tensor that describes the current response in direction $\hat{e}_\alpha$ to an externally applied electric field in direction $\hat{e}_\beta$. Here we anticipated the long-time limit where the conductivity tensor only depends on the time difference $t-t'$, as discussed in the previous section. In this case, the corresponding expression in frequency domain reads

$$J_e^\alpha(\mathbf{r},\omega) = \int d\mathbf{r}' \sum_\beta \sigma^{\alpha\beta}(\mathbf{r},\mathbf{r}',\omega)\, E^\beta_{\mathrm{ext}}(\mathbf{r}',\omega). \tag{6.16}$$

The external electric field $\mathbf{E}$ is given by the electric potential ϕ_{ext} and the vector potential $\mathbf{A}_{\mathrm{ext}}$

$$\mathbf{E}_{\mathrm{ext}}(\mathbf{r},t) = -\nabla_\mathbf{r}\phi_{\mathrm{ext}}(\mathbf{r},t) - \partial_t \mathbf{A}_{\mathrm{ext}}(\mathbf{r},t). \tag{6.17}$$

The electric current is $\mathbf{J}_e = -e\langle \mathbf{J}_{\mathrm{tot}}\rangle$, where $\mathbf{J}_{\mathrm{tot}}$ is the total current density operator. For simplicity, we assume only one kind of particles, say electrons, but generalization to several branches of charge carriers is straightforward.

After application of the perturbation that drives the electrical current the total vector potential is

$$\mathbf{A}_{\mathrm{tot}} = \mathbf{A} + \mathbf{A}_{\mathrm{ext}}, \tag{6.18}$$

where $\mathbf{A}$ denotes the vector potential in *equilibrium*, i.e. prior to the onset of the perturbation $\mathbf{A}_{\mathrm{ext}}$. The Hamiltonian describing the coupling to a vector potential was introduced in Sec. 1.4.3, see Eq. (1.97). Inserting the total vector potential (6.18) in the Hamiltonian (1.97) and keeping only *linear* terms in $\mathbf{A}_{\mathrm{ext}}$ (which is in Eq. (Eq. (1.96)), we obtain the vector potential contribution to the perturbation. Also adding the potential energy due to the external electrical potential, we arrive at the total perturbation Hamiltonian

$$H'(t) = -e\int d\mathbf{r}\, \rho(\mathbf{r})\phi_{\mathrm{ext}}(\mathbf{r},t) + e\int d\mathbf{r}\, \mathbf{J}(\mathbf{r})\cdot \mathbf{A}_{\mathrm{ext}}(\mathbf{r},t), \tag{6.19}$$

where the equilibrium current density operator $\mathbf{J}$ is given in Eq. (1.98a). It has two terms: a paramagnetic term and a diamagnetic term proportional to the equilibrium vector potential $\mathbf{A}$. The total current operator follows from Eq. (1.98a) with the vector potential given by the *total* vector potential Eq. (6.18):

$$\mathbf{J}_{\text{tot}}(\mathbf{r}, t) = \mathbf{J}(\mathbf{r}) + \mathbf{J}_{\text{ext}}(\mathbf{r}) = \mathbf{J}(\mathbf{r}) + \frac{e}{m}\mathbf{A}_{\text{ext}}(\mathbf{r}, t)\rho(\mathbf{r}). \tag{6.20}$$

With the perturbation Hamiltonian (6.19) and the total current operator (6.20), we are now ready to use the linear response theory to calculate the conductivity. It is somewhat unusual, because part of the perturbation also appears in the operator of which we want to calculate the expectation value and we therefore get two contributions to $\mathbf{J}_e = -e\langle\mathbf{J}_{\text{tot}}\rangle$, namely

$$\mathbf{J}_e(\mathbf{r}, t) = -e\langle\mathbf{J}\rangle(t) - \frac{e^2}{m}\mathbf{A}_{\text{ext}}(\mathbf{r}, t)\langle\rho(\mathbf{r})\rangle_0. \tag{6.21}$$

The last term is proportional to $\mathbf{A}_{\text{ext}}$ and therefore linear response result is found by taking the expectation value of the density operator with respect to the equilibrium Hamiltonian. To find the linear component of the first term in (6.20) we use the general Kubo formula in Eq. (6.6). For this purpose it is convenient to choose a specific gauge, namely one where the external electrical potential is zero $\phi_{\text{ext}} = 0$. This is always possible by a suitable choice of $\mathbf{A}_{\text{ext}}(\mathbf{r}, t)$ as seen in Eq. (6.17). The final result does not, of course, depend on the choice of gauge. With this gauge choice the linear response result for $\langle\mathbf{J}\rangle$ due to the perturbation (6.19) becomes

$$\langle\mathbf{J}(\mathbf{r})\rangle(t) = e \int d\mathbf{r}' \int dt' \sum_\beta C^R_{\mathbf{J}(\mathbf{r})J^\beta(\mathbf{r}')}(t - t')A^\beta_{\text{ext}}(\mathbf{r}', t'), \tag{6.22}$$

where we used that the equilibrium state does not carry any current[21], i.e.$\langle\mathbf{J}\rangle_0 = 0$. Inserting (6.22) back into Eq. (6.20) we then have the final linear response for the current. Converting this to frequency domain, we have

$$\mathbf{J}_e(\mathbf{r})(\omega) = -e^2 \int d\mathbf{r}' \sum_\beta C^R_{\mathbf{J}(\mathbf{r})J^\beta(\mathbf{r}')}(\omega)A^\beta_{\text{ext}}(\mathbf{r}', \omega) - \frac{e^2}{m}\mathbf{A}_{\text{ext}}(\mathbf{r}, \omega)\langle\rho(\mathbf{r})\rangle_0. \tag{6.23}$$

Finally, to express the response in terms of the electric field, we Fourier transform the relation between the external vector potential and the external electric field: $\mathbf{A}_{\text{ext}}(\omega) = \mathbf{E}_{\text{ext}}(\mathbf{r}, \omega)/i\omega$. When inserting this into Eq. (6.23) and comparing the result with Eq. (6.15), we arrive at the final expression for the linear response formula for the conductivity tensor

$$\sigma^{\alpha\beta}(\mathbf{r}, \mathbf{r}', \omega) = \frac{ie^2}{\omega}\Pi^R_{\alpha\beta}(\mathbf{r}, \mathbf{r}', \omega) + \frac{ie^2 n(\mathbf{r})}{\omega m}\delta(\mathbf{r} - \mathbf{r}')\delta_{\alpha\beta}, \tag{6.24}$$

where we have used the symbol $\Pi^R = C^R_{J_0 J_0}$ for the retarded current-current correlation function and where $n(\mathbf{r}) = \langle\rho(\mathbf{r})\rangle_0$. In the time domain, the retarded current-current correlation function is given by

$$\Pi^R_{\alpha\beta}(\mathbf{r}, \mathbf{r}', t - t') = C^R_{J_0^\alpha(\mathbf{r})J_0^\beta(\mathbf{r}')}(t - t') = -i\theta(t - t')\left\langle\left[\hat{J}_0^\alpha(\mathbf{r}, t), \hat{J}_0^\beta(\mathbf{r}', t')\right]\right\rangle_0. \tag{6.25}$$

---

[21] In fact, this statement is not true for superconductors, see Section 18.6.2.

Finding the conductivity of a given system has thus been reduced to finding the retarded current-current correlation function. This formula will be used extensively in Chap. 14.

6.3 Kubo formula for conductance

For a given material the conductivity σ is the proportionality coefficient between the electric field $\mathbf{E}$ and the current density $\mathbf{J}$. The conductivity thus describes an intrinsic property of the material. The conductance on the other hand is the proportionality coefficient between the current I through a given sample and the voltage V applied to it, i.e., it is a sample-specific quantity. The conductance G is defined by the usual Ohm's law

$$I = GV. \tag{6.26}$$

For a material where the conductivity can be assumed to be local in space one can find the conductance of a specific sample by the relation

$$G = \frac{W}{L}\sigma, \tag{6.27}$$

where L is the length of the sample, and W the area of the cross-section. For samples which are inhomogeneous such that this simple relation is not applicable, one must use the Kubo formula for conductance rather than that for conductivity. One example is the so-called mesoscopic conductors, which are systems smaller than a typical thermalization or equilibration length, whereby a local description is inadequate.

The current passing through the sample is equal to the integrated current density through a cross-section. Here we are interested in the DC-response only (or in frequencies where the corresponding wavelength is much longer than the sample size). Because of current conservation we can, of course, choose any cross-section, and it is convenient to choose an equipotential surface and to define a coordinate system $(\xi, \mathbf{a}_\xi)$, where ξ is a coordinate parallel to the field line and where $\mathbf{a}_\xi$ are coordinates on the plane perpendicular to the ξ-direction (see Figure 6.2). In this coordinate system the external electric field is directed along the $\hat{\xi}$-direction, $\mathbf{E}_{\text{ext}}(\mathbf{r}) = \hat{\xi}E_{\text{ext}}(\xi)$. The current I is

$$I_e = \int da_\xi\, \hat{\xi} \cdot \mathbf{J}_\mathbf{e}(\xi, \mathbf{a}_\xi) = \int da_\xi \int d\mathbf{r}'\, \hat{\xi} \cdot \sigma(\mathbf{r}, \mathbf{r}'; \omega = 0)\mathbf{E}_{\text{ext}}(\mathbf{r}')$$

$$= \int da_\xi \int da_{\xi'} \int d\xi'\, \hat{\xi} \cdot \sigma(\xi, \mathbf{a}_\xi, \xi', \mathbf{a}_{\xi'}; \omega = 0) \cdot \hat{\xi}' E_{\text{ext}}(\xi'), \tag{6.28}$$

where $\hat{\xi}$ is a unit vector normal to the surface element da_ξ and σ is the conductivity tensor. In order to get the DC-response we should take the limit $\omega \to 0$ of this expression. If we furthermore take the real part of (6.28) we see that what determines the DC-current is the real part of the first term in Eq. (6.24) and hence the retarded correlation function of the current densities. Since the total particle current at the coordinate ξ is given by $I(\xi) = \int da_\xi\, \hat{\xi} \cdot \mathbf{J}_e$, the current can instead be written as

$$I_e(\xi) = \lim_{\omega \to 0} \int d\xi'\, \text{Re}\left[\frac{ie^2}{\omega}C^R_{I(\xi)I(\xi')}(\omega)\right] E(\xi') \equiv \int d\xi'\, G(\xi, \xi')E_{\text{ext}}(\xi'), \tag{6.29}$$

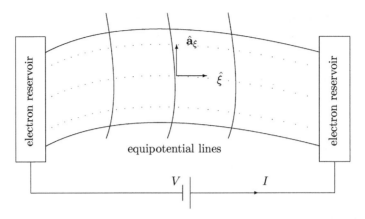

FIG. 6.2. The principle of a conductance measurement, which, in contrast to the conductivity, is a sample-specific quantity. In the Kubo formula derivation we use a coordinate system given by the equipotential lines, which together with use of current conservation allows a simple derivation.

where $C^R_{I(\xi)I(\xi')}$ is the correlation function between total currents. Because of current conservation the DC-current may be calculated at any point ξ and thus the result cannot depend on ξ. Consequently, the function inside the square brackets in Eq. (6.29) cannot depend on ξ. Furthermore, since the zero frequency conductance function $G(\xi, \xi')$ can be shown to be a symmetric function, it cannot depend on ξ' either; see Exercise 6.5. This simplification is the reason for choosing the skew coordinate system defined by the field lines. We can therefore perform the integration over ξ' which is just the voltage difference $V = -\int d\xi' E_{\text{ext}}(\xi') = \phi_{\text{ext}}(\infty) - \phi_{\text{ext}}(-\infty)$, and finally we arrive at the for linear response formula for the conductance

$$G = \lim_{\omega \to 0} \text{Re} \left[\frac{ie^2}{\omega} C^R_{II}(\omega) \right]. \tag{6.30}$$

Here C^R_{II} is the retarded current-current function. In the time domain it is

$$C^R_{II}(t - t') = -i\theta(t - t')\langle [\hat{I}(t), \hat{I}(t')] \rangle, \tag{6.31}$$

where the current operator I denotes the current through an arbitrary cross-section along the sample.

6.4   Kubo formula for the dielectric function

When dealing with systems containing charged particles, as for example the electron gas, one is often interested in the dielectric properties of the system, and in particular the linear response properties. When such a system is subjected to an external electromagnetic perturbation, the charge is redistributed and the system gets polarized. This in turn affects the measurement, an effect known as *screening*. The typical experiment is to exert an external potential ϕ_{ext} and measure the resulting total potential

ϕ_{tot}. The total potential is the sum of the external one and the potential ϕ_{ind} created by the induced polarization,

$$\phi_{\text{tot}} = \phi_{\text{ext}} + \phi_{\text{ind}}. \tag{6.32}$$

Alternatively, to use the potentials, we can work with electric fields or charges. The charges are related to the potentials through a set of Poisson equations[22]

$$\rho_{\text{tot}} = \rho_{\text{ext}} + \rho_{\text{ind}}, \quad \begin{cases} \nabla^2 \phi_{\text{tot}} = -\frac{1}{\varepsilon_0} \rho_{e,\text{tot}}, \\ \nabla^2 \phi_{\text{ext}} = -\frac{1}{\varepsilon_0} \rho_{e,\text{ext}}, \\ \nabla^2 \phi_{\text{ind}} = -\frac{1}{\varepsilon_0} \rho_{e,\text{ind}}, \end{cases} \tag{6.33}$$

and likewise for electric fields, $E_{\text{tot}}, E_{\text{ext}}$, and E_{ind}, which are related to the corresponding charges by a set of Gauss laws, $\nabla \cdot \mathbf{E} = \frac{1}{\varepsilon_0} \rho_e$. Here we have used the symbols ρ_e for the charge density, whereas ρ as defined in Chap. 1 defines particle densities.

The ratio between the external and the total potential is the dielectric response function, also called the relative permittivity ε

$$\phi_{\text{tot}} = \varepsilon^{-1} \phi_{\text{ext}}, \tag{6.34}$$

which is well known from classical electrodynamics.[23] However, in reality, the permittivity is non-local both in time and space, and the general relations between the total and the external potentials are

$$\phi_{\text{tot}}(\mathbf{r}, t) = \int d\mathbf{r}' \int dt' \, \varepsilon^{-1}(\mathbf{r}t, \mathbf{r}'t') \, \phi_{\text{ext}}(\mathbf{r}', t'), \tag{6.35a}$$

$$\phi_{\text{ext}}(\mathbf{r}, t) = \int d\mathbf{r}' \int dt' \, \varepsilon(\mathbf{r}t, \mathbf{r}'t') \, \phi_{\text{tot}}(\mathbf{r}', t'). \tag{6.35b}$$

Our present task is to find the dielectric function $\varepsilon(\mathbf{r}t, \mathbf{r}'t')$, or rather its inverse $\varepsilon^{-1}(\mathbf{r}t, \mathbf{r}'t')$, assuming linear response theory and for this purpose the induced potential is needed.

The external perturbation is represented by the following term in the Hamiltonian,

$$H' = \int d\mathbf{r} \, \rho_e(\mathbf{r}) \, \phi_{\text{ext}}(\mathbf{r}, t). \tag{6.36}$$

The induced charge density follows from linear response theory, and since the induced charge is defined as the deviation from the equilibrium charge distribution, i.e., $\rho_{e,\text{ind}} = \langle \rho_e \rangle - \langle \rho_e \rangle_0$, we have

$$\rho_{e,\text{ind}}(\mathbf{r}, t) = \int d\mathbf{r}' \int_{t_0}^{\infty} dt' C_{\rho_e(\mathbf{r})\rho_e(\mathbf{r}')}^{R}(t, t') \, \phi_{\text{ext}}(\mathbf{r}', t'), \tag{6.37a}$$

$$C_{\rho_e(\mathbf{r})\rho_e(\mathbf{r}')}^{R}(t, t') \equiv \chi_e^{R}(\mathbf{r}t, \mathbf{r}'t') = -i\theta(t - t')\langle [\hat{\rho}_e(\mathbf{r}, t), \hat{\rho}_e(\mathbf{r}', t')] \rangle_0. \tag{6.37b}$$

[22]Here we do not include the the equilibrium background charges, e.g., from the static charges of the lattice.

[23]In electrodynamics, the permittivity is defined as the proportionality constant between the electric displacement field, $\mathbf{D}$, and the electric field, $\mathbf{D} = \varepsilon_0 \varepsilon \mathbf{E}$. In the present formulation, $\mathbf{E}_{\text{ext}}$ plays the role of the $\mathbf{D}$-field, that is, $\mathbf{D} = \varepsilon_0 \mathbf{E}_{\text{ext}}$, while $\mathbf{E}_{\text{tot}}$ is the $\mathbf{E}$-field.

The retarded charge-charge correlation function χ_e^R is called the polarizability function, and it is an important function, which we will encounter many times. Once the induced charge is known, the potential follows from the Coulomb interaction $V_c(\mathbf{r} - \mathbf{r}') = 1/(4\pi\epsilon_0|\mathbf{r} - \mathbf{r}'|)$ as

$$\phi_{\text{ind}}(\mathbf{r}) = \int d\mathbf{r}' \, V_c(\mathbf{r} - \mathbf{r}') \rho_{e,\text{ind}}(\mathbf{r}'), \tag{6.38}$$

and hence

$$\phi_{\text{tot}}(\mathbf{r}, t) = \phi_{\text{ext}}(\mathbf{r}, t) + \int d\mathbf{r}' \int d\mathbf{r}'' \int_{t_0}^{\infty} dt' V_c(\mathbf{r} - \mathbf{r}') \chi_e^R(\mathbf{r}'t, \mathbf{r}''t') \, \phi_{\text{ext}}(\mathbf{r}'', t'). \tag{6.39}$$

From this expression we read off the inverse of the dielectric function as

$$\varepsilon^{-1}(\mathbf{rt}, \mathbf{r}'t') = \delta(\mathbf{r} - \mathbf{r}')\delta(t - t') + \int d\mathbf{r}'' V_c(\mathbf{r} - \mathbf{r}'') \chi_e^R(\mathbf{r}''t, \mathbf{r}'t'), \tag{6.40}$$

which ends our derivation. In later chapters we will make extensive use of the dielectric function ε and the polarizability χ. The dielectric function expressed in Eq. (6.40) includes all correlation effects, but often we must use some approximation to compute the polarizability.

6.4.1 Dielectric function for translation-invariant system

In the translation-invariant case the polarizability can only depend on the differences of the arguments, i.e. $\chi_e^R(\mathbf{rt}, \mathbf{r}'t') = \chi_e^R(\mathbf{r} - \mathbf{r}'; t - t')$, and therefore the problem is considerably simplified by the use of frequency and momentum space, where both Eqs. (6.35) have the form of convolutions. After Fourier transformation they become

$$\phi_{\text{tot}}(\mathbf{q},\omega) = \varepsilon^{-1}(\mathbf{q}, \omega)\phi_{\text{ext}}(\mathbf{q}, \omega), \quad \text{or} \quad \phi_{\text{ext}}(\mathbf{q},\omega) = \varepsilon(\mathbf{q}, \omega)\phi_{\text{tot}}(\mathbf{q}, \omega), \tag{6.41}$$

with the dielectric function being

$$\varepsilon^{-1}(\mathbf{q}, \omega) = 1 + V_c(\mathbf{q})\chi_e^R(\mathbf{q}, \omega). \tag{6.42}$$

6.4.2 Relation between dielectric function and conductivity

Both the dielectric function ε and the conductivity σ give the response of a system to an applied electromagnetic field, and one would therefore expect that they are related, and of course they are. Here we consider again the translational-invariant case, and using the definition of conductivity

$$\mathbf{J}_e(\mathbf{q}, \omega) = \sigma(\mathbf{q}, \omega)\mathbf{E}_{\text{ext}}(\mathbf{q}, \omega) = -i\sigma(\mathbf{q}, \omega)\mathbf{q}\phi_{\text{ext}}(\mathbf{q}, \omega), \tag{6.43}$$

and the continuity equation,

$$-i\omega\rho_e(\mathbf{q}, \omega) + i\mathbf{q} \cdot \mathbf{J}_e(\mathbf{q}, \omega) = 0, \tag{6.44}$$

we obtain

$$-i\mathbf{q} \cdot \boldsymbol{\sigma}(\mathbf{q},\omega)\mathbf{q}\phi_{\text{ext}}(\mathbf{q},\omega) = \omega\rho_e(\mathbf{q},\omega) = \omega\chi_e^R(\mathbf{q},\omega)\,\phi_{\text{ext}}(\mathbf{q},\omega). \qquad (6.45)$$

Finally, using Eq. (6.42) and knowing that for an isotropic system, the conductivity tensor is diagonal, we arrive at the relation

$$\varepsilon^{-1}(\mathbf{q},\omega) = 1 - i\frac{q^2}{\omega}V_c(\mathbf{q})\sigma(\mathbf{q},\omega). \qquad (6.46)$$

So if we know the conductivity, we can find the dielectric response and *vice versa*.

6.5   Summary and outlook

We have developed a general method for calculating the response to weak perturbations. This method, called linear response theory, is widely used because many experimental investigations are done in the linear response regime. In this regime the lack of equilibrium is not important, and one can think of this as probing the individual excitations of the systems. Because the perturbation is weak, it is not necessary to include interactions between these excitations.

The general formula,

$$\langle A(t)\rangle - \langle A\rangle_0 = \int_{t_0}^{\infty} dt'\, C_{AH'}^R(t,t'),$$

where

$$C_{AH'}^R(t,t') = -i\theta(t-t')\left\langle\left[\hat{A}(t),\hat{H}'(t')\right]\right\rangle_0,$$

is a retarded correlation function of the quantity $A(t)$ that we measure and the quantity $H'(t)$, to which the weak external perturbation couples. In the case of conductivity we saw that C^R was the current-current correlation function, while the dielectric response reduces to a charge-charge correlation. These two correlation functions will be used later in Chapters 12 and 14, but already in the following two chapters we will make use of the linear response result. In Chapter 7 we apply the Kubo formalism for single-particle transport in mesoscopic systems, and in Chapter 8.4.1 we use it to calculate the tunneling current between two conductors.

7

TRANSPORT IN MESOSCOPIC SYSTEMS

Electron transport in mesoscopic systems is a relatively new field of condensed matter physics. Since its beginning in the early 1980s many interesting and fundamental results have emerged from this fascinating research area bordering to the macroscopic classical world on one side and to the microscopic quantum world on the other. Here, we combine this brief introduction to mesoscopic physics with our first in-depth use of the Kubo formalism given in the previous chapter.

The physics of mesoscopic systems is a vast field and in this chapter we will narrow the scope and focus on the Landauer–Büttiker single-particle formalism for the conductance of nano-scale coherent systems. In Chapter 10 we will include correlation effects to the discussion. By coherent we mean that the quantum-mechanical coherence length is longer than the sample size. The phenomena to be discussed in this chapter all rely on quantum effects, all being clear manifestations of the propagation of electron waves through the structures.

The field of mesoscopic transport is interesting in that it combines physics on many length scales. The important length scales are the coherence length ℓ_ϕ, the energy relaxation length ℓ_{in}, the elastic mean free path ℓ_0, the Fermi wave length λ_F of the electron, the atomic Bohr radius a_0, and of course the sample size $\mathcal{L}$.

The typical mesoscopic structures that we have in mind are fabricated on semiconductor chips, such as the electrostatically confined 2D electron gases described in Section 2.3.2, or molecular systems, such as the carbon nanotube placed between two metallic contacts shown in Fig. 2.12. At low temperatures (e.g. between 50 mK and 4 K), the length scales of these system often fulfil the following inequalities defining the mesoscopic regime,

$$a_0 \ll \lambda_F \lesssim \ell_0 < \mathcal{L} < \ell_\phi \lesssim \ell_{in}. \tag{7.1}$$

Metallic systems are more difficult to bring into the mesoscopic regime because of their small Fermi wavelength, $\lambda_F \approx a_0$. However, there is one relatively simple experiment involving a narrow metallic wire subject to an externally applied stress where, as a clear signature of quantum transport, the conductance of the wire decreases in pronounced steps of size $2e^2/h$ just before the wire breaks. This can even be observed at room temperature, whereas the more sophisticated high-tech devices based on semiconductor nanostructures only show quantum effects at low temperatures as shown in Fig. 7.2.

This chapter deals with the physics of single-particle quantum transport which can be understood by invoking the Fermi liquid picture of non-interacting electrons. This picture will be discussed in Chapter 15. When interactions are important, another rich field of physics appears of which some topics are treated in Chapter 10.

7.1 The S-matrix and scattering states

We consider the standard setup in the field of mesoscopics, namely a mesoscopic sample connected to electron reservoirs (or contacts) in the form of macroscopic metal contacts. In the mesoscopic regime, the condition (7.1) is fulfilled per definition, and this implies that we can consider the electron motion to be quantum mechanically coherent in the entire sample. Furthermore, since the reservoirs are macroscopic conductors, much larger than the entrance to the mesoscopic region, we can safely assume that electrons entering the reservoir will be thermalized at the temperature and chemical potential of the contact before returning to the mesoscopic sample. The contact is thus required to be reflectionless, meaning that an electron impinging on a contact will be fully absorbed and thermalized by the contact before being re-emitted into the sample. Fig. 7.1 illustrates how a contact formed as a "horn" can give a reflectionless contact reservoir. This way of treating conductance in mesoscopic systems is denoted the two-probe Landauer–Büttiker formalism.

7.1.1 Definition of the S-matrix

In the following, we solve for the eigenstates of energy E in a geometry similar to Fig. 7.1. Since the system is open, any real value of E is possible. The system is divided into five regions: the left reservoir, left lead L, the mesoscopic sample M, the right lead R, and the right reservoir. The role of the reservoir is just to provide thermalized electrons to occupy the quantum states of the mesoscopic sample. For simplicity, the leads are taken to be perfect straight segments, with the same constant cross-section Ω and impenetrable walls at the boundary $\partial\Omega$.[24] We use the coordinate vector $\mathbf{r}_\perp$ for the perpendicular coordinates in the yz plane such that $(x, y, z) = (x, \mathbf{r}_\perp)$, where $\mathbf{r}_\perp \in \Omega$ and x is the coordinate in the direction of a lead. The leads are denoted by $\alpha = L, R$. For each lead, the Hamiltonian H_α and the associated eigenstates $\phi_{\alpha n E}^{\pm}(x, \mathbf{r}_\perp)$ with energy E and transverse mode quantum number $n = 1, 2, \ldots, N$, are given by

$$H_\alpha = \frac{1}{2m}p_x^2 + \frac{1}{2m}p_\perp^2, \tag{7.2a}$$

$$\phi_{\alpha n E}^{\pm}(x, \mathbf{r}_\perp) = \frac{1}{\sqrt{k_n(E)}}\, \chi_n(\mathbf{r}_\perp)\, e^{\pm i k_n(E) x}, \tag{7.2b}$$

$$\frac{1}{2m}p_\perp^2\, \chi_n(\mathbf{r}_\perp) = \varepsilon_n\, \chi_n(\mathbf{r}_\perp), \tag{7.2c}$$

$$\chi_n(\mathbf{r}_\perp) = 0, \quad \text{for } \mathbf{r}_\perp \in \partial\Omega \tag{7.2d}$$

$$E = \frac{1}{2m}k_n^2 + \varepsilon_n, \quad \text{i.e.,} \quad k_n(E) = \sqrt{2m(E - \varepsilon_n)}. \tag{7.2e}$$

The quantum number "$+/-$" represents right/left moving states with the wavenumber $k_n(E)$, and $\chi_n(\mathbf{r}_\perp)$ are the transverse eigenfunctions of the transverse Hamiltonian $\frac{1}{2m}p_\perp^2$ with eigenenergies ε_n. We have assumed a straight lead segment, so that $\chi_n(\mathbf{r}_\perp)$ is independent of x, but we could have an x-dependent cross section $\Omega = \Omega(x)$, as

---

[24]For specific choices of the shape of the cross-section Ω see Exercise 7.1.

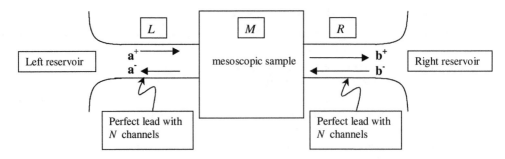

FIG. 7.1. The geometry considered in the derivation of the Landauer formula. Two
    reflectionless contacts each with N channels connect to a mesoscopic region. The
    wavefunction is written as a superposition of incoming and outgoing wave at the
    two entrances. When solving the Schrödinger equation, the system is separated in
    three regions: L, R and M.

long as the change is adiabatic and the derivatives $\partial_x \chi_n$ can be neglected. For a
cross section with some symmetry, the transverse eigenenergies ε_n can have degeneracies, but we choose the basis so that the transverse wavefunctions are orthogonal:
$\langle \chi_n | \chi_m \rangle = \delta_{nm}$. For calculations it will often be assumed that only a finite number
of transverse modes contribute for energetic reasons and therefore a cut-off value N
is chosen for the quantum numbers n. In summary, each lead has a set of eigenstates
with quantum numbers (n, η, E), where $\eta = \pm$ gives the direction of the current
carried by the state.

The wavefunctions $\phi_\lambda^\pm$ have been normalized in a particular manner so that they
all carry the same absolute probability current in a given cross section:

$$\int_\Omega d\mathbf{r}_\perp \left(\phi_\lambda^\pm(x, \mathbf{r}_\perp) \right)^* \frac{p_x}{m} \left(\phi_\lambda^\pm(x, \mathbf{r}_\perp) \right) = \pm \frac{1}{m}. \tag{7.3}$$

Because of this normalization, it is more natural to label the states in terms of their
energy E rather than, as usual, their k values. The transformation from a discrete to a
continuous set of energy levels looks a bit different in the two cases. In the following,
$\tilde{\phi}_k = \frac{1}{\sqrt{\mathcal{L}'}} e^{ikx}$ is a state with the usual normalization, while $\phi_k = \frac{1}{\sqrt{k}} e^{ikx}$. The
length $\mathcal{L}'$, not to be confused with the length $\mathcal{L}$ of the mesoscopic region, describes a
normalization length associated with the plane waves in the leads,

$$\sum_{k>0} \langle \tilde{\phi}_k | A | \tilde{\phi}_k \rangle \to \mathcal{L}' \int_0^\infty \frac{dk}{2\pi} \langle \tilde{\phi}_k | A | \tilde{\phi}_k \rangle$$

$$= \int_0^\infty \frac{dk}{2\pi} k \langle \phi_k | A | \phi_k \rangle$$

$$= \int_0^\infty \frac{dE}{2\pi} \frac{k}{dE/dk} \langle \phi_k | A | \phi_k \rangle$$

$$= \frac{m}{2\pi} \int_0^\infty dE \langle \phi_k | A | \phi_k \rangle. \tag{7.4}$$

As we shall see in detail later, the formulation of conductance in terms of the transmission stems from the cancellation, seen in Eq. (7.4), of the k dependence of the velocity, $\propto k$, and that of the density of states, $\propto dk/dE$; a feature particular of one spatial dimension.

The eigenfunctions in the complex middle region M are in general not easy to find, but fortunately, the general conductance formula can be established without specifying them. All we will need is the transmission coefficients, relating the amplitudes of the incoming and outgoing electron waves. Let us therefore introduce the so-called scattering matrix or S-matrix formalism.

A given eigenstate $\psi_E(x, \mathbf{r}_\perp)$ with energy E must be matched together by a linear combination of $\phi_{LnE}^\pm$ in L, another linear combination of $\phi_{RnE}^\pm$ in R, and an unknown complicated function $\psi_{M,E}$ in M. We can therefore write $\psi_E(x, \mathbf{r}_\perp)$ as

$$\psi_E(x, \mathbf{r}_\perp) = \begin{cases} \sum_n a_n^+ \phi_{LnE}^+(x, \mathbf{r}_\perp) + \sum_n a_n^- \phi_{LnE}^-(x, \mathbf{r}_\perp), & (x, \mathbf{r}_\perp) \in L, \\ \psi_{M,E}(x, \mathbf{r}_\perp), & (x, \mathbf{r}_\perp) \in M, \qquad (7.5) \\ \sum_n b_n^+ \phi_{RnE}^+(x, \mathbf{r}_\perp) + \sum_n b_n^- \phi_{RnE}^-(x, \mathbf{r}_\perp), & (x, \mathbf{r}_\perp) \in R, \end{cases}$$

where $a_n^\pm$ and $b_n^\pm$ are sets of some unknown coefficients to be determined. In vector form they are written as $\mathbf{a}^+ = (a_1^+, a_2^+, \ldots)$ and similarly for $\mathbf{a}^-$ and $\mathbf{b}^\pm$. As usual, the wavefunction and its derivative must be continuous. For a given $\psi_{M,E}$ in M this condition gives $4 \times N$ linearly independent equations to determine $a_n^\pm$ and $b_n^\pm$,

$$a_n^+ + a_n^- = \sqrt{k_n(E)} \int_\Omega d\mathbf{r}_\perp \, \chi_n^*(\mathbf{r}_\perp) \psi_{M,E}(0, \mathbf{r}_\perp), \qquad (7.6a)$$

$$b_n^+ e^{ik_n(E)L} + b_n^- e^{-ik_n(E)L} = \sqrt{k_n(E)} \int_\Omega d\mathbf{r}_\perp \, \chi_n^*(\mathbf{r}_\perp) \psi_{M,E}(L, \mathbf{r}_\perp), \qquad (7.6b)$$

$$a_n^+ - a_n^- = \frac{1}{i\sqrt{k_n(E)}} \int_\Omega d\mathbf{r}_\perp \, \chi_n^*(\mathbf{r}_\perp) \big(\partial_x \psi_{M,E}(x, \mathbf{r}_\perp)\big)_{x=0}, \qquad (7.6c)$$

$$b_n^+ e^{ik_n(E)L} - b_n^- e^{-ik_n(E)L} = \frac{1}{i\sqrt{k_n(E)}} \int_\Omega d\mathbf{r}_\perp \, \chi_n^*(\mathbf{r}_\perp) \big(\partial_x \psi_{M,E}(x, \mathbf{r}_\perp)\big)_{x=L}. \qquad (7.6d)$$

Fortunately, we need not solve this equation system, unless we want an explicit expression for the wavefunction. It is merely written down to illustrate the linear dependence of the coefficients $\{a_n^\pm\}$ and $\{b_n^\pm\}$. A particular useful way of representing the linear dependence is through the so-called scattering matrix, or S-matrix, which relates the amplitudes of outgoing waves, ϕ_L^- and ϕ_R^+, to incoming waves, ϕ_L^+ and ϕ_R^-,

$$\mathbf{c}_{\text{out}} \equiv \begin{pmatrix} \mathbf{a}^- \\ \mathbf{b}^+ \end{pmatrix} = \begin{pmatrix} \mathbf{r} & \mathbf{t}' \\ \mathbf{t} & \mathbf{r}' \end{pmatrix} \begin{pmatrix} \mathbf{a}^+ \\ \mathbf{b}^- \end{pmatrix} \equiv \mathbf{S} \begin{pmatrix} \mathbf{a}^+ \\ \mathbf{b}^- \end{pmatrix} \equiv \mathbf{S} \, \mathbf{c}_{\text{in}}. \qquad (7.7)$$

Here we have defined the important dependent S-matrix to be a matrix of size $2N \times 2N$ with the $N \times N$ reflection matrices $\mathbf{r}$ and $\mathbf{r}'$ and transmission matrices $\mathbf{t}$ and $\mathbf{t}'$ as block elements,

$$\mathbf{S}(E) = \begin{pmatrix} \mathbf{r}(E) & \mathbf{t}'(E) \\ \mathbf{t}(E) & \mathbf{r}'(E) \end{pmatrix}. \qquad (7.8)$$

The matrix element $t_{nn'}$ represents the transmission amplitude for an incoming wave from the left in state n' to be transmitted into state n on the right hand side. The amplitude for transmission in the opposite direction is given by $t'_{nn'}$. Similarly the element $r_{nn'}$ gives the amplitude for being reflected back into the left lead in state n. Due to Eqs. (7.6a)–(7.6d) the coefficients of the scattering matrix are clearly energy dependent. Most of the time, however, we suppress this dependence in the notation.

7.1.2 Definition of the scattering states

We now define the so-called scattering states $\psi_\lambda = \psi_{\alpha n E}$ (not to be confused with $\phi^\pm_{\alpha n E}$), which are states with an incoming wave in one particular lead state αn, i.e., $\mathbf{c}_{\text{in}} = (0, \dots, 0, 1, 0, \dots)$. The lead index α refers to the lead from which the incoming wave arrives. For the left lead $\alpha = L$, the scattering states are

$$\psi_{LnE}(x, \mathbf{r}_\perp) = \begin{cases} \phi^+_{LnE}(x, \mathbf{r}_\perp) + \sum_{n'} r_{n'n} \phi^-_{Ln'E}(x, \mathbf{r}_\perp), & (x, \mathbf{r}_\perp) \in L, \\ \psi_{M,E}(x, \mathbf{r}_\perp) & (x, \mathbf{r}_\perp) \in M, \\ \sum_{n'} t_{n'n} \phi^+_{Rn'E}(x, \mathbf{r}_\perp), & (x, \mathbf{r}_\perp) \in R. \end{cases} \quad (7.9)$$

and for the right lead $\alpha = R$ they are

$$\psi_{RnE}(x, \mathbf{r}_\perp) = \begin{cases} \sum_{n'} t'_{n'n} \phi^-_{Ln'E}(x, \mathbf{r}_\perp), & (x, \mathbf{r}_\perp) \in L. \\ \psi_{M,E}(x, \mathbf{r}_\perp) & (x, \mathbf{r}_\perp) \in M, \\ \phi^-_{RnE}(x, \mathbf{r}_\perp) + \sum_{n'} r'_{n'n} \phi^+_{Rn'E}(x, \mathbf{r}_\perp), & (x, \mathbf{r}_\perp) \in R, \end{cases} \quad (7.10)$$

To determine the conductance of the mesoscopic system one needs to find $\psi_{M,E}(x, \mathbf{r}_\perp)$, which in general involves tedious, often numerical, work. However, as we shall see in the following, it is possible to derive some general results in terms of the the S-matrix without knowledge of the exact form of $\psi_{M,E}(x, \mathbf{r}_\perp)$.

7.1.3 Unitarity of the S-matrix

Before we calculate the transport properties of a mesoscopic system, let us look at some properties of the S-matrix. First of all, it must be unitary, i.e., $\mathbf{S}^{-1} = \mathbf{S}^\dagger$. This is a consequence of probability current conservation. The incoming electron flux $\sum_n |c_{\text{in}}|^2 = |\mathbf{c}_{\text{in}}|^2$ must equal the outgoing flux $\sum_n |c_{\text{out}}|^2 = |\mathbf{c}_{\text{out}}|^2$ and therefore

$$\mathbf{c}^\dagger_{\text{out}} \mathbf{c}_{\text{out}} = \mathbf{c}^\dagger_{\text{in}} \mathbf{c}_{\text{in}} \quad \Rightarrow \quad \mathbf{c}^\dagger_{\text{in}} (1 - \mathbf{S}^\dagger \mathbf{S}) \mathbf{c}_{\text{in}} = 0, \quad (7.11)$$

and hence $\mathbf{S}^\dagger = \mathbf{S}^{-1}$. From the unitarity follows some properties of $\mathbf{r}$, $\mathbf{r}'$, $\mathbf{t}$, and $\mathbf{t}'$, which we will make use of below,

$$\mathbf{S}^\dagger \mathbf{S} = 1 \quad \Leftrightarrow \quad \begin{cases} 1 = \mathbf{r}^\dagger \mathbf{r} + \mathbf{t}^\dagger \mathbf{t} = \mathbf{r}'^\dagger \mathbf{r}' + \mathbf{t}'^\dagger \mathbf{t}', \\ 0 = \mathbf{r}^\dagger \mathbf{t}' + \mathbf{t}^\dagger \mathbf{r}' = \mathbf{t}'^\dagger \mathbf{r} + \mathbf{r}'^\dagger \mathbf{t}, \end{cases} \quad (7.12)$$

and furthermore

$$\mathbf{S} \mathbf{S}^\dagger = 1 \quad \Leftrightarrow \quad \begin{cases} 1 = \mathbf{r}' \mathbf{r}'^\dagger + \mathbf{t} \mathbf{t}^\dagger = \mathbf{r} \mathbf{r}^\dagger + \mathbf{t}' \mathbf{t}'^\dagger, \\ 0 = \mathbf{r} \mathbf{t}^\dagger + \mathbf{t}' \mathbf{r}'^\dagger = \mathbf{t} \mathbf{r}^\dagger + \mathbf{r}' \mathbf{t}'^\dagger. \end{cases} \quad (7.13)$$

We also show the unitarity in a bit more explicit way by calculating the currents on the left- and right-hand sides of the system. This we do because we will need the currents later on anyway. The particle current through a cross-section for a given wavefunction $\Psi(x, \mathbf{r}_\perp)$ is, cf. Eq. (1.98b),

$$I(x) = \int_\Omega d\mathbf{r}_\perp \; \Psi^*(x, \mathbf{r}_\perp) \overset{\leftrightarrow}{J}_x \Psi(x, \mathbf{r}_\perp), \quad \overset{\leftrightarrow}{J}_x = \frac{1}{2mi}\left(\overset{\rightarrow}{\partial}_x - \overset{\leftarrow}{\partial}_x\right), \qquad (7.14)$$

where the arrows indicate to which side the differential operators are acting. For a stationary state, i.e. an eigenstate with energy E, the continuity equation gives $\partial_x J = -\dot{\rho} = 0$, and thus $I(x)$ cannot depend on x. Let us compute $I(x)$ for a state with incoming coefficients $\mathbf{c}_{\text{in}} = (\mathbf{a}^+, \mathbf{b}^-)$. First calculate the current $I_L(x)$ entering from the left contact by using Eq. (7.14) with $\Psi = \psi_E$ for $(x, \mathbf{r}_\perp) \in L$ from Eq. (7.5),

$$I_L(x) = \int_\Omega d\mathbf{r}_\perp \left(\mathbf{a}^+ \cdot \boldsymbol{\phi}^+_{L,E} + \mathbf{a}^- \cdot \boldsymbol{\phi}^-_{L,E}\right)^* \overset{\leftrightarrow}{J}_x \left(\mathbf{a}^+ \cdot \boldsymbol{\phi}^+_{L,E} + \mathbf{a}^- \cdot \boldsymbol{\phi}^-_{L,E}\right)$$

$$= \frac{1}{m}\left(|\mathbf{a}^+|^2 - |\mathbf{r}\mathbf{a}^+ + \mathbf{t}'\mathbf{b}^-|^2\right), \qquad (7.15)$$

where $\boldsymbol{\phi}^+_{L,E} = (\phi^+_{L,1E}, \phi^+_{L,2E}, \dots)$ and $\boldsymbol{\phi}^-_{L,E} = (\phi^-_{L,1E}, \phi^-_{L,2E}, \dots)$, and where we have used Eq. (7.7) to express the outgoing amplitudes $\mathbf{a}^-$ by the incoming ones $\mathbf{r}\mathbf{a}^+ + \mathbf{t}'\mathbf{b}^-$. Similarly, but now taking $(x, \mathbf{r}_\perp) \in R$ in Eq. (7.5) and later replacing the outgoing amplitudes $\mathbf{b}^+$ by the incoming ones $\mathbf{t}\mathbf{a}^+ + \mathbf{r}'\mathbf{b}^-$ from Eq. (7.7), we obtain the current I_R entering in the right contact

$$I_R(x) = \frac{1}{m}\left(-|\mathbf{b}^-|^2 + |\mathbf{t}\mathbf{a}^+ + \mathbf{r}'\mathbf{b}^-|^2\right). \qquad (7.16)$$

Expansion of the squared terms yields

$$I_L = \frac{1}{m}\left\{(\mathbf{a}^+)^\dagger\left(1 - \mathbf{r}^\dagger\mathbf{r}\right)\cdot\mathbf{a}^+ - (\mathbf{b}^-)^\dagger\left(\mathbf{t}'^\dagger\mathbf{t}'\right)\mathbf{b}^- - 2\,\mathrm{Re}\left[(\mathbf{a}^+)^\dagger\mathbf{r}^\dagger\mathbf{t}'\mathbf{b}^-\right]\right\}, \quad (7.17a)$$

$$I_R = \frac{1}{m}\left\{(\mathbf{b}^-)^\dagger\left(-1 + \mathbf{r}'^\dagger\mathbf{r}'\right)\mathbf{b}^- + (\mathbf{a}^+)^\dagger\left(\mathbf{t}^\dagger\mathbf{t}\right)\mathbf{a}^+ + 2\,\mathrm{Re}\left[(\mathbf{a}^+)^\dagger\mathbf{t}^\dagger\mathbf{r}'\mathbf{b}^-\right]\right\}. \quad (7.17b)$$

The continuity equation demands that $I_L = -I_R$ and for this to be true, for any value of the coefficient matrices $\mathbf{a}^+$ and $\mathbf{b}^-$, the terms in Eqs. (7.17) must match term by term. This leads to Eq. (7.12) and hence $\mathbf{S}$ is unitary.

7.1.4   Time-reversal symmetry

Time-reversal symmetry means that $H = H^*$, because if $\Psi(\mathbf{r}, t)$ is a solution to the Schrödinger equation so is $\Psi^*(\mathbf{r}, -t)$. We shall show that in this case the scattering matrix is not only unitary but also symmetric, $\mathbf{S} = \mathbf{S}^T$. This will be of importance for the study in Chapter 16 of disordered systems with and without an applied magnetic field.

A non-zero magnetic field $\mathbf{B} = \nabla \times \mathbf{A}$ breaks time-reversal symmetry, and in this case, according to Eq. (1.90), the Schrödinger equation for an electron in steady state is

$$H_\mathbf{B}\,\Psi_\mathbf{B}(\mathbf{r}) \;=\; \left[-\frac{1}{2m}\,(\nabla_\mathbf{r}+ie\mathbf{A})^2 + V(\mathbf{r})\right]\Psi_\mathbf{B}(\mathbf{r}) \;=\; E\,\Psi_\mathbf{B}(\mathbf{r}). \qquad (7.18)$$

It is seen that $H_\mathbf{B} = H^*_{-\mathbf{B}}$, so it follows that

$$H_\mathbf{B}\,\Psi_\mathbf{B}(\mathbf{r}) \;=\; E\,\Psi_\mathbf{B}(\mathbf{r}) \quad \Leftrightarrow \quad H^*_{-\mathbf{B}}\,\Psi^*_{-\mathbf{B}}(\mathbf{r}) \;=\; E\,\Psi^*_{-\mathbf{B}}(\mathbf{r}), \qquad (7.19)$$

or in short: if $\Psi_\mathbf{B}(\mathbf{r})$ is a solution so is $\Psi^*_{-\mathbf{B}}(\mathbf{r})$. We can therefore construct new eigenstates by complex conjugation followed by reversal of the magnetic field. This is perhaps not surprising, given the fact that magnetic fields are generated by currents. A complete treatment would take these currents into account, and the sign change of $\mathbf{B}$ would follow automatically from the sign change of the currents. In our treatment the magnetic field is taken as an external entity, and we must ensure the proper behavior under time reversal by changing the sign of $\mathbf{B}$ by hand.

Suppose we have an eigenstate which is a linear combination of incoming and outgoing waves $\Psi_\mathbf{B}(\mathbf{r}) = (\mathbf{c}_{\text{in}}\phi_{\text{in}},\,\mathbf{c}_{\text{out}}\phi_{\text{out}})$, then we can make a new eigenstate by $\Psi^{\text{new}}_{-\mathbf{B}}(\mathbf{r}) = \Psi^*_\mathbf{B}(\mathbf{r})$, which is a solution for $-\mathbf{B}$. However, because complex conjugation reverses the direction of propagation, the new incoming and outgoing wavefunctions are $\mathbf{c}^{\text{new}}_{\text{in}} = \mathbf{c}^*_{\text{out}}$, and $\mathbf{c}^{\text{new}}_{\text{out}} = \mathbf{c}^*_{\text{in}}$. Since Ψ^{new} is a solution for $-\mathbf{B}$, we have

$$\mathbf{c}^{\text{new}}_{\text{out}} = \mathbf{S}_{-\mathbf{B}}\mathbf{c}^{\text{new}}_{\text{in}} \quad \Rightarrow \quad \mathbf{c}^*_{\text{in}} = \mathbf{S}_{-\mathbf{B}}\mathbf{c}^*_{\text{out}} = \mathbf{S}_{-\mathbf{B}}\mathbf{S}^*_\mathbf{B}\mathbf{c}^*_{\text{in}}, \qquad (7.20)$$

which shows that

$$\mathbf{S}_{-\mathbf{B}}\mathbf{S}^*_\mathbf{B} = 1 \quad \Rightarrow \quad \mathbf{S}^*_{-\mathbf{B}} = \mathbf{S}^\dagger_\mathbf{B} \quad \Rightarrow \quad \mathbf{S}_\mathbf{B} = \mathbf{S}^T_{-\mathbf{B}}. \qquad (7.21)$$

In case of time-reversal symmetry $\mathbf{B} = \mathbf{0}$, the scattering matrix therefore has an additional symmetry besides unitary: it is also a symmetric matrix, $\mathbf{S}_{\mathbf{B}=0} = \mathbf{S}^T_{\mathbf{B}=0}$.

7.2   Conductance and transmission coefficients

Next, we calculate the conductance. This will be done in two different ways. First, we will argue on physical grounds that the population of the scattering states is given by the equilibrium distribution functions of the reservoirs, which allows us to calculate the current directly. Secondly, we calculate the conductance using linear response theory, and, fortunately, we find the same result. While the first method is more physically appealing, one could get in doubt whether the Pauli principle is treated correctly. The linear response result shows that indeed the first method gave the right answer, at least in the linear response limit.

The answer we find, the celebrated Landauer formula, is very simple and physically sensible: the conductance of a mesoscopic sample is given by the sum of all the transmission possibilities an electron has when propagating with an energy equal to the chemical potential, $E = \mu$,

$$G(\mu) = \frac{2e^2}{h}\sum_n \mathcal{T}_n(\mu) = \frac{2e^2}{h}\,\text{Tr}\big[\mathbf{t}(\mu)^\dagger\mathbf{t}(\mu)\big], \qquad (7.22)$$

where, although often suppressed, the energy dependence is written explicitly, where h has been reintroduced, and where the eigenvalues $\mathcal{T}_n$ of the matrix $\mathbf{t}^\dagger\mathbf{t}$ are introduced.

The eigenvalues $\mathcal{T}_n$ should not be confused with the transmission probabilities, i.e. the probability that an electron in a given incoming state n ends up on the other side. This probability is $T_n = \left(\mathbf{t}^\dagger \mathbf{t}\right)_{nn}$, but the invariance of the trace in fact leads to $\sum_n \mathcal{T}_n = \sum_n T_n$. So we can write Eq. (7.22) in terms of $\mathcal{T}_n$ or T_n as we please.

The Landauer formula tells us that the conductance of a mesoscopic sample is quantized in units of $2e^2/h$. The number of quanta will be the number of transmission possibilities, denoted channels, connecting the two sides. However, since $\mathcal{T}_n$ is a number between 0 and 1 one expect this quantization to show up only for some special geometries where $\mathcal{T}_n$ is either 0 or 1. This is in fact what happens for the quantum point contact, which is discussed below in Sec. 7.3.1. There, a particular smooth interface between the two reservoirs which ensures that $\mathcal{T}_n$ changes in a well-controlled manner between 0 and 1. However, there are other examples where the conductance quantum e^2/h shows up, namely in the fluctuations of conductance. These fluctuations are universal in the sense that they have an amplitude of the order e^2/h independent of the average conductance. This is discussed in Section 16.6.3.

7.2.1   The Landauer formula, heuristic derivation

We argued above that if the reservoirs are much wider than the mesoscopic region and its leads, then we can assume reflectionless transmission from the leads to the reservoirs, i.e., the electrons entering the reservoir from the sample are always completely thermalized before returning. Thus all electrons coming from any given reservoir α have a Fermi–Dirac energy distribution n_F of that particular reservoir, characterized by the corresponding chemical μ_α. Furthermore, since the mesoscopic region is defined to be phase coherent, no energy relaxation takes place there, and consequently electrons originating from reservoir α maintain the distribution function of that reservoir. Therefore it is natural to express the occupation of the scattering eigenstates $\psi_{\alpha n E}$ in lead α by the distribution function f_α and the chemical potential μ_α of the reservoir connected to that lead,

$$f_\alpha(\varepsilon) = n_F(\varepsilon - \mu_\alpha), \quad \alpha = L, R. \tag{7.23}$$

If we use the scattering states Eqs. (7.9) and (7.10) as our basis states, the calculation of the current through the system becomes rather easy. By construction the scattering states have very simple incoming coefficients $(\mathbf{a}^+, \mathbf{b}^-)$. For example for ψ_{LnE} we have

$$\left(\mathbf{a}^+(\psi_{LnE})\right)_{n'} = \delta_{nn'}, \qquad \left(\mathbf{b}^-(\psi_{LnE})\right)_{n'} = 0. \tag{7.24}$$

Although ψ_{LnE} contains many other coefficients than the incoming ones, the latter ones suffice according to Eqs. (7.17a) and (7.17b) to calculate the current I_{LnE} carried by ψ_{LnE} if it is occupied by an electron. Due to current conservation, this current is the same at any cross-section in the mesoscopic sample. Combining Eq. (7.24) with Eqs. (7.17a) and (7.17b), we find expressions for I_{LnE} at the left and right lead, respectively. A completely analogous calculation yields the current I_{RnE} for the scattering state ψ_{RnE} coming from the right lead. The final result is

$$I_{LnE} = \frac{1}{m}\left[1 - \left(\mathbf{r}^\dagger(E)\mathbf{r}(E)\right)_{nn}\right] = \frac{1}{m}\left(\mathbf{t}^\dagger(E)\mathbf{t}(E)\right)_{nn}, \tag{7.25}$$

$$I_{RnE} = -\frac{1}{m}\left(\mathbf{t}'^\dagger(E)\mathbf{t}'(E)\right)_{nn} = \frac{1}{m}\left[-1 + \left(\mathbf{r}'^\dagger(E)\mathbf{r}'(E)\right)_{nn}\right]. \tag{7.26}$$

The electrical current I_e is now obtained by multiplying the occupancy factors (7.23) (including a factor of 2 for spin) with the current carried by an occupied scattering state Eqs. (7.25) and (7.26) and then summing over all scattering states,

$$I_e = -2e\sum_{\alpha n E} I_{\alpha n E}\, f_\alpha(E). \tag{7.27}$$

Transforming from a sum to an energy integral as in Eq. (7.4), the electric current becomes

$$I_e = \frac{-2e}{2\pi}\sum_n \int_{\varepsilon_n}^\infty dE\left[\left(\mathbf{t}^\dagger\mathbf{t}\right)_{nn} n_F(E - \mu_L) - \left(\mathbf{t}'^\dagger\mathbf{t}'\right)_{nn} n_F(E - \mu_R)\right]. \tag{7.28}$$

The sum over diagonal elements of $\left(\mathbf{t}^\dagger\mathbf{t}\right)$ is simply the trace. Taking traces of the unitarity conditions Eqs. (7.12) and (7.13) leads to $\mathrm{Tr}[\mathbf{t}'^\dagger\mathbf{t}'] = \mathrm{Tr}[\mathbf{t}^\dagger\mathbf{t}]$, and we obtain

$$I_e(V_L, V_R) = \frac{-e}{\pi}\int_0^\infty dE\,\mathrm{Tr}[\mathbf{t}^\dagger(E)\mathbf{t}(E)]\left[n_F(E - \mu + eV_L) - n_F(E - \mu + eV_R)\right], \tag{7.29}$$

where we have introduced the applied voltages as shifts of the equilibrium chemical potential μ as $\mu_L = \mu - eV_L$ and $\mu_R = \mu - eV_R$. To arrive at the linear response result, we perform a Taylor expansion of I_e in eV around μ, which then gives (after reinserting $\hbar$)

$$I_e(V_L, V_R) \approx \frac{2e^2}{h}\sum_n \int_0^\infty dE\,\left(\mathbf{t}^\dagger(E)\mathbf{t}(E)\right)_{nn}\left(-\frac{\partial n_F}{\partial E}\right)(V_L - V_R), \tag{7.30}$$

which then yields the conductance $G = I/(V_L - V_R)$ as

$$G(\mu, T) = \frac{2e^2}{h}\int_0^\infty dE\,\mathrm{Tr}[\mathbf{t}^\dagger(E)\mathbf{t}(E)]\left(-\frac{\partial n_F}{\partial E}\right) = \frac{2e^2}{h}\sum_n \int_0^\infty dE\,\mathcal{T}_n(E)\left(-\frac{\partial n_F}{\partial E}\right). \tag{7.31}$$

Furthermore, for low temperatures $(-\partial n_F/\partial E)(E) \approx \delta(E - \mu)$,

$$G(\mu, 0) = \frac{2e^2}{h}\mathrm{Tr}\left[\mathbf{t}^\dagger(\mu)\mathbf{t}(\mu)\right] = \frac{2e^2}{h}\sum_n \mathcal{T}_n(\mu). \tag{7.32}$$

Thus, we have derived the Landauer formula. It has been assumed that the spin degrees of freedom are degenerate which gives rise to a simple factor of 2. If they are not degenerate, the trace must also include a trace over the spin degrees of freedom.

The expression Eq. (7.28) for current relies on the fact that the scattering states are eigenstates of the system, which means that we should not include any kind of blocking factors $(1-n_F)$ to ensure that the final state is empty, as one would normally do in a Boltzmann equation. A scattering state extends over the entire sample, so once it is occupied in one lead, it is automatically also occupied in the other. Thus we are not talking about a scattering event from one reservoir to the other, but rather about the thermal population of eigenmodes. In order to dismiss any concern about this point, the next section is devoted to a derivation of Eq. (7.22) from first principles using the linear response formalism of Chap. 6.

7.2.2 The Landauer formula, linear response derivation

Our starting point is Eq. (6.30) expressing the conductance G in terms of the current-current correlation function,

$$G(\omega) = -\frac{2e^2}{\omega} \, \mathrm{Im} \int_{-\infty}^{\infty} dt \, e^{i(\omega+i\eta)t}(-i)\theta(t) \, \langle [I(x,t), I(x,0)] \rangle_0 \,, \qquad (7.33)$$

where the current operator $I(x)$ due to current conservation can be evaluated at any cross-section x in the perfect leads, rendering G independent of x, for example we could chose to calculate the current in the left lead. Again we consider the spin degenerate case which is the reason for the factor of 2.

In second quantization the current operator is given by

$$I(x) = \sum_{\lambda\lambda'} j_{\lambda\lambda'}(x) \, c_\lambda^\dagger c_{\lambda'}\,, \qquad (7.34)$$

$$j_{\lambda\lambda'}(x) = \frac{1}{2mi} \int_\Omega d\mathbf{r}_\perp \, \psi_\lambda^*(x,\mathbf{r}_\perp) \left(\overrightarrow{\partial}_x - \overleftarrow{\partial}_x \right) \psi_{\lambda'}(x,\mathbf{r}_\perp), \qquad (7.35)$$

where $\{\psi_\lambda\}$ is a complete set of eigenstates, and $j_{\lambda\lambda'}$ is a matrix element of the current operator in this basis. We will, of course, use the scattering states of Section 7.1.2 enumerated by the combined quantum number $\lambda = (\alpha, n, E)$.

We start by calculating the commutator in Eq. (7.33),

$$\langle [I(x',t), I(x',0)] \rangle_0 = \sum_{\nu\nu'} j_{\nu\nu'}(x') \sum_{\lambda\lambda'} j_{\lambda\lambda'}(x') e^{i(E_\lambda - E_{\lambda'})t} \left\langle \left[c_\lambda^\dagger c_{\lambda'}, c_\nu^\dagger c_{\nu'} \right] \right\rangle_0$$

$$= \sum_{\lambda\lambda'} |j_{\lambda\lambda'}(x')|^2 \, e^{i(E_\lambda - E_{\lambda'})t} \left[n_F(E_\lambda) - n_F(E_{\lambda'}) \right], \qquad (7.36)$$

where we used that $\langle c_\lambda^\dagger c_{\lambda'} \rangle_0 = \delta_{\lambda\lambda'} n_F(E_\lambda)$, and that $j_{\lambda\lambda'}(x') = (j_{\lambda'\lambda}(x'))^*$. Inserting this into Eq. (7.33) yields

$$G(\omega) = \frac{2e^2}{\omega} \, \mathrm{Im} \sum_{\lambda\lambda'} \frac{|j_{\lambda\lambda'}(x')|^2}{(\omega + i\eta + E_\lambda - E_{\lambda'})} \left[n_F(E_\lambda) - n_F(E_{\lambda'}) \right], \qquad (7.37)$$

and in the DC-limit, $\omega \to 0$, one has

$$G(0) = 2e^2 \pi \sum_{\lambda\lambda'} |j_{\lambda\lambda'}(x')|^2 \left(-\frac{\partial n_F(E_\lambda)}{\partial E_\lambda} \right) \delta \left(E_\lambda - E_{\lambda'} \right). \qquad (7.38)$$

Using the scattering states labeled by $\lambda = (\alpha, n, E)$, changing the sum over these eigenstates to integrals over energy, i.e., $\sum_\lambda \to \sum_{nn} \frac{m}{2\pi} \int dE$, and setting $T = 0$ such that $(-\partial n_F(E)/\partial E) = \delta(E - \mu)$, the conductance becomes

$$G(0) = 2e^2 \pi \left(\frac{m}{2\pi}\right)^2 \sum_{nn',\alpha\alpha'} |j_{\alpha n\mu,\alpha'n'\mu}(x')|^2. \qquad (7.39)$$

According to the continuity equation, the current matrix elements $j_{\alpha n\mu,\alpha'n'\mu}(x')$ are independent of the x'-coordinate,[25] and we evaluate them in the L or R region at our convenience

$$j_{\alpha n\mu,\alpha'n'\mu}(x') = \frac{1}{m} \begin{pmatrix} (\mathbf{t}^\dagger \mathbf{t})_{nn'} & (\mathbf{t}^\dagger \mathbf{r}')_{nn'} \\ -(\mathbf{t}'^\dagger \mathbf{r})_{nn'} & (-\mathbf{t}'^\dagger \mathbf{t}')_{nn'} \end{pmatrix} \equiv \frac{1}{m}\mathbf{j}, \qquad (7.40)$$

where the rows and columns correspond to $\alpha = L, R$, respectively, and where scattering matrices are evaluated at the energy μ. Hence we obtain

$$\sum_{nn',\alpha\alpha'} |j_{\alpha n\mu,\alpha'n'\mu}(x')|^2 = \left(\frac{1}{m}\right)^2 \mathrm{Tr}\left[\mathbf{j}^\dagger \mathbf{j}\right]$$

$$= \left(\frac{1}{m}\right)^2 \mathrm{Tr}\left[(\mathbf{t}^\dagger \mathbf{t})^2 + (\mathbf{t}'^\dagger \mathbf{t}')^2 + \mathbf{r}'^\dagger \mathbf{t}\mathbf{t}^\dagger \mathbf{r}' + \mathbf{r}^\dagger \mathbf{t}'\mathbf{t}'^\dagger \mathbf{r}\right]$$

$$= 2\left(\frac{1}{m}\right)^2 \mathrm{Tr}\left[\mathbf{t}^\dagger \mathbf{t}\right], \qquad (7.41)$$

after using the result Eq. (7.13). Now inserting Eq. (7.41) into Eq. (7.39) yields

$$G(0) = \frac{2e^2}{h} \mathrm{Tr}\left[\mathbf{t}^\dagger \mathbf{t}\right], \qquad (7.42)$$

which again is the Landauer formula. We have thus seen that it can be derived microscopically, and any doubt about the validity of the treatment of the occupation factor in the heuristic derivation, has been removed.

7.2.3   The Landauer–Büttiker formalism for multiprobe systems

In the previous sections, our study was restricted to two-probe systems. The lead index α could only take the values L and R, but this can straightforwardly be extended to multiprobe systems, such that α can take values corresponding to each of the leads in the system. We will not go through the derivation in detail, since the reader should be able to fill in the gaps.

Consider first the the two-probe formula (7.30) for the current I_L in the left lead, but now rewritten by use of Eq. (7.28) and linearization in the applied voltages, V_L and V_R:

$$I_L(V_L, V_R) = \frac{2e^2}{h} \int_0^\infty dE \left(-\frac{\partial n_F}{\partial E}\right) \left(\mathrm{Tr}\left[\mathbf{t}^\dagger \mathbf{t}\right] V_L - \mathrm{Tr}\left[\mathbf{t}'^\dagger \mathbf{t}'\right] V_R\right). \qquad (7.43)$$

The first term contains transmission processes starting in lead L and ending in lead R, while the second term contains transmission in the opposite direction. The trace

---

[25] $\langle\lambda|\nabla\cdot\mathbf{J}|\lambda'\rangle = -\langle\lambda|\dot{\rho}|\lambda'\rangle = -i\langle\lambda|[H,\rho]|\lambda'\rangle = -i(E_\lambda - E_{\lambda'})\langle\lambda|\rho|\lambda'\rangle = 0$ when $E_\lambda = E_{\lambda'} = \mu$.

is written as $\text{Tr}[t^\dagger t] = \sum_{nn'} t^*_{n'n} t_{nn'} = \sum_{nn'} |t_{nn'}|^2$, and the current can therefore be written as

$$I_L(V_L, V_R) = \frac{2e^2}{h} \sum_{nn'} \int_0^\infty dE \left(-\frac{\partial n_F}{\partial E}\right) \left(|t_{n'n,RL}|^2 V_L - |t_{nn',LR}|^2 V_R\right). \qquad (7.44)$$

The current in the left lead is thus seen to be the sum of the transmission probabilities of going from any mode n in lead L to any mode n' in lead R minus the probabilities for the opposite processes.

This result is generally valid for the multiprobe case. The current in lead α is found by adding in a similar way the probability for all possible transitions between this lead and any of the other leads α' each with chemical potentials given by $\mu_{\alpha'} = \mu - eV_{\alpha'}$,

$$I_\alpha = \frac{2e^2}{h} \sum_{\alpha' \neq \alpha} \int_0^\infty dE \left(-\frac{\partial n_F}{\partial E}\right) (T_{\alpha'\alpha} V_\alpha - T_{\alpha\alpha'} V_{\alpha'}), \qquad (7.45)$$

where the total transmission probabilities $T_{\alpha\alpha'} = \sum_{nn'} |t_{n'n,\alpha\alpha'}|^2$ is introduced. This formula constitutes the multiprobe Landauer–Büttiker formulism. It gives the linear relation between the applied voltages V_α and the resulting currents I_α, the coefficients being given in terms of the transmission probabilities. The formula is widely used in the analysis of coherent transport in mesoscopic structures, but we will not go further into this here, but instead proceed to study examples of two-probe geometries.

7.3   Electron wave guides

7.3.1   *Quantum point contact and conductance quantization*

One of the most striking consequences of the Landauer formula for conductance is that the conductance of a perfect channel is $2e^2/h$, and if there are N "perfect" channels it is $N2e^2/h$. This has been experimentally tested in numerous experiments, and it is now a well established fact. The first experiments showing this were done by groups in Delft (Holland) and Cambridge (England) in 1988. The technique they used was a so-called splitgate geometry, where a set of metallic gate electrodes was put on top of a 2D electron gas such that a narrow contact between the two sides of the 2DEG was formed; see Fig. 2.10(b). By applying voltage to the gates, the width of the constriction could be controlled very accurately. As the width decreases, quantum channels are squeezed out one by one, leading to a staircase of conductance, each step being of height $2e^2/h$ (see Fig. 7.2). We will now see how this nice effect can happen.

Suppose there is a smooth constriction between two electron reservoirs. Smooth here means a horn-like shape, where the curvature at all points is large compared to the wavelength of the wave which is going to be transmitted through the horn. The relevant wave equation for an electron horn is of course the Schrödinger equation, but there is in principle no difference between the electron wave guide and horn wave guides used in loud speakers, water waves, or other wave phenomena. So the quantized conductance is nothing but a manifestation of the wave nature of a quantum particle, but a very striking one.

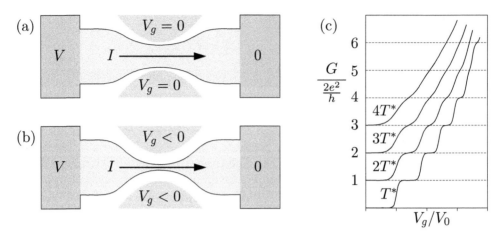

FIG. 7.2. (a) A sketch of a narrow semiconductor quantum wire in a 2D electron gas. A bias voltage V along the wire leads to the current I. (b) The gate-electrodes on the sides are biased by the negative gate-voltage V_g, which results in an *in situ* change of the width of the wire: the electrons are simply repelled by the gate-electrodes. Large negative values of V_g can result in pinch-off. (c) Quantized conductance in the trace of G (in units of $2e^2/h$) for the wire versus V_g. At the lowest temperature T^*, the conductance exhibits clear steps at integer values of $2e^2/h$. As the temperature is increased, the conductance steps disappear gradually. In actual experiments T^* can be as high as 10 K while V_g is swept in intervals of the order of 500 mV.

The Schrödinger equation for the quantum point contact geometry in 2D is

$$\left[-\frac{1}{2m} \left(\partial_x^2 + \partial_y^2 \right) + V_{\text{conf}}(x, y) \right] \Psi(x, y) = E\, \Psi(x, y), \tag{7.46}$$

where $V_{\text{conf}}(x, y)$ is the confinement potential. Because the change along the x-direction is assumed to be smooth, we try to separate the motion in longitudinal and transverse motion. Inspired by that, we expand the wavefunction in terms of the transverse eigenstates $\chi_{nx}(y)$ which however are x-dependent now, as are the expansion coefficients $\phi_n(x)$,

$$\Psi(x, y) = \sum_n \phi_n(x) \chi_{nx}(y). \tag{7.47}$$

This is always possible at any given fixed x since, being solutions of the transverse Schrödinger equation, $\{\chi_n(x)\}$ forms a complete set,

$$\left[-\frac{1}{2m} \partial_y^2 + V_{\text{conf}}(x, y) \right] \chi_{nx}(y) = \varepsilon_n(x) \chi_{nx}(y). \tag{7.48}$$

Inserting Eq. (7.47) into Eq. (7.46) and multiplying from the left with $\chi_{nx}^*(y)$ followed by integration over the transverse direction, y, one finds

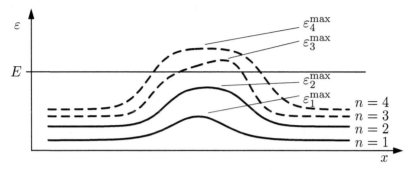

FIG. 7.3. Illustration of the effective 1D barrier created in the adiabatic wire Fig. 7.2. If the energy E of an incident electron in channel n is larger than the maximum transverse kinetic energy $\varepsilon_n^{\max}$, the result is perfect transmission (open channels, full line) as for $n = 1$ and $n = 2$ in the figure. Lower values of E yield complete reflection (closed channels, dashed lines) as for $n = 3$ and $n = 4$. The height of the effective barriers $\varepsilon_n(x)$ is controlled by the gate-voltage V_g.

$$\left[-\frac{1}{2m} \partial_x^2 + \varepsilon_n(x) \right] \phi_n(x) = E\phi_n(x) + \delta_n, \tag{7.49}$$

where

$$\delta_n = \frac{1}{m} \sum_{n'} \int dy \chi_{nx}^*(y) \left[(\partial_x \phi_{n'}(x)) (\partial_x \chi_{n'x}(y)) + \frac{1}{2} \phi_{n'}(x) \partial_x^2 \chi_{n'x}(y) \right]. \tag{7.50}$$

As mentioned, the fundamental approximation we wanted to impose was the smooth geometry approximation, often referred to as the adiabatic approximation. It means that the derivative of the transverse mode with respect to longitudinal direction is neglected, i.e., $\partial_x \chi_{n'x}(y) \approx 0$. In the case of hard walls placed at $y = \pm d(x)/2$,

$$V_{\mathrm{conf}}(x,y) = \begin{cases} 0, & \text{for } y \in [-d(x)/2, d(x)/2], \\ \infty, & \text{otherwise}, \end{cases} \tag{7.51}$$

the transverse wavefunctions are the well-known wavefunctions for a particle in a box

$$\chi_{nx}(y) = \sqrt{\frac{2}{d(x)}} \sin\left(n\pi \left[\frac{y}{d(x)} + \frac{1}{2} \right] \right), \tag{7.52}$$

with the corresponding eigenenergies

$$\varepsilon_n(x) = \frac{\pi^2}{2m \left[d(x) \right]^2} n^2. \tag{7.53}$$

Taking the derivative $\partial_x \chi_{n'x}(y)$, will give something proportional to $d'(x)$. The essence of the adiabatic approximation is that $d'(x) \ll 1$, such that $\partial_x \chi$ can be neglected and we end up with an effective 1D problem of decoupled modes ϕ_n, which obey the 1D Schrödinger equation with an energy barrier $\varepsilon_n(x)$,

$$\left[-\frac{1}{2m} \partial_x^2 + \varepsilon_n(x) \right] \phi_n(x) = E\phi_n(x). \tag{7.54}$$

The transverse direction has thus been translated into an effective 1D barrier. The barrier is there because the part $\varepsilon_n(x)$ of the total kinetic energy E is bound into the transverse motion. Let $x = 0$ be the position in the constriction where this is most narrow, i.e., $d_{\min} = d(0)$. At this position the transverse kinetic energy reaches its maximum, $\varepsilon_n^{\max} \equiv \varepsilon_n(0)$. If $E < \varepsilon_n^{\max}$ the mode cannot transmit (neglecting tunneling through the barrier, of course). If, however, $E > \varepsilon_n^{\max}$ the mode has sufficient energy to pass over the barrier and get through the constriction, this is illustrated in Fig. 7.3.

For smooth barriers, we can use the WKB approximation result for the wavefunction

$$\phi_n(x) \approx \phi_n^{\text{WKB}}(x) = \frac{1}{\sqrt{p(x)}} \exp\left(i \int_{-\infty}^x dx' p(x') \right), \quad p(x) = \sqrt{2m(E - \varepsilon_n(x))}, \tag{7.55}$$

which is a solution to Eq. (7.54) if $|p'(x)/p^2(x)| \ll 1$ and $|p''(x)/p^3(x)| \ll 1$. In this case, we can directly read off the transmission amplitude because in the notation used for the scattering states, we have $r = 0$ and hence $|t| = 1$. The conductance is therefore

$$G = \frac{2e^2}{h} \sum_n \Theta(E_F - \varepsilon_n^{\max}). \tag{7.56}$$

All sub-bands with energy smaller than E_F contribute with one conductance quantum, which results in a step structure of the conductance as a function of $\varepsilon_n^{\max}$. This is roughly what is seen experimentally, where $\varepsilon_n^{\max}$ is changed by changing the width of the constriction through the voltage of the gate electrodes.

Obviously the WKB approximation breaks down if $p(x)$ is too small. Right at the point where a new channel opens, which happens when $E_F = \varepsilon_n(0)$, we would expect some smearing of the step. The shape of the smearing will, in general, depend on the geometry of the constriction and is, in contrast to the step heights, not universal. A useful model is the so-called saddle point model for the constriction, where the confinement potential is modelled by

$$V_{\text{conf}}(x, y) = \frac{1}{2} m\omega_y^2 y^2 - \frac{1}{2} m\omega_x^2 x^2 + V_0, \tag{7.57}$$

where V_0 is a constant. The saddle point model can be thought of as a quadratic expansion of the confinement potential near its maximum. Using this potential, it can be shown that the transmission probability has a particular simple form (Büttiker 1990), namely

$$T_n(E) = \frac{1}{\exp\left[-\pi \left(E - V_0 - (n + \frac{1}{2})\omega_y \right) / \omega_x \right] + 1}. \tag{7.58}$$

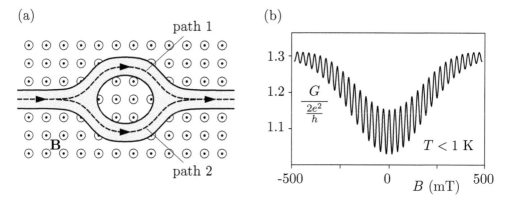

FIG. 7.4. (a) A ring-shaped device exhibiting the Aharonov–Bohm (AB) effect due
    to interference between electron path 1 and electron path 2. The interference is
    modulated by the magnetic flux enclosed by the paths. (b) A sketch of a typical
    low-temperature trace of conductance versus B-field for the ring. The amplitude
    of the AB-oscillations can be of the order 20%. The overall dip of the conductance
    near zero field is due to weak localization (see Section 16.5).

For this model the smearing of the conductance steps thus has the form of a Fermi
function. Experiments using the splitgate geometry indeed show that the conductance
traces (meaning conductance versus gate voltage) are well described by Eq. (7.58).

7.3.2   The Aharonov–Bohm effect

A particular nice example of interference effects in mesoscopic systems is the Aharonov–
Bohm effect, where an applied magnetic field $\mathbf{B}$ is used to control the phase of two
interfering paths. The geometry is illustrated in Fig. 7.4. Each of the arms in the ring
could be an adiabatic wave guide, where the wavefunction can be assumed to be of
the form in Eq. (7.55). Because of the applied $\mathbf{B}$-field we must add a vector potential
$\mathbf{A}$ to the Schrödinger equation (7.46) as in Eq. (7.18).

At small magnetic fields, we can neglect the orbital changes induced by $\mathbf{B}$ in the
narrow arms of the ring and the effect of the magnetic field thus only enters through
the flux enclosed by the ring. This particular limit is relatively easy to analyze, because
for the electron paths in the ring ending at $\mathbf{r}$, the phase factor $\exp\left(-ie \int_{\text{path}}^{\mathbf{r}} d\mathbf{l} \cdot \mathbf{A}\right)$
is unique.[26] For any function $f(\mathbf{r})$ it is straightforward to prove that

$$(\nabla_{\mathbf{r}} + ie\mathbf{A})\left[f(\mathbf{r}) \exp\left(-ie \int_{\text{path}}^{\mathbf{r}} d\mathbf{l} \cdot \mathbf{A}\right)\right] = \exp\left(-ie \int_{\text{path}}^{\mathbf{r}} d\mathbf{l} \cdot \mathbf{A}\right)\left[\nabla_{\mathbf{r}} f(\mathbf{r})\right]. \quad (7.59)$$

Using this formula twice on the wavefunction

$$\Psi_{\mathbf{B}}(\mathbf{r}) \equiv \Psi_0(\mathbf{r}) \exp\left(-ie \int^{\mathbf{r}} d\mathbf{l} \cdot \mathbf{A}\right), \quad (7.60)$$

[26] Because inside the region where $\mathbf{B} = 0$ the line integral is independent of path

where $\Psi_0(\mathbf{r})$ is an eigenstate of the field-free Hamiltonian with energy E,

$$-\frac{1}{2m}\nabla_{\mathbf{r}}^2\Psi_0(\mathbf{r}) = E\,\Psi_0(\mathbf{r}), \tag{7.61}$$

it is easily proven that $\Psi_{\mathbf{B}}(\mathbf{r})$ is an eigenstate of the Hamiltonian with a magnetic field $\mathbf{B} = \nabla \times \mathbf{A}$,

$$-\frac{1}{2m}(\nabla_{\mathbf{r}} + ie\mathbf{A})^2\Psi_{\mathbf{B}}(\mathbf{r}) = E\,\Psi_{\mathbf{B}}(\mathbf{r}). \tag{7.62}$$

In the weak B-field limit, we can thus approximate the line integral by an integral following the center of the wave guides. Furthermore, we assume ideal adiabatic arms, i.e., no backscattering. In that case, the transmission coefficient is given by a sum corresponding to the two paths

$$t \propto \exp\left(-ie\int_{\text{path 1}}^{\mathbf{r}} d\mathbf{l}\cdot\mathbf{A}\right) + e^{i\phi_0}\exp\left(-ie\int_{\text{path 2}}^{\mathbf{r}} d\mathbf{l}\cdot\mathbf{A}\right), \tag{7.63}$$

where ϕ_0 is some phase shift due to different length of the two arms. The transmission probability now becomes

$$|t|^2 \propto 1 + \cos\left(\phi_0 - e\int_{\text{path 1+2}}^{\mathbf{r}} d\mathbf{l}\cdot\mathbf{A}\right) = 1 + \cos\left(\phi_0 - 2\pi\frac{\Phi}{\Phi_0}\right), \tag{7.64}$$

where Φ is the flux enclosed and $\Phi_0 = h/e$ is the one-electron flux quantum. The conductance will oscillate as a function of applied magnetic field, a signature of quantum interference, which indeed can be verified experimentally as illustrated in Fig. 7.4. Note that the effect persists even if there is no magnetic field along the electron trajectories, which is a manifestation of the non-locality of quantum mechanics.

7.4 Summary and outlook

A system is said to be mesoscopic if the length scales involved, the coherence length ℓ_ϕ, the energy relaxation length ℓ_{in}, the elastic mean free path ℓ_0, the Fermi wavelength λ_F of the electron, the atomic Bohr radius a_0, and of the sample size $\mathcal{L}$, fulfil the inequality

$$a_0 \ll \lambda_F \lesssim \ell_0 < \mathcal{L} < \ell_\phi \lesssim \ell_{\text{in}}.$$

In a mesoscopic system, the electron motion is quantum mechanically coherent in the entire sample. The transport is therefore analyzed as a wave scattering problem, where the scattering states $\psi_{\alpha n E}$, which for two-probe systems are defined as

$$\psi_{LnE}(x,\mathbf{r}_\perp) = \begin{cases} \phi_{LnE}^+(x,\mathbf{r}_\perp) + \sum_{n'} r_{n'n}\phi_{Ln'E}^-(x,\mathbf{r}_\perp), & (x,\mathbf{r}_\perp) \in L, \\ \psi_{M,E}(x,\mathbf{r}_\perp), & (x,\mathbf{r}_\perp) \in M, \\ \sum_{n'} t_{n'n}\phi_{Rn'E}^+(x,\mathbf{r}_\perp), & (x,\mathbf{r}_\perp) \in R, \end{cases}$$

are populated with electrons from the metal contacts (the electron reservoirs). The coefficients r and t appearing in the scattering states comes from the S-matrix, which relates the amplitudes of incoming and outgoing waves,

$$\mathbf{c}_{\text{out}} \equiv \begin{pmatrix} \mathbf{a}^- \\ \mathbf{b}^+ \end{pmatrix} = \begin{pmatrix} \mathbf{r} & \mathbf{t}' \\ \mathbf{t} & \mathbf{r}' \end{pmatrix} \begin{pmatrix} \mathbf{a}^+ \\ \mathbf{b}^- \end{pmatrix} \equiv \mathbf{S} \begin{pmatrix} \mathbf{a}^+ \\ \mathbf{b}^- \end{pmatrix} \equiv \mathbf{S}\, \mathbf{c}_{\text{in}}.$$

The conductance $G(\mu, T)$ is calculated from the retarded current-current correlation function using linear response theory. The result is the Landauer formula

$$G(\mu, T) = \frac{2e^2}{h} \int_0^\infty dE \ \text{Tr}[\mathbf{t}^\dagger(E)\mathbf{t}(E)] \left(-\frac{\partial n_{\text{F}}}{\partial E} \right) = \frac{2e^2}{h} \sum_n \int_0^\infty dE \ T_n(E) \left(-\frac{\partial n_{\text{F}}}{\partial E} \right),$$

where the fundamental quantum unit of conductance, $2e^2/h$, appears.

We shall study more aspects of mesoscopic systems later in the book. In Chapter 10 interactions are taken into account, and in Section 16.6 quantum fluctuation effects are dealt with.

More material about mesoscopic physics can be found in the books by Altshuler, Lee, and Webb (1991), Ferry and Goodnick (1999), and Datta (1997), Imry (1997) as well as in the review papers by Beenakker and van Houten (1991) and van Ruitenbeek (1999).

GREEN'S FUNCTIONS

8.1 "Classical" Green's functions

The Green's function method is very useful in the theory of ordinary and partial differential equations. It has a long history with numerous applications.

To illustrate the idea of the method, let us consider the familiar problem of finding the electrical potential ϕ given a fixed charge distribution ρ_e, i.e., we want to solve Poisson's equation

$$\nabla_{\mathbf{r}}^2 \phi(\mathbf{r}) = -\frac{1}{\varepsilon_0}\rho_e(\mathbf{r}). \tag{8.1}$$

It turns out to be a good idea instead to look for the solution G of a related but simpler differential equation

$$\nabla_{\mathbf{r}}^2 G(\mathbf{r}) = \delta(\mathbf{r}), \tag{8.2}$$

where $\delta(\mathbf{r})$ is the Dirac delta function. $G(\mathbf{r})$ is called the Green's function for the Laplace operator $\nabla_{\mathbf{r}}^2$. This is a good idea, because once we have found $G(\mathbf{r})$, the electrical potential follows as

$$\phi(\mathbf{r}) = -\frac{1}{\varepsilon_0}\int d\mathbf{r}'\, G(\mathbf{r}-\mathbf{r}')\rho_e(\mathbf{r}'). \tag{8.3}$$

That this is a solution to Eq. (8.1) is easily verified by letting $\nabla_{\mathbf{r}}^2$ act directly on the integrand and then use Eq. (8.2).

The easiest way to find $G(\mathbf{r})$ is by Fourier transformation, which immediately gives

$$-k^2 G(\mathbf{k}) = 1 \quad \Rightarrow \quad G(\mathbf{k}) = -\frac{1}{k^2}, \tag{8.4}$$

and hence

$$G(\mathbf{r}) = \int\frac{d\mathbf{k}}{(2\pi)^3}e^{i\mathbf{k}\cdot\mathbf{r}}G(\mathbf{k}) = -\int\frac{d\mathbf{k}}{(2\pi)^3}\frac{e^{i\mathbf{k}\cdot\mathbf{r}}}{k^2} = -\frac{1}{4\pi r}. \tag{8.5}$$

When inserting this into Eq. (8.3), we obtain the well-known potential created by a charge distribution

$$\phi(\mathbf{r}) = \frac{1}{4\pi\varepsilon_0}\int d\mathbf{r}'\frac{\rho_e(\mathbf{r}')}{|\mathbf{r}-\mathbf{r}'|}. \tag{8.6}$$

8.2 Green's function for the one-particle Schrödinger equation

Green's functions are particularly useful for problems where one looks for perturbation theory solutions. Consider, e.g., the Schrödinger equation

$$[H_0(\mathbf{r}) + V(\mathbf{r})]\,\Psi_E = E\Psi_E, \tag{8.7}$$

where we know the eigenstates of H_0, and where we want to treat V as a perturbation. Here, we consider the case of an open system, i.e., there is a continuum of states and hence we are free to choose any E. This situation is relevant for scattering problems where a flux of incoming particles (described by H_0) interacts with a system (described by V). The interaction induces transitions from the incoming state to different outgoing states. The procedure outlined below is then a systematic way of calculating the effect of the interaction between the "beam" and the "target" on the outgoing states.

In order to solve the Schrödinger equation, we define the corresponding Green's function by the differential equation

$$\Big[E - H_0(\mathbf{r})\Big]G_0(\mathbf{r}, \mathbf{r}', E) = \delta(\mathbf{r} - \mathbf{r}'), \tag{8.8}$$

with the boundary condition $G_0(\mathbf{r}, \mathbf{r}') = G_0(\mathbf{r}', \mathbf{r})$. It is natural to identify the operator $[E - H_0(\mathbf{r})]$ as the inverse of $G_0(\mathbf{r}, \mathbf{r}')$, and therefore we write[27]

$$G_0^{-1}(\mathbf{r}, E) = E - H_0(\mathbf{r}) \quad \text{or} \quad G_0^{-1}(\mathbf{r}, E)\, G_0(\mathbf{r}, \mathbf{r}', E) = \delta(\mathbf{r} - \mathbf{r}'). \tag{8.9}$$

Now, the Schrödinger equation can be rewritten as

$$\Big[G_0^{-1}(\mathbf{r}, E) - V(\mathbf{r})\Big]\Psi_E = 0, \tag{8.10}$$

and by inspection we see that the solution may be written as an integral equation

$$\Psi_E(\mathbf{r}) = \Psi_E^0(\mathbf{r}) + \int d\mathbf{r}'\, G_0(\mathbf{r}, \mathbf{r}', E)V(\mathbf{r}')\Psi_E(\mathbf{r}'), \tag{8.11}$$

where Ψ_E^0 is an eigenstate to H_0 with eigenenergy E. This is verified by inserting Ψ_E from Eq. (8.11) into the $G_0^{-1}\Psi_E$-term of Eq. (8.10) and then using Eq. (8.9).

One can then solve the integral equation Eq. (8.11) by iteration, and up to first order in V the solution is

$$\Psi_E(\mathbf{r}) = \Psi_E^0(\mathbf{r}) + \int d\mathbf{r}'\, G_0(\mathbf{r}, \mathbf{r}', E)V(\mathbf{r}')\Psi_E^0(\mathbf{r}') + \mathcal{O}\left(V^2\right), \tag{8.12}$$

What we have generated by the iteration procedure is nothing but the ordinary (non-degenerate) perturbation theory. The next leading terms are also easily found by continuing the iteration procedure. The Green's function method is thus useful for this kind of iterative calculations, and one can regard the Green's functions G_0 of the unperturbed system as simple building blocks, from which the solutions of more complicated problems can be build.

Before we introduce the many-body Green's function in the next section, we continue to study the case of non-interacting particles, but now including time dependence. Again we consider the case where the Hamiltonian has a free-particle part H_0 and a perturbation V, $H = H_0 + V$. The time-dependent Schrödinger equation is

$$\Big[i\partial_t - H_0(\mathbf{r}) - V(\mathbf{r})\Big]\Psi(\mathbf{r}, t) = 0. \tag{8.13}$$

---

[27]In order to emphasize the matrix structure, we could have written this as $\int d\mathbf{r}''\ G_0^{-1}(\mathbf{r}, \mathbf{r}'')\, G_0(\mathbf{r}'', \mathbf{r}') = \delta(\mathbf{r} - \mathbf{r}')$, where the inverse Green's function is a function of two arguments. But in the $\mathbf{r}$-representation it is in fact diagonal $G_0^{-1}(\mathbf{r}, \mathbf{r}') = [E - H_0(\mathbf{r})]\delta(\mathbf{r} - \mathbf{r}')$.

Similar to Eq. (8.8) we define the Green's functions by

$$[i\partial_t - H_0(\mathbf{r})]G_0(\mathbf{r}t, \mathbf{r}'t') = \delta(\mathbf{r} - \mathbf{r}')\delta(t - t'). \qquad (8.14a)$$

$$[i\partial_t - H_0(\mathbf{r}) - V(\mathbf{r})]G(\mathbf{r}t, \mathbf{r}'t') = \delta(\mathbf{r} - \mathbf{r}')\delta(t - t'). \qquad (8.14b)$$

The inverse of the Green's functions are thus

$$G_0^{-1}(\mathbf{r}, t) = i\partial_t - H_0(\mathbf{r}) \qquad (8.15a)$$

$$G^{-1}(\mathbf{r}, t) = i\partial_t - H_0(\mathbf{r}) - V(\mathbf{r}). \qquad (8.15b)$$

From these building blocks we easily build the solution of the time-dependent Schrödinger equation. First we observe that the following expression is a solution to Eq. (8.13):

$$\Psi(\mathbf{r}, t) = \Psi^0(\mathbf{r}, t) + \int d\mathbf{r}' \int dt' \, G_0(\mathbf{r}t, \mathbf{r}'t')V(\mathbf{r}')\Psi(\mathbf{r}', t'), \qquad (8.16)$$

or in terms of the full Green's function

$$\Psi(\mathbf{r}, t) = \Psi^0(\mathbf{r}, t) + \int d\mathbf{r}' \int dt' \, G(\mathbf{r}, \mathbf{r}'; t, t')V(\mathbf{r}')\Psi^0(\mathbf{r}', t'), \qquad (8.17)$$

both of which can be shown by inspection (see Exercise 8.1). As for the static case in Eq. (8.11) we can iterate the solution and get

$$\Psi = \Psi^0 + G_0V\Psi^0 + G_0VG_0V\Psi^0 + G_0VG_0VG_0V\Psi^0 + \cdots$$

$$= \Psi^0 + (G_0 + G_0VG_0 + G_0VG_0VG_0 + \cdots)V\Psi^0, \qquad (8.18)$$

where the integration variables have been suppressed. By comparison with Eq. (8.17), we see that the full Green's function G is given by

$$G = G_0 + G_0VG_0 + G_0VG_0VG_0 + \cdots$$

$$= G_0 + G_0V(G_0 + G_0VG_0 + \cdots). \qquad (8.19)$$

Noting that the last parenthesis is nothing but G itself, we have derived the so-called Dyson equation

$$G = G_0 + G_0VG. \qquad (8.20)$$

This equation will play an important role, when we introduce the Feynman diagrams later in the book. The Dyson equation can also be derived directly from Eqs. (8.14) by multiplying Eq. (8.14b) with G_0 from the left.

The Green's function $G(\mathbf{r}t, \mathbf{r}'t')$ defined here is the non-interacting version of the retarded single-particle Green's function that will be introduced in the following section. It is also often called a propagator because it propagates the wavefunction: if the wavefunction is known at some time t', then the wavefunction at a later time t is given by

$$\Psi(\mathbf{r}, t) = \int d\mathbf{r}' \, G(\mathbf{r}t, \mathbf{r}'t')\Psi(\mathbf{r}', t'), \qquad (8.21)$$

which can be checked by inserting Eq. (8.21) into the Schrödinger equation and using the definition Eq. (8.14b). That the Green's function is nothing but a propagator is immediately clear when we write it as[28]

---

[28]Another way to write Eq. (8.21) is simply: $\int d\mathbf{r}' \langle \mathbf{r}|e^{-iH(t-t')}|\mathbf{r}'\rangle \langle \mathbf{r}'|\Psi(t')\rangle = \langle \mathbf{r}|\Psi(t)\rangle$.

$$G(\mathbf{r}t, \mathbf{r}'t') = G^R(\mathbf{r}t, \mathbf{r}'t') = -i\theta(t - t')\langle\mathbf{r}|e^{-iH(t-t')}|\mathbf{r}'\rangle, \qquad (8.22)$$

which indeed is a solution of the partial differential equation defining the Green's function, Eq. (8.14b), the proof being left as an exercise (see Exercise 8.2). Another solution is

$$G(\mathbf{r}t, \mathbf{r}'t') = G^A(\mathbf{r}t, \mathbf{r}'t') = i\theta(t' - t)\langle\mathbf{r}|e^{-iH(t-t')}|\mathbf{r}'\rangle. \qquad (8.23)$$

We have here labelled the two solutions R and A. Here R means *retarded*, because the presence of the particle at $\mathbf{r}$ at time t depends on its position $\mathbf{r}'$ at an earlier time t', whereas *advanced* envokes a dependence on the postion $\mathbf{r}'$ of the particle at a future time t'.

The retarded Green's function (8.22) thus expresses the amplitude for the particle to be in state $|\mathbf{r}\rangle$ at time t, given that it was in the state $|\mathbf{r}'\rangle$ at time t'. We could, of course, calculate the propagator in a different basis, e.g., suppose the particle initially is in state $|\phi_{n'}\rangle$ at time t'; then the propagator for ending in state $|\phi_n\rangle$ is

$$G^R(nt, n't') = -i\theta(t - t')\langle\phi_n|e^{-iH(t-t')}|\phi_{n'}\rangle. \qquad (8.24)$$

The completeness of the basis states ensures that the Green's functions Eqs. (8.22) and (8.24) are related by a simple change of basis

$$G^R(\mathbf{r}t, \mathbf{r}'t') = \sum_{nn'}\langle\mathbf{r}|\phi_n\rangle G^R(nt, n't')\langle\phi_{n'}|\mathbf{r}'\rangle. \qquad (8.25)$$

If we choose the eigenstates $|\phi_n\rangle$ of the Hamiltonian as the basis states, the Green's function becomes

$$G^R(\mathbf{r}t, \mathbf{r}'t') = -i\theta(t - t')\sum_n\langle\mathbf{r}|\phi_n\rangle\langle\phi_n|\mathbf{r}'\rangle e^{-iE_n(t-t')}. \qquad (8.26)$$

8.2.1   Example: from the S-matrix to the Green's function

Propagation from one point to another in quantum mechanics is generally expressed in terms of transmission amplitudes, as in Chapter 7. In fact, there is a general relation that links the S-matrix introduced in Chapter 7 and the asymptotic behavior of the propagator or Green's function defined in Eq. (8.22).

Consider a simple example of a typical 1D scattering problem, where an electron incident from the left interacts with a barrier, located between $x > 0$ and $x < W$. As in Eq. (7.9), the incoming wave is given by $\exp(ikx) + r_k\exp(-ikx)$ for $x < 0$, while the outgoing wave on the other side, $x > W$, is given by $t_k\exp(ikx)$, with k being positive. Similarly, we define eigenstates for wave incoming from the right, with reflection and transmission amplitudes denoted r'_k and t'_k, respectively. In short, we have a one-dimensional version of the scattering states in Eqs. (7.9) and (7.10), with eigenenergy $E = k^2/2m$. Inserting these eigenstates into Eq. (8.26), one finds in the asymptotic limit far from the scattering region (see Exercise 8.7 for details) that for $x' < 0$ and $x > W$

$$G^R(x, x', \omega) = t_{k_\omega} G_0^R(x, x', \omega), \quad (x - x')k_\omega \gg 1, \qquad (8.27)$$

where G_0 is the Green's function in the absence of the scattering potential and $k_\omega = \sqrt{2m\omega}$. From Eq. (8.27) it is evident that the Green's function contains information about the transmission amplitudes for the particle. See also Exercise 12.2 for a specific example. The generalization of Eq. (8.27) to the multi-channel case is in principle straightforward. Instead of a single transmission amplitude, we have a matrix that contains amplitudes for transitions between different channels, as explained in the previous chapter. However, for now we will end the discussion on the Green's function for the single-particle Scrödinger equation, and instead move on and introduce the many-body Green's functions.

8.3   Single-particle Green's functions of many-body systems

In many-particle physics we adopt the Green's function philosophy and define some simple building blocks, also called Green's functions, from which we obtain solutions to our problems. The Green's functions contain only part of the full information carried by the wavefunctions of the systems but they include the relevant information for the given problem. When we define the many-body Green's functions, it is not immediately clear that they are solutions to differential equations as for the Schrödinger-equation Green's functions defined above. But as we will see later, they are in fact solutions of equations of motions with similar structure, and this is the justification for denoting them Green's functions. Let us simply carry on and define the different types of Green's functions that we will be working with.

There are various types of single-particle Green's functions. The retarded Green's function is defined as[29]

$$G^R(\mathbf{r}\sigma t, \mathbf{r}'\sigma't') = -i\theta(t-t')\left\langle[\Psi_\sigma(\mathbf{r}t), \Psi^\dagger_{\sigma'}(\mathbf{r}'t')]_{B,F}\right\rangle, \quad \begin{cases} B : \text{bosons}, \\ F : \text{fermions}, \end{cases} \quad (8.28)$$

where the (anti-) commutator $[\cdots,\cdots]_{B,F}$ is defined as

$$\begin{aligned} [A,B]_B &= [A,B] = AB - BA, \\ [A,B]_F &= \{A,B\} = AB + BA. \end{aligned} \quad (8.29)$$

Similarly, we define a advanced Green's function as

$$G^A(\mathbf{r}\sigma t, \mathbf{r}'\sigma't') = i\theta(t'-t)\langle[\Psi_\sigma(\mathbf{r}t), \Psi^\dagger_{\sigma'}(\mathbf{r}'t')]_{B,F}\rangle, \quad (8.30)$$

Notice the similarity between the many-body Green's function Eq. (8.28) and the one for the propagator for the single-particle wavefunction, in Eq. (8.22). For non-interacting particles they are indeed identical.

The second type of single-particle Green's functions is the so-called greater and lesser Green's functions

$$G^>(\mathbf{r}\sigma t, \mathbf{r}'\sigma't') = -i\langle\Psi_\sigma(\mathbf{r}t)\Psi^\dagger_{\sigma'}(\mathbf{r}'t')\rangle, \quad (8.31a)$$

$$G^<(\mathbf{r}\sigma t, \mathbf{r}'\sigma't') = -i\,(\pm1)\,\langle\Psi^\dagger_{\sigma'}(\mathbf{r}'t')\Psi_\sigma(\mathbf{r}t)\rangle. \quad (8.31b)$$

We see that the retarded and advanced Green's function can be written in terms of these two functions as

---

[29] Recall the definition of the thermal average given in Eq. (1.118).

$$G^R(\mathbf{r}\sigma t, \mathbf{r}'\sigma' t') = \theta(t - t') \left[G^>(\mathbf{r}\sigma t, \mathbf{r}'\sigma' t') - G^<(\mathbf{r}\sigma t, \mathbf{r}'\sigma' t') \right], \qquad (8.32a)$$

$$G^A(\mathbf{r}\sigma t, \mathbf{r}'\sigma' t') = \theta(t' - t) \left[G^<(\mathbf{r}\sigma t, \mathbf{r}'\sigma' t') - G^>(\mathbf{r}\sigma t, \mathbf{r}'\sigma' t') \right]. \qquad (8.32b)$$

Even though we call these Green's functions for "single-particle Green's functions", they are truly many-body objects because they describe the propagation of single particles governed by the full many-body Hamiltonian. Therefore the single-particle functions can include all sorts of correlation effects.

The Green's functions in Eqs. (8.28), (8.31a), and (8.31b) are often referred to as propagators. The reason is that they give the amplitude of a particle inserted in point $\mathbf{r}'$ at time t' to propagate to position $\mathbf{r}$ at time t. In this sense G^R has its name "retarded" because it is required that $t > t'$. The relation between the real-space retarded Green's function and the corresponding one in a general $|\psi_\nu\rangle$-basis as defined in Eq. (1.69) is analogously to Eq. (8.25) and given by

$$G^R(\mathbf{r}\sigma t, \mathbf{r}'\sigma' t') = \sum_{\nu\nu'} \langle \mathbf{r}|\psi_\nu\rangle G^R(\nu\sigma t, \nu'\sigma' t')\langle \psi'_\nu|\mathbf{r}'\rangle, \qquad (8.33)$$

where

$$G^R(\nu\sigma t, \nu'\sigma' t') = -i\theta(t - t') \langle [a_{\nu\sigma}(t), a^\dagger_{\nu'\sigma'}(t')]_{B,F} \rangle, \qquad (8.34)$$

and similar expressions for $G^>$, $G^<$, and G^A.

8.3.1   Green's function of translation-invariant systems

For a system with translational invariance $G(\mathbf{r}, \mathbf{r}')$ can only depend on the difference $\mathbf{r} - \mathbf{r}'$, and the $\mathbf{k}$-representation becomes a natural basis:

$$G^R(\mathbf{r} - \mathbf{r}', \sigma t, \sigma' t') = \frac{1}{\mathcal{V}} \sum_{\mathbf{k}\mathbf{k}'} e^{i\mathbf{k}\cdot\mathbf{r}} G^R(\mathbf{k}\sigma t, \mathbf{k}'\sigma' t') e^{-i\mathbf{k}'\cdot\mathbf{r}'},$$

$$= \frac{1}{\mathcal{V}} \sum_{\mathbf{k}\mathbf{k}'} e^{i\mathbf{k}\cdot(\mathbf{r}-\mathbf{r}')} G^R(\mathbf{k}\sigma t, \mathbf{k}'\sigma' t') e^{i(\mathbf{k}-\mathbf{k}')\cdot\mathbf{r}'}. \qquad (8.35)$$

Since the right-hand side cannot explicitly depend on neither the origin nor on $\mathbf{r}'$, it follows that $G(\mathbf{k},\mathbf{k}') = \delta_{\mathbf{k},\mathbf{k}'} G(\mathbf{k})$, allowing us to write

$$G^R(\mathbf{r} - \mathbf{r}', \sigma t, \sigma' t') = \frac{1}{\mathcal{V}} \sum_{\mathbf{k}} e^{i\mathbf{k}\cdot(\mathbf{r}-\mathbf{r}')} G^R(\mathbf{k}, \sigma t, \sigma' t'), \qquad (8.36a)$$

$$G^R(\mathbf{k}, \sigma t, \sigma' t') = -i\theta(t - t') \langle [a_{\mathbf{k}\sigma}(t), a^\dagger_{\mathbf{k}\sigma'}(t')]_{B,F} \rangle. \qquad (8.36b)$$

The other types of Green's functions have similar forms.

8.3.2   Green's function of free electrons

A particular case often encountered in the theory of quantum liquids, is the simple case of free particles. Consider therefore the Hamiltonian for free electrons (or other fermions)

$$H = \sum_{\mathbf{k}\sigma} \xi_{\mathbf{k}\sigma} c^{\dagger}_{\mathbf{k}\sigma} c_{\mathbf{k}\sigma}, \tag{8.37}$$

and the corresponding greater function in $\mathbf{k}$-space, which we denote $G_0^>$ to indicate that it is the propagator of free electrons. Because the Hamiltonian is diagonal in the quantum numbers $\mathbf{k}$ and σ, so is the Green's function, and therefore

$$G_0^> (\mathbf{k}\sigma, t - t') = -i \left\langle c_{\mathbf{k}\sigma}(t) c^{\dagger}_{\mathbf{k}\sigma}(t') \right\rangle. \tag{8.38}$$

Because of the simple form of the Hamiltonian we are able to find the time dependence of the c-operators (see Eq. (5.24)),

$$c_{\mathbf{k}\sigma}(t) = e^{iHt} c_{\mathbf{k}\sigma} e^{-iHt} = c_{\mathbf{k}\sigma} e^{-i\xi_{\mathbf{k}} t}, \quad \text{and likewise} \quad c^{\dagger}_{\mathbf{k}}(t) = c^{\dagger}_{\mathbf{k}} e^{i\xi_{\mathbf{k}} t}. \tag{8.39}$$

Now $G^>$ becomes

$$G_0^> (\mathbf{k}\sigma; t - t') = -i \langle c_{\mathbf{k}\sigma} c^{\dagger}_{\mathbf{k}\sigma} \rangle e^{-i\xi_{\mathbf{k}}(t-t')}, \tag{8.40}$$

and because the Hamiltonian is diagonal in $\mathbf{k}$ and the occupation of free electrons is given by the Fermi–Dirac distribution, we of course have $\langle c_{\mathbf{k}\sigma} c^{\dagger}_{\mathbf{k}\sigma} \rangle = 1 - n_F(\xi_{\mathbf{k}})$. In exactly the same way, we can evaluate $G_0^<, G_0^R$, and finally G_0^A,

$$G_0^> (\mathbf{k}\sigma, t - t') = -i(1 - n_F(\xi_{\mathbf{k}})) e^{-i\xi_{\mathbf{k}}(t-t')}, \tag{8.41a}$$

$$G_0^< (\mathbf{k}\sigma, t - t') = i n_F(\xi_{\mathbf{k}}) e^{-i\xi_{\mathbf{k}}(t-t')}, \tag{8.41b}$$

$$G_0^R (\mathbf{k}\sigma, t - t') = -i\theta(t - t') e^{-i\xi_{\mathbf{k}}(t-t')}. \tag{8.41c}$$

$$G_0^A (\mathbf{k}\sigma, t - t') = i\theta(t' - t) e^{-i\xi_{\mathbf{k}}(t-t')}. \tag{8.41d}$$

We see that $G^>$ gives the propagation of electrons, because it requires an empty state while $G^<$ gives the propagation of holes, because it is proportional to the number of electrons. This is perhaps more clearly seen, if we write the definition at $T = 0$ of, for example, $G_0^>$,

$$G_0^> (\mathbf{k}, \mathbf{k}', t - t') = -i\langle G| c_{\mathbf{k}}(t) c^{\dagger}_{\mathbf{k}'}(t') |G\rangle = -i\langle G| c_{\mathbf{k}} e^{-iH(t-t')} c^{\dagger}_{\mathbf{k}'} |G\rangle e^{iE_0(t-t')}, \tag{8.42}$$

which precisely is the overlap between a state with an added electron in state $\mathbf{k}'$ and with a state with an added electron in $\mathbf{k}$ and allowing time to evolve from t' to t. Here $|G\rangle$ denotes the groundstate of the free electrons, i.e., the filled Fermi sea, $|G\rangle = |\text{FS}\rangle$.

By Fourier transforming from the time domain to the frequency domain, we obtain information about the possible energies of the propagating particle. This is intuitively clear from Eqs. (8.41), because the propagators evolve periodically in time with the period given by the energy of the electron. For example in the frequency domain, the electron propagator is

$$G_0^> (\mathbf{k}\sigma, \omega) = -2\pi i \left[1 - n_F(\xi_{\mathbf{k}}) \right] \delta(\xi_{\mathbf{k}} - \omega). \tag{8.43}$$

The corresponding $\mathbf{r}$-dependent propagator, which expresses propagation of a particle in real space, is given by

$$\frac{G_0^>(\mathbf{r}-\mathbf{r}',\omega)}{-2\pi i} = \int \frac{d\mathbf{k}}{(2\pi)^3}\left[1-n_F(\xi_{\mathbf{k}})\right]e^{i\mathbf{k}\cdot(\mathbf{r}-\mathbf{r}')}\delta\left(\xi_{\mathbf{k}}-\omega\right)$$

$$= d(\omega)\left[1-n_F(\omega)\right]\frac{\sin(k_\omega\rho)}{k_\omega\rho},\quad \frac{k_\omega^2}{2m}=\omega,\quad \rho=|\mathbf{r}-\mathbf{r}'|, \qquad (8.44)$$

where $d(\varepsilon) = \frac{1}{\pi^2}m^{3/2}\sqrt{\frac{1}{2}\varepsilon}$ is the density of states per spin in three dimensions, see also Eq. (2.31). The propagation from point $\mathbf{r}'$ to $\mathbf{r}$ of a particle with energy ω is thus determined by the density of states d, the availability of an empty state $(1-n_F)$, and the interference function $\frac{1}{x}\sin(x)$ that gives the amplitude of a spherical wave spreading out from the point $\mathbf{r}'$. See also Exercise 8.3.

8.3.3 The Lehmann representation

A method we will often be using when proving formal results is the so-called Lehmann representation, which is just another name for using the set of eigenstates, $\{|n\rangle\}$, of the full Hamiltonian H as basis set. Let us for example study the diagonal Green's function $G^>(\nu t,\nu t')$, which we write as

$$G^>(\nu;t,t') = -i\langle c_\nu(t)c_\nu^\dagger(t')\rangle = -i\frac{1}{Z}\sum_n\langle n|e^{-\beta H}c_\nu(t)c_\nu^\dagger(t')|n\rangle, \qquad (8.45\text{a})$$

and then insert $1 = \sum_{n'}|n'\rangle\langle n'|$ to get

$$G^>(\nu;t,t') = -i\frac{1}{Z}\sum_{nn'}e^{-\beta E_n}\langle n|c_\nu|n'\rangle\langle n'|c_\nu^\dagger|n\rangle e^{i(E_n-E_{n'})(t-t')}. \qquad (8.45\text{b})$$

In the frequency domain, we then obtain

$$G^>(\nu;\omega) = \frac{-2\pi i}{Z}\sum_{nn'}e^{-\beta E_n}\langle n|c_\nu|n'\rangle\langle n'|c_\nu^\dagger|n\rangle\delta(E_n-E_{n'}+\omega). \qquad (8.46)$$

In the same way we have (for fermions, c)

$$G^<(\nu;\omega) = \frac{2\pi i}{Z}\sum_{nn'}e^{-\beta E_n}\langle n|c_\nu^\dagger|n'\rangle\langle n'|c_\nu|n\rangle\delta(E_n-E_{n'}-\omega)$$

$$= \frac{2\pi i}{Z}\sum_{nn'}e^{-\beta E_{n'}}\langle n'|c_\nu^\dagger|n\rangle\langle n|c_\nu|n'\rangle\delta(E_{n'}-E_n-\omega)$$

$$= \frac{2\pi i}{Z}\sum_{nn'}e^{-\beta(E_n+\omega)}\langle n'|c_\nu^\dagger|n\rangle\langle n|c_\nu|n'\rangle\delta(E_{n'}-E_n-\omega)$$

$$= -G^>(\nu;\omega)e^{-\beta\omega}. \qquad (8.47)$$

When remembering the positive infinitesimal η discussed in Section 5.8, the retarded Green's function G^R (for fermions) becomes

$$G^R(\nu,\omega) = -i \int_0^\infty dt\, e^{i(\omega+i\eta)t} \frac{1}{Z} \sum_{nn'} e^{-\beta E_n} \left(\langle n|c_\nu|n'\rangle\langle n'|c_\nu^\dagger|n\rangle e^{i(E_n-E_{n'})t} \right.$$

$$\left. + \langle n|c_\nu^\dagger|n'\rangle\langle n'|c_\nu|n\rangle e^{-i(E_n-E_{n'})t} \right)$$

$$= \frac{1}{Z} \sum_{nn'} e^{-\beta E_n} \left(\frac{\langle n|c_\nu|n'\rangle\langle n'|c_\nu^\dagger|n\rangle}{\omega + E_n - E_{n'} + i\eta} + \frac{\langle n|c_\nu^\dagger|n'\rangle\langle n'|c_\nu|n\rangle}{\omega - E_n + E_{n'} + i\eta} \right)$$

$$= \frac{1}{Z} \sum_{nn'} \frac{\langle n|c_\nu|n'\rangle\langle n'|c_\nu^\dagger|n\rangle}{\omega + E_n - E_{n'} + i\eta} \left(e^{-\beta E_n} + e^{-\beta E_{n'}} \right). \tag{8.48}$$

Using $(\omega+i\eta)^{-1} = \mathcal{P}\frac{1}{\omega} - i\pi\delta(\omega)$ and taking the imaginary part, we obtain

$$2\,\mathrm{Im}\,G^R(\nu,\omega) = -\frac{2\pi}{Z} \sum_{nn'} \langle n|c_\nu|n'\rangle\langle n'|c_\nu^\dagger|n\rangle \left(e^{-\beta E_n} + e^{-\beta E_{n'}} \right) \delta(\omega + E_n - E_{n'})$$

$$= -\frac{2\pi}{Z} \sum_{nn'} \langle n|c_\nu|n'\rangle\langle n'|c_\nu^\dagger|n\rangle e^{-\beta E_n} (1 + e^{-\beta\omega}) \delta(\omega + E_n - E_{n'})$$

$$= -i(1 + e^{-\beta\omega})G^>(\nu,\omega). \tag{8.49}$$

Defining the spectral function A as
$$A(\nu,\omega) = -2\,\mathrm{Im}\,G^R(\nu,\omega), \tag{8.50}$$

we have derived the important general relations

$$iG^>(\nu,\omega) = A(\nu,\omega)\left[1 - n_F(\omega)\right], \tag{8.51a}$$
$$-iG^<(\nu,\omega) = A(\nu,\omega)n_F(\omega). \tag{8.51b}$$

These relations, known as the fluctuation-dissipation theorem, relate the occupations to the spectral functions. The name of the theorem is better understood in the case of bosons, where similar relations hold, see Exercise 8.4. Take, say, electromagnetic modes described as bosons and consider the current-current Green's function function. In this case, the current fluctuations are given by the occupation number and the dissipation is related to the retarded correlation, as shown in Chapter 6.

The Lehmann representation also gives the following general and useful relations,

$$G^R(\nu,\omega) = \int \frac{d\omega'}{2\pi} \frac{A(\nu,\omega')}{\omega - \omega' + i\eta}, \tag{8.52a}$$

$$G^A(\nu,\omega) = \int \frac{d\omega'}{2\pi} \frac{A(\nu,\omega')}{\omega - \omega' - i\eta}. \tag{8.52b}$$

The first one is easily derived from Eq. (8.48) by inserting $1 = \int d\omega'\, \delta(\omega' + E_n - E_{n'})$ inside the sum on the right hand and by comparing with the first line of Eq. (8.49). We also note from these two relations that the retarded and advanced Green's function are simply the complex conjugate of one and another:

$$G^R(\nu,\omega) = \left[G^A(\nu,\omega)\right]^*, \tag{8.53}$$

or in the case of non-diagonal Green's functions the corresponding relation is

$$G^R(\nu, \nu', \omega) = \left[G^A(\nu', \nu, \omega) \right]^*, \tag{8.54}$$

which is proven by using the definition of the retarded and advanced Green's functions (see Exercise 8.8).

8.3.4 *The spectral function*

The spectral function $A(\nu, \omega)$ can be thought of as either the quantum state resolution of a particle with given energy ω or as the energy resolution for a particle in a given quantum number ν. It gives an indication of how well the excitation created by adding a particle in state ν can be described by a free non-interacting particle. For example the retarded propagator (8.41c) for free electrons,

$$G_0^R(\mathbf{k}\sigma, \omega) = -i \int_{-\infty}^{\infty} dt\, \theta(t - t')\, e^{i\omega(t-t')} e^{-i\xi_\mathbf{k}(t-t') - \eta(t-t')} = \frac{1}{\omega - \xi_\mathbf{k} + i\eta}, \tag{8.55}$$

has the corresponding spectral function

$$A_0(\mathbf{k}\sigma, \omega) = -2\,\mathrm{Im}\, G_0^R(\mathbf{k}\sigma, \omega) = 2\pi\delta(\omega - \xi_\mathbf{k}). \tag{8.56}$$

Thus for the idealized case of non-interacting free electrons, the spectral function is a delta function, which tells us that an excitation with energy ω can only happen by adding an electron to the state $\mathbf{k}$ given by $\xi_\mathbf{k} = \omega$, as expected.

This result is true for any quadratic Hamiltonian, i.e., for non-interacting systems. If we have for example

$$H_0 = \sum_\nu \xi_\nu c_\nu^\dagger c_\nu, \tag{8.57}$$

where ν labels the eigenstates of the system, the spectral function is again given by a simple delta function

$$A_0(\nu, \omega) = 2\pi\delta(\omega - \xi_\nu). \tag{8.58}$$

Generally, due to interactions, the spectral function differs from a delta function, but it may still be a peaked function, which then indicates that the non-interacting approximation is not too far from the truth. In Chapter 15 this is discussed in much more detail.

We will now show that the spectral function is a probability distribution. First, as one must require, it is always positive. This follows from Eq. (8.48), the definition Eq. (8.50) of the spectral function, and the fact that $\langle n|c_\nu|n'\rangle\langle n'|c_\nu^\dagger|n\rangle = |\langle n|c_\nu|n'\rangle|^2$. Second, it obeys the sum rule

$$\int_{-\infty}^{\infty} \frac{d\omega}{2\pi} A(\nu, \omega) = 1. \tag{8.59}$$

This formula is easily derived by considering the Lehmann representation of $-2\,\mathrm{Im}\, G^R$ in Eq. (8.49),

$$\int_{-\infty}^{\infty} \frac{d\omega}{2\pi} A(\nu, \omega) = -\int_{-\infty}^{\infty} \frac{d\omega}{2\pi} 2 \operatorname{Im} G^R(\nu, \omega)$$

$$= \int_{-\infty}^{\infty} d\omega \frac{1}{Z} \sum_{nn'} \langle n|c_\nu|n'\rangle\langle n'|c_\nu^\dagger|n\rangle \left(e^{-\beta E_n} + e^{-\beta E_{n'}}\right) \delta(\omega + E_n - E_{n'})$$

$$= \frac{1}{Z} \sum_{nn'} \langle n|c_\nu|n'\rangle\langle n'|c_\nu^\dagger|n\rangle \left(e^{-\beta E_n} + e^{-\beta E_{n'}}\right)$$

$$= \langle c_\nu c_\nu^\dagger \rangle + \langle c_\nu^\dagger c_\nu \rangle = \langle c_\nu c_\nu^\dagger + c_\nu^\dagger c_\nu \rangle = 1, \qquad (8.60)$$

where the last equality follows from the Fermi operator anti-commutation relations.

Alternatively, one can show Eq. (8.60) by using that for any Fourier transform $f(\omega)$ in the frequency domain we have $\int \frac{d\omega}{2\pi} f(\omega) = f(t{=}0)$ in the time domain, and hence

$$\int_{-\infty}^{\infty} \frac{d\omega}{2\pi} G^R(\nu, \omega) = G^R(\nu, t{=}0) = -i\theta(0)\langle\{c_\nu, c_\nu^\dagger\}\rangle = -i \times \frac{1}{2} \times 1, \qquad (8.61)$$

from which Eq. (8.60) follows. Of course, setting $\theta(0) = \frac{1}{2}$, should be done with some care, because it $\theta(x)$ is not well-defined for $x \to 0$, and therefore the derivation Eq. (8.61) only serves as an easy way to remember the sum rule.

Remarkably, the spectral function is similar to the density of states at a given energy. This is evident since for fermions the occupation n_ν of a given state ν is given by Eq. (8.51b)

$$\bar{n}_\nu = \langle c_\nu^\dagger c_\nu \rangle = -iG^<(\nu, t = 0)$$

$$= -i \int_{-\infty}^{\infty} \frac{d\omega}{2\pi} G^<(\nu, \omega)$$

$$= \int_{-\infty}^{\infty} \frac{d\omega}{2\pi} A(\nu, \omega) n_F(\omega). \qquad (8.62)$$

The physical interpretation is that the occupation of the quantum state $|\nu\rangle$ is an energy integral of the spectral density projected onto the state $|\nu\rangle$ weighted by the occupation at the given energy. For a Fermi gas at low temperatures, $T \ll T_F$, we expect that if the state $|\nu\rangle$ is far below the Fermi surface, e.g., $\varepsilon_\nu \ll E_F$, then $\langle c_\nu^\dagger c_\nu \rangle \approx 1$. This in fact follows from the sum rule, because if $\varepsilon_\nu \ll E_F$ and the width of $A(\nu, \omega)$ is also small compared to E_F then the Fermi function in Eq. (8.62) is approximately unity and since $A(\nu, \omega)$ integrates to 2π the expected result follows.

8.3.5 *Broadening of the spectral function*

When interactions are present, the spectral function changes from the ideal delta function to a broadened profile. One possible mechanism of broadening in a metal is by e.g., electron-phonon interaction, which redistributes the spectral weight because of energy exchange between the electron and the phonon system. Another mechanism for broadening is the electron–electron interaction. See Chapter 15.

As a simple example we consider a Green's function which decays in time due to processes that scatter the particle out of the state ν. In this situation the retarded Green's function becomes

$$G^R(\nu, t) \approx -i\theta(t)e^{-i\xi_\nu t}e^{-t/\tau}, \tag{8.63}$$

where τ is the characteristic decay time. Such a decaying Green's function corresponds to a finite width of the spectral function

$$A(\nu, \omega) = -2\,\mathrm{Im}\int_{-\infty}^{\infty} dt\, e^{i\omega t} G^R(\nu, t) \approx 2\,\mathrm{Im}\,i \int_0^\infty dt\, e^{i\omega t} e^{-i\xi_\nu t} e^{-t/\tau}$$

$$= \frac{2/\tau}{(\omega - \xi_\nu)^2 + (1/\tau)^2}. \tag{8.64}$$

Thus the width in energy space is given by τ^{-1}.

The simple notion of single electron propagators becomes less well defined for interacting systems, which is reflected in a broadening of the spectral function. Amazingly, the free electron picture is still a good description in many cases and in particular for metals, which is quite surprising since the Coulomb interaction between the electrons is a rather strong interaction. The reason for this will be discussed later in the Chapter 15 on Fermi liquid theory. In other cases, the single-particle approximation is not justified at all. One such example is discussed in Chapter 19, where it shown that one-dimensional systems does not fall into the class of systems for which Fermi liquid theory applies.

8.4 Measuring the single-particle spectral function

In order to probe the single-particle properties of a many-body system, say a solid-state sample, one must have a way of measuring how the electrons propagate as a function of energy. In practice, this means taking out or inserting a particle with definite energy. There are not too many ways of doing this, because most experiments measure density or other two-particle properties. For example the response to an electromagnetic field couples to the charge or current, which, as we saw in Chapter 6, measures charge-charge or current-current correlation functions, both being two-particle propagators.

In principle, there is only one way to measure the single-particle properties, which is to insert/remove a single electron into/out of a many-body system. This can be achieved by a so-called tunnel junction device or by subjecting the sample to a beam of electrons. However, in some cases also optical experiments approximately measures the single-particle density of states. For example, when a photon is absorbed and an electron is kicked out from an occupied state to, e.g., a freely propagating state outside the material as in the photo-electric effect.

In the following, we study in detail the tunneling case where an electron tunnels from one material to the other and show how the tunneling current is expressed in terms of the spectral functions and thus provides a direct measurement of these.

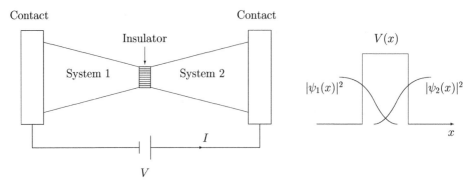

FIG. 8.1. Measurement setup for the tunnel experiment. Two systems are brought into close contact, separated by an insulating material, e.g., an oxide or for the so-called scanning tunneling microscope (STM) simply vacuum. The right panel illustrates the electron wavefunctions in the two subsystems which have a small overlap in the insulator region. In the tunneling Hamiltonian this is modeled by the matrix element $T_{\nu\nu'}$.

8.4.1 Tunneling spectroscopy

The tunnel experiment setup consists of two conducting materials brought into close contact, such that electrons can tunnel from one to the other. This is illustrated in Fig. 8.1. Systems 1 and 2 are described by their respective Hamiltonians, H_1 and H_2, involving electron operators, $c_{1,\nu}$ and $c_{2,\mu}$, where the single-particle states $|\nu\rangle$ and $|\mu\rangle$ are complete sets for system 1 and 2, respectively.

The coupling between the two sides of the junction is due to the finite overlap of the wavefunctions, which gives rise to a term in the Hamiltonian of the form

$$H_T = \sum_{\nu\mu} \left(T_{\nu\mu} c_{1,\nu}^\dagger c_{2,\mu} + T_{\nu\mu}^* c_{2,\mu}^\dagger c_{1,\nu} \right). \tag{8.65}$$

This is the most general one-particle operator which couples the two systems. The tunnel matrix element is defined as

$$T_{\nu\mu} = \int d\mathbf{r}\, \psi_\nu^*(\mathbf{r}) H(\mathbf{r}) \psi_\mu(\mathbf{r}), \tag{8.66}$$

with $H(\mathbf{r})$ being the (first quantization) one-particle Hamiltonian.

The current through the device is defined by the rate of change of particles, $I_e = -e\langle I\rangle$, where $I = \dot{N}_1$, and hence

$$I = i[H, N_1] = i[H_T, N_1] = i \sum_{\nu\mu} \sum_{\nu'} \left[\left(T_{\nu\mu} c_{1,\nu}^\dagger c_{2,\mu} + T_{\nu\mu}^* c_{2,\mu}^\dagger c_{1,\nu} \right), c_{1,\nu'}^\dagger c_{1,\nu'} \right]$$

$$= -i \sum_{\nu\mu} \left(T_{\nu\mu} c_{1,\nu}^\dagger c_{2,\mu} - T_{\nu\mu}^* c_{2,\mu}^\dagger c_{1,\nu} \right) \equiv -i(L - L^\dagger). \tag{8.67}$$

The current passing from 1 to 2 is driven by a shift of chemical potential difference, which means that $\mu_1 \neq \mu_2$. The coupling between the two systems is assumed to be very weak, since the tunnel matrix element is exponentially suppressed with distance between the two systems. Therefore, we calculate the current to lowest order in the coupling. The current operator itself is already linear in $T_{\nu\mu}$ and therefore we need only one more order, which means that linear response theory is applicable. According to the general Kubo formula derived in Chapter 6 the particle current is to first order in H_T given by

$$\langle I \rangle(t) = \int_{-\infty}^{\infty} dt' C^R_{I_p H_T}(t, t'), \tag{8.68a}$$

$$C^R_{I_p H_T}(t - t') = -i\theta(t - t')\langle [\hat{I}_p(t), \hat{H}_T(t')] \rangle_0, \tag{8.68b}$$

where the time development is governed by $H = H_1 + H_2$. The correlation function $C_{I H_T}$ can be simplified a bit as

$$C^R_{I_p H_T}(t - t') = -\theta(t - t') \left\langle \left[\hat{L}(t) - \hat{L}^\dagger(t), \hat{L}(t') + \hat{L}^\dagger(t') \right] \right\rangle_0$$

$$= -\theta(t - t') \left[\left\langle \left[\hat{L}(t), \hat{L}(t') \right] \right\rangle_0 - \left\langle \left[\hat{L}^\dagger(t), \hat{L}(t') \right] \right\rangle_0 + \text{c.c.} \right]. \tag{8.69}$$

Now the combination $\left\langle \left[\hat{L}(t), \hat{L}(t') \right] \right\rangle$ involves terms of the form

$$\left\langle \left(c^\dagger_{1,\nu} c_{2,\mu} \right)(t) \left(c^\dagger_{1,\nu} c_{2,\mu} \right)(t') \right\rangle_0,$$

with two electrons created in system 1 and two electrons annihilated in system 2 and therefore is does not conserve the number of particles in each system. Naturally, the number of particles is a conserved quantity and matrix elements of this type must vanish.[30] We are therefore left with

$$I_p(t) = 2\,\mathrm{Re} \int_{-\infty}^{\infty} dt' \theta(t - t') \left\langle \left[\hat{L}^\dagger(t), \hat{L}(t') \right] \right\rangle_0$$

$$= 2\,\mathrm{Re} \int_{-\infty}^{\infty} dt' \theta(t - t') \sum_{\nu\mu} \sum_{\nu'\mu'} T^*_{\nu\mu} T_{\nu'\mu'} \left\langle \left[\hat{c}^\dagger_{2,\mu}(t)\hat{c}_{1,\nu}(t), \hat{c}^\dagger_{1,\nu'}(t')\hat{c}_{2,\mu'}(t') \right] \right\rangle_0$$

$$= 2\,\mathrm{Re} \int_{-\infty}^{\infty} dt' \theta(t - t') \sum_{\nu\mu} \sum_{\nu'\mu'} T^*_{\nu\mu} T_{\nu'\mu'} \left(\left\langle \hat{c}_{1,\nu}(t)\hat{c}^\dagger_{1,\nu'}(t') \right\rangle_0 \left\langle \hat{c}^\dagger_{2,\mu}(t)\hat{c}_{2,\mu'}(t') \right\rangle_0 \right.$$

$$\left. - \left\langle \hat{c}^\dagger_{1,\nu'}(t')\hat{c}_{1,\nu}(t) \right\rangle_0 \left\langle \hat{c}_{2,\mu'}(t')\hat{c}^\dagger_{2,\mu}(t) \right\rangle_0 \right). \tag{8.70}$$

Now the time dependence due to the shift in energy by the applied voltages is explicitly pulled out such that

$$\hat{c}_1(t) = \tilde{c}_1(t)e^{-i(-e)V_1 t}, \tag{8.71a}$$

$$\hat{c}_2(t) = \tilde{c}_2(t)e^{-i(-e)V_2 t}, \tag{8.71b}$$

with the time dependence of $\tilde{c}$ being given by the Hamiltonian with a common chemical potential μ. Furthermore, we are of course allowed to choose a basis set

---

[30]This is in fact not true for superconductors which are characterized by having a spontaneous breaking of the symmetry corresponding to the conservation of particles and therefore such two-particle tunnel processes are allowed and give rise to the so-called Josephson current (see Chapter 18).

where the Green's function of the decoupled system (i.e., without H_T) is diagonal, $G^>_{\nu\nu'} = \delta_{\nu\nu'} G^>_\nu$. The particle current then becomes (after change of variable $t' \to t'+t$)

$$I_p = 2\mathrm{Re} \int_{-\infty}^{0} dt' \sum_{\nu\mu} |T_{\nu\mu}|^2 e^{i(-e)(V_1-V_2)t'} \left[G^>_1(\nu;-t') G^<_2(\mu;t') - G^<_1(\nu;-t') G^>_2(\mu;t') \right].$$

$$(8.72)$$

After Fourier transformation (and reinsertion of the convergence factor $e^{\eta t'}$) this expression becomes

$$I_p = \int_{-\infty}^{\infty} \frac{d\omega}{2\pi} \sum_{\nu\mu} |T_{\nu\mu}|^2 \left[G^>_1(\nu;\omega) G^<_2(\mu;\omega-eV) - G^<_1(\nu;\omega) G^>_2(\mu;\omega-eV) \right],$$

$$(8.73)$$

with the voltage given by $V = V_1 - V_2$. The lesser and greater Green's functions are now written in terms of the spectral function, see Eq. (8.51), and we finally arrive at

$$I_p = \int_{-\infty}^{\infty} \frac{d\omega}{2\pi} \sum_{\nu\mu} |T_{\nu\mu}|^2 A_1(\nu,\omega) A_2(\mu,\omega-eV) [n_F(\omega+eV) - n_F(\omega)]. \qquad (8.74)$$

In Eq. (8.74) we see that the current is determined by two factors: the availability of states, given by the difference of occupation functions, and by the density of states at a given energy. Therefore by sweeping the voltage across the junction one gets information about $A(\nu,\omega)$. This is a widely used spectroscopic principle in, e.g., the study of superconductors, where it was used to verify the famous prediction of the BCS theory of superconductivity that there is an excitation gap in the superconductor, and that the density of states peaks near the gap (see Section 18.4.3). Also it is used to study small structures such as quantum dots, where the individual quantum levels become visible due to size quantization.

The tunnel spectroscopy technique amounts to a sweep of an external voltage which controls the chemical potential while measuring the differential conductance dI/dV. If the other material is a simple metal where one can assume the density of states to be more or less constant, i.e.,

$$\sum_{\mu} |T_{\nu\mu}|^2 A_2(\mu,\omega+eV) \approx \text{const.} \qquad (8.75)$$

then

$$\frac{dI}{dV} \propto \int_{-\infty}^{\infty} d\omega \left(-\frac{\partial n_F(\omega+eV)}{\partial \omega} \right) \sum_{\nu} A_1(\nu,\omega). \qquad (8.76)$$

At low temperatures where the derivative of the Fermi function tends to a delta function Eq. (8.76) becomes

$$\frac{dI}{dV} \propto \sum_{\nu} A_1(\nu,-eV). \qquad (8.77)$$

So the spectral function can in fact be measured in a rather direct way, which is illustrated in Fig. 8.2.

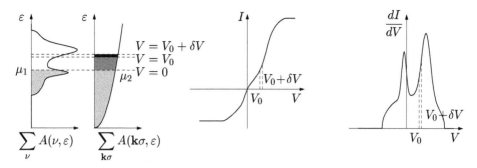

FIG. 8.2. The principle used in tunneling spectroscopy. The left panel shows the two density of states in the two materials. The right one is a metal, where there is little variations with energy and the experiment can therefore be used to get information about the density of states of the left material. The two right-most panels show the resulting current and the differential conductance trace. It is seen how the differential conductance is a direct measure of $\sum_\nu A_1(\nu, \omega)$.

8.5 Two-particle correlation functions of many-body systems

While the single-particle Green's functions defined above measure the properties of individual particles, the higher order Green's functions give the response of the quantum system to processes involving several particles. One important type of higher order Green's functions are the retarded correlation functions, which were encountered in the linear response chapter. For example, we saw that the response to electromagnetic radiation was determined by the retarded auto-correlation function of the charge and current densities. Typical functions that we will encounter are of the type

$$C_{AA}(t, t') = -i\theta(t - t') \left\langle \left[A(t), A(t') \right] \right\rangle, \tag{8.78}$$

where A is some two-particle operator.

In order to treat a specific case, we evaluate the polarization function $\chi = C_{\rho\rho}$ for a non-interacting electron gas, see Eq. (6.37b). This function gives for example information about the dissipation due to an applied field, because the dissipation, which is the real part of the conductivity,[31] is according to Eq. (6.46) given by (take for simplicity the translation-invariant case)

$$\operatorname{Re} \sigma \left(\mathbf{q}, \omega \right) = -\frac{\omega e^2}{q^2} \operatorname{Im} \chi^R(\mathbf{q}, \omega). \tag{8.79}$$

---

[31]Because the power dissipated at any given point in space and time is $P(\mathbf{r},t) = \mathbf{J}_e(\mathbf{r}, t) \cdot \mathbf{E}(\mathbf{r}, \mathbf{t})$, the total energy being dissipated is

$$W = \int d\mathbf{r} dt\, \mathbf{E}(\mathbf{r}, \mathbf{t}) \cdot \mathbf{J}_e(\mathbf{r}, t) = \int \frac{d\omega}{2\pi} \frac{1}{\mathcal{V}} \sum_q \mathbf{E}^*(\mathbf{q}, \omega) \cdot \mathbf{J}_e(\mathbf{q}, \omega) = \int \frac{d\omega}{2\pi} \frac{1}{\mathcal{V}} \sum_q |\mathbf{E}(\mathbf{q}, \omega)|^2 \operatorname{Re} \sigma(\mathbf{q}, \omega),$$

where it was used that $\sigma^*(\mathbf{q}, \omega) = \sigma(-\mathbf{q}, -\omega)$, which can be proven from the Kubo formula 6.24.

In momentum space, the polarization is given by the Fourier transform of $\chi(\mathbf{r}, \mathbf{r}')$, which can only depend on the difference $\mathbf{r} - \mathbf{r}'$. We therefore choose an arbitrary $\mathbf{r}'$ and write

$$\chi^R(\mathbf{q}, t - t') = \int d\mathbf{r}\, \chi(\mathbf{r} - \mathbf{r}', t - t') e^{-i\mathbf{q}\cdot(\mathbf{r}-\mathbf{r}')}$$

$$= -i\theta(t - t') \int d\mathbf{r}\, \langle [\rho(\mathbf{r}, t), \rho(\mathbf{r}', t')] \rangle\, e^{-i\mathbf{q}\cdot(\mathbf{r}-\mathbf{r}')}$$

$$= -i\theta(t - t') \int d\mathbf{r}\, \frac{1}{V^2} \sum_{\mathbf{q}_1 \mathbf{q}_2} \langle [\rho(\mathbf{q}_1, t), \rho(\mathbf{q}_2, t')] \rangle\, e^{i\mathbf{q}_1\cdot\mathbf{r} + i\mathbf{q}_2\cdot\mathbf{r}'} e^{-i\mathbf{q}\cdot(\mathbf{r}-\mathbf{r}')}$$

$$= -i\theta(t - t') \frac{1}{V} \sum_{\mathbf{q}_2} \langle [\rho(\mathbf{q}, t), \rho(\mathbf{q}_2, t')] \rangle\, e^{i(\mathbf{q}_2 + \mathbf{q})\cdot\mathbf{r}'}. \tag{8.80}$$

Now this cannot depend on $\mathbf{r}'$ and one sees that $\mathbf{q}_2 = -\mathbf{q}$ (or formally one can integrate over $\mathbf{r}'$ and divide by volume to get a delta function, $\delta_{\mathbf{q}_2 + \mathbf{q}, 0}$) and thus

$$\chi^R(\mathbf{q}, t - t') = -i\theta(t - t') \frac{1}{V} \Big\langle [\rho(\mathbf{q}, t), \rho(-\mathbf{q}, t')] \Big\rangle. \tag{8.81}$$

The Fourier transform of the charge operator was derived in Eq. (1.95). For free electrons, the time dependence is given by, see Eq. (8.39),

$$\rho(\mathbf{q}, t) = \sum_{\mathbf{k}\sigma} c_{\mathbf{k}\sigma}^\dagger c_{\mathbf{k}+\mathbf{q}\sigma} e^{i(\xi_\mathbf{k} - \xi_{\mathbf{k}+\mathbf{q}})t}, \tag{8.82}$$

which, when inserted into Eq. (8.81), yields

$$\chi_0^R(\mathbf{q}, t - t') = -i\theta(t - t') \frac{1}{V} \sum_{\mathbf{k}\mathbf{k}'\sigma\sigma'} \langle [c_{\mathbf{k}\sigma}^\dagger c_{\mathbf{k}+\mathbf{q}\sigma}, c_{\mathbf{k}'\sigma'}^\dagger c_{\mathbf{k}'-\mathbf{q}\sigma'}] \rangle_0 e^{i(\xi_\mathbf{k} - \xi_{\mathbf{k}+\mathbf{q}})t} e^{i(\xi_{\mathbf{k}'} - \xi_{\mathbf{k}'-\mathbf{q}})t'},$$

$$\tag{8.83}$$

where the subindex "0" indicates that we are using the free electron approximation. Using the formula $[c_\nu^\dagger c_\mu, c_{\nu'}^\dagger c_{\mu'}] = c_\nu^\dagger c_{\mu'} \delta_{\mu,\nu'} - c_{\nu'}^\dagger c_\mu \delta_{\nu,\mu'}$, the commutator is easily evaluated and we find

$$\chi_0^R(\mathbf{q}, t - t') = -i\theta(t - t') \frac{1}{V} \sum_{\mathbf{k}\sigma} [n_F(\xi_\mathbf{k}) - n_F(\xi_{\mathbf{k}+\mathbf{q}})]\, e^{i(\xi_\mathbf{k} - \xi_{\mathbf{k}+\mathbf{q}})(t-t')}, \tag{8.84}$$

because $\langle c_k^\dagger c_k \rangle = n_F(\xi_k)$. In frequency space, we find when using the rule in Section 5.8 on how to transform retarded functions

$$\chi_0^R(\mathbf{q}, \omega) = -i \int_{t'}^{\infty} dt\, e^{i\omega(t-t')} \frac{1}{V} \sum_{\mathbf{k}\sigma} [n_F(\xi_\mathbf{k}) - n_F(\xi_{\mathbf{k}+\mathbf{q}})]\, e^{i(\xi_\mathbf{k} - \xi_{\mathbf{k}+\mathbf{q}})(t-t')} e^{-\eta(t-t')}$$

$$= \frac{1}{V} \sum_{\mathbf{k}\sigma} \frac{n_F(\xi_\mathbf{k}) - n_F(\xi_{\mathbf{k}+\mathbf{q}})}{\xi_\mathbf{k} - \xi_{\mathbf{k}+\mathbf{q}} + \omega + i\eta}. \tag{8.85}$$

This function is known as the Lindhard function, and later on, when discussing the elementary excitations of the electron gas, we will study it in much more detail.

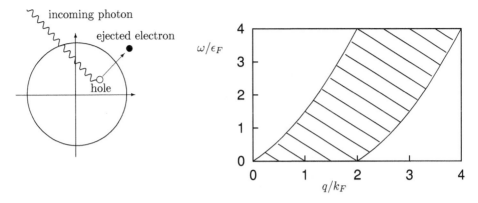

FIG. 8.3. Absorption of a photon creates an electron-hole pair excitation in the free electron gas (left panel). The possible range of q and ω is given by the dashed area (right panel). The strength of the interaction depends on the imaginary part of the polarization function (see Eq. (8.86)).

Within the non-interacting approximation and according to Eq. (8.79) the dissipation of the electron gas is proportional to Im χ_0^R, where

$$- \operatorname{Im} \chi_0^R(\mathbf{q},\omega) = \frac{\pi}{V} \sum_{\mathbf{k}\sigma} \left[n_F(\xi_{\mathbf{k}}) - n_F(\xi_{\mathbf{k}+\mathbf{q}}) \right] \delta(\xi_{\mathbf{k}} - \xi_{\mathbf{k}+\mathbf{q}} + \omega). \qquad (8.86)$$

We can now analyze for what $\mathbf{q}$ and ω excitations are possible, i.e., for which $(\mathbf{q},\omega)$ Eq. (8.86) is non-zero. Let us take $T = 0$ where n_F is either zero or one, which means that $n_F(\xi_{\mathbf{k}}) - n_F(\xi_{\mathbf{k}+\mathbf{q}})$ is only non-zero if $(k > k_F$ and $|\mathbf{k} + \mathbf{q}| < k_F)$ or $(k < k_F$ and $|\mathbf{k} + \mathbf{q}| > k_F)$. The first case corresponds to $\omega < 0$, while the latter corresponds to $\omega > 0$. However, because of the symmetry Im $\chi_0^R(\mathbf{q}, \omega) = -\text{Im } \chi_0^R(-\mathbf{q}, -\omega)$, which is easily seen from Eq. (8.85), we need only study one case, for example $\omega > 0$. The delta function together with the second condition thus imply

$$0 < \omega = q^2 \frac{1}{2m} + \mathbf{k} \cdot \mathbf{q} \frac{1}{m} \quad \Rightarrow \quad \begin{cases} \omega_{\max} = \frac{1}{2m}q^2 + v_F q \\ \omega_{\min} = \frac{1}{2m}q^2 - v_F q, \ q > 2k_F. \end{cases} \qquad (8.87)$$

The possible range of excitations in $(\mathbf{q},\omega)$-space is shown in Fig. 8.3. The excitations which give rise to the dissipation are electron-hole pair excitations, where an electron within the Fermi sea is excited to a state outside the Fermi sea. There is a continuum of such excitations given by conditions in Eq. (8.87).

While the electron-hole pair excitations are the only possible source of dissipation in the non-interacting electron gas, this is certainly not true for the interacting case which is more complicated. There is one particular type of excitation which is immensely important, namely the plasmon excitation. This we study in great detail later in this book.

The excitation of the electrons gas can be measured by, e.g., inelastic light scattering (Raman scattering), where the change of momentum and energy of an incoming photon is measured. The process discussed here where an electron within the Fermi sea is scattering to an empty state outside the Fermi sea, is illustrated in the left hand side of Fig. 8.3.

8.6 Summary and outlook

The concept of Green's functions in many-body physics has been introduced in this chapter, and we will use Green's functions in practically all discussions in the remaining part of the course. The Green's functions describe the dynamical properties of excitations. We have so far seen two examples of this: the density of states is related to the spectral function, and it can be measured for example in a tunneling experiment, and second, the absorption of electromagnetic radiation is given by the charge-charge retarded correlation function.

The physical picture to remember is that the Green's function $G\left(\mathbf{r}\sigma t, \mathbf{r}'\sigma't'\right)$ gives the amplitude for propagation from the space-time point $\mathbf{r}'t'$ to $\mathbf{r}t$, with initial spin σ' and final spin σ.

In this chapter, we have defined the following many-body Green's functions

$$G^R(\mathbf{r}\sigma t, \mathbf{r}'\sigma't') = -i\theta(t-t')\langle[\Psi_\sigma(\mathbf{r}t), \Psi_{\sigma'}^\dagger(\mathbf{r}'t')]_{B,F}\rangle, \text{ retarded Green's function,}$$

$$G^A(\mathbf{r}\sigma t, \mathbf{r}'\sigma't') = i\theta(t'-t)\langle[\Psi_\sigma(\mathbf{r}t), \Psi_{\sigma'}^\dagger(\mathbf{r}'t')]_{B,F}\rangle, \quad \text{advanced Green's function,}$$

$$G^>(\mathbf{r}\sigma t, \sigma'\mathbf{r}'t') = -i\langle\Psi_\sigma(\mathbf{r}t)\Psi_{\sigma'}^\dagger(\mathbf{r}'t')\rangle, \qquad\qquad \text{greater Green's function,}$$

$$G^<(\mathbf{r}\sigma t, \sigma'\mathbf{r}'t') = -i\left(\pm 1\right)\langle\Psi_{\sigma'}^\dagger(\mathbf{r}'t')\Psi_\sigma(\mathbf{r}t)\rangle, \qquad \text{lesser Green's function}$$

and their corresponding Fourier transforms. The important spectral function is in the frequency domain and in a diagonal basis given by

$$A(\nu, \omega) = -2\operatorname{Im} G^R(\nu, \omega), \quad \text{spectral function.}$$

The spectral function is related to the density of states. For non-interacting electrons the spectral function is given by a Dirac delta function

$$A_0(\nu, \omega) = 2\pi\delta(\xi_\nu - \omega), \quad \text{non-interacting case.}$$

We have also seen examples of how the Green's functions enter physical observables. One example was tunneling experiments where the tunneling density of states is given by the spectral function of the system. Also when calculating two-particle correlation functions, we saw that the Green's function enter as natural building blocks. This is precisely the idea of the Green's function, namely that, for example, in perturbation theory the Green's functions come in as building blocks from which the theory is built. This will be used extensively later in the book.

9

EQUATION OF MOTION THEORY

In the previous chapters, we saw how various physical observables can be expressed in terms of retarded Green's functions. In many cases we need to calculate the time dependence of these functions. There are several ways of attacking this problem, one of which is the equation of motion technique. The basic idea of this method is to generate a series of coupled differential equations by differentiating the function at hand a number of times. If these equations close, the problem is in principle solvable; and if not, one needs to invoke physical arguments to truncate the set of equations in a reasonable fashion. For example, one can neglect certain correlations. We shall study examples of both situations in this chapter.

9.1 The single-particle Green's function

Let us consider the retarded Green's function G^R (8.28) for either fermions or bosons,

$$G^R(\mathbf{r}t, \mathbf{r}'t') = -i\theta(t - t')\langle[\Psi(\mathbf{r}t), \Psi^\dagger(\mathbf{r}'t')]_{B,F}\rangle. \tag{9.1}$$

We find the equation of motion for G^R as the derivative with respect to the first time argument

$$
\begin{aligned}
i\partial_t G^R(\mathbf{r}t, \mathbf{r}'t') &= (-i)\left[i\partial_t\theta(t - t')\right]\langle[\Psi(\mathbf{r}t), \Psi^\dagger(\mathbf{r}'t')]_{B,F}\rangle \\
&\quad + (-i)\,\theta(t - t')\langle[i\partial_t\Psi(\mathbf{r}t), \Psi^\dagger(\mathbf{r}'t')]_{B,F}\rangle, \\
&= \delta(t - t')\delta(\mathbf{r} - \mathbf{r}') + \\
&\quad + (-i)\,\theta(t - t')\langle[i\partial_t\Psi(\mathbf{r}t), \Psi^\dagger(\mathbf{r}'t')]_{B,F}\rangle. \tag{9.2}
\end{aligned}
$$

Here, we used that the derivative of a step function is a delta function and the commutation relations for field operators at equal times $\left[\Psi(\mathbf{r}), \Psi^\dagger(\mathbf{r}')\right]_{B,F} = \delta(\mathbf{r} - \mathbf{r}')$. Next, let us study the time-derivative of the annihilation operator (throughout this chapter we assume that H is time independent),

$$i\partial_t\Psi(\mathbf{r}t) = -\left[H, \Psi(\mathbf{r})\right](t) = -[H_0, \Psi(\mathbf{r})](t) - [V_{\text{int}}, \Psi(\mathbf{r})](t), \tag{9.3}$$

where the interaction part of the Hamiltonian includes all the interactions in the given problem, while H_0 describes the quadratic part of the Hamiltonian, for example the kinetic energy. If H_0 is the usual kinetic energy Hamiltonian of free particles, we have

$$
\begin{aligned}
-[H_0, \Psi(\mathbf{r})] &= \frac{1}{2m}\int d\mathbf{r}'\left[\Psi^\dagger(\mathbf{r}')\nabla^2_{\mathbf{r}'}\Psi(\mathbf{r}'), \Psi(\mathbf{r})\right] \\
&= -\frac{1}{2m}\nabla^2_{\mathbf{r}}\Psi(\mathbf{r}). \tag{9.4}
\end{aligned}
$$

In this case, the equation of motion becomes

$$\left(i\partial_t + \frac{1}{2m}\nabla_{\mathbf{r}}^2\right)G^R(\mathbf{r}t,\mathbf{r}'t') = \delta(t-t')\delta(\mathbf{r}-\mathbf{r}') + D^R(\mathbf{r}t,\mathbf{r}'t'), \tag{9.5a}$$

$$D^R(\mathbf{r}t,\mathbf{r}'t') = -i\theta(t-t')\left\langle\left[-[V_{\text{int}},\Psi(\mathbf{r})](t),\Psi^\dagger(\mathbf{r}'t')\right]_{B,F}\right\rangle. \tag{9.5b}$$

The function D^R thus equals the corrections to the free-particle Green's function. After evaluating $[V_{\text{int}},\Psi(\mathbf{r})]$ we can, as in Section 5.5, continue the generation of differential equations. It is now evident why the many-body functions G^R are called Green's functions. The equation in (9.5a) has the structure of the classical Green's function, that we saw in Section 8.1, where the Green's function of a differential operator, L, was defined as "LG equals a delta function".

Often, it is convenient to work in some other basis, say $\{\nu\}$. The Hamiltonian is again written as $H = H_0 + V_{\text{int}}$, where the quadratic part of the Hamiltonian is

$$H_0 = \sum_{\nu\nu'} t_{\nu'\nu}a_{\nu'}^\dagger a_\nu. \tag{9.6}$$

Of course, the Hamiltonian H_0 can include both the kinetic energy contribution and the single-particle potential energy, which are both quadratic in creation and annihilation operators. The differential equation for the Green's function in this basis,

$$G^R(\nu t,\nu't') = -i\theta(t-t')\langle[a_\nu(t),a_{\nu'}^\dagger(t')]_{B,F}\rangle, \tag{9.7}$$

is found in exactly the same way as above. By differentiation, the commutator with H_0 is generated,

$$-[H_0,a_\nu] = \sum_{\nu''} t_{\nu\nu''}a_{\nu''}, \tag{9.8}$$

and hence

$$\sum_{\nu''}(i\delta_{\nu\nu''}\partial_t - t_{\nu\nu''})\,G^R(\nu''t,\nu't') = \delta(t-t')\delta_{\nu\nu'} + D^R(\nu t,\nu't'), \tag{9.9a}$$

$$D^R(\nu t,\nu't') = -i\theta(t-t')\left\langle\left[-[V_{\text{int}},a_\nu](t),a_{\nu'}^\dagger(t')\right]_{B,F}\right\rangle. \tag{9.9b}$$

In this book, we will mainly deal with problems where the Hamiltonian does not depend explicitly on time (linear response was an exception, but even there the time-dependent problem was transformed into a retarded correlation function of a time-independent problem). Therefore, the Green's function can only depend on the time difference $t-t'$, and in this case it is always useful to work with the Fourier transforms. Recall, that when performing the Fourier transformation, the derivative ∂_t becomes $-i\omega$, and the delta function becomes unity, $\delta(t) \to 1$, so we can write the equation of motion in frequency domain as,

$$\sum_{\nu''} [\delta_{\nu\nu''}(\omega + i\eta) - t_{\nu\nu''}] G^R(\nu''\nu';\omega) = \delta_{\nu\nu'} + D^R(\nu,\nu';\omega), \qquad (9.10\text{a})$$

$$D^R(\nu,\nu';\omega) = -i \int_{-\infty}^{\infty} dt e^{i(\omega+i\eta)(t-t')} \theta(t-t') \left\langle \left[-[V_{\text{int}}, a_\nu](t), a_{\nu'}^\dagger(t') \right]_{B,F} \right\rangle.$$
$$(9.10\text{b})$$

Here it is important to remember that the frequency of the retarded functions must carry a small positive imaginary part, η, to ensure proper convergence (see Section 5.8).

9.1.1 Non-interacting particles

For non-interacting particles, which means that the Hamiltonian is bilinear in annihilation or creation operators, we can in fact solve for the Green's function.[32] In this case we have

$$\sum_{\nu''} (\delta_{\nu\nu''}(\omega + i\eta) - t_{\nu\nu''}) G_0^R(\nu''\nu';\omega) = \delta_{\nu\nu'}, \qquad (9.11)$$

where the subindex 0 on G_0^R indicates that it is the Green's function corresponding to a non-interacting Hamiltonian. As in Section 8.1 we define the inverse Green's function as

$$\left(G_0^R\right)^{-1}(\nu\nu';\omega) = \delta_{\nu\nu'}(\omega + i\eta) - t_{\nu\nu'} \equiv \left(\mathbf{G}_0^R\right)^{-1}_{\nu\nu'}, \qquad (9.12)$$

and in matrix notation, Eq. (9.11) becomes

$$\left(\mathbf{G}_0^R\right)^{-1}\mathbf{G}_0^R = \mathbf{1}. \qquad (9.13)$$

Therefore, in order to find the Green's function, all we need to do is to invert the matrix $\left(\mathbf{G}_0^R\right)^{-1}_{\nu\nu'}$. For a diagonal basis in which $t_{\nu\nu'} = \varepsilon_\nu \, \delta_{\nu\nu'}$, the solution is

$$\left(\mathbf{G}_0^R\right)_{\nu\nu'} = G_0^R(\nu,\omega) \, \delta_{\nu\nu'} = \frac{1}{\omega - \varepsilon_\nu + i\eta} \, \delta_{\nu\nu'}, \qquad (9.14)$$

which, of course, agrees with the result found in Eq. (8.55).

9.2 Single level coupled to a continuum

To illustrate how the equation of motion theory can be used to determine the Green's function, we now study a model, which we will use several times in the book, namely

$$H = H_0 + H_{\text{hyb}} + H_l, \qquad (9.15)$$

where H_0 describes a non-interacting electron gas,

$$H_0 = \sum_\nu \xi_\nu c_\nu^\dagger c_\nu, \qquad (9.16)$$

---

[32] Here we only consider terms of the form $c^\dagger c$ but also anomalous terms like cc could be included. Such a term is indeed relevant for superconductors. We return to this in Chapter 18.

H_l is the Hamiltonian of a single localized level,

$$H_l = \xi_0 c_l^\dagger c_l, \tag{9.17}$$

and H_{hyb} describes the hybridization or tunneling between the localized level and the continuum of states,

$$H_{\text{hyb}} = \sum_\nu (t_\nu^* c_\nu^\dagger c_l + t_\nu c_l^\dagger c_\nu). \tag{9.18}$$

We are interested in finding the diagonal Green's function for the local state. To do that, we need the following two retarded Green's functions,

$$G^R(l, l, t - t') = -i\theta(t - t')\langle \{c_l(t), c_l^\dagger(t')\} \rangle, \tag{9.19a}$$

$$G^R(\nu, l, t - t') = -i\theta(t - t')\langle \{c_\nu(t), c_l^\dagger(t')\} \rangle. \tag{9.19b}$$

appropriate for fermions.

The equations of motion are now found by letting ν and ν'' in Eq. (9.11) run over both the continuum states ν and the localized state l, and we obtain the coupled equations

$$(\omega + i\eta - \xi_0)\, G^R(l, l, \omega) - \sum_\nu t_\nu G^R(\nu, l, \omega) = 1, \tag{9.20a}$$

$$(\omega + i\eta - \xi_\nu)\, G^R(\nu, l, \omega) - t_\nu^* G^R(l, l, \omega) = 0. \tag{9.20b}$$

It is now a simple matter to solve the last equation and insert it into the first. We obtain

$$G^R(l, l, \omega) = \frac{1}{\omega - \xi_0 - \Sigma^R(\omega)}, \tag{9.21}$$

where

$$\Sigma^R(\omega) = \sum_\nu \frac{|t_\nu|^2}{\omega - \xi_\nu + i\eta}. \tag{9.22}$$

The function $\Sigma^R(\omega)$ is our first encounter with the concept known as "self-energy." The self-energy changes the pole of $G^R(l, l)$ and furthermore gives some broadening to the spectral function. In the time domain, the imaginary part translates into a lifetime. It arises because the coupling to the continuum states introduces off-diagonal terms in the Hamiltonian, so that it is no longer diagonal in the l-operator. The true diagonal modes are instead superpositions of ν- and l-states.

9.3   Anderson's model for magnetic impurities

In order to exemplify the usefulness of the equation of motion technique, we proceed by applying it to a famous model for the appearance of a magnetic moment of impurities of certain magnetic ions embedded in a non-magnetic host metal. The host metal,

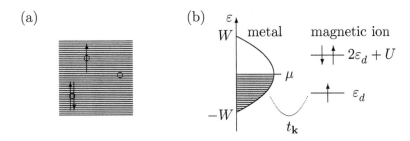

FIG. 9.1. The Anderson model describing magnetic impurities embedded in a homogeneous host metal. (a) The electrons in the conduction band of the non-magnetic host metal (horizontal lines) couple to the magnetic impurity ions (circles), each having zero, one, or two valence electrons (vertical arrows). (b) The bare onsite energy of the state on the magnetic ion is ϵ_d. But the energy of electrons residing on the impurity ion also depends on whether it is doubly occupied or not, therefore the state with two electrons residing on the ion has energy $2\epsilon_d + U$.

e.g., Nb or Mo, has a conduction band, which can be described by an effective non-interacting model

$$H_c = \sum_{\mathbf{k}\sigma} (\varepsilon_{\mathbf{k}} - \mu) \, c_{\mathbf{k}\sigma}^\dagger c_{\mathbf{k}\sigma}. \tag{9.23}$$

For the impurity ion we assume that it has only one spin-degenerate state in the active shell, which is typically the d shell. In addition to the bare energy cost for an electron to reside in the d-state, there is an interaction energy that depends on the state being doubly occupied or not. The Hamiltonian $H_d + H_U$ for the impurity ion d is thus modelled as

$$H_d + H_U = \sum_{\sigma} (\varepsilon_d - \mu) \, c_{d\sigma}^\dagger c_{d\sigma} + U n_{d\uparrow} n_{d\downarrow}, \tag{9.24}$$

where $n_{d\sigma} = c_{d\sigma}^\dagger c_{d\sigma}$ is the number operator for d-electrons on the ion. The crucial input here is the correlation between electrons on the impurity ion, because the interaction in the narrower d-shell of a magnetic ion is particular strong, and this is in fact the reason for the magnetism. The states forming the conduction band are primarily s-states that are more extended in space, and hence interactions are less important for those.

The electrons occupying the conduction band couple to the outermost electrons of the magnetic impurity ions, e.g., the d-shell of an Fe ion. The coupling occurs because the d-orbital and the conduction band states overlap spatially and also lie close in energy, giving rise to a "hybridization" between the two. The overlapping orbitals lead to a non-diagonal matrix element of the Hamiltonian,

$$H_{\text{hyb}} = \sum_{\mathbf{k}\sigma} t_{\mathbf{k}} c_{d\sigma}^\dagger c_{\mathbf{k}\sigma} + \sum_{\mathbf{k}\sigma} t_{\mathbf{k}}^* c_{\mathbf{k}\sigma}^\dagger c_{d\sigma}. \tag{9.25}$$

The bare d-electron energy ε_d is below the chemical potential, and from the kinetic energy point of view it is favorable to fill the orbital by two electrons. However, this costs the potential energy U, and it is not possible if $2\varepsilon_d + U > 2\mu$. Furthermore, the system gains further kinetic energy by the hybridization, which on the other hand is complicated by the fact that the hopping in and out of the impurity orbital with, say, spin-up electrons, depends on the occupation of spin-down electrons. The hybridization therefore seems to randomize the spin on the magnetic ion.

The Hamiltonian H formed by the sum of these three energy contributions,

$$H = H_c + H_d + H_U + H_{\text{hyb}}, \tag{9.26}$$

is known as the Anderson model (see Fig. 9.1 for an illustration). Although the Anderson model looks simple, its full solution is very complicated, and in fact the model has a very rich phase diagram. The Anderson model has been used to describe numerous effects in the physics of strongly correlated electron systems.

It turns out that, for certain values of the parameters, it is energetically favorable for the system to have a magnetic moment (and thus minimizing the on-site interaction energy) while for other values there is no magnetic moment (thus gaining maximum hybridization energy). The physical question we try to answer here is: under which circumstances is the material magnetic?

9.3.1 The equation of motion for the Anderson model

The magnetization in the z-direction is given by the expectation value of the difference $n_\uparrow - n_\downarrow$ between spin-up and spin-down occupancy. The occupation of a quantum state was found in Eq. (8.62) in terms of the spectral function. For the d-electron occupation we therefore have

$$\langle n_{d\sigma} \rangle = \int \frac{d\omega}{2\pi} A(d\sigma, \omega) n_{\text{F}}(\omega), \tag{9.27}$$

where $A(d\sigma, \omega)$ is the spectral function, which follows from the retarded Green's function, G^R (see Eq. (8.50)). All we need to find is then

$$G^R(d\sigma; t - t') = -i\theta(t - t') \left\langle \left\{ c_{d\sigma}(t), c_{d\sigma}^\dagger(t') \right\} \right\rangle. \tag{9.28}$$

Let us write the equation of motion of this function using Eq. (9.10). Due to the hybridization term, the Hamiltonian is not diagonal in the d-operators, and the equations of motion will involve another Green's function, namely

$$G^R(\mathbf{k}\sigma, d\sigma, t - t') = -i\theta(t - t') \left\langle \left\{ c_{\mathbf{k}\sigma}(t), c_{d\sigma}^\dagger(t') \right\} \right\rangle. \tag{9.29}$$

The equations of motion are thus found by letting ν'' in Eq. (9.10) run over both d and k and we obtain the coupled equations

$$\left(\omega + i\eta - \varepsilon_d + \mu\right) G^R(d\sigma, \omega) - \sum_{\mathbf{k}} t_{\mathbf{k}} G^R(\mathbf{k}\sigma, d\sigma, \omega) = 1 + U D^R(d\sigma, \omega), \tag{9.30}$$

$$\left(\omega + i\eta - \varepsilon_{\mathbf{k}} + \mu\right) G^R(\mathbf{k}\sigma, d\sigma, \omega) - t_{\mathbf{k}}^* G^R(d\sigma, \omega) = 0, \tag{9.31}$$

where

$$D^R(d\sigma, t - t') = -i\theta(t - t') \left\langle \left\{ -[n_{d\uparrow} n_{d\downarrow}, c_{d\sigma}](t), c_{d\sigma}^\dagger(t') \right\} \right\rangle. \tag{9.32}$$

For $\sigma = \uparrow$, the commutator in this expression is

$$[n_{d\uparrow} n_{d\downarrow}, c_{d\uparrow}] = n_{d\downarrow}[n_{d\uparrow}, c_{d\uparrow}] = -n_{d\downarrow} c_{d\uparrow}, \tag{9.33}$$

and likewise, we find the commutator for spin-down by interchanging up and down. We thus face the following more complicated Green's function,

$$D^R(d\uparrow, t - t') = -i\theta(t - t')\langle\{n_{d\downarrow}(t)\, c_{d\uparrow}(t), c_{d\uparrow}^\dagger(t')\}\rangle. \tag{9.34}$$

9.3.2   Mean-field approximation for the Anderson model

Differentiating the function in Eq. (9.34) with respect to time would generate yet another function to be determined, and the set of equations does not close. This is investigated further in Chapter 10, where the same model is used to describe transport in interacting mesoscopic systems. Here, however, we use the model to describe a very different physical situation, where many (weakly interacting) magnetic impurities reside in a host metal. For this situation, a mean-field approximation grasps the important physics that the spin-up electron population depends on the spin-down population, and therefore we replace the interaction part H_U by its mean-field version

$$H_U^{MF} = U\langle n_{d\uparrow}\rangle\, n_{d\downarrow} + U\langle n_{d\downarrow}\rangle\, n_{d\uparrow} - U\langle n_{d\uparrow}\rangle\langle n_{d\downarrow}\rangle. \tag{9.35}$$

When writing the mean-field decoupling in this way, it is important to realize that we have implicitly assumed that the system might spontaneously break the symmetry, if we allow for solutions with $\langle n_{d\uparrow}\rangle \neq \langle n_{d\downarrow}\rangle$. Naturally, this cannot occur for a single impurity, because the Hamiltonian is symmetric with respect to interchanging up and down spin. Therefore, it takes a macroscopic sample to break the symmetry, as discussed in Section 4.4, which is the situation we have in mind here.

With this approximation, the function D^R becomes

$$D^R(d\uparrow, t - t') = -i\theta(t - t')\langle n_{d\downarrow}\rangle\Big\langle\{c_{d\uparrow}(t), c_{d\uparrow}^\dagger(t')\}\Big\rangle$$
$$= \langle n_{d\downarrow}\rangle\, G^R(d\uparrow, t - t'). \tag{9.36}$$

In other words, since the mean-field approximation makes the Hamiltonian quadratic we can include $U\langle n_{d\downarrow}(t)\rangle$ to the energy of the spin-up d-electrons in our equation of motion. Inserting (9.36) in Eq. (9.30) and solving Eq. (9.31) for $G^R(d\uparrow, \omega)$ gives

$$\left(\omega + i\eta - \varepsilon_d + \mu - U\langle n_{d\downarrow}\rangle\right)G^R(d\uparrow, \omega) - \sum_{\mathbf{k}}\frac{|t_{\mathbf{k}}|^2}{\omega - \varepsilon_{\mathbf{k}} + \mu + i\eta}\,G^R(d\uparrow, \omega) = 1, \tag{9.37}$$

and likewise for the spin-down Green's function. The final answer is

$$G^R(d\uparrow, \omega) = \frac{1}{\omega - \varepsilon_d + \mu - U\langle n_{d\downarrow}\rangle - \Sigma^R(\omega)}, \tag{9.38a}$$

$$\Sigma^R(\omega) = \sum_{\mathbf{k}}\frac{|t_{\mathbf{k}}|^2}{\omega - \varepsilon_{\mathbf{k}} + \mu + i\eta}. \tag{9.38b}$$

The function $\Sigma^R(\omega)$ is recognized as the self-energy in Eq. (9.22). As mentioned, the self-energy changes the pole and broadens G^R. The "bare" d-electron energy ε_d is thus seen to be renormalized by two effects: first, the energy is shifted by $U \langle n_{d\downarrow} \rangle$ due to the interaction with the averaged density of electrons having opposite spin, and second, the coupling to the conduction band electrons gives through $\Sigma(\omega)$ an energy shift and most importantly an imaginary part.

Assuming that the coupling $t_{\mathbf{k}}$ only depends on the length of $\mathbf{k}$ and thus on $\varepsilon_{\mathbf{k}}$, the self-energy Σ is

$$
\begin{aligned}
\Sigma^R(\omega) &= \int d\varepsilon \, d(\varepsilon) \frac{|t(\varepsilon)|^2}{\omega - \varepsilon + \mu + i\eta} \\
&= \mathcal{P} \int d\varepsilon \, d(\varepsilon) \frac{|t(\varepsilon)|^2}{\omega - \varepsilon + \mu} - i\pi d(\omega + \mu) \left| t(\omega + \mu) \right|^2.
\end{aligned}
\tag{9.39}
$$

The density of states $d(\varepsilon)$ and the coupling matrix element $t(\varepsilon)$ depend on the details of the material, but fortunately it is not important for the present considerations. Let us assume that the product $d(\varepsilon) |t(\varepsilon)|^2$ is constant within the band limits, given by $-D < \varepsilon < D$, such that we can write

$$
2\pi d(\varepsilon) |t(\varepsilon)|^2 = \begin{cases} \Gamma, & \text{for } -D < \varepsilon < D, \\ 0, & \text{for } D < |\varepsilon|. \end{cases}
\tag{9.40}
$$

This approximation is good, if the width of the Green's function (which we shall see shortly is given by Γ) turns out to be small compared to the scale on which $d(\varepsilon) |t(\varepsilon)|^2$ typically changes. Since in practice $\Gamma \ll \varepsilon_F$, the approximation is indeed valid. For $\omega + \mu \in [-D, D]$ we obtain

$$
\begin{aligned}
\Sigma^R(\omega) &\approx \frac{\Gamma}{2\pi} \int_{-D}^{D} \frac{d\varepsilon}{\omega - \varepsilon + \mu} - i\Gamma/2 \\
&= \frac{\Gamma}{2\pi} \ln \left| \frac{D + \omega + \mu}{D - \omega - \mu} \right| - i\Gamma/2.
\end{aligned}
\tag{9.41}
$$

The real part gives a shift of energy, and since it is a slowly varying function, we simply include it as a shift of ε_d and define the new onsite energy $\tilde{\varepsilon} = \varepsilon_d + \operatorname{Re} \Sigma^R$.

The spectral function hence becomes

$$
\begin{aligned}
A(d\uparrow, \omega) &= -2 \operatorname{Im} G^R(d\uparrow, \omega) \\
&= \frac{\Gamma}{\left(\omega - \tilde{\varepsilon} + \mu - U \langle n_{d\downarrow} \rangle \right)^2 + (\Gamma/2)^2},
\end{aligned}
\tag{9.42}
$$

and Γ is thus the width of the spectral function. Note that the spectral function derived here is an example of the Lorentzian form discussed in Section 8.3.5.

Now, the self-consistent mean-field equation for $\langle n_{d\uparrow} \rangle$ follows as

$$
\begin{aligned}
\langle n_{d\uparrow} \rangle &= \int \frac{d\omega}{2\pi} n_F(\omega) A(d\uparrow, \omega) \\
&= \int \frac{d\omega}{2\pi} n_F(\omega) \frac{\Gamma}{\left(\omega - \tilde{\varepsilon} + \mu - U \langle n_{d\downarrow} \rangle \right)^2 + (\Gamma/2)^2}.
\end{aligned}
\tag{9.43}
$$

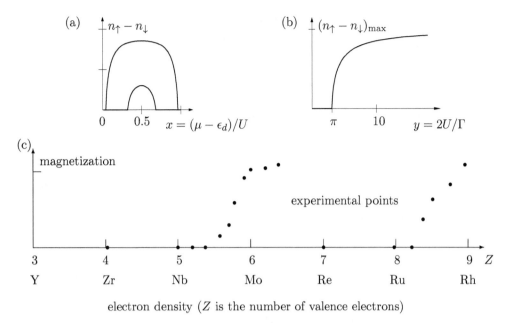

FIG. 9.2. (a) The mean-field solution of the Anderson model: the magnetization $n_\uparrow - n_\downarrow$ as a function of electron density n_{el}, i.e., the chemical potential μ, for two different values of Γ. (b) The maximum magnetization as function of the correlation energy. There is a critical density and a critical U/Γ above which the magnetization sets in. The latter means that too strong hybridization destroys the magnetization. (c) Experimental results (Clogston et $al.$ 1962) for the magnetic moment of Fe-atoms embedded in transition metals. The electron density and hence μ is varied by changing Z using different alloys. For $4 < Z < 8$ the magnetization curve is seen to be quite similar to the prediction of the model. For $Z > 8$ the effect of having more than two d-orbitals in the Fe atoms becomes important and the simple model is no longer adequate.

If we neglect the finite bandwidth, which is justified because $\Gamma \ll D$, and if we furthermore consider low temperatures, $T = 0$, we obtain

$$\langle n_{d\uparrow} \rangle \approx \int_{-\infty}^{0} \frac{d\omega}{2\pi} \frac{\Gamma}{(\omega - \tilde{\varepsilon} + \mu - U \langle n_{d\downarrow} \rangle)^2 + (\Gamma/2)^2}$$

$$= \frac{1}{2} - \frac{1}{\pi} \tan^{-1} \left(\frac{\tilde{\varepsilon} - \mu + U \langle n_{d\downarrow} \rangle}{\Gamma/2} \right). \tag{9.44}$$

Doing the same for the spin-down occupation, we finally arrive at the following two coupled equations

$$\cot(\pi n_\uparrow) = (n_\downarrow - x)\, y, \quad x = -(\tilde{\varepsilon} - \mu)/U, \tag{9.45a}$$
$$\cot(\pi n_\downarrow) = (n_\uparrow - x)\, y, \quad y = 2U/\Gamma. \tag{9.45b}$$

The solution of these equations gives the occupation of the d-orbital and in particular tells us whether there is a finite magnetization, i.e., whether there exists a solution $n_\downarrow \neq n_\uparrow$, different from the trivial solution $n_\downarrow = n_\uparrow$.[33] In Fig. 9.2 solutions of these equations are shown together with experimental data. As is evident there, the model describes the observed behavior, at least qualitatively.

9.4   The two-particle correlation function

The two-particle retarded correlation functions, such as the density-density correlation, was in Chapter 6 shown to give the linear response properties. Also for this quantity, one can generate a set of equation of motions, and as for the single-particle Green's function they are not solvable in general. But even so, they may provide a good starting point for various approximation schemes.

Consider for example the retarded charge-charge correlation function

$$\chi^R(\mathbf{r}t, \mathbf{r}'t') = -i\theta(t - t')\Big\langle \big[\rho(\mathbf{r}t), \rho(\mathbf{r}'t')\big]\Big\rangle. \tag{9.46}$$

In Chapter 6 it was shown that this function is related to the dielectric response function and therefore tells about the screening properties of the material.

9.4.1   *The random phase approximation*

A commonly used approximation scheme for correlation functions is the so-called random phase approximation (RPA). For the case of the electron gas, which is one of our main topics in this book, RPA is exact in some limits, but also in general gives a decent description of the interacting electron gas. In Chapter 13, RPA is derived using Feynman diagrams, but here we derive it using the equation of motion technique. The two derivations give complementary insight into the physical content of the approximation.

We will for simplicity work with the translation-invariant electron gas and with the Hamiltonian given by the usual kinetic energy plus interaction energy (here, we disregard for the moment the spin degree of freedom because it is not important for the discussion),

$$H = \sum_{\mathbf{k}} \xi_{\mathbf{k}} c_{\mathbf{k}}^\dagger c_{\mathbf{k}} + \frac{1}{2\mathcal{V}} \sum_{\mathbf{k}\mathbf{k}'\mathbf{q}\neq 0} V(q) c_{\mathbf{k}+\mathbf{q}}^\dagger c_{\mathbf{k}'-\mathbf{q}}^\dagger c_{\mathbf{k}'} c_{\mathbf{k}} = H_0 + V_{\text{int}}. \tag{9.47}$$

Furthermore, the $\mathbf{q} = \mathbf{0}$ component is canceled by the positively charged background (see Section 2.2). The charge-charge correlation function is

$$\chi^R(\mathbf{q}, t - t') = -i\theta(t - t')\frac{1}{\mathcal{V}}\Big\langle \big[\rho(\mathbf{q},t), \rho(-\mathbf{q},t')\big]\Big\rangle, \quad \rho(\mathbf{q}) = \sum_{\mathbf{k}} c_{\mathbf{k}}^\dagger c_{\mathbf{k}+\mathbf{q}}. \tag{9.48}$$

However, it turns out to be better to work with the function

$$\chi^R(\mathbf{k}\mathbf{q}, t - t') = -i\theta(t - t')\Big\langle \big[(c_{\mathbf{k}}^\dagger c_{\mathbf{k}+\mathbf{q}})(t), \rho(-\mathbf{q},t')\big]\Big\rangle, \tag{9.49}$$

---

[33]It must be checked that the magnetic solution has lower energy, which in fact it does have.

from which we easily obtain $\chi(\mathbf{q})$ by summing over $\mathbf{k}$ as $\chi^R(\mathbf{q}) = \frac{1}{\mathcal{V}} \sum_{\mathbf{k}} \chi^R(\mathbf{kq})$. Let us find the equation of motion for χ^R,

$$i\partial_t \chi^R(\mathbf{kq}, t - t') = \delta(t - t')\langle[(c_\mathbf{k}^\dagger c_{\mathbf{k}+\mathbf{q}})(t), \rho(-\mathbf{q},t')]\rangle$$
$$- i\theta(t - t')\langle[-[H, c_\mathbf{k}^\dagger c_{\mathbf{k}+\mathbf{q}}](t), \rho(-\mathbf{q}, t')]\rangle. \qquad (9.50)$$

For this purpose we need the following commutators

$$\left[c_\mathbf{k}^\dagger c_{\mathbf{k}+\mathbf{q}}, \rho(-\mathbf{q})\right] = \sum_{\mathbf{k}'} \left[c_\mathbf{k}^\dagger c_{\mathbf{k}+\mathbf{q}}, c_{\mathbf{k}'}^\dagger c_{\mathbf{k}'-\mathbf{q}}\right] = c_\mathbf{k}^\dagger c_\mathbf{k} - c_{\mathbf{k}+\mathbf{q}}^\dagger c_{\mathbf{k}+\mathbf{q}}, \qquad (9.51)$$

$$[H_0, c_\mathbf{k}^\dagger c_{\mathbf{k}+\mathbf{q}}] = (\xi_\mathbf{k} - \xi_{\mathbf{k}+\mathbf{q}}) c_\mathbf{k}^\dagger c_{\mathbf{k}+\mathbf{q}}, \qquad (9.52)$$

$$[V_{\text{int}}, c_\mathbf{k}^\dagger c_{\mathbf{k}+\mathbf{q}}] = \frac{1}{2\mathcal{V}} \sum_{\mathbf{k}',\mathbf{q}'\neq 0} V(q') \left\{ c_{\mathbf{k}+\mathbf{q}'}^\dagger c_{\mathbf{k}'-\mathbf{q}'}^\dagger c_{\mathbf{k}'} c_{\mathbf{k}+\mathbf{q}} + c_{\mathbf{k}'+\mathbf{q}'}^\dagger c_{\mathbf{k}-\mathbf{q}'}^\dagger c_{\mathbf{k}+\mathbf{q}} c_{\mathbf{k}'} \right.$$
$$\left. - c_{\mathbf{k}'+\mathbf{q}'}^\dagger c_\mathbf{k}^\dagger c_{\mathbf{k}+\mathbf{q}+\mathbf{q}'} c_{\mathbf{k}'} - c_\mathbf{k}^\dagger c_{\mathbf{k}'-\mathbf{q}'}^\dagger c_{\mathbf{k}'} c_{\mathbf{k}+\mathbf{q}-\mathbf{q}'} \right\}. \qquad (9.53)$$

When this is inserted into Eq. (9.50) a new six-particle Green's function is generated. Likewise for each level of the equation of motion, a new Green's function with two additional electron operators pops up. Here, at this stage we truncate the series by the random phase approximation, which says that the right-hand side of (9.53) is replaced by a mean-field expression where pairs of operators are replaced by their average values. Using the recipe from Chapter 4, we get

$$[V_{\text{int}}, c_\mathbf{k}^\dagger c_{\mathbf{k}+\mathbf{q}}] \approx \frac{1}{2\mathcal{V}} \sum_{\mathbf{k}'\mathbf{q}'\neq 0} V(q') \left\{ c_{\mathbf{k}+\mathbf{q}'}^\dagger c_{\mathbf{k}+\mathbf{q}} \left\langle c_{\mathbf{k}'-\mathbf{q}'}^\dagger c_{\mathbf{k}'} \right\rangle + \left\langle c_{\mathbf{k}+\mathbf{q}'}^\dagger c_{\mathbf{k}+\mathbf{q}} \right\rangle c_{\mathbf{k}'-\mathbf{q}'}^\dagger c_{\mathbf{k}'} \right.$$
$$+ \left\langle c_{\mathbf{k}-\mathbf{q}'}^\dagger c_{\mathbf{k}+\mathbf{q}} \right\rangle c_{\mathbf{k}'+\mathbf{q}'}^\dagger c_{\mathbf{k}'} + c_{\mathbf{k}-\mathbf{q}'}^\dagger c_{\mathbf{k}+\mathbf{q}} \left\langle c_{\mathbf{k}'+\mathbf{q}'}^\dagger c_{\mathbf{k}'} \right\rangle$$
$$- c_{\mathbf{k}'+\mathbf{q}'}^\dagger c_{\mathbf{k}'} \left\langle c_\mathbf{k}^\dagger c_{\mathbf{k}+\mathbf{q}+\mathbf{q}'} \right\rangle - \left\langle c_{\mathbf{k}'+\mathbf{q}'}^\dagger c_{\mathbf{k}'} \right\rangle c_\mathbf{k}^\dagger c_{\mathbf{k}+\mathbf{q}+\mathbf{q}'}$$
$$\left. - c_\mathbf{k}^\dagger c_{\mathbf{k}+\mathbf{q}-\mathbf{q}'} \left\langle c_{\mathbf{k}'-\mathbf{q}'}^\dagger c_{\mathbf{k}'} \right\rangle - \left\langle c_\mathbf{k}^\dagger c_{\mathbf{k}+\mathbf{q}-\mathbf{q}'} \right\rangle c_{\mathbf{k}'-\mathbf{q}'}^\dagger c_{\mathbf{k}'} \right\}$$
$$= \frac{V(q)}{\mathcal{V}} \left(\langle n_{\mathbf{k}+\mathbf{q}}\rangle - \langle n_\mathbf{k}\rangle\right) \sum_{\mathbf{k}'} c_{\mathbf{k}'-\mathbf{q}}^\dagger c_{\mathbf{k}'}, \qquad (9.54)$$

where we used that $\langle c_\mathbf{k}^\dagger c_{\mathbf{k}'}\rangle = \langle n_\mathbf{k}\rangle \delta_{\mathbf{k},\mathbf{k}'}$. Note that the exchange pairings, which we included in the Hartree–Fock approximation, is not included here.

Collecting everything and going to the frequency domain, the equation of motion becomes,

$$(\omega + i\eta + \xi_\mathbf{k} - \xi_{\mathbf{k}+\mathbf{q}}) \chi^R(\mathbf{kq}, \omega) \approx -\left(\langle n_{\mathbf{k}+\mathbf{q}}\rangle - \langle n_\mathbf{k}\rangle\right) \left(1 + \frac{V(q)}{\mathcal{V}} \sum_{\mathbf{k}'} \chi^R(\mathbf{k}'\mathbf{q}, \omega)\right),$$
$$(9.55)$$

which, when summed over $\mathbf{k}$, allows us to find an approximate equation for $\chi^R(\mathbf{q}, \omega)$

$$\chi^R(\mathbf{q}, \omega) = \frac{1}{\mathcal{V}} \sum_\mathbf{k} \chi^R(\mathbf{kq}, \omega) \approx \frac{1}{\mathcal{V}} \sum_\mathbf{k} \frac{\langle n_\mathbf{k}\rangle - \langle n_{\mathbf{k}+\mathbf{q}}\rangle}{\omega + \xi_\mathbf{k} - \xi_{\mathbf{k}+\mathbf{q}} + i\eta} [1 + V(q)\chi^R(\mathbf{q}, \omega)], \quad (9.56)$$

and hence

$$\chi^R(\mathbf{q},\omega) \approx \frac{\chi_0^R(\mathbf{q},\omega)}{1 - V(q)\chi_0^R(\mathbf{q},\omega)} \equiv \chi^{R,\mathrm{RPA}}(\mathbf{q},\omega). \tag{9.57}$$

This is the RPA result for the polarizability function. The free particle polarizability $\chi_0^R(\mathbf{q},\omega)$ was derived in Section 8.5. According to Eq. (6.46), the RPA dielectric function becomes

$$\varepsilon^{\mathrm{RPA}}(\mathbf{q},\omega) = \left[1 + V(q)\chi^R(\mathbf{q},\omega)\right]^{-1} = 1 - V(q)\,\chi_0^R(\mathbf{q},\omega). \tag{9.58}$$

Replacing the expectation values $\langle n_\mathbf{k} \rangle$ and $\langle n_{\mathbf{k+q}} \rangle$ by the Fermi–Dirac distribution function, we recognize the Lindhard function studied in Section 8.5. There we studied a non-interacting electron gas and found that $\chi^R(\mathbf{q},\omega)$ indeed was equal to the numerator in (9.57), and the two results therefore agree nicely.

Also in Section 8.5 did we analyze the excitation of the non-interacting electrons gas, and that analysis is basically still correct. The excitations, shown in Section 8.5 to be related to the imaginary part of $\chi^R(\mathbf{q},\omega)$ and thus related to the structure of the electron-hole excitations of the non-interacting gas (depicted in Fig. 8.3), is preserved here, but of course the strength is now modified by the real part of the denominator of (9.57).

However, the interactions add other fundamental excitations, namely collective modes, and in the case of a charge liquid these modes are the plasmon modes. These additional collective modes are given by the part where the imaginary part of $\chi_0^R(\mathbf{q},\omega)$ is zero because then there is a possibility of a pole in the polarizability. If we for brevity set $\operatorname{Im}\chi_0^R(\mathbf{q},\omega) = -\delta$, we can write

$$-\operatorname{Im}\chi^R(\mathbf{q},\omega) = \frac{\delta}{\left[1 - V(q)\operatorname{Re}\chi_0^R(\mathbf{q},\omega)\right]^2 + \delta^2}$$
$$= \pi\delta\left(1 - V(q)\operatorname{Re}\chi_0^R(\mathbf{q},\omega)\right). \tag{9.59}$$

Thus, there is a well-defined mode when $1 - V(q)\operatorname{Re}\chi_0^R(\mathbf{q},\omega) = 0$. This is the plasma oscillation mode, also called a plasmon, which is treated further in Chapter 14

9.5   Summary and outlook

In this chapter, we have seen a method to deal with the dynamical aspects of interacting many-body systems, namely the equation of motion method applied to the Green's functions. The set of differential equation is not soluble in general, and in fact only a very small set of Hamiltonians describing interacting systems can be solved exactly. Therefore approximations are necessary, and we saw particular examples of this, namely the mean-field solution of Anderson's model for a magnetic impurity embedded in a metallic host, and the RPA approximation for the charge auto correlation function. The Anderson model will be studied further in Chapter 10, where we will focus more on transport in interacting mesoscopic systems.

In Chapter 11, we use the equation of motion to derive the Green's functions in the imaginary time formalism and to derive the famous Wick's theorem. Wick's theorem will then pave the way for introducing the Feynman diagrams.

10

TRANSPORT IN INTERACTING MESOSCOPIC SYSTEMS

In Chapter 7 we studied transport of electrons through phase coherent mesoscopic systems. The underlying assumption was that the charge carriers could be viewed as non-interacting particles. This assumption is generally valid for extended systems[34] at low temperatures. In the case of mesoscopic systems or nanostructures, "extended" should be thought of as many channels or at least a few almost open channels that connect the system to the electron reservoirs of the leads. When this is no longer the case, interactions play a dominant role for the transport properties. This is easily understood physically, because if the charge tends to be localized on the nanostructure, the flow of electrons becomes correlated due to the Coulomb interactions. The degree of correlation is controlled by the Coulomb energy cost for adding or removing electrons. This energy can become considerable, of order of meV for small systems such as semiconductor quantum dots, or even as large as eV for molecular devices. Because of this charging effect, it is not permissible to think about the particles as independent entities, and hence a many-body approach is needed. In this chapter, we shall study a few examples that illustrate the importance of many-body physics, and discuss the different theoretical tools available.

10.1 Model Hamiltonians

As in Chapter 7, we study a system consisting of a middle region connected by leads to large electron reservoirs. The leads themselves are assumed to be metallic systems, in which the electrons can be regarded as non-interacting. In contrast, the middle region is a nanostructure where interaction effects may be important. The interactions can be either electron-electron interactions, or, if the charge on the structure couples to mechanical vibrations, it could be electron-phonon interactions. The connection between lead and interacting region is given by a tunneling term that allows electrons to be exchanged. The system is illustrated in Fig. 10.1 . The Hamiltonian we are going to study throughout this chapter is thus given by

$$H = H_L + H_R + H_D + H_T, \tag{10.1}$$

where H_L and H_R are the Hamiltonians for the left and right leads, respectively, H_D is the Hamiltonian describing the "dot" region and, finally, H_T is the tunneling Hamiltonian for the two the junctions, see Section 8.4.1. The states in the uncoupled dot, the left lead, and the right lead are defined by the fermion operators $c^\dagger_{\nu_D}$, $c^\dagger_{\nu_L}$ and $c^\dagger_{\nu_R}$, respectively. The tunneling Hamiltonian H_T coupling these three subsystems is given by

---

[34]Here, "extended" means 3D metals, where Fermi liquid theory is applicable, see Chapter 15.

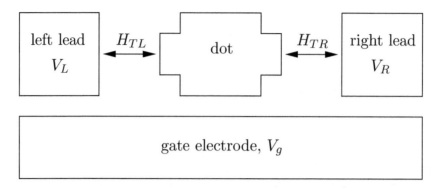

FIG. 10.1.   Schematic illustration of the nanostructure or mesoscopic "quantum dot" system connected to metallic electrodes. The left and right electrodes, with voltages V_L and V_R, are assumed to be described by non-interacting electrons, whereas in the middle region ("dot") interactions may be present. The three regions are connected by tunneling processes described by tunneling Hamiltonians H_{TL} and H_{TR}. In addition there is a gate electrode, which by its voltage V_g is used to change the electrostatic potential of the nanostructure. Such systems are realized in semiconductor quantum dots or in molecular electronics devices. An example of such a nanoelectronic device is shown in Fig. 2.12(c).

$$H_T = H_{TL} + H_{TR}, \tag{10.2}$$

with

$$H_{TL} = \sum_{\nu_L \nu_D} \left(t_{L,\nu_L,\nu_D} c^\dagger_{\nu_L} c_{\nu_D} + t^*_{L,\nu_L,\nu_D} c^\dagger_{\nu_D} c_{\nu_L} \right), \tag{10.3a}$$

$$H_{TR} = \sum_{\nu_R \nu_D} \left(t_{R,\nu_R,\nu_D} c^\dagger_{\nu_R} c_{\nu_D} + t^*_{R,\nu_R,\nu_D} c^\dagger_{\nu_D} c_{\nu_R} \right). \tag{10.3b}$$

The "dot" Hamiltonian includes both a single particle part and an interaction part,

$$H_D = H_{D0} + H_{\text{int}}, \tag{10.4}$$

where the non-interacting part is as usual

$$H_{D0} = \sum_{\nu_D} \xi_{\nu_D} c^\dagger_{\nu_D} c_{\nu_D}, \tag{10.5}$$

and where the interaction part H_{int} is to be specified. One example of the quantum dot region could be a dot or molecule with a number of single particle levels, ν_D, and Coulomb interactions between the electrons occupying these, i.e., as in Eq. (1.62),

$$H_{\text{int}} = \frac{1}{2} \sum_{\substack{\nu_1 \nu_2 \\ \nu_3 \nu_4}} V_{\nu_1 \nu_2, \nu_3 \nu_4} c^\dagger_{\nu_1} c^\dagger_{\nu_2} c_{\nu_4} c_{\nu_3}. \tag{10.6}$$

To simplify our discussion, we restrict ourselves to two interaction models. The first model is the so-called constant-interaction model or capacitor model,

$$H_{int}^{CI} = E_C \sum_{\nu\nu'} n_\nu n_{\nu'} = E_C \left(\sum_\nu n_\nu \right)^2, \qquad (10.7)$$

which is nothing but the charging energy for a capacitor with charge $Q = -e \sum_\nu n_\nu$, and where $E_C = e^2/2C$ is the energy associated with charging a capacitor C by a single electron. This model is most relevant when there are many single particle levels within the energy window of interest, or when the system is sufficiently disordered, such that the charge distribution can be approximated by a smeared distribution similar to electrostatic energy minimum configuration.

The second interaction model to be studied is an Anderson-type model, similar to the one discussed in Section 9.3, the only difference being that "the magnetic impurity", which is now our nanostructure, is connected to two metallic electron reservoirs with separate chemical potentials. For this model, the dot Hamiltonian is

$$H_D = \sum_{\sigma=\uparrow\downarrow} \xi_{d\sigma} c_{d\sigma}^\dagger c_{d\sigma} + U n_{d\downarrow} n_{d\uparrow}, \qquad (10.8)$$

where $c_{d\sigma}$ is the operator associated with the dot-level.

In the following, we discuss different approaches for calculating transport properties of these model systems.

10.2   Sequential tunneling: the Coulomb blockade regime

The sequential tunneling regime, which is equivalent to the weak tunneling regime, assumes that time spent on the dot between tunneling event is assumed to be the largest time scale in the problem, larger than coherence times in the mesoscopic systems. As a consequence, coherence between successive tunneling events is not possible. This means that we can describe the mesoscopic system by a distribution function that gives the probability of finding the system in a particular state α. The distribution function is denoted $P(\alpha)$. In equilibrium this is, of course, nothing but the Boltzmann distribution function in Eq. (1.113). However, the out-of-equilibrium conditions become important when a voltage bias is applied across the system, and we must determine the non-equilibrium distribution function. In order to do so, we look at transitions between the various α-states, and since we consider the weak tunneling limit, we can use Fermi's golden rule with the tunneling Hamiltonian as a perturbation, see Section 5.6. The transition between two states α and β due to tunneling through the left junction is

$$\Gamma_{\beta\alpha}^L = 2\pi \sum_{f_\beta i_\alpha} |\langle f_\beta | H_{TL} | i_\alpha \rangle|^2 W_{i_\alpha} \, \delta(E_{f_\beta} - E_{i_\alpha}), \qquad (10.9)$$

where the sum over initial states runs over all configurations of the internal degrees of freedom i_α that give the state α, each weighted by a thermal distribution function W_{i_α}. Similarly, we sum over configurations of the final states that give the final state β. The meaning of this will become clear in the two examples considered below.

Once the transition rates are known, we can set up our kinetic equations, also called master equations, for the dynamical behavior of the distribution function $P(\alpha)$,

$$\frac{d}{dt}P(\alpha) = \underbrace{-\sum_{\beta}\Gamma_{\beta\alpha}P(\alpha)}_{\text{Tunneling out of state }\alpha} + \underbrace{\sum_{\beta}\Gamma_{\alpha\beta}P(\beta)}_{\text{Tunneling into state }\alpha} , \qquad (10.10)$$

where the Γ-coefficients originate from tunneling through either of the contacts.

The master equations are rather self-explanatory: the first term on the right-hand side gives the rate at which the state α decays. This is proportional to the probability that the system is in state α in the first place, multiplied by the rates for transitions from α to any other state β. The second term gives the rate for the opposite process, i.e., the rate at which the system ends up in state α given that is started in some other state β. The master equations are to be understood as semi-classical equations, because we assume that in between tunneling events, the state of the system can be described by classical probabilities, $P(\alpha)$. The approach is similar in spirit to the Boltzmann equation discussed in Chapter 15. Later in this chapter, we go beyond the semi-classical approach and discuss the situation where the tunneling rate is comparable to energy differences of the mesoscopic system.

The only situation we study here is steady state where $dP(\alpha)/dt = 0$, and thus

$$0 = -\sum_{\beta}\Gamma_{\beta\alpha}P(\alpha) + \sum_{\beta}\Gamma_{\alpha\beta}P(\beta). \qquad (10.11)$$

From this set of linear equations for $P(\alpha)$ in combination with the normalization condition $\sum_{\alpha}P(\alpha) = 1$, one can determine the distribution function.

10.2.1 Coulomb blockade for a metallic dot

We start by looking at the classical model for Coulomb blockade in a metallic dot (the results are summarized in Fig. 10.2). Here, "metallic" means that there is a continuous density of electronic states in the dot. Between the rare tunneling events, the occupancy of the states in the dot are assumed to relax to a Fermi–Dirac distribution. The only unknown variable of the mesoscopic system is therefore the number N of electrons on the dot. The Hamiltonian is given by the constant-interaction model,

$$H_D^{CB} = \sum_{\nu}\xi_{\nu}c_{\nu}^{\dagger}c_{\nu} + E(N), \qquad (10.12)$$

$$E(N) = E_C N^2 - eV_g N, \qquad (10.13)$$

where we have included the gate voltage V_G explicitly. The gate voltage gives rise to an electric field that increases the energy of the electrons residing on the dot.[35]

Tunneling onto the dot will increase the number of electrons on the dot by one. The rates for these processes are calculated using Eq. (10.9) with $i_{\alpha} = i_N$ and $f_{\beta} = f_{N+1}$.

---

[35] In the actual experimental set-up, eV_G is a combination of the source, drain, and gate voltages together with the respective mutual capacitances. Here we lump all this into one gate voltage dependence.

The final state is $|f_{N+1}\rangle = c_{\nu_D}^\dagger c_{\nu_L} |i_N\rangle$, and we should thus sum over the states ν_D and ν_L to span the possible configurations after the tunneling event. We have

$$\Gamma_{N+1,N}^L = 2\pi \sum_{\nu_L \nu_D} \sum_{i_N} \left| \langle i_N | c_{\nu_L}^\dagger c_{\nu_D} H_{TL} | i_N \rangle \right|^2 W_{i_N} \, \delta\left[E(N+1) - E(N) + \varepsilon_{\nu_D} - \varepsilon_{\nu_L} \right]$$

$$= 2\pi \sum_{\nu_L \nu_D} \sum_{i_N} |t_{L,\nu_L,\nu_D}|^2 \left| \langle i_N | c_{\nu_L}^\dagger c_{\nu_D} c_{\nu_D}^\dagger c_{\nu_L} | i_N \rangle \right|^2 W_{i_N}$$

$$\times \delta\left[E(N+1) - E(N) + \varepsilon_{\nu_D} - \varepsilon_{\nu_L} \right]. \tag{10.14}$$

Using that the lead and dot states are independent, so that $W_{i_N} = W_{i_{ND}} W_{i_L}$, where $|i_{ND}\rangle$ and $|i_L\rangle$ is the initial state of dot and the left lead, respectively, we write

$$\Gamma_{N+1,N}^L = 2\pi \sum_{\nu_L \nu_D} \sum_{i_{ND} i_L} |t_{L,\nu_L,\nu_D}|^2 \left| \langle i_L | c_{\nu_L}^\dagger c_{\nu_L} | i_L \rangle \langle i_{ND} | c_{\nu_D} c_{\nu_D}^\dagger | i_{ND} \rangle \right|^2$$

$$\times W_{i_{ND}} W_{i_L} \, \delta\left[E(N+1) - E(N) + \varepsilon_{\nu_D} - \varepsilon_{\nu_L} \right]. \tag{10.15}$$

Here $\langle i_L | c_{\nu_L}^\dagger c_{\nu_L} | i_L \rangle$ gives the occupation of the state ν_L for the initial state $|i_L\rangle$, and it is therefore either zero or one (which means that the square can be omitted). The occupations and the sum over initial states in the lead can be replaced by the definitions of the Fermi-Dirac distribution function, see Eq. (1.125),

$$\sum_{i_L} W_{i_L} \langle i_L | c_{\nu_L}^\dagger c_{\nu_L} | i_L \rangle = n_F(\varepsilon_{\nu_L} - \mu_L). \tag{10.16a}$$

However, the sum over occupation of the dot state is restricted to a given number of electrons on the dot and therefore it is strictly speaking not equal to the thermodynamical Fermi-Dirac distribution defined in the grand canonical ensemble. Here, we assume that the Fermi energy of the dot is large compare to temperature and voltage, so that it is still approximately true that

$$\sum_{i_{ND}} W_{i_{ND}} \langle i_{ND} | c_{\nu_D} c_{\nu_D}^\dagger | i_{ND} \rangle \approx 1 - n_F(\varepsilon_{\nu_D} - \mu_D). \tag{10.16b}$$

Furthermore, because of the continuous density of states, we replace the sums over ν_L and ν_D by integrals over energies ε_L and ε_D as follows:

$$2\pi \sum_{\nu_L \nu_D} |t_{L,\nu_L,\nu_D}|^2 \quad \rightarrow \quad 2\pi \int_{-\infty}^{\infty} d\varepsilon_L d\varepsilon_D |t_L|^2 d_D d_L \approx \gamma^L \int_{-\infty}^{\infty} d\varepsilon_L d\varepsilon_D, \tag{10.17}$$

where the last step assumes that the combined tunneling density of states $|t|^2 d_D d_L$ is constant in the relevant energy range. This approximation is similar to Eq. (8.75). We then collect everything and obtain

$$\Gamma_{N+1,N}^L = \gamma^L \int_{-\infty}^{\infty} d\varepsilon_L d\varepsilon_D \, n_F(\varepsilon_L - \mu_L)$$

$$\times \left[1 - n_F(\varepsilon_D - \mu_D) \right] \delta\left[E(N+1) - E(N) + \varepsilon_D - \varepsilon_L \right]. \tag{10.18}$$

Using the relations

$$n_{\rm F}(\varepsilon_1)\left[1 - n_{\rm F}(\varepsilon_2)\right] = n_{\rm B}(\varepsilon_1 - \varepsilon_2)\left[n_{\rm F}(\varepsilon_2) - n_{\rm F}(\varepsilon_1)\right], \qquad (10.19{\rm a})$$

$$\int_{-\infty}^{\infty} d\varepsilon \left[n_{\rm F}(\varepsilon) - n_{\rm F}(\varepsilon + \omega)\right] = \omega, \qquad (10.19{\rm b})$$

we arrive at

$$\Gamma_{N+1,N}^{\alpha} = \gamma^{\alpha} f\left[E(N+1) - E(N) + \mu_D - \mu_\alpha\right], \qquad (10.20)$$

$$f(E) = \frac{E}{e^{\beta E} - 1}, \qquad (10.21)$$

where α is either L or R. In the same way, we find the tunneling rates for decreasing the number of particles on the dot,

$$\Gamma_{N-1,N}^{\alpha} = \gamma^{\alpha} f\left[E(N-1) - E(N) - \mu_D + \mu_\alpha\right]. \qquad (10.22)$$

If we, furthermore, define the total rates as the sum of left and right contributions,

$$\Gamma_{N\mp1,N} = \Gamma_{N\mp1,N}^{L} + \Gamma_{N\mp1,N}^{R}, \qquad (10.23)$$

then the master equation finally becomes

$$\frac{d}{dt} P(N) = -(\Gamma_{N+1,N} + \Gamma_{N-1,N})P(N) + \Gamma_{N,N+1}P(N+1) + \Gamma_{N,N-1}P(N-1) = 0. \qquad (10.24)$$

The solution of this set of equations is, by inspection, seen to be given by the following recursive relation:

$$\Gamma_{N-1,N}\, P(N) = \Gamma_{N,N-1}\, P(N-1). \qquad (10.25)$$

Once $P(N)$ is determined, the current through the device can be found as the current through, say, the left junction

$$I = (-e) \sum_{N} \left(\Gamma_{N+1,N}^{L} - \Gamma_{N-1,N}^{L}\right) P(N), \qquad (10.26)$$

where the first term gives the rate for electrons entering the dot through the left tunnel junction, while the last term correspondingly gives the rate at which electrons leave the dot through the same contact.

Coulomb blockade for a metallic dot results in an oscillatory dependence as a function of gate voltage. The optimum number of particles follows from Eq. (10.13): $\mathcal{N}_{\rm opt} = eV_g/2E_C$. When this is an integer there is an energy gap for adding electrons, whereas when $\mathcal{N}_{\rm opt}$ is a half-integer two charge states are degenerate and current can flow. The solution of the Coulomb model Eqs. (10.25) and (10.26) for a metallic dot is shown in Fig. 10.2. The occupations of the different states near the optimum number are shown in the leftmost panel, with N_0 being an integer. The middle panel shows the conductance of the system, which is seen to peak when $\mathcal{N}_{\rm opt}$ is a half integer. The rightmost figure shows a density plot of the differential conductance dI/dV, where the Coulomb gap is seen as dark areas. This is a plot often used to characterize the addition spectrum of such single electron transistors (see also Fig. 10.6).

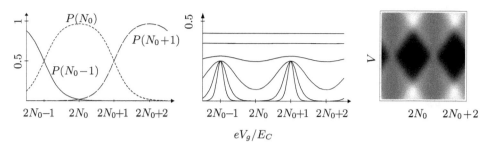

FIG. 10.2. Different ways to plot the consequences of Coulomb blockade. The left figure shows how the occupations change when the gate voltage is scanned, while the middle panel shows the conductance for different temperatures, $kT/E_C = 0.05, 0.1, 0.25, 0.5, 1$ and 2. Note that N_0 is an integer. The density plot to the right shows how the differential conductance dI/dV changes when changing gate voltage and bias voltage simultaneously. The dark and light regions are regions with low and high conductance, respectively. The diamond-shaped figure is a clear feature of the Coulomb blockade physics.

10.2.2 Coulomb blockade for a quantum dot

When the dot is so small that the single particle levels inside are resolved, one cannot use the result from the previous section, where a metallic dot was considered. Instead, one has to define a probability function $P(\alpha)$ for each configuration of occupations of the single particle levels inside the dot. Here, $|\alpha\rangle$ means a specific many-body state of the dot, so for each charge N it could be the groundstate or one of the excited states of the N-particle system. For a dot with a discrete set of eigenstates, the tunneling rates are different from the metallic dot case, because we do not assume that the states are distributed according to a Fermi–Dirac distribution. Instead, we have to find the distribution functions $P(\alpha)$ from the master equation.

The tunneling rate for tunneling between dot states $|\alpha_{D,N}\rangle$ and $|\beta_{D,N+1}\rangle$ through the left barrier, follows from Eq. (10.9). Here we have indicated the number of electrons in the dot as a subscript to the dot states. The final state is $c_{\nu_L}|i_L\rangle|\beta_{D,N+1}\rangle$, where $|i_L\rangle$ is the initial lead state and $|\beta_{D,N+1}\rangle$ the final dot state. We have

$$\Gamma^L_{\beta_{N+1}\alpha_N} = 2\pi \sum_{i_L \nu_L} \left| \langle \beta_{D,N+1} | \langle i_L | c^\dagger_{\nu_L} H_{TL} | \alpha_{D,N} \rangle | i_L \rangle \right|^2 W_{i_L}\, \delta(E_{\beta_{N+1}} - \varepsilon_{\nu_L} - E_{\alpha_N}).$$

$$(10.27)$$

Inserting the tunneling Hamiltonian from Eq. (10.3a), we find (see Exercise 10.1)

$$\Gamma^L_{\beta_{N+1}\alpha_N} = 2\pi \sum_{\nu_L} \left| \langle \beta_{D,N+1} | \sum_{\nu_D} t^*_{L,\nu_L,\nu_D} c^\dagger_{\nu_D} | \alpha_{D,N} \rangle \right|^2 n_F(\varepsilon_{\nu_L} - \mu_L) \delta(E_{\beta_{N+1}} - \varepsilon_{\nu_L} - E_{\alpha_N}).$$

$$(10.28)$$

Often, one makes the assumption, as we did in the previous section, that the combination $\gamma^L = 2\pi |t_{L,\nu_L,\nu_D}|^2 d_L(\nu_L)$ of the tunneling matrix element and the density of states is constant. With this simplification the transition rate $\Gamma^L_{\beta_{N+1}\alpha_N}$ becomes

$$\Gamma^L_{\beta_{N+1}\alpha_N} = \gamma^L \left| \langle \beta_{D,N+1} | \sum_{\nu_D} c^\dagger_{\nu_D} | \alpha_{D,N} \rangle \right|^2 n_{\mathrm{F}}(E_{\beta_{N+1}} - E_{\alpha_N} - \mu_L). \qquad (10.29)$$

Similarly, we have for the transition rate for decreasing the number of particles as

$$\Gamma^L_{\beta_{N-1}\alpha_N} = \gamma^L \left| \langle \beta_{D,N-1} | \sum_{\nu_D} c_{\nu_D} | \alpha_{D,N} \rangle \right|^2 \left[1 - n_{\mathrm{F}}(E_{\alpha_N} - E_{\beta_{N-1}} - \mu_L) \right]. \qquad (10.30)$$

With this result, we are now able to solve the master equation for a given problem. In principle, the calculation follows the method used in the previous metallic case, but we will not go into specific details here. We do, however, recommend the reader to do Exercise 10.2, where an example of a dot with a single resolved level is considered. See also the original paper by Beenakker (1991).

10.3   Coherent many-body transport phenomena

Until now, we have dealt with the regime where the tunneling in and out of the dot is incoherent. However, for stronger coupling to the leads this is no longer a correct description, because when the tunneling coupling energy becomes larger than the thermal energy and the bias voltage, the delocalization of the charge due to quantum processes cannot be disregarded. For very strong coupling the Coulomb blockade itself may be strongly modified and even disappear completely. To illustrate the importance of higher order tunneling processes, we begin by discussing the so-called cotunneling processes. A cotunneling process is a higher order tunneling process, where the electron is transferred from one side of the structure to the other. Since the intermediate state, where the electron is on the dot itself, has an energy which is larger than the initial energy, this is a virtual process. Using second order perturbation theory, such processes give a contribution of order t^2/U, where t is the tunneling amplitude for each tunneling and U is the energy of the intermediate state.[36]

10.3.1   Cotunneling

Above we used Fermi's golden rule, which is valid to second order in the tunneling Hamiltonian. To higher orders, the tunneling Hamiltonian is replaced by the T-matrix introduced in Section 5.7. The T-matrix is given self-consistently by

$$T = H_T + H_T \frac{1}{E_i - H_0 + i\eta} T, \qquad (10.31)$$

where E_i is the energy of the initial state. To find the leading correction to Fermi's golden rule, we thus replace Eq. (10.9) by

$$\Gamma^{(2)}_{\beta\alpha} = 2\pi \sum_{f_\beta, i_\alpha} \left| \langle f_\beta | H_T \frac{1}{E_{i_\alpha} - H_0} H_T | i_\alpha \rangle \right|^2 W_{i_\alpha} \, \delta(E_{f_\beta} - E_{i_\alpha}). \qquad (10.32)$$

The cotunneling process is illustrated in Fig. 10.3. There are two different ways that an electron can get from, say, left to right: (1) first an electron is transferred from left

---

[36] In other contexts these types of processes are often called super-exchange processes.

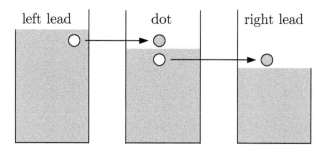

FIG. 10.3. Illustration of the coherent, inelastic cotunneling process. Shaded areas illustrate filled electron states, and in the cotunneling process an electron is transferred from the high energy side to the central dot region, while at the same time an electron is moved from dot to right. The virtual intermediate state has an excess charge and hence a higher energy due to the charging energy, while the charging energy of the final state is the same as for the initial state.

lead to the dot, *then* an electron is transferred from dot to right lead, or (2) first an electron is taken from the dot to the right lead followed by a tunneling through the left junction. The rate for transferring an electron from left to right is thus

$$\Gamma_{RL}^{(2)} = 2\pi \sum_{f_\beta, i_\alpha} \left| \langle f_\beta | \left(H_{TR} \frac{1}{E_{i_\alpha} - H_0} H_{TL} + H_{TL} \frac{1}{E_{i_\alpha} - H_0} H_{TR} \right) | i_\alpha \rangle \right|^2$$
$$\times W_{i_\alpha} \, \delta(E_{f_\beta} - E_{i_\alpha}). \tag{10.33}$$

Furthermore, the cotunneling processes are either elastic or inelastic. Elastic means that the energies of the initial and final electron state in the leads are the same, while inelastic means that these energies are different. In the latter case the energy of the dot has been changed by the process, e.g., by the appearance of an electron-hole pair (see Fig. 10.3). A special type of elastic cotunneling involves a spin flip during the process. This process will be discussed thoroughly in connection with the Kondo effect in Section 10.5.

10.3.2 *Inelastic cotunneling for a metallic dot*

We start by looking at cotunneling in metallic dots. Because metallic dots by definition have a continuum of states, there is a large phase space for inelastic processes (Averin and Nazarov 1990). The process of inelastic cotunneling is illustrated in Fig. 10.3. The final state is an electron-hole pair on the dot, plus one electron transferred from the left to the right electrode, i.e.,

$$|f\rangle = c_{\nu_R}^\dagger c_{\nu_L} c_{\nu_D}^\dagger c_{\nu_D'} |i\rangle. \tag{10.34}$$

Using the same set of approximations as was used in deriving the tunneling rates, Γ, to fourth order in H_T, we have (see Exercise 10.3)

$$\Gamma^{(2)}_{RL,\text{in}}(N) = \frac{\gamma^L \gamma^R}{2\pi} \int_{-\infty}^{\infty} d\varepsilon_L \, d\varepsilon_R \, d\varepsilon_D \, d\varepsilon'_D \, \delta(\varepsilon_R - \varepsilon_L + \varepsilon_D - \varepsilon'_D)$$

$$\times \left(\frac{1}{E(N) - E(N+1) + \varepsilon_L - \varepsilon_D} + \frac{1}{E(N) - E(N-1) - \varepsilon_R + \varepsilon'_D} \right)^2$$

$$\times n_F(\varepsilon_L - \mu_L) \left[1 - n_F(\varepsilon_R - \mu_R) \right] n_F(\varepsilon'_D - \mu_D) \left[1 - n_F(\varepsilon_D - \mu_D) \right]. \tag{10.35}$$

For small source-drain bias voltages, $eV \ll E(N) - E(N \pm 1)$, we find at zero temperature (see Exercise 10.4)

$$\Gamma^{(2)}_{RL,\text{in}}(N) \approx (eV)^3 \frac{\gamma^L \gamma^R}{12\pi} \left(\frac{1}{E(N) - E(N+1)} + \frac{1}{E(N) - E(N-1)} \right)^2. \tag{10.36}$$

As is evident from this expression, the current is a non-linear function of voltage. For the linear response, one can thus neglect the cotunneling effect. The prediction for the cotunneling current at small values of Γ has been verified in great detail experimentally, and only at the degeneracy points, where one of the denominators in Eq. (10.36) is zero, does the present approach break down, and one must go to higher orders in the tunneling Hamiltonian. This interesting many-body problem is, however, not discussed further here.

10.3.3   Elastic cotunneling for a quantum dot

For a non-metallic dot, i.e., for a mesoscopic system with discrete eigenstates, inelastic cotunneling is of course also possible. However, because the spectrum is discrete there is in this case a threshold voltage for leaving the quantum dot in an excited state. We will not go further into this here, but instead turn to elastic processes, which are allowed even in linear response.

Above we considered the case, when an electron-hole excitation was created in the dot. If, however, it is the same electron level in the dot that is emptied and filled, no energy is dissipated in the dot, and the cotunneling is thus "elastic." This possibility is not relevant for a metallic dot,[37] but for a dot with a discrete spectrum it can become relevant. To illustrate this, we take our other extreme model — the single dot-level model — with dot-level ν_d having energy ε_d. For the elastic case the final state is then

$$|f\rangle = c^\dagger_{\nu_R} c_{\nu_L} |i\rangle. \tag{10.37}$$

The rate for the elastic process is again calculated starting from Eq. (10.33). Given that the dot is empty, the rate for a process where an electron is transferred from left to right is

$$\Gamma^{(2)}_{RL,\text{elas}}(0) = 2\pi \sum_{\nu_L, \nu_R} \sum_{i_0} W_{i_0} |t_R|^2 |t_L|^2 \, \delta(\varepsilon_{\nu_R} - \varepsilon_{\nu_L})$$

$$\times \left| \langle f_0 | c^\dagger_{\nu_R} c_{\nu_d} \frac{1}{\varepsilon_{\nu_L} - \varepsilon_d} c^\dagger_{\nu_d} c_{\nu_L} | i_0 \rangle \right|^2. \tag{10.38}$$

---

[37]Because this possibility has an infinitesimal phase space, being only one out of infinitely many processes.

In contrast to Eq. (10.33) there is only one term here, because the process where the electron is first moved from dot to lead cannot happen in this case, simply because the dot is empty to begin with. When inserting Eq. (10.37) we obtain

$$
\Gamma^{(2)}_{RL,\text{elas}}(0) = 2\pi \sum_{\nu_L,\nu_R} |t_R|^2 |t_L|^2 \, \delta(\varepsilon_{\nu_R} - \varepsilon_{\nu_L}) \left(\frac{1}{\varepsilon_{\nu_L} - \varepsilon_d} \right)^2
$$
$$
\times \sum_{i_0} W_{i_0} \left| \langle i_0 | c_{\nu_R} c^\dagger_{\nu_R} | i_0 \rangle \langle i_0 | c^\dagger_{\nu_L} c_{\nu_L} | i_0 \rangle \right|^2. \tag{10.39}
$$

Since $\langle i_0 | c_{\nu_R} c^\dagger_{\nu_R} | i_0 \rangle$ is either one or zero, we can remove the square, $|\cdot|^2$, in Eq. (10.39) and we are left with the definition of the Fermi–Dirac distribution function. If we then convert the sums to integrals we arrive at

$$
\Gamma^{(2)}_{RL,\text{elas}}(0) = \frac{\Gamma^L \Gamma^R}{2\pi} \int d\varepsilon_L \int d\varepsilon_R \, \delta(\varepsilon_R - \varepsilon_L) \left(\frac{1}{\varepsilon_L - \varepsilon_d} \right)^2
$$
$$
\times n_F(\varepsilon_L - \mu_L) \left[1 - n_F(\varepsilon_R - \mu_R) \right], \tag{10.40}
$$

where $\Gamma^\alpha = 2\pi |t_\alpha|^2 d_\alpha$. Let $\mu_L = eV/2$ and $\mu_R = -eV/2$, and furthermore assume that $\varepsilon_d \gg eV$, so that we can neglect ε_L in the denominator, we find for $T = 0$ that

$$
\Gamma^{(2)}_{RL,\text{elas}}(0) = \theta(V) \frac{eV}{\varepsilon_d^2} \frac{\Gamma^L \Gamma^R}{2\pi}. \tag{10.41}
$$

The important points to notice here are that the cotunneling current is linear in the bias voltage and that there is current even when it should be blocked by the Coulomb blockade. In this sense, the cotunneling effect is a leakage current. However, when the tunneling coupling becomes even stronger the perturbative treatment in Γ^L and Γ^R is no longer a good starting point and new methods have to applied. These methods are developed in the remaining part of the chapter. See also Fig. 10.8 where the various regimes for transport through a quantum dot are shown.

10.4 The conductance for Anderson-type models

We now turn to the Anderson-type model for the quantum dot system. The dot Hamiltonian was written in Eq. (10.8), and the left and right tunneling Hamiltonians are

$$
H_{TL} = \sum_{\nu_L \sigma} \left(t_L c^\dagger_{\nu_L \sigma} c_{d\sigma} + t_L^* c^\dagger_{d\sigma} c_{\nu_L \sigma} \right), \quad H_{TR} = \sum_{\nu_R \sigma} \left(t_R c^\dagger_{\nu_R \sigma} c_{d\sigma} + t_R^* c^\dagger_{d\sigma} c_{\nu_R \sigma} \right),
$$
$$
\tag{10.42}
$$

where we, furthermore, make the simplification that the tunneling amplitudes t_L and t_R are independent of energy and spin. With these simplifications, it is convenient to make the following rotation in the space of L- and R-electron operators,

$$
\begin{pmatrix} c_{\nu_e \sigma} \\ c_{\nu_o \sigma} \end{pmatrix} = \frac{1}{\sqrt{|t_L|^2 + |t_R|^2}} \begin{pmatrix} t_L^* & t_R^* \\ -t_R & t_L \end{pmatrix} \begin{pmatrix} c_{\nu_L \sigma} \\ c_{\nu_R \sigma} \end{pmatrix}. \tag{10.43}
$$

The rotation is thus done for each pair ν_L and ν_R, and each pair is labelled by the quantum numbers, ν and the even or odd labels e or o, respectively. If we for simplicity let the two electrodes have identical dispersion relations $\xi_{\nu_L} = \xi_{\nu_R}$, the lead Hamiltonians transform as

$$
\begin{aligned}
H_{LR} &= \sum_{\nu\sigma} \left(c^\dagger_{\nu_L\sigma} \; c^\dagger_{\nu_R\sigma} \right) \begin{pmatrix} \xi_\nu & 0 \\ 0 & \xi_\nu \end{pmatrix} \begin{pmatrix} c_{\nu_L\sigma} \\ c_{\nu_R\sigma} \end{pmatrix} \\
&= \sum_{\nu\sigma} \left(c^\dagger_{\nu_e\sigma} \; c^\dagger_{\nu_o\sigma} \right) \begin{pmatrix} \xi_\nu & 0 \\ 0 & \xi_\nu \end{pmatrix} \begin{pmatrix} c_{\nu_e\sigma} \\ c_{\nu_o\sigma} \end{pmatrix},
\end{aligned}
\tag{10.44}
$$

which easily follows from the unitary of the transformation in Eq. (10.43). The reason for introducing the transformation in the first place is that the tunneling Hamiltonian now only involves the even sector of the lead electrons, i.e.,

$$
H_T = \sum_{\nu\sigma} \sqrt{|t_L|^2 + |t_R|^2} \left(c^\dagger_{\nu_e\sigma} c_{d\sigma} + c^\dagger_{d\sigma} c_{\nu_e\sigma} \right).
\tag{10.45}
$$

What has been achieved is that the Hamiltonian has been separated into two independent sectors: one which involves only a non-interacting electron gas of the odd combination of lead electrons, and one which is a single level coupled to a single electron reservoir made out of the even states. The latter is then nothing but the well-known Anderson model with a *single* electron reservoir.

10.4.1   The conductance in linear response

The next step is to calculate the conductance of the structure. Here, we concentrate on the linear response limit, and later we briefly comment on the generalization to the non-linear transport regime. In Section 6.3, the conductance was shown to be given by the retarded current-current correlation function C^R_{II}. In Eq. (8.67) the current operator for a tunneling system was found. We define particle-current operators for the left and right leads by $I_L = \dot{N}_L$ and $I_R = \dot{N}_R$, and they follow from Eq. (8.67)

$$
I_L = i \sum_{\nu_L\sigma} \left(t_L c^\dagger_{\nu_L\sigma} c_{d\sigma} - t_L^* c^\dagger_{d\sigma} c_{\nu_L\sigma} \right), \quad I_R = i \sum_{\nu_R\sigma} \left(t_R c^\dagger_{\nu_R\sigma} c_{d\sigma} - t_R^* c^\dagger_{d\sigma} c_{\nu_R\sigma} \right).
\tag{10.46}
$$

In the case of a time-independent voltage, we can calculate the expectation value of the current in any of the leads, or we can use any linear combination of the two, i.e., $I \equiv \alpha I_L - (1-\alpha) I_R$. By a clever choice of α, namely $\alpha = |t_R|^2/\{|t_L|^2 + |t_R|^2\}$, the current becomes a function of the odd electron combination only,

$$
I = \frac{-i}{\sqrt{|t_L|^2 + |t_R|^2}} \sum_{\nu_o\sigma} \left(t_L t_R c^\dagger_{\nu_o\sigma} c_{d\sigma} - t_L^* t_R^* c^\dagger_{d\sigma} c_{\nu_o\sigma} \right).
\tag{10.47}
$$

The current-current correlation function now simplifies greatly, because the c_{ν_o} and the dot operators belong to separate parts of the Hamiltonian. In the same spirit as in Section 8.4.1, we write the current operators as $I = -i(L - L^\dagger)$, and we obtain

$$C_{II}^R(t) = -i\theta(t)\left\langle [I(t), I(0)]\right\rangle = i\theta(t)\left\langle [L(t) - L^\dagger(t), L(0) - L^\dagger(0)]\right\rangle, \qquad (10.48)$$

where the expectation value is with respect to the full Hamiltonian, but without the voltage term. Since $\langle L(t)L(0)\rangle = 0$, because it does not conserve the number of particles, we reduce Eq. (10.48) to

$$C_{II}^R(t) = -i\theta(t)\left(\left\langle [L(t), L^\dagger(0)]\right\rangle + \left\langle [L^\dagger(t), L(0)]\right\rangle\right). \qquad (10.49)$$

The expectation value in the first term becomes

$$\left\langle [L(t), L^\dagger(0)]\right\rangle$$

$$= \sum_{\nu_o \nu_o' \sigma\sigma'} \frac{|t_L|^2|t_R|^2}{|t_L|^2 + |t_R|^2}\left[\left\langle c_{\nu_o\sigma}^\dagger(t)c_{d\sigma}(t)c_{d\sigma'}^\dagger c_{\nu_o'\sigma'}\right\rangle - \left\langle c_{d\sigma'}^\dagger c_{\nu_o'\sigma'}c_{\nu_o\sigma}^\dagger(t)c_{d\sigma}(t)\right\rangle\right]$$

$$= \sum_{\nu_o \nu_o' \sigma\sigma'} \frac{|t_L|^2|t_R|^2}{|t_L|^2 + |t_R|^2}\left[\left\langle c_{\nu_o\sigma}^\dagger(t)c_{\nu_o'\sigma'}\right\rangle\left\langle c_{d\sigma}(t)c_{d\sigma'}^\dagger\right\rangle - \left\langle c_{d\sigma'}^\dagger c_{d\sigma}(t)\right\rangle\left\langle c_{\nu_o'\sigma'}c_{\nu_o\sigma}^\dagger(t)\right\rangle\right]$$

$$= \sum_{\nu_o \sigma} \frac{|t_L|^2|t_R|^2}{|t_L|^2 + |t_R|^2}\left[G^<(\nu_o\sigma, -t)G^>(d\sigma, t) - G^>(\nu_o\sigma, -t)G^<(d\sigma, t)\right]. \qquad (10.50)$$

The other term in Eq. (10.49) follows from Eq. (10.50) by a change of sign and $t \to -t$. In the frequency domain we thus find

$$C_{II}^R(\omega) = -i\int_0^\infty dt\, e^{i\omega t}\sum_{\nu_o\sigma}\frac{|t_L|^2|t_R|^2}{|t_L|^2 + |t_R|^2}$$

$$\times \left[G^<(\nu_o\sigma, -t)G^>(d\sigma, t) - G^>(\nu_o\sigma, -t)G^<(d\sigma, t)\right.$$

$$\left. - G^<(\nu_o\sigma, t)G^>(d\sigma, -t) + G^>(\nu_o\sigma, t)G^<(d\sigma, -t)\right]. \qquad (10.51)$$

When using that $(G^>(\nu, t))^* = (-i\langle c_\nu(t)c_\nu^\dagger\rangle)^* = i\langle c_\nu c_\nu^\dagger(t)\rangle = -G^>(\nu, -t)$ and similarly $(G^<(\nu, t))^* = -G^<(\nu, -t)$, we see that the expression in the square parenthesis is purely imaginary and also an odd function of t. These observations are used to write the imaginary part of $C_{II}^R(\omega)$ as

$$\text{Im}\,C_{II}^R(\omega) = -\frac{1}{2}\int_{-\infty}^\infty dt\, e^{i\omega t}\sum_{\nu_o\sigma}\frac{|t_L|^2|t_R|^2}{|t_L|^2 + |t_R|^2}$$

$$\times \left[G^<(\nu_o\sigma, -t)G^>(d\sigma, t) - G^>(\nu_o\sigma, -t)G^<(d\sigma, t)\right.$$

$$\left. - G^<(\nu_o\sigma, t)G^>(d\sigma, -t) + G^>(\nu_o\sigma, t)G^<(d\sigma, -t)\right]. \qquad (10.52)$$

The Fourier transform of a product is $\int dt e^{i\omega t}f(t)g(-t) = \int (d\omega'/2\pi)f(\omega + \omega')g(\omega')$, see Eq. (A.16), and we hence obtain

$$\text{Im}\, C_{II}^R(\omega) = -\frac{1}{2} \int_{-\infty}^{\infty} \frac{d\omega'}{2\pi} \sum_{\nu_o \sigma} \frac{|t_L|^2 |t_R|^2}{|t_L|^2 + |t_R|^2}$$

$$\times \left[G^>(d\sigma, \omega') \{ G^<(\nu_o \sigma, \omega' + \omega) - G^<(\nu_o \sigma, \omega' - \omega) \} \right.$$

$$\left. - G^<(d\sigma, \omega') \{ G^>(\nu_o \sigma, \omega' + \omega) - G^>(\nu_o \sigma, \omega' - \omega) \} \right]. \quad (10.53)$$

Similar to Eq. (8.74), we use the relation Eq. (8.51) to substitute the Green's functions $G^>$ and $G^<$ by the spectral function A. Thereby $\text{Im}\, C_{II}^R(\omega)$ becomes

$$\text{Im}\, C_{II}^R(\omega) = \frac{1}{2} \int_{-\infty}^{\infty} \frac{d\omega'}{2\pi} \sum_{\nu_o \sigma} \frac{|t_L|^2 |t_R|^2}{|t_L|^2 + |t_R|^2}$$

$$\times \left[A(d\sigma, \omega') A(\nu_o \sigma, \omega' + \omega) \left\{ n_F(\omega + \omega') - n_F(\omega') \right\} \right.$$

$$\left. - A(d\sigma, \omega') A(\nu_o \sigma, \omega' - \omega) \left\{ n_F(\omega - \omega') - n_F(\omega') \right\} \right]. \quad (10.54)$$

Since we should insert $\text{Im}\, C_{II}$ into the Kubo formula (6.30) for conductance and take the limit $\omega \to 0$, we expand Eq. (10.54) to first order in ω. Furthermore, we also use that the decoupled lead electrons are described by a non-interacting electron gas, i.e., $A(\nu_o \sigma, \omega) = 2\pi \delta(\omega - \xi_{\nu_o})$. With these two steps we arrive at

$$G = e^2 \sum_{\nu_o \sigma} \frac{|t_L|^2 |t_R|^2}{|t_L|^2 + |t_R|^2} A(d\sigma, \xi_{\nu_o}) \left(-\frac{\partial n_F(\xi_{\nu_o})}{\partial \xi_{\nu_o}} \right). \quad (10.55)$$

When converting the sum over lead states to an integral $\sum_{\nu_o} \to \int d\xi d(\xi)$, and, furthermore, including the density of states into the definition of the level width functions, $\Gamma^\alpha = 2\pi |t_\alpha|^2 d$, we obtain the final result for the conductance G through a single level in the dot including Coulomb interactions,

$$G = e^2 \sum_{\sigma} \int \frac{d\xi}{2\pi} \frac{\Gamma^L \Gamma^R}{\Gamma^L + \Gamma^R} A(d\sigma, \xi) \left(-\frac{\partial n_F(\xi)}{\partial \xi} \right). \quad (10.56)$$

This is a quite remarkable result because it directly relates the spectral function of the mesoscopic system to the conductance. Note that no approximations have been made apart from the initial one that the tunneling amplitudes are constant and spin independent and the left and right leads have identical dispersion relations. Therefore, the spectral function should be calculated in the presence of tunnel couplings to the leads, and the expression in Eq. (10.56) thus enables us to study the conductance through a mesoscopic system to any order in hybridization term H_T.

Above we have calculated the linear response conductance, but it can in fact be generalized to finite bias voltage as well. This was done by Meir and Wingreen (1992), who showed that the following very intuitive result,

$$I = e \sum_{\sigma} \int \frac{d\xi}{2\pi} \frac{\Gamma^L \Gamma^R}{\Gamma^L + \Gamma^R} A(d\sigma, \xi) \left[n_F(\xi - \mu_L) - n_F(\xi - \mu_R) \right], \quad (10.57)$$

gives the current at any voltage bias. Here μ_L and μ_R are the chemical potentials of the left and right reservoirs, respectively. However, in order to derive this result one needs non-equilibrium Green's function formalism which is outside the scope of this book. Interested readers can consult, e.g., the book by Haug and Jauho (Haug and Jauho 1996).

10.4.2   Calculation of Coulomb blockade peaks

In order to calculate the conductance G using Eq. (10.56), it is necessary to determine the spectral function A. To do this we must first find the retarded Green's function

$$G^R(d\sigma, t) = -i\theta(t) \left\langle \left\{ c_{d\sigma}(t), c_{d\sigma}^\dagger \right\} \right\rangle. \tag{10.58}$$

Let us start from the equation of motion theory in Section 9.3.1 and Eqs. (9.30) and (9.31). But instead of performing the mean-field approximation in Eq. (9.36) we take it one step further. The mean-field approach is not appropriate here, because the current through the dot depends crucially on the correlations. Clearly, if U is large (compared to Γ), and the mean occupation is close to unity, the motion of the electrons hopping through the dot will be correlated so that there is never more than one electron at any given time on the dot. This type of correlated motion cannot be captured by the mean-field approximation, where the electrons only interact with the mean occupation. The situation here, where we are studying a single dot, is thus very different from the case studied in Section 9.3, where we considered an ensemble of magnetic impurities. We therefore iterate the equations of motion one step further and look at the equation of motion for the function D^R, defined in Eq. (9.34),

$$i\partial_t D^R(d\uparrow, t) = \delta(t)\left\langle \left\{ n_{d\downarrow}c_{d\uparrow}, c_{d\uparrow}^\dagger \right\} \right\rangle - i\theta(t)\left\langle \left\{ -[H, n_{d\downarrow}c_{d\uparrow}](t), c_{d\uparrow}^\dagger \right\} \right\rangle. \tag{10.59}$$

The commutator in the last term on the right-hand side results in two terms

$$[H, n_{d\downarrow}c_{d\uparrow}] = [H_D, n_{d\downarrow}c_{d\uparrow}] + [H_T, n_{d\downarrow}c_{d\uparrow}], \tag{10.60a}$$

with

$$[H_D, n_{d\downarrow}c_{d\uparrow}] = n_{d\downarrow}[H_D, c_{d\uparrow}] = (-\xi_{d\uparrow} - U)n_{d\downarrow}c_{d\uparrow}, \tag{10.60b}$$

$$[H_T, n_{d\downarrow}c_{d\uparrow}] = -n_{d\downarrow}\sum_{\nu_e}\tilde{t}c_{\nu_e\uparrow} + \sum_{\nu_e}\tilde{t}\left(c_{\nu_e\downarrow}^\dagger c_{d\downarrow} - c_{d\downarrow}^\dagger c_{\nu_e\downarrow}\right)c_{d\uparrow}, \tag{10.60c}$$

where H_D was defined in Eq. (10.8) and H_T in Eq. (10.45), and where we now define $\tilde{t} = \sqrt{|t_L|^2 + |t_R|^2}$. The commutator in Eq. (10.60b) is proportional to $n_{d\downarrow}c_{d\uparrow}$ itself, whereas the commutator in Eq. (10.60c) generates two new terms. The first of these leads to the definition of the function

$$F^R(\nu\uparrow, t) = -i\theta(t)\left\langle \left\{ (n_{d\downarrow}c_{\nu_e\uparrow})(t), c_{d\uparrow}^\dagger \right\} \right\rangle, \tag{10.61}$$

while the second term gives rise to a new type of correlation. In fact, it corresponds to higher-order correlations, where the spin on the quantum dot flips during the tunneling process. For now these more intricate correlation functions will be omitted, but in Section 10.5 we shall deal with it when discussing the Kondo effect in quantum dots. Here, we are thus left with F^R, and the equation of motion of this function leads to the following commutators

$$[H, n_{d\downarrow}c_{\nu_e\uparrow}] = -\xi_{\nu_e}c_{\nu_e\uparrow} + [H_T, c_{\nu_e\uparrow}], \tag{10.62a}$$

$$[H_T, n_{d\downarrow}c_{\nu_e\uparrow}] = n_{d\downarrow}[H_T, c_{\nu_e\uparrow}] + [H_T, n_{d\downarrow}]c_{\nu_e\uparrow}$$

$$= -\tilde{t}n_{d\downarrow}c_{d\uparrow} + \sum_{\nu_e}\tilde{t}\left(c_{\nu_e\downarrow}^\dagger c_{d\downarrow} - c_{d\downarrow}^\dagger c_{\nu_e\downarrow}\right)c_{\nu_e\uparrow}. \tag{10.62b}$$

Again, we neglect the last term, which corresponds to spin-flip processes, and in fact the set of equations is seen to close. We obtain

$$(i\partial_t - \xi_{d\uparrow} - U)D^R(d\uparrow, t) = \delta(t)\langle n_{d\downarrow}\rangle + \sum_\nu \tilde{t}F^R(\nu\uparrow, t), \tag{10.63}$$

$$(i\partial_t - \xi_{\nu\uparrow})F^R(\nu\uparrow, t) = \tilde{t}D^R(d\uparrow, t). \tag{10.64}$$

Of course, we also have a similar equation for the corresponding spin-down functions. After Fourier transformation and insertion of the result for F^R into the first equation, we obtain

$$\left[\omega - \xi_{d\uparrow} - U - \Sigma^R(\omega)\right]D^R(d\uparrow, \omega) = \langle n_{d\downarrow}\rangle, \tag{10.65}$$

where Σ^R is the same self-energy function $\Sigma^R(\omega)$ as in Eq. (9.38b),

$$\Sigma^R(\omega) = \sum_\nu \frac{|\tilde{t}|^2}{\omega - \xi_\nu + i\eta} = \int \frac{d\xi}{2\pi} \frac{\Gamma}{\omega - \xi + i\eta}, \tag{10.66}$$

and where we have defined the total level width by $\Gamma = \Gamma^L + \Gamma^R$. The corresponding expression for G^R follows from Eqs. (9.30) and (9.31) as

$$\left[\omega - \xi_{d\uparrow} - \Sigma^R(\omega)\right]G^R(d\uparrow, \omega) = 1 + UD^R(d\uparrow, \omega). \tag{10.67}$$

We are now in position to determine $G^R(d\uparrow, \omega)$ by combining Eqs. (10.65) and (10.67). The result is

$$G^R(d\uparrow, \omega) = \frac{1 - \langle n_{d\downarrow}\rangle}{\omega - \xi_{d\uparrow} - \Sigma^R(\omega)} + \frac{\langle n_{d\downarrow}\rangle}{\omega - \xi_{d\uparrow} - U - \Sigma^R(\omega)}. \tag{10.68}$$

In the limit where we can regard Γ as a constant (energy independent) level width, the spectral function is thus simply given by the sum of two Lorentzians,

$$A(d\uparrow, \omega) = \frac{(1 - \langle n_{d\downarrow}\rangle)\Gamma}{(\omega - \xi_{d\uparrow}^2) + (\Gamma/2)^2} + \frac{\langle n_{d\downarrow}\rangle\Gamma}{(\omega - \xi_{d\uparrow} - U)^2 + (\Gamma/2)^2}. \tag{10.69}$$

The conductance now follows by inserting Eq. (10.69) into Eq. (10.56). The interpretation of the two-peak structure is straightforward: the first term corresponds to the density of states for adding an electron with spin up when the dot is empty, while

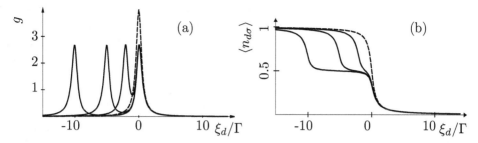

FIG. 10.4. The conductance and occupation of the quantum dot Anderson model.
Each curve is calculated on the basis of Eqs. (10.56) and (10.69) for a given value
of U/Γ; the dashed line is $U/\Gamma = 0$, while the others are $U/\Gamma = 2$, 5, and 10,
respectively. (a) The dimensionless conductance $g = G(h/2e^2)(\Gamma^L + \Gamma^R)/\Gamma^L\Gamma^R$
plotted versus ε_d/Γ. At $U/\Gamma = 0$ as single peak is seen, but for increasing values of
U/Γ a two-peak structure develops with an increasing peak spacing. (b) The corresponding occupations $\langle n_{d\sigma} \rangle$ exhibiting an increased smearing as U/Γ increases.
We have used $T = 0$ in the plot.

the second term is the density of states for adding electrons when the dot is already
occupied by an electron with spin down. In the first situation the spectral function
is centered around the single particle energy, $\xi_{d\uparrow}$, while in the second situation the
Coulomb interaction energy U is added. The result is thus seen to be an improvement
of the mean-field expression in Eq. (9.42), which has a single peak. Only when $\Gamma \gg U$
do the two results become the same.

The two peaks in the spectral function correspond to the Coulomb-blockade physics
discussed previously. The first peak reflects the presence of a possible channel for conductance when the empty state and the singly occupied states are degenerate, whereas
the second peak comes from the second possible channel which is open when the singly
and doubly occupied states are degenerate. Experimentally, one can move between
the peaks by adjusting ξ_d, which in practice is done through the gate voltage. The
Lorentzian shape of the peaks then gives the tunneling broadening that we were interested in, and which is beyond the master-equation approach. The conductance based
on the tunnel-broadened spectral function is shown in Fig. 10.4.

The result for the same model using the master-equation approach is considered in
Exercise 10.2. Here it is shown that any given conductance peak is broadened by the
temperature only, and that its height increases linearly with the inverse temperature.
This of course cannot be correct at very low temperatures $kT \ll \Gamma$, where the peak
height must be limited by Γ. But we see a nice correspondence between the two
approaches: in the weak tunneling limit, $\Gamma \ll kT$, the more general result derived
in this section reduces to the master-equation result given in Exercise 10.2. This
correspondence is the topic of Exercise 10.6.

Finally, note that one should also solve the occupation which enters in Eq. (10.69)
self-consistently. The occupation is given by

$$\langle n_{d\sigma} \rangle = \int \frac{d\omega}{2\pi}\, n_{\rm F}(\omega) A(d\sigma, \omega), \qquad (10.70)$$

which therefore constitutes two coupled linear equations, one for each spin. In Fig. 10.4, we show the results for the conductance using the formulae derived in this section; see also Exercise 10.7. On the figure it is seen how the conductance with its two-peak structure deviates from the mean-field approximation, which always leads to only one peak.

10.5 The Kondo effect in quantum dots

In Section 10.3.1 we discussed cotunneling processes for both metallic dots and quantum dots. We saw that higher order tunneling gave rise to a "leakage" current, which was present even when the semiclassical theory predicted that the current was blocked. This physics was further developed in Section 10.4.2, where we studied tunneling broadening of the Coulomb blockade peaks. There is, however, one very interesting phenomenon, which is not covered by any of the approaches we have considered so far, namely the so-called Kondo resonance. This occurs in the case when the quantum dot has an average occupation close to an odd integer number of electrons. The state with an odd integer of electrons is degenerate due to spin degeneracy, and the fluctuations between the two degenerate states cause the resonance. The Kondo resonance or Kondo effect is a well-known effect from the study of resistivity of metals due to scattering off magnetic impurities. There the scattering has an anomalous effect on the low-temperature resistivity: it causes the resistivity to increase with decreasing temperatures. This effect was explained by Kondo in 1964, using a model for the interaction between the conduction electron and the localized magnetic moment; a model which is now known as the Kondo model. The anomalous behavior turns out to be an intriguing many-body effect. For a review of the Kondo effect in metals, see, e.g., the book by Hewson (1993).

The quantum dot with a single electron occupying the top-most quantum state is, in fact, similar to the spin one-half magnetic impurity. But in contrast to the increased resistivity for the magnetic impurities, the increased scattering in the mesoscopic devices gives rise to an increased conductance. In the following we establish explicitly the correspondence between the scattering against magnetic impurities and tunneling through a quantum dot. We also show, how the anomalous scattering arises by performing a perturbation theory calculation to third order in the exchange coupling. Furthermore, we briefly direct the reader to more advanced treatments, i.e., beyond the perturbation theory, of the Kondo problem.

10.5.1 From the Anderson model to the Kondo model

In order to study the physics in the region, where the quantum dot is occupied by an odd number of electrons, we start from the Anderson model in the range of parameters, where the energy of the singly-occupied state is much lower than the energies of both the empty and doubly-occupied states. In this case, we use an approach due to Schrieffer and Wolff (1966), who performed a canonical transformation that eliminates the two high-energy states: the empty state $|0\rangle$ and the doubly-occupied state $|2\rangle$.

The idea of the canonical transformation is to transform the Hamiltonian so that it is projected onto a subspace with only one electron on the dot, and excursions to the doubly occupied or empty states occur only virtually. In order to achieve this goal, the tunneling term should not be present in the transformed Hamiltonian, since it couples states with different number of electrons on the dot. This cannot be done exactly, but we can at least do it to linear order in the tunneling term, which then means that we seek a canonical transformation

$$H_S \equiv e^{iS} H e^{-iS} = e^{iS} (H_D + H_T + H_{LR}) e^{-iS}, \tag{10.71}$$

that does not have a linear term in H_T. If S is chosen to be linear in the tunneling amplitudes, we can arrange that H_T cancels by it requiring that to *linear order* in S or H_T, one should have

$$e^{iS}(H_D + H_{LR} + H_T)e^{-iS} \approx H_D + H_{LR} + H_T + i[S, H_D + H_{LR}] = H_D + H_{LR}, \tag{10.72}$$

which implies

$$i[S, H_D + H_{LR}] = -H_T. \tag{10.73}$$

The problem is then to find the operator S that does this trick, and this is precisely what was done by Schrieffer and Wolff. They found

$$S = S^+ + S^- = (S^-)^\dagger + S^-, \tag{10.74}$$

where

$$S^- = -i \sum_{\alpha=L,R} \sum_{\nu_\alpha \sigma} \left(\frac{t_\alpha}{\xi_{\nu_\alpha} - E_2 + E_1} n_{d\bar\sigma} c^\dagger_{\nu_\alpha \sigma} c_{d\sigma} + \frac{t_\alpha}{\xi_{\nu_\alpha} + E_0 - E_1} (1 - n_{d\bar\sigma}) c^\dagger_{\nu_\alpha \sigma} c_{d\sigma} \right). \tag{10.75}$$

Here the first term $S^+ = (S^-)^\dagger$ takes care of the virtual transitions to the doubly-occupied state, while the second term S^- deals with the fluctuations due to transitions to the empty state. Let us look at the commutator in Eq. (10.73) explicitly, and let us denote the first term in Eq. (10.75) by S_1^- and the second term by S_2^-, so that $S^- = S_1^- + S_2^-$. We have for the commutator between S_1^- and $H_0 = H_{LR} + H_D$ that

$$[S_1^-, H_0] = -i \sum_{\alpha=L,R} \sum_{\nu_\alpha \sigma} \frac{t_\alpha}{\xi_{\nu_\alpha} - E_2 + E_1}$$

$$\times \left[n_{d\bar\sigma} c^\dagger_{\nu_\alpha \sigma} c_{d\sigma}, \sum_{\beta=L,R} \sum_{\nu_\beta \sigma'} \xi_{\nu_\beta} n_{\nu_\beta \sigma'} + \sum_{\sigma'} \xi_d n_{d\sigma'} + U n_{d\uparrow} n_{d\downarrow} \right]$$

$$= -i \sum_{\alpha=L,R} \sum_{\nu_\alpha \sigma} \frac{t_\alpha}{\xi_{\nu_\alpha} - E_2 + E_1} \left[n_{d\bar\sigma}^2 U + n_{d\bar\sigma}(-\xi_{\nu_\alpha} + \xi_d) \right] c^\dagger_{\nu_\alpha \sigma} c_{d\sigma}. \tag{10.76}$$

Now using that $n_{d\bar\sigma}^2 = n_{d\bar\sigma}$ and that $E_2 - E_1 = U + \xi_d$, this becomes

$$[S_1^-, H_0] = i \sum_{\alpha=L,R} \sum_{\nu_\alpha \sigma} t_\alpha n_{d\bar\sigma} c^\dagger_{\nu_\alpha \sigma} c_{d\sigma}. \tag{10.77}$$

Similarly, we get from the commutator involving S_2^-

$$[S_2^-, H_0] = \sum_{\alpha=L,R} \sum_{\nu_\alpha \sigma} \frac{-it_\alpha}{\xi_{\nu_\alpha} + E_0 - E_1} \left[(1 - n_{d\bar\sigma}) c_{\nu_\alpha \sigma}^\dagger c_{d\sigma}, H_0 \right]$$

$$= i \sum_{\alpha=L,R} \sum_{\nu_\alpha \sigma} t_\alpha \left[1 - n_{d\bar\sigma} \right] c_{\nu_\alpha \sigma}^\dagger c_{d\sigma}. \tag{10.78}$$

When adding the two and using that $[S, H] = [S^-, H] + [S^+, H] = [S^-, H] - [S^-, H]^\dagger$, we are able to conclude that Eq. (10.73) is indeed fulfilled.

We have now successfully cancelled out the term linear in the tunneling, but of course all the higher order terms are still there, i.e.,

$$H_S = H_D + H_{LR} + i[S, H_T] - \frac{1}{2!}[S, [S, H]] + \frac{i}{3!}\left[S, [S, [S, H]] \right] \cdots , \tag{10.79}$$

In general, this cannot be evaluated, but the idea is now to keep the next leading order correction, i.e., the terms to second order in S or H_T. The Schrieffer–Wolff transformation thus deals with the limit where the tunneling is weak, but it still allows for virtual transitions to lowest order in the dot-lead coupling. The Hamiltonian is then approximated by

$$H_S \approx H_D + H_{LR} + i[S, H_T] - \frac{1}{2}[S, [S, H_D + H_{LR}]], \tag{10.80}$$

and because of Eq. (10.73) this also equals

$$H_S \approx H_D + H_{LR} + \frac{i}{2}[S, H_T]. \tag{10.81}$$

Now we write H_T as the sum of a term H_T^+ that corresponds to tunneling onto the dot and a term H_T^- that corresponds to tunneling out of the dot, i.e.,

$$H_T = H_T^+ + H_T^-, \tag{10.82}$$

where

$$H_T^+ = \sum_{\alpha=L,R} \sum_{\nu_\alpha \sigma} t_\alpha^* c_{d\sigma}^\dagger c_{\nu_\alpha \sigma}, \quad H_T^- = \sum_{\alpha=L,R} \sum_{\nu_\alpha \sigma} t_\alpha c_{\nu_\alpha \sigma}^\dagger c_{d\sigma}. \tag{10.83}$$

When inserting Eqs. (10.74) and (10.82) into Eq. (10.81), we find two kinds of terms. One type is the two-particle tunneling terms $[S^+, H^+]$ and $[S^-, H^-]$, which have two quantum dot annihilation or creation operators, i.e., $c_{d\uparrow}^\dagger c_{d\downarrow}^\dagger$ or $c_{d\uparrow} c_{d\downarrow}$, and these terms thus correspond to processes, where the state of the quantum dot changes from empty to doubly occupied or vice versa. Since we are interested in deriving an effective Hamiltonian for the subspace with singly occupancy, we can omit these two terms all together. The remaining terms (last term in Eq. (10.81)) are,

$$H_S^{(2)} = \frac{i}{2} \left([S^-, H_T^+] + [S^+, H_T^-] \right). \tag{10.84}$$

The first commutator in this expression becomes

$$[S^-, H_T^+] = -i \sum_{\alpha,\beta=L,R} \sum_{\nu_\alpha \nu_\beta \sigma} \left(\frac{t_\alpha t_\beta^*}{\xi_{\nu_\alpha} - E_2 + E_1} - \frac{t_\alpha t_\beta^*}{\xi_{\nu_\alpha} + E_0 - E_1} \right)$$

$$\times \left[n_{d\bar\sigma} \left(c_{\nu_\alpha \sigma}^\dagger c_{\nu_\beta \sigma} - n_{d\sigma} \delta_{\nu_\alpha \nu_\beta} \right) - c_{d\bar\sigma}^\dagger c_{d\sigma} c_{\nu_\alpha \sigma}^\dagger c_{\nu_\beta \bar\sigma} \right]$$

$$- i \sum_{\alpha,\beta=L,R} \sum_{\nu_\alpha \nu_\beta \sigma} \frac{t_\alpha t_\beta^*}{\xi_{\nu_\alpha} + E_0 - E_1} \left(c_{\nu_\alpha \sigma}^\dagger c_{\nu_\beta \sigma} - n_\sigma \delta_{\nu_\alpha \nu_\beta} \right). \tag{10.85}$$

At this point, we make use of the fact that we are interested in low temperature behavior. Since our approximate expression is not valid for energies larger than the energy $E_2 - E_1$ for adding or the energy $E_0 - E_1$ for removing an electron, we might as well neglect the dependence on ξ_{ν_α} in the denominators of the above expression. We have limited ourselves to the case when the occupation is one, i.e., $\sum_\sigma n_{d\sigma} = 1$, and hence the last term in the last parenthesis reduces to a simple constant shift of the energy which is unimportant here. Also the term $n_{d\sigma} n_{d\bar\sigma}$ vanishes in this case. With these simplifications, we obtain

$$[S^-, H_T^+] \approx i \sum_{\alpha,\beta=L,R} \sum_{\nu_\alpha \nu_\beta \sigma} \left(\frac{t_\alpha t_\beta^*}{E_2 - E_1} - \frac{t_\alpha t_\beta^*}{E_1 - E_0} \right) \left(n_{d\bar\sigma} c_{\nu_\alpha \sigma}^\dagger c_{\nu_\beta \sigma} - c_{d\bar\sigma}^\dagger c_{d\sigma} c_{\nu_\alpha \sigma}^\dagger c_{\nu_\beta \bar\sigma} \right)$$

$$- i \sum_{\alpha,\beta=L,R} \sum_{\nu_\alpha \nu_\beta \sigma} \frac{t_\alpha t_\beta^*}{E_0 - E_1} c_{\nu_\alpha \sigma}^\dagger c_{\nu_\beta \sigma}. \tag{10.86}$$

Because we have a fixed number of electrons on the dot, it is very instructive to express the effective Hamiltonian, which we are in process of deriving, in terms of the difference $n_{d\sigma} - n_{d\bar\sigma}$ between up and down occupation operator. Therefore, we write $n_{d\sigma} = (n_{d\sigma} + n_{d\bar\sigma} + n_{d\sigma} - n_{d\bar\sigma})/2 = 1/2 + (n_{d\sigma} - n_{d\bar\sigma})/2$, and Eq. (10.86) becomes

$$[S^-, H_T^+] \approx \frac{i}{2} \sum_{\alpha,\beta=L,R} \sum_{\nu_\alpha \nu_\beta \sigma} \left(\frac{t_\alpha t_\beta^*}{E_2 - E_1} - \frac{t_\alpha t_\beta^*}{E_1 - E_0} \right)$$

$$\times \left((n_{d\bar\sigma} - n_{d\sigma}) c_{\nu_\alpha \sigma}^\dagger c_{\nu_\beta \sigma} - 2 c_{d\bar\sigma}^\dagger c_{d\sigma} c_{\nu_\alpha \sigma}^\dagger c_{\nu_\beta \bar\sigma} \right)$$

$$+ \frac{i}{2} \sum_{\alpha,\beta=L,R} \sum_{\nu_\alpha \nu_\beta \sigma} \left(\frac{t_\alpha t_{\alpha'}^*}{E_2 - E_1} + \frac{t_\alpha t_{\alpha'}^*}{E_1 - E_0} \right) c_{\nu_\alpha \sigma}^\dagger c_{\nu_\beta \sigma}. \tag{10.87}$$

The first term clearly involves spin-flip processes and therefore resembles a spin-spin interaction between the spin of the impurity and the spin of the scattering lead electrons. Indeed, if we write such a spin-spin interaction in a basis-independent form using the spinor notation introduced in Eq. (1.92)

$$\mathbf{S}_d \cdot \mathbf{S}_{\nu_\alpha, \nu_\beta} = \frac{1}{4} \sum_{i=x,y,z} \sum_{\sigma\sigma'\sigma_1\sigma_1'} (c_{d\sigma}^\dagger \tau_{\sigma\sigma'}^i c_{d\sigma'})(c_{\nu_\alpha,\sigma_1}^\dagger \tau_{\sigma_1\sigma_1'}^i c_{\nu_\beta\sigma_1'}), \tag{10.88}$$

where $\tau_{\sigma\sigma'}^i$, $i = x, y, z$ are the Pauli spin matrices defined in Eq. (1.91), one obtains after some rewriting

$$\mathbf{S}_d \cdot \mathbf{S}_{\nu_\alpha, \nu_\beta} = \frac{1}{2} \sum_\sigma c^\dagger_{d\bar\sigma} c_{d\sigma} c^\dagger_{\nu_\alpha\sigma} c_{\nu_\beta\bar\sigma} + \frac{1}{4} \sum_\sigma \left([c^\dagger_{d\bar\sigma} c_{d\sigma} - c^\dagger_{d\bar\sigma} c_{d\bar\sigma}] c^\dagger_{\nu_\alpha\sigma} c_{\nu_\beta\sigma} \right), \quad (10.89)$$

and comparing this with Eq. (10.86), we see that

$$[S^-, H_T^+] \approx -i \sum_{\alpha,\beta=L,R} \sum_{\nu_\alpha \nu_\beta} J_{\alpha\beta} \mathbf{S}_d \cdot \mathbf{S}_{\nu_\alpha, \nu_\beta} - i \sum_{\alpha,\beta=L,R} \sum_{\nu_\alpha \nu_\beta \sigma} W_{\alpha\beta} c^\dagger_{\nu_\alpha\sigma} c_{\nu_\beta\sigma}. \quad (10.90)$$

Here, we have defined the energy for exchange scattering

$$J_{\alpha\beta} = 2 \left(\frac{t_\alpha t_\beta^*}{E_2 - E_1} + \frac{t_\alpha t_\beta^*}{E_0 - E_1} \right) = \frac{2U t_\alpha t_\beta^*}{(\xi_d + U)(-\xi_d)}, \quad (10.91a)$$

and the energy for potential scattering

$$W_{\alpha\beta} = -\frac{1}{2} \left(\frac{t_\alpha t_\beta^*}{E_2 - E_1} + \frac{t_\alpha t_\beta^*}{E_1 - E_0} \right) = \frac{(2\xi_d + U) t_\alpha t_\beta^*}{2(\xi_d + U)(-\xi_d)}. \quad (10.91b)$$

It is instructive to express the exchange and potential scattering amplitudes in terms of the a dimensionless parameter $x = 1 + 2\xi_d/U$, which is zero at the particle-hole symmetric point(i.e., for $E_2 - E_1 = E_1 - E_0$ which is in the middle of an odd Coulomb diamond, see Fig. 10.6) and ± 1 at the charge degeneracy points given by $E_0 = E_1$ and $E_1 = E_2$, respectively. Inserting x into (10.91) gives

$$J_{\alpha\beta} = \frac{8}{1 - x^2} \frac{t_\alpha t_\beta^*}{U}, \quad W_{\alpha\beta} = \frac{2x}{1 - x^2} \frac{t_\alpha t_\beta^*}{U}. \quad (10.92)$$

We see that the exchange coupling $J_{\alpha\beta}$ is maximal at the particle-hole symmetric point and diverges at the charge degeneracy points. In contrast, $W_{\alpha\beta}$ is zero for $x = 0$. Note that the elements of the exchange energy are related as $J_{LR} J_{RL} = J_{LL} J_{RR}$ and similarly for $W_{\alpha\beta}$.

The final expression for the virtual scattering processes is now obtained by inserting Eq. (10.90) into Eq. (10.84), while using that $[S^+, H_T^-] = -[S^-, H_T^+]^\dagger$, and we find

$$H_S^{(2)} = \sum_{\alpha,\beta=L,R} \sum_{\nu_\alpha \nu_\beta} J_{\alpha\beta} \mathbf{S}_d \cdot \mathbf{S}_{\nu\alpha, \nu_\beta} + \sum_{\alpha,\beta=L,R} \sum_{\nu_\alpha \nu_\beta \sigma} W_{\alpha\beta} c^\dagger_{\nu_\alpha\sigma} c_{\nu_\beta\sigma}. \quad (10.93)$$

The first term is the famous Kondo Hamiltonian. Note that the spin-spin interaction is anti-ferromagnetic, which means that the energy is lowered by having anti-parallel spins. Both terms express scattering on the dot due to virtual scattering in and out of the dot, with the energies of the virtual processes appearing in the dominators of $J_{\alpha\alpha'}$ and $W_{\alpha\alpha'}$. While the Kondo Hamiltonian describes the fact that the spin of lead electrons can change, if at the same time the spin of the dot changes, the second term describes the "potential scattering" contribution, which is independent of the spin component. The two processes are illustrated in Fig. 10.5. It is important to realize that the Kondo Hamiltonian is a simplification of the Anderson model, even though the latter seems simpler. The reason for this little paradox is that the dot has four electron states $|0\rangle, |\uparrow\rangle, |\downarrow\rangle$, and $|2\rangle$ that are coupled to the lead electrons, whereas the Kondo model only involves a single spin and hence two states in the dot on which the lead electrons may scatter.

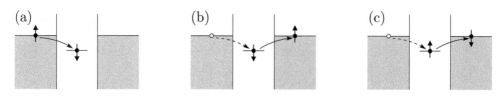

FIG. 10.5. Illustration of the scattering processes that give rise to the Kondo model and the potential scattering terms in Eq. (10.93). (a) shows the initial state with an up spin electron in the left lead and the dot occupied by a single spin down electron. The two possible final states are shown in (b) and (c), where (b) shows a scattering process that conserves the spin of the localized level, while (c) involves an exchange of the spins of transmitted electron and the localized electron. See also Fig. 10.7.

10.5.2 Comparing the Kondo effect in metals and quantum dots

The Kondo model was, as mentioned above, first used in the context of magnetic impurities in metals, where the scattering of electrons on the localized spins gives a contribution to the resistivity. In Chapter 16 we shall see that the *resistivity* of metals is related to the imaginary part of the self-energy, while for the *conductance* of a quantum dot we have seen in Section 10.4 that it is related to the spectral function of the dot states. Therefore, the scattering, which in the case of metals gives rise to an enhancement of the resistivity, in the quantum dot case gives rise to an enhancement of the conductance. To see the relation between scattering on the "impurity" and the current see also Exercise 10.9. Below, we shall see that scattering in the case of conductance in fact gives rise to an anomalous contribution to the conductance. In Section 10.5.5 the essence of the effect is explained qualitatively. Furthermore, in Section 13.7 the self-energy of the lead electrons is calculated for the Kondo-model.

10.5.3 Kondo-model conductance to second order in $H_S^{(2)}$

In this and the following section we calculate the current through a quantum dot up to third order in the exchange couplings, $J_{\alpha\alpha'}$. This will reveal the divergence that is the precursor of the Kondo resonance. In the same spirit as our approach to elastic cotunneling in Section 10.3.3, we consider the amplitude for tunneling an electron from left to right, i.e., the final state of the electron system is $|f^e_{\nu_R\sigma\nu_L\sigma'}\rangle = c^\dagger_{\nu_R\sigma}c_{\nu_L\sigma'}|i^e\rangle$, where $|i^e\rangle$ denotes the initial state of lead electrons. In addition, also the spin of the dot spin may have changed in the process, and since neither the final state nor the initial state is measured, we must sum over all possible initial and final spin states of the dot spin. These are denoted $|\sigma_i\rangle$ and $|\sigma_f\rangle$, respectively. The rate for transferring electrons from left to right is then

$$\Gamma_{RL} = 2\pi \sum_{i\nu_R\nu_L\sigma\sigma'\sigma_i\sigma_f} \frac{1}{2}P_i \left| t^{fi(1)}_{\nu_R\sigma,\nu_L\sigma'} + t^{fi(2)}_{\nu_R\sigma,\nu_L\sigma'} + \cdots \right|^2 \delta(\xi_{\nu_L} - \xi_{\nu_R}), \qquad (10.94)$$

where P_i is the distribution function of the initial electronic state, the factor of $1/2$ is normalization for the average over the two initial spin state, and $t^{fi(n)}_{\nu_R\sigma,\nu_L\sigma'}$ are the transmission amplitudes to order n in $H_S^{(2)}$ in Eq. (10.84).

To first order in $H_S^{(2)}$ the amplitude for the electron transfer from left to right is

$$t_{\nu_R\sigma,\nu_L\sigma'}^{fi(1)} = \langle f_{\nu_R\sigma,\nu_L\sigma'}^e | \langle \sigma_f | H_S^{(2)} | \sigma_i \rangle | i^e \rangle \tag{10.95}$$

$$= \left(W_{RL}\delta_{\sigma\sigma'}\delta_{\sigma_i\sigma_f} + \frac{J_{RL}}{2} \sum_{j=x,y,z} \langle \sigma_f | S_d^j | \sigma_i \rangle \tau_{\sigma\sigma'}^j \right) n_{\nu_L\sigma'}^i [1 - n_{\nu_R\sigma}^i],$$

where $n_{\nu_\alpha\sigma}^i$ is the occupation of the single particle state $\nu_\alpha\sigma$ in lead α in the initial state $|i^e\rangle$. If we were to take the absolute square now to obtain the transmission probability, we would derive the cotunneling contribution, since we have from Section 10.3.3 that cotunneling is second order in Γ, which here corresponds to second order in J or W. The lowest-order contribution to the tunnel rate thus is

$$\Gamma_{RL}^{(2)} \approx 2\pi d_0^2 \int d\xi \, n_F(\xi - \mu_L) \left[1 - n_F(\xi - \mu_R) \right] \left(W_{RL}^2 + \frac{3J_{RL}^2}{8} \right), \tag{10.96}$$

where d_0 is the density of states per spin direction. Here we have used that

$$\sum_{\sigma\sigma'\sigma_i\sigma_f} \left| \sum_j \langle \sigma_f | S_d^j | \sigma_i \rangle \tau_{\sigma\sigma'}^j \right|^2 = \frac{1}{4} \sum_{\sigma\sigma'\sigma_i\sigma_f} \sum_{jk} \tau_{\sigma_i\sigma_f}^j \tau_{\sigma_f\sigma_i}^k \tau_{\sigma\sigma'}^j \tau_{\sigma'\sigma}^k$$

$$= \frac{1}{4} \sum_{jk} \left(\text{Tr} \left[\tau^j \tau^k \right] \right)^2 = 3, \tag{10.97}$$

since $\text{Tr} \left[\tau^j \tau^k \right] = 2\delta_{jk}$.

The second-order contribution Eq. (10.96) corresponds to the elastic cotunneling theory Eq. (10.40).

10.5.4   *Kondo-model conductance to third order in $H_S^{(2)}$*

The peculiarity of the Kondo effect appears in the next order in perturbation theory. The next leading order for the tunneling rate Γ_{RL} is third order in J and W, and we have from Eq. (10.94) that

$$\Gamma_{RL}^{(3)} \approx 4\pi \sum_{\nu_R\nu_L i\sigma\sigma'\sigma_i\sigma_f} \frac{1}{2} P_i \, \text{Re} \left(\left[t_{\nu_R\sigma,\nu_L\sigma'}^{fi(1)} \right]^* t_{\nu_R\sigma,\nu_L\sigma'}^{fi(2)} \right) \delta(\xi_{\nu_R} - \xi_{\nu_L}). \tag{10.98}$$

In order to find the third-order contribution, we move on to calculate the amplitude t_{RL} to second order in $H_S^{(2)}$. To this end, one needs the T-matrix, which was given in Eq. (10.31) (see Section 5.7 for a derivation), and we have

$$t_{\nu_R\sigma,\nu_L\sigma'}^{fi(2)} = \langle f_{\nu_R\sigma,\nu_L\sigma'}^e | \langle \sigma_f | H_S^{(2)} \frac{1}{E_i - H_{LR} + i\eta} H_S^{(2)} | \sigma_i \rangle | i^e \rangle. \tag{10.99}$$

This second-order term must involve the factor J_{RL} and one factor of J_{RR} or J_{LL} in order for $t^{(2)}$ to be non-zero. To study this, we first write $H_S^{(2)}$ as

$$H_S^{(2)} = \sum_{\alpha\beta=L,R} H_{S\alpha\beta}^{(2)}, \tag{10.100a}$$

$$H_{S\alpha\beta}^{(2)} = \sum_{\nu_\alpha\sigma,\nu_\beta\sigma'} c_{\nu_\alpha\sigma}^\dagger c_{\nu_\beta\sigma'} L_{\alpha\alpha',\sigma\sigma'}, \tag{10.100b}$$

$$L_{\alpha\beta,\sigma\sigma'} = W_{\alpha\beta}\delta_{\sigma\sigma'} + \frac{J_{\alpha\beta}}{2} \sum_j S_d^j \tau_{\sigma\sigma'}^j, \tag{10.100c}$$

and then write $t_{\nu_R\sigma,\nu_L\sigma'}^{fi(2)}$ as

$$t_{\nu_R\sigma,\nu_L\sigma'}^{fi(2)} = \sum_{\beta=L,R} \left[\langle i^e | c_{\nu_L\sigma'}^\dagger c_{\nu_R\sigma} \langle \sigma_f | H_{SRL}^{(2)} \frac{1}{E_i - H_{LR} + i\eta} H_{S\beta\beta}^{(2)} | \sigma_i \rangle | i^e \rangle \right.$$

$$\left. + \langle i^e | c_{\nu_L\sigma'}^\dagger c_{\nu_R\sigma} \langle \sigma_f | H_{S\beta\beta}^{(2)} \frac{1}{E_i - H_{LR} + i\eta} H_{SRL}^{(2)} | \sigma_i \rangle | i^e \rangle \right]. \tag{10.101}$$

Inserting Eq. (10.100b) and noting that we can separate the expectation value $\langle i^e | \cdot | i^e \rangle$ in two parts corresponding to the left and right electrodes, we obtain for the term with, say, $\beta = R$,

$$t_{\nu_R\sigma,\nu_L\sigma'}^{fi(2),R} = \langle i_L^e | c_{\nu_L\sigma'}^\dagger c_{\nu_L\sigma'} | i_L^e \rangle \sum_{\nu_1\nu_2\nu_3,\sigma_1\sigma_2\sigma_3}$$

$$\times \left[\langle i_R^e | c_{\nu_R\sigma} c_{\nu_1\sigma_1}^\dagger \frac{\langle \sigma_f | L_{RL,\sigma_1\sigma'} L_{RR,\sigma_2\sigma_3} | \sigma_i \rangle}{\xi_{\nu_3} - \xi_{\nu_2} + i\eta} c_{\nu_2\sigma_2}^\dagger c_{\nu_3\sigma_3} | i_R^e \rangle \right.$$

$$\left. + \langle i_R^e | c_{\nu_R\sigma} c_{\nu_2\sigma_2}^\dagger c_{\nu_3\sigma_3} \frac{\langle \sigma_f | L_{RR,\sigma_2\sigma_3} L_{RL,\sigma_1\sigma'} | \sigma_i \rangle}{\xi_{\nu_L} - \xi_{\nu_1} + i\eta} c_{\nu_1\sigma_1}^\dagger | i_R^e \rangle \right], \tag{10.102}$$

where ν_1, ν_2 and ν_3 are quantum number in the right lead. In this expression we see that $\nu_R\sigma = \nu_2\sigma_2$ and $\nu_1\sigma_1 = \nu_3\sigma_3$ and furthermore, because of the delta function part of Eq. (10.94), one can replace ξ_{ν_L} by ξ_{ν_R}, and the terms combine to

$$t_{\nu_R\sigma,\nu_L\sigma'}^{fi(2),R} = -n_{\nu_L}^i [1 - n_{\nu_R}^i] \sum_{\nu_1,\sigma_1} \left[\frac{\langle \sigma_f | L_{RL,\sigma_1\sigma'} L_{RR,\sigma\sigma_1} | \sigma_i \rangle \, n_{\nu_1\sigma_1}^i}{\xi_{\nu_1} - \xi_{\nu_R} + i\eta} \right.$$

$$\left. + \frac{\langle \sigma_f | L_{RR,\sigma\sigma_1} L_{RL,\sigma_1\sigma'} | \sigma_i \rangle [1 - n_{\nu_1\sigma_1}^i]}{\xi_{\nu_1} - \xi_{\nu_R} - i\eta} \right]. \tag{10.103}$$

Here, we see that the occupation $n_{\nu_1\sigma_1}^i$ of the intermediate state appears in the scattering amplitude. For ordinary potential scattering, i.e., scattering on an external potential, we do not expect the occupation factors to enter the amplitude, because it is essentially a single-particle property. Indeed, below it is shown that the occupation factor is only important for the spin-dependent scattering, i.e., for the first term of $H_S^{(2)}$ in Eq. (10.93), whereas the second term that corresponds to a simple one-body scattering on an external potential (see Eq. (1.61)) does not have such a factor.

The amplitude in Eq. (10.103) has two different contributions: one from the energy conserving scattering, i.e., the delta function part of $1/(\xi_1 - \xi + i\eta)$ and one from the principal part. The first term corresponds to a classical scattering event where the intermediate state conserves energy, while the latter is the non-conserving virtual scattering contribution. The latter term is also the important term for our calculation because it could lead to a large contribution when $\xi_{\nu_1} = \xi_{\nu_R}$, and therefore we concentrate our effort on this term. Now, in order to perform the sum over the intermediate energy ξ_{ν_1}, we assume for simplicity the density of states to be a constant d_0 from a lower cut-off $-D$ to an upper cut-off $+D$. For the term in Eq. (10.103) that does not depend on $n^i_{\nu_1 \sigma_1}$, we are therefore left with

$$-n^i_{\nu_L}[1 - n^i_{\nu_R}] d_0 \mathcal{P} \int_{-D}^{D} d\xi_1 \frac{1}{\xi_1 - \xi_{\nu_R}} \sum_{\sigma_1} \langle \sigma_f | L_{RR,\sigma\sigma_1} L_{RL,\sigma_1\sigma'} | \sigma_i \rangle \approx 0, \quad (10.104)$$

because the integral for $\xi_{\nu_R} \ll D$ is zero.[38] For the other terms, where the occupation $n^i_{\nu_1 \sigma_1}$ of the intermediate state enter, it is quite a different story. For this part, we have two terms in Eq. (10.103) which can be combined into a commutator as:

$$t^{fi(2),R}_{\nu R\sigma, \nu' L\sigma'} = -n^i_{\nu_L}[1 - n^i_{\nu_R}] d_0 \mathcal{P} \int_{-D}^{D} d\xi_{\nu_1} \frac{n^i_{\nu_1}}{\xi_{\nu_1} - \xi_{\nu_R}} \underbrace{\sum_{\sigma_1} \langle \sigma_f | [L_{RL,\sigma_1\sigma'}, L_{RR,\sigma\sigma_1}] | \sigma_i \rangle}_{=c}.$$

$$(10.105)$$

Now it is clear that the potential scattering part of L, see Eq. (10.100c), gives no contribution, because the commutator is zero, while the spin scattering term does. The result we find below thus results from the fact that spin-scattering processes are non-commuting while potential scattering processes commute. More explicitly, we obtain

$$c = \sum_{\sigma_1} \langle \sigma_f | [L_{RL,\sigma_1\sigma'}, L_{RR,\sigma\sigma_1}] | \sigma_i \rangle$$

$$= \sum_{\sigma_1} \langle \sigma_f | \left[\left(W_{RL}\delta_{\sigma_1\sigma'} + \frac{J_{RL}}{2} \sum_j S^j_d \tau^j_{\sigma_1\sigma'} \right), \left(W_{RR}\delta_{\sigma\sigma_1} + \frac{J_{RR}}{2} \sum_j S^j_d \tau^j_{\sigma\sigma_1} \right) \right] | \sigma_i \rangle$$

$$= \frac{J_{RL}J_{RR}}{4} \sum_{jk} \langle \sigma_f | [S^j_d, S^k_d] | \sigma_i \rangle \sum_{\sigma_1} \tau^k_{\sigma\sigma_1} \tau^j_{\sigma_1\sigma'}, \quad (10.106)$$

where have used that $W_{RL}J_{RR} - J_{RL}W_{RR} = 0$, which follows from the definitions of W and J in Eqs. (10.91a) and (10.91b). Therefore, when multiplying $t^{(2)}$ by $t^{(1)}$ in Eq. (10.95), we must evaluate the following two expressions

$$\sum_{\sigma\sigma'\sigma_i\sigma_f} \delta_{\sigma\sigma'}\delta_{\sigma_i\sigma_f} \sum_{jk} \langle \sigma_f | [S^j_d, S^k_d] | \sigma_i \rangle \sum_{\sigma_1} \tau^k_{\sigma\sigma_1} \tau^j_{\sigma_1\sigma'} = 0, \quad (10.107a)$$

$$\sum_{l\sigma\sigma'\sigma_i\sigma_f} \langle \sigma_i | S^l_d | \sigma_f \rangle \tau^l_{\sigma'\sigma} \sum_{jk} \langle \sigma_f | [S^j_d, S^k_d] | \sigma_i \rangle \sum_{\sigma_1} \tau^k_{\sigma\sigma_1} \tau^j_{\sigma_1\sigma'} = a. \quad (10.107b)$$

[38] For a non-symmetric band there is of course a small convergent correction to this result.

To verify these equations we write the spin operators as $S_d^j = \tau^j/2$ and utilize the Pauli-matrix product rule

$$\tau^j \tau^k = \delta_{jk} + i \sum_l \epsilon_{jkl}\, \tau^l, \tag{10.108}$$

where the Levi–Civita symbol ϵ_{ijk} from Eq. (1.11) is used. Eq. (10.107a) is valid because $\sigma_i = \sigma_f$ which leaves us with a commutator of two Pauli matrices which by Eq. (10.108) is a third Pauli matrix, the trace of which is zero. The other term, a, becomes

$$a = \sum \langle \sigma_i | S_d^l | \sigma_f \rangle \langle \sigma_f | \left[S_d^j, S_d^k \right] | \sigma_i \rangle \tau^l_{\sigma'\sigma} \tau^k_{\sigma\sigma_1} \tau^j_{\sigma_1\sigma'} = \frac{1}{8} \sum_{jkl} \mathrm{Tr}\left(\tau^l \left[\tau^j, \tau^k \right] \right) \mathrm{Tr}\left(\tau^l \tau^k \tau^j \right), \tag{10.109}$$

and by Eq. (10.108) the trace of a product of three Pauli matrices is only non-zero if they are all different, and since $\tau^j \tau^k = -\tau^k \tau^j$ for $j \neq k$, we have

$$a = \frac{1}{4} \sum_{jkl} \mathrm{Tr}\left(\tau^l \tau^j \tau^k \right) \mathrm{Tr}\left(\tau^k \tau^l \tau^j \right) = -\sum_{jkl} \epsilon_{ljk}\epsilon_{klj} = -\sum_{jkl} \epsilon_{ljk}^2 = -6. \tag{10.110}$$

where we have used Eq. (10.108) to write $\mathrm{Tr}\left(\tau^l \tau^k \tau^j \right) = 2i\epsilon_{lkj}$.

When combining all the parts that go into $\Gamma_{RL}^{(3)}$, i.e., starting from Eq. (10.98), inserting Eq. (10.95) and Eq. (10.105), which according to Eq. (10.107) gives only one non-zero term, one finally arrives at

$$\Gamma_{RL}^{(3)} \approx -\frac{3\pi}{2}(J_{RL}d_0)^2 \int d\xi\, n_\mathrm{F}(\xi - \mu_L) \left[1 - n_\mathrm{F}(\xi - \mu_R) \right]$$
$$\times (J_{LL} + J_{RR})d_0 \mathcal{P} \int_{-D}^{D} d\xi_1 \frac{n_\mathrm{F}(\xi_1)}{\xi_1 - \xi}, \tag{10.111}$$

where we furthermore inserted the term proportional to J_{LL}, because we remembered that in Eq. (10.101), we put $\beta = R$, but we could instead have used $\beta = L$ in which case J_{RR} is replaced by J_{LL}. The factor $\frac{3\pi}{2}$ comes from: 2π (Eq. (10.98)) $\times$ 6 (Eq. (10.110)) $\times \frac{1}{2}$ (Eq. (10.95)) $\times \frac{1}{4}$ (Eq. (10.106)).

Adding the two contributions in Eqs. (10.96) and (10.111) we obtain the current running from left to right. Similar expression can of course be derived from the scatterings from right to left and after subtracting the two, we obtain for the current and the conductance (reinserting $\hbar$)

$$I = \frac{e}{h} \int d\xi\, [n_\mathrm{F}(\xi - \mu_L) - n_\mathrm{F}(\xi - \mu_R)]\, T(\xi), \tag{10.112a}$$

$$G = \frac{e^2}{h} \int d\xi \left(-\frac{dn_\mathrm{F}(\xi)}{d\xi} \right) T_{\mu_L = \mu_R = 0}(\xi), \tag{10.112b}$$

where

$$T(\xi) \approx 4\pi^2 (W_{RL} d_0)^2 + \frac{3}{2}\pi^2 (J_{RL} d_0)^2 \left(1 - 2(J_{LL} + J_{RR}) d_0 \mathcal{P} \int_{-D}^{D} d\xi_1 \frac{n_F(\xi_1)}{\xi_1 - \xi} \right).$$
(10.113)

This expression for the transmission coefficient is one of the main result of our discussion of the Kondo effect in quantum dots. It shows that there is a peak in the conductance at low temperatures, because for $T \ll D$ the integral in Eq. (10.113) has a logarithmic divergence, as we will see below. As a side remark, we can now make a connection to the spectrum function for the Anderson model. The relationship between conductance and the local spectral function was derived in Eq. (10.56). Here, we have also derived an equation for the conductance for the Anderson model projected onto the singly occupied space. In this limit, we can therefore identify $T(\xi)$ and $A(d\sigma, \xi)$.

Let us now focus on the conductance for the particle-hole symmetric point where $W_{LR} = 0$. Inserting Eq. (10.113) into Eq. (10.112b), we find the perturbative result for the conductance:

$$G = \frac{3\pi^2 e^2}{2h} g_{LR}^2 \left(1 + 4gf(D/T) \right),$$
(10.114a)

$$f(x) = \ln(x) + 0.568 \ldots \approx \ln(x) \text{ for } x \ll 1.$$
(10.114b)

where $g_{LR}^2 = J_{LL} J_{RR} d_0^2$ and $g = (J_{LL} + J_{RR}) d_0 / 2$. Here, energy integrations was done by first rewriting the integral in Eq. (10.112b) as

$$- \int_{-\infty}^{\infty} d\xi \int_{-D}^{D} d\xi_1 \left(-\frac{dn_F(\xi)}{d\xi} \right) \frac{n_F(\xi_1)}{\xi_1 - \xi} = \int_{-\infty}^{\infty} d\xi \int_0^D d\xi_1 \left(-\frac{dn_F(\xi)}{d\xi} \right) \left(\frac{2\xi n_F(\xi_1)}{\xi^2 - \xi_1^2} + \frac{1}{\xi + \xi_1} \right),$$

such that the first term vanishes by antisymmetry of the integrand leaving the last term which can be integrated over ξ_1 to give

$$\int_{-\infty}^{\infty} d\xi \left(-\frac{dn_F(\xi)}{d\xi} \right) \ln \left| \frac{D}{\xi} \right| = \ln \left(\frac{D}{T} \right) - \int_{-\infty}^{\infty} dx \frac{e^x}{(e^x + 1)^2} \ln |x| = f(D/T). \quad (10.115)$$

We have now reached the important conclusion that the perturbative results for the conductance diverges logarithmically at low temperatures. Therefore at sufficiently low temperatures the perturbation theory stops to be valid, and the system crosses over to a strong coupling regime, where one has to use other methods. These methods are more advanced many-particle methods which are outside the scope of this book.

However, we can give a simple estimate of the temperature at which the Kondo resonance starts to dominate transport: when the last term in the parenthesis in Eq. (10.113) is comparable to the first term. This defines the so-called Kondo temperature, T_K. Based on this definition, the Kondo temperature is

$$T_K = D \exp(-1/4g) = \exp \left(-\frac{1}{2(J_{LL} + J_{RR}) d_0} \right).$$
(10.116)

Below the Kondo temperature we thus expect perturbation theory to break down and there the so-called Kondo resonance evolves. In typical experimental realizations in

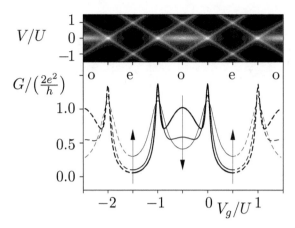

FIG. 10.6. Illustration of the conductance properties of a mesoscopic Kondo system with discrete levels. Top panel: gray scale plot of the differential conductance dI/dV as in Fig. 10.2. The bright slanted lines originate from classically allowed transitions between charge states N and $N+1$. The center of each "diamond" corresponds to a state with an integer number of electrons. For odd numbers (o) the Kondo effect generates an enhanced conductance (bright horizontal lines), whereas for even numbers (e) the conductance remains small. Bottom panel: the linear-response conductance for low (thick line), medium (normal line), and high (thin line) temperature. The arrows indicate the increase and decrease of the conductance as a function of temperature for the even and odd number occupations, respectively. In this section, we have only dealt with a single-level model restricting the allowed occupations to 0, 1, and 2 (the full lines). For multi-level systems the pattern is repeated (dashed lines).

quantum dot systems, the Kondo temperature is in the range of a few kelvin. The subtle and interesting Kondo resonance has been observed in many experiments in quantum dots and in various scenarios for creating the doubly degenerate groundstate of the dot. In Fig. 10.6, we show an illustration of the Kondo effect in a quantum dot. The Kondo peak is clearly seen as an enhancement in the differential conductance for odd-number occupations.

10.5.5 *Origin of the logarithmic divergence*

In this section, we pin-point the important scattering processes that lead to the logarithmic divergence. The primary difference between scattering off a spin and scattering off a static potential is that the first is a many-body effect, while the latter is a single-particle effect. In Chapter 7 we saw how, e.g., the conductance of non-interacting particles is fully determined by the one-particle transmission coefficients and the fact that there are other electrons only enters through the distribution function of the initial state. Therefore, only processes where the occupation factors of the intermediate states enter are true many-body processes. One such example is the second-order contribution to the transmission amplitude $t^{(2)}$ calculated in Section 10.5.4. In Fig. 10.7,

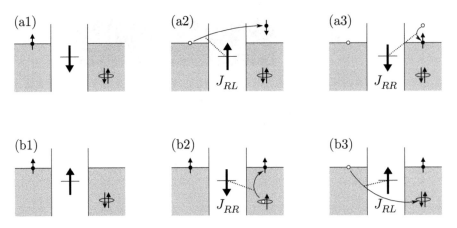

FIG. 10.7. Illustration of the two scattering processes that add up and give a log-arithmically diverging contribution to the scattering. It is second order in the exchange coupling and involves two spin flip processes. In (a) an electron is first scattered from left to right into an empty state and then in a second interaction event scattered down to a state near the Fermi energy. In (b) the same final state is reached through a slightly different path: first an electron-hole pair is created on the right hand side, then the hole is filled by an electron from the left. In (a) and (b) the initial state of the localized spin need not be the same, and because the non-commuting spin flips processes of the localized spin is reversed, the many possible intermediate add up. This is in stark contrast to a potential scattering process where the two scattering commute and the many-body contribution in fact cancels. See Section 10.5.5 for a discussion.

we show the processes that involve a scattering of a spin-up electron from left to right, via an intermediate state where the spin has been reversed. This can only happen if the localized spin also flips. The interaction between the two can be written as $\mathbf{s} \cdot \mathbf{S} = s^+ S^- + s^- S^+ + s_z S_z$ and the sum of the two different paths then is

$$t_{RL}^{(2)} \propto J_{RR} J_{RL} \sum_\nu \left(\frac{S^- S^+ (1 - n_\nu)}{-\xi_\nu} - \frac{S^+ S^- n_\nu}{\xi_\nu} \right). \tag{10.117}$$

The minus sign appears because the lead electrons appear in different order in the two terms: the first term has a factor $s_{RR}^+ s_{RL}^- = c_{R\uparrow}^\dagger c_{R\downarrow} c_{R\downarrow}^\dagger c_{L\uparrow}$, whereas the second term has a factor $s_{RL}^- s_{RR}^+ = c_{R\downarrow}^\dagger c_{L\uparrow} c_{R\uparrow}^\dagger c_{R\downarrow}$, and since the down-spin operators should pair up to give $1 - n$ and n, respectively, we see that the order of $c_{R\uparrow}^\dagger$ and $c_{L\downarrow}$ is indeed reversed.

The conclusion is now clear: because S^+ and S^- do not commute, $[S^+, S^-] = S_z$, the occupation factors for the intermediate state do not cancel. The logarithmic divergence appears because there is a large number of many-body intermediate states that can mediate the process and because the weight of these (assuming a constant density of states) falls off as $1/\xi$. Thus we have

$$t_{RL}^{(2)} \propto S_z \int_0^D d\xi \, \frac{1}{\xi}. \tag{10.118}$$

Combining this with the first order expression for the same process $t_{RL}^{(1)} \propto S_z$, we conclude that the many-body effect survives and they can potentially give a large contribution to the scattering coefficient, in agreement with the more detailed calculation performed in the preceding sections.

10.5.6 The Kondo problem beyond perturbation theory

From the form of the Kondo temperature in Eq. (10.116), we see that the result is in fact non-perturbative in $J_{\beta\beta}$. Therefore, we cannot expect to gain much by continuing our perturbation expansion and, as mentioned above, other methods and indeed non-perturbative ones, have to be utilized. For the equilibrium properties of the Kondo or Anderson models, one can use the numerical renormalization group methods developed by Wilson (1975). This method has since been extended to calculations of response functions, such as, e.g., the conductance. Alternatively, the Anderson model can be solved using the Bethe ansatz (Tsvelick and Wiegmann 1983).

An alternative approach to the Kondo problem is the so-called "poor man's scaling" developed by Anderson (see, e.g., the book by Hewson). The principle behind the procedure is to relate Kondo Hamiltonians with different exchange couplings and bandwidth with each other, such that by decreasing the bandwidth but increasing the exchange coupling the spectrum remains the same. The result of such a scaling analysis is that if one sums up the higher terms and keep the most divergent terms, the logarithmic term in Eq. (10.114a) is the first term of a geometric series which can be summed up to give

$$G \approx \frac{3\pi^2 e^2}{2h} \frac{g_{LR}^2}{1 - 4g \ln(D/T)} = \frac{3\pi^2 e^2}{h} \frac{g_{LR}^2}{g_{LL}^2 + g_{RR}^2} \frac{1}{\ln(T/T_K)}, \tag{10.119}$$

where the Kondo temperature was given in Eq. (10.116). This result describes the high temperature limit $T > T_K$, whereas for low temperatures, one most use other non-perturbative methods as discussed above. What remains true, however, is that the conductance is a function of a single parameter, namely T/T_K, and the Kondo temperature therefore appears as a new characteristic energy scale of the problem. Moreover, it turns out that the conductance at the Kondo resonance, i.e., when the dot is occupied by one electron, increases all the way to the non-interacting limit, $2e^2/h$ (for the symmetric case $t_L = t_R$), which is indeed a very remarkable result. The conductance of a quantum dot in the case with an odd number of electrons occupying the dot is sketched in Fig. 10.8 as a function of temperature. At temperatures larger than $T > \Delta/k_B$, where Δ is the energy for adding or removing an electron, the Coulomb blockade is destroyed by thermal fluctuations.[39] At lower temperatures, $T_K < T < \Delta/k_B$, the conductance drops due to Coulomb blockade and the background conductance is given by either cotunneling or thermally activated tunneling.

---

[39]If the spacing between single-particle energy levels, δ is smaller than Δ there is an additional regime $\delta < T < \Delta$

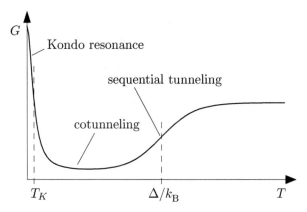

FIG. 10.8. Sketch of the temperature dependence of the conductance of a quantum dot ranging from the high temperature sequential tunneling result, $k_{\mathrm{B}}T \approx \Delta$ to the low temperature Kondo resonance behavior, $T < T_K$.

Finally, at temperatures below the Kondo temperature the strong coupling Kondo resonance sets in and the conductance approaches $2e^2/h$. We refer the interested reader to the review by Pustilnik and Glazman (2004), where the transition between the various regimes is further reviewed. See also the review by Aleiner *et al.* (2002). Fig. 10.8 shows schematically this temperature dependence.

10.6   Summary and outlook

The chapter has been devoted to calculating transport properties of mesoscopic systems using the many-body formalism. Many of the results have been derived starting from Fermi's golden rule, which is valid in the weak tunneling limits. In the sequential tunneling limit the current could be found from a kinetic equation approach which determined the occupations of the quantum states of the dot. The steady state solution was determined from

$$0 = -\sum_{\beta} \Gamma_{\beta\alpha} P(\alpha) + \sum_{\beta} \Gamma_{\alpha\beta} P(\beta),$$

where the tunneling rates $\Gamma_{\beta\alpha}$ are determined by Fermi's golden rule. In this approach the current is always linear in Γ coefficients and the current is blocked at low temperatures and low voltages unless two charge states are generate.

However, to higher orders in the tunneling coupling the electrons can tunnel through the classically forbidden charge states; a process known as cotunneling or super-exchange. This process is second order in the Γ coefficients, and in order to derive an expression for the current we had to resort to the generalized Fermi's golden using the T-matrix. The general form of the cotunneling contribution was found to be $I_{\mathrm{cotun}} \propto V\Gamma_L\Gamma_R/E$, where E is the energy of the intermediate forbidden state.

For even stronger tunneling coupling, the Green's function technique turned out to be useful, and we expressed the conductance in terms of the spectral function. For the Anderson-type model, we found that

$$G = e^2 \sum_\sigma \int \frac{d\xi}{2\pi} \frac{\Gamma^L \Gamma^R}{\Gamma^L + \Gamma^R} A(d\sigma, \xi) \left(-\frac{\partial n_F(\xi)}{\partial \xi} \right).$$

Unfortunately, it is in general not possible to solve for the Green's function due to the combination of interactions in the dot and the coupling to the continuum of state in the leads. In this book, we have so far seen two suggestions how to find the spectral function for an Anderson model. In Section 9.3 we saw that a mean-field approximation leads to a Lorentzian form for the spectral function, while in Section 10.4.2 in was shown that a better approximation leads to a two-peak structure of the spectral function. Only when the tunneling-broadened energy Γ is much larger than the interaction energy U is the mean-field result recovered. The two-peak structure of the spectral function reflects the broadening of the Coulomb blockade, obtained in the rate equation approach, due to finite tunneling life time.

For even stronger tunneling coupling and at low temperature yet another interesting effect sets in, namely the Kondo effect. In this regime, the spectral function develops a peak at the Fermi energy which corresponds to the resonant spin-flip scattering and the conductance therefore increases again.

We have seen, how the different techniques in many-body theory had to be used in order to understand the current-voltage characteristic of mesoscopic systems. In the following chapters of this book, we will expand on the tools available, e.g., the Feynman diagram technique which is explored in Chapters 12 and 13. Finally, we return to the physics of mesoscopic systems in Chapter 16, where we discuss the effects of disorder, and in Chapter 19 where electron correlations in one-dimension are discussed.

For further reading and references, the reader can consult some of the reviews (Averin and Likharev (1990), Aleiner et al. (2002), Pustilnik and Glazman (2004)) and books (Grabert and Devoret (1992), Ferry and Goodnick (1999) Sohn et al. (1997)) on interaction effects in general, and more specialized books on Kondo physics (Hewson 1993) and kinetic equations (Breuer and Petruccione (2002) and (Weiss 1999)).

11

IMAGINARY-TIME GREEN'S FUNCTIONS

We have seen that physical observables often have the form of Green's functions, or that they can be derived in a simple way from the Green's functions. In all the situations we have so far studied the physical observables have been related to the retarded Green's functions. In analogy with Eqs. (8.28) and (8.29), a more general class of functions containing the Green's functions can be defined as

$$C_{AB}^R(t,t') = -i\theta(t-t')\left\langle [A(t), B(t')]_{B,F} \right\rangle, \quad \begin{cases} B: & \text{for bosons} \\ F: & \text{for fermions} \end{cases}. \tag{11.1}$$

When A and B are single-particle annihilation and creation operators, it is the single particle Green's function defined in Eq. (8.28), from which one could derive the density of states. When A and B are two-particle operators, e.g., the density or current operators, C^R has the form of a retarded correlation function that was shown to give the linear response results of Chapter 6. In Eq. (11.1) boson operators mean either single-particle operators like b or $b^\dagger$ or an even number of fermion operators such as $c^\dagger c$ appearing in, e.g., the density operator ρ. The important feature that distinguishes the boson case from the fermion case is the sign change obtained upon interchange.

In this chapter, we introduce a mathematical method to work out the retarded Green's functions. For technical reasons it is convenient to use a mapping to a more general Green's function, where the time and frequency arguments are imaginary quantities. This has no real physical meaning, and is only a clever mathematical trick, which we need to learn. This is much like treating electrical circuit theory with complex numbers even though all currents and voltages are real. The present chapter concentrates on the mathematical details of the technique, while applications are left for later. The imaginary time formalism is particularly useful when we want to perform perturbation theory, and this will eventually lead us to the Feynman diagrams.

Let us for example look at the definition of the following correlation function:

$$C_{AB}(t,t') = -\left\langle A(t)B(t') \right\rangle, \tag{11.2}$$

from which we can find the retarded function as $C^R = i\theta(t-t')(C_{AB} \mp C_{BA})$. By definition we have

$$C_{AB}(t,t') = -\frac{1}{Z}\text{Tr}\left[e^{-\beta H} A(t)B(t') \right]. \tag{11.3}$$

Suppose the Hamiltonian is $H = H_0 + V$, where V is the perturbation. Then we saw in Chapter 5 that the interaction picture provides a systematic way of expanding in powers of V. We could try to utilize this and write C_{AB} as

$$C_{AB}(t,t') = -\frac{1}{Z}\text{Tr}\left[e^{-\beta H}\hat{U}(0,t)\hat{A}(t)\hat{U}(t,t')\hat{B}(t')\hat{U}(t',0) \right]. \tag{11.4}$$

In Eq. (5.18) we saw also how a single $\hat{U}$ operator could be expanded as a time-ordered exponential. This would in Eq. (11.4) result in three time-ordered exponentials, which could be collected into a single time-ordered exponential. But the trouble arises for the density matrix $e^{-\beta H}$, which should also be expanded in powers of the interaction. To make a long story short: this is a mess, and a new idea is therefore needed. One solution to this problem is to use imaginary times instead of real times, but bear in mind that this is just a mathematical trick without physical contents.

To employ imaginary time is not as far fetched as it might look, because both the density operator $\rho = e^{-\beta H}/Z$ from Eq. (1.116) and the time-evolution operator $U(t) = e^{-iHt}$ are exponential functions of the Hamiltonian.[40] They therefore satisfy similar differential equations: U satisfies the Schrödinger equation, $i\partial_t U = HU$, while ρ is the solution to $\partial_\beta \rho = -H\rho$, known as the Bloch equation. In order to treat both U and ρ in one go, the time argument is replaced by an imaginary quantity $t \to -i\tau$, where τ is real and has the dimension time. In the end this means that both U and ρ can be treated in just one expansion in powers of V. Furthermore, we will see that there is a well-defined method to obtain the physically relevant quantity, i.e., to go back to physical real times from the imaginary-time function.

As for real time we can define an imaginary-time Heisenberg picture by substituting it by τ. We define

$$A(\tau) = e^{\tau H} A e^{-\tau H}, \quad \tau \text{ a Greek letter.} \tag{11.5}$$

In this notation, the imaginary-time definitions are used when the time argument is a Greek letter . The usual definition is used when the times are written with roman letters,

$$A(t) = e^{itH} A e^{-itH}, \quad t \text{ a Roman letter.} \tag{11.6}$$

Similar to the interaction picture defined for real times, we can define the interaction picture for imaginary times as

$$\hat{A}(\tau) = e^{\tau H_0} A e^{-\tau H_0}. \tag{11.7}$$

Letting $H = H_0 + V$, the relation between the Heisenberg and the interaction picture in imaginary time follows the arguments in Chapter 5. If we consider a product of operators $A(\tau)B(\tau')$ and write it in terms of the corresponding operators in the interaction representation, we get

$$A(\tau)B(\tau') = \hat{U}(0,\tau)\hat{A}(\tau)\hat{U}(\tau,\tau')\hat{B}(\tau')\hat{U}(\tau',0), \tag{11.8}$$

where, like in Eq. (5.12), the time-evolution operator $\hat{U}$ in the interaction picture is

$$\hat{U}(\tau,\tau') = e^{\tau H_0} e^{-(\tau-\tau')H} e^{-\tau' H_0}, \tag{11.9}$$

from which it follows directly that

---

[40]Note that we consider only time-independent Hamiltonians in this section. If they are time dependent, one cannot use the ordinary equilibrium statistical mechanics but instead one must use a non-equilibrium formalism. This we did in the linear response limit in Chapter 6, but we will not cover the more general case of nonlinear time-dependent response in this book.

$$\hat{U}(\tau, \tau'')\hat{U}(\tau'', \tau') = \hat{U}(\tau, \tau'). \tag{11.10}$$

An explicit expression for $U(\tau, \tau')$ is found in analogy with the derivation of Eq. (5.18). First we differentiate Eq. (11.9) with respect to τ and find

$$\partial_\tau \hat{U}(\tau, \tau') = e^{\tau H_0}(H_0 - H)e^{-(\tau - \tau')H}e^{-\tau' H_0} = -\hat{V}(\tau)\hat{U}(\tau, \tau'). \tag{11.11}$$

This is analogous to Eq. (5.13) and the boundary condition, $\hat{U}(\tau, \tau) = 1$, is of course the same. Now the same iterative procedure is applied, and we end with

$$\hat{U}(\tau, \tau') = \sum_{n=0}^{\infty} \frac{1}{n!}(-1)^n \int_{\tau'}^{\tau} d\tau_1 \cdots \int_{\tau'}^{\tau} d\tau_n \, T_\tau \left(\hat{V}(\tau_1) \cdots \hat{V}(\tau_n) \right)$$

$$= T_\tau \exp\left(-\int_{\tau'}^{\tau} d\tau_1 \hat{V}(\tau_1) \right). \tag{11.12}$$

The time ordering is again the same as defined in Section 5.3, i.e., the operators are ordered such that $T_\tau[A(\tau)B(\tau')]$ is equal to $A(\tau)B(\tau')$ for $\tau > \tau'$ and $B(\tau')A(\tau)$ when $\tau' > \tau$. Above it was argued that the density operator naturally can be treated within the imaginary time formalism, and indeed it can, because by combining Eqs. (11.9) and (11.12) we obtain

$$e^{-\beta H} = e^{-\beta H_0}\hat{U}(\beta, 0) = e^{-\beta H_0}T_\tau \exp\left(-\int_0^\beta d\tau_1 \hat{V}(\tau_1) \right). \tag{11.13}$$

Below, we use this property together with the properties of the time-ordering to write expectation values in a very compact and useful way.

Consider the time-ordered expectation value of the pair of operators in Eq. (11.8)

$$\langle T_\tau A(\tau)B(\tau') \rangle = \frac{1}{Z}\text{Tr}\left\{ e^{-\beta H}T_\tau[A(\tau)B(\tau')] \right\}, \tag{11.14}$$

and let us first study the case $\beta > \tau > \tau' > 0$. In this case, the operators $A(\tau)$ and $B(\tau)$ are ordered as in Eq. (11.8). Inserting Eqs. (11.8) and (11.13), we have

$$\langle T_\tau A(\tau)B(\tau') \rangle = \frac{1}{Z}\text{Tr}\left[e^{-\beta H_0}\hat{U}(\beta, 0)\hat{U}(0, \tau)\hat{A}(\tau)\hat{U}(\tau, \tau')\hat{B}(\tau')\hat{U}(\tau', 0) \right]$$

$$= \frac{1}{Z}\text{Tr}\left[e^{-\beta H_0}\hat{U}(\beta, \tau)\hat{A}(\tau)\hat{U}(\tau, \tau')\hat{B}(\tau')\hat{U}(\tau', 0) \right], \tag{11.15}$$

where Eq. (11.10) has been used to combine $\hat{U}(\beta, 0)$ and $\hat{U}(0, \tau)$. Now the time-ordering is reintroduced. First we observe that when writing $\hat{U}(\tau, \tau')$ as in Eq. (11.12) only interactions $\hat{V}(\tau'')$ with time arguments τ'' between τ and τ' enter. Therefore, by definition of the time-ordering operator we may write

$$\hat{U}(\beta, \tau)\hat{A}(\tau)\hat{U}(\tau, \tau')\hat{B}(\tau')\hat{U}(\tau', 0) = T_\tau \left(\hat{U}(\beta, 0)\hat{A}(\tau)\hat{B}(\tau') \right). \tag{11.16}$$

This was derived for the case $\tau > \tau'$, and in the opposite case $\tau < \tau'$ the same line of arguments are used. First write

$$T_\tau \left(\hat{A}(\tau)\hat{B}(\tau') \right) = \pm \hat{B}(\tau')\hat{A}(\tau), \tag{11.17}$$

and after the same steps as above, we have similarly to Eq. (11.15) that

$$\pm \hat{U}(\beta,\tau)\hat{B}(\tau')\hat{U}(\tau,\tau')\hat{A}(\tau)\hat{U}(\tau',0) = T_\tau \left(\hat{U}(\beta,0)\hat{A}(\tau)\hat{B}(\tau') \right), \tag{11.18}$$

where the sign change was reabsorbed into time-ordering operator. We now see that Eq. (11.16) and Eq. (11.18) are identical and therefore we have the result for both $\tau > \tau'$ and $\tau < \tau'$.

Our final result for the expectation value in Eq. (11.14) is

$$\begin{aligned} \langle T_\tau \left(A(\tau)B(\tau') \right) \rangle &= \frac{1}{Z}\mathrm{Tr}\left[e^{-\beta H_0} T_\tau \left(\hat{U}(\beta,0)\hat{A}(\tau)\hat{B}(\tau') \right) \right] \\ &= \frac{\left\langle T_\tau \left(\hat{U}(\beta,0)\hat{A}(\tau)\hat{B}(\tau') \right) \right\rangle_0}{\left\langle \hat{U}(\beta,0) \right\rangle_0}, \end{aligned} \tag{11.19}$$

where we have used that $Z = \mathrm{Tr}\left[e^{-\beta H} \right] = \mathrm{Tr}\left[e^{-\beta H_0}\hat{U}(\beta,0) \right]$, and where the averages $\langle \cdots \rangle_0$ appear after we divide both numerator and denominator by $Z_0 = \mathrm{Tr}\left[e^{-\beta H_0} \right]$.

The result in Eq. (11.19) demonstrates that the trick of using imaginary time indeed allows for a systematic expansion of the complicated looking expression in Eq. (11.4). However, before we can see the usefulness fully, we need to relate the correlation functions written in imaginary time and the correlation function with real-time arguments.

11.1 Definitions of Matsubara Green's functions

The imaginary-time Green's functions, also called Matsubara Green's functions, are defined in the following way

$$\mathcal{C}_{AB}(\tau,\tau') \equiv -\left\langle T_\tau \left[A(\tau)B(\tau') \right] \right\rangle, \tag{11.20}$$

where the time-ordering symbol in imaginary time has been introduced. It means that operators are ordered according to history and just like the time-ordering operator seen in Chapter 5 with the later "times" to the left,

$$T_\tau \left[A(\tau)B(\tau') \right] = \theta(\tau - \tau')A(\tau)B(\tau') \pm \theta\left(\tau' - \tau \right) B(\tau')A(\tau), \quad \begin{cases} + & \text{for bosons,} \\ - & \text{for fermions.} \end{cases} \tag{11.21}$$

The next question is: What values can τ have? From the definition in Eq. (11.20) three things are clear. First, $\mathcal{C}_{AB}(\tau,\tau')$ is a function of the time difference only, i.e.,

$\mathcal{C}_{AB}(\tau, \tau') = \mathcal{C}_{AB}(\tau - \tau')$. This follows from the cyclic properties of the trace. We have for $\tau > \tau'$

$$
\begin{aligned}
\mathcal{C}_{AB}(\tau, \tau') &= \frac{-1}{Z} \operatorname{Tr}\left[e^{-\beta H} e^{\tau H} A e^{-\tau H} e^{\tau' H} B e^{-\tau' H} \right] \\
&= \frac{-1}{Z} \operatorname{Tr}\left[e^{-\beta H} e^{-\tau' H} e^{\tau H} A e^{-\tau H} e^{\tau' H} B \right] \\
&= \frac{-1}{Z} \operatorname{Tr}\left[e^{-\beta H} e^{(\tau - \tau') H} A e^{-(\tau - \tau') H} B \right] \\
&= \mathcal{C}_{AB}(\tau - \tau'),
\end{aligned}
\tag{11.22}
$$

and of course likewise for $\tau' > \tau$. Second, convergence of $\mathcal{C}_{AB}(\tau, \tau')$ is guaranteed only if $-\beta < \tau - \tau' < \beta$. For $\tau > \tau'$ the equality $\tau - \tau' < \beta$ is clearly seen if one uses the Lehmann representation in Eq. (11.22) to get a factor $\exp\left(-\left[\beta - \tau + \tau'\right] E_n\right)$, and likewise, the second equality is obtained if $\tau < \tau'$. Thirdly, we have the property

$$
\mathcal{C}_{AB}(\tau) = \pm \mathcal{C}_{AB}(\tau + \beta), \quad \text{for } \tau < 0,
\tag{11.23}
$$

which again follows from the cyclic properties of the trace. The proof of Eq. (11.23) for $\tau < 0$ is

$$
\begin{aligned}
\mathcal{C}_{AB}(\tau + \beta) &= \frac{-1}{Z} \operatorname{Tr}\left[e^{-\beta H} e^{(\tau + \beta) H} A e^{-(\tau + \beta) H} B \right] \\
&= \frac{-1}{Z} \operatorname{Tr}\left[e^{\tau H} A e^{-\tau H} e^{-\beta H} B \right] \\
&= \frac{-1}{Z} \operatorname{Tr}\left[e^{-\beta H} B e^{\tau H} A e^{-\tau H} \right] \\
&= \frac{-1}{Z} \operatorname{Tr}\left[e^{-\beta H} B A(\tau) \right] \\
&= \pm \frac{-1}{Z} \operatorname{Tr}\left[e^{-\beta H} T_\tau \left(A(\tau) B \right) \right] \\
&= \pm \mathcal{C}_{AB}(\tau).
\end{aligned}
\tag{11.24}
$$

11.1.1 *Fourier transform of Matsubara Green's functions*

Next we wish to find the Fourier transforms with respect to the "time" argument τ. Because of the properties above, we take $\mathcal{C}_{AB}(\tau)$ to be defined in the interval $-\beta < \tau < \beta$, and thus according to the theory of Fourier transformations we have a discrete Fourier series on that interval given by

$$
\mathcal{C}_{AB}(n) \equiv \frac{1}{2} \int_{-\beta}^{\beta} d\tau \, e^{i\pi n \tau / \beta} \mathcal{C}_{AB}(\tau),
\tag{11.25a}
$$

$$
\mathcal{C}_{AB}(\tau) = \frac{1}{\beta} \sum_{n=-\infty}^{\infty} e^{-i\pi n \tau / \beta} \mathcal{C}_{AB}(n).
\tag{11.25b}
$$

However, due to the symmetry property (11.24) this can be simplified as

$$
\begin{aligned}
\mathcal{C}_{AB}(n) &= \frac{1}{2} \int_0^\beta d\tau\, e^{i\pi n \tau/\beta} \mathcal{C}_{AB}(\tau) + \frac{1}{2} \int_{-\beta}^0 d\tau\, e^{i\pi n \tau/\beta} \mathcal{C}_{AB}(\tau), \\
&= \frac{1}{2} \int_0^\beta d\tau\, e^{i\pi n \tau/\beta} \mathcal{C}_{AB}(\tau) + e^{-i\pi n} \frac{1}{2} \int_0^\beta d\tau\, e^{i\pi n \tau/\beta} \mathcal{C}_{AB}(\tau - \beta), \\
&= \frac{1}{2}\left(1 \pm e^{-i\pi n}\right) \int_0^\beta d\tau\, e^{i\pi n \tau/\beta} \mathcal{C}_{AB}(\tau),
\end{aligned}
\tag{11.26}
$$

and since the factor $\left(1 \pm e^{-i\pi n}\right)$ is zero for plus sign and n odd, or for minus sign and n even, and 2 otherwise, we obtain

$$
\mathcal{C}_{AB}(n) = \int_0^\beta d\tau\, e^{i\pi n \tau/\beta} \mathcal{C}_{AB}(\tau), \quad
\begin{cases}
n \text{ is even for bosons}, \\
n \text{ is odd for fermions}.
\end{cases}
\tag{11.27}
$$

From now on we use the following notation for the Fourier transforms of the Matsubara Green's functions

$$
\mathcal{C}_{AB}(i\omega_n) = \int_0^\beta d\tau\, e^{i\omega_n \tau} \mathcal{C}_{AB}(\tau), \quad
\begin{cases}
\omega_n = \frac{2n\pi}{\beta}, & \text{for bosons}, \\
\omega_n = \frac{(2n+1)\pi}{\beta}, & \text{for fermions}.
\end{cases}
\tag{11.28}
$$

The frequency variable ω_n is denoted a Matsubara frequency. Note how the information about the temperature is contained in the Matsubara frequencies through $\beta = 1/k_{\mathrm{B}}T$.

Finally, we remark that the boundaries of the integral $\int_0^\beta d\tau$ in Eq. (11.28) leads to a minor ambiguity of how to treat the boundary $\tau = 0$, e.g., if $\mathcal{C}_{AB}(\tau)$ includes a delta function $\delta(\tau)$. A consistent choice is always to move the time argument into the interior of the interval $[0, \beta]$, e.g., replace $\delta(\tau)$ by $\delta(\tau - 0^+)$.

11.2 Connection between Matsubara and retarded functions

We shall now see why the Matsubara Green's functions have been introduced at all. In the frequency domain they are in fact the same analytic function as the usual real times Green's functions. In other words, there exists an analytic function $\mathcal{C}_{AB}(z)$, where z is a complex frequency argument in the upper half plane, that equals $\mathcal{C}_{AB}(i\omega_n)$ on the imaginary axis and $C_{AB}^R(\omega)$ on the real axis. This means that once we have one of the two, the other one follows by analytic continuation. Since it is in many cases much easier to compute the Matsubara function $\mathcal{C}_{AB}(i\omega_n)$, this is a powerful method for finding the corresponding retarded function. Indeed we shall now show that the appropriate analytic continuation is $C_{AB}^R(\omega) = \mathcal{C}_{AB}(i\omega_n \to \omega + i\eta)$, where η is a positive infinitesimal.

The relation between the two functions $\mathcal{C}_{AB}$ and C_{AB}^R is proven by use of the Lehmann representation. In Section 8.3.3 we calculated the retarded single-particle

Green's function, and the result (8.48) can be carried over for fermions. In the general case we obtain

$$C_{AB}^R(\omega) = \frac{1}{Z} \sum_{nn'} \frac{\langle n|A|n'\rangle \langle n'|B|n\rangle}{\omega + E_n - E_{n'} + i\eta} \left(e^{-\beta E_n} - (\pm) e^{-\beta E_{n'}} \right). \tag{11.29}$$

Here, it is important to note that it is assumed that the grand-canonical ensemble is being used, because the complete set of states includes states with any number of particles. Therefore the connection between imaginary time functions and retarded real-time functions derived below is only valid in this ensemble.

The Matsubara function is calculated in a similar way. For $\tau > 0$, we have

$$\mathcal{C}_{AB}(\tau) = \frac{-1}{Z} \operatorname{Tr} \left[e^{-\beta H} e^{\tau H} A e^{-\tau H} B \right]$$

$$= \frac{-1}{Z} \sum_{nn'} e^{-\beta E_n} \langle n|A|n'\rangle \langle n'|B|n\rangle e^{\tau(E_n - E_{n'})}, \tag{11.30}$$

and hence

$$\mathcal{C}_{AB}(i\omega_n) = \int_0^\beta d\tau\, e^{i\omega_n \tau} \frac{-1}{Z} \sum_{nn'} e^{-\beta E_n} \langle n|A|n'\rangle \langle n'|B|n\rangle e^{\tau(E_n - E_{n'})},$$

$$= \frac{-1}{Z} \sum_{nn'} e^{-\beta E_n} \frac{\langle n|A|n'\rangle \langle n'|B|n\rangle}{i\omega_n + E_n - E_{n'}} \left(e^{i\omega_n \beta} e^{\beta(E_n - E_{n'})} - 1 \right),$$

$$= \frac{-1}{Z} \sum_{nn'} e^{-\beta E_n} \frac{\langle n|A|n'\rangle \langle n'|B|n\rangle}{i\omega_n + E_n - E_{n'}} \left(\pm e^{\beta(E_n - E_{n'})} - 1 \right)$$

$$= \frac{1}{Z} \sum_{nn'} \frac{\langle n|A|n'\rangle \langle n'|B|n\rangle}{i\omega_n + E_n - E_{n'}} \left(e^{-\beta E_n} - (\pm) e^{-\beta E_{n'}} \right), \tag{11.31}$$

Eqs. (11.29) and (11.31) show that $\mathcal{C}_{AB}(i\omega_n)$ and $C_{AB}^R(\omega)$ coincide and that they are just special cases of the same function, because we can generate both $\mathcal{C}_{AB}(i\omega_n)$ and $C_{AB}^R(\omega)$ from the following function, defined in the entire complex plane except for the real axis,

$$C_{AB}(z) = \frac{1}{Z} \sum_{nn'} \frac{\langle n|A|n'\rangle \langle n'|B|n\rangle}{z + E_n - E_{n'}} \left(e^{-\beta E_n} - (\pm) e^{-\beta E_{n'}} \right). \tag{11.32}$$

This function is analytic in the upper (or lower) half plane, but has a series of poles at $E_{n'} - E_n$ along the real axis. According to the theory of analytic functions: if two functions coincide in an infinite set of points then they are fully identical functions within the entire domain, where at least one of them is an analytic function, and furthermore, there is only one such common function. This means that if we know $\mathcal{C}_{AB}(i\omega_n)$ we can find $C_{AB}^R(\omega)$ by analytic continuation:

$$C_{AB}^R(\omega) = \mathcal{C}_{AB}(i\omega_n \to \omega + i\eta). \tag{11.33}$$

Warning: this way of performing the analytic continuation is only true when $\mathcal{C}_{AB}(i\omega_n)$ is written as a rational function which is analytic in the upper half plane. If not, it is

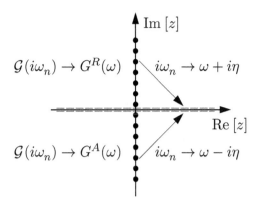

FIG. 11.1. The analytic continuation procedure in the complex z-plane. The Matsubara function originally defined only for $z = i\omega_n$ goes to the retarded or advanced Green's functions defined infinitesimally close to real axis by letting $i\omega_n \to \omega \pm i\eta$.

not obvious how to perform the continuation. For example look at the definition in Eq. (11.28). If we naïvely insert $i\omega_n \to \omega + i\eta$ before doing the integral, the answer is completely different and of course wrong. Later we shall see examples of how to perform the analytic continuation correctly.

To summarize: Using the Lehmann representation we have shown that there exists a function $C_{AB}(z)$ which is analytic for z not purely real, and which coincides with the Matsubara function, i.e., $C_{AB}(z = i\omega_n) = C_{AB}(i\omega_n)$. Approaching the real axis from positive imaginary values, this function is identical to the retarded function, i.e., $C_{AB}(z = \omega + i0^+) = C_{AB}^R(\omega)$. However, it is not a simple task to determine $C_{AB}(z)$ unless it has been reduced to a rational function as in Eq. (11.31), where it is evident that the replacement in (11.33) $i\omega_n \to z \to \omega + i\eta$ gives the right analytic function. This is illustrated in Fig. 11.1.

11.2.1 *Advanced functions*

The function $C_{AB}(z)$ is analytic for all z away from the real axis. Therefore instead of the continuation in the upper half plane, we could do the same thing in the lower half plane $i\omega_n \to z \to \omega - i\eta$, which gives the so-called advanced Green's function,

$$C_{AB}^A(\omega) = C_{AB}(i\omega_n \to \omega - i\eta). \tag{11.34}$$

The advanced Green's function is in the time domain defined as

$$C_{AB}^A(t, t') = i\theta(t' - t)\left\langle [A(t), B(t')]_{B,F} \right\rangle. \tag{11.35}$$

The term "advanced" means that it gives the state of the system at previous times based on the state of system at present times. The retarded one, as was explained in Chapter 8, gives the present state of the system as it has evolved from the state at previous times, i.e., the effect of retardation.

11.3   Single-particle Matsubara Green's function

An important type of Matsubara functions are the single-particle Green's function $\mathcal{G}$. They are defined as

$$\mathcal{G}(\mathbf{r}\sigma\tau, \mathbf{r}'\sigma\tau') = -\left\langle T_\tau\left(\Psi_\sigma(\mathbf{r},\tau)\Psi_\sigma^\dagger(\mathbf{r}',\tau')\right)\right\rangle, \quad \text{real space,} \tag{11.36a}$$

$$\mathcal{G}(\nu\tau, \nu'\tau') = -\left\langle T_\tau\left(c_\nu(\tau)c_{\nu'}^\dagger(\tau')\right)\right\rangle, \quad \{\nu\} \text{ representation.} \tag{11.36b}$$

11.3.1   Matsubara Green's function for non-interacting particles

For non-interacting particles the Matsubara Green's functions can be evaluated in the same way that we found the retarded Green's function in Section 8.3.2. Suppose the Hamiltonian is diagonal in the ν-quantum numbers,

$$H_0 = \sum_\nu \xi_\nu c_\nu^\dagger c_\nu, \tag{11.37}$$

so that

$$c_\nu(\tau) = e^{\tau H_0} c_\nu e^{-\tau H_0} = e^{-\xi_\nu \tau} c_\nu, \quad c_\nu^\dagger(\tau) = e^{\tau H_0} c_\nu^\dagger e^{-\tau H_0} = e^{\xi_\nu \tau} c_\nu^\dagger, \tag{11.38}$$

which gives

$$\begin{aligned}
\mathcal{G}_0(\nu, \tau - \tau') &= -\left\langle T_\tau\left(c_\nu(\tau)c_\nu^\dagger(\tau')\right)\right\rangle, \\
&= -\theta(\tau - \tau')\langle c_\nu(\tau)c_\nu^\dagger(\tau')\rangle - (\pm)\,\theta(\tau' - \tau)\langle c_\nu^\dagger(\tau')c_\nu(\tau)\rangle \\
&= -\left[\theta(\tau - \tau')\langle c_\nu c_\nu^\dagger\rangle(\pm)\theta(\tau' - \tau)\langle c_\nu^\dagger c_\nu\rangle\right]e^{-\xi_\nu(\tau-\tau')}.
\end{aligned} \tag{11.39}$$

For fermions this is

$$\mathcal{G}_{0,F}(\nu, \tau - \tau') = -\left[\theta(\tau - \tau')(1 - n_F(\xi_\nu)) - \theta(\tau' - \tau)n_F(\xi_\nu)\right]e^{-\xi_\nu(\tau-\tau')}, \tag{11.40}$$

while the bosonic free-particle Green's function reads

$$\mathcal{G}_{0,B}(\nu, \tau - \tau') = -\left[\theta(\tau - \tau')\left(1 + n_B(\xi_\nu)\right) + \theta(\tau' - \tau)n_B(\xi_\nu)\right]e^{-\xi_\nu(\tau-\tau')}. \tag{11.41}$$

In the frequency representation, the fermionic Green's function is

$$\begin{aligned}
\mathcal{G}_{0,F}(\nu, ik_n) &= \int_0^\beta d\tau\, e^{ik_n\tau}\mathcal{G}_{0,F}(\nu,\tau), \quad k_n = (2n+1)\frac{\pi}{\beta} \\
&= -\left(1 - n_F(\xi_\nu)\right)\int_0^\beta d\tau\, e^{ik_n\tau}e^{-\xi_\nu\tau}, \\
&= -\left(1 - n_F(\xi_\nu)\right)\frac{1}{ik_n - \xi_\nu}\left(e^{ik_n\beta}e^{-\xi_\nu\beta} - 1\right), \\
&= \frac{1}{ik_n - \xi_\nu},
\end{aligned} \tag{11.42}$$

because $e^{ik_n\beta} = -1$ and $1 - n_F(\varepsilon) = \left(e^{-\beta\varepsilon} + 1\right)^{-1}$, while the bosonic one becomes

$$\mathcal{G}_{0,B}(\nu, iq_n) = \int_0^\beta d\tau\, e^{iq_n\tau} \mathcal{G}_{0,B}(\nu,\tau), \quad q_n = 2n\frac{\pi}{\beta}$$

$$= -(1 + n_B(\xi_\nu)) \int_0^\beta d\tau\, e^{iq_n\tau} e^{-\xi_\nu\tau},$$

$$= -(1 + n_B(\xi_\nu)) \frac{1}{iq_n - \xi_\nu} \left(e^{iq_n\beta} e^{-\xi_\nu\beta} - 1\right),$$

$$= \frac{1}{iq_n - \xi_\nu}, \tag{11.43}$$

because $e^{iq_n\beta} = 1$ and $1 + n_B(\varepsilon) = -\left(e^{-\beta\varepsilon} - 1\right)^{-1}$. Here, we have anticipated the notation that is used later: Matsubara frequencies ik_n and ip_n are used for fermion frequencies, while iq_n and $i\omega_n$ are used for boson frequencies.

According to our recipe Eq. (11.33), the retarded free particles Green's functions are for both fermions and bosons

$$G_0^R(\nu, \omega) = \frac{1}{\omega - \xi_\nu + i\eta}, \tag{11.44}$$

in agreement with Eq. (8.55).

11.4 Evaluation of Matsubara sums

When working with Matsubara Green's functions, we will often encounter sums over Matsubara frequencies, similar to integrals over frequencies in the real-time formalism. For example sums of the type

$$\mathcal{S}_1(\nu, \tau) = \frac{1}{\beta} \sum_{ik_n} \mathcal{G}(\nu, ik_n) e^{ik_n\tau}, \quad \tau > 0, \tag{11.45}$$

or sums with products of Green's functions. The imaginary time formalism has been introduced, because it will be used to perform perturbation expansions, and therefore the types of sums that we will encounter, are often products of such free Green's functions, e.g.,

$$\mathcal{S}_2(\nu_1, \nu_2, i\omega_n, \tau) = \frac{1}{\beta} \sum_{ik_n} \mathcal{G}_0(\nu_1, ik_n) \mathcal{G}_0(\nu_2, ik_n + i\omega_n) e^{ik_n\tau}, \quad \tau > 0. \tag{11.46}$$

This section is devoted to the mathematical techniques for evaluating such sums. To be more general, we define the two generic sums

$$\mathcal{S}^F(\tau) = \frac{1}{\beta} \sum_{ik_n} g(ik_n) e^{ik_n\tau}, \quad ik_n \text{ fermion frequency}, \tag{11.47a}$$

$$\mathcal{S}^B(\tau) = \frac{1}{\beta} \sum_{i\omega_n} g(i\omega_n) e^{i\omega_n\tau}, \quad i\omega_n \text{ boson frequency}, \tag{11.47b}$$

and study them for $\tau > 0$.

To evaluate these, the trick is to rewrite them as integrals over a complex variable and to use residue theory. For this we need two functions, $n_F(z)$ and $n_B(z)$, which have poles at $z = ik_n$ and $z = i\omega_n$, respectively. These functions turn out to be the well-known Fermi and Bose distribution functions,

$$n_F(z) = \frac{1}{e^{\beta z} + 1}, \quad \text{poles for } z = i(2n+1)\frac{\pi}{\beta}, \tag{11.48a}$$

$$n_B(z) = \frac{1}{e^{\beta z} - 1}, \quad \text{poles for } z = i(2n)\frac{\pi}{\beta}. \tag{11.48b}$$

The residues at these values are

$$\operatorname*{Res}_{z=ik_n} [n_F(z)] = \lim_{z \to ik_n} \frac{(z - ik_n)}{e^{\beta z} + 1} = \lim_{\delta \to 0} \frac{\delta}{e^{\beta ik_n} e^{\beta \delta} + 1} = -\frac{1}{\beta}, \tag{11.49a}$$

$$\operatorname*{Res}_{z=i\omega_n} [n_B(z)] = \lim_{z \to i\omega_n} \frac{(z - i\omega_n)}{e^{\beta z} - 1} = \lim_{\delta \to 0} \frac{\delta}{e^{\beta i\omega_n} e^{\beta \delta} - 1} = +\frac{1}{\beta}. \tag{11.49b}$$

According to the theory of analytic functions, the contour integral which encloses one of these points, but no singularity of $g(z)$, is given by

$$\oint dz\, n_F(z)g(z) = 2\pi i \operatorname*{Res}_{z=ik_n} [n_F(z)g(z)] = -\frac{2\pi i}{\beta} g(ik_n), \tag{11.50}$$

for fermions and similarly for boson frequencies

$$\oint dz\, n_B(z)g(z) = 2\pi i \operatorname*{Res}_{z=i\omega_n} [n_B(z)g(z)] = \frac{2\pi i}{\beta} g(i\omega_n). \tag{11.51}$$

If we therefore define contours $\mathcal{C}$, which enclose all point $z = ik_n$ in the fermionic case and all points $z = i\omega_n$ in the bosonic case, but only regions where $g(z)$ is analytic, we can write

$$S^F = -\int_{\mathcal{C}} \frac{dz}{2\pi i} n_F(z)g(z)e^{z\tau}, \tag{11.52a}$$

$$S^B = +\int_{\mathcal{C}} \frac{dz}{2\pi i} n_B(z)g(z)e^{z\tau}. \tag{11.52b}$$

In the following two subsections, we use the contour integration technique in two special cases.

11.4.1　Summations over functions with simple poles

Consider a Matsubara frequency sum like Eq. (11.46), but let us take a slightly more general function, which could include more free Green's function. Let us therefore consider the sum

$$S_0^F(\tau) = \frac{1}{\beta} \sum_{ik_n} g_0(ik_n)e^{ik_n\tau}, \quad \tau > 0, \tag{11.53}$$

where $g_0(z)$ has a number of known simple poles, e.g., in the form of non-interacting Green's functions like (11.46)

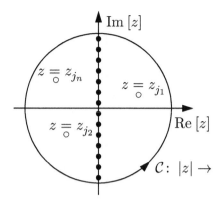

FIG. 11.2. The contour $\mathcal{C}$ used to perform the Matsubara sum for a function with known poles z_j. The contribution from the contour goes to zero as $|z| \to \infty$, and hence the contributions from the (black) poles at the Matsubara frequencies $z = ik_n$ and the (white) poles at $z = z_j$ add up to zero.

$$g_0(z) = \prod_j \frac{1}{z - z_j}, \tag{11.54}$$

where $\{z_j\}$ is the set of known poles and hence $g_0(z)$ is analytic elsewhere in the z-plane. Because we know the poles of g_0, a good choice for a contour is one that encompasses the entire complex plane $\mathcal{C}_\infty$: $z = R e^{i\theta}$ where $R \to \infty$ (see Fig. 11.2). Such a contour would give us the contribution for poles of $n_F(z)$ plus the contributions from poles of $g_0(z)$. Furthermore, the contour integral itself gives zero because the integrand goes to zero exponentially for $z \in \mathcal{C}_\infty$ (remember $0 < \tau < \beta$)

$$n_F(z)e^{\tau z} = \frac{e^{\tau z}}{e^{\beta z} + 1} \propto \begin{cases} e^{(\tau - \beta)\,\mathrm{Re}\,z} \to 0, & \text{for } \mathrm{Re}\,z > 0, \\ e^{\tau\,\mathrm{Re}\,z} \to 0, & \text{for } \mathrm{Re}\,z < 0. \end{cases} \tag{11.55}$$

Hence

$$\begin{aligned}
0 &= \int_{\mathcal{C}_\infty} \frac{dz}{2\pi i} n_F(z) g_0(z) e^{z\tau} \\
&= -\frac{1}{\beta} \sum_{ik_n} g_0(ik_n) e^{ik_n \tau} + \sum_j \operatorname*{Res}_{z=z_j} [g_0(z)]\, n_F(z_j) e^{z_j \tau},
\end{aligned} \tag{11.56}$$

and thus

$$S_0^F(\tau) = \sum_j \operatorname*{Res}_{z=z_j} [g_0(z)]\, n_F(z_j) e^{z_j \tau}. \tag{11.57}$$

The Matsubara sum has thus been simplified considerably, and we shall use this formula several times in the following parts of the book. For bosons the derivation is almost identical, and we obtain

$$\frac{1}{\beta} \sum_{i\omega_n} g_0(i\omega_n) e^{i\omega_n \tau} = S_0^B(\tau) = -\sum_j \operatorname*{Res}_{z=z_j} [g_0(z)]\, n_B(z_j) e^{z_j \tau}. \tag{11.58}$$

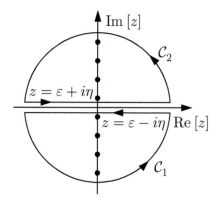

FIG. 11.3. The contour $\mathcal{C}$ consisting of two parts, $\mathcal{C}_1$ and $\mathcal{C}_2$, used to perform the Matsubara sum for a function with known branch cuts. Such a function is analytic in the entire complex plane except on the branch cuts. The contribution from the circular parts of the contour goes to zero as $|z| \to \infty$, and hence only the paths parallel to the cut (here the real axis) contribute.

11.4.2 Summations over functions with known branch cuts

The second type of sums we will meet are of the form in Eq. (11.45). If it is the full Green's function, including for example the influence of interaction, we do not know the poles of the Green's function, but we do know that it is analytic for z not on the real axis. This general property of the Green's function was shown in Section 11.2.

In general, consider the sum

$$S(\tau) = \frac{1}{\beta} \sum_{ik_n} g(ik_n) e^{ik_n \tau}, \quad \tau > 0, \tag{11.59}$$

where it is known that $g(z)$ is analytic in the entire complex plane except on the real axis. A contour, which includes all points $z = ik_n$ and no singularities of g, is therefore $\mathcal{C} = \mathcal{C}_1 + \mathcal{C}_2$ depicted in Fig. 11.3. As in the example studied in the previous section (see Eq. (11.55)), the part where $|z| \to \infty$ does not contribute to the integral, and we are left with the parts of the contour running parallel to the real axis. They are shifted by an infinitesimal amount η away from the real axis on either side

$$\begin{aligned}
\mathcal{S}(\tau) &= -\int_{\mathcal{C}_1 + \mathcal{C}_2} \frac{dz}{2\pi i} n_F(z) g(z) e^{z\tau} \\
&= -\frac{1}{2\pi i} \int_{-\infty}^{\infty} d\varepsilon\, n_F(\varepsilon) \left[g(\varepsilon + i\eta) - g(\varepsilon - i\eta) \right] e^{\varepsilon \tau}.
\end{aligned} \tag{11.60}$$

As an example of the use of this method, the sum in Eq. (11.45) is evaluated as

$$\mathcal{S}_1(\nu,\tau) = -\frac{1}{2\pi i} \int_{-\infty}^{\infty} d\varepsilon\, n_F(\varepsilon)\left[\mathcal{G}(\nu,\varepsilon+i\eta) - \mathcal{G}(\nu,\varepsilon-i\eta)\right] e^{\varepsilon\tau}$$

$$= -\frac{1}{2\pi i} \int_{-\infty}^{\infty} d\varepsilon\, n_F(\varepsilon) 2i\,\mathrm{Im}\left[G^R(\nu,\varepsilon)\right] e^{\varepsilon\tau}$$

$$= \int_{-\infty}^{\infty} \frac{d\varepsilon}{2\pi}\, n_F(\varepsilon) A(\nu,\varepsilon) e^{\varepsilon\tau}, \tag{11.61}$$

according to the definition of the spectral function in Eq. (8.50). In the second equality we used that $\mathcal{G}(\varepsilon - i\eta) = [\mathcal{G}(\varepsilon + i\eta)]^*$, which follows from Eq. (11.32) with $A = c_\nu$ and $B = c_\nu^\dagger$. Now, setting the time argument in the single particle imaginary time Green's function, Eq. (11.36b), to a negative infinitesimal 0^-, we are in fact led to an expression for the expectation value of the occupation:

$$\langle c_\nu^\dagger c_\nu \rangle = \mathcal{G}(\nu, 0^-)$$

$$= \frac{1}{\beta} \sum_{ik_n} \mathcal{G}(\nu, ik_n) e^{-ik_n 0^-} = \mathcal{S}_1(\nu, 0^+)$$

$$= \int_{-\infty}^{\infty} \frac{d\varepsilon}{2\pi}\, n_F(\varepsilon) A(\nu, \varepsilon). \tag{11.62}$$

This result agrees with our previous finding, Eq. (8.62).

11.5   Equation of motion

The equation of motion technique, used in Chapter 9 to find various Green's functions, can also be used for the Matsubara functions. In the imaginary time formalism, the time derivative of an operator is

$$\partial_\tau A(\tau) = \partial_\tau \left(e^{\tau H} A e^{-\tau H}\right) = e^{\tau H}[H, A]e^{-\tau H} = [H, A](\tau). \tag{11.63}$$

If we differentiate the Matsubara function Eq. (11.20) with respect to τ, we obtain

$$-\partial_\tau \mathcal{C}_{AB}(\tau - \tau') = \frac{\partial}{\partial\tau}\left[\theta(\tau - \tau')\langle A(\tau)B(\tau')\rangle \pm \theta(\tau' - \tau)\langle B(\tau')A(\tau)\rangle\right]$$

$$= \delta(\tau - \tau')\langle AB - (\pm)BA\rangle + \left\langle T_\tau\{[H, A](\tau)B(\tau')\}\right\rangle,$$

where as usual the minus sign in $\pm$ is for fermion operators, whereas the plus sign should be used for boson operators.

For the single-particle Green's functions defined in Eqs. (11.36), we then obtain for both fermion and boson Green's functions

$$-\partial_\tau \mathcal{G}(\mathbf{r}\tau, \mathbf{r}'\tau') = \delta(\tau - \tau')\delta(\mathbf{r} - \mathbf{r}') + \left\langle T_\tau\{[H, \Psi(\mathbf{r})](\tau)\,\Psi^\dagger(\mathbf{r}', \tau')\}\right\rangle, \tag{11.64a}$$

$$-\partial_\tau \mathcal{G}(\nu\tau, \nu'\tau') = \delta(\tau - \tau')\delta_{\nu\nu'} + \left\langle T_\tau\{[H, c_\nu](\tau)\, c_{\nu'}^\dagger(\tau')\}\right\rangle. \tag{11.64b}$$

For non-interacting electrons the Hamiltonian is quadratic, i.e., of the general form

$$H_0 = \int d\mathbf{r} \int d\mathbf{r}'\, \Psi^\dagger(\mathbf{r}) h_0(\mathbf{r},\mathbf{r}') \Psi(\mathbf{r}') \qquad \text{or} \qquad H_0 = \sum_{\nu\nu'} h_{0,\nu\nu'} c_\nu^\dagger c_{\nu'}. \qquad (11.65)$$

In these cases, the equations of motion reduce to

$$-\partial_\tau \mathcal{G}_0(\mathbf{r}\tau, \mathbf{r}'\tau') - \int d\mathbf{r}''\, h_0(\mathbf{r},\mathbf{r}'') \mathcal{G}_0(\mathbf{r}''\tau, \mathbf{r}'\tau') = \delta(\tau-\tau')\delta(\mathbf{r}-\mathbf{r}'), \qquad (11.66a)$$

or

$$-\partial_\tau \mathcal{G}_0(\nu\tau, \nu'\tau') - \sum_{\nu''} h_{0,\nu\nu''} \mathcal{G}_0(\nu''\tau, \nu'\tau') = \delta(\tau-\tau')\delta_{\nu\nu'}. \qquad (11.66b)$$

In matrix form, Eq. (11.66) reads

$$\mathcal{G}_0^{-1}\mathcal{G}_0 = 1 \quad \text{with} \quad \mathcal{G}_0^{-1} = -\partial_\tau - h_0. \qquad (11.67)$$

This equation and the boundary condition $\mathcal{G}(\tau) = \pm\mathcal{G}(\tau+\beta)$ determine the solution. For example, for a diagonal Hamiltonian those given in Eqs. (11.40) and (11.41).

11.6   Wick's theorem

We end this rather technical part by proving an extremely useful theorem, which we will need later when doing perturbation theory, and which is used in the example ending this chapter. The theorem — the celebrated Wick's theorem — states that for non-interacting particles, i.e., when the Hamiltonian is quadratic, higher-order Green's function involving more than one particle can be factorized into products of single-particle Green's functions.

Consider an n-particle Green's function defined as

$$\mathcal{G}_0^{(n)}(\nu_1\tau_1,\ldots,\nu_n\tau_n; \nu_1'\tau_1',\ldots,\nu_n'\tau_n')$$
$$= (-1)^n \left\langle T_\tau \left[\hat{c}_{\nu_1}(\tau_1)\cdots\hat{c}_{\nu_n}(\tau_n)\hat{c}_{\nu_n'}^\dagger(\tau_n')\cdots\hat{c}_{\nu_1'}^\dagger(\tau_1') \right] \right\rangle_0. \qquad (11.68)$$

The average is taken with respect to a non-interacting Hamiltonian H_0, such as the one of Eq. (11.65), and this is indicated by the subscript 0. The time evolution is also with respect to H_0, and it is given by

$$\hat{c}(\tau) = e^{\tau H_0} c\, e^{-\tau H_0}. \qquad (11.69)$$

The expression in (11.68) is indeed quite complicated to look at, if we write out all the possible orderings and the conditions for that particular ordering. For example if $n = 2$, there are 4 time arguments, which can be ordered in 4! different ways. Let us simplify the writing by defining one operator symbol for both creation and annihilation operators

$$d_j(\sigma_j) = \begin{cases} \hat{c}_{\nu_j}(\tau_j), & j \in [1,n], \\ \hat{c}_{\nu'_{(2n+1-j)}}^\dagger(\tau'_{(2n+1-j)}), & j \in [n+1, 2n], \end{cases} \qquad (11.70)$$

and furthermore define the permutations of the $2n$ operators as

$$P\big(d_1(\sigma_1)\cdots d_{2n}(\sigma_{2n})\big) = d_{P_1}(\sigma_{P_1})\cdots d_{P_{2n}}(\sigma_{P_{2n}}), \qquad (11.71)$$

where P_j denotes the jth variable in the permutation P, e.g., define the list (a,b,c) and the permutation (c,a,b) then $P = (3,1,2)$. Which permutation is the correct one of course depends on how the time arguments in (11.68) are really ordered. Therefore if we sum over all permutations (the set of all permutations of N elements is denoted S_N) and include the corresponding conditions, we can rewrite $\mathcal{G}_0^{(n)}$ as

$$\mathcal{G}_0^{(n)}(j_1, \ldots, j_{2n}) = (-1)^n \sum_{P \in S_{2n}} (\pm 1)^P \, \theta(\sigma_{P_1} - \sigma_{P_2}) \cdots \theta(\sigma_{P_{n-1}} - \sigma_{P_n})$$
$$\times \left\langle d_{P_1}(\sigma_{P_1}) \cdots d_{P_{2n}}(\sigma_{P_{2n}}) \right\rangle_0, \tag{11.72}$$

where the factor $(\pm 1)^P$ takes into account that for fermions (minus sign) it costs a sign change every time a pair of operators are commuted.

The easiest way to prove Wick's theorem is through the equation of motion for the n-particle Green's function. Thus we differentiate $\mathcal{G}_0^{(n)}$ with respect to one of the time arguments, $\tau_1, \ldots, \tau_n$. This gives two kinds of contributions: the terms coming from the derivative of the theta functions and one term from the derivative of the expectation value itself. For example for τ_1, the last one gives

$$\left[-\frac{\partial}{\partial \tau_1} \mathcal{G}_0^{(n)} \right]_{\text{not } \theta\text{-functions}}$$
$$= -(-1)^n \left\langle T_\tau \left[[\hat{H}_0, \hat{c}_{\nu_1}](\tau_1) \cdots \hat{c}_{\nu_n}(\tau_n) \hat{c}_{\nu'_n}^\dagger(\tau'_n) \cdots \hat{c}_{\nu'_1}^\dagger(\tau'_1) \right] \right\rangle_0, \tag{11.73}$$

which is similar to the derivation that lead to Eqs. (11.66) and (11.67), so therefore we have

$$\mathcal{G}_{0i}^{-1} \mathcal{G}_0^{(n)} = -\partial_{\tau_i}^\theta \, \mathcal{G}_0^{(n)}, \tag{11.74}$$

where $\mathcal{G}_{0i}^{-1}$ means that it works on the coordinates ν_i, τ_i. On the right-hand side, the derivative only acts on the theta functions in Eq. (11.72).

Take now for example the case when τ_i is next to τ'_j. There are two such terms in (11.72), corresponding to τ_i being either smaller or larger than τ'_j, and they will have different order of the permutation. In this case, $\mathcal{G}^{(n)}$ has the structure

$$\mathcal{G}_0^{(n)} = \quad [\cdots \theta(\tau_i - \tau'_j) \cdots] \left\langle \cdots \hat{c}_{\nu_i}(\tau_i) \hat{c}_{\nu'_j}^\dagger(\tau'_j) \cdots \right\rangle_0$$
$$\pm [\cdots \theta(\tau'_j - \tau_i) \cdots] \left\langle \cdots \hat{c}_{\nu'_j}^\dagger(\tau'_j) \hat{c}_{\nu_i}(\tau_i) \cdots \right\rangle_0, \tag{11.75}$$

and when it is differentiated with respect to τ_i, it results in two delta functions, and consequently

$$-\partial_{\tau_i}^\theta \, \mathcal{G}_0^{(n)} = \left([\cdots] \left\langle \cdots \hat{c}_{\nu_i}(\tau_i) \hat{c}_{\nu'_j}^\dagger(\tau'_j) \cdots \right\rangle_0 \mp [\cdots] \left\langle \cdots \hat{c}_{\nu'_j}^\dagger(\tau'_j) \hat{c}_{\nu_i}(\tau_i) \cdots \right\rangle_0 \right) \delta(\tau_i - \tau'_j). \tag{11.76}$$

We can pull out the equal time commutator or anti-commutator for boson or fermions, respectively

$$\left[\hat{c}_{\nu_i}(\tau_i), \hat{c}_{\nu'_j}^\dagger(\tau_i) \right]_{B,F} = \delta_{\nu_i, \nu'_j}. \tag{11.77}$$

If τ_i is next to τ_j instead of τ'_j, we get in the same manner the (anti-)commutator

$$\left[\hat{c}_{\nu_i}(\tau_i), \hat{c}_{\nu_j}(\tau_i)\right]_{B,F} = 0, \tag{11.78}$$

which therefore does not contribute. The number of creation and annihilation operators has thus both been reduced by one, and it leaves a Green's function which is no longer an n-particle Green's function but an $(n-1)$-Green's function. In fact, we saw a special case of this in Eq. (11.66) where a one-particle Green's function was reduced to a zero-particle Green's function, i.e., a constant. However, we have not determined the sign of the new $(n-1)$-Green's function, which we will denote by $(-1)^x$. The sign will (for fermions) depend on the τ'_j in question. Including this undetermined sign, our equation of motion (11.74) now looks like

$$\mathcal{G}_{0i}^{-1}\mathcal{G}_0^{(n)} = \sum_{j=1}^{n} \delta_{\nu_i,\nu'_j}\delta(\tau_i - \tau'_j)(-1)^x \,\mathcal{G}_0^{(n-1)}(\underbrace{\nu_1\tau_1,\ldots,\nu_n\tau_n}_{\text{without } i}; \underbrace{\nu'_1\tau'_1,\ldots,\nu'_n\tau'_n}_{\text{without } j}). \tag{11.79}$$

In order to determine x, let us collect the signs that go into $(-1)^x$. There is a (-1) from $(-\partial_\tau)$, a $(-1)^n$ from the definition in (11.68), $[(-1)^{n-1}]^{-1}$ from the definition of $\mathcal{G}^{(n-1)}$, and for fermions a factor $(-1)^{n-i+n-j}$ from moving $\hat{c}_{\nu'_j}^\dagger$ next to $\hat{c}_{\nu_i}$. Hence

$$\text{fermions:} \quad (-1)^x = -(-1)^n(-1)^{1-n}(-1)^{2n-i-j} = (-1)^{j+i}, \tag{11.80a}$$

$$\text{bosons:} \quad (-1)^x = -(-1)^n(-1)^{1-n} = 1. \tag{11.80b}$$

Now Eq. (11.79) can be integrated, and because $\mathcal{G}_0^{(n)}$ has the same boundary conditions as $\mathcal{G}_0$, the solution is the same as for the differential equation in Eq. (11.66), namely the single-particle Green's function. As a result

$$\mathcal{G}_0^{(n)}(\nu_1\tau_1,\ldots,\nu_n\tau_n;\nu'_1\tau'_1,\ldots,\nu'_n\tau'_n)$$
$$= \sum_{j=1}^{n} (\pm)^{j+i} \,\mathcal{G}_0\left(\nu_i\tau_i,\nu'_j\tau'_j\right)\mathcal{G}_0^{(n-1)}(\underbrace{\nu_1\tau_1,\ldots,\nu_n\tau_n}_{\text{without } i}; \underbrace{\nu'_1\tau'_1,\ldots,\nu'_n\tau'_n}_{\text{without } j}). \tag{11.81}$$

By recalling the definition of determinants and permanents, Eqs. (1.24) and (1.25), this iteration formula is immediately recognized as the determinant in case of the minus sign, and the permanent in the case of the plus sign. We therefore arrive at

$$\mathcal{G}_0^{(n)}(1,\ldots,n;1',\ldots,n') = \begin{vmatrix} \mathcal{G}_0(1,1') & \cdots & \mathcal{G}_0(1,n') \\ \vdots & \ddots & \vdots \\ \mathcal{G}_0(n,1') & \cdots & \mathcal{G}_0(n,n') \end{vmatrix}_{B,F}, \quad i \equiv (\nu_i,\tau_i) \tag{11.82}$$

where we used a shorthand notation with the orbital and the time arguments being collected into one variable, and where the determinant $|\cdot|_{B,F}$ means that for fermions it is the usual determinant, while for bosons it should be understood as a permanent where all have terms come with a plus sign; this is Wick's theorem.

11.7   Example: polarizability of free electrons

In Section 8.5 we calculated the polarizability of *non-interacting free electrons*. In order to illustrate the working principle of the imaginary time formalism, we do it again here.

The starting point is the physical quantity which is needed: the frequency-dependent retarded charge–charge correlation function $\chi^R(\mathbf{q}, \omega)$, which follows from the corresponding Matsubara function by

$$\chi^R(\mathbf{q}, \omega) = \chi(\mathbf{q}, iq_n \to \omega + i\eta). \tag{11.83}$$

To find $\chi(\mathbf{q}, iq_n)$, we begin from the time-dependent χ,

$$\chi_0(\mathbf{q}, \tau) = -\frac{1}{\mathcal{V}} \langle T_\tau (\hat{\rho}(\mathbf{q}, \tau) \hat{\rho}(-\mathbf{q})) \rangle_0, \tag{11.84}$$

and expresses it as a two-particle Green's function

$$\chi_0(\mathbf{q}, \tau) = -\frac{1}{\mathcal{V}} \sum_{\mathbf{k},\mathbf{k'}\sigma\sigma'} \left\langle T_\tau \left(\hat{c}_{\mathbf{k}\sigma}^\dagger(\tau) \hat{c}_{\mathbf{k}+\mathbf{q}\sigma}(\tau) \hat{c}_{\mathbf{k'}\sigma'}^\dagger \hat{c}_{\mathbf{k'}-\mathbf{q}\sigma'} \right) \right\rangle_0. \tag{11.85}$$

Note that the time dependence of the *c*-operators is given by the full Hamiltonian. However, because we are considering a noninteracting, free electron gas, the full Hamiltonian and the unperturbed Hamiltonian are the same. In order to be consistent with the notation in this chapter, we write the time-dependent operators with a hat.

By Wick's theorem, this is given by a product of single-particle Green's functions with all possible pairings and with the sign given by the number of times we interchange two fermion operators, i.e.,

$\chi_0(\mathbf{q}, \tau)$

$$= \frac{1}{\mathcal{V}} \left[\sum_{\mathbf{k},\mathbf{k'},\sigma,\sigma'} \left\langle T_\tau \left(\hat{c}_{\mathbf{k}+\mathbf{q}\sigma}(\tau) \hat{c}_{\mathbf{k'}\sigma'}^\dagger \right) \right\rangle_0 \left\langle T_\tau \left(\hat{c}_{\mathbf{k'}-\mathbf{q}\sigma'} \hat{c}_{\mathbf{k}\sigma}^\dagger(\tau) \right) \right\rangle_0 - \overbrace{\langle \hat{\rho}(\mathbf{q}) \rangle_0 \langle \hat{\rho}(-\mathbf{q}) \rangle_0}^{=0 \text{ for } q \neq 0} \right]$$

$$= \frac{1}{\mathcal{V}} \sum_{\mathbf{k}\sigma} \mathcal{G}_0(\mathbf{k} + \mathbf{q}\sigma, \tau) \mathcal{G}_0(\mathbf{k}\sigma, -\tau). \tag{11.86}$$

where we consider only $q \neq 0$ and use that $\mathcal{G}_0(\mathbf{k}, \mathbf{k'}) \propto \delta_{\mathbf{k},\mathbf{k'}}$.

The next step is to calculate the frequency-dependent function, which amounts to the Fourier transform of the sum of products in (11.86). The Fourier transformation of a product in the time domain is a convolution in the frequency domain. Because one function has argument τ while the other has argument $-\tau$, the internal frequencies in the two Fourier transforms come with the same sign,

$$\chi_0(\mathbf{q}, iq_n) = \frac{1}{\beta} \sum_{ik_n} \frac{1}{\mathcal{V}} \sum_{\mathbf{k}\sigma} \mathcal{G}_0(\mathbf{k} + \mathbf{q}\sigma, ik_n + iq_n) \mathcal{G}_0(\mathbf{k}\sigma, ik_n). \tag{11.87}$$

The sum over Matsubara frequencies ik_n in Eq. (11.87) has exactly the form studied in Section 11.4.1. Remembering that $\mathcal{G}_0(\mathbf{k}\sigma, ik_n) = 1/(ik_n - \xi_{\mathbf{k}})$, we can read off the answer from Eq. (11.57) by inserting the poles $z = \xi_{\mathbf{k+q}} - iq_n$ and $z = \xi_{\mathbf{k}}$ of the Green's functions $\mathcal{G}_0(\mathbf{k} + \mathbf{q}\sigma, ik_n + iq_n)$ and $\mathcal{G}_0(\mathbf{k}\sigma, ik_n)$, respectively,

$$
\begin{aligned}
\chi_0(\mathbf{q}, iq_n) &= \frac{1}{\mathcal{V}} \sum_{\mathbf{k}\sigma} \left\{ n_F(\xi_{\mathbf{k}}) \mathcal{G}_0(\mathbf{k+q}\sigma, \xi_{\mathbf{k}} + iq_n) + n_F(\xi_{\mathbf{k+q}} - iq_n) \mathcal{G}_0(\mathbf{k}\sigma, \xi_{\mathbf{k+q}} - iq_n) \right\} \\
&= \frac{1}{\mathcal{V}} \sum_{\mathbf{k}\sigma} \frac{n_F(\xi_{\mathbf{k}}) - n_F(\xi_{\mathbf{k+q}})}{iq_n + \xi_{\mathbf{k}} - \xi_{\mathbf{k+q}}}.
\end{aligned} \tag{11.88}
$$

Here, we used that

$$
n_F(\xi_{\mathbf{k+q}} - iq_n) = \frac{1}{e^{\beta \xi_{\mathbf{k+q}}} e^{-\beta iq_n} + 1} = \frac{1}{e^{\beta \xi_{\mathbf{k+q}}} + 1}, \tag{11.89}
$$

because iq_n is a bosonic Matsubara frequency. After the substitution of Eq. (11.83), Eq. (11.88) gives the result we found in Eq. (8.85).

11.8 Summary and outlook

When performing calculations of physical quantities at finite temperatures, it turns out that the easiest way to find the "real time", introduced in Chapter 8, is often to go via the imaginary-time formalism. This formalism has been introduced in this chapter, and in the following chapters on Feynman diagrams it is a necessary tool. There, we will see, why it is more natural to use the imaginary time Green's function, also called Matsubara Green's function. The reason is that the time evolution operator and the Boltzmann weight factor can be treated on an equal footing and one single perturbation expansion suffices. In the real time formalism there is no simple way of doing this.

We have also derived some very useful relations concerning sums over Matsubara frequencies. The main points to remember are the following:

1. Non-interacting particle Green's function, valid for both bosons and fermions,

$$
\mathcal{G}_0(\nu, i\omega_n) = \frac{1}{i\omega_n - \xi_\nu}.
$$

2. Matsubara frequency sum over products of non-interacting Green's functions, valid for $\tau > 0$,

$$
S^F(\tau) = \frac{1}{\beta} \sum_{ik_n} g_0(ik_n) e^{ik_n \tau} = \sum_j \operatorname{Res}\left(g_0(z_j)\right) n_F(z_j) e^{z_j \tau}, \quad ik_n \text{ fermion frequency,}
$$

$$
S^B(\tau) = \frac{1}{\beta} \sum_{i\omega_n} g_0(i\omega_n) e^{i\omega_n \tau} = -\sum_j \operatorname{Res}\left(g_0(z_j)\right) n_B(z_j) e^{z_j \tau}, \quad i\omega_n \text{ boson frequency,}
$$

with $g_0(z) = \prod_i 1/(z - \xi_i)$. If we perform a sum over functions, for which the poles are unknown, but where the branch cuts are known, we can use the contour depicted in Fig. 11.3. For example if $g(ik_n)$ is known to be analytic everywhere but on the real axis, we obtain

$$S^F(\tau) = \frac{1}{\beta} \sum_{ik_n} g(ik_n) e^{ik_n \tau}$$

$$= -\int_{-\infty}^{\infty} \frac{d\varepsilon}{2\pi i} n_F(\varepsilon) \left[g(\varepsilon + i\eta) - g(\varepsilon - i\eta) \right]$$

$$= -\int_{-\infty}^{\infty} \frac{d\varepsilon}{2\pi i} n_F(\varepsilon) \left[g^R(\varepsilon) - g^A(\varepsilon) \right].$$

3. Finally, we proved the important Wick's theorem, which states that for non-interacting systems, i.e., quadratic Hamiltonians, an n-particle Green's function is equal to a sum of products of single-particle Green's functions, where all possible pairings should be included in the sum. For fermions we must furthermore keep track of the number of factors -1, because each time we interchange two fermion operators we must include such a factor. The end result was

$$\mathcal{G}_0^{(n)}(1,\ldots,n;1',\ldots,n') = \begin{vmatrix} \mathcal{G}_0(1,1') & \cdots & \mathcal{G}_0(1,n') \\ \vdots & \ddots & \vdots \\ \mathcal{G}_0(n,1') & \cdots & \mathcal{G}_0(n,n') \end{vmatrix}_{B,F}, \quad i \equiv (\nu_i, \tau_i),$$

where

$$\mathcal{G}_0^{(n)}(1,\ldots,n;1',\ldots,n') = (-1)^n \left\langle T_\tau \left[\hat{c}(1) \cdots \hat{c}(n) \hat{c}^\dagger(n') \cdots \hat{c}^\dagger(1') \right] \right\rangle_0.$$

12

FEYNMAN DIAGRAMS AND EXTERNAL POTENTIALS

From the previous chapters on linear response theory and Green's functions, it is clear that complete calculations of thermal averages of time-dependent phenomena in quantum field theory are a rather formidable task. Even the basic imaginary time-evolution operator $\hat{U}(\tau)$ itself is an infinite series to all orders in the interaction $\hat{V}(\mathbf{r}, \tau)$. One simply faces the problem of getting lost in the myriads of integrals, and not being able to maintain a good physical intuition of which terms are important. In 1948 Feynman solved this problem as part of his seminal work on quantum electrodynamics by inventing the ingenious diagrams that today bear his name. The Feynman diagrams are both an exact mathematical representation of perturbation theory to infinite order and a powerful pictorial method that elucidate the physical content of the complicated expressions. In this chapter, we introduce the Feynman diagrams for the case of non-interacting particles in an external potential. Our main example of their use will be the analysis of electron-impurity scattering in disordered metals.

12.1 Non-interacting particles in external potentials

Consider a time-independent Hamiltonian H in the space representation describing non-interacting fermions in an external spin-diagonal single-particle potential $V_\sigma(\mathbf{r})$,

$$H = H_0 + V = \sum_\sigma \int d\mathbf{r}\, \Psi_\sigma^\dagger(\mathbf{r})H_0(\mathbf{r})\Psi_\sigma(\mathbf{r}) + \sum_\sigma \int d\mathbf{r}\, \Psi_\sigma^\dagger(\mathbf{r})V_\sigma(\mathbf{r})\Psi_\sigma(\mathbf{r}). \quad (12.1)$$

As usual, we assume that the unperturbed system described by the time-independent Hamiltonian H_0 is solvable, and that we know the corresponding eigenstates $|\nu\rangle$ and Green's functions $\mathcal{G}_\nu^0$. In the following, it will prove helpful to introduce the short-hand notation

$$(\mathbf{r}_1, \sigma_1, \tau_1) \equiv (1) \quad \text{and} \quad \int d1 \equiv \sum_{\sigma_1} \int d\mathbf{r}_1 \int_0^\beta d\tau_1 \quad (12.2)$$

for points and integrals in space-time.

We want to study the full Green's function, $\mathcal{G}(b, a) = -\langle T_\tau\, \Psi(b)\Psi^\dagger(a)\rangle$, governed by H, and the bare one, $\mathcal{G}^0(b, a) = -\langle T_\tau\, \hat{\Psi}(b)\hat{\Psi}^\dagger(a)\rangle_0$, governed by H_0. We note that since no particle-particle interaction is present in Eq. (12.1), both the full Hamiltonian H and the bare H_0 have the simple form of Eq. (11.65), and the equations of motion for the two Green's functions have the same form as Eq. (11.66),

$$\left[-\partial_{\tau_b} - H_0(b)\right] \mathcal{G}^0(b, a) = \delta(b-a) \Leftrightarrow \left[-\partial_{\tau_b} - H(b) + V(b)\right] \mathcal{G}^0(b, a) = \delta(b-a), \quad (12.3a)$$

$$\left[-\partial_{\tau_b} - H(b)\right] \mathcal{G}(b, a) = \delta(b-a) \Leftrightarrow \mathcal{G}(b, a) = \left[-\partial_{\tau_b} - H(b)\right]^{-1} \delta(b-a), \quad (12.3b)$$

204

where we have also given the formal solution of $\mathcal{G}$, which is helpful in acquiring the actual solution for $\mathcal{G}$. Substituting $\delta(b - a)$ in Eq. (12.3b) by the expression from Eq. (12.3a) yields,

$$
\begin{aligned}
[-\partial_{\tau_b} - H(b)]\,\mathcal{G}(b, a) &= [-\partial_{\tau_b} - H(b) + V(b)]\,\mathcal{G}^0(b, a) \\
&= [-\partial_{\tau_b} - H(b)]\,\mathcal{G}^0(b, a) + V(b)\,\mathcal{G}^0(b, a) \\
&= [-\partial_{\tau_b} - H(b)]\,\mathcal{G}^0(b, a) + \int d1\,\delta(b - 1)\,V(1)\,\mathcal{G}^0(1, a).
\end{aligned}
\tag{12.4}
$$

Acting from the left with $[-\partial_{\tau_b} - H(b)]^{-1}$ gives an integral equation for $\mathcal{G}$, the so-called Dyson equation,

$$
\mathcal{G}(b, a) = \mathcal{G}^0(b, a) + \int d1\,\mathcal{G}(b, 1)\,V(1)\,\mathcal{G}^0(1, a),
\tag{12.5}
$$

where we have used the second expression in Eq. (12.3b) to introduce $\mathcal{G}$ in the integrand. By iteratively inserting $\mathcal{G}$ itself in the integrand on the left-hand side, we obtain the infinite perturbation series

$$
\begin{aligned}
\mathcal{G}(b, a) = \mathcal{G}^0(b, a) &+ \int d1\,\mathcal{G}^0(b, 1)\,V(1)\,\mathcal{G}^0(1, a) \\
&+ \int d1 \int d2\,\mathcal{G}^0(b, 1)\,V(1)\,\mathcal{G}^0(1, 2)\,V(2)\,\mathcal{G}^0(2, a) \\
&+ \int d1 \int d2 \int d3\,\mathcal{G}^0(b, 1)\,V(1)\,\mathcal{G}^0(1, 2)\,V(2)\,\mathcal{G}^0(2, 3)\,V(3)\,\mathcal{G}^0(3, a) + \cdots.
\end{aligned}
\tag{12.6}
$$

The solutions Eqs. (12.5) and (12.6) for $\mathcal{G}$ are easy to interpret. The propagator $\mathcal{G}$ of a fermion in an external potential is given as the sum of all possible processes involving unperturbed propagation, described by $\mathcal{G}^0$, intersected by any number of scattering events V. So in this simple case there is really no need for further elucidation, but we will anyway proceed by introducing the corresponding Feynman diagrams.

The first step is to define the basic graphical vocabulary, i.e., to define the pictograms representing the basic quantities $\mathcal{G}$, $\mathcal{G}^0$, and V of the problem. This vocabulary is known as the Feynman rules,

$$
\mathcal{G}(b, a) = \quad , \qquad \mathcal{G}^0(b, a) = \quad , \qquad \int d1\,V(1)\ldots \;=\; \text{✹}\,1.
\tag{12.7}
$$

Note how the fermion lines point from the points of creation, e.g., $\Psi^\dagger(a)$, to the points of annihilation, e.g., $\Psi(b)$. Using the Feynman rules, the infinite perturbation series Eq. (12.6) becomes

In this form we clearly see how the full propagator from a to b is the sum over all possible ways to connect a and b with bare propagators via any number of scattering events. We can also perform calculations by manipulating the diagrams. Let us for example derive an integral form equivalent to Eq. (12.5) from Eq. (12.8),

which by using the Feynman rules can be written as

$$\mathcal{G}(b,a) = \mathcal{G}^0(b,a) + \int d1\, \mathcal{G}^0(b,1)\, V(1)\, \mathcal{G}(1,a). \tag{12.10}$$

The former integral equation (12.5) for $\mathcal{G}$ is obtained by pulling out the bottom part $V(n)\,\mathcal{G}^0(n,a)$ of every diagram on the right-hand side of Eq. (12.8), thereby exchanging the arrow and the double arrow in the last diagram of Eq. (12.9).

This is a first demonstration of the compactness of the Feynman diagram technique, and how visual clarity is obtained without loss of mathematical rigor.

12.2   Elastic scattering and Matsubara frequencies

When a fermion system interacts with a static external potential, no energy is transferred between the two systems, a situation referred to as elastic scattering. The lack of energy transfer in elastic scattering is naturally reflected in a particularly simple form of the single-particle Green's function $\mathcal{G}(ik_n)$ in Matsubara frequency space. In the following, the spin index σ is left out since the same answer is obtained for the two spin directions.

First, we note that since the Hamiltonian H in Eq. (12.1) is time independent for static potentials, we know from Eq. (11.22) that $\mathcal{G}(\mathbf{r}\tau, \mathbf{r}'\tau')$ depends only on the time difference $\tau - \tau'$. According to Eqs. (11.25b) and (11.28) it can therefore be expressed in terms of a Fourier transform with just one fermionic Matsubara frequency ik_n,

$$\mathcal{G}(\mathbf{r}\tau, \mathbf{r}'\tau') = \frac{1}{\beta} \sum_n \mathcal{G}(\mathbf{r}, \mathbf{r}'; ik_n)\, e^{-ik_n(\tau-\tau')},$$

$$\mathcal{G}(\mathbf{r}, \mathbf{r}'; ik_n) = \int_0^\beta d(\tau - \tau')\mathcal{G}(\mathbf{r}\tau, \mathbf{r}'\tau')\, e^{ik_n(\tau-\tau')}. \tag{12.11}$$

The Fourier transform of the time convolution $\int d\tau_1\, \mathcal{G}^0(\tau_b - \tau_1)V\mathcal{G}(\tau_1 - \tau_a)$ appearing in the integral equation of $\mathcal{G}$, is the product $\mathcal{G}^0(ik_n)V\mathcal{G}(ik_n)$. The elastic scattering, i.e., the time-independent V, cannot change the frequencies of the propagators. In Matsubara frequency space the Dyson equation Eq. (12.10) takes the form

$$\mathcal{G}(\mathbf{r}_b, \mathbf{r}_a; ik_n) = \mathcal{G}^0(\mathbf{r}_b, \mathbf{r}_a; ik_n) + \int d\mathbf{r}_1\, \mathcal{G}^0(\mathbf{r}_b, \mathbf{r}_1; ik_n)\, V(1)\, \mathcal{G}(\mathbf{r}_1, \mathbf{r}_a; ik_n). \tag{12.12}$$

As seen previously, the expressions are simplified by transforming from the $|\mathbf{r}\rangle$-basis to the basis $|\nu\rangle$ which diagonalizes H_0. We define the transformed Green's function in this basis as,

$$\mathcal{G}_{\nu\nu'} \equiv \int d\mathbf{r} d\mathbf{r}'\, \langle\nu|\mathbf{r}\rangle\mathcal{G}(\mathbf{r}, \mathbf{r}')\langle\mathbf{r}'|\nu'\rangle \quad \Leftrightarrow \quad \mathcal{G}(\mathbf{r}, \mathbf{r}') = \sum_{\nu\nu'}\langle\mathbf{r}|\nu\rangle\mathcal{G}_{\nu\nu'}\langle\nu'|\mathbf{r}'\rangle. \tag{12.13}$$

In a similar way, we define the $|\nu\rangle$-transform of $V(\mathbf{r})$ as $V_{\nu\nu'} \equiv \int d\mathbf{r}\, \langle\nu|\mathbf{r}\rangle V(\mathbf{r})\langle\mathbf{r}|\nu'\rangle$. In the $|\nu, ik_n\rangle$ representation, the equation of motion Eq. (12.3b) for $\mathcal{G}$ is a matrix equation,

$$\sum_{\nu''}[(ik_n - \xi_\nu)\delta_{\nu\nu''} - V_{\nu,\nu''}]\mathcal{G}_{\nu'',\nu'}(ik_n) = \delta_{\nu\nu'} \quad \text{or} \quad [ik_n\bar{\bar{1}} - \bar{\bar{E}}_0 - \bar{\bar{V}}]\bar{\bar{\mathcal{G}}}(ik_n) = \bar{\bar{1}}, \tag{12.14}$$

where $\bar{\bar{E}}_0$ is a diagonal matrix with the eigenenergies $\xi_\nu = \varepsilon_\nu - \mu$ along the diagonal. We have thus reduced the problem of finding the full Green's function to a matrix inversion problem. We note in particular that in accordance with Eq. (11.43) the bare propagator $\mathcal{G}^0$ has the simple diagonal form

$$\sum_{\nu''}(ik_n - \xi_\nu)\delta_{\nu\nu''}\, \mathcal{G}^0_{\nu'',\nu'}(ik_n) = \delta_{\nu\nu'} \quad \Rightarrow \quad \mathcal{G}^0_{\nu,\nu'}(ik_n) = \frac{1}{ik_n - \xi_\nu}\, \delta_{\nu\nu'}. \tag{12.15}$$

We can utilize this to rewrite the integral equation Eq. (12.12) as a simple matrix equation,

$$\mathcal{G}(\nu_b\nu_a; ik_n) = \delta_{\nu_b\nu_a}\, \mathcal{G}^0(\nu_a\nu_a; ik_n) + \sum_{\nu_c}\mathcal{G}^0(\nu_b\nu_b; ik_n)\, V_{\nu_b\nu_c}\, \mathcal{G}(\nu_c\nu_a; ik_n). \tag{12.16}$$

We can also formulate Feynman rules in (ν, ik_n)-space. We note that $\bar{\bar{\mathcal{G}}}_0$ is diagonal in ν, while $\bar{\bar{V}}$ is a general matrix. To obtain the sum of all possible quantum processes one must sum over all matrix indices different from the externally given ν_a and ν_b.

The frequency argument is suppressed, since the Green's functions are diagonal in ik_n,

$$
\mathcal{G}_{\nu_b\nu_a} = \quad , \qquad \mathcal{G}^0_{\nu_b,\nu_a} = \delta_{\nu_b\nu_a} \quad = \frac{\delta_{\nu_a\nu_b}}{ik_n - \xi_{\nu_a}}, \qquad V_{\nu\nu'} = \begin{matrix}\nu \\ \nu'\end{matrix}. \tag{12.17}
$$

Using these Feynman rules in (ν, ik_n)-space we can express Dyson's equation Eq. (12.16) diagrammatically as,

$$
\quad = \delta_{\nu_b\nu_a} \quad + \quad . \tag{12.18}
$$

12.3   Random impurities in disordered metals

An important example of elastic scattering by external potentials is the case of random impurities in a disordered metal. One well-controlled experimental realization of this is provided by a perfect metal Cu lattice with Mg^{II} ions substituting a small number of randomly chosen Cu^I ions. The valence of the impurity ions is one higher than the host ions, and as a first approximation an impurity ion at site $\mathbf{P}_j$ gives rise to a simple screened monocharge Coulomb potential $u_j(\mathbf{r}) = -(e_0^2/|\mathbf{r} - \mathbf{P}_j|)\,e^{-|\mathbf{r}-\mathbf{P}_j|/a}$. The screening is due to the electrons in the system trying to neutralize the impurity charge, and as a result the range of the potential is finite, given by the so-called screening length a. This will be discussed in detail in Chapter 14.

In Fig. 12.1(a) a number of randomly positioned impurities in an otherwise perfect metal lattice is shown. The presence of the impurities can be detected by measuring the (longitudinal) resistivity ρ_{xx} of the metal as a function of temperature. At high temperature, the resistivity is mainly due to electron-phonon scattering, and since the vibrational energy $\hbar\omega\,(n+\tfrac{1}{2})$ in thermal equilibrium is proportional to $k_B T$, the number n of phonons, and hence the electron-phonon scattering rate, is also proportional to T (see e.g., Exercise 3.2). At lower temperature the phonon degrees of freedom begin to freeze out, and the phase space available for scattering also shrinks, and consequently the resistivity becomes proportional to some power α of T. Finally, at the lowest temperatures, typically a few kelvin, only the electron-impurity scattering is left preventing the Bloch electrons in moving unhindered through the crystal. As a result, the resistivity levels off at some value ρ_0, the so-called residual resistivity. The temperature behavior of the resistivity is depicted in Fig. 12.1(b).

We postpone the calculation of the resistivity, and in this section just concentrate on studying the electron Matsubara Green's function $\mathcal{G}$ for electrons moving in such a disordered metal. We use the plane wave states $|\mathbf{k}\sigma\rangle$ from the effective mass approximation Eq. (2.16) as the unperturbed basis $|\nu\rangle$.

Now consider N_{imp} identical impurities situated at the randomly distributed but fixed positions $\mathbf{P}_j$. The elastic scattering potential $V(\mathbf{r})$ then acquires the form

(a)

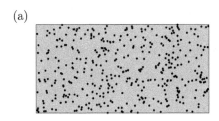

(b)

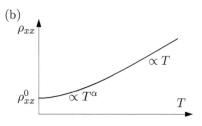

FIG. 12.1. (a) A disordered metal consisting of an otherwise perfect metal lattice with a number of randomly positioned impurities giving rise to elastic electron-impurity scattering. (b) The electrical resistivity $\rho_{xx}(T)$ of the disordered metal as a function of temperature. At high T the electron-phonon scattering dominates giving rise to a linear behavior, while at low T only the electron-impurity scattering is effective and gives rise to the non-zero value ρ_{xx}^0 of ρ_{xx} at $T = 0$.

$$V(\mathbf{r}) = \sum_{j=1}^{N_{\text{imp}}} u(\mathbf{r} - \mathbf{P}_j), \quad \mathbf{P}_j \text{ is randomly distributed.} \tag{12.19}$$

Two small dimensionless parameters of the system serve as guides to obtain good approximative solutions. One is stating that the ratio between the impurity density, $n_{\text{imp}} = N_{\text{imp}}/\mathcal{V}$, and the electron density n_{el} is much smaller than unity:

$$\frac{n_{\text{imp}}}{n_{\text{el}}} \ll 1. \tag{12.20}$$

The other small parameter is stating that the strength of the scattering potential is weak. We assume that the scattering potential $u(\mathbf{r} - \mathbf{P}_j)$ differs only significantly from 0 for $|\mathbf{r} - \mathbf{P}_j| < a$, and that the characteristic value in that region is $\tilde{u}$. Weak scattering means that $\tilde{u}$ is much smaller than some characteristic level spacing $\hbar^2/ma^2$ as follows,[41]

$$\tilde{u} \frac{ma^2}{\hbar^2} \ll \min\{1, k_{\text{F}}a\}. \tag{12.21}$$

12.3.1 Feynman diagrams for the impurity scattering

With the random potential Eq. (12.19), the Dyson equation (12.12) becomes

$$\mathcal{G}(\mathbf{r}_b, \mathbf{r}_a; ik_n) = \mathcal{G}^0(\mathbf{r}_b - \mathbf{r}_a; ik_n) + \sum_{j=1}^{N_{\text{imp}}} \int d\mathbf{r}_1 \, \mathcal{G}^0(\mathbf{r}_b - \mathbf{r}_1; ik_n) \, u(\mathbf{r}_1 - \mathbf{P}_j) \, \mathcal{G}(\mathbf{r}_1, \mathbf{r}_a; ik_n),$$

$$\tag{12.22}$$

where we have used the fact that the unperturbed system is translation invariant and hence that $\mathcal{G}^0(\mathbf{r}_1, \mathbf{r}_a; ik_n) = \mathcal{G}^0(\mathbf{r}_1 - \mathbf{r}_a; ik_n)$. We now want to deduce the Feynman

---

[41] Assume that u is only important in a sphere of radius a around the scattering center. The level spacing for non-perturbed states in that sphere is near the ground state given by the size quantization $\hbar^2/ma^2$. For high energies around, say $\varepsilon = p^2/2m$, the level spacing is $(\partial\varepsilon/\partial p)\,\Delta p = (p/m)\,(\hbar/a) = ka\,\hbar^2/ma^2$, where $p = \hbar k$ has been used. Thus u is weak if it is smaller than the smallest of these level spacings.

rules for constructing diagrams in this situation. First expand the Dyson equation Eq. (12.22) in orders n of the scattering potential $u(\mathbf{r}-\mathbf{P}_j)$, and obtain $\mathcal{G}(\mathbf{r}_b, \mathbf{r}_a) = \sum_{n=0}^{\infty} \mathcal{G}^{(n)}(\mathbf{r}_b, \mathbf{r}_a)$, where the frequency argument ik_n has been suppressed. The nth-order term $\mathcal{G}^{(n)}$ is

$$\mathcal{G}^{(n)}(\mathbf{r}_b, \mathbf{r}_a) = \sum_{j_1}^{N_{\text{imp}}} \cdots \sum_{j_n}^{N_{\text{imp}}} \int d\mathbf{r}_1 \cdots \int d\mathbf{r}_n \tag{12.23}$$
$$\times \mathcal{G}^0(\mathbf{r}_b - \mathbf{r}_n)\, u(\mathbf{r}_n - \mathbf{P}_{j_n}) \cdots u(\mathbf{r}_2 - \mathbf{P}_{j_2})\, \mathcal{G}^0(\mathbf{r}_2 - \mathbf{r}_1)\, u(\mathbf{r}_1 - \mathbf{P}_{j_1})\, \mathcal{G}^0(\mathbf{r}_1 - \mathbf{r}_a).$$

This nth-order contribution can be interpreted as the sum over all processes involving n scattering events in all possible combination of impurities. Naturally, we can never hope to solve this problem exactly. Not only is it for all practical purposes impossible to know where all the impurities in a given metallic sample *de facto* are situated, but even if we did, no simple solution for the Green's function could be found. However, if we are satisfied with the answer to the less-ambitious and more practical question of what is the average behavior, then we shall soon find an answer. To this end we reformulate Dyson's equation in $\mathbf{k}$ space since according to Eq. (12.15) $\mathcal{G}_{\mathbf{k}}^0$ of the impurity-free, and therefore translation-invariant, problem has the simple form,

$$\mathcal{G}_{\mathbf{k}}^0(ik_n) = \frac{1}{ik_n - \xi_{\mathbf{k}}}, \qquad \mathcal{G}^0(\mathbf{r}-\mathbf{r}'; ik_n) = \frac{1}{V} \sum_{\mathbf{k}} \mathcal{G}_{\mathbf{k}}^0(ik_n)\, e^{i\mathbf{k}\cdot(\mathbf{r}-\mathbf{r}')}. \tag{12.24}$$

The Fourier transform of the impurity potential $u(\mathbf{r}-\mathbf{P}_j)$ is

$$u(\mathbf{r}-\mathbf{P}_j) = \frac{1}{V} \sum_{\mathbf{q}} u_{\mathbf{q}}\, e^{i\mathbf{q}\cdot(\mathbf{r}-\mathbf{P}_j)} = \frac{1}{V} \sum_{\mathbf{q}} e^{-i\mathbf{q}\cdot\mathbf{P}_j}\, u_{\mathbf{q}}\, e^{i\mathbf{q}\cdot\mathbf{r}}, \tag{12.25}$$

while the Fourier expansion of $\mathcal{G}^{(n)}(\mathbf{r}_b, \mathbf{r}_a; ik_n)$ in Eq. (12.23) is

$$\mathcal{G}^{(n)}(\mathbf{r}_b, \mathbf{r}_a) = \sum_{j_1 \cdots j_n}^{N_{\text{imp}}} \frac{1}{V^n} \sum_{\mathbf{q}_1 \cdots \mathbf{q}_n} \frac{1}{V^2} \sum_{\mathbf{k}_a \mathbf{k}_b} \frac{1}{V^{n-1}} \sum_{\mathbf{k}_1 \cdots \mathbf{k}_{n-1}} \int d\mathbf{r}_1 \cdots \int d\mathbf{r}_n \tag{12.26}$$
$$\times \mathcal{G}_{\mathbf{k}_b}^0 u_{\mathbf{q}_n} \mathcal{G}_{\mathbf{k}_{n-1}}^0 u_{\mathbf{q}_{n-1}} \cdots u_{\mathbf{q}_2} \mathcal{G}_{\mathbf{k}_1}^0 u_{\mathbf{q}_1} \mathcal{G}_{\mathbf{k}_a}^0\, e^{-i(\mathbf{q}_n \cdot \mathbf{P}_{j_n} + \cdots + \mathbf{q}_2 \cdot \mathbf{P}_{j_2} + \mathbf{q}_1 \cdot \mathbf{P}_{j_1})}$$
$$\times e^{i\mathbf{k}_b \cdot (\mathbf{r}_b - \mathbf{r}_n)} e^{i\mathbf{q}_n \cdot \mathbf{r}_n} e^{i\mathbf{k}_{n-1} \cdot (\mathbf{r}_n - \mathbf{r}_{n-1})} \cdots e^{i\mathbf{q}_2 \cdot \mathbf{r}_2} e^{i\mathbf{k}_1 \cdot (\mathbf{r}_2 - \mathbf{r}_1)} e^{i\mathbf{q}_1 \cdot \mathbf{r}_1} e^{i\mathbf{k}_a \cdot (\mathbf{r}_1 - \mathbf{r}_a)}.$$

This complicated expression can be simplified significantly by performing the n spatial integrals, $\int d\mathbf{r}_j\, e^{i(\mathbf{k}_j - \mathbf{k}_{j-1} - \mathbf{q}_j)\cdot\mathbf{r}_j} = V\, \delta_{\mathbf{k}_j, \mathbf{k}_{j-1} + \mathbf{q}_j}$, which may be interpreted as momentum conservation in each electron-impurity scattering: the change of the electron momentum is absorbed by the impurity. Utilizing these delta functions in the n $\mathbf{q}$-sums leads to

$$\mathcal{G}^{(n)}(\mathbf{r}_b \mathbf{r}_a) = \frac{1}{V^2} \sum_{\mathbf{k}_a \mathbf{k}_b} e^{i\mathbf{k}_b \cdot \mathbf{r}_b} e^{-i\mathbf{k}_a \cdot \mathbf{r}_a} \sum_{j_1 \cdots j_n}^{N_{\text{imp}}} \sum_{\mathbf{k}_1 \cdots \mathbf{k}_{n-1}} \frac{1}{V^{n-1}} \tag{12.27}$$
$$\times \mathcal{G}_{\mathbf{k}_b}^0 u_{\mathbf{k}_b - \mathbf{k}_{n-1}} \mathcal{G}_{\mathbf{k}_{n-1}}^0 \cdots u_{\mathbf{k}_2 - \mathbf{k}_1} \mathcal{G}_{\mathbf{k}_1}^0 u_{\mathbf{k}_1 - \mathbf{k}_a} \mathcal{G}_{\mathbf{k}_a}^0\, e^{-i[(\mathbf{k}_b - \mathbf{k}_{n-1}) \cdot \mathbf{P}_{j_n} + \cdots + (\mathbf{k}_1 - \mathbf{k}_a) \cdot \mathbf{P}_{j_1}]}.$$

Introducing the Fourier transform $\mathcal{G}^{(n)}_{\mathbf{k}_b \mathbf{k}_a}$ of $\mathcal{G}^{(n)}(\mathbf{r}_b, \mathbf{r}_a)$ as

$$
\mathcal{G}^{(n)}(\mathbf{r}_b, \mathbf{r}_a) = \frac{1}{\mathcal{V}^2} \sum_{\mathbf{k}_a \mathbf{k}_b} e^{i\mathbf{k}_b \cdot \mathbf{r}_b} e^{-i\mathbf{k}_a \cdot \mathbf{r}_a} \, \mathcal{G}^{(n)}_{\mathbf{k}_b \mathbf{k}_a}, \tag{12.28}
$$

with

$$
\mathcal{G}^{(n)}_{\mathbf{k}_b \mathbf{k}_a} = \sum_{j_1 \cdots j_n}^{N_{\mathrm{imp}}} \frac{1}{\mathcal{V}^{n-1}} \sum_{\mathbf{k}_1 \cdots \mathbf{k}_{n-1}} e^{-i[(\mathbf{k}_b - \mathbf{k}_{n-1}) \cdot \mathbf{P}_{j_n} + \cdots + (\mathbf{k}_1 - \mathbf{k}_a) \cdot \mathbf{P}_{j_1}]}
$$
$$
\times \, \mathcal{G}^0_{\mathbf{k}_b} u_{\mathbf{k}_b - \mathbf{k}_{n-1}} \mathcal{G}^0_{\mathbf{k}_{n-1}} \cdots u_{\mathbf{k}_2 - \mathbf{k}_1} \mathcal{G}^0_{\mathbf{k}_1} \cdots u_{\mathbf{k}_1 - \mathbf{k}_a} \mathcal{G}^0_{\mathbf{k}_a}. \tag{12.29}
$$

We can now easily deduce the Feynman rules for the diagrams corresponding to $\mathcal{G}^{(n)}_{\mathbf{k}_b \mathbf{k}_a}$:

---

(1) Let dashed arrows $j \bullet \text{-- -} \blacktriangleleft \text{-- -} \bigstar \mathbf{q}, \mathbf{P}_j$ denote a scattering event $u_{\mathbf{q}} e^{-i\mathbf{q} \cdot \mathbf{P}_j}$.

(2) Draw n scattering events.

(3) Let straight arrows $\text{———} \blacktriangleleft \text{———} \mathbf{k}$ denote $\mathcal{G}^0_{\mathbf{k}}$.

(4) Let $\mathcal{G}^0_{\mathbf{k}_a}$ go into vertex $\bullet 1$ and $\mathcal{G}^0_{\mathbf{k}_b}$ away from vertex $\bullet n$.

(5) Let $\mathcal{G}^0_{\mathbf{k}_j}$ go from vertex j to vertex $j + 1$.

(6) Maintain momentum conservation at each vertex.

(7) Perform the sums $\frac{1}{\mathcal{V}} \sum_{\mathbf{k}_j}$ over all internal momenta $\mathbf{k}_j$, and $\sum_{j_1 \cdots j_n}^{N_{\mathrm{imp}}}$ over $\mathbf{P}_{j_l}$.

$$\tag{12.30}$$

The diagram corresponding to Eq. (12.29) is

$$\tag{12.31}$$

This diagram is very suggestive. One can see how an incoming electron with momentum $\mathbf{k}_a$ is scattered n times under momentum conservation with the impurities and leaves the system with momentum $\mathbf{k}_b$. However, as mentioned above, it is not possible to continue the study of impurity scattering on general grounds without further assumptions. We therefore begin to consider the possibility of performing an average over the random positions $\mathbf{P}_j$ of the impurities.

12.4 Impurity self-average

If the electron wavefunctions are completely coherent throughout the entire disordered metal, each true electronic eigenfunction exhibit an extremely complex diffraction pattern spawned by the randomly positioned scatterers. If one imagines changing some external parameter, e.g., the average electron density or an external magnetic field,

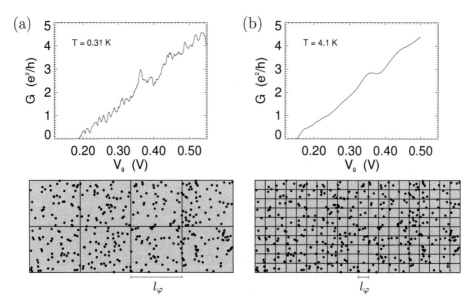

FIG. 12.2. (a) The measured conductance of a disordered GaAs sample at $T = 0.31$ K displaying random but reproducible quantum fluctuations as a function of a gate voltage V_g controlling the electron density. The fluctuations are due to phase coherent scattering against randomly positioned impurities. Below is indicated that the phase coherence length l_φ is comparable to the size of the sample. (b) The same system at $T = 4.1$ K. The fluctuations are almost gone due to the smallness of l_φ at this temperature. The sample now contains a large number of independent but phase-coherent subsystems of size l_φ, as illustrated below the plot. As a result a substantial self-averaging occurs, which suppresses the quantum fluctuations.

each individual diffraction pattern will of course change drastically due to the sensitivity of the scattering phases of the wavefunctions. Significant quantum fluctuations must therefore occur in any observable at sufficiently low temperatures.

Using modern nanotechnology to fabricate small (but still macroscopic) samples, and standard cryogenic equipment to cool down these samples to ultralow temperatures, one can in fact obtain an experimental situation, where the electrons can traverse the sample without loosing their quantum-mechanical phase coherence. In Fig. 12.2(a) is shown the conductance trace of a GaAs nano-device, such as the one shown in Fig. 2.10, at 0.31 K as a function of the electron density. This density can be controlled by applying a gate voltage V_g on an external electrode. The conductance G is seen to fluctuate strongly for minute changes of V_g. These fluctuations turn out to be perfectly reproducible as V_g is swept up and down several times.

As the temperature of a given sample is raised, the amount of electron-electron and electron-phonon scattering increases because of an increased phase space for scattering and an increased number of phonons. The quantum mechanical phase of each individual electron is changed by a small random amount at each inelastic scattering event, and as a result the coherence length l_φ for the electrons diminishes. At sufficiently high temperature (e.g., 4.1 K) l_φ is much smaller than the size of the device, and we can think of the device as being composed as a number of phase-independent small phase-coherent subsystems. Therefore, when one measures an observable the result is in fact an incoherent average of all these subsystems. Note that this average is imposed by the physical properties of the system itself, and this effective averaging is consequently denoted *self-averaging*. This effect is illustrated in Fig. 12.2(b) where the conductance trace at 4.1 K is seen to be much smoother than the one at 0.31 K, and where the many small phase-coherent subsystems of the sample are indicated below the experimental graph.

For very large (mm-sized) macroscopic samples, l_φ is much smaller than the sample size at all experimental realizable temperatures ($T > 10$ mK for electron gases in metals and semiconductors), and we are in the impurity self-averaging case. Mathematically, the impurity average is performed by summing over all the phase-independent coherent subsystems and dividing by their number N_{sys}. But due to the random distribution of the impurities, this average is the same as an average over the impurity position within a single subsystem, as can be seen from Fig. 12.2. However, even on the rather small length scale l_φ the system is already homogeneous, and one can as well perform the position average over the entire volume of the sample. Thus in the following, we average over all possible uncorrelated positions $\mathbf{P}_j$ of the N_{imp} impurities for the entire system,

$$
\frac{1}{\mathcal{V}} \langle \mathcal{G}_{\mathbf{k}_b \mathbf{k}_a} \rangle_{\text{imp}} \equiv \delta_{\mathbf{k}_b \mathbf{k}_a} \bar{\mathcal{G}}_{\mathbf{k}_a} \equiv \frac{\delta_{\mathbf{k}_b \mathbf{k}_a}}{N_{\text{sys}}} \sum_{i=1}^{N_{\text{sys}}} \mathcal{G}_{\mathbf{k}_a}^{\text{sys}_i}
$$

$$
\sim \delta_{\mathbf{k}_b \mathbf{k}_a} \frac{1}{\mathcal{V}} \int d\mathbf{P}_1 \frac{1}{\mathcal{V}} \int d\mathbf{P}_2 \cdots \frac{1}{\mathcal{V}} \int d\mathbf{P}_{N_{\text{imp}}} \mathcal{G}_{\mathbf{k}_a}. \tag{12.32}
$$

Here, we have anticipated that the impurity-averaged Green's function is diagonal in $\mathbf{k}$ due to the restoring of translation invariance upon average. Some care must be taken regarding the average over the impurity positions $\mathbf{P}_j$. Any nth-order contribution to $\mathcal{G}_{\mathbf{k}}$ contains n scattering events, but they need not be on n different scatterers. In fact, the number of scatterers involved could be any number between 1 and n. We must therefore carefully sort out all possible ways to scatter on p different impurities.

As mentioned in Eq. (12.20), we work in the limit of small impurity densities n_{imp}. For a given fixed number n of scattering events the most important contribution therefore comes from processes involving just one impurity. Then, down by the small factor $n_{\text{imp}}/n_{\text{el}}$, follow processes involving two impurities, etc. We note that in Eq. (12.29) the only reference to the impurity positions is the exponential $\exp[i(\mathbf{q}_1 \cdot \mathbf{P}_{j_1} + \mathbf{q}_2 \cdot \mathbf{P}_{j_2} + \cdots + \mathbf{q}_n \cdot \mathbf{P}_{j_n})]$, with the scattering vectors $\mathbf{q}_i = \mathbf{k}_i - \mathbf{k}_{i-1}$. The sum in Eq. (12.29) over impurity positions in this exponential is now ordered according to how many impurities are involved,

$$\sum_{j_1,\ldots,j_n}^{N_{\text{imp}}} e^{i\sum_{l=1}^n \mathbf{q}_l\cdot\mathbf{P}_{j_l}} = \sum_{h_1}^{N_{\text{imp}}} e^{i(\sum_{\mathbf{q}_{j_1}\in Q}\mathbf{q}_{j_1})\cdot\mathbf{P}_{h_1}}$$

$$+\sum_{Q_1\cup Q_2=Q}\sum_{h_1}^{N_{\text{imp}}}\sum_{h_2}^{N_{\text{imp}}} e^{i(\sum_{\mathbf{q}_{l_1}\in Q_1}\mathbf{q}_{l_1})\cdot\mathbf{P}_{h_1}}\,e^{i(\sum_{\mathbf{q}_{l_2}\in Q_2}\mathbf{q}_{l_2})\cdot\mathbf{P}_{h_2}}$$

$$+\sum_{Q_1\cup Q_2\cup Q_3=Q}\sum_{h_1}^{N_{\text{imp}}}\sum_{h_2}^{N_{\text{imp}}}\sum_{h_3}^{N_{\text{imp}}} e^{i(\sum_{\mathbf{q}_{l_1}\in Q_1}\mathbf{q}_{l_1})\cdot\mathbf{P}_{h_1}}\,e^{i(\sum_{\mathbf{q}_{l_2}\in Q_2}\mathbf{q}_{l_2})\cdot\mathbf{P}_{h_2}}\,e^{i(\sum_{\mathbf{q}_{l_3}\in Q_3}\mathbf{q}_{l_3})\cdot\mathbf{P}_{h_3}}$$

$$+\cdots \tag{12.33}$$

Here, $Q = \{\mathbf{q}_1, \mathbf{q}_2, \ldots, \mathbf{q}_n\}$ is the set of the n scattering vectors, while $Q_1 \cup Q_2 \cup \ldots \cup Q_p = Q$ denotes all possible unions of non-empty disjoint subsets spanning Q. By definition all the scattering vectors in one particular subset Q_i are connected to the same impurity $\mathbf{P}_{h_i}$. The first term on the right-hand side of Eq. (12.33) thus corresponds to all scattering being on the same impurity, the second term that only two impurities are involved *et cetera*. Note that, strictly speaking, two different impurities cannot occupy the same position. Nevertheless, in Eq. (12.33) we let the h-sums run unrestricted. This introduces a small error of the order $1/N_{\text{imp}}$ for the important terms in the low impurity density limit involving only a few impurities.[42]

Since all the p positions $\mathbf{P}_h$ now are manifestly different we can perform the impurity average indicated in Eq. (12.32) over each exponential factor independently. The detailed calculation is straightforward but somewhat cumbersome; the result may perhaps be easier to understand than the derivation. As depicted in Eq. (12.38), the impurity-averaged Green's function is a sum over scattering processes on the position-averaged impurities. Since translation invariance is restored by the averaging, the sum of all scattering momenta on the same impurity must be zero, cf. Fig. 12.3. But let us see how these conclusions are reached.

The impurity average indicated in Eq. (12.32) over each independent exponential factor results in some Kronecker delta's, meaning that all scattering vectors $\mathbf{q}_{h_i}$ connected to the same impurity must add up to zero,

$$\left\langle e^{i(\sum_{\mathbf{q}_{h_i}\in Q_i}\mathbf{q}_{h_i})\cdot\mathbf{P}_{h_i}} \right\rangle_{\text{imp}} = \frac{1}{\mathcal{V}}\int d\mathbf{P}_{h_i}\, e^{i(\sum_{\mathbf{q}_{h_i}\in Q_h}\mathbf{q}_{h_i})\cdot\mathbf{P}_{h_i}} = \delta_{0,\sum_{\mathbf{q}_{h_i}\in Q_h}\mathbf{q}_{h_i}}. \tag{12.34}$$

This of course no longer depends on the p impurity positions $\mathbf{P}_{h_i}$; the averaging has restored translation invariance. The result of the impurity averaging can now be written as

$$\left\langle \sum_{j_1,\ldots,j_n}^{N_{\text{imp}}} e^{i\sum_{l=1}^n \mathbf{q}_l\cdot\mathbf{P}_{j_l}} \right\rangle_{\text{imp}} = \sum_{p=1}^n \left[\sum_{Q_1\cup\ldots\cup Q_p=Q} \prod_{h=1}^p \left(N_{\text{imp}}\,\delta_{0,\sum_{\mathbf{q}_{h_i}\in Q_h}\mathbf{q}_{h_i}} \right) \right], \tag{12.35}$$

[42]This error occurs since our approximation amounts to saying that the $(p+1)$-st impurity can occupy any of the N_{imp} impurity sites, and not just the $N_{\text{imp}} - p$ available sites. For the important terms we have $p \ll N_{\text{imp}}$, and hence the error is of order $p/N_{\text{imp}} \ll 1$.

which, when inserted in Eq. (12.29), leads to

$$\langle \mathcal{G}_{\mathbf{k}}^{(n)} \rangle_{\text{imp}} = \frac{1}{\mathcal{V}^{n-1}} \sum_{\mathbf{k}_1 \ldots \mathbf{k}_{n-1}} \sum_{p=1}^{n} \sum_{Q_1 \cup \ldots \cup Q_p = Q} \prod_{h=1}^{p} \left(N_{\text{imp}} \delta_{0, \sum_{Q_h} (\mathbf{k}_{h_i} - \mathbf{k}_{(h_i - 1)})} \right)$$

$$\times \mathcal{G}_{\mathbf{k}}^0 u_{\mathbf{k} - \mathbf{k}_1} \mathcal{G}_{\mathbf{k}_1}^0 u_{\mathbf{k}_1 - \mathbf{k}_2} \mathcal{G}_{\mathbf{k}_2}^0 \cdots u_{\mathbf{k}_{n-1} - \mathbf{k}} \mathcal{G}_{\mathbf{k}}^0. \tag{12.36}$$

We note that due to the p factors containing δ-functions there are in fact only $n - 1 - p$ free momentum sums $\frac{1}{\mathcal{V}} \sum_{\mathbf{k}'}$ to perform. The remaining p volume prefactors are combined with N_{imp} to yield p impurity density factors $n_{\text{imp}} = N_{\text{imp}}/\mathcal{V}$.

The Feynman rules for constructing the nth-order contribution $\langle \mathcal{G}_{\mathbf{k}}^{(n)} \rangle_{\text{imp}}$ to the impurity-averaged Green's function $\langle \mathcal{G}_{\mathbf{k}} \rangle_{\text{imp}}$ are now easy to establish:

---

(1) Let scattering lines $- - \blacktriangleleft - - - \mathbf{q}$ denote the scattering amplitude $u_{\mathbf{q}}$.

(2) Let ✹ denote a momentum-conserving impurity-averaged factor $n_{\text{imp}} \delta_{0 \sum \mathbf{q}}$.

(3) Let fermion lines ——◄—— $\mathbf{k}$ denote unperturbed Green's functions $\mathcal{G}_{\mathbf{k}}^0$.

(4) Draw p impurity stars. Let n_1 scattering lines go out from star 1, n_2 from star 2, etc., so the total number $n_1 + n_2 + \cdots + n_p$ of scattering lines is n.

(5) Draw all topologically different diagrams containing an unbroken chain of $n+1$ fermion lines and connect the end points of the n first fermion lines to one of the n scattering lines.

(6) Let the first and last fermion line be $\mathcal{G}_{\mathbf{k}}^0$.

(7) Maintain momentum conservation at each vertex.

(8) Make sure that the sum of all momenta leaving an impurity star is zero.

(9) Perform the sum $\frac{1}{\mathcal{V}} \sum_{\mathbf{k}_j}$ over all free internal momenta $\mathbf{k}_j$.

(10) Sum over all orders n of scattering and over p, with $1 \leq p \leq n$.

$$\tag{12.37}$$

The diagrammatic expansion of $\langle \mathcal{G}_{\mathbf{k}} \rangle_{\text{imp}}$ has a direct intuitive appeal,

$$\tag{12.38}$$

In this expression, showing all diagrams up to third order and three diagrams of fourth order, we have for visual clarity suppressed all momentum labels and even the arrows of the scattering lines. For each order the diagrams are arranged after powers

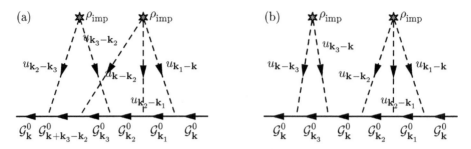

FIG. 12.3. Two fully labelled fifth-order diagrams both with two impurity scatterers. Diagram (a) is a so-called irreducible diagram, i.e., it cannot be cut into two pieces by cutting one internal fermion line. In contrast, diagram (b) is reducible. It consists of two irreducible parts.

of n_{imp}, i.e., the number of impurity stars. In Fig. 12.3 two diagrams with complete labels are shown. In the following section we gain further insight in the solution of $\langle \mathcal{G}_{\mathbf{k}} \rangle_{\text{imp}}$ by rearranging the terms in the diagrammatic expansion, a procedure known as resummation.

12.5 Self-energy for impurity scattered electrons

In Fig. 12.3 we introduce the concept of irreducible diagrams, i.e., diagrams in the expansion of $\langle \mathcal{G}_{\mathbf{k}} \rangle_{\text{imp}}$ that cannot be cut into two pieces by cutting a single internal fermion line. We now use this concept to resum the diagrammatic expansion Eq. (12.38) for $\langle \mathcal{G}_{\mathbf{k}} \rangle_{\text{imp}}$. We remind the reader that this resummation is correct only in the limit of low impurity density. First we define the so-called self-energy $\Sigma_{\mathbf{k}}$ by

$$\Sigma_{\mathbf{k}} \equiv \left\{ \begin{array}{l} \text{The sum of all irreducible diagrams in } \langle \mathcal{G}_{\mathbf{k}} \rangle_{\text{imp}} \\ \text{without the two external fermion lines } \mathcal{G}_{\mathbf{k}}^0 \end{array} \right\}$$

$$= \; ! \; + \; \triangle \; + \left(\triangle + \triangle \right) + \left(\triangle + \cdots \right) + \cdots$$

$$= \; \bigoplus \tag{12.39}$$

Using $\Sigma_{\mathbf{k}}$ and the product form of $\langle \mathcal{G}_{\mathbf{k}} \rangle_{\text{imp}}$ in Fourier space, Eq. (12.38) becomes

$$\langle \mathcal{G}_{\mathbf{k}} \rangle_{\text{imp}} = \; \longleftarrow + \longleftarrow \bigoplus \longleftarrow + \longleftarrow \bigoplus \longleftarrow \bigoplus \longleftarrow + \ldots$$

$$= \; \longleftarrow + \longleftarrow \bigoplus \times \left(\longleftarrow + \longleftarrow \bigoplus \longleftarrow + \cdots \right)$$

$$= \mathcal{G}_{\mathbf{k}}^0 + \mathcal{G}_{\mathbf{k}}^0 \, \Sigma_{\mathbf{k}} \, \langle \mathcal{G}_{\mathbf{k}} \rangle_{\text{imp}}. \tag{12.40}$$

This algebraic Dyson equation, equivalent to Eqs. (12.9) and (12.18), is readily solved,

$$\langle \mathcal{G}_{\mathbf{k}}(ik_n) \rangle_{\text{imp}} = \frac{\mathcal{G}_{\mathbf{k}}^0}{1 - \mathcal{G}_{\mathbf{k}}^0 \, \Sigma_{\mathbf{k}}} = \frac{1}{(\mathcal{G}_{\mathbf{k}}^0)^{-1} - \Sigma_{\mathbf{k}}} = \frac{1}{ik_n - \xi_{\mathbf{k}} - \Sigma_{\mathbf{k}}(ik_n)}. \tag{12.41}$$

From this solution we immediately learn that $\Sigma_{\mathbf{k}}$ enters $\langle \mathcal{G}_{\mathbf{k}} \rangle_{\mathrm{imp}}$ as an additive correction to the original unperturbed energy, $\xi_{\mathbf{k}} \to \xi_{\mathbf{k}} + \Sigma_{\mathbf{k}}$, hence the name self-energy. The problem of finding $\langle \mathcal{G}_{\mathbf{k}} \rangle_{\mathrm{imp}}$ is thus reduced to a calculation of $\Sigma_{\mathbf{k}}$. In the following we go through various approximations for $\Sigma_{\mathbf{k}}$.

12.5.1 Lowest-order approximation

One marvellous feature of the self-energy $\Sigma_{\mathbf{k}}$ is that even if it is approximated by a finite number of diagrams, the Dyson equation (12.40) actually ensures that some diagrams of all orders are included in the perturbation series for $\langle \mathcal{G}_{\mathbf{k}} \rangle_{\mathrm{imp}}$. This allows for essential changes in $\langle \mathcal{G}_{\mathbf{k}} \rangle_{\mathrm{imp}}$, notably one can move the poles of $\langle \mathcal{G}_{\mathbf{k}} \rangle_{\mathrm{imp}}$ and hence change the excitation energies. This would not be possible if only a finite number of diagrams were used in the expansion of $\langle \mathcal{G}_{\mathbf{k}} \rangle_{\mathrm{imp}}$ itself.

Bearing in mind the inequalities in Eqs. (12.20) and (12.21), the lowest-order approximation $\Sigma_{\mathbf{k}}^{\mathrm{LOA}}$ to $\Sigma_{\mathbf{k}}$ is obtained by including only the diagram with the fewest number of impurity stars and scattering lines,

$$\Sigma_{\mathbf{k}}^{\mathrm{LOA}}(ik_n) \equiv \begin{array}{c} \star \\ | \\ \bullet \end{array} = n_{\mathrm{imp}} u_0 = n_{\mathrm{imp}} \int d\mathbf{r}\, u(\mathbf{r}), \qquad (12.42)$$

i.e., a constant, which upon insertion into Dyson's equation (12.41) yields

$$\mathcal{G}_{\mathbf{k}}^{\mathrm{LOA}}(ik_n) = \frac{1}{ik_n - (\xi_{\mathbf{k}} + n_{\mathrm{imp}} u_0)}. \qquad (12.43)$$

But this just reveals a simple constant shift of all the energy levels with the amount $n_{\mathrm{imp}} u_0$. This shift constitutes a redefinition of the origin of the energy axis with no dynamical consequences. In the following, it is absorbed into the definition of the chemical potential and will therefore not appear in the equations.

12.5.2 First-order Born approximation

The simplest non-trivial low-order approximation to the self-energy is the so-called first-order Born approximation given by the "wigwam"-diagram

$$\Sigma_{\mathbf{k}}^{\mathrm{1BA}}(ik_n) \equiv \quad \begin{array}{c} \star \\ \mathbf{k\!-\!k'} \diagup\!\!\!\diagdown \mathbf{k'\!-\!k} \\ \hline \mathbf{k'} \end{array} \quad = \frac{n_{\mathrm{imp}}}{\mathcal{V}} \sum_{\mathbf{k'}} |u_{\mathbf{k-k'}}|^2 \frac{1}{ik_n - \xi_{\mathbf{k'}}}, \qquad (12.44)$$

where we have used that $u_{-\mathbf{k}} = u_{\mathbf{k}}^*$ since $u(\mathbf{r})$ is real. We shall see shortly that $\Sigma_{\mathbf{k}}^{\mathrm{1BA}} = \mathrm{Re}\,\Sigma_{\mathbf{k}}^{\mathrm{1BA}} + i\,\mathrm{Im}\,\Sigma_{\mathbf{k}}^{\mathrm{1BA}}$ moves the poles of $\langle \mathcal{G}_{\mathbf{k}} \rangle_{\mathrm{imp}} = \!\!=\!\!\Longleftarrow$ away from the real axis, i.e., the propagator acquires a finite lifetime. By Eq. (12.40) we see that $\mathcal{G}_{\mathbf{k}}^{\mathrm{1BA}}$ is the sum of propagations with any number of sequential wigwam-diagrams:

$$\begin{array}{c} \underset{\mathrm{1BA}}{=\!=\!=} = \quad \longleftarrow \quad + \quad \overset{\star}{\underset{\blacktriangleleft\!-\!\blacktriangleleft}{\diagup\diagdown}} \quad + \quad \overset{\star}{\underset{\blacktriangleleft\!-\!\blacktriangleleft\!-\!\blacktriangleleft}{\diagup\diagdown}}\,\overset{\star}{\diagup\diagdown} \quad + \quad \overset{\star}{\underset{\blacktriangleleft\!-\!\blacktriangleleft\!-\!\blacktriangleleft\!-\!\blacktriangleleft}{\diagup\diagdown}}\,\overset{\star}{\diagup\diagdown}\,\overset{\star}{\diagup\diagdown} \quad + \cdots \end{array}$$
$$\hspace{12cm} (12.45)$$

In the evaluation of $\Sigma_{\mathbf{k}}^{\mathrm{1BA}}$ we shall rely on our physical insight to facilitate the math. We know that for the electron gas in a typical metal $\varepsilon_{\mathrm{F}} \sim 7\ \mathrm{eV} \sim 80\,000\ \mathrm{K}$, so as usual only electrons with an energy $\varepsilon_{\mathbf{k}}$ in a narrow shell around $\varepsilon_{\mathrm{F}} \approx \mu$ play a

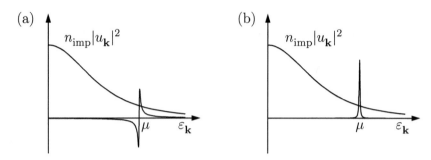

FIG. 12.4.   (a) The functions $n_{\mathrm{imp}}|u_{\mathbf{k}}|^2$ and $(\omega - \varepsilon_{\mathbf{k}} + \mu)/[(\omega - \varepsilon_{\mathbf{k}} + \mu)^2 + \eta^2]$ appearing in the expression for Re $\Sigma_{\mathbf{k}}^{\mathrm{1BA}}(ik_n)$. (b) The functions $n_{\mathrm{imp}}|u_{\mathbf{k}}|^2$ and $|k_n|/[(\omega - \varepsilon_{\mathbf{k}} + \mu)^2 + \eta^2]$ appearing in the expression for Im $\Sigma_{\mathbf{k}}^{\mathrm{1BA}}(ik_n)$.

role. For $T < 800$ K we have $k_{\mathrm{B}}T/\varepsilon_{\mathrm{F}} < 10^{-2}$, and for applied voltage drops V_{ext} less than 70 mV over the coherence length $l_{\varphi} < 10^{-5}$ m (the typical size we are looking at), i.e., applied electrical fields less than 7000 V/m, we have $eV_{\mathrm{ext}}/\varepsilon_{\mathrm{F}} < 10^{-2}$. Thus we are only interested in $\Sigma_{\mathbf{k}}^{\mathrm{1BA}}(ik_n)$ for

$$|\mathbf{k}| \sim k_{\mathrm{F}} \quad \text{and} \quad |ik_n \to \omega + i\,\mathrm{sgn}(k_n)\eta| \ll \varepsilon_{\mathrm{F}}. \tag{12.46}$$

Here, we have also anticipated that at the end of the calculation, as sketched in Fig. 11.1, we need to perform an analytical continuation down to the real axis, either from the upper half-plane, where $k_n > 0$, as $ik_n \to \omega + i\eta$, or from the lower half-plane, where $k_n < 0$, as $ik_n \to \omega - i\eta$.

Furthermore, as we shall study in great detail later, the electron gas redistributes itself to screen out the external charges from the impurities, and as a result $u_{\mathbf{k}-\mathbf{k}'}$ varies in a smooth and gentle way for $0 < |\mathbf{k} - \mathbf{k}'| < 2k_{\mathrm{F}}$.

With this physical input in mind we continue:

$$\Sigma_{\mathbf{k}}^{\mathrm{1BA}}(\omega + i\,\mathrm{sgn}(k_n)\eta) = \frac{n_{\mathrm{imp}}}{V}\sum_{\mathbf{k}'}|u_{\mathbf{k}-\mathbf{k}'}|^2 \frac{1}{(\omega - \xi_{\mathbf{k}'}) + i\,\mathrm{sgn}(k_n)\eta} \tag{12.47}$$

$$= \frac{n_{\mathrm{imp}}}{V}\sum_{\mathbf{k}'}|u_{\mathbf{k}-\mathbf{k}'}|^2 \left[\frac{\omega - \xi_{\mathbf{k}'}}{(\omega - \xi_{\mathbf{k}'})^2 + \eta^2} - i\,\mathrm{sgn}(k_n)\,\pi\delta(\omega - \xi_{\mathbf{k}'})\right].$$

Since $|u_{\mathbf{k}-\mathbf{k}'}|^2$ varies smoothly and $|\omega - \xi_{\mathbf{k}'}| \ll \varepsilon_{\mathrm{F}} \approx \mu$ we get the functional behavior shown in Fig. 12.4. Since $(\omega - \xi_{\mathbf{k}'})/((\omega - \xi_{\mathbf{k}'})^2 + \eta^2)$ is an odd function of $\omega - \xi_{\mathbf{k}'}$ and the width η is small, we have[43] Re $\Sigma_{\mathbf{k}}^{\mathrm{1BA}}(ik_n) \approx 0$; for the imaginary part of Σ^{1BA} we obtain the usual delta function for $\eta \to 0$. Finally, we assume that the spectral function for the unperturbed system forces ω to equal $\xi_{\mathbf{k}}$, and then check for consistency at the end of the calculation. We obtain:

[43]Strictly speaking, we only get vanishing real part if the slope of $|u_{\mathbf{k}-\mathbf{k}'}|^2$ is zero near μ. If this is not the case we do get a non-zero real part. However, since $|u_{\mathbf{k}-\mathbf{k}'}|^2$ is slowly varying near μ we get the same real part for all $\mathbf{k}$ and $\mathbf{k}'$ near k_{F}. This contribution is absorbed into the definition of μ.

$$\Sigma_{\mathbf{k}}^{1\mathrm{BA}}(ik_n) = -i\pi\,\mathrm{sgn}(k_n)\,\frac{n_{\mathrm{imp}}}{\mathcal{V}}\sum_{\mathbf{k}'}|u_{\mathbf{k}-\mathbf{k}'}|^2\,\delta(\xi_{\mathbf{k}} - \xi_{\mathbf{k}'}) = -i\,\mathrm{sgn}(k_n)\,\frac{1}{2\tau_{\mathbf{k}}}, \quad (12.48)$$

where we have introduced the impurity scattering time $\tau_{\mathbf{k}}$ defined as

$$\frac{1}{\tau_{\mathbf{k}}} \equiv 2\pi\,\frac{n_{\mathrm{imp}}}{\mathcal{V}}\sum_{\mathbf{k}'}|u_{\mathbf{k}-\mathbf{k}'}|^2\,\delta(\xi_{\mathbf{k}} - \xi_{\mathbf{k}'}). \quad (12.49)$$

This result can also be found using Fermi's golden rule. Now we have obtained the first-order Born approximation for $\mathcal{G}_{\mathbf{k}}(ik_n)$ in Eq. (12.41) and the analytic continuation $ik_n \to z$ thereof into the entire complex plane,

$$\mathcal{G}_{\mathbf{k}}^{1\mathrm{BA}}(ik_n) = \frac{1}{ik_n - \xi_{\mathbf{k}} + i\frac{\mathrm{sgn}(k_n)}{2\tau_{\mathbf{k}}}} \xrightarrow[ik_n \to z]{} \mathcal{G}_{\mathbf{k}}^{1\mathrm{BA}}(z) = \begin{cases} \frac{1}{z - \xi_{\mathbf{k}} + \frac{i}{2\tau_{\mathbf{k}}}}, & \mathrm{Im}\,z > 0 \\[2mm] \frac{1}{z - \xi_{\mathbf{k}} - \frac{i}{2\tau_{\mathbf{k}}}}, & \mathrm{Im}\,z < 0. \end{cases} \quad (12.50)$$

We see that $\mathcal{G}_{\mathbf{k}}^{1\mathrm{BA}}(z)$ has a branch cut along the real axis, but that it is analytic separately in the upper and the lower half-plane. This is a property that will play an important role later, when we calculate the electrical resistivity of disordered metals. Note that this behavior is in accordance with the general results obtained in Sec. 9.2 concerning the analytic properties of Matsubara Green's functions.

We close this section by remarking three properties summarized in Fig. 12.5 related to the retarded Green's function $G^{\mathrm{R},1\mathrm{BA}}(\omega) = \mathcal{G}^{1\mathrm{BA}}(\omega + i\eta)$. First, it is seen by Fourier transforming to the time domain that $G_{\mathbf{k}}^{\mathrm{R},1\mathrm{BA}}(t)$ decays exponentially in time with $\tau_{\mathbf{k}}$ as the typical time scale,

$$G_{\mathbf{k}}^{\mathrm{R},1\mathrm{BA}}(t) \equiv \int \frac{d\omega}{2\pi}\,\frac{e^{-i(\omega + i\eta)t}}{\omega - \xi_{\mathbf{k}} + i/2\tau_{\mathbf{k}}} = -i\,\theta(t)\,e^{-i\xi_{\mathbf{k}}t}\,e^{-t/2\tau_{\mathbf{k}}}. \quad (12.51)$$

Second, exploiting that $\omega, \tau_{\mathbf{k}}^{-1} \ll \varepsilon_{\mathrm{F}}$, it is seen by Fourier transforming back to real space that $G^{\mathrm{R},1\mathrm{BA}}(r,\omega)$ decays exponentially in space with $l_{\mathbf{k}} \equiv v_{\mathrm{F}}\tau_{\mathbf{k}}$ as the typical length scale,

$$G^{\mathrm{R},1\mathrm{BA}}(r,\omega) \equiv \int \frac{d\mathbf{k}}{(2\pi)^3}\,\frac{e^{i\mathbf{k}\cdot\mathbf{r}}}{\omega - \xi_{\mathbf{k}} + i/2\tau_{\mathbf{k}}} = \frac{\pi d(\varepsilon_{\mathrm{F}})}{k_{\mathrm{F}}|\mathbf{r}|}\,e^{ik_{\mathrm{F}}|\mathbf{r}|}\,e^{-|\mathbf{r}|/2l_{\mathbf{k}}}. \quad (12.52)$$

Thirdly, the spectral function $A_{\mathbf{k}}^{1\mathrm{BA}}(\omega)$ is a Lorentzian of width $2\tau_{\mathbf{k}}$:

$$A_{\mathbf{k}}^{1\mathrm{BA}}(\omega) \equiv -2\,\mathrm{Im}\,\mathcal{G}_{\mathbf{k}}^{1\mathrm{BA}}(\omega + i\eta) = \frac{1/\tau_{\mathbf{k}}}{(\omega - \xi_{\mathbf{k}})^2 + 1/4\tau_{\mathbf{k}}^2} \quad (12.53)$$

In conclusion, the impurity-averaged first Born approximation has resulted in a self-energy with a non-zero imaginary part. The poles of the Matsubara Green's function $\mathcal{G}_{\mathbf{k}}^{1\mathrm{BA}}(ik_n)$ are therefore shifted away from the real axis, resulting in a both temporal and spatial exponential decay of the retarded Green's function. This is interpreted as follows: the impurity scattering transforms the free electrons into quasi-particles with a finite lifetime given by the scattering time $\tau_{\mathbf{k}}$ and a finite mean free

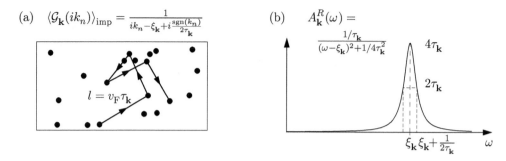

(a) $\langle \mathcal{G}_{\mathbf{k}}(ik_n) \rangle_{\mathrm{imp}} = \dfrac{1}{ik_n - \xi_{\mathbf{k}} + i\frac{\mathrm{sgn}(k_n)}{2\tau_{\mathbf{k}}}}$

$l = v_F \tau_{\mathbf{k}}$

(b) $A_{\mathbf{k}}^R(\omega) =$

$\dfrac{1/\tau_{\mathbf{k}}}{(\omega - \xi_{\mathbf{k}})^2 + 1/4\tau_{\mathbf{k}}^2}$

$4\tau_{\mathbf{k}}$

$2\tau_{\mathbf{k}}$

$\xi_{\mathbf{k}} \quad \xi_{\mathbf{k}} + \frac{1}{2\tau_{\mathbf{k}}} \qquad \omega$

FIG. 12.5.   (a) The impurity-averaged Green's function $\langle \mathcal{G}_{\mathbf{k}}(ik_n) \rangle_{\mathrm{imp}}$. The imaginary part of the self-energy is related to the scattering time $\tau_{\mathbf{k}}$ and hence also to the elastic scattering length $l = v_F \tau_{\mathbf{k}}$. (b) In the Born approximation the spectral function $A_{\mathbf{k}}(\omega)$ is a Lorentzian centered around $\xi_{\mathbf{k}} = 0$ with a width $1/2\tau_{\mathbf{k}}$.

path given by $l_{\mathbf{k}} = v_F \tau_{\mathbf{k}}$. The finite lifetime of the quasiparticles is also reflected in the broadening of the spectral function. The characteristic sharp δ-function for free electrons, $A_{\mathbf{k}}(\omega) = 2\pi\delta(\omega - \xi_{\mathbf{k}})$, is broadened into a Lorentzian of width $1/2\tau_{\mathbf{k}}$. This means that a particle with momentum $\mathbf{k}$ can have an energy ω that differs from $\xi_{\mathbf{k}}$ with an amount $\hbar/2\tau_{\mathbf{k}}$.

This calculation of self-averaged impurity scattering constitutes a first and very important example of what can happen in a many-particle system. Note in particular the important role played by the self-energy, and the fact that it can have a non-zero imaginary part. The results are obtained in the first-order Born approximation, where the self-energy is approximated by a single diagram. But what happens if we take more diagrams into account? The surprising answer is that in the low impurity density limit, $n_{\mathrm{imp}} \ll n_{\mathrm{el}}$ no qualitative difference arises by taking more diagrams into account. Only at higher impurity densities where scattering events from different impurities begin to interfere new physical effects, such as weak localization, appear. Let us see how this conclusion is reached.

12.5.3   The full Born approximation

A natural extension of the first Born approximation is the full Born approximation, which is exact to lowest order in n_{imp}. It is defined by the following self-energy $\Sigma_{\mathbf{k}}^{\mathrm{FBA}}(ik_n)$, where any number of scattering on the same impurity is taken into account, i.e., more dashed lines on the wigwam-diagram,

$$\Sigma_{\mathbf{k}}^{\mathrm{FBA}} \equiv \; \raisebox{-1ex}{\vdots}_{\mathbf{k}\,\mathbf{k}} + \raisebox{-1ex}{$\triangle$}_{\mathbf{k}\quad\mathbf{k}} + \raisebox{-1ex}{$\triangle$}_{\mathbf{k}\quad\mathbf{k}} + \raisebox{-1ex}{$\triangle$}_{\mathbf{k}\quad\mathbf{k}} + \cdots$$

$$= \; \raisebox{-1ex}{\vdots}_{\mathbf{k}\,\mathbf{k}} + \raisebox{-1ex}{$\nearrow$}^{\mathbf{k}-\mathbf{k}'}_{\mathbf{k}\,\mathbf{k}\quad\mathbf{k}'} \times \left(\raisebox{-1ex}{\vdots}_{\delta_{\mathbf{k}'\mathbf{k}}} + \raisebox{-1ex}{$\triangle$}_{\mathbf{k}'\quad\mathbf{k}} + \raisebox{-1ex}{$\triangle$}_{\mathbf{k}'\quad\mathbf{k}} + \cdots \right). \qquad (12.54)$$

In the parenthesis at the end of the second line we find a factor, which we denote $t_{\mathbf{k}',\mathbf{k}}$, that is not diagonal in $\mathbf{k}$ but with a diagonal that equals the self-energy $t_{\mathbf{k},\mathbf{k}} = \Sigma_{\mathbf{k}}^{\mathrm{FBA}}$. In scattering theory $t_{\mathbf{k}',\mathbf{k}}$ is known as the transition matrix. When this matrix is known, all consequences of the complete scattering sequence can be calculated. An integral equation for the transition matrix is derived diagrammatically:

$$= n_{\mathrm{imp}} u_0 \delta_{\mathbf{k}_1 \mathbf{k}_2} + \sum_{\mathbf{k}'} u_{\mathbf{k}_1 - \mathbf{k}'} \, \mathcal{G}_{\mathbf{k}'}^0 \, t_{\mathbf{k}',\mathbf{k}_2}. \tag{12.55}$$

This equation can in many cases be solved numerically. As before, the task is simplified by the fact that we are only interested at electrons moving at the Fermi surface. The real part of the diagonal element $t_{\mathbf{k},\mathbf{k}}(ik_n)$, the one yielding the self-energy, is almost constant for $|\mathbf{k}| \sim k_{\mathrm{F}}$ and is absorbed into the definition of the chemical potential μ. We are then left with $\mathrm{Im}\, t_{\mathbf{k},\mathbf{k}}(ik_n)$, and by applying the optical theorem,[44] $\mathrm{Im}\, t_{\mathbf{k},\mathbf{k}} = \mathrm{Im} \sum_{\mathbf{k}'} t_{\mathbf{k},\mathbf{k}'}^\dagger \mathcal{G}_{\mathbf{k}'}^0 t_{\mathbf{k}',\mathbf{k}}$, we obtain

$$\mathrm{Im}\, \Sigma_{\mathbf{k}}^{\mathrm{FBA}}(ik_n) = \mathrm{Im}\, t_{\mathbf{k},\mathbf{k}}(ik_n) = \mathrm{Im} \sum_{\mathbf{k}'} \frac{|t_{\mathbf{k},\mathbf{k}'}|^2}{ik_n - \xi_{\mathbf{k}'}}$$

$$\xrightarrow[ik_n \to \omega + i\,\mathrm{sgn}(k_n)\eta]{} -\mathrm{sgn}(k_n) \pi \sum_{\mathbf{k}'} |t_{\mathbf{k},\mathbf{k}'}|^2 \, \delta(\omega - \xi_{\mathbf{k}'}). \tag{12.56}$$

This has the same form as Eq. (12.48) with $|t_{\mathbf{k},\mathbf{k}'}|^2$ instead of $n_{\mathrm{imp}} |u_{\mathbf{k}-\mathbf{k}'}|^2$, and we write

$$\Sigma_{\mathbf{k}}^{\mathrm{FBA}}(ik_n) = -i\,\mathrm{sgn}(k_n)\frac{1}{2\tau_{\mathbf{k}}}, \quad \text{with} \quad \frac{1}{\tau_{\mathbf{k}}} \equiv 2\pi \sum_{\mathbf{k}'} |t_{\mathbf{k},\mathbf{k}'}|^2 \, \delta(\xi_{\mathbf{k}} - \xi_{\mathbf{k}'}). \tag{12.57}$$

By iteration of Dyson's equation we find that $\mathcal{G}^{\mathrm{FBA}}$ is the sum of propagations with any number and any type of sequential wigwam-diagrams:

$$\tag{12.58}$$

---

[44]Eq. (12.55) states that (i): $t = u + u\mathcal{G}^0 t$. Since $u^\dagger = u$ we find the Hermitian conjugate of (i) to be (ii): $u = -t^\dagger(\mathcal{G}^0)^\dagger u + t^\dagger$. Insert (ii) into (i): $t = u + (t^\dagger \mathcal{G}^0 t - t^\dagger(\mathcal{G}^0)^\dagger u \mathcal{G}^0 t)$. Both u and $t^\dagger(\mathcal{G}^0)^\dagger u \mathcal{G}^0 t$ are Hermitian so $\mathrm{Im}\, t_{\mathbf{k},\mathbf{k}} = \mathrm{Im}\, \langle \mathbf{k}|t^\dagger \mathcal{G}^0 t|\mathbf{k}\rangle = \mathrm{Im} \sum_{\mathbf{k}'} t_{\mathbf{k},\mathbf{k}'}^\dagger \mathcal{G}_{\mathbf{k}'}^0 t_{\mathbf{k}',\mathbf{k}}$.

12.5.4 *The self-consistent full Born approximation and beyond*

Many more diagrams can be taken into account using the self-consistent full Born approximation defined by substituting the bare $\mathcal{G}^0$ with the full Born approximation $\mathcal{G}$. With this approach Eqs. (12.54) and (12.55) yield

$$t_{\mathbf{kk}}^{\mathrm{SCBA}} \equiv \quad + \quad + \quad + \quad + \cdots$$

$$= n_{\mathrm{imp}} u_0 \delta_{\mathbf{kk}} + \sum_{\mathbf{k'}} u_{\mathbf{k}-\mathbf{k'}}\, \mathcal{G}_{\mathbf{k'}}\, t_{\mathbf{k',k}}^{\mathrm{SCBA}}; \tag{12.59}$$

a self-consistent equation in $\Sigma_{\mathbf{k}}^{\mathrm{SCBA}}$ since $\mathcal{G}_{\mathbf{k'}} = (ik_n - \xi_{\mathbf{k'}} - \Sigma_{\mathbf{k'}}^{\mathrm{SCBA}})^{-1}$. We again utilize that $t_{\mathbf{k,k}}$ is only weakly dependent on energy for $|\mathbf{k}| \approx k_{\mathrm{F}}$ and $\omega \ll \varepsilon_{\mathrm{F}}$, and if furthermore the scattering strength is moderate, $|\Sigma_{\mathbf{k}}^{\mathrm{SCBA}}| \ll \varepsilon_{\mathrm{F}}$, we obtain almost the same result as in Eq. (12.56). Only the imaginary part $\Sigma_{\mathbf{k}}^i$ of $\Sigma_{\mathbf{k}}^{\mathrm{SCBA}} = \Sigma_{\mathbf{k}}^R + i\Sigma_{\mathbf{k}}^i$ plays a role, since the small real part $\Sigma_{\mathbf{k}}^R$ can be absorbed into μ,

$$\Sigma_{\mathbf{k}}^i = \mathrm{Im} \sum_{\mathbf{k'}} \frac{|t_{\mathbf{k,k'}}^{\mathrm{SCBA}}|^2}{ik_n - \xi_{\mathbf{k'}} - i\Sigma_{\mathbf{k'}}^i} \tag{12.60}$$

For the retarded self-energy we thus get

$$\Sigma_{\mathbf{k}}^{\mathrm{R,SCBA}}(\omega) = -i\mathrm{sgn}(\omega)\, \pi \sum_{\mathbf{k'}} |t_{\mathbf{k,k'}}|^2\, \delta(\omega - \xi_{\mathbf{k'}}). = -i\, \mathrm{sgn}(\omega)\, \frac{1}{2\tau_{\mathbf{k}}}, \tag{12.61}$$

with

$$\frac{1}{\tau_{\mathbf{k}}} \approx 2\pi \sum_{\mathbf{k'}} |t_{\mathbf{k,k'}}|^2\, \delta(\omega - \xi_{\mathbf{k'}}). \tag{12.62}$$

Here we have done several steps. First we note that the denominator has the fastest variation with energy, hence we can approximate it with the usual delta function. Then we note that the only self-consistency requirement is thus connected with the sign of the imaginary part. But this requirement is fulfilled by taking $\mathrm{Im}\Sigma^{\mathrm{SCBA}}(ik_n) \propto -\mathrm{sgn}(\omega)$ as seen by direct substitution. Finally, we note that the only difference between the full Born and the self-consistent Born approximation is in the case of strong scattering, where the limiting δ-function in Eq. (12.60) may acquire a small renormalization. When all the dust has settled, we have the simple result for the self-energy given by Eq. (12.61).

By iteration of Dyson's equation we find that $\mathcal{G}^{\mathrm{SCBA}}$ is the sum of propagations with any number and any type of sequential wigwam-diagrams inside wigwam-diagrams but without crossings of any scattering lines,

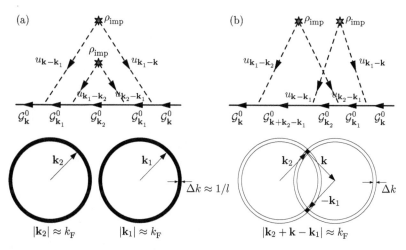

FIG. 12.6. (a) The non-crossing wigwam diagrams, one inside the other, where $\mathbf{k}_1$ and $\mathbf{k}_2$ can take any value on the spherical shell of radius k_F and thickness $\Delta k \approx 1/l$. The phase space is $\Omega_a \propto (4\pi k_F^2 \Delta k)^2$. (b) The crossing wigwam diagram has the same restrictions for $\mathbf{k}_1$ and $\mathbf{k}_2$ as in (a) plus the constraint that $|\mathbf{k} + \mathbf{k}_2 - \mathbf{k}_1| \approx k_F$. For fixed $\mathbf{k}_2$ the variation of $\mathbf{k}_1$ within its Fermi shell is restricted to the intersection between this shell and the Fermi shell of $\mathbf{k} + \mathbf{k}_2 - \mathbf{k}_1$, i.e., to a ring with cross-section $1/l^2$ and radius $\approx k_F$. The phase space is now $\Omega_b \propto (4\pi k_F^2 \Delta k)(2\pi k_F \Delta k^2)$. Thus the crossing diagram (b) is suppressed relative to the non-crossing diagram (a) with a factor $1/k_F l$.

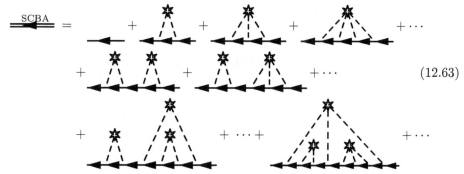

$$(12.63)$$

We have now resummed most of the diagrams in the diagrammatic expansion of $\langle \mathcal{G}_\mathbf{k} \rangle_{\text{imp}}$ with the exception of wigwam-diagrams with crossing lines. In Fig. 12.6 are shown two different types of irreducible diagrams of the same order in both n_{imp} and $u_\mathbf{k}$. Also sketched is the phase space Ω available for the internal momenta $\mathbf{k}_1$ and $\mathbf{k}_2$ in the two cases. At zero temperature the energy broadening around the Fermi energy ε_F is given by $|\Sigma| \approx \hbar/\tau$ which relaxes $|\mathbf{k}_1|, |\mathbf{k}_2| = k_F$ a bit. In $\mathbf{k}$-space the broadening Δk is given by $\hbar^2 (k_F + \Delta k)^2 / 2m \approx \varepsilon_F + \hbar/\tau$ which gives $\Delta k \approx 1/v_F \tau = 1/l$, i.e., the inverse scattering length. This means that $\mathbf{k}_1$ and $\mathbf{k}_2$ are both confined to a thin spherical shell of thickness $1/l$ and radius k_F.

In Fig. 12.6(a), where no crossing of scattering lines occurs, no further restrictions applies, so the volume of the available phase space is $\Omega_a = (4\pi k_{\mathrm{F}}^2/l)^2$. In Fig. 12.6(b), where the scattering lines cross, the Feynman rules dictate one further constraint, namely $|\mathbf{k} + \mathbf{k}_1 - \mathbf{k}_2| \approx k_{\mathrm{F}}$. Thus only one of the two internal momenta are free to be anywhere on the Fermi shell, the other is bound to the intersection between two Fermi shells, i.e., on a ring with radius $\sim k_{\mathrm{F}}$ and a cross-section $1/l^2$ as indicated in Fig. 12.6(b). So $\Omega_b = (4\pi k_{\mathrm{F}}^2/l)(2\pi k_{\mathrm{F}}/l^2)$. Thus by studying the phase space available for the non-crossed and the crossed processes we have found that the crossed ones are suppressed by a factor $\Omega_b/\Omega_a \approx 1/(k_{\mathrm{F}}l)$. Such a suppression factor enters the calculation for each crossing of scattering lines in a diagram. Since for metals $1/k_{\mathrm{F}} \sim 1$ Å we find that

$$\frac{1}{k_{\mathrm{F}}l} \ll 1, \quad \text{for } l \gg 1 \text{ Å}. \tag{12.64}$$

In conclusion: all cases where the scattering length l is greater than 1 Å we have by the various Born approximations indeed resummed the perturbation series for $\langle \mathcal{G}_{\mathbf{k}}(ik_n) \rangle_{\mathrm{imp}}$ taking all relevant diagrams into account and obtained $\Sigma_{\mathbf{k}}(ik_n) = -i\frac{\mathrm{sgn}(k_n)}{2\tau_{\mathbf{k}}}$. It is interesting to note that in e.g., doped semiconductors it is possible to obtain a degenerate electron gas with a very low density. In these systems $1/k_{\mathrm{F}}$ or the Fermi wavelength is much larger than in metals, and the condition in Eq. (12.64) is violated. In this case one may therefore observe deviations from the simple theory presented here. One example is the observation of weak localization, which is an increase in the resistivity due to quantum interference between scattering events involving several impurities at the same time. The weak localization effect is studied in Section 16.5.

12.6 Summary and outlook

In this chapter we have introduced the Feynman diagrams for elastic impurity scattering. We have applied the diagrammatic technique to an analysis of the single-particle Matsubara Green's function for electron propagation in disordered metals. The main result was the determination of the self-energy $\Sigma_{\mathbf{k}}(ik_n)$ in terms of the scattering time $\tau_{\mathbf{k}}$,

$$\Sigma_{\mathbf{k}}^{\mathrm{FBA}}(ik_n) = -i\,\mathrm{sgn}(k_n)\frac{1}{2\tau_{\mathbf{k}}}, \quad \text{with } \frac{1}{\tau_{\mathbf{k}}} \equiv 2\pi \sum_{\mathbf{k}'} |t_{\mathbf{k},\mathbf{k}'}|^2\, \delta(\xi_{\mathbf{k}} - \xi_{\mathbf{k}'}),$$

and the scattering-time broadened spectral function

$$A_{\mathbf{k}}^{\mathrm{1BA}}(\omega) = \frac{1/\tau_{\mathbf{k}}}{(\omega - \xi_{\mathbf{k}})^2 + 1/4\tau_{\mathbf{k}}^2}.$$

The structure in the complex plane of the Green's function was found to be:

$$\mathcal{G}_{\mathbf{k}}^{\mathrm{1BA}}(ik_n) = \frac{1}{ik_n - \xi_{\mathbf{k}} + i\frac{\mathrm{sgn}(k_n)}{2\tau_{\mathbf{k}}}} \xrightarrow{ik_n \to z} \mathcal{G}_{\mathbf{k}}^{\mathrm{1BA}}(z) = \begin{cases} \frac{1}{z - \xi_{\mathbf{k}} + \frac{i}{2\tau_{\mathbf{k}}}}, & \mathrm{Im}\, z > 0 \\ \frac{1}{z - \xi_{\mathbf{k}} - \frac{i}{2\tau_{\mathbf{k}}}}, & \mathrm{Im}\, z < 0. \end{cases}$$

These results will be employed in Chap. 16 in the study of the residual resistivity of metals.

The theory presented here provides in combination with the Kubo formalism the foundation for a microscopic quantum theory of resistivity. The technique can be extended to the study of quantum effects like weak localization (see Section 16.5) and universal conductance fluctuations (see Fig. 12.2). These more subtle quantum effects are fundamental parts of the modern research field known as mesoscopic physics. They can be explained within the theoretical framework presented here, by taking higher-order correlations into account. For example is weak localization explained by treating crossed diagrams like the one in Fig. 12.6(b), which was neglected in calculation presented in this chapter.

13

FEYNMAN DIAGRAMS AND PAIR INTERACTIONS

It is in the case of interacting particles and fields that the power of quantum field theory and Feynman diagrams really comes into play. In this chapter, we develop the Feynman diagram technique for a system of fermions with pair interactions. The primary example we will work with is a gas of fermions interacting through Coulomb interactions, but the technique is derived in a more general framework. In general, the Hamiltonian has two parts $H = H_0 + W$.

One part represents the single particle energy, i.e., the kinetic energy and the single particle potential. In the real-space basis it reads

$$H_0 = \sum_\sigma \int d\mathbf{r} \, \Psi_\sigma^\dagger(\mathbf{r}) h_0(\mathbf{r}) \Psi_\sigma(\mathbf{r}), \tag{13.1}$$

where as usual $h_0(\mathbf{r})$ is the single-particle (first quantization) Hamiltonian. Using a different basis as in Eq. (1.69), the single-particle part of the Hamiltonian becomes

$$H_0 = \sum_{\nu_1 \nu_2} c_{\nu_1 \sigma}^\dagger h_{0,\nu_1\nu_2} c_{\nu_2 \sigma}, \tag{13.2}$$

where $h_{0,\nu_1\nu_2} = \langle \nu_1 | h_0 | \nu_2 \rangle$.

The second part is the interaction part. In the real-space basis it reads

$$W = \frac{1}{2} \sum_{\sigma_1,\sigma_2} \int d\mathbf{r}_1 d\mathbf{r}_2 \, \Psi^\dagger(\sigma_1,\mathbf{r}_1)\Psi^\dagger(\sigma_2,\mathbf{r}_2) \, W(\sigma_2,\mathbf{r}_2; \sigma_1,\mathbf{r}_1) \, \Psi(\sigma_2,\mathbf{r}_2)\Psi(\sigma_1,\mathbf{r}_1),$$

$$\tag{13.3}$$

where we have specialized to the case where no spin flip processes occur at the vertices, this being the case for our coming main examples: electron-electron interactions by direct Coulomb or phonon-mediated interactions. However, at the end of the chapter we take an example where the spin flip processes are essential, namely the Kondo model discussed in Section 10.5. Again, we could express the interactions as in the ν basis, which would transform Eq. (13.3) into

$$W = \frac{1}{2} \sum_{\nu_1 \nu_2 \nu_3 \nu_4} V_{\nu_1 \nu_2, \nu_3 \nu_4} \, a_{\nu_1 \sigma_1}^\dagger a_{\nu_2 \sigma_2}^\dagger a_{\nu_4 \sigma_2} a_{\nu_3 \sigma_1}. \tag{13.4}$$

The main goal of this chapter is to derive the Feynman rules for the diagrammatic expansion in orders of W of the full single-particle Matsubara Green's function (11.36a),

$$\mathcal{G}(\sigma_b,\mathbf{r}_b,\tau_b; \sigma_a,\mathbf{r}_a,\tau_a) \equiv -\left\langle T_\tau \Psi(\sigma_b,\mathbf{r}_b,\tau_b)\Psi^\dagger(\sigma_a,\mathbf{r}_a,\tau_a) \right\rangle, \tag{13.5}$$

226

or equivalently in the ν basis,

$$\mathcal{G}(\sigma_b, \nu_b, \tau_b; \sigma_a, \nu_a, \tau_a) \equiv -\Big\langle T_\tau \Psi(\sigma_b, \nu_b, \tau_b) \Psi^\dagger(\sigma_a, \nu_a, \tau_a) \Big\rangle. \qquad (13.6)$$

The diagrammatic technique derived below applies equally well to the two bases. Which basis to use, naturally depends on the problem at hand, the most obvious choice being the basis that diagonalizes the non-interacting part of the Hamiltonian. For example, in Section 13.4 we take the translational-invariant case for which the plane wave basis is used. However, because the interaction vertex $W(\sigma_2, \mathbf{r}_2; \sigma_1, \mathbf{r}_1)$ in Eq. (13.3) depends on two coordinates only, whereas $V_{\nu_1 \nu_2, \nu_3 \nu_4}$ in Eq. (13.6) depends on four variables, we will use the former, when developing the diagrammatical technique below.

13.1 The perturbation series for $\mathcal{G}$

The field operators in Eq. (13.5) defining $\mathcal{G}$ are of course given in the Heisenberg picture, but using Eq. (11.19) we can immediately transform the expression for $\mathcal{G}$ into the interaction picture. With the short-hand notation $(\sigma_1, \mathbf{r}_1, \tau_1) = (1)$, we obtain

$$\mathcal{G}(b, a) = -\frac{\text{Tr}\left(e^{-\beta H} T_\tau \Psi(b) \Psi^\dagger(a)\right)}{\text{Tr}\left(e^{-\beta H}\right)} = -\frac{\Big\langle T_\tau\left[\hat{U}(\beta, 0)\,\hat{\Psi}(b)\,\hat{\Psi}^\dagger(a)\right]\Big\rangle_0}{\Big\langle \hat{U}(\beta, 0)\Big\rangle_0}.$$

$$(13.7)$$

The subscript 0 indicates that the averages in Eq. (13.7) are with respect to $e^{-\beta H_0}$ rather than $e^{-\beta H}$ as in Eq. (13.5). The expansion Eq. (11.12) for $\hat{U}$ is now inserted into Eq. (13.7),

$$\mathcal{G}(b, a) = -\frac{\displaystyle\sum_{n=0}^\infty \frac{(-1)^n}{n!} \int_0^\beta d\tau_1 \cdots \int_0^\beta d\tau_n \Big\langle T_\tau\left[\hat{W}(\tau_1) \cdots \hat{W}(\tau_n) \hat{\Psi}(b)\,\hat{\Psi}^\dagger(a)\right]\Big\rangle_0}{\displaystyle\sum_{n=0}^\infty \frac{(-1)^n}{n!} \int_0^\beta d\tau_1 \cdots \int_0^\beta d\tau_n \Big\langle T_\tau\left[\hat{W}(\tau_1) \cdots \hat{W}(\tau_n)\right]\Big\rangle_0}.$$

$$(13.8)$$

Here we need to calculate τ-integral of $\hat{W}(\tau)$. But one precaution must be taken regarding the ordering of the four operators in the basic two-particle interaction operator. According to Eq. (13.3), the two creation operators must always be to the left of the two annihilation operators. To make sure of that we add an infinitesimal time $\eta = 0^+$ to the time-arguments of $\Psi^\dagger(1)$ and $\Psi^\dagger(2)$, which gives the right ordering when the time-ordering operator T_τ of Eq. (13.7) acts. The τ-integrals of $\hat{W}(\tau)$ is therefore

$$\int_0^\beta d\tau_j\, \hat{W}(\tau_j) = \frac{1}{2} \int dj \int dj'\, \hat{\Psi}^\dagger(j_+) \hat{\Psi}^\dagger(j'_+)\, W_{j,j'}\, \hat{\Psi}(j') \hat{\Psi}(j), \qquad (13.9)$$

where we have defined j_+, $\int dj$, and $W_{j,j'}$ as

$$j_+ \equiv (\sigma_j, \mathbf{r}_j, \tau_j + \eta), \quad \int dj \equiv \sum_{\sigma_j} \int d\mathbf{r}_j \int_0^\beta d\tau_j, \quad W_{j,j'} \equiv W(\mathbf{r}_j, \mathbf{r}_{j'})\, \delta(\tau_j - \tau_{j'}).$$

$$(13.10)$$

It is only in expressions where the initial and final times coincide that the infinitesimal shift in time of $\Psi^\dagger$ plays a role. Next, we insert Eq. (13.9) for $\hat{W}$ into Eq. (13.8) for $\mathcal{G}$,

$$\mathcal{G}(b,a) = \qquad\qquad (13.11)$$

$$\frac{-\displaystyle\sum_{n=0}^{\infty} \frac{(-\frac{1}{2})^n}{n!} \int d1d1'..dndn'\, W_{1,1'}..W_{n,n'} \left\langle T_\tau\left[\hat{\Psi}_1^\dagger\hat{\Psi}_{1'}^\dagger\hat{\Psi}_{1'}\hat{\Psi}_1..\hat{\Psi}_n^\dagger\hat{\Psi}_{n'}^\dagger\hat{\Psi}_{n'}\hat{\Psi}_n\ \hat{\Psi}_b\hat{\Psi}_a^\dagger\right]\right\rangle_0}{\displaystyle\sum_{n=0}^{\infty} \frac{(-\frac{1}{2})^n}{n!} \int d1d1'..dndn'\, W_{1,1'}..W_{n,n'} \left\langle T_\tau\left[\hat{\Psi}_1^\dagger\hat{\Psi}_{1'}^\dagger\hat{\Psi}_{1'}\hat{\Psi}_1..\hat{\Psi}_n^\dagger\hat{\Psi}_{n'}^\dagger\hat{\Psi}_{n'}\hat{\Psi}_n\right]\right\rangle_0}.$$

The great advantage of Eq. (13.11) is that the average of the field operators now involves bare propagation and thermal average both with respect to H_0. In fact, using Eq. (11.68), we recognize that the average of the products of field operators in the numerator is the bare $(2n+1)$-particle Green's function $\mathcal{G}_0^{(2n+1)}(b,1,1',..,n';a,1,1',..,n')$ times $(-1)^{2n+1} = -1$, while in the denominator it is the bare $(2n)$-particle Green's function $\mathcal{G}_0^{(2n)}(1,1',..,n';1,1',..,n')$ times $(-1)^{2n} = 1$. The resulting sign, -1, thus cancels the sign in Eq. (13.11). Now is the time for our main use of Wick's theorem Eq. (11.82): the bare many-particle Green's functions in the expression for the full single-particle Green's function are written in terms of determinants containing the bare single-particle Green's functions $\mathcal{G}^0(l,j)$,

$$\mathcal{G}(b,a) = \qquad\qquad (13.12)$$

$$\frac{\displaystyle\sum_{n=0}^{\infty} \frac{(-\frac{1}{2})^n}{n!} \int d1d1'..dndn'\, W_{1,1'}..W_{n,n'} \begin{vmatrix} \mathcal{G}^0(b,a) & \mathcal{G}^0(b,1) & \mathcal{G}^0(b,1') & \cdots & \mathcal{G}^0(b,n') \\ \mathcal{G}^0(1,a) & \mathcal{G}^0(1,1) & \mathcal{G}^0(1,1') & \cdots & \mathcal{G}^0(1,n') \\ \mathcal{G}^0(1',a) & \mathcal{G}^0(1',1) & \mathcal{G}^0(1',1') & \cdots & \mathcal{G}^0(1',n') \\ \vdots & & & \ddots & \vdots \\ \mathcal{G}^0(n',a) & \mathcal{G}^0(n',1) & \mathcal{G}^0(n',1') & \cdots & \mathcal{G}^0(n',n') \end{vmatrix}}{\displaystyle\sum_{n=0}^{\infty} \frac{(-\frac{1}{2})^n}{n!} \int d1d1'..dndn'\, W_{1,1'}..W_{n,n'} \begin{vmatrix} \mathcal{G}^0(1,1) & \mathcal{G}^0(1,1') & \cdots & \mathcal{G}^0(1,n') \\ \mathcal{G}^0(1',1) & \mathcal{G}^0(1',1') & \cdots & \mathcal{G}^0(1',n') \\ \vdots & & \ddots & \vdots \\ \mathcal{G}^0(n',1) & \mathcal{G}^0(n',1') & \cdots & \mathcal{G}^0(n',n') \end{vmatrix}}.$$

This voluminous formula is the starting point for defining the Feynman rules for the diagrammatic expansion of $\mathcal{G}$ in terms of the pair interaction W. We have suppressed, but not forgotten, the fact that, according to Eqs. (13.9) and (13.10), the initial time τ_j in $\mathcal{G}^0(l,j)$ is to be shifted infinitesimally to $\tau_j + \eta$.

13.2   The Feynman rules for pair interactions

We formulate first a number of basic Feynman rules that are derived directly from Eq. (13.12). However, these basic rules can be used to prove that the denominator cancels out. This in turn leads to the formulation of the final Feynman rules to be used in all later calculations.

13.2.1   Feynman rules for the denominator of $\mathcal{G}(b,a)$

The basic Feynman rules for the nth-order term in the denominator of $\mathcal{G}(b,a)$ are

---

(1) Fermion lines: $j_2 \bullet\!\!\blacktriangleleft\!\!\bullet j_1 \equiv \mathcal{G}^0(j_2, j_1)$ with $\tau_1 \to \tau_1 + \eta$.

(2) Interaction lines: $j \bullet\!\!\sim\!\!\bullet j' \equiv W_{j,j'}$.

(3) Vertices: $j \bullet \equiv \int dj\, \delta_{\sigma_j^{\mathrm{in}}, \sigma_j^{\mathrm{out}}}$, i.e., sum over internal variables, no spin flip.

(4) Draw $(2n)!$ sets of n interaction lines $j \bullet\!\!\sim\!\!\bullet j'$.

(5) For each set connect the $2n$ vertices with $2n$ fermion lines: one entering and one leaving each vertex. This can be done in $(2n)!$ ways.

$$(13.13\mathrm{a})$$

However, also the sign of each term in the expansion of the determinant must be found. Here, the concept of fermion loops enters the game. A fermion loop is an uninterrupted sequence of fermion lines starting at some vertex j and ending there again after connecting to other vertices, e.g., $j_1 \bigcirc$, $j_1 \bigcirc j_2$, or $j_1 \triangle j_2$. The overall sign coming from the determinant is $(-1)^F$, where F is the number of fermion loops in the given diagram. An outline of the proof is as follows. The product of the diagonal terms in the determinant is per definition positive, and diagramatically it consist of n factors $\bigcirc\!\!j\!\!\sim\!\!j'\!\!\bigcirc$, and thus $F = 2n$ is even. All other diagrams can be constructed one by one simply by pairwise interchange of the end-points of the fermion lines. This changes the sign from the determinant-expansion of the given product, since $\mathrm{sgn}[..\mathcal{G}^0(j_1, j_1')..\mathcal{G}^0(j_2, j_2')..] = -\mathrm{sgn}[..\mathcal{G}^0(j_1, j_2')..\mathcal{G}^0(j_2, j_1')..]$, and at the same time it changes the number of fermion loops by 1, e.g. $\triangle$ becomes $\oslash\ \bigcirc$. Thus we obtain the last Feynman rule,

---

(6) Multiply by $\frac{1}{n!}(-\frac{1}{2})^n (-1)^F$, F being the number of fermion loops, and add the resulting diagrams of order n.

$$(13.13\mathrm{b})$$

For all n there are $(2n)!$ terms or diagrams of order n in the expansion of the determinant in the denominator $\langle \hat{U}(\beta, 0)\rangle_0$ of $\mathcal{G}(b,a)$ in Eq. (13.12). Suppressing the labels, but indicating the number of diagrams of each order, this expansion takes the following form using Feynman diagrams,

$$\left\langle \hat{U}(\beta,0)\right\rangle_0 = 1 + \left[\ \text{}\ \right]_{\text{2 terms}}$$

$$+ \cdots \tag{13.14}$$

13.2.2 Feynman rules for the numerator of $\mathcal{G}(b,a)$

The numerator $\langle T_\tau [\hat{U}(\beta,0)\hat{\Psi}(b)\hat{\Psi}^\dagger(a)]\rangle_0$ of $\mathcal{G}(b,a)$ differs from the denominator by the presence of the two external field operators $\hat{\Psi}(b)$ and $\hat{\Psi}^\dagger(a)$ that act at the external space-time points (b) and (a). This raises the dimension of the nth-order determinant from $2n$ to $2n+1$. Consequently, only Feynman rules (4) and (5) given for the denominator have to be changed to give the rules for the numerator,

(4') Draw $(2n+1)!$ sets of n lines $j\text{\textbullet}\!\!\sim\!\!\!\text{\textbullet}j'$ and two external vertices $\bullet a$ and $\bullet b$.
(5') For each set connect the $2n+2$ vertices with $2n+1$ fermion lines: one leaving a, one entering b, and one entering and leaving each internal vertex j.

$$\tag{13.15}$$

Using these rules we obtain the diagrammatic expansion of the numerator,

$$\left\langle T_\tau [\hat{U}(\beta,0)\hat{\Psi}(b)\hat{\Psi}^\dagger(a)]\right\rangle_0 =$$

$$+ \cdots \tag{13.16}$$

13.2.3   The cancellation of disconnected Feynman diagrams

It looks like we are drowning in diagrams, but in fact there is a major reduction at hand. We note that in Eq. (13.16) two classes of diagrams appear: those being connected into one piece with the external vertices a and b, the so-called connected diagrams (e.g., the last second-order diagram), and those consisting of two or more pieces, the so-called disconnected diagrams (e.g., the first second-order diagram). Furthermore, we note that the parts of the diagrams in Eq. (13.16) disconnected from the external vertices are the same as the diagrams appearing in Eq. (13.14) order by order. We also note that a diagram containing two or more disconnected parts can be written as a product containing one factor for each disconnected part. A detailed combinatorial analysis (given at the end of this section and in Section 13.6) reveals that the denominator in $\mathcal{G}$ cancels exactly the disconnected parts of the diagrams in the numerator leaving only the connected ones,

Being left with only the connected diagrams, we find that since now all lines in the diagram are connected in a specific way to the external points a and b, the combinatorics of the permutations of the internal vertex indices is particularly simple. There are $n!$ ways to choose the enumeration j of the n interaction lines $j \bullet\!\!\!\sim\!\!\!\bullet j'$, and for each line there are two ways to put a given pair of labels j and j'. We conclude that all $2^n \, n!$ diagrams with the same topology relative to the external points give the same value. Except for the sign this factor cancels the prefactor $\frac{1}{n!}(-\frac{1}{2})^n$, i.e., we are left with a factor of (-1) for each of the n interaction lines. In conclusion, for pair interactions the final version of the Feynman rules for expanding $\mathcal{G}$ diagrammatically is

(1) Fermion lines: $j_2 \bullet\!\!\!-\!\!\!\bullet j_1 \equiv \mathcal{G}^0(j_2, j_1)$ with $\tau_1 \to \tau_1 + \eta$.
(2) Interaction lines: $j \bullet\!\!\sim\!\!\sim\!\!\bullet j' \equiv -W_{j,j'}$.
(3) Vertices: $j \bullet \equiv \int dj\, \delta_{\sigma_j^{in}, \sigma_j^{out}}$, i.e., sum over internal variables, no spin flip.
(4) At order n draw all topologically different, connected diagrams containing n interaction lines $j \bullet\!\!\sim\!\!\sim\!\!\bullet j'$, 2 vertices $\bullet a$ and $\bullet b$, and $2n+1$ fermion lines, so that one leaves $\bullet a$, one enters $\bullet b$, and one enters and leaves each vertex $\bullet j$.
(5) Multiply each diagram by $(-1)^F$, F being the number of fermion loops.
(6) Sum over all the topologically different diagrams.

$$\text{(13.18)}$$

Pay attention to the fact that only the topology of the diagrams is mentioned. Thus they can be stretched, mirror inverted and otherwise deformed. No notion of a time-axis is implied in the imaginary time version of the Feynman diagrams.

For completeness, we give the following proof of the cancellation of the disconnected diagrams, but the reader may skip it, since the essential conclusion has already been given above. The proof, which is a special case of the more general proof given in Section 13.6, contains eight steps. We study the numerator of Eq. (13.12). (1) Since all internal vertices have one incoming and one outgoing fermion line, the external vertices a and b are always connected. (2) If vertex j somehow is connected to a, so is j' due to the interaction line $W_{j,j'}$. (3) In a diagram of order n, a is connected with r W-lines, where $0 \le r \le n$. The number of disconnected W-lines is denoted m, i.e., $m = n - r$. (4) In all terms of the expanded numerator, the integral factorizes into a product of two integrals, one over the $2r$ variables connected to a and one over the $2m$ variables disconnected from a. (5) The r pairs of vertex variables j and j' connected to a can be chosen out of the available n pairs in $n!/[r!(n-r)!]$ ways, each choice yielding the same value of the total integral. (6) The structure of the sum is now:

$$\sum_{n=0}^{\infty} \frac{1}{n!} \left(\frac{-1}{2}\right)^n I[1, 1', .., n, n'] \tag{13.19}$$

$$= \sum_{n=0}^{\infty} \frac{1}{n!} \left(\frac{-1}{2}\right)^n \sum_{r=0}^{n} \frac{n!}{r!(n-r)!} I[1, 1', .., r, r']_{\text{con}} \, I[r+1, (r+1)', .., n, n']_{\text{discon}}$$

$$= \sum_{r=0}^{\infty} \frac{1}{r!} \left(\frac{-1}{2}\right)^r I[1, 1', .., r, r']_{\text{con}} \sum_{m=0}^{\infty} \frac{1}{m!} \left(\frac{-1}{2}\right)^m I[r+1, (r+1)', .., (r+m), (r+m)']_{\text{discon}}.$$

(7) In the connected part all $r!$ permutations of the vertex variable pairs (j, j') yield the same result, and so do all the 2^n ways of ordering each pair, if as usual $W_{j,j'} = W_{j',j}$. (8) The disconnected part is seen to be $\langle \hat{U}(\beta, 0) \rangle_0$. We thus reach the conclusion

$$\left\langle T_\tau \hat{U}(\beta, 0) \, \hat{\Psi}(b) \, \hat{\Psi}^\dagger(a) \right\rangle_0 = \left\langle \hat{U}(\beta, 0) \right\rangle_0 \sum_{r=0}^{\infty} [-W(1, 1')]...[-W(r, r')] \, \text{Det}\left[\mathcal{G}^0\right]_{\substack{\text{connected} \\ \text{topological diff.}}}^{(2r+1)\times(2r+1)}.$$

$$\text{(13.20)}$$

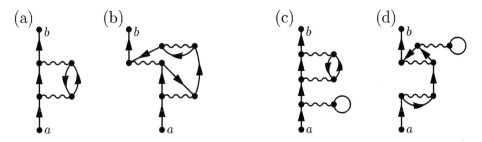

FIG. 13.1. Examples of irreducible, (a) and (b), and reducible, (c) and (d), Feynman diagrams in the expansion of $\mathcal{G}(b,a)$ in the presence of pair interactions.

13.3 Self-energy and Dyson's equation

In complete analogy with Fig. 12.3 for impurity-scattering, we can now define the concept of irreducible diagrams in $\mathcal{G}(b,a)$ in the case of pair interactions based on Eq. (13.17). As depicted in Fig. 13.1, such diagrams are the ones that cannot be cut into two pieces by cutting a single fermion line. Continuing the analogy with the impurity-scattering case, we can also define the self-energy $\Sigma(l,j)$ as

$$\Sigma(l,j) \equiv \left\{ \begin{array}{l} \text{The sum of all irreducible diagrams in } \mathcal{G}(b,a) \\ \text{without the two external fermion lines } \mathcal{G}^0(j,a) \text{ and } \mathcal{G}^0(b,l) \end{array} \right\}$$

$$= \overset{\delta_{lj}}{\bullet\!\!\!\frown\!\!\!\frown\!\!\bigcirc} + \; \overset{l\qquad j}{\bullet\!\!\frown\!\!\frown\!\!\bullet} + \; \overset{l}{}\underset{j}{} + \; \overset{l}{}\underset{j}{} + \; \cdots$$

$$= \overset{l}{}\underset{j}{} \tag{13.21}$$

From Eqs. (13.17) and (13.21) we obtain Dyson's equation for $\mathcal{G}(b,a)$

$$\mathcal{G}(b,a) = \overset{b\qquad a}{\bullet\!\!\blacktriangleleft\!\!\bullet}$$

$$= \overset{b\quad a}{\bullet\!\!\blacktriangleleft\!\!\bullet} + \overset{b\quad l\quad j\quad a}{\bullet\!\!\blacktriangleleft\!\!\bullet} + \overset{b\quad l\quad j\quad a}{\bullet\!\!\blacktriangleleft\!\!\bullet\!\!\blacktriangleleft\!\!\bullet} + \cdots$$

$$= \overset{b\quad a}{\bullet\!\!\blacktriangleleft\!\!\bullet} + \overset{b\quad l\quad j}{\bullet\!\!\blacktriangleleft\!\!\bullet} \times \left(\overset{j\quad a}{\bullet\!\!\blacktriangleleft\!\!\bullet} + \overset{j\qquad a}{\bullet\!\!\blacktriangleleft\!\!\bullet\!\!\blacktriangleleft\!\!\bullet} + \cdots \right)$$

$$= \overset{b\quad a}{\bullet\!\!\blacktriangleleft\!\!\bullet} + \overset{b\quad l\quad j\quad a}{\bullet\!\!\blacktriangleleft\!\!\bullet}$$

$$= \mathcal{G}^0(b,a) \quad + \quad \int dl \int dj \; \mathcal{G}^0(b,l) \, \Sigma(l,j) \, \mathcal{G}(j,a). \tag{13.22}$$

Note how Dyson's equation in this case is an integral equation. We shall shortly see that for a translation-invariant system it becomes an algebraic equation in $\mathbf{k}$-space.

13.4 The Feynman rules in Fourier space

For the special case where H_0 describes a translation-invariant system and where the interaction $W_{j,j'}$ only depends on the coordinate differences $\mathbf{r}_j - \mathbf{r}'_j$ and $\tau_j - \tau'_j$, it is a great advantage to Fourier transform the representation from $(\mathbf{r}, \tau)$-space to $(\mathbf{q}, iq_n)$-space. Our main example of such a system is the jellium model for Coulomb-interacting electrons studied in Section 2.2. In terms of the Fourier transformed interaction $W(\mathbf{q}) = 4\pi e_0^2/q^2$, the Coulomb interaction $W(\mathbf{r}\tau; \mathbf{r}', \tau')$ is written as

$$W(\mathbf{r}\tau; \mathbf{r}'\tau') = \frac{1}{\beta \mathcal{V}} \sum_{\mathbf{q}, iq_n} W(q) \, e^{[i\mathbf{q}\cdot(\mathbf{r}-\mathbf{r}')-iq_n(\tau-\tau')]}. \qquad (13.23)$$

It is important to realize that the Matsubara frequency iq_n is bosonic since the Coulomb interaction is bosonic in nature: *two* fermions are annihilated and *two* fermions are created by the interaction, i.e., *one* boson object is annihilated and *one* is created. Furthermore, we note that due to the factor $\delta(\tau - \tau')$ in Eq. (13.10), the Matsubara frequency iq_n appears only in the argument of the exponential function.

Likewise, using Eq. (11.42), the electron Green's function $\mathcal{G}_\sigma^0(\mathbf{r}\tau, \mathbf{r}'\tau')$ for spin σ in $(\mathbf{k}, ik_n)$-space can be expressed as

$$\mathcal{G}_\sigma^0(\mathbf{r}\tau; \mathbf{r}'\tau') = \frac{1}{\beta \mathcal{V}} \sum_{\mathbf{k}, ik_n} \mathcal{G}_\sigma^0(\mathbf{k}, ik_n) \, e^{[i\mathbf{k}\cdot(\mathbf{r}-\mathbf{r}')-ik_n(\tau-\tau')]}, \qquad (13.24)$$

where $\mathcal{G}_\sigma^0(\mathbf{k}, ik_n) = 1/(ik_n - \xi_{\mathbf{k}})$ depends on $\mathbf{k}$ and ik_n, but not on σ, and $\xi_{\mathbf{k}} \equiv \varepsilon_{\mathbf{k}} - \mu$. In the case of the Coulomb-interacting electron gas in the jellium model, we thus see that the Green's function $\mathcal{G}_\sigma^0$ and the interaction W both depend only on the space and imaginary time differences $\mathbf{r}-\mathbf{r}'$ and $\tau-\tau'$. It follows from Eqs. (13.23) and (13.24) that it saves some writing to introduce the four-vector notation $\tilde{k} \equiv (\mathbf{k}, ik_n)$, $\tilde{r} \equiv (\mathbf{r}, \tau)$, and $i\tilde{k}\cdot\tilde{r} \equiv i\mathbf{k}\cdot\mathbf{r} - ik_n\tau$. Using this notation we analyze the Fourier transform of the basic Coulomb-scattering vertex,

$$\int d\tilde{r} \, \mathcal{G}_\sigma^0(\tilde{r}_2, \tilde{r}) \, \mathcal{G}_\sigma^0(\tilde{r}, \tilde{r}_1) \, W(\tilde{r}_3; \tilde{r}) \quad = \quad \qquad (13.25)$$

where the $(\mathbf{r}, \tau)$-space points $\tilde{r}_1$, $\tilde{r}_2$, $\tilde{r}_3$, and $\tilde{r}$ are indicated, as are the wave vectors $\tilde{k}$, $\tilde{p}$, and $\tilde{q}$ to be used in the Fourier transform. On top of their usual meaning, the arrows now also indicate the choice of sign for the four-momentum vectors: $\tilde{k}$ flows from $\tilde{r}_1$ to $\tilde{r}$, $\tilde{p}$ from $\tilde{r}$ to $\tilde{r}_2$, and $\tilde{q}$ from $\tilde{r}$ to $\tilde{r}_3$. Inserting the Fourier transforms of Eqs. (13.23) and (13.24) into Eq. (13.25), we obtain with this sign convention

$$\int d\tilde{r}\, \mathcal{G}^0_\sigma(\tilde{r}_2, \tilde{r})\, \mathcal{G}^0_\sigma(\tilde{r}, \tilde{r}_1)\, W(\tilde{r}_3; \tilde{r})$$

$$= \int d\tilde{r}\, \frac{1}{(\beta\mathcal{V})^3} \sum_{\tilde{k}\tilde{p}\tilde{q}} \mathcal{G}^0_\sigma(\tilde{p})\, \mathcal{G}^0_\sigma(\tilde{k})\, W(\tilde{q})\, e^{i[\tilde{p}\cdot(\tilde{r}_2-\tilde{r})+\tilde{k}\cdot(\tilde{r}-\tilde{r}_1)+\tilde{q}\cdot(\tilde{r}_3-\tilde{r})]}$$

$$= \frac{1}{(\beta\mathcal{V})^3} \sum_{\tilde{k}\tilde{p}\tilde{q}} \mathcal{G}^0_\sigma(\tilde{p})\, \mathcal{G}^0_\sigma(\tilde{k})\, W(\tilde{q})\, e^{i[\tilde{p}\cdot\tilde{r}_2-\tilde{k}\cdot\tilde{r}_1+\tilde{q}\cdot\tilde{r}_3]} \int d\tilde{r}\, e^{-i(\tilde{p}-\tilde{k}+\tilde{q})\cdot\tilde{r}}$$

$$= \frac{1}{(\beta\mathcal{V})^2} \sum_{\tilde{k}\tilde{q}} \mathcal{G}^0_\sigma(\tilde{k}-\tilde{q})\, \mathcal{G}^0_\sigma(\tilde{k})\, W(\tilde{q})\, e^{i[\tilde{k}\cdot(\tilde{r}_2-\tilde{r}_1)+\tilde{q}\cdot(\tilde{r}_3-\tilde{r}_2)]}. \qquad (13.26)$$

From this follows that in Fourier space the four-momentum $(\mathbf{k}, ik_n)$ is conserved at each Coulomb-scattering vertex: $\tilde{k} = \tilde{p} + \tilde{q}$. Since each vertex consists of two fermion lines and one interaction line, the momentum conservation combined with the odd values of the fermion Matsubara frequencies leads, in agreement with our previous remarks, to even values for the Matsubara frequencies of the interaction lines. The momentum conservation rule for each of the $2n$ vertices also leads to $2n$ delta function constraints on the $2n$ internal fermion momenta and the n interaction line momenta, and whence the number of independent internal momenta equals n, i.e., the order of the diagram. For each independent momentum a factor $1/\beta\mathcal{V}$ remains from the corresponding Fourier transform. The topology of the diagram in $(\mathbf{r}, \tau)$-space is not changed by the Fourier transform. We therefore end up with the following Feynman rules for the n-order diagrams in the expansion of $\mathcal{G}_\sigma(\mathbf{k}, ik_n)$, where $(\mathbf{k}, ik_n)$ is to be interpreted as the externally given four-vector momentum.

(1) Fermion lines with four-momentum orientation: $\equiv \mathcal{G}^0_\sigma(\mathbf{k}, ik_n)$.

(2) Interaction lines with four-momentum orientation: $\equiv -W(q)$.

(3) Conserve the spin and four-momentum at each vertex, i.e., incoming momenta must equal the outgoing, and no spin flipping.

(4) At order n draw all topologically different connected diagrams containing n oriented interaction lines $W(\tilde{q})$, two external fermion lines $\mathcal{G}^0_\sigma(\mathbf{k}, ik_n)$, and $2n - 1$ internal fermion lines $\mathcal{G}^0_\sigma(\mathbf{p}_j, ip_j)$. All vertices must contain an incoming and an outgoing fermion line as well as an interaction line.

(5) Multiply each diagram by $(-1)^F$, F being the number of fermion loops.

(6) Multiply $\mathcal{G}^0_\sigma(\mathbf{k}, ik_n)$ in the "same-time" diagrams , by $e^{ik_n\eta}$.

(7) Multiply by $\frac{1}{\beta\mathcal{V}}$ for each internal four-momentum $\tilde{p}$; perform the sum $\sum_{\tilde{p}\sigma'}$.

$$(13.27)$$

Note how the two "same-time" diagrams in rule (6) are the only ones where it is relevant to take explicitly into account the infinitesimal shift $\tau_j \to \tau_j + \eta$ mentioned in Eqs. (13.9) and (13.10). The factor $e^{ik_n\eta}$ follows directly from the Fourier transform when this shift is included.

In $(\mathbf{k}, ik_n)$-space the third Feynman rule concerning the conservation of four-momentum at the scattering vertices simplifies many calculations. Most noteworthy is the fact that Dyson's equation becomes an algebraic equation. Due to four-momentum conservation, a four-momentum $\tilde{k}_j$ entering a self-energy diagram, such as the ones shown in Eq. (13.21), must also exit it, i.e., $\tilde{k}_l = \tilde{k}_j$. The self-energy (with spin σ) is thus diagonal in $\mathbf{k}$-space,

$$\Sigma_\sigma(\tilde{k}, \tilde{k}') = \delta_{\tilde{k}\tilde{k}'} \Sigma_\sigma(\tilde{k}), \qquad \Sigma_\sigma(\tilde{k}) \equiv \Sigma_\sigma(\tilde{k}, \tilde{k}). \tag{13.28}$$

Dyson's equation (13.22) is therefore an algebraic equation,

$$\mathcal{G}_\sigma(\tilde{k}) = \mathcal{G}_\sigma^0(\tilde{k}) + \mathcal{G}_\sigma^0(\tilde{k}) \, \Sigma_\sigma(\tilde{k}) \, \mathcal{G}_\sigma(\tilde{k}),$$

$$\tag{13.29}$$

with the solution

$$\mathcal{G}_\sigma(\mathbf{k}, ik_n) = \frac{\mathcal{G}_\sigma^0(\mathbf{k}, ik_n)}{1 - \mathcal{G}_\sigma^0(\mathbf{k}, ik_n) \, \Sigma_\sigma(\mathbf{k}, ik_n)} = \frac{1}{ik_n - \xi_\mathbf{k} - \Sigma_\sigma(\mathbf{k}, ik_n)}. \tag{13.30}$$

As in Eq. (12.41), the self-energy $\Sigma_\sigma(\mathbf{k}, ik_n)$, induced here by the Coulomb interaction W, appears as a direct additive renormalization of the bare energy $\xi_\mathbf{k} = \varepsilon_\mathbf{k} - \mu$.

13.5 Examples of how to evaluate Feynman diagrams

The Feynman diagrams are an extremely useful tool to gain an overview of the very complicated infinite-order perturbation calculation, and they allow one to identify the important processes for a given physical problem. When this part of the analysis is done one is (hopefully) left with only a few important diagrams that then need to be evaluated. We end this chapter by studying the explicit evaluation of three simple Feynman diagrams in Fourier space using the Feynman rules (13.27).

13.5.1   The Hartree self-energy diagram

To evaluate a given diagram the first task is to label the fermion and interaction lines with four-momenta and spin obeying the conservation rules at each vertex, rule (3) in Eq. (13.27). We start with the so-called Hartree diagram $\mathcal{G}_\sigma^H$ (which is zero in the presence of a charge-compensating background), where we in accordance with Eq. (13.21) strip off the two external fermion lines to obtain the self-energy Σ^H:

$$\tag{13.31}$$

The four-momentum transfer along the interaction line is zero, while the four-momentum $(\mathbf{p}, ip_n)$ and the spin σ' in the fermion loop are free to take any value. The self-energy diagram is a first-order diagram, i.e., $n = 1$. It contains one internal four-momentum, $(\mathbf{p}, ip_n)$, yielding a factor of $1/\beta\mathcal{V}$, one internal spin, σ', and one fermion loop, i.e.,

$F = 1$. The Feynman rules therefore lead to the following expression for the Hartree self-energy diagram Eq. (13.31),

$$
\Sigma_\sigma^H(\mathbf{k}, ik_n) \equiv \ \rule{1.5cm}{0pt} = \frac{-1}{\beta \mathcal{V}} \sum_{\sigma'} \sum_{\mathbf{p}} \sum_{ip_n} \left[-W(0) \right] \mathcal{G}_{\sigma'}^0(\mathbf{p}, ip_n)\, e^{ip_n \eta}
$$

$$
= \frac{2W(0)}{\beta} \int \frac{d\mathbf{p}}{(2\pi)^3} \sum_{ip_n} \frac{e^{ip_n \eta}}{ip_n - \xi_{\mathbf{p}}}
$$

$$
= 2W(0) \int \frac{d\mathbf{p}}{(2\pi)^3}\, n_{\mathrm{F}}(\xi_{\mathbf{p}}) \quad = \quad W(0)\, \frac{N}{\mathcal{V}}. \qquad (13.32)
$$

Note the need for Feynman rule Eq. (13.27)(6) for evaluating this specific diagram. The spin sum turns into a simple factor 2. The Matsubara sum can easily be carried out using the method of Section 11.4.1. The evaluation of the $\mathbf{p}$-integral is elementary and yields $N/2$.

According to Eq. (13.30) the self-energy is the interaction-induced renormalization of the non-interacting single-particle energy. This renormalization we have calculated by completely different means in Sec. 4.2 using the Hartree–Fock mean field approximation. We see that the diagrammatic result Eq. (13.32) exactly equals the Hartree part of the mean field energy in Eq. (4.24b). In other words we have shown that the "tadpole" self-energy diagram is the diagrammatic equivalent of the Hartree mean-field approximation.

13.5.2   *The Fock self-energy diagram*

We treat the Fock diagram $\mathcal{G}_\sigma^F$ and Fock self-energy Σ_σ^F similarly:

$$
\mathcal{G}_\sigma^F(\mathbf{k}, ik_n) = \ \rule{2cm}{0pt} \ = \ \mathcal{G}_\sigma^0(\mathbf{k}, ik_n) \ \rule{2cm}{0pt} \ . \qquad (13.33)
$$

$$
\begin{array}{c} \mathbf{k} - \mathbf{p} \\ ik_n - ip_n \end{array} \quad \mathcal{G}_\sigma^0(\mathbf{k}, ik_n)
$$

$$
\mathbf{p}, ip_n, \sigma'
$$

Once more, the external fermion lines are written explicitly as two factors $\mathcal{G}_\sigma^0(\mathbf{k}, ik_n)$, leaving the Fock self-energy Σ_σ^F to be determined. The four-momentum transferred by the interaction line is $(\mathbf{k} - \mathbf{p}, ik_n - ip_n)$. This diagram is a first-order diagram, i.e., $n = 1$. It contains one internal four-momentum, $(\mathbf{p}, ip_n)$, yielding a factor $1/\beta\mathcal{V}$.

However, in contrast to Eq. (13.31), the internal spin σ' is now forced to be equal to the external spin σ. Finally, no fermion loops are present, i.e., $F = 0$. The Feynman rules therefore lead to the following expression for the Fock self-energy diagram Eq. (13.33):

$$\Sigma_\sigma^F(\mathbf{k}, ik_n) \equiv \text{} = \frac{1}{\beta \mathcal{V}} \sum_{\sigma'} \sum_{\mathbf{p}} \sum_{ip_n} \left[-W(\mathbf{k}-\mathbf{p}) \right] \delta_{\sigma\sigma'} \mathcal{G}_{\sigma'}^0(\mathbf{p}, ip_n) \, e^{ip_n\eta}$$

$$= \frac{-1}{\beta} \int \frac{d\mathbf{p}}{(2\pi)^3} \, W(\mathbf{k}-\mathbf{p}) \sum_{ip_n} \frac{e^{ip_n\eta}}{ip_n - \xi_{\mathbf{p}}}$$

$$= -\int \frac{d\mathbf{p}}{(2\pi)^3} \, W(\mathbf{k}-\mathbf{p}) \, n_F(\xi_{\mathbf{p}}). \tag{13.34}$$

Note that also for this specific diagram we have used Feynman rule (6). The spin sum turned into a simple factor 1. The Matsubara sum can easily be carried out using the method of Section 11.4.1. The evaluation of the $\mathbf{p}$-integral is in principle elementary. We see that this self-energy diagram exactly equals the Fock part of the energy in Eq. (4.24b) calculated using the Hartree–Fock mean field approximation. We have thus shown that the "half-oyster" self-energy diagram[45] is the diagrammatic equivalent of the Fock mean field approximation.

13.5.3   *The pair-bubble self-energy diagram*

Our last example is the pair-bubble diagram $\mathcal{G}_\sigma^P$, which, as we shall see in Chapter 14, plays a central role in studies of the electron gas. We proceed as in the previous examples:

$$\mathcal{G}_\sigma^P(\mathbf{k}, ik_n) \equiv \text{}$$

$$\tag{13.35}$$

Removing the two external fermion lines $\mathcal{G}_\sigma^0(\mathbf{k}, ik_n)$, leaves us with the pair-bubble self-energy diagram Σ_σ^P. We immediately note that this diagram is of second order, i.e., $n = 2$, containing one fermion loop, i.e., $F = 1$. At the first vertex the incoming momentum $(\mathbf{k}, ik_n)$ is split, sending $(\mathbf{q}, iq_n)$ out through the interaction line, while the remainder $(\mathbf{k}-\mathbf{q}, ik_n-iq_n)$ continues in the fermion line. At the fermion loop, $(\mathbf{q}, iq_n)$ is joined by the internal fermion momentum $(\mathbf{p}, ip_n)$ and continues in a new fermion line as $(\mathbf{p}+\mathbf{q}, ip_n+iq_n)$. At the top of the loop, the momentum $(\mathbf{q}, iq_n)$ is sent out through the interaction line, where it ultimately recombines with the former fermion momentum $(\mathbf{k}-\mathbf{q}, ik_n-iq_n)$. We have thereby ensured that the exit momentum equals that of the entrance: $(\mathbf{k}, ik_n)$. The internal degrees of freedom are $(\mathbf{q}, iq_n)$, $(\mathbf{p}, ip_n)$, and σ', the former two yielding a prefactor $1/(\beta\mathcal{V})^2$. The Feynman rules lead to the following expression for the pair-bubble self-energy Eq. (13.35):

---

[45] A full "oyster" diagram can be seen in Eq. (13.14)

$\Sigma_\sigma^P(\mathbf{k}, ik_n)$

$$= \frac{-1}{(\beta V)^2} \sum_{\sigma'\mathbf{pq}} \sum_{ip_n iq_n} \left[-W(\mathbf{q})\right]^2 \mathcal{G}_{\sigma'}^0(\mathbf{p}, ip_n)\, \mathcal{G}_{\sigma'}^0(\mathbf{p}+\mathbf{q}, ip_n+iq_n)\, \mathcal{G}_\sigma^0(\mathbf{k}-\mathbf{q}, ik_n-iq_n)$$

$$= \frac{1}{\beta} \sum_{iq_n} \int \frac{d\mathbf{q}}{(2\pi)^3} W(\mathbf{q})^2\, \Pi^0(\mathbf{q}, iq_n)\, \mathcal{G}_\sigma^0(\mathbf{k}-\mathbf{q}, ik_n-iq_n), \tag{13.36}$$

where we have separated out the contribution $\Pi^0(\mathbf{q}, iq_n)$ from the fermion loop,

$$\Pi^0(\mathbf{q}, iq_n) \equiv \quad \text{\scriptsize(loop)} \quad = \frac{-2}{\beta} \sum_{ip_n} \int \frac{d\mathbf{p}}{(2\pi)^3} \frac{1}{(ip_n + iq_n - \xi_{\mathbf{p+q}})} \frac{1}{(ip_n - \xi_{\mathbf{p}})}. \tag{13.37}$$

The loop contribution $\Pi^0(\mathbf{q}, iq_n)$ is traditionally denoted the pair-bubble, and we shall study it in more detail in the coming chapters. Here we just note that the spin sum becomes a factor 2, and that the Matsubara sum over ip_n can easily be carried out using the method of Section 11.4.1. The evaluation of the $\mathbf{p}$-integral is in principle elementary. Inserting the result for $\Pi^0(\mathbf{q}, iq_n)$ into the pair-bubble self-energy diagram (13.35) leads to a bit more involved Matsubara frequency summation over iq_n and momentum integration over $\mathbf{q}$. However, the calculation can be performed, and we shall return to it later.

13.6 Cancellation of disconnected diagrams, general case

Now we show the cancellation of disconnected diagrams for a general Hamiltonian of the form $H_0 + W$, where W is the perturbation. The proof is similar to the one in Eq. (13.19). We look for the expectation value of some operator $A(\tau)B(\tau')$. The Green's function studied so far in this chapter is just $A(\tau)B(\tau') = -\hat{\Psi}(b)\,\hat{\Psi}^\dagger(a)$. If we, furthermore, absorb the minus sign appearing in Eq. (13.8) by defining $V = -W$, we have

$$\langle T_\tau \hat{A}(\tau)\hat{B}(\tau')\rangle = \frac{\displaystyle\sum_{n=0}^\infty \frac{1}{n!} \int_0^\beta d\tau_1 \cdots \int_0^\beta d\tau_n \left\langle T_\tau\left[\hat{V}(\tau_1)\cdots\hat{V}(\tau_n)\hat{A}(\tau)\hat{B}(\tau')\right]\right\rangle_0}{\displaystyle\sum_{n=0}^\infty \frac{1}{n!} \int_0^\beta d\tau_1 \cdots \int_0^\beta d\tau_n \left\langle T_\tau\left[\hat{V}(\tau_1)\cdots\hat{V}(\tau_n)\right]\right\rangle_0}.$$

$$\tag{13.38}$$

Consider the nth term $V^{(n)}$ in the numerator

$$V^{(n)} = \frac{1}{n!} \int_0^\beta d\tau_1 \cdots \int_0^\beta d\tau_n \left\langle T_\tau\left[\hat{V}(\tau_1)\cdots\hat{V}(\tau_n)\hat{A}(\tau)\hat{B}(\tau')\right]\right\rangle_0. \tag{13.39}$$

When applying Wick's theorem, this will have both connected and non-connected terms. We define "connected" in the following way. When applying Wick's theorem to a given average $\langle T_\tau \hat{V}_1 \cdots \hat{V}_n \hat{A}\hat{B}\rangle$ the determinant (or permanent for bosons) generates a number of terms; a term is said to be disconnected if it can be factorized as $\langle T_\tau \hat{V}_1 \cdots \hat{V}_j \hat{A}\hat{B}\rangle\langle T_\tau \hat{V}_j \cdots \hat{V}_n\rangle$. The connected diagram to the order n in $\hat{V}$ is thus the average $\langle T_\tau \hat{V}_1 \cdots \hat{V}_n \hat{A}\hat{B}\rangle$ minus all the possible factorizations. We write this condition in the following way, where all disconnected terms are explicitly subtracted,

$$\left\langle T_\tau\left[\hat{V}(\tau_1)\cdots\hat{V}(\tau_n)\hat{A}(\tau)\hat{B}(\tau')\right]\right\rangle_{0,\text{con}} \equiv \left\langle T_\tau\left[\hat{V}(\tau_1)\cdots\hat{V}(\tau_n)\hat{A}(\tau)\hat{B}(\tau')\right]\right\rangle_0$$

$$- \sum_{P\in S_n}\sum_{j=0}^{n-1}\left\langle T_\tau\left[\hat{V}(\tau_{P_1})\cdots\hat{V}(\tau_{P_j})\hat{A}(\tau)\hat{B}(\tau')\right]\right\rangle_{0,\text{con}}\left\langle T_\tau\left[\hat{V}(\tau_{P_{j+1}})\cdots\hat{V}(\tau_{P_n})\right]\right\rangle_0,$$

$$(13.40)$$

where S_n is the permutation group as in Eq. (1.25). Note that the last factor can be decomposed further into disconnected parts. Next, we want to find all the possible ways that a connected diagram of order m can be generated in the numerator of Eq. (13.38). Clearly, the first term appears when $n = m$ in the sum, but also higher n can give a connected diagram of order m. For example, the $n = m + 1$ term will give n terms of the form $\langle T_\tau \hat{V}_1 \cdots \hat{V}_m \hat{A}\hat{B}\rangle_{0,\text{con.}}\langle\hat{V}\rangle_0$ because there are n different $\hat{V}$'s that can be factorized out. The term with $n = m+2$ will have $n(n-1)/2$ terms of the form $\langle T_\tau \hat{V}_1 \cdots \hat{V}_m \hat{A}\hat{B}\rangle_{0,\text{con.}}\langle\hat{V}_{m+1}\hat{V}_n\rangle_0$ etc. Because we can relabel the time arguments as we wish, these $n(n-1)/2$ are all identical. Continuing this argument, we now write all the terms containing a connected diagram of order m:

$$\int_0^\beta d\tau_1 \cdots \int_0^\beta d\tau_m \left\langle T_\tau\left[\hat{V}(\tau_1)\cdots\hat{V}(\tau_m)\hat{A}(\tau)\hat{B}(\tau')\right]\right\rangle_{0,\text{con}} \quad (13.41)$$

$$\times\left(\frac{1}{m!} + \frac{m+1}{(m+1)!1!}\int_0^\beta d\tau_{m+1}\left\langle T_\tau\left[\hat{V}(\tau_1)\right]\right\rangle_0\right.$$

$$\left.+ \frac{(m+2)(m+1)}{(m+2)!2!}\int_0^\beta d\tau_{m+1}\int_0^\beta d\tau_{m+2}\left\langle T_\tau\left[\hat{V}(\tau_1)\hat{V}(\tau_2)\right]\right\rangle + \cdots\right).$$

A factor $1/m!$ can be pulled out of this, and we recognize the parenthesis as the denominator in Eq. (13.38), which therefore cancels out! We have thus proved that Eq. (13.38) can be written as

$$\langle T_\tau \hat{A}(\tau)\hat{B}(\tau')\rangle = \sum_{n=0}^\infty \frac{1}{n!}\int_0^\beta d\tau_1 \cdots \int_0^\beta d\tau_n \left\langle T_\tau\left[\hat{V}(\tau_1)\cdots\hat{V}(\tau_n)\hat{A}(\tau)\hat{B}(\tau')\right]\right\rangle_{0,\text{con}}$$

$$(13.42)$$

As a final step, we can also get rid of the factorial factor, because, as for the pair interaction discussed above, one can argue that each diagram generated by Wick's theorem are generated in $n!$ topologically identical variants. Two diagrams are said to be topologically identical if they can be made identical after permuting their internal time arguments. The final expression for the general way to construct diagrams for the expectation value $\langle T_\tau \hat{A}(\tau)\hat{B}(\tau')\rangle$ is therefore

$$\langle T_\tau \hat{A}(\tau)\hat{B}(\tau')\rangle = \sum_{n=0}^{\infty} \int_0^\beta d\tau_1 \cdots \int_0^\beta d\tau_n \left\langle T_\tau \left[\hat{V}(\tau_1)\cdots\hat{V}(\tau_n)\hat{A}(\tau)\hat{B}(\tau') \right] \right\rangle_{0,\,\text{con-diff}}$$
(13.43)

where the subscript "con-diff" means all connected, different diagrams.

13.7 Feynman diagrams for the Kondo model

So far we have derived the diagram technique for two-particle interactions or for interactions, where we can use Wick's theorem. Here we take a different example, namely the Kondo model, which was introduced in Section 10.5. In Chapter 10 we studied the model in the context of interacting mesoscopic systems. The reason for studying this case here is also to alert the reader to the fact that when dealing with spin operators, the diagram technique has to be revised. The Kondo model reads

$$H = H_0 + H_J,$$
(13.44)

where H_0 describes the non-interacting electron gas,

$$H_0 = \sum_{\mathbf{k}\sigma} \xi_{\mathbf{k}} c_{\mathbf{k}\sigma}^\dagger c_{\mathbf{k}\sigma},$$
(13.45)

while H_J represents the interaction with the localized spin degree of freedom,

$$H_J = J \sum_{\mathbf{k}\mathbf{k}'} \mathbf{S}_d \cdot \mathbf{S}_{\mathbf{k},\mathbf{k}'}.$$
(13.46)

Here $\mathbf{S}_d$ is the spin operator for the local spin, and

$$\mathbf{S}_{\mathbf{k},\mathbf{k}'} = \sum_{\sigma\sigma'} c_{\mathbf{k}\sigma}^\dagger \boldsymbol{\tau}_{\sigma\sigma'} c_{\mathbf{k}'\sigma'}$$
(13.47)

is the operator for the electron gas spin that couples to the localized spin. Remember that $\boldsymbol{\tau} = \frac{1}{2}(\tau^x,\tau^y,\tau^z)$ contains the three Pauli spin matrices defined in Eq. (1.91). The Hamiltonian that governs the impurity spin is here equal to zero, because the system is isotropic. However, if an external magnetic field is applied, we must add a term $\mathbf{S}_d \cdot \mathbf{B}$ to the Hamiltonian. Also the electron gas is isotropic in spin, because the energy $\xi_{\mathbf{k}}$ does not depend on spin at all.

As for the pair interaction in Eq. (13.8), our strategy is to expand the Green's function

$$\mathcal{G}(\mathbf{k}\sigma\tau,\mathbf{k}'\sigma'\tau') = -\left\langle T_\tau \left[c_{\mathbf{k}\sigma}(\tau) c_{\mathbf{k}'\sigma'}^\dagger(\tau') \right] \right\rangle$$
(13.48)

in powers of the interaction H_J in analogy to Eq. (13.38). We could try to utilize the general principles derived in Section 13.6. There is, however, one caveat, namely that Wick's theorem does not apply to the spin operators. This means, e.g., that an average $\langle \hat{S}^i(\tau_1)\hat{S}^j(\tau_2)\hat{S}^k(\tau_3)\hat{S}^l(\tau_4)\rangle$ does not factorize into averages of pairs. Consequently, we must derive a new set of rules for this kind of interaction. The number of disconnected diagrams that are cancelled out is thus reduced to the ones where the electron operators defining the Green's function, i.e., $c_{\mathbf{k}\sigma}(\tau)$ and $c_{\mathbf{k}'\sigma'}^\dagger(\tau')$ are fully disconnected from the rest. Also the concept of "topologically different" has to be revisited. It only works if we can argue that the average

$$\left\langle T_\tau \left[\hat{S}^{i_1}(\tau_1) \cdots \hat{S}^{i_n}(\tau_n) \right] \right\rangle, \tag{13.49}$$

does not change under a permutation of the time arguments. Of course, since the time ordering determines the order of the operators, the only concern is possible sign changes caused by the relabelling of the time arguments. Because the spin operators conserves the number of fermions, they behave like bosons in this respect, which for example can be seen from the fermion representation in Eq. (1.92). Thus, we only have to count topologically identical electron diagrams once and omit the $1/n!$ factor. In total we now have for the expansion

$$\mathcal{G}(\mathbf{k}\sigma\tau, \mathbf{k}'\sigma'\tau') = \sum_{n=0}^{\infty} (-J)^n \int_0^\beta d\tau_1 \cdots \int_0^\beta d\tau_n \sum_{i_1 \cdots i_n} \left\langle T_\tau \left[\hat{S}^{i_1}(\tau_1) \cdots \hat{S}^{i_n}(\tau_n) \right] \right\rangle_0 \times$$

$$\sum_{\substack{\sigma_1 \cdots \sigma_n \\ \sigma_1' \cdots \sigma_n'}} \left\langle T_\tau \left[\hat{C}_{\sigma_1}^\dagger(\tau_1) \tau_{\sigma_1\sigma_1'}^{i_1} \hat{C}_{\sigma_1'}(\tau_1) \cdots \hat{C}_{\sigma_n}^\dagger(\tau_n) \tau_{\sigma_n\sigma_n'}^{i_n} \hat{C}_{\sigma_n'}(\tau_n) c_{\mathbf{k}\sigma}(\tau) c_{\mathbf{k}'\sigma'}^\dagger(\tau') \right] \right\rangle_{0,\,\mathrm{con-diff}},$$

$$\tag{13.50}$$

where we for convenience define

$$C_\sigma = \sum_{\mathbf{k}} c_{\mathbf{k}\sigma}. \tag{13.51}$$

The different diagrams up to third order in J are shown in Fig. 13.2. A number of these vanish because of the isotropic assumption, i.e., no magnetic field and no spin polarization. First of all, the linear term in J is zero because $\langle S \rangle = 0$. Second, the diagrams which have fermion loops are zero for the following reasons. A loop with a single fermion line forces the Pauli spin matrix in the vertex to be diagonal and after summation over spin, one has a trace: $\mathrm{Tr}[\tau^i] = 0$. A loop with two fermion lines that appears in the third order diagram gives rise to a trace over two Pauli spin matrices, $\mathrm{Tr}[\tau^i\tau^j] = \delta_{ij}$, and because the trace over the three spin operators gives $\mathrm{Tr}[\tau^i\tau^j\tau^k] = 2i\epsilon_{ijk}$ (see Section 13.7.2), we end up with a factor $\epsilon_{ijk}\delta_{ij} = 0$, where ϵ_{ijk} is the Levi-Civita symbol defined in Eq. (1.11). Thus we can supplement the rules in Eq. (13.48) by the condition that all fermion lines must connect to the end point of the diagram.

It is now clear from Fig. 13.2 that we can collect all terms in the expansion of the Green's function as

$$\mathcal{G}(\mathbf{k}\sigma\tau, \mathbf{k}'\sigma'\tau') = \mathcal{G}^0(\mathbf{k}\sigma, \tau - \tau')\delta_{\mathbf{k},\mathbf{k}'}\delta\sigma, \sigma' \tag{13.52}$$

$$+ \int_0^\beta d\tau_1 \int_0^\beta d\tau_2\, \mathcal{G}^0(\mathbf{k}\sigma, \tau - \tau_1)\Sigma(\mathbf{k}\sigma\tau_1, \mathbf{k}'\sigma'\tau_2)\mathcal{G}^0(\mathbf{k}'\sigma', \tau_2 - \tau').$$

Below we include up to third order, because we learned in Chapter 10, that the interesting Kondo effect shows up in third order perturbation theory.

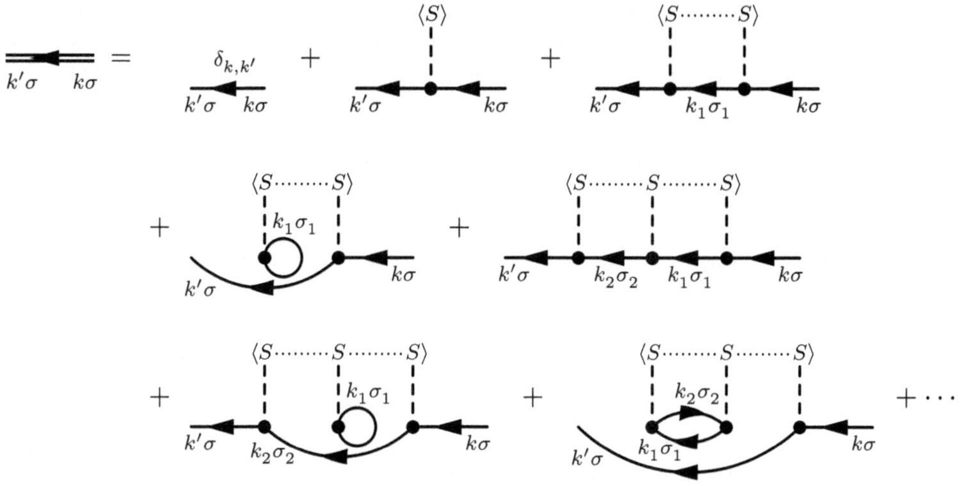

FIG. 13.2. The topologically different diagrams up to third order in J for the Kondo model. In the text it shown that the diagrams which contain a fermion loop are zero, which leaves only one non-zero diagram to each order in J. Note that the diagram involves an average over S-operators, which cannot be drawn as usual diagrams because the average of S-operators does not obey Wick's theorem.

13.7.1 Kondo model self-energy, second order in J

We have for the second-order term that

$$\Sigma^{(2)}(\mathbf{k}\sigma\tau_1, \mathbf{k}'\sigma'\tau_2) = (-J)^2 \sum_{\mathbf{k}_1, \sigma_1, i, j} \mathcal{G}^0(\mathbf{k}_1\sigma_1, \tau_1 - \tau_2) \left\langle T_\tau \hat{S}^i(\tau_1)\hat{S}^j(\tau_2) \right\rangle_0 \frac{\tau^i_{\sigma\sigma_1}}{2} \frac{\tau^j_{\sigma_1\sigma'}}{2}$$

$$= \delta_{\sigma\sigma'} \frac{3J^2}{16} \sum_{\mathbf{k}_1} \mathcal{G}^0(\mathbf{k}_1\sigma_1, \tau_1 - \tau_2), \tag{13.53}$$

where it was used that $\langle T_\tau[\hat{S}^i_d(\tau_1)\hat{S}^j_d(\tau_2)]\rangle_0 = \langle S^i_d S^j_d\rangle_0 = \frac{1}{8}\text{Tr}\left[\tau^i\tau^j\right] = \frac{1}{4}\delta_{ij}$, and that $\mathcal{G}^0(\mathbf{k}\sigma_1)$ does not depend on σ_1. In the frequency domain Eq. (13.53) becomes

$$\Sigma^{(2)}(\mathbf{k}\sigma, \mathbf{k}'\sigma', ik_n) = \delta_{\sigma\sigma'} \frac{3J^2}{16} \sum_{\mathbf{k}_1} \frac{1}{ik_n - \xi_{\mathbf{k}_1}} = -i\delta_{\sigma\sigma'} \frac{3J^2\pi d_\sigma(\varepsilon_F)}{16} \text{sgn}(k_n), \tag{13.54}$$

where we assumed a constant density of states, $d_\sigma(\varepsilon_F)$ (per spin direction). The important information about the scattering lies in the imaginary part of the retarded self-energy,

$$-\text{Im}\,\Sigma^{R,(2)}(\mathbf{k}\sigma, \mathbf{k}'\sigma', \omega) = \delta_{\sigma\sigma'} \frac{3J^2\pi d_\sigma(\varepsilon_F)}{16}. \tag{13.55}$$

When these expressions are compared with the first Born approximation for usual impurity-scattering in Eqs. (12.44) and (12.48), we see that the scattering on magnetic impurities gives a contribution similar to the scattering on non-magnetic impurities.

Of course, in the case when many such impurities are present the above result is multiplied by the density of impurities as in Section 12.5. Only in the next leading order does the internal structure of the magnetic impurities show up. This is the shown in the following.

13.7.2 Kondo model self-energy, third order in J

The third-order contribution reads

$$\Sigma^{(3)}(\mathbf{k}\sigma\tau_1, \mathbf{k}'\sigma'\tau_2) = (-J)^3 \sum_{\substack{\mathbf{k}_1\mathbf{k}_2 \\ \sigma_1\sigma_2}} \int_0^\beta d\tau_3\, \mathcal{G}^0(\mathbf{k}_1\sigma_1, \tau_1 - \tau_3)\mathcal{G}^0(\mathbf{k}_2\sigma_2, \tau_3 - \tau_2)$$

$$\times \left\langle T_\tau \hat{S}^i(\tau_1)\hat{S}^j(\tau_3)\hat{S}^k(\tau_2)\right\rangle_0 \frac{\tau^i_{\sigma\sigma_1}}{2}\frac{\tau^j_{\sigma_1\sigma_2}}{2}\frac{\tau^k_{\sigma_2\sigma'}}{2}. \tag{13.56}$$

Contrary to the second-order term, the average involving the spin operators depends on the time arguments, because the spin operators do not commute. In fact, it is straightforward to show that $\operatorname{Tr}\left[\tau^i\tau^j\tau^k\right] = 2i\epsilon_{ijk}$. Since the self-energy only depends on the time difference $\tau_1 - \tau_2$, we set $\tau_2 = 0$ and separate the two possibilities $\tau_1 > \tau_3$ and $\tau_3 > \tau_1$:

$$\theta(\tau_1 - \tau_3)\left\langle S^iS^jS^k\right\rangle_0 + \theta(\tau_3 - \tau_1)\left\langle S^jS^iS^k\right\rangle_0 = \frac{2i}{8}\operatorname{sgn}(\tau_1 - \tau_3)\epsilon_{ijk} \tag{13.57}$$

where we used that $\epsilon_{ijk} = -\epsilon_{jik}$. Furthermore, since the unperturbed Green's functions in Eq. (13.56) are spin independent, we can perform the σ_1 and σ_2 sums, and because Eq. (13.57) forces i, j and k to be different we see that only $\sigma = \sigma'$ contributes (because $\tau^x\tau^y\tau^z$ equals the identity matrix). The sums over i, j and k hence reduce to $\sum_{ijk}\epsilon_{ijk}^2 = 6$, and thus

$$\Sigma^{(3)}(\mathbf{k}\sigma\tau_1, \mathbf{k}'\sigma'0) = -\delta_{\sigma\sigma'}\frac{3J^3}{16}\sum_{\mathbf{k}_1\mathbf{k}_2}\int_0^\beta d\tau_3\, \mathcal{G}^0(\mathbf{k}_1\sigma_1, \tau_1 - \tau_3)\mathcal{G}^0(\mathbf{k}_2\sigma_2, \tau_3)\operatorname{sgn}(\tau_1 - \tau_3).$$

$$\tag{13.58}$$

We proceed by performing the τ_3 integral, followed by a Fourier transformation. These steps are done in Exercise 13.4, and the result is

$$\Sigma^{(3)}(\mathbf{k}\sigma, \mathbf{k}'\sigma', ik_n) = \delta_{\sigma\sigma'}\frac{3J^3}{16}\sum_{\mathbf{k}_1,\mathbf{k}_2}\frac{1 - 2n_F(\xi_{\mathbf{k}_2})}{\xi_{\mathbf{k}_1} - \xi_{\mathbf{k}_2}}\frac{1}{ik_n - \xi_{\mathbf{k}_1}}. \tag{13.59}$$

Let us now calculate the imaginary part of the retarded self-energy and again assume a constant density of states, but with specified cut-offs at $\pm D$ as in Section 10.5.4. After performing one of the energy integrals we have

$$-\operatorname{Im}\Sigma^{R,(3)}(\mathbf{k}\sigma, \mathbf{k}'\sigma', \omega) = \delta_{\sigma\sigma'}\frac{3J[Jd_\sigma(\varepsilon_F)]^2\pi}{16}\int_{-D}^{D}d\xi\,\frac{1 - 2n_F(\xi)}{\omega - \xi}. \tag{13.60}$$

This is logarithmically divergent when the temperature or the frequency goes to zero. This is the famous result obtained originally by Kondo.

Collecting the second and third-order terms, one now finds in the low temperature case

$$-\mathrm{Im}\,\Sigma^R(\mathbf{k}\sigma, \mathbf{k}'\sigma', \omega) \approx \delta_{\sigma\sigma'} \frac{3J^2 d_\sigma(\varepsilon_\mathrm{F})\pi}{16}\left[1 + 2Jd_\sigma(\varepsilon_\mathrm{F})\ln\left(\frac{D}{\sqrt{\omega^2+(k_\mathrm{B}T)^2}}\right)+\cdots\right].$$
(13.61)

We have thus reached the same conclusion as in Chapter 10, that there is an enhanced scattering in third order of the exchange coupling.[46] To learn about the physics of the Kondo model, see for example Hewson (1993) or Mattuck (1976).

13.8   Summary and outlook

In this chapter, we have established the Feynman rules for writing down the Feynman diagrams constituting the infinite-order perturbation expansion of the full single-particle Green's functions. Our main example was the Coulomb interaction, but we also showed how other types of interactions can be analyzed by the diagrammatic technique. Later in Chapter 17, we derive the Feynman rule for yet another type of interactions, namely the electron-phonon interactions.

The Feynman diagram technique enables a systematic analysis of the infinitely many terms that need to be taken into account in a given calculation. Using the Feynman diagrammatic analysis one can, as we shall see in the following chapters, identify which subclasses of diagrams that give the most important contributions. We have already given explicit examples of how to evaluate some of the diagrams that are going to play an important role. Indeed, we show in Chapter 14 that the diagrams analyzed in Eqs. (13.34) and (13.37) are the ones that dominate the physics of the interacting electron gas in the high-density limit. We shall learn how these diagrams determine the groundstate energy of the system as well as its dielectric properties such as static and dynamic screening.

For other books treating the diagram technique, see for example Mahan (1990), Mattuck (1976), Fetter and Walecka (1971), Doniach and Sondheimer (1974), Rickayzen (1991), and Schrieffer (1983).

---

[46]In fact, when deriving the Kondo model from the Anderson model as in Chapter 10, one can relate the spectral function in Eq. (10.56) to the self-energy calculated here and therefore obtain Eq. (10.113) from Eq. (13.61) (see also Exercise 10.8).

14

THE INTERACTING ELECTRON GAS

In Section 2.2 we studied the Coulomb interaction as a perturbation to the non-interacting electron gas in the jellium model. This was expected to be a valid procedure in the high density limit, where according to Eq. (2.35) the interaction energy is negligible. Nevertheless, the second-order perturbation analysis of Section 2.2.2 revealed a divergence in the contribution $E_{\text{dir}}^{(2)}$ from the direct processes (see Eq. (2.49)).

In this chapter, we reanalyze the Coulomb-interacting electron gas in the jellium model using the Feynman diagram technique, and we show how a meaningful finite groundstate energy can be found. To ensure well-behaved finite integrals during our analysis we work with the Yukawa-potential with an artificial range $1/\alpha$ instead of the pure long-range Coulomb potential (see Eq. (1.102) and the associated footnote),

$$W(\mathbf{r} - \mathbf{r}') = \frac{e_0^2}{|\mathbf{r} - \mathbf{r}'|}\, e^{-\alpha|\mathbf{r} - \mathbf{r}'|}, \quad W(\mathbf{q}) = \frac{4\pi e_0^2}{q^2 + \alpha^2}. \tag{14.1}$$

The range $1/\alpha$ has no physical origin. At the end at the calculation we take the limit $\alpha \to 0$ to recover the Coulomb interaction. For example, with the Yukawa potential we can obtain a finite value for $E_{\text{dir}}^{(2)}$ in Eq. (2.49) if α is finite, but the divergence reappears as soon as we take the limit $\alpha \to 0$,

$$E_{\text{dir}}^{(2)} \propto \int_0^\infty dq\, q^2 \frac{1}{(q^2 + \alpha^2)^2} \frac{1}{q}\, q\, q \sim -\ln(\alpha) \xrightarrow[\alpha \to 0]{} \infty. \tag{14.2}$$

The main result of the following diagrammatic calculation is that the dynamics of the interacting system by itself creates a renormalization of the pure Coulomb interaction into a Yukawa-like potential independent of the value of α, which then without problems can be taken to zero. The starting point of the theory is the self-energy $\Sigma_\sigma(\mathbf{k}, ik_n)$.

14.1 The self-energy in the random phase approximation

To construct the diagrammatic expansion of the self-energy $\Sigma_\sigma(\mathbf{k}, ik_n)$ in $(\mathbf{k}, ik_n)$-space, we use the Feynman rules (13.27). In analogy with Eq. (13.21), the self-energy is given by the sum of all the irreducible diagrams in $\mathcal{G}_\sigma(\mathbf{k}, ik_n)$ removing the two external fermion lines $\mathcal{G}_\sigma^0(\mathbf{k}, ik_n)$. We recall that due to the charge-compensating background in the jellium model, the Hartree self-energy diagram $\Sigma_\sigma^H(\mathbf{k}, ik_n) = $ vanishes, as it is exactly compensated by the positive background. Thus,

$$\Sigma_\sigma(\mathbf{k}, ik_n) = \quad \text{} \quad + \cdots \tag{14.3}$$

For each order of W we want to identify the most important terms, and then resum the infinite series taking only these terms into account. This is achieved by noting that each diagram in the expansion is characterized by its density dependence through the dimensionless electron distance parameter r_s of Eq. (2.37) and its degree of divergence in the cut-off parameter α.

14.1.1   The density dependence of self-energy diagrams

Consider a typical self-energy diagram $\Sigma_\sigma^{(n)}(\mathbf{k}, ik_n)$ of order n,

$$
\Sigma_\sigma^{(n)}(\mathbf{k}, ik_n) = \quad \propto \quad \underbrace{\int d\tilde{k}_1 \cdots \int d\tilde{k}_n}_{n \text{ internal momenta}} \overbrace{W()\cdots W()}^{n \text{ interaction terms}} \underbrace{\mathcal{G}^0()\cdots\mathcal{G}^0()}_{2n-1 \text{ Green's fcts}} .
$$
(14.4)

We then make the integral dimensionless by measuring momenta and frequencies in powers of the Fermi momentum k_F and pulling out the corresponding factors of k_F. The contributions from kinematics are

$$
k \propto k_F, \qquad \varepsilon \propto k_F^2, \qquad \text{and} \qquad \frac{1}{\beta} \propto k_F^2.
$$
(14.5a)

The contribution related to phase space in the form of summation over internal momenta yields

$$
\int d\tilde{k}_1 = \frac{1}{\beta} \sum_{ik_n} \int \frac{d\mathbf{k}}{(2\pi)^3} \propto k_F^{2+3} = k_F^5,
$$
(14.5b)

and finally the contributions from dynamics are

$$
W(\mathbf{q}) \propto \frac{1}{q^2 + \alpha^2} \propto k_F^{-2},
$$
(14.5c)

and

$$
\mathcal{G}_\sigma^0(\tilde{k}) = \frac{1}{ik_n - \varepsilon_\mathbf{k}} \propto k_F^{-2}.
$$
(14.5d)

The self-energy diagram therefore has the following dependence on k_F and r_s:

$$
\Sigma_\sigma^{(n)}(\mathbf{k}, ik_n) \quad \propto \quad \left(k_F^5\right)^n \left(k_F^{-2}\right)^n \left(k_F^{-2}\right)^{2n-1} = k_F^{-(n-2)} \quad \propto \quad r_s^{n-2},
$$
(14.6)

where in the last proportionality we have used $r_s = (9\pi/4)^{\frac{1}{3}}/(a_0 k_F)$ from Eq. (2.37). We can conclude that for two different orders n and n' in the high density limit, $r_s \to 0$, we have

$$
n < n' \quad \Rightarrow \quad \left|\Sigma_\sigma^{(n)}(\mathbf{k}, ik_n)\right| \gg \left|\Sigma_\sigma^{(n')}(\mathbf{k}, ik_n)\right|, \quad \text{for } r_s \to 0.
$$
(14.7)

Eqs. (14.6) and (14.7) are the precise statements of how to identify the most important self-energy diagrams in the high density limit.

14.1.2 The divergence number of self-energy diagrams

The singular nature of the Yukawa-modified Coulomb potential Eq. (14.1) in the limit of small q and α leads to a divergent behavior of the self-energy integrals. The more interaction lines carrying the same momentum there are in a given diagram, the more divergent is this diagram. For example (taking $\alpha = 0$), two lines with the momentum $\mathbf{q}$ contributes with $W(\mathbf{q})^2$ which diverges as q^{-4} for $q \to 0$ independent of the behavior of any other internal momentum $\mathbf{p}$ in the diagram. In contrast, two lines with different momenta $\mathbf{q}$ and $\mathbf{q} - \mathbf{p}$ contributes with $W(\mathbf{q})W(\mathbf{q} - \mathbf{p})$, which diverges as q^{-4} only when both $q \to 0$ and $p \to 0$ at the same time, i.e., in a set of measure zero in the integral over $\mathbf{q}$ and $\mathbf{p}$.

In view of this discussion it is natural to define a divergence number $\delta_\sigma^{(n)}$ of the self-energy diagram $\Sigma_\sigma^{(n)}(\mathbf{k}, ik_n)$ as

$$\delta_\sigma^{(n)} \equiv \begin{cases} \text{the largest number of interaction lines in} \\ \Sigma_\sigma^{(n)}(\mathbf{k}, ik_n) \text{ having the same momentum } \mathbf{q}. \end{cases} \tag{14.8}$$

Consider two diagrams $\Sigma_\sigma^{(n,1)}$ and $\Sigma_\sigma^{(n,2)}$ of the same order n. With one notable exception, it is in general not possible to determine which diagram is the larger based alone on knowledge of the divergence number. The exception involves the diagram with the maximal divergence number, i.e., when all n momenta in the diagram are the same. In the limit $\alpha \to 0$, this diagram is the largest:

$$\delta_\sigma^{(n,1)} = n \quad \Rightarrow \quad \left|\Sigma_\sigma^{(n,1)}(\mathbf{k}, ik_n)\right| \gg \left|\Sigma_\sigma^{(n,2)}(\mathbf{k}, ik_n)\right|, \\ \begin{cases} \text{for } \alpha \to 0 \text{ and} \\ \text{any } n\text{-order diagram } \Sigma^{(n,2)}. \end{cases} \tag{14.9}$$

14.1.3 RPA resummation of the self-energy

Using the order n and the divergence number δ, we now order the self-energy diagrams in a (n, δ)-table. According to Eqs. (14.7) and (14.9), the most important terms are those in the diagonal in this table where $\delta = n$. Examples of some of the entries in the (n,d)-table are the following diagrams with $1 \leq \delta \leq n = 4$, shown in Eq. (14.10) without arrows on the interaction lines for graphical clarity:

$\Sigma_\sigma(\tilde{k})$	$n=1$	$n=2$	$n=3$	$n=4$
$\delta = 1$				
$\delta = 2$	–			
$\delta = 3$	–	–		
$\delta = 4$	–	–	–	

$$(14.10)$$

It is clear that the most important diagrams in the high-density limit are those having a low order. For each given order, the diagrams with the highest divergence number are the most important. The self-energy in the random phase approximation (RPA) is an infinite sum containing diagrams of all orders n, but only the most divergent one for each n,

$$\Sigma_\sigma^{\text{RPA}}(\tilde{k}) \equiv \quad + \quad + \quad + \quad + \cdots$$

$$(14.11)$$

Below we are going to analyze parts of the diagrams individually. This is straightforward to do, since the Feynman rules Eq. (13.27) are still valid for each part. An important part of the self-energy diagrams in Eq. (14.11) is clearly the pair-bubble $\Pi^0(\mathbf{q}, iq_n) \equiv$ already introduced in Section 13.5.3. It plays a crucial role, because it ensures that all interaction lines $W(\mathbf{q})$ carry the same momentum $\mathbf{q}$. To make the fermion-loop sign from the pair-bubble appear explicitly, we prefer to work with $\chi_0 \equiv -\Pi^0$, i.e.,

$$\equiv -\chi_0(\mathbf{q}, iq_n), \qquad \chi_0(\mathbf{q}, iq_n) = \frac{2}{\beta} \sum_{ip_n} \int \frac{d\mathbf{p}}{(2\pi)^3} \frac{1}{(ip_n + iq_n - \xi_{\mathbf{p}+\mathbf{q}})} \frac{1}{(ip_n - \xi_{\mathbf{p}})}.$$

$$(14.12)$$

In fact, this χ_0 is the same correlation function as the one introduced for other reasons in Section 11.7.

By introducing a renormalized interaction line $-W^{\mathrm{RPA}}(\tilde{q}) = $ we can, omitting interaction line arrows, rewrite the RPA self-energy as

$$\tag{14.13}$$

or, by pulling out the convergent Fock self-energy , as

$$\tag{14.14}$$

In the following we study the properties of the renormalized Coulomb interaction $W^{\mathrm{RPA}}(\tilde{q})$.

14.2 The renormalized Coulomb interaction in RPA

In Eqs. (14.13) and (14.14) we introduced the renormalized Coulomb interaction $W^{\mathrm{RPA}}(\mathbf{q}, iq_n)$. It can be found using a Dyson equation approach as follows:

$$-W^{\mathrm{RPA}}(\mathbf{q}, iq_n) \equiv$$

$$\tag{14.15}$$

In $(\mathbf{q}, iq_n)$-space this is an algebraic equation with the solution

$$-W^{\mathrm{RPA}}(\mathbf{q}, iq_n) = \text{} = \frac{-W(\mathbf{q})}{1 - W(\mathbf{q})\,\chi_0(\mathbf{q}, iq_n)}. \tag{14.16}$$

Note the cancellation of the explicit signs from W and χ_0 in the denominator. We can now insert the specific form Eq. (14.1) for the Yukawa-modified Coulomb interaction, and let the artificial cut-off parameter α tend to zero. The final result is

$$W^{\text{RPA}}(\mathbf{q}, iq_n) \xrightarrow[\alpha \to 0]{} \frac{4\pi e_0^2}{q^2 - 4\pi e_0^2 \, \chi_0(\mathbf{q}, iq_n)}. \tag{14.17}$$

$W^{\text{RPA}}(\mathbf{q}, iq_n)$ thus has a form similar to $W(\mathbf{q})$, but with the important difference that the artificially introduced parameter α in the latter has been replaced with the pair-bubble function $-4\pi e_0^2 \chi_0(\mathbf{q}, iq_n)$ having its origin in the dynamics of the interacting electron gas. Note that the pair-bubble is a function of both momentum and frequency. From now on, we no longer need a finite value of α, and it is put to zero in the following.

In the static, long-wave limit, $\mathbf{q} \to 0$ and $iq_n = 0 + i\eta$, we find that W^{RPA} appears in a form identical with the Yukawa-modified Coulomb interaction, i.e., a screened Coulomb interaction

$$W^{\text{RPA}}(\mathbf{q}, 0) \xrightarrow[q \to 0]{} \frac{4\pi e_0^2}{q^2 + k_s^2}, \tag{14.18}$$

where the so-called Thomas–Fermi screening wavenumber k_s has been introduced,

$$k_s^2 \equiv -4\pi e_0^2 \, \chi_0(0, 0). \tag{14.19}$$

In the extreme long wave limit we have

$$W^{\text{RPA}}(0, 0) = \frac{-1}{\chi_0(0, 0)}. \tag{14.20}$$

In the following section we calculate the pair-bubble $\chi_0(\mathbf{q}, iq_n)$, find the value of the Thomas–Fermi screening wavenumber k_s, and discuss a physical interpretation of the random phase approximation.

14.2.1   Calculation of the pair-bubble

In Eq. (13.37) the pair-bubble diagram is given in terms of a $\mathbf{p}$-integral and a Matsubara frequency sum. The sum is carried out in Eq. (11.88) using the recipe Eq. (11.57),

$$\chi_0(\mathbf{q}, iq_n) = 2 \int \frac{d\mathbf{p}}{(2\pi)^3} \frac{n_F(\xi_\mathbf{p}) - n_F(\xi_{\mathbf{p}+\mathbf{q}})}{iq_n + \xi_\mathbf{p} - \xi_{\mathbf{p}+\mathbf{q}}}. \tag{14.21}$$

The frequency dependence of the retarded pair-bubble χ_0^R can now be found by the usual analytical continuation $iq_n \to \omega + i\eta$. We still have to perform the rather involved $\mathbf{p}$-integral. However, it is a simple matter to obtain the static, long-wave limit $q \to 0$ and $iq_n = 0$, and thus determine $\chi_0^R(\mathbf{q}, 0)$. In this limiting case $\xi_{\mathbf{p}+\mathbf{q}} \to \xi_\mathbf{p}$, and we can perform a Taylor expansion in energy,

$$\chi_0^R(\mathbf{q}, 0) \xrightarrow[q \to 0]{} 2 \int \frac{d\mathbf{p}}{(2\pi)^3} \frac{(\xi_{\mathbf{p}+\mathbf{q}} - \xi_\mathbf{p}) \frac{\partial n_F}{\partial \xi_\mathbf{p}}}{\xi_{\mathbf{p}+\mathbf{q}} - \xi_\mathbf{p}} = -\int d\xi_\mathbf{p} \, d(\mu + \xi_\mathbf{p}) \left[-\frac{\partial n_F}{\partial \xi_\mathbf{p}} \right]$$

$$\simeq -d(\varepsilon_F), \quad \text{for } k_B T \ll \varepsilon_F. \tag{14.22}$$

In the static, long-wave limit at low temperatures, $\chi_0^R(\mathbf{q}, 0)$ is simply minus the density of states at the Fermi level, and consequently, according to Eq. (14.20), we find that $W^{\mathrm{RPA}}(q \to 0, 0)$ becomes

$$W^{\mathrm{RPA}}(q \to 0, 0) = \frac{1}{d(\varepsilon_{\mathrm{F}})}. \tag{14.23}$$

The Thomas–Fermi screening wavenumber k_s is found by combining Eq. (14.19) with Eqs. (2.31) and (2.36),

$$k_s^2 = -4\pi e_0^2 \, \chi_0^R(0, 0) = 4\pi \, e_0^2 \, d(\varepsilon_{\mathrm{F}}) = \frac{4}{\pi} \frac{k_{\mathrm{F}}}{a_0}, \tag{14.24}$$

a_0 being the Bohr radius. This result is very important, because it relates the screening length $1/k_s$ to microscopic parameters of the electron gas. It is therefore useful for numerous applications. For metals $k_s \approx 0.1 \text{ nm}^{-1}$.

We now turn to the more general case, but limit the calculation of $\chi_0(\mathbf{q}, \omega + i\eta)$ to the low temperature regime $k_{\mathrm{B}}T \ll \varepsilon_{\mathrm{F}}$. Finite temperature effects can be obtained by using the Sommerfeld expansion or by numerical integration. In the low temperature limit, an analytical expression is obtained by a straightforward but rather tedious calculation. In the $\mathbf{p}$-integral the only angular dependence of the integrand is through $\cos\theta$, and we have

$$\lambda \equiv \cos\theta, \quad \int \frac{d\mathbf{p}}{(2\pi)^3} = \int_0^\infty \frac{dp}{4\pi^2} p^2 \int_{-1}^1 d\lambda, \quad \xi_{\mathbf{p}-\mathbf{q}} - \xi_{\mathbf{p}} = \frac{1}{2m}(q^2 - 2pq\lambda). \tag{14.25}$$

In the low temperature limit the Fermi–Dirac distribution is a step-function, and the real part of χ_0 is most easily calculated by splitting Eq. (14.21) in two terms, substituting $\mathbf{p}$ by $\mathbf{p} - \mathbf{q}$ in the first term, and collecting the terms again,

$$\mathrm{Re}\,\chi_0(\mathbf{q}, \omega + i\eta) = -\mathcal{P}\int_0^{k_{\mathrm{F}}} \frac{dp}{2\pi^2} p^2 \int_{-1}^1 d\lambda \left[\frac{1}{\frac{1}{2m}(q^2 - 2pq\lambda) + \omega} + \frac{1}{\frac{1}{2m}(q^2 + 2pq\lambda) - \omega} \right]. \tag{14.26}$$

The integrand is now made dimensionless by measuring all momenta in units of k_{F} and all frequencies and energies in units of ε_{F}, such as

$$x \equiv \frac{q}{2k_{\mathrm{F}}} \quad \text{and} \quad x_0 \equiv \frac{\omega}{4\varepsilon_{\mathrm{F}}}. \tag{14.27}$$

Then the λ-integral, followed by the p-integral, is carried out using standard logarithmic integrals.[47] The final result for the retarded function χ_0^R is

$$\mathrm{Re}\,\chi_0^R(\mathbf{q}, \omega) = -2\,d(\varepsilon_{\mathrm{F}}) \left(\frac{1}{2} + \frac{f(x, x_0) + f(x, -x_0)}{8x} \right), \tag{14.28a}$$

where

[47] Useful integrals are $\int dx \frac{1}{ax+b} = \frac{1}{a}\ln(ax+b)$ and $\int dx \ln(ax+b) = \frac{1}{a}[(ax+b)\ln(ax+b) - ax]$.

$$f(x, x_0) \equiv \left[1 - \left(\frac{x_0}{x} - x \right)^2 \right] \ln \left| \frac{x + x^2 - x_0}{x - x^2 + x_0} \right|. \tag{14.28b}$$

The imaginary part of χ_0 in Eq. (14.21) is

$$\mathrm{Im}\, \chi_0(\mathbf{q}, \omega + i\eta) = \int_0^{k_F} \frac{dp}{2\pi} p^2 \int_{-1}^1 d\lambda \, [n_F(\xi_{\mathbf{p+q}}) - n_F(\xi_{\mathbf{p}})] \, \delta(\xi_{\mathbf{p+q}} - \xi_{\mathbf{p}} - \omega). \tag{14.29}$$

Using $\delta(f[x]) = \sum_{x_0} \delta(x - x_0)/|f'[x_0]|$, where x_0 are the zeros of $f[x]$, the λ-integral can be performed. A careful analysis of when the delta function and the theta functions are non-zero leads to

$$\mathrm{Im}\, \chi_0^R(\mathbf{q}, \omega) = -d(\varepsilon_F) \begin{cases} \frac{\pi}{8x} \left[1 - \left(\frac{x_0}{x} - x \right)^2 \right], & \text{for } |x - x^2| < x_0 < x + x^2, \\ \frac{\pi}{2} \frac{x_0}{x}, & \text{for } 0 < x_0 < x - x^2, \\ 0, & \text{for other } x_0 \geq 0. \end{cases} \tag{14.30}$$

14.2.2   The electron-hole pair interpretation of RPA

We have learned above that the RPA results in a screened Coulomb interaction. To gain some physical insight into the nature of this renormalization, we study the pair-bubble diagram a little closer in the $(\mathbf{q}, \tau)$-representation. Choosing $\tau > 0$ in Eq. (11.86), we arrive at

$$\chi_0(\mathbf{q}, \tau > 0) = -\sum_{\sigma} \int \frac{d\mathbf{p}}{(2\pi)^3} \, \langle c_{\mathbf{p}\sigma}(\tau) c_{\mathbf{p}\sigma}^\dagger \rangle_0 \, \langle c_{\mathbf{p+q}\sigma}^\dagger(\tau) c_{\mathbf{p+q}\sigma} \rangle_0. \tag{14.31}$$

Consequently, we can interpret $\chi_0(\mathbf{q}, \tau > 0)$ as the sum of all processes of the following type: at $\tau = 0$ an electron is created in the state $|\mathbf{p}\sigma\rangle$ and a hole in the state $|\mathbf{p+q}\sigma\rangle$, which correspond to an electron jumping from the latter state to the former. At the later time τ the process is reversed, and the electron falls back into the hole state. In the time interval from 0 to τ an electron-hole pair is thus present, but this corresponds to a polarization of the electron gas, and we now see the origin of the renormalization of the Coulomb interaction. The RPA scheme takes interaction processes into account thus changing the dielectric properties of the non-interacting electron gas. The imaginary part of $\chi_0^R(\mathbf{q}, \omega)$, describes the corresponding dissipative processes, where momentum $\mathbf{q}$ and energy ω is absorbed by the electron gas (see also the discussion in Section 8.5).

In the remaining sections of the chapter we study how the effective RPA interaction influences the groundstate energy and the dielectric properties (in linear response) of the electron gas.

14.3   The groundstate energy of the electron gas

We first show how to express the groundstate energy in terms of the single-particle Green's functions $\mathcal{G}(\mathbf{k}, ik_n)$. That this is at all possible is perhaps surprising due to the presence of the two-particle Coulomb interaction. But using the equation of motion technique combined with an "integration over the coupling constant" method, we obtain the result.

Let λ be a real number $0 \leq \lambda \leq 1$, and define

$$H_\lambda \equiv H_0 - \mu N + \lambda W, \tag{14.32}$$

where H_0 is the kinetic energy and W the Coulomb interaction Eq. (2.34). For $\lambda = 0$ we have the non-interacting electron gas while for $\lambda = 1$ we retrieve the full Coulomb-interacting electron gas. According to the definition in Eq. (1.121b), the thermodynamic potential $\Omega = U - TS - \mu N$ is given by

$$\Omega(\lambda) = -\frac{1}{\beta} \ln \mathrm{Tr} \left[e^{-\beta(H_0 - \mu N + \lambda W)} \right]. \tag{14.33}$$

By differentiating with respect to λ, we find

$$\frac{\partial \Omega}{\partial \lambda} = -\frac{1}{\beta} \frac{\mathrm{Tr} \left[-\beta W \, e^{-\beta(H_0 - \mu N + \lambda W)} \right]}{\mathrm{Tr} \left[e^{-\beta(H_0 - \mu N + \lambda W)} \right]} = \langle W \rangle_\lambda. \tag{14.34}$$

By integration over λ from 0 to 1, the change in Ω due to the interactions is found,

$$\Omega(1) - \Omega(0) = \int_0^1 \frac{d\lambda}{\lambda} \langle \lambda W \rangle_\lambda. \tag{14.35}$$

The subscript λ refers to averaging with respect to H_λ. At $T = 0$ the thermodynamical potential reduces to the groundstate energy, which thus can be calculated as

$$E = E_0 + \lim_{T \to 0} \int_0^1 \frac{d\lambda}{\lambda} \langle \lambda W \rangle_\lambda, \tag{14.36}$$

where E_0 is the groundstate energy without interactions. The expectation value $\langle \lambda W \rangle_\lambda$ can be related to $\mathcal{G}_\sigma^\lambda(\mathbf{k}, ik_n)$ through the equation of motion for $\mathcal{G}_\sigma^\lambda(\mathbf{k}, \tau)$ using Eqs. (5.31) and (11.64b)

$$- \partial_\tau \mathcal{G}_\sigma^\lambda(\mathbf{k}, \tau) = \delta(\tau) + \left\langle T_\tau [H_\lambda, c_{\mathbf{k}\sigma}](\tau) \, c_{\mathbf{k}\sigma}^\dagger \right\rangle_\lambda$$

$$= \delta(\tau) + \left[\xi_{\mathbf{k}} \mathcal{G}_\sigma^\lambda(\mathbf{k}, \tau) - 2 \sum_{\mathbf{k}'\sigma'\mathbf{q}} \frac{\lambda}{2} W(q) \langle T_\tau c_{\mathbf{k}'\sigma'}^\dagger(\tau) \, c_{\mathbf{k}'+\mathbf{q}\sigma'}(\tau) \, c_{\mathbf{k}-\mathbf{q}\sigma}(\tau) \, c_{\mathbf{k}\sigma}^\dagger \rangle_\lambda \right]. \tag{14.37}$$

We now let $\tau = 0^- = -\eta$, and note that the last term is nothing but the interaction part $\langle \lambda W \rangle_\lambda$ of the Hamiltonian. Furthermore, using Fourier transforms we can at $\tau = -\eta$ write $\mathcal{G}_\sigma^\lambda(\mathbf{k}, -\eta) = \frac{1}{\beta} \sum_{ik_n} \mathcal{G}_\sigma^\lambda(\mathbf{k}, ik_n) \, e^{ik_n\eta}$ and $\delta(-\eta) = \frac{1}{\beta} \sum_{ik_n} e^{ik_n\eta}$. We therefore arrive at the following compact expression,

$$\frac{1}{\beta \mathcal{V}} \sum_{ik_n \mathbf{k}\sigma} (ik_n - \xi_{\mathbf{k}}) \, \mathcal{G}_\sigma^\lambda(\mathbf{k}, ik_n) \, e^{ik_n\eta} = \frac{1}{\beta \mathcal{V}} \sum_{ik_n \mathbf{k}\sigma} e^{ik_n\eta} + 2\langle \lambda W \rangle_\lambda. \tag{14.38}$$

Collecting the sums on the left-hand side yields

$$\frac{1}{\beta \mathcal{V}} \sum_{ik_n \mathbf{k}\sigma} \left[(ik_n - \xi_{\mathbf{k}})\mathcal{G}_\sigma^\lambda(\mathbf{k}, ik_n) - 1 \right] e^{ik_n\eta} = 2\langle \lambda W \rangle_\lambda. \tag{14.39}$$

We now utilize that $1 = [\mathcal{G}_\sigma^\lambda]^{-1}\mathcal{G}_\sigma^\lambda$ and furthermore that $[\mathcal{G}_\sigma^\lambda]^{-1} = ik_n - \varepsilon_{\mathbf{k}} - \Sigma_\sigma^\lambda$ to obtain

$$\langle \lambda W \rangle_\lambda = \frac{1}{2\beta \mathcal{V}} \sum_{ik_n} \sum_{\mathbf{k}\sigma} \Sigma_\sigma^\lambda(\mathbf{k}, ik_n) \, \mathcal{G}_\sigma^\lambda(\mathbf{k}, ik_n) \, e^{ik_n\eta}, \tag{14.40}$$

and when this is inserted in Eq. (14.36) we finally arrive at the expression for the groundstate energy

$$E = E^0 + \lim_{T \to 0} \frac{1}{2\beta \mathcal{V}} \sum_{ik_n} \sum_{\mathbf{k}\sigma} \int_0^1 \frac{d\lambda}{\lambda} \Sigma_\sigma^\lambda(\mathbf{k}, ik_n) \, \mathcal{G}_\sigma^\lambda(\mathbf{k}, ik_n) \, e^{ik_n\eta}. \tag{14.41}$$

This expression allows for a diagrammatic calculation with the additional Feynman rule that $\lim_{T\to 0} \int_0^1 \frac{d\lambda}{\lambda}$ must be performed at the end of the calculation. Moreover, it is a remarkable result, because it relates the groundstate energy of the interacting system to the single-particle Green's function and the related self-energy.

To improve the high-density, second-order perturbation theory of Section 2.2, we include in Eq. (14.41) all diagrams up to second order and, through RPA, the most divergent diagram of each of the higher orders. Since the self-energy Σ contains diagrams from first order and up, we do not have to expand the Green's function $\mathcal{G}$ beyond first order:

$$\mathcal{G}_\sigma^\lambda(\mathbf{k}, ik_n) \approx \text{(diagrams)} \tag{14.43}$$

Note that only the second diagram in Eq. (14.42) needs to be renormalized. This is because only this diagram is divergent without renormalization. Combining Eq. (14.41) with Eqs. (14.43) and (14.42) we obtain to (renormalized) second order:[48]

Note the similarity between the three diagrams in this expression for $E - E^0$ and the ones depicted in Figs. 2.6 and 2.8 . We will not go through the calculation of these

[48]Note that the last diagram in the first line of Eq. (14.44) is zero for $T = 0$.

diagrams. The techniques are similar to those employed in the calculation of the pair-bubble diagram in Section 14.2.1. The RPA renormalization of the interaction line in the second diagram in Eq. (14.44) renders the diagram finite. Since the Thomas–Fermi wavenumber k_s replaced α as a cut-off, we know from Eq. (14.2) that this diagram must be proportional to $\log k_s$ and hence to $\log r_s$. We are now in a position to continue the expansion Eq. (2.43) of E/N in terms of the dimensionless distance parameter r_s,

$$\frac{E}{N} \xrightarrow[r_s \to 0]{} \left(\frac{2.211}{r_s^2} - \frac{0.916}{r_s} + 0.0622 \log r_s - 0.094 \right) \text{Ry.} \tag{14.45}$$

This expression ends the discussion of the groundstate energy of the interacting electron gas in the jellium model. By employing the powerful quantum field theoretic method, in casu resummation of the Feynman diagram series for the single-electron self-energy and Green's function, we could finally solve the problem posed by the failed second-order perturbation theory.

Having achieved this solution, we will also be able to study other aspects of the interacting electron gas. In the following we focus on the dielectric properties of the system.

14.4 The dielectric function and screening

Already from Eq. (14.16) it is clear that the internal dynamics of the interacting electron gas lead to a screening of the pure Coulomb interaction. One suspects that also external potentials ϕ_{ext} will be screened similarly; and indeed, as we shall see below, this is in fact the case. As in Section 6.4, we study the linear response of the interacting system due to the perturbation H' caused by ϕ_{ext},

$$H' = \int d\mathbf{r} \; [-e\, \rho(\mathbf{r})] \; \phi_{\text{ext}}(\mathbf{r}, t), \tag{14.46}$$

where $\rho(\mathbf{r})$ is the particle density and not, as in Section 6.4, the charge density. Since the unperturbed system even with its Coulomb-interacting electrons is translation invariant, we write all expressions in Fourier $(\mathbf{q}, \omega)$-space. The external potential $\phi_{\text{ext}}(\mathbf{q}, \omega)$ creates an induced charge density $-e\rho_{\text{ind}}(\mathbf{q}, \omega)$. Through the Coulomb interaction this in turn corresponds to an induced potential

$$\phi_{\text{ind}}(\mathbf{r}, t) = \int d\mathbf{r}' \frac{-e\rho_{\text{ind}}(\mathbf{r}', t)}{4\pi\epsilon_0 |\mathbf{r} - \mathbf{r}'|} \quad \Rightarrow \quad \phi_{\text{ind}}(\mathbf{q}, \omega) = \frac{1}{e^2} W(\mathbf{q}) \left[-e\, \rho_{\text{ind}}(\mathbf{q}, \omega) \right]. \tag{14.47}$$

We divide by e^2, since $W(\mathbf{q})$ by definition contains this factor. Next step is to use the Kubo formula, which relates $[-e\, \rho_{\text{ind}}(\mathbf{r}, \omega)]$ with the external potential and with the retarded density-density correlator,

$$[-e\, \rho_{\text{ind}}(\mathbf{q}, \omega)] = (-e)^2 \, C^R_{\rho\rho}(\mathbf{q}, -\mathbf{q}, \omega) \, \phi_{\text{ext}}(\mathbf{q}, \omega) \equiv e^2 \, \chi^R(\mathbf{q}, \omega) \, \phi_{\text{ext}}(\mathbf{q}, \omega). \tag{14.48}$$

Collecting our partial results, we have

$$\phi_{\text{ind}}(\mathbf{q}, \omega) = W(\mathbf{q}) \, \chi^R(\mathbf{q}, \omega) \, \phi_{\text{ext}}(\mathbf{q}, \omega), \tag{14.49}$$

where $\chi^R(\mathbf{q}, \omega)$ is the Fourier transform of the retarded Kubo density-density correlation function $\chi^R(\mathbf{q}, t - t')$, see Eqs. (8.80) and (8.81),

$$\chi^R(\mathbf{q}, t - t') \equiv C_{\rho\rho}^R(\mathbf{q}t, -\mathbf{q}t') = -i\theta(t - t')\frac{1}{\mathcal{V}}\left\langle\left[\rho(\mathbf{q}t), \rho(-\mathbf{q}t')\right]\right\rangle_{\text{eq}}. \qquad (14.50)$$

Here the subscript "eq" refers to averaging in equilibrium, i.e., with respect to $H = H_0 + W$ omitting H'. Using Eq. (14.49) the total potential $\phi_{\text{tot}}(\mathbf{q}, \omega)$ can be written in terms of the polarization function χ^R,

$$\phi_{\text{tot}}(\mathbf{q}, \omega) = \phi_{\text{ext}}(\mathbf{q}, \omega) + \phi_{\text{ind}}(\mathbf{q}, \omega) = \left[1 + W(\mathbf{q})\,\chi^R(\mathbf{q}, \omega)\right]\phi_{\text{ext}}(\mathbf{q}, \omega). \qquad (14.51)$$

When recalling that ϕ_{tot} corresponds to the electric field $\mathbf{E}$, and ϕ_{ext} to the displacement field $\mathbf{D} = \epsilon_0\varepsilon\mathbf{E}$, we see that the following expression for the dielectric function or electrical permittivity ε has been derived:

$$\frac{1}{\varepsilon(\mathbf{q}, \omega)} = 1 + W(\mathbf{q})\,\chi^R(\mathbf{q}, \omega). \qquad (14.52)$$

So upon calculating $\chi^R(\mathbf{q}, \omega)$, we can determine $\varepsilon(\mathbf{q}, \omega)$. But according to Eq. (11.33) and the specific calculation in Section 11.7, we can obtain $\chi^R(\mathbf{q}, \omega)$ by analytic continuation of the corresponding Matsubara Green's function χ^R,

$$\chi^R(\mathbf{q}, \omega) = \chi(\mathbf{q}, iq_n \to \omega + i\eta), \qquad (14.53)$$

where $\chi(\mathbf{q}, iq_n)$ is the imaginary-time Fourier transform of $\chi(\mathbf{q}, \tau)$ given by Eq. (11.84),

$$\chi(\mathbf{q}, \tau) = -\frac{1}{\mathcal{V}}\left\langle T_\tau\,\rho(\mathbf{q}, \tau)\,\rho(-\mathbf{q}, 0)\right\rangle_{\text{eq}}. \qquad (14.54)$$

We will calculate the latter Green's function using the Feynman diagram technique. From Eq. (1.94) we can read off the Fourier transform $\rho(\pm\mathbf{q})$:

$$\rho(\mathbf{q}) = \sum_{\mathbf{p}\sigma'}c_{\mathbf{p}\sigma'}^{\dagger}c_{\mathbf{p}+\mathbf{q}\sigma'}, \qquad \rho(-\mathbf{q}) = \sum_{\mathbf{k}\sigma}c_{\mathbf{k}+\mathbf{q}\sigma}^{\dagger}c_{\mathbf{k}\sigma}. \qquad (14.55)$$

Hence, $\chi(\mathbf{q}, \tau)$ is seen to be a two-particle Green's function of the form

$$\chi(\mathbf{q}, \tau) = -\frac{1}{\mathcal{V}}\left\langle T_\tau\sum_{\mathbf{p}\sigma'\mathbf{k}\sigma}c_{\mathbf{p}\sigma'}^{\dagger}(\tau)c_{\mathbf{p}+\mathbf{q}\sigma'}(\tau)\,c_{\mathbf{k}+\mathbf{q}\sigma}^{\dagger}c_{\mathbf{k}\sigma}\right\rangle_{\text{eq}}$$

$$= \frac{1}{\mathcal{V}}\left[-\left\langle\rho_{q=0}\right\rangle_{\text{eq}}\left\langle\rho_{q=0}\right\rangle_{\text{eq}} + \sum_{\mathbf{p}\sigma'\mathbf{k}\sigma}\left\langle T_\tau\,c_{\mathbf{p}+\mathbf{q}\sigma'}(\tau)c_{\mathbf{k}\sigma}c_{\mathbf{p}\sigma'}^{\dagger}(\tau + \eta)c_{\mathbf{k}+\mathbf{q}\sigma}^{\dagger}(\eta)\right\rangle_{\text{eq}}^{\text{con}}\right].$$

$$(14.56)$$

Here, as in Eq. (13.9), $\eta = 0^+$ has been inserted to ensure correct ordering, and we have divided the contributions to χ into two parts. One part where the two density operators are disconnected from one another, and one "connected" part where they mix. The disconnected part is zero since the expectation of the charge density in the neutralized and homogeneous jellium model is zero. The second term has a structure similar to the simple pair-bubble diagram with an external momentum $\mathbf{q}$ flowing through it.

It is now possible to apply the Feynman rules Eq. (13.27) directly and to write the diagrammatic expansion in $(\mathbf{q}, iq_n)$-space of $\chi(\mathbf{q}, iq_n) = \chi(\tilde{q})$. We only have to pay special attention to rule (4), where for the single-particle Green's function, it is stated that the diagrams must contain two Green's functions with the external momentum $\mathbf{k}$. This rule was a direct consequence of the definition of $\mathcal{G}(\mathbf{k}, \tau)$,

$$\mathcal{G}(\mathbf{k}, \tau) = -\langle T_\tau\, c_{\mathbf{k}\sigma}(\tau)\, c_{\mathbf{k}\sigma}^\dagger \rangle \quad \rightarrow \quad \tag{14.57}$$

Likewise for $\chi(\mathbf{q}, \tau)$, except this is a two-particle Green's function with two operators at each of the external vertices instead of just one. One straightforwardly obtains the following vertices corresponding to $\rho(\mathbf{q})$ and $\rho(-\mathbf{q})$,

$$\chi(\mathbf{q}, \tau) \sim \langle\, T_\tau\, c_{\mathbf{p}\sigma'}^\dagger(\tau) c_{\mathbf{p}+\mathbf{q}\sigma'}(\tau)\, c_{\mathbf{k}\sigma} c_{\mathbf{k}+\mathbf{q}\sigma}^\dagger \rangle \quad \rightarrow \tag{14.58}$$

The initial (right) vertex absorbs an external four-momentum $\tilde{q}$, while the final (left) vertex reemits $\tilde{q}$. We must then have that $\chi(\tilde{q})$ is the sum of all possible diagrams that connect the two ρ-vertices and that involve any number of Coulomb interaction lines.

$$-\chi(\tilde{q}) \equiv \tag{14.59}$$

$$+ \cdots,$$

where we have shown some repesentative diagrams, but certainly not all.

In analogy with the self-energy diagrams in Section 13.3, we define the irreducible diagrams in the χ-sum as the ones that cannot be cut into two pieces by cutting any single interaction line and obtain

$$-\chi^{\text{irr}}(\tilde{q}) \equiv \boxed{\text{the sum of all irreducible diagrams in } -\chi(\tilde{q})}$$

$$\equiv \qquad (14.60)$$

Hence we can resum $\chi(\tilde{q})$ in terms of $\chi^{\text{irr}}(\tilde{q})$ and obtain a Dyson equation for it,

$$= \quad -\chi^{\text{irr}}(\tilde{q}) - \chi^{\text{irr}}(\tilde{q})\, W(\tilde{q})\, \chi(\tilde{q})\,,$$
$$(14.61)$$

with the solution

$$-\chi(\tilde{q}) = \quad = \quad = \quad \frac{-\chi^{\text{irr}}(\tilde{q})}{1 - W(\tilde{q})\,\chi^{\text{irr}}(\tilde{q})}\,. \qquad (14.62)$$

With this result for $\chi(\mathbf{q}, iq_n)$ we can determine the dielectric function,

$$\frac{1}{\varepsilon(\mathbf{q}, iq_n)} = 1 + W(\mathbf{q})\,\frac{\chi^{\text{irr}}(\mathbf{q}, iq_n)}{1 - W(\mathbf{q})\,\chi^{\text{irr}}(\mathbf{q}, iq_n)} = \frac{1}{1 - W(\mathbf{q})\,\chi^{\text{irr}}(\mathbf{q}, iq_n)}, \qquad (14.63)$$

or more directly

$$\varepsilon(\mathbf{q}, iq_n) = 1 - W(\mathbf{q})\,\chi^{\text{irr}}(\mathbf{q}, iq_n) = 1 - \frac{e^2}{\epsilon_0 q^2}\,\chi^{\text{irr}}(\mathbf{q}, iq_n). \qquad (14.64)$$

Note it is e^2 and not e_0^2 that appears in the last expression. In RPA, $\chi^{\mathrm{irr}}(\mathbf{q}, iq_n)$ is approximated by the simple pair-bubble,

$$-\chi^{\mathrm{irr}}(\mathbf{q}, iq_n) = \boxed{\ } \quad \longrightarrow \quad -\chi^{\mathrm{irr}}_{RPA}(\mathbf{q}, iq_n) = \boxed{\ } = -\chi_0(\mathbf{q}, iq_n),$$

(14.65)

and the full correlation function $\chi(\mathbf{q}, iq_n)$ is approximated by $\chi^{\mathrm{RPA}}(\mathbf{q}, iq_n)$,

$$-\chi^{\mathrm{RPA}}(\mathbf{q}, iq_n) = \boxed{\ } = \frac{\bigcirc}{1 - \bigcirc\!\!\sim} = \frac{-\chi_0(\mathbf{q}, iq_n)}{1 - W(\tilde{q})\,\chi_0(\mathbf{q}, iq_n)}. \quad (14.66)$$

This results in the RPA dielectric function $\varepsilon^{\mathrm{RPA}}(\mathbf{q}, iq_n)$,

$$\varepsilon^{\mathrm{RPA}}(\mathbf{q}, iq_n) \;=\; 1 - W(\mathbf{q})\,\chi_0(\mathbf{q}, iq_n) \;=\; 1 - \frac{e^2}{\epsilon_0 q^2}\,\chi_0(\mathbf{q}, iq_n). \quad (14.67)$$

The entire analysis presented in this section leads to the conclusion that the external potentials treated in linear response theory are renormalized (or screened) in exactly the same way as the internal Coulomb interactions of the previous section,

$$\phi_{\mathrm{tot}}^{\mathrm{RPA}}(\mathbf{q}, iq_n) \;=\; \frac{1}{\varepsilon^{\mathrm{RPA}}(\mathbf{q}, iq_n)}\,\phi_{\mathrm{ext}}(\mathbf{q}, iq_n) \;=\; \frac{\phi_{\mathrm{ext}}(\mathbf{q}, iq_n)}{1 - \frac{e^2}{\epsilon_0 q^2}\,\chi_0(\mathbf{q}, iq_n)}. \quad (14.68)$$

This conclusion can be summarized in the following two diagrammatic expansions. One is the internal electron-electron interaction potential represented by the screened Coulomb-interaction line W^{RPA}. The other is the external impurity potential given by Eqs. (12.25) and (12.30) and represented by the screened electron-impurity line u^{RPA},

$$-W^{\mathrm{RPA}}(\mathbf{q}, iq_n) = \sim\!\!\sim + \sim\!\!\bigcirc\!\!\sim + \sim\!\!\bigcirc\!\!\sim\!\!\bigcirc\!\!\sim + \cdots$$

(14.69)

$$u^{\mathrm{RPA}}(\mathbf{q}) = \bullet =\!=\!=\!=\!\star = \bullet\!-\!-\!\star + \sim\!\!\bigcirc\!\!-\!\star + \sim\!\!\bigcirc\!\!\sim\!\!-\!\star + \cdots$$

(14.70)

14.5   Plasma oscillations and Landau damping

We now leave the static case and turn on an external potential with frequency ω. The goal of this section is to investigate the frequency dependence of the dielectric function $\varepsilon(\mathbf{q}, \omega)$. We could choose to study the full case described through $\chi_0^R(\mathbf{q}, \omega)$ by Eqs. (14.28a) and (14.30), but to draw some clear-cut physical conclusions, we confine the discussion to the case of high frequencies, long wavelengths and low temperatures, all defined by the conditions

$$v_F q \ll \omega \ (\text{or } x \ll x_0), \qquad q \ll k_F \ (\text{or } x \ll 1), \qquad k_B T \ll \varepsilon_F. \qquad (14.71)$$

In this regime we see from Eq. (14.30) that $\operatorname{Im} \chi_0^R = 0$. To proceed we adopt the following notation

$$\lambda \equiv \cos \theta, \qquad \int \frac{d\mathbf{p}}{(2\pi)^3} = \int_0^\infty \frac{dp}{4\pi^2} p^2 \int_{-1}^1 d\lambda, \qquad \xi_{\mathbf{p}+\mathbf{q}} - \xi_{\mathbf{p}} \approx v_{\mathbf{p}} q \lambda. \qquad (14.72)$$

Utilizing this in Eq. (14.21) and Taylor expanding n_F as in Eq. (14.22), we obtain

$$\operatorname{Re} \chi_0^R(\mathbf{q}, \omega) \approx \frac{1}{2\pi^2} \int dp \, p^2 \, \delta(\varepsilon_p - \varepsilon_F) \int_{-1}^1 d\lambda \, \frac{v_{\mathbf{p}} q \lambda}{\omega - v_{\mathbf{p}} q \lambda}. \qquad (14.73)$$

We rewrite the delta function in energy-space to one in k-space, and furthermore we introduce a small dimensionless variable z:

$$\delta(\varepsilon_{\mathbf{p}} - \varepsilon_F) = \frac{\delta(p - k_F)}{v_F}, \qquad p \to k_F, \qquad v_p \to v_F, \qquad z \equiv \frac{q v_F}{\omega} \lambda \ll 1. \qquad (14.74)$$

This is inserted in Eq. (14.73). The variable λ is substituted by z, and the smallness of this new variable permits the Taylor expansion $z/(1-z) \approx z + z^2 + z^3 + z^4$.

$$\operatorname{Re} \chi_0^R(\mathbf{q}, \omega) \approx \frac{1}{2\pi^2} k_F^2 \frac{1}{v_F} \frac{\omega}{q v_F} \int_{-q v_F/\omega}^{q v_F/\omega} dz \, \frac{z}{1-z}$$

$$\approx \frac{1}{2\pi^2} \frac{k_F^2 \omega}{q v_F^2} \left[\frac{1}{2} z^2 + \frac{1}{3} z^3 + \frac{1}{4} z^4 + \frac{1}{5} z^5 \right]_{-q v_F/\omega}^{+q v_F/\omega}$$

$$= \frac{n}{m} \frac{q^2}{\omega^2} \left[1 + \frac{3}{5} \left(\frac{q v_F}{\omega} \right)^2 \right], \qquad (14.75)$$

where in the last line we used $v_F = k_F/m$ and $3\pi^2 n = k_F^3$. Combining Eqs. (14.67) and (14.75), we find the RPA dielectric function in the high-frequency and long-wavelength limit to be

$$\varepsilon^{\text{RPA}}(\mathbf{q}, \omega) = 1 - \frac{\omega_p^2}{\omega^2} \left[1 + \frac{3}{5} \left(\frac{q v_F}{\omega} \right)^2 \right], \qquad (14.76)$$

where the characteristic frequency ω_p, well known as the electronic plasma frequency, has been introduced

$$\omega_p \equiv \sqrt{\frac{n e^2}{m \epsilon_0}}. \qquad (14.77)$$

14.5.1 *Plasma oscillations and plasmons*

The plasma frequency is an important parameter of the interacting electron gas setting the energy scale for several processes, e.g., it marks the limit below which metals
reflect incoming electromagnetic radiation, and above which they become transparent.
Typical values are $\omega \approx 10^{16}$ Hz, putting it in the ultraviolet part of the electromagnetic spectrum. It is determined by the electron density n and the effective band
mass m of Eq. (2.16). The former parameter can be found by Hall-effect measurements, while the latter can be determined from de Haas–van Alphen effect.[49] Using
the observed parameters for aluminum, $n = 1.81 \times 10^{29}$ m^{-2} and $m = 1.115\, m_0$, we
obtain $\omega_p^{\text{Al}} = 2.27 \times 10^{16}$ Hz $= 15.0$ eV.

A very direct manifestation of the plasmon frequency is the existence of the collective charge density oscillations, the plasma oscillations. Theoretically, the existence
of these oscillations is proved as follows. Consider the relation $\mathbf{D} = \varepsilon\, \epsilon_0 \mathbf{E}$ or the
related one, $\phi_{\text{ext}}(\mathbf{q}, \omega) = \varepsilon(\mathbf{q}, \omega)\, \phi_{\text{tot}}(\mathbf{q}, \omega)$. Note that $\varepsilon(\mathbf{q}, \omega) = 0$ in fact allows for
a situation where the total potential varies in space and time in the absence of any
external potential driving these variations. We are thus about to identify an oscillatory eigenmode for the electron gas. Let us calculate its properties in RPA from
Eq. (14.76),

$$
\varepsilon^{\text{RPA}}(\mathbf{q}, \omega) = 0 \quad \Rightarrow \quad \omega^2 \approx \omega_p^2 + \frac{3}{5}(q v_{\text{F}})^2 \quad \Rightarrow \quad \omega(q) \approx \omega_p + \frac{3}{10}\frac{v_{\text{F}}^2}{\omega_p} q^2. \quad (14.78)
$$

Recall that in the high frequency regime Im χ_0^R and consequently Im ε is zero, so no
damping occurs. Thus by Eq. (14.78) it is indeed possible to find oscillatory eigenmodes, the plasma oscillations. They have a simple quadratic dispersion relation $\omega(q)$
starting out from ω_p for $q = 0$ and then going up as q is increased.

But how could one be convinced of the existence of these oscillations? One beau-
tiful and very direct verification is the experiment discussed in Fig. 14.1. If some
eigenmodes exist with a frequency $\sim \omega_p$, then, as is the case with any harmonic oscillator, they must be quantized leading to oscillator quanta, denoted plasmons, with
a characteristic energy of ω_p. In the experiment, high energy electrons with an initial
energy $E_i = 20$ keV are shot through a 258 nm wide aluminum foil. The final energy,
E_f, is measured on the other side of the foil, and the energy loss $E_i - E_f$ can be plotted. The result of the measurement is shown in Fig. 14.1(a). The energy loss clearly
happens in quanta of size ΔE. Some electrons traverse the foil without exciting any
plasmons (the first peak), others excite one or more as sketched in Fig. 14.1(b). On
the plot, electrons exciting as many as seven plasmons are clearly seen. Note that
the most probable process is not the plasmon-free traversal, but instead a traversal
during which two plasmons are excited. The value of the energy loss quantum was
measured to be $\Delta E = 14.8$ eV in very good agreement with the value of the plasma
frequency of 15.0 eV for bulk aluminum.

---

[49]The de Haas–van Alphen effect is oscillations in the magnetization of a system as the function
of an applied external magnetic field. The Fermi surface can be mapped out using this technique as
described in Ashcroft and Mermin (1981).

(a)   (b)

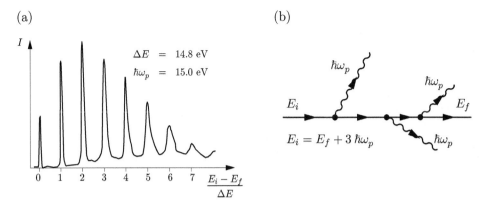

FIG. 14.1. (a) A sketch of an experiment allowing for direct observation of plasmons. The technique is high-energy electron transmission spectroscopy on a thin (258 nm wide) aluminum foil. For the original data see Marton *et al. Phys. Rev.* **126**, 182 (1962). The initial energy is $E_i = 20$ keV, and the final energy E_f is measured at zero scattering angle on the other side of the foil. The energy loss $E_i - E_f$ clearly reveals loss in quanta of ΔE. The energy quantum ΔE was found to be 14.8 eV in good agreement with the plasma frequency determined by other methods to be 15.0 eV. (b) A diagram representing a typical microscopic process, here the emission of three plasmons during the traversal.

14.5.2 *Landau damping*

Finally, we discuss the damping of excitations, which is described by the imaginary part $\mathrm{Im}\,\chi_0^R(\mathbf{q},\omega)$. The pure plasma oscillations discussed above are examples of undamped or long-lived excitations. This can be elucidated by going to the retarded functions in Eq. (14.68)

$$\phi_{\text{tot}}^{\text{RPA,R}}(\mathbf{q},\omega) \;=\; \frac{\phi_{\text{ext}}(\mathbf{q},\omega)}{1 - \frac{e^2}{\epsilon_0 q^2}\,\chi_0(\mathbf{q},\omega + i\eta)}. \tag{14.79}$$

In regions of the (q,ω) plan where the imaginary part of χ_0 vanishes, we can have a pole in Eq. (14.79) with a real frequency. However, in regions where $\mathrm{Im}\,\chi_0^R \neq 0$ we end up with a broadened (Lorentzian-type) peak as a function of ω, signaling a temporal decay of the total potential with a decay time proportional to $\mathrm{Im}\,\chi_0^R$,

$$\phi_{\text{tot}}^{\text{RPA,R}}(\mathbf{q},\omega) \;=\; \frac{\phi_{\text{ext}}(\mathbf{q},\omega)}{1 - \frac{e^2}{\epsilon_0 q^2}\,\mathrm{Re}\,\chi_0^R(\mathbf{q},\omega) - i\frac{e^2}{\epsilon_0 q^2}\,\mathrm{Im}\,\chi_0^R(\mathbf{q},\omega)}. \tag{14.80}$$

In Eq. (14.30), we have within RPA calculated the region in the (q,ω)-plane of non-vanishing $\mathrm{Im}\,\chi_0^R$, and this region is shown in Fig. 14.2. The physical origin of the non-zero imaginary part is the ability for the electron gas to absorb incoming energy by generating electron-hole pairs. Outside the appropriate area in (q,ω)-space, energy and momentum constraints prohibit the excitation of electron-hole pairs, and the electron gas cannot absorb energy by that mechanism.

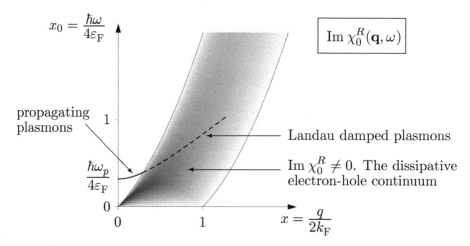

FIG. 14.2. A gray-scale plot of $\mathrm{Im}\chi_0^R(\mathbf{q},\omega)$. The darker a shade the higher the value. The variables are rescaled according to Eq. (14.27): $x = q/2k_\mathrm{F}$ and $x_0 = \omega/4\varepsilon_\mathrm{F}$. Note that $\mathrm{Im}\,\chi_0^R(\mathbf{q},\omega) \neq 0$ only in the gray-scaled area, which is bounded by the constraint functions given in Eq. (14.30). Also shown is the plasmon branch with its propagating and damped parts. The parameters chosen for this branch are those of aluminum, $\varepsilon_\mathrm{F} = 11.7$ eV and $\omega_p = 15.0$ eV.

Another way to understand the effect of a non-vanishing $\mathrm{Im}\,\chi_0^R$ is to link it to the conductivity σ of the electron gas. It is well known that the real part of σ is associated with dissipation (Joule heating), when a current $\mathbf{J}$ is flowing. But from Eq. (6.45) it follows that

$$e^2 \,\mathrm{Im}\,\chi_0^R = -\frac{1}{\omega}\,\mathbf{q}\cdot(\mathrm{Re}\,\sigma)\cdot\mathbf{q}, \qquad (14.81)$$

whereby it is explicitly confirmed that a non-vanishing $\mathrm{Im}\,\chi_0^R$ is associated with the ability of the system to dissipate energy.

Finally, we remark that in Fig. 14.2 is shown the dispersion relation for the plasmon excitation. It starts out as a bona fide excitation in the region of the (q,ω)-space where the RPA dissipation is zero. Hence the plasmons have infinite life times for small q. However, at some point the dispersion curve crosses into the dissipative $\mathrm{Im}\,\chi_0 \neq 0$ area, and there the plasmon acquires a finite life time. In other words, for high q-values the plasmonic excitations are not exact eigenmodes of the system, and they are damped out as a function of time. In the literature, this damping mechanism is called Landau damping.

14.6   Summary and outlook

In this chapter we have used the Feynman rules for pairwise interacting particles to analyze the Coulomb-interacting electron gas in the jellium model. The main result was that the RPA resummation of diagrams to all orders in perturbation theory were valid in the high-density limit. In particular we found the self-energy

$$\Sigma_\sigma^{\mathrm{RPA}}(\mathbf{k}, ik_n) = \quad \times \left[\quad + \quad + \quad + \quad + \cdots \right] = \quad .$$

This result was used to calculate the groundstate energy of the electron gas,

$$\frac{E-E^0}{N} = \quad + \quad + \quad = \left(\frac{2.211}{r_s^2} - \frac{0.916}{r_s} + 0.0622 \log r_s - 0.094 \right) \mathrm{Ry}.$$

We also used the RPA analysis to study the dielectric properties of the electron gas. One main result was finding the screening of the Coulomb interaction both for the internal interaction and for external potentials, here expressed by their Dyson's equations

$$-W^{\mathrm{RPA}}(\mathbf{q}, iq_n) = \quad = \quad + \quad .$$

$$u^{\mathrm{RPA}}(\mathbf{q}) = \quad = \quad + \quad .$$

Explicit expressions for the dielectric function $\varepsilon(\mathbf{q}, \omega)$ was found in two cases, (i) the static, long-wave limit and (ii) the high-frequency, long-wave limit,

$$\varepsilon^{\mathrm{RPA}}(\mathbf{q}, 0) = 1 + \frac{k_s^2}{q^2}, \quad \text{where} \quad k_s^2 = \frac{4}{\pi} \frac{k_{\mathrm{F}}}{a_0},$$

$$\varepsilon^{\mathrm{RPA}}(\mathbf{q}, \omega \gg qv_{\mathrm{F}}) = 1 - \frac{\omega_p^2}{\omega^2} \left[1 + \frac{3}{5} \left(\frac{qv_{\mathrm{F}}}{\omega} \right)^2 \right].$$

Finally, we studied the plasma oscillations of the electron gas found from the condition $\varepsilon^{\mathrm{RPA}}(\mathbf{q}, \omega) = 0$, and found the dispersion relation involving the plasma frequency ω_p,

$$\omega(q) = \omega_p + \frac{3}{10} \frac{v_{\mathrm{F}}^2}{\omega_p} q^2, \quad \text{where} \quad \omega_p \equiv \sqrt{\frac{ne^2}{m\epsilon_0}}.$$

The RPA analysis has already given us a good insight in some central physical properties of the electron gas. Moreover, it plays a crucial role in the studies of electron-impurity scattering, electron-phonon interaction, superconductivity, and of many other physical phenomena involving the electron gas.

15

FERMI LIQUID THEORY

The concept of Fermi liquid theory was developed by Landau in 1957–59 and later refined by others.[50] The basic conclusion is that a gas of interacting particles can be described by a system of almost non-interacting "quasiparticles." These quasiparticles are approximate excitations of the system at sufficiently short time scales. What we mean by "sufficiently short" of course has to be quantified, and this condition will set the limits for the applicability of the theory.

The Fermi liquid theory is conceptually extremely important, because it explains why the apparently immensely complicated system of, e.g., interacting electrons in a metal can be regarded as a gas of non-interacting particles. This is an enormous simplification, and it gives the theoretical explanation of why all the results that one obtains from the widely used free-electron model work so well.

The quasiparticle concept furthermore gives the theoretical foundation of the semi-classical description. The quasiparticle distribution function satisfies a kinetic equation, which may include scattering from one state to another for example due to impurity scattering. This equation is known as the Landau transport equation, and is equivalent to the well-known Boltzmann equation from kinetic gas theory. In this description the potential is allowed to vary in space due to some external perturbation or due to interactions with the inhomogeneous density of quasiparticles. Using the Landau transport equation we shall see that the collective modes derived in the previous chapter also come out naturally from a semi-classical description and, furthermore, that also the conductivity, which is calculated from microscopic considerations in Chapter 16, can be understood in terms of scattering of quasiparticles.

15.1 Adiabatic continuity

The Fermi liquid theory is based on the assumption that starting from the non-interacting system of particles, one can analyze the interacting case by applying perturbation theory. This is, in fact, a rather stringent criterion, because it means that one cannot cross a phase boundary line. Take, for instance, the ferromagnetic transition described in Section 4.5: starting from the paramagnetic phase it is impossible by perturbation theory to reach the ferromagnetic phase.[51]

If the excitations of the non-interacting system are connected to the excitations of the interacting system by a one-to-one correspondence (at least on short timescales as explained below), the two cases are said to be connected by "adiabatic continuity".

---

[50]See the collection of reprints in the book: D. Pines *The Many-body problem*, Addison–Wesley (1961, 1997).

[51]This fact can be understood from the concept of broken symmetry explained in Section 4.4. The phase with broken symmetry can only occur if the ensemble of states in the statistical average is truncated.

If we imagine that we start from the non-interacting system excited in some state and then turn on the interaction adiabatically, i.e., so slowly that the occupation numbers are not changed, then we would end up in a corresponding excited state of the interacting system.

15.1.1 Example: one-dimensional well

As a simple example of adiabatic continuity, we consider a particle trapped in a 1D potential. The 1D potential will have a number of bound states with discrete eigenenergies and a continuum of eigenenergies corresponding to the delocalized states. We now imagine changing the potential slowly. As an example, consider a potential of the form

$$V(x,t) = -V_0(t) \, \exp\left(-\frac{x^2}{2x_0^2}\right). \tag{15.1}$$

The depth of the well is time dependent, and let us suppose that it is changing from an initial value V_{01} to a final value V_{02}. If this change is slow, the solution of the Schrödinger equation

$$i\partial_t \psi(x,t) = H(t)\psi(x,t) = \left(\frac{p^2}{2m} + V(x,t)\right)\psi(x,t), \tag{15.2}$$

can be approximated by the adiabatic solution

$$\psi_{\text{adia}}(x,t) \approx \psi_{V_0(t)}(x) \exp\left(-iE_{V_0(t)}t\right), \tag{15.3}$$

where $\psi_{V_0(t)}(x)$ is the solution of the static (or instantaneous) Schrödinger equation with energy $E_{V_0(t)}$,

$$H(t)\psi_{V_0(t)}(x) = E_{V_0(t)}\psi_{V_0(t)}(x). \tag{15.4}$$

Note that both $\psi_{V_0(t)}(x)$ and $E_{V_0(t)}$ depend parametrically on the time through $V_0(t)$. The accuracy of the solution in Eq. (15.3) is estimated by inserting Eq. (15.3) into Eq. (15.2), which yields

$$i\partial_t \psi_{\text{adia}}(x,t) = E_{V_0(t)}\psi_{\text{adia}}(x,t) + i\left(\frac{\partial\psi_{\text{adia}}(x,t)}{\partial V_0(t)}\right)\left(\frac{\partial V_0(t)}{\partial t}\right) = H\psi_{\text{adia}}(x,t). \tag{15.5}$$

Thus, we have an approximate solution if the first term dominates over the second term.

Apparently our conclusion is that if the rate of change of $V_0(t)$ is small enough, the solution for the new value of $V_0 = V_{02}$ can be found by starting from the solution with the old value of $V_0 = V_{01}$ and "adiabatically" changing it to its new value. For example, if the first excited state is a bound state, it will change to a somewhat modified state with a somewhat modified energy, but most importantly it is still the first excited state and is still a bound state. This may sound completely trivial, but it is not, and it is not always true. For example if the real solution during this change of V_0 from V_{01} to V_{02} changes from a bound state to an unbound state (if V_{02} is small enough there is only one bound state), then it does not matter how slowly we change V_0. The two states can simply not be connected through small changes of V_0, because one is a decaying function and the other is an oscillatory function. This is an example where perturbation theory to any order would never give the right answer. The important message is, however, that if we avoid these transitions between different kinds of states, adiabatic continuity does work. In the following this idea is applied to the case of interacting particles.

15.1.2   *The quasiparticle concept and conserved quantities*

The principle of adiabatic continuity is now applied to study a system of interacting particles. It is used to bring the excitations of the interacting case back to the well-known excitations of the non-interacting case, thus making computation possible. In doing so, we gain the fundamental understanding that the interacting and the non-interacting cases have a lot in common at least under some restricting circumstances. This turns out to be realized in many systems. The following arguments are not meant to be the full theoretical explanation for the applicability of Fermi liquid theory, but rather to give a physical intuition for the reason why the quasiparticle picture is valid.

When calculating physical quantities, such as spectral functions, response functions or occupation numbers, we are facing matrix elements between different states, for instance, between states with added particles or added particle-hole pairs. Since we are dealing with the low-energy properties of the system, let us consider states with single particles or single electron-hole pairs added to the groundstate

$$|(\mathbf{k}\sigma)_p\rangle = c_{\mathbf{k}\sigma}^\dagger |G\rangle, \quad |(\mathbf{k}\sigma)_p; (\mathbf{k'}\sigma')_h\rangle = c_{\mathbf{k}\sigma}^\dagger c_{\mathbf{k'}\sigma'} |G\rangle, \quad etc., \qquad (15.6)$$

where $|G\rangle$ is the groundstate of the interacting system. The first term inserts a particle while the second term creates both a particle and hole.

Imagine letting time evolve according to a Hamiltonian where the interaction is gradually *switched off* at a rate ζ

$$H_\zeta = H_0 + H_{\text{int}} e^{-\zeta t}, \quad t > 0. \qquad (15.7)$$

If, under the conditions of adiabaticity, we can bring the states Eq. (15.6) all the way back to the non-interacting case, then the matrix elements are identical to those of the non-interacting case. There are two important assumptions built into this construction:

---

1. The adiabatic procedure is valid when the energy of the state is large compared to the rate of change, i.e., $\varepsilon_{k\sigma} \gg \zeta$, or, since typical excitation energies are of order of the temperature, this is equivalent to assuming $k_B T \gg \zeta$.
2. The interactions do not induce transitions among the states in question, or in other words the lifetime τ_{life} of the state is long compared to ζ^{-1}, that is $\tau_{\text{life}} \gg \zeta^{-1}$.

This apparently leaves an energy window where the idea makes sense, namely when we can choose a switch-off rate ζ such that

$$\tau_{\text{life}}^{-1} \ll \zeta \ll k_B T. \tag{15.8}$$

The last condition can in principle always be met at high enough temperatures, whereas the first one is not necessarily possible. Below we shall see that it is indeed possible to make the approximations consistent,[52] because the lifetime turns out to be inversely proportional to the square of the temperature, $\tau_{\text{life}}^{-1} \propto T^2$. Thus there is always a temperature range at low temperature where Eq. (15.8) is fulfilled.

Next, we discuss the properties of the state with an added particle, $|(\mathbf{k}\sigma)_p\rangle$. It is clear that the state where the interaction is switched off has a number of properties in common with the initial state, namely those that are conserved by the Hamiltonian: (i) it has an excess charge e (compared to the groundstate), (ii) it carries current $-e\hbar\mathbf{k}/m$, and (iii) it has excess spin σ. These properties are all conserved quantities because the corresponding operators (i) the total charge $Q = -eN$, (ii) the total current $\mathbf{J}_e = -e\sum_{\mathbf{k}\sigma} \mathbf{v_k} n_{\mathbf{k}\sigma}$, and (3) the total spin $\mathbf{S} = \sum_{\mathbf{k}\sigma} \sigma n_{\mathbf{k}\sigma}$ all commute with the Hamiltonian.

Most importantly, the adiabatic continuity principle can also be used to relate the distribution function to the distribution function of non-interacting particles, i.e., the Fermi–Dirac distribution function. This leads us to the definition of quasiparticles:

Quasiparticles describe excitations of the interacting system corresponding to the creation or annihilation of particles (e.g., particle-hole pair state $|(\mathbf{k}\sigma)_p; (\mathbf{k}'\sigma')_h\rangle$) in an intermediate time window: short enough that the state does not decay due to electron-electron scattering, but long enough to ensure a well-defined energy of the quasiparticle. The quasiparticles are thus not to be thought of as the exact eigenstates.

The quasiparticles can be labelled by the same quantum numbers as in the non-interacting case, provided that the corresponding operators commute with the Hamiltonian. For a translation-invariant system of electrons interacting through the Coulomb interaction, the quasiparticle quantum numbers are thus $\mathbf{k}$ and σ and they carry charge $-e$ and velocity $\mathbf{v_k} = \hbar\mathbf{k}/m$.

At low temperatures there are only a few weakly interacting quasiparticles. In equilibrium these are distributed according to the Fermi–Dirac distribution function.

In the following we make use of the quasiparticle concept to calculate the screening and the transport properties of an electron gas.

15.2   Semi-classical treatment of screening and plasmons

In Chapter 14 we saw how the collective modes of a charged Fermi gas came out of a rigorous diagrammatical analysis. Here, we shall rederive some of this using a less rigorous but maybe physically more appealing approach. Consider a uniform electron gas which is subject to an external potential $\phi_{\text{ext}}(\mathbf{r}, t)$. We can include the external

---

[52]This is true in three dimensions, but not necessary in lower dimensions, see Chapter 19.

potential as a local change of the chemical potential of the charged quasiparticles.[53] Now, if the local potential of the quasiparticles is space and time dependent, so is then the density of quasiparticles. This in turn changes the electrical potential, because the quasiparticles are charged, and therefore the total potential ϕ_{tot} is given by the sum of the external potential ϕ_{ext} and the induced potential ϕ_{ind}. The induced potential is caused by the excess or deficit of quasiparticles. Thus we write the resulting local potential $\phi_{\text{tot}}(\mathbf{r}, t)$ as

$$\phi_{\text{tot}}(\mathbf{r}, t) = \phi_{\text{ext}}(\mathbf{r}, t) + \phi_{\text{ind}}(\mathbf{r}, t). \tag{15.9}$$

The induced potential ϕ_{ind} created by the induced density ρ_{ind}, which in turn depends on the total potential, must be determined self-consistently.

15.2.1 Static screening

First we consider linear static screening. To linear order in the local total potential and at low temperatures, the induced charge density is given by

$$\rho_{\text{ind}}(\mathbf{r}) = \frac{2}{\mathcal{V}} \sum_{\mathbf{k}} \left[n_F\left(\xi_{\mathbf{k}} + (-e)\phi_{\text{tot}}(\mathbf{r})\right) - n_F\left(\xi_{\mathbf{k}}\right) \right]$$

$$\approx -(-e)\phi_{\text{tot}}(\mathbf{r}) \frac{2}{\mathcal{V}} \sum_{\mathbf{k}} \left(-\frac{\partial n_F(\xi_{\mathbf{k}})}{\partial \xi_{\mathbf{k}}} \right) \approx -(-e)\phi_{\text{tot}}(\mathbf{r}) d(\varepsilon_F), \tag{15.10}$$

where $\xi_{\mathbf{k}}$ is the quasiparticle energy measured relative to the equilibrium chemical potential, and $d(\varepsilon_F)$ is the density of states at the Fermi level. From this we obtain the induced potential in real space

$$\phi_{\text{ind}}(\mathbf{r}) = \frac{1}{-e} \int d\mathbf{r}' W(\mathbf{r} - \mathbf{r}') \rho_{\text{ind}}(\mathbf{r}), \tag{15.11}$$

and in $\mathbf{q}$-space

$$\phi_{\text{ind}}(\mathbf{q}) = \frac{1}{-e} W(\mathbf{q}) \rho_{\text{ind}}(\mathbf{q}) = -W(\mathbf{q})\phi_{\text{tot}}(\mathbf{q}) d(\varepsilon_F). \tag{15.12}$$

When inserted into Eq. (15.9), this yields a solution for ϕ_{tot},

$$\phi_{\text{tot}}(\mathbf{q}) = \frac{\phi_{\text{ext}}(\mathbf{q})}{1 + W(\mathbf{q}) d(\varepsilon_F)}, \tag{15.13}$$

and hence

$$\varepsilon(\mathbf{q}) = 1 + W(\mathbf{q}) d(\varepsilon_F), \tag{15.14}$$

in full agreement with the conclusions of the RPA results (14.67) and (14.68) using $\chi_0^R = -d(\varepsilon_F)$ from Eq. (14.22).

---

[53]Note that we are here invoking a new concept, namely local equilibrium, because otherwise we could not talk about a local potential. Clearly, this only makes sense on length scales larger than a typical thermalization length. The thermalization length is the length scale on which thermal equilibrium is established.

15.2.2 *Dynamical screening*

In the dynamical case, we expect to find collective excitations similar to the plasmons found in Section 14.5. In order to treat this case we need to refine the analysis a bit to allow for the time it takes the charge to adjust to the varying external potential. Consequently, the induced charge density at point $\mathbf{r}$ at time t now depends on the total potential at some other point $\mathbf{r}'$ and at some other (previous) time t'. The way to describe this, is to look at the deviation of the distribution function $n_{\mathbf{k}}$ of a quasiparticle with a given momentum $\mathbf{p} = \hbar\mathbf{k}$ (below we as usual use $\hbar = 1$). This depends on both $\mathbf{r}$ and t, so that

$$n_{\mathbf{k}} = n_{\mathbf{k(t)}}(\mathbf{r}, t). \tag{15.15}$$

The dynamics are controlled by two things: the conservation of charge and the change of momentum with time. The first dependence arises from the flow of the distribution function. Because we are interested in times shorter than the lifetime of the quasiparticles, *the number of quasiparticles in each* $\mathbf{k}$ *state is conserved.* The conservation of particles in state $\mathbf{k}$ is expressed through the continuity equation

$$\dot{n}_{\mathbf{k}} + \nabla_{\mathbf{r}} \cdot \mathbf{j}_{\mathbf{k}} = 0, \tag{15.16}$$

where the current carried of quasiparticles in state $\mathbf{k}$ is given by $\mathbf{j}_{\mathbf{k}} = \mathbf{v}_{\mathbf{k}}n_{\mathbf{k}} = (\hbar\mathbf{k}/m)n_{\mathbf{k}}$, and hence we obtain

$$\partial_t(n_{\mathbf{k}}) + \dot{\mathbf{k}} \cdot \nabla_{\mathbf{k}}n_{\mathbf{k}} + \mathbf{v}_{\mathbf{k}} \cdot \nabla_r n_{\mathbf{k}} = 0, \tag{15.17}$$

which is known as the collision-free Boltzmann equation.[54]
   The second dependence follows from how a negatively charged particle is accelerated in a field, i.e., simply from Newton's law

$$\dot{\mathbf{p}} = -(-e)\nabla_{\mathbf{r}}\phi_{\text{tot}}(\mathbf{r}, t). \tag{15.18}$$

Again, it is convenient to use Fourier space and introducing the Fourier transform $n_{\mathbf{k}}(\mathbf{q}, \omega)$. Using Eqs. (15.17) and (15.18), we find

$$(-i\omega + i\mathbf{q} \cdot \mathbf{v}_{\mathbf{k}})n_{\mathbf{k}}(\mathbf{q}, \omega) = -ie\,(\mathbf{q} \cdot \nabla_{\mathbf{k}}n_{\mathbf{k}})\,\phi_{\text{tot}}(\mathbf{q}, \omega) = ie\,(\mathbf{q}\cdot\nabla_{\mathbf{k}}\xi_{\mathbf{k}})\left(-\frac{\partial n_{\mathbf{k}}}{\partial \xi_{\mathbf{k}}}\right)\phi_{\text{tot}}(\mathbf{q}, \omega). \tag{15.19}$$

To linear order in the potential ϕ_{tot} we can replace the $n_{\mathbf{k}}$ on the right-hand side by the equilibrium distribution $n_{\mathbf{k}}^0 = n_F(\xi_{\mathbf{k}})$, and hence we find

$$n_{\mathbf{k}}(\mathbf{q}, \omega) = \frac{\mathbf{q}\cdot\nabla_{\mathbf{k}}\xi_{\mathbf{k}}}{\omega - \mathbf{q} \cdot \mathbf{v}_{\mathbf{k}}}\left(-\frac{\partial n_F(\xi_{\mathbf{k}})}{\partial \xi_{\mathbf{k}}}\right)\left[-e\phi_{\text{tot}}(\mathbf{q}, \omega)\right]. \tag{15.20}$$

From this expression we easily obtain the induced density by summation over $\mathbf{k}$,

---

[54]Here, as in fluid dynamics, $\mathbf{r}$ and t are independent space and time coordinates of the fields in question in contrast to mechanics where $\mathbf{r} = \mathbf{r}(t)$ follows the particle motion.

$$\rho_{\text{ind}}(\mathbf{q}, \omega) = \frac{2}{V} \sum_{\mathbf{k}} \frac{\mathbf{q} \cdot \nabla_{\mathbf{k}} \xi_{\mathbf{k}}}{\omega - \mathbf{q} \cdot \mathbf{v}_{\mathbf{k}}} \left(-\frac{\partial n_F(\xi_{\mathbf{k}})}{\partial \xi_{\mathbf{k}}} \right) \left[-e\phi_{\text{tot}}(\mathbf{q}, \omega) \right], \qquad (15.21)$$

where the factor 2 comes from to spin degeneracy. This is inserted into Eqs. (15.12) and (15.9) and we obtain the dielectric function $\varepsilon = \phi_{\text{ext}}/\phi_{\text{tot}}$ in the dynamical case

$$\varepsilon(\mathbf{q}) = 1 - W(\mathbf{q}) \frac{2}{V} \sum_{\mathbf{k}} \frac{\mathbf{q} \cdot \nabla_{\mathbf{k}} \xi_{\mathbf{k}}}{\omega - \mathbf{q} \cdot \mathbf{v}_{\mathbf{k}}} \left(-\frac{\partial n_F(\xi_{\mathbf{k}})}{\partial \xi_{\mathbf{k}}} \right). \qquad (15.22)$$

At $\omega = 0$ we recover the static case in Eq. (15.14), because $\nabla_{\mathbf{k}} \xi_{\mathbf{k}} = v_{\mathbf{k}}$. At long wavelengths or large frequencies $qv \ll \omega$, we find by expanding in powers of q that

$$\varepsilon(\mathbf{q}) \approx 1 - \frac{W(\mathbf{q})}{\omega^2} \frac{2}{V} \sum_{\mathbf{k}} (\mathbf{q} \cdot \mathbf{v}_{\mathbf{k}})^2 \left(-\frac{\partial n_F(\xi_{\mathbf{k}})}{\partial \xi_{\mathbf{k}}} \right) = 1 - \left(\frac{\omega_p}{\omega} \right)^2, \qquad (15.23)$$

which agrees with Eq. (14.76).

We have thus shown that in the long wavelength limit the semi-classical treatment, which relies on the Fermi liquid theory, gives the same result as the fully microscopic theory, based on renormalization by summation of the most important diagrams. We have also gained some physical understanding of this renormalization, because we saw explicitly how it was due to the screening of the external potential by the mobile quasiparticles.

15.3 Semi-classical transport equation

Our last application of the semi-classical approach is the calculation of conductivity of a uniform electron gas with some embedded impurities. This will in fact lead us to the famous Drude formula. Historically, the Drude formula was first derived in an incorrect way, namely by assuming that the charge carriers form a classical gas. We know now that they follow a Fermi–Dirac distribution, but amazingly the result turns out to be the same. In Section 16 we will furthermore see, how the very same result can be derived in a microscopic quantum theory starting from the Kubo formula and using a diagrammatic approach.

As explained in Chapter 10, the finite resistivity of metals at low temperatures is due to scattering against impurities or other imperfections in the crystal structure. These collisions take momentum out of the electron system, thus introducing a mechanism for momentum relaxation and hence resistivity. A simple-minded approach to conductivity would be to say that the forces acting on a small volume of charge is the sum of the external force and a friction force that is taken to be proportional to the velocity of the fluid at the given point. In steady state these forces balance and hence

$$-(-e)\mathbf{E} + \frac{m\mathbf{v}}{\tau_{\text{p-relax}}} = 0 \Rightarrow \mathbf{J} = -en\mathbf{v} = \frac{e^2 n \tau_{\text{p-relax}}}{m} \mathbf{E} \Rightarrow \sigma = \frac{ne^2 \tau_{\text{p-relax}}}{m}, \qquad (15.24)$$

where σ is the conductivity and $\tau_{\text{p-relax}}$ is the momentum relaxation time.

Microscopically, the momentum relaxation corresponds to scattering of quasiparticles from one state $|\mathbf{k}\sigma\rangle$ with momentum $\hbar\mathbf{k}$ to another state $|\mathbf{k}'\sigma'\rangle$ with momentum

$\hbar \mathbf{k}'$. For non-magnetic impurities, the ones considered here, the spin is conserved and thus $\sigma = \sigma'$. The new scattering process thus introduced means that the number of quasiparticles in a given $\mathbf{k}$-state is no longer conserved and we have to modify Eq. (15.16) to take into account the processes that change the occupation number $n_{\mathbf{k}}$. The rate of change is given by the rate, $\Gamma(\mathbf{k}'\sigma \leftarrow \mathbf{k}\sigma)$, at which scattering from the state $|\mathbf{k}\sigma\rangle$ to some other state $|\mathbf{k}'\sigma'\rangle$ occurs. It can be found from Fermi's golden rule

$$\Gamma(\mathbf{k}'\sigma \leftarrow \mathbf{k}\sigma) = 2\pi \left| \langle \mathbf{k}'\sigma | V_{\text{imp}} | \mathbf{k}\sigma \rangle \right|^2 \delta(\xi_{\mathbf{k}} - \xi_{\mathbf{k}'}), \qquad (15.25)$$

where V_{imp} is the impurity potential. The fact that the scattering on an external potential is an elastic scattering process, is reflected in the energy-conserving delta function. The total impurity potential is a sum over single impurity potentials situated at positions $\mathbf{R}_i$, see Chapter 12,

$$V_{\text{imp}}(\mathbf{r}) = \sum_i u(\mathbf{r} - \mathbf{R}_i). \qquad (15.26)$$

We can then find the rate Γ by the adiabatic procedure, where the matrix element $\langle \mathbf{k}'\sigma | V_{\text{imp}} | \mathbf{k}\sigma \rangle$ is identified with non-interacting counterpart $\langle \mathbf{k}'\sigma | V_{\text{imp}} | \mathbf{k}\sigma \rangle_0$, where $|\mathbf{k}\sigma\rangle_0 = e^{i\mathbf{k}\cdot\mathbf{r}}/\sqrt{\mathcal{V}}$, and we obtain

$$\Gamma(\mathbf{k}'\sigma \leftarrow \mathbf{k}\sigma) = \Gamma_{\mathbf{k}'\sigma,\mathbf{k}\sigma} = \frac{2\pi}{\mathcal{V}^2} \left| \sum_j \int d\mathbf{r}\, e^{-i\mathbf{k}'\cdot\mathbf{r}} u(\mathbf{r} - \mathbf{R}_j) e^{+i\mathbf{k}\cdot\mathbf{r}} \right|^2 \delta(\xi_{\mathbf{k}} - \xi_{\mathbf{k}'}). \quad (15.27)$$

Of course, we do not know the location of the impurities exactly and therefore we perform a positional average. The average is done assuming only lowest order scattering, i.e., leaving out interference between scattering on different impurities. Therefore we can simply replace the sum over impurities by the number of scattering centers, $N_{\text{imp}} = n_{\text{imp}}\mathcal{V}$, and multiplied by the impurity potential for a single impurity $u(\mathbf{r})$. We obtain

$$\Gamma_{\mathbf{k}',\mathbf{k}} = 2\pi \frac{n_{\text{imp}}}{\mathcal{V}} \left| \int d\mathbf{r}\, e^{i(\mathbf{k}-\mathbf{k}')\cdot\mathbf{r}} u(\mathbf{r}) \right|^2 \delta(\xi_{\mathbf{k}} - \xi_{\mathbf{k}'}) \equiv \frac{n_{\text{imp}}}{\mathcal{V}} W_{\mathbf{k}',\mathbf{k}}. \qquad (15.28)$$

Now, the change of $n_{\mathbf{k}}$ due to collisions is included in the differential equation Eq. (15.16) as an additional term. The time derivative of $n_{\mathbf{k}}$ becomes

$$\dot{n}_{\mathbf{k}(t)}(\mathbf{r},t) = \left(\frac{d}{dt} n_{\mathbf{k}} \right)_{\text{flow}-\text{force}} - \left(\frac{\partial}{\partial t} n_{\mathbf{k}} \right)_{\text{collisions}}, \qquad (15.29)$$

where the change due to "flow and force" is given by the left-hand side in Eq. (15.17). The new collision term is not a derivative but an integral functional of $n_{\mathbf{k}}$

$$\left(\frac{\partial}{\partial t} n_{\mathbf{k}} \right)_{\text{collisions}} = -\frac{n_{\text{imp}}}{\mathcal{V}} \sum_{\mathbf{k}'} \left[n_{\mathbf{k}}(1 - n_{\mathbf{k}'}) W_{\mathbf{k}',\mathbf{k}} - n_{\mathbf{k}'}(1 - n_{\mathbf{k}}) W_{\mathbf{k},\mathbf{k}'} \right]. \qquad (15.30)$$

The first term in the sum represents the rate for being scattered out of the state $\mathbf{k}$, and the second term represents the rate for being scattered into to state $\mathbf{k}$ from the state $\mathbf{k}'$. The total rate is obtained from Eq. (15.28) times the probability for the initial state to be filled and the final state to be empty. Because $W_{\mathbf{k},\mathbf{k}'} = W_{\mathbf{k}',\mathbf{k}}$, we have

$$\left(\frac{\partial}{\partial t} n_{\mathbf{k}}\right)_{\text{collisions}} = -\frac{n_{\text{imp}}}{V} \sum_{\mathbf{k}'} W_{\mathbf{k}',\mathbf{k}} \left(n_{\mathbf{k}} - n_{\mathbf{k}'}\right), \qquad (15.31)$$

and the full Boltzmann transport equation in the presence of impurity scattering now reads

$$\partial_t(n_{\mathbf{k}}) + \dot{\mathbf{k}} \cdot \nabla_{\mathbf{k}} n_{\mathbf{k}} + \mathbf{v}_{\mathbf{k}} \cdot \nabla_{\mathbf{r}} n_{\mathbf{k}} = -\frac{n_{\text{imp}}}{V} \sum_{\mathbf{k}'} W_{\mathbf{k}',\mathbf{k}} \left(n_{\mathbf{k}} - n_{\mathbf{k}'}\right). \qquad (15.32)$$

The Boltzmann equation for impurity scattering is rather easily solved in the linear response regime. First we note that $\dot{\mathbf{p}} = -e\mathbf{E}$, and therefore to linear order in $\mathbf{E}$ the term $\nabla_{\mathbf{k}} n_{\mathbf{k}}$ multiplying $\dot{\mathbf{k}}$ can be replaced by the equilibrium occupation, which at zero temperature becomes $\nabla_{\mathbf{k}} n_{\mathbf{k}}^0 = \nabla_{\mathbf{k}} \theta(k_F - k) = -\hat{\mathbf{k}}\delta(k_F - k)$, where $\hat{\mathbf{k}}$ is a unit vector oriented along $\mathbf{k}$. Let us furthermore concentrate on the long wavelength limit such that $\nabla_{\mathbf{r}} n_{\mathbf{k}} \approx 0$. By going to the frequency domain, we obtain

$$-i\omega n_{\mathbf{k}} + e\mathbf{E} \cdot \hat{\mathbf{k}}\,\delta(k_F - k) = -\frac{n_{\text{imp}}}{V} \sum_{\mathbf{k}'} W_{\mathbf{k}',\mathbf{k}} \left(n_{\mathbf{k}} - n_{\mathbf{k}'}\right). \qquad (15.33)$$

Without the $n_{\mathbf{k}'}$-term on the right-hand side this equation is simple to solve, because the right-hand side is then of the form $\tau^{-1} n_{\mathbf{k}}$ similar to $-i\omega n_{\mathbf{k}}$ on the left-hand side. This hints that we can obtain the full solution by some imaginary shift of ω, so let us try the ansatz

$$n_{\mathbf{k}}(\omega) = \frac{1}{i\omega - 1/\tau^{\text{tr}}} e\mathbf{E} \cdot \hat{\mathbf{k}}\delta\left(k_F - k\right), \qquad (15.34)$$

where the relaxation time τ^{tr} needs to be determined. That this in fact is a solution is seen by substitution

$$\frac{-i\omega}{i\omega - 1/\tau^{\text{tr}}} e\mathbf{E} \cdot \hat{\mathbf{k}}\delta(k_F - k) + e\mathbf{E} \cdot \hat{\mathbf{k}}\,\delta(k_F - k)$$

$$= \frac{-e}{i\omega - 1/\tau^{\text{tr}}} \frac{n_{\text{imp}}}{V} \sum_{|\mathbf{k}'|} W_{\mathbf{k}',\mathbf{k}} \left(\hat{\mathbf{k}}\delta(k_F - k) - \hat{\mathbf{k}}'\delta(k_F - k')\right) \cdot \mathbf{E}. \qquad (15.35)$$

Since $W_{\mathbf{k}',\mathbf{k}}$ includes an energy conserving delta function, we can set $k = k' = k_F$ and remove the common factor $\delta(k_F - k)$ to obtain

$$\frac{-i\omega}{i\omega - 1/\tau^{\text{tr}}} e\mathbf{E} \cdot \hat{\mathbf{k}} + e\mathbf{E} \cdot \hat{\mathbf{k}} = \frac{-e}{i\omega - 1/\tau^{\text{tr}}} \frac{n_{\text{imp}}}{V} \sum_{|\mathbf{k}'|=k_F} W_{\mathbf{k}',\mathbf{k}}(\hat{\mathbf{k}} - \hat{\mathbf{k}}') \cdot \mathbf{E}. \qquad (15.36)$$

which is solved by

$$\frac{1}{\tau^{\text{tr}}} \cos\theta_{\mathbf{k}} = \frac{n_{\text{imp}}}{V} \sum_{k'=k_F} W_{\mathbf{k}',\mathbf{k}}(\cos\theta_{\mathbf{k}} - \cos\theta_{\mathbf{k}'}). \qquad (15.37)$$

Here, $\theta_{\mathbf{k}}$ is the angle between the vector $\mathbf{k}$ and the electric field $\mathbf{E}$. For a uniform system, the scattering does not depend on direction and the only angular dependence of $W_{\mathbf{k},\mathbf{k}'}$ is the angle between $\mathbf{k}$ and $\mathbf{k}'$. This means that we can choose $\mathbf{E}$ parallel to $\mathbf{k}$, and obtain

$$\frac{1}{\tau^{\text{tr}}} = \frac{n_{\text{imp}}}{\mathcal{V}} \sum_{k'=k_F} W_{\mathbf{k}',\mathbf{k}} (1 - \cos \theta_{\mathbf{k},\mathbf{k}'}). \tag{15.38}$$

The time τ^{tr} is known as the transport time, because it is the time that enters the expression for the conductivity, as we see by calculating the current density

$$\begin{aligned}
\mathbf{J} &= -\frac{2e}{\mathcal{V}} \sum_{\mathbf{k}} n_{\mathbf{k}} \mathbf{v}_{\mathbf{k}} \\
&= -\frac{2e}{\mathcal{V}} \sum_{\mathbf{k}} \left[\frac{e\delta(k_F - k)}{i\omega - 1/\tau^{\text{tr}}} \hat{\mathbf{k}} \cdot \mathbf{E} \right] \frac{\mathbf{k}}{m} \\
&= \frac{2e^2}{(2\pi)^2} \frac{\mathbf{E}}{-i\omega + 1/\tau^{\text{tr}}} \frac{1}{m} \int_0^\infty dk\, k^3 \delta(k_F - k) \int_{-1}^1 d(\cos\theta) \cos^2\theta \\
&= \frac{2e^2 \mathbf{E}}{-i\omega + 1/\tau^{\text{tr}}} \frac{1}{(2\pi)^2 m} k_F^3 \frac{2}{3} = \frac{e^2 n}{(-i\omega + 1/\tau^{\text{tr}})\, m} \mathbf{E}, \tag{15.39}
\end{aligned}$$

where we have used the relation between density and k_F, $n = k_F^3/3\pi^2$. The result for the conductivity is

$$\sigma = \sigma_0 \frac{1}{1 - i\omega\tau^{\text{tr}}}; \quad \sigma_0 = \frac{n e^2 \tau^{\text{tr}}}{m}, \tag{15.40}$$

which agrees with Eq. (15.24) found by the simplified analysis. The reason that the two approaches give the same result is that we can treat the quasiparticle as being independent, and the analysis that was applied in the fluid dynamical picture in Eq. (15.24) is indeed applicable to each quasiparticle separately.

Often one uses an even simpler approximation for the collision term, namely the so-called relaxation time approximation. In this approximation the collision is replaced by

$$\left(\frac{\partial}{\partial t} n_{\mathbf{k}} \right)_{\text{collisions}} = -\frac{n_{\mathbf{k}} - n_{\mathbf{k}}^0}{\tau_0}, \tag{15.41}$$

where $n_{\mathbf{k}}^0$ is the equilibrium distribution function, and τ_0 is the relaxation time. This approximation in fact gives the correct answer if the relaxation time is identified with the transport time $\tau_0 = \tau^{\text{tr}}$. At first sight, it is tempting to think of the τ_0 as the time for scattering out of the state $\mathbf{k}$, i.e., the lifetime of the state $\mathbf{k}$. This would, however, only give the first term in the right-hand side of Eq. (15.31) and it is therefore incorrect. The lifetime, which was also calculated in Eq. (12.49), is given by the first Born approximation,

$$\frac{1}{\tau_{\text{life}}} = \frac{n_{\text{imp}}}{\mathcal{V}} \sum_{\mathbf{k}'} W_{\mathbf{k}',\mathbf{k}}. \tag{15.42}$$

This time expresses the rate for scattering out of a given state $\mathbf{k}$, but it does not tell us how much the momentum is degraded by the scattering process. This is precisely what the additional cosine term in Eq. (15.38) accounts for. If the quasiparticle scatters forward, i.e., $\cos\theta \approx 1$, the state $\mathbf{k}$ is destroyed but the momentum is almost conserved and such a process therefore does not effect the conductivity. On the contrary when the particle is scattered backward, i.e., $\cos\theta \approx -1$, there is a large change in momentum, corresponding to a large momentum relaxation. Therefore the transport time is precisely the momentum-relaxation time defined in the simple fluid dynamical picture in Eq. (15.24).

15.3.1 *Finite lifetime of the quasiparticles*

Above we first assumed that the quasiparticles have an infinite lifetime. Then we included some finite lifetime induced by scattering against impurities. But we never included scattering of quasiparticle on other quasiparticles. Here we investigate the validity of this approach by studying the rate of quasiparticle-quasiparticle scattering. Clearly, there is a mechanism for quasiparticle scattering against quasiparticles because they are charged and therefore interact through the Coulomb interaction. The interaction between the particles is screened by the other particles and we should use the RPA result for the interaction. The Coulomb interactions thus introduces a two-particle scattering where momentum and energy are exchanged, but total momentum and total energy are conserved in the scattering event. If two particles in states $|\mathbf{k}\sigma; \mathbf{k}'\sigma'\rangle$ scatter, the final state will be a state $|\mathbf{k}+\mathbf{q}\sigma; \mathbf{k}'-\mathbf{q}\sigma'\rangle$, such that the initial and the final energies are the same $\varepsilon_\mathbf{k}+\varepsilon_{\mathbf{k}'} = \varepsilon_{\mathbf{k}+\mathbf{q}}+\varepsilon_{\mathbf{k}'-\mathbf{q}}$ or counting from the chemical potential $\xi_\mathbf{k}+\xi_{\mathbf{k}'} = \xi_{\mathbf{k}+\mathbf{q}}+\xi_{\mathbf{k}'-q}$. The rate for quasiparticle-quasiparticle scattering can be calculated using Fermi's golden rule,

$$\Gamma_{\mathbf{k}+\mathbf{q}\sigma,\mathbf{k}'-\mathbf{q}\sigma';\mathbf{k}'\sigma',\mathbf{k}\sigma} = 2\pi \left|\langle \mathbf{k}+\mathbf{q}\sigma, \mathbf{k}'-\mathbf{q}\sigma'|W^{\mathrm{RPA}}(\mathbf{q})|\mathbf{k}'\sigma',\mathbf{k}\sigma\rangle\right|^2$$
$$\times \delta(\xi_\mathbf{k}+\xi_{\mathbf{k}'}-\xi_{\mathbf{k}+\mathbf{q}}-\xi_{\mathbf{k}'-\mathbf{q}}), \tag{15.43}$$

where $W^{\mathrm{RPA}}(\mathbf{q})$ is the RPA-screened Coulomb interaction. From this rate we can obtain the total rate for changing the state of a given quasiparticle in state $|\mathbf{k}\sigma\rangle$ by the Coulomb interaction. The total rate is found by multiplying Γ with the probability that the state $|\mathbf{k}'\sigma'\rangle$ is occupied and that the final states are unoccupied and then summing over all possible $\mathbf{k}'$ and $\mathbf{q}$. So the result for this "lifetime" $\tau_\mathbf{k}$ of the state $|\mathbf{k}\sigma\rangle$ is given by

$$\frac{1}{\tau_\mathbf{k}} = \overset{\text{spin}}{2} \frac{2\pi}{V^2} \sum_{\mathbf{k}'\mathbf{q}} \overset{\text{screened interaction}}{\left|\frac{W(\mathbf{q})}{\varepsilon^{\mathrm{RPA}}(\mathbf{q},0)}\right|^2} \delta\left(\xi_\mathbf{k} + \xi_{\mathbf{k}'} - \xi_{\mathbf{k}+\mathbf{q}} - \xi_{\mathbf{k}'-\mathbf{q}}\right)$$

$$\times \Bigg\{ \underbrace{n_\mathbf{k}n_{\mathbf{k}'}\left[1 - n_{\mathbf{k}+\mathbf{q}}\right]\left[1 - n_{\mathbf{k}'-\mathbf{q}}\right]}_{\text{scattering out of state } \mathbf{k}}$$

$$- \underbrace{\left[1 - n_\mathbf{k}\right]\left[1 - n_{\mathbf{k}'}\right]n_{\mathbf{k}+\mathbf{q}}n_{\mathbf{k}'-\mathbf{q}}}_{\text{scattering into state } \mathbf{k}} \Bigg\}. \tag{15.44}$$

The expression (15.44) can be evaluated explicitly for a particle in state $\mathbf{k}$ added to a filled Fermi sea, i.e., $n_\mathbf{k} = 1$ and $n_\mathbf{p} = \theta(k_F - p)$ for $\mathbf{p}$ equal to $\mathbf{k}', \mathbf{k}' - \mathbf{q}$, or (a)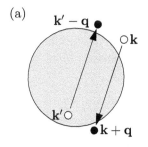

(b)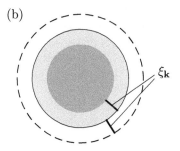

FIG. 15.1. (a) A **k**-space representation of the two-particle scattering event near the Fermi surface that gives rise to a finite lifetime of the quasiparticles. (b) Both momentum and energy have to be conserved. This together with the Pauli principle cause the phase space available for the scattering to be very limited as illustrated. The dashed circle indicates the energy of the initial state. Since the particle can only loose energy, the other particle which is scattered out of state **k′** can only gain energy. Furthermore, because of the Pauli principle the final states of both particles have to lie outside the Fermi surface and therefore the phase space volume for the final state **k + q** (white area) and for the initial state **k′** (light gray area) both scale with $\xi_{\mathbf{k}}$ giving rise to a maximum phase space $(4\pi k_F^2 \xi_{\mathbf{k}})^2 \propto \xi_{\mathbf{k}}^2$.

k + q. But for now we just want the energy dependence of the lifetime. A simple phase space argument gives the answer (see also Fig. 15.1). We look at the situation with a particle above the Fermi surface $\xi_{\mathbf{k}} > 0$. Suppose then we have integrated out the angle dependence, which takes care of the delta function. At $T = 0$ this gives the condition that $\xi_{\mathbf{k}} + \xi_{\mathbf{k}'} - \xi_{\mathbf{k+q}} > 0$. Then we are left with two energy integrals over $\xi_{\mathbf{k}'} \equiv \xi' < 0$ and $\xi_{\mathbf{k}'-\mathbf{q}} \equiv \xi'' > 0$. We then have

$$\frac{1}{\tau_{\mathbf{k}}} \sim |W|^2 \left[d(\varepsilon_F)\right]^3 \int_{-\infty}^{0} d\xi' \int_{0}^{\infty} d\xi'' \Theta(\xi_{\mathbf{k}} + \xi' - \xi'')$$

$$= |W|^2 \left[d(\varepsilon_F)\right]^3 \int_{-\infty}^{0} d\xi' \, (\xi_{\mathbf{k}} + \xi') \, \Theta(\xi_{\mathbf{k}} + \xi')$$

$$= |W|^2 \left[d(\varepsilon_F)\right]^3 \frac{\xi_{\mathbf{k}}^2}{2}, \quad \text{for } T < \xi_k. \tag{15.45}$$

This is a very important result because it tells us that the lifetime of the quasiparticles diverges as we approach the Fermi level and thus the notion of quasiparticles is a consistent picture. At finite temperature, the typical excitation energy is $k_B T$ and $\xi_{\mathbf{k}}$ is replaced by $k_B T$

$$\frac{1}{\tau_{\mathbf{k}}} \propto T^2, \quad \text{for } T > \xi_k. \tag{15.46}$$

The conclusion from the analysis is that the lifetime of the quasiparticles based on Fermi's golden rule diverges at low temperatures and therefore the condition for the adiabatic approach expressed in Eq. (15.8) holds as long the temperature is much smaller than the Fermi energy. Because the Fermi energy in, e.g., a metal in general is a fairly large energy scale, Eq. (2.27), the condition in fact holds for even moderately elevated temperatures. As illustrated in Fig. 15.1 the physical reason for the smallness of the scattering rate is that although the Coulomb scattering matrix elements are big, the Pauli principle severely restricts the phase space available for scattering.

15.4 Microscopic basis of the Fermi liquid theory

15.4.1 *Renormalization of the single particle Green's function*

The Fermi liquid theory relies on the assumption that the excitation created by adding a particle to the system can be described by a free particle, a quasiparticle, with a long lifetime. The function that measures precisely the density of states for adding particles is the retarded Green's function G^R. If the retarded Green's function of the interacting system turns out to be similar to that of free particles, the quasiparticle picture will have a real physical meaning. This is what we are going to show in this section and thereby give a microscopic foundation of the Fermi liquid theory.

We consider the one-particle retarded Green's function, which in general can be written as

$$G^R(\mathbf{k}\sigma, \omega) = \frac{1}{\omega - \xi_{\mathbf{k}} - \Sigma^R(\mathbf{k}\sigma, \omega)}, \qquad (15.47)$$

where $\xi_k = k^2/2m - \mu$ is the free electron energy measured with the respect to the chemical potential μ, and where $\Sigma^R(\mathbf{k}, \omega)$ is the irreducible retarded self-energy. To calculate the self-energy we should in principle include all possible diagrams, which of course is not possible in the general case. Fortunately, important conclusions can be drawn from the first non-trivial approximation, namely the RPA which in Chapter 12 was shown to give the exact answer in the high-density limit. Let us first write the general form of G^R by separating the self-energy in real and imaginary parts,

$$G^R(\mathbf{k}\sigma, \omega) = \frac{1}{\omega - \left[\xi_{\mathbf{k}} + \operatorname{Re}\Sigma^R(\mathbf{k}, \omega)\right] - i\operatorname{Im}\Sigma^R(\mathbf{k}, \omega)}. \qquad (15.48)$$

We then anticipate the quasiparticle picture by looking at $\mathbf{k}$-values close to the $\tilde{k}_F$, meaning close to the renormalized Fermi energy. The renormalized Fermi wave number $\tilde{k}_F$ is defined by the condition that the real part of the energy vanishes $\xi_{\tilde{k}_F} + \operatorname{Re}\Sigma(\tilde{k}_F, 0) = 0$. At small energies and for k close to $\tilde{k}_F$, we can expand $(G^R)^{-1}$ in powers of $k - \tilde{k}_F$ and ω, which leads to

$$G^R(\mathbf{k}, \omega) \approx \left[\omega - \omega\partial_\omega \operatorname{Re}\Sigma^R - (k - \tilde{k}_F)\partial_k(\xi + \operatorname{Re}\Sigma^R) - i\operatorname{Im}\Sigma^R\right]^{-1}$$

$$\equiv Z\left[\omega - \tilde{\xi}_{\mathbf{k}} + \frac{i}{2\tilde{\tau}_{\mathbf{k}}(\omega)}\right]^{-1}, \qquad (15.49)$$

where

$$Z^{-1} = 1 - \frac{\partial}{\partial \omega} \operatorname{Re} \Sigma(\tilde{k}_F, \omega) \Big|_{\omega=0}, \tag{15.50}$$

$$\tilde{\xi}_{\mathbf{k}} = (k - \tilde{k}_F) Z \frac{\partial}{\partial \mathbf{k}} (\xi_{\mathbf{k}} + \operatorname{Re} \Sigma(\mathbf{k}, 0))_{k=\tilde{k}_F}, \tag{15.51}$$

$$\frac{1}{\tilde{\tau}_{\mathbf{k}}(\omega)} = -2Z \operatorname{Im} \Sigma^R(\mathbf{k}, \omega). \tag{15.52}$$

The imaginary part of $\Sigma^R(\mathbf{k}, \omega)$ is not expanded, because we look at its form later. The effective energy $\tilde{\xi}_{\mathbf{k}}$ is usually expressed as

$$\tilde{\xi}_{\mathbf{k}} = \frac{1}{m^*} (k - \tilde{k}_F)\tilde{k}_F, \tag{15.53}$$

where the effective mass by Eq. (15.51) is seen to be

$$\frac{m}{m^*} = Z \left(1 + \frac{m}{\tilde{k}_F} \frac{\partial}{\partial k} \operatorname{Re} \Sigma(k, 0) \Big|_{k=\tilde{k}_F} \right). \tag{15.54}$$

In Section 15.3.1 we saw that the lifetime goes to infinity at low temperatures. If this also holds here the spectral function therefore has a Lorentzian shape near $k = \tilde{k}_F$. For a very small imaginary part we could namely approximate $\operatorname{Im} \Sigma^R \approx -\eta$, and hence Eq. (15.49) gives

$$A(\mathbf{k}, \omega) = -2\operatorname{Im} G^R(\mathbf{k}, \omega) \approx 2\pi Z \delta(\omega - \tilde{\xi}_{\mathbf{k}}). \tag{15.55}$$

This shows that with a small imaginary part, the Green's function and the spectral function has a sharp peak at $\omega = \tilde{\xi}_{\mathbf{k}}$. The peaked spectral function therefore resembles that of a free gas and the pole is identified as the quasiparticle that was defined in the Fermi liquid theory. However, because the general sum rule

$$\int_{-\infty}^{\infty} \frac{d\omega}{2\pi} A(\mathbf{k}, \omega) = 1 \tag{15.56}$$

is not fulfilled by Eq. (15.55), the integral only amounts to Z, the quasiparticle peak cannot be the whole story. There must be another part of the spectral function, which we denote A', that has an integrated weight given by $1 - Z$. See Fig. 15.2. Therefore we instead write

$$A(\mathbf{k}, \omega) = 2\pi Z \, \delta(\omega - \tilde{\xi}_{\mathbf{k}}) + A'(\mathbf{k}, \omega), \tag{15.57}$$

where the remaining contribution A', not associated with the pole, contains more complicated many-body excitations not describable by a free-electron like peak. The constant Z is called the renormalization constant and it is a measure of the quasiparticle weight. Typically Z is found from experiments to be between 0.7 and 1 for $r_s < 3$, where $r_s = (3/4\pi a_0^3 n)^{-1/3}$ is the density parameter of the electron gas, see Eq. (2.37). The renormalization constant shows up, e.g., in the distribution function $n(\mathbf{k})$, where the jump at the Fermi level is a direct measure of Z. For a discussion on the measurements of Z using Compton scattering (see, e.g., the book by Mahan (1990)).

We still need to show that the assumption of a large $\tau_{\mathbf{k}}$ is valid, so we now turn to evaluating the imaginary part of the self-energy.

$$A(\mathbf{k}, \omega) = A_{\rm qp}(\mathbf{k}, \omega) + A'(\mathbf{k}, \omega)$$

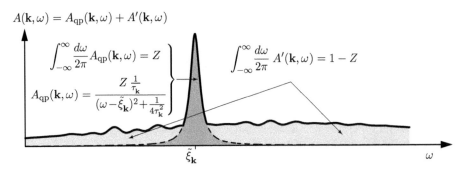

$$\int_{-\infty}^{\infty} \frac{d\omega}{2\pi} A_{\rm qp}(\mathbf{k}, \omega) = Z$$

$$\int_{-\infty}^{\infty} \frac{d\omega}{2\pi} A'(\mathbf{k}, \omega) = 1 - Z$$

$$A_{\rm qp}(\mathbf{k}, \omega) = \frac{Z \frac{1}{\tau_{\mathbf{k}}}}{(\omega - \tilde{\xi}_{\mathbf{k}})^2 + \frac{1}{4\tau_{\mathbf{k}}^2}}$$

FIG. 15.2. The spectral function $A(\mathbf{k}, \omega)$ as resulting from the analysis of the RPA approximation. It contains a distinct peak, which is identified with the quasiparticle. This part called $A_{\rm qp}$, however, only carries part of the integrated spectral weight and the rest must therefore be contained in the background function A' stemming from other types of excitations.

15.4.2 Imaginary part of the single-particle Green's function

We base our analysis on the most important diagram, the RPA self-energy , in Eq. (14.13). In the Matsubara frequency domain it is given by

$$\Sigma_{\sigma}^{\rm RPA}(\mathbf{k}\sigma, ik_n) = -\frac{1}{\beta} \sum_{i\omega_n} \frac{1}{\mathcal{V}} \sum_{\mathbf{q}} \frac{W(\mathbf{q})}{\varepsilon^{\rm RPA}(\mathbf{q}, i\omega_n)} \mathcal{G}_0(\mathbf{k}+\mathbf{q}, \sigma; ik_n + i\omega_n), \qquad (15.58)$$

where $W/\varepsilon^{\rm RPA}$ is the screened interaction. As usual, we perform the Matsubara summation by a contour integration

$$\Sigma_{\sigma}^{\rm RPA}(\mathbf{k}\sigma, ik_n) = -\int_{\mathcal{C}} \frac{dz}{2\pi i} n_B(z) \frac{1}{\mathcal{V}} \sum_{\mathbf{q}} \frac{W(\mathbf{q})}{\varepsilon^{\rm RPA}(\mathbf{q}, z)} \mathcal{G}_0(\mathbf{k}+\mathbf{q}, \sigma; ik_n + z), \qquad (15.59)$$

where $\mathcal{C}$ is a suitable contour that encloses all the bosonic Matsubara frequencies $z = i\omega_n$. The integrand is analytic everywhere except in $z = \xi_{\mathbf{k}+\mathbf{q}} - ik_n$ and for z purely real. If we use a contour like $\mathcal{C} = \mathcal{C}_1 + \mathcal{C}_2$ in Fig. 11.3, we will include all the Matsubara frequencies except the one in origin (note that the points shown in Fig. 11.3) are the fermionic Matsubara frequencies). Therefore, we add a loop $\mathcal{C}_3$ around the origin, so that the contour $\mathcal{C} = \mathcal{C}_1 + \mathcal{C}_2 + \mathcal{C}_3$ shown in Fig. 15.3 includes all boson Matsubara frequencies $z = i\omega_n$. The small loop $\mathcal{C}_3$ shown in Fig. 15.3 is now seen to cancel parts of the counters $\mathcal{C}_1$ and $\mathcal{C}_2$ so that they are modified to run between $]-\infty, -\delta]$ and $[\delta, \infty[$ only, and this is equivalent to stating that the integration are replaced by the principal part, when letting $\delta \to 0^+$. As seen in Fig. 15.3 we, however, also enclose the pole in $z = \xi_{\mathbf{k}+\mathbf{q}} - ik_n$, which we therefore have to subtract again. Now , we obtain

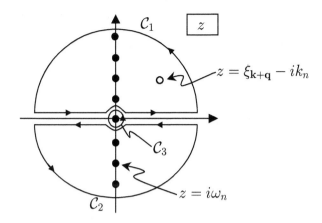

FIG. 15.3. The contour $\mathcal{C} = \mathcal{C}_1 + \mathcal{C}_2 + \mathcal{C}_3$ used for integration for a the Matsubara sum that enters the RPA self-energy in Eq. (15.59). The poles from the boson frequencies are shown by black dots, while that of $\mathcal{G}_0$ is the white dot. The contour $\mathcal{C}_3$ which picks up the contribution from the pole $z = 0$ cancels the parts of $\mathcal{C}_1$ and $\mathcal{C}_2$ given by the small loops.

$$\Sigma^{\text{RPA}}(\mathbf{k}\sigma, ik_n) = -\frac{1}{\mathcal{V}} \sum_{\mathbf{q}} \mathcal{P} \int_{-\infty}^{\infty} \frac{d\omega}{2\pi i} n_B(\omega)$$

$$\times \left[\frac{W(\mathbf{q})}{\varepsilon^{\text{RPA}}(\mathbf{q}, \omega + i\eta)} \mathcal{G}_0(\mathbf{k} + \mathbf{q}, \sigma; ik_n + \omega) - (\eta \to -\eta) \right]$$

$$+ \frac{1}{\mathcal{V}} \sum_{\mathbf{q}} n_B(\xi_{\mathbf{k}+\mathbf{q}} - ik_n) \left[\frac{W(\mathbf{q})}{\varepsilon^{\text{RPA}}(\mathbf{q}, \xi_{\mathbf{k}+\mathbf{q}} - ik_n)} \right]. \tag{15.60}$$

In the last term we should use that $n_B(\xi_{\mathbf{k}+\mathbf{q}} - ik_n) = -n_F(\xi_{\mathbf{k}+\mathbf{q}})$ because ik_n is a fermion frequency. Now, that we have performed the Matsubara sum, we are allowed to get the retarded self-energy by the substitution $ik_n \to \varepsilon + i\eta$, which leads to

$$\Sigma^{\text{RPA},R}(\mathbf{k}\sigma, \varepsilon) = -\frac{1}{\mathcal{V}} \sum_{\mathbf{q}} \mathcal{P} \int_{-\infty}^{\infty} \frac{d\omega}{2\pi i} n_B(\omega)$$

$$\times (2i) \, \text{Im} \left[\frac{1}{\varepsilon^{\text{RPA}}(\mathbf{q}, \omega + i\eta)} \right] W(\mathbf{q}) G_0^R(\mathbf{k} + \mathbf{q}, \sigma; \varepsilon + \omega)$$

$$- \frac{1}{\mathcal{V}} \sum_{\mathbf{q}} n_F(\xi_{\mathbf{k}+\mathbf{q}}) \left[\frac{W(\mathbf{q})}{\varepsilon^{\text{RPA}}(\mathbf{q}, \xi_{\mathbf{k}+\mathbf{q}} - \varepsilon - i\eta)} \right], \tag{15.61}$$

because $\left[\varepsilon^{\text{RPA}}(\mathbf{q}, \omega + i\eta) \right] = \left[\varepsilon^{\text{RPA}}(\mathbf{q}, \omega - i\eta) \right]^*$. The imaginary part of the self-energy becomes

$$\text{Im} \, \Sigma^{\text{RPA}}(\mathbf{k}\sigma, \varepsilon) = \frac{1}{\mathcal{V}} \sum_{\mathbf{q}} [n_B(\xi_{\mathbf{k}+\mathbf{q}} - \varepsilon) + n_F(\xi_{\mathbf{k}+\mathbf{q}})] \, \text{Im} \left[\frac{W(\mathbf{q})}{\varepsilon^{\text{RPA}}(\mathbf{q}, \xi_{\mathbf{k}+\mathbf{q}} - \varepsilon + i\eta)} \right],$$

$$\tag{15.62}$$

where we used that $-\operatorname{Im} G_0^R(\mathbf{k}+\mathbf{q}, \sigma; \varepsilon+\omega) = \pi\delta(\varepsilon+\omega-\xi_{\mathbf{k}+\mathbf{q}})$ and then performed the ω-integration. Since we are interested in the case where a particle with $\xi_{\mathbf{k}}$ is scattered, we evaluate the imaginary part in $\varepsilon = \xi_{\mathbf{k}}$ and find

$$\frac{1}{\tau_{\mathbf{k}}} = -2\operatorname{Im}\Sigma^{\text{RPA}}(\mathbf{k}\sigma, \xi_{\mathbf{k}})$$

$$= \frac{-2}{\mathcal{V}}\sum_{\mathbf{q}}\left[n_B(\xi_{\mathbf{k}+\mathbf{q}} - \xi_{\mathbf{k}}) + n_F(\xi_{\mathbf{k}+\mathbf{q}})\right]\left|\frac{W(q)}{\varepsilon^{\text{RPA},R}(\mathbf{q}, \xi_{\mathbf{k}+\mathbf{q}} - \xi_{\mathbf{k}})}\right|^2$$

$$\times \operatorname{Im}\chi_0^R(\mathbf{q}, \xi_{\mathbf{k}+\mathbf{q}} - \xi_{\mathbf{k}}). \tag{15.63}$$

The imaginary part of the polarization function follows from Eq. (14.21)

$$\operatorname{Im}\chi_0^R(\mathbf{q}, \xi_{\mathbf{k}+\mathbf{q}}-\xi_{\mathbf{k}}) = \frac{2\pi}{\mathcal{V}}\sum_{\mathbf{k}'}\left[n_F(\xi_{\mathbf{k}'}) - n_F(\xi_{\mathbf{k}'-\mathbf{q}})\right]\delta(\xi_{\mathbf{k}'}-\xi_{\mathbf{k}'-\mathbf{q}}-\xi_{\mathbf{k}+\mathbf{q}}+\xi_{\mathbf{k}}) \tag{15.64}$$

here we have shifted $\mathbf{k}' \to \mathbf{k}' - \mathbf{q}$ as compared with Eq. (14.21), and when this is inserted back into (15.63), we obtain

$$\frac{1}{\tau_{\mathbf{k}}} = -2\operatorname{Im}\Sigma^{\text{RPA}}(\mathbf{k}\sigma, \varepsilon)$$

$$= -\frac{4\pi}{\mathcal{V}^2}\sum_{\mathbf{q}\mathbf{k}'}\left[n_B(\xi_{\mathbf{k}+\mathbf{q}} - \xi_{\mathbf{k}}) + n_F(\xi_{\mathbf{k}+\mathbf{q}})\right]\left[n_F(\xi_{\mathbf{k}'}) - n_F(\xi_{\mathbf{k}'-\mathbf{q}})\right]$$

$$\times \left|\frac{W(q)}{\varepsilon^{\text{RPA},R}(\mathbf{q}, \xi_{\mathbf{k}+\mathbf{q}} - \xi_{\mathbf{k}})}\right|^2 \delta(\xi_{\mathbf{k}'} - \xi_{\mathbf{k}'-\mathbf{q}} - \xi_{\mathbf{k}+\mathbf{q}} + \xi_{\mathbf{k}}). \tag{15.65}$$

Let us study the occupation factors in this expression and compare with the Fermi's golden rule expression Eq. (15.44). For the first term in the first parenthesis, we use the identity

$$n_B(\epsilon_1 - \epsilon_2)\left[n_F(\epsilon_2) - n_F(\epsilon_1)\right] = n_F(\epsilon_1)\left[1 - n_F(\epsilon_2)\right], \tag{15.66a}$$

combined with $\xi_{\mathbf{k}} - \xi_{\mathbf{k}+\mathbf{q}} = -\xi_{\mathbf{k}'} + \xi_{\mathbf{k}'-\mathbf{q}}$. For the second term, we use the obvious identity

$$n_F(\epsilon_1)\left[1 - n_F(\epsilon_2)\right] - n_F(\epsilon_2)\left[1 - n_F(\epsilon_1)\right] = n_F(\epsilon_1) - n_F(\epsilon_2) \tag{15.66b}$$

and $n_F(-\epsilon) = 1 - n_F(\epsilon)$. Altogether this allows us to write the occupation factors in Eq. (15.65) as

$$-n_F(\xi_{\mathbf{k}'})[1 - n_F(\xi_{\mathbf{k}+\mathbf{q}})]\left[1 - n_F(\xi_{\mathbf{k}'-\mathbf{q}'})\right] - [1 - n_F(\xi_{\mathbf{k}'})]n_F(\xi_{\mathbf{k}+\mathbf{q}})n_F(\xi_{\mathbf{k}'-\mathbf{q}}) \tag{15.67}$$

At low temperature, the first term is due to the energy conservation condition non-zero for $\xi_{\mathbf{k}} > 0$, while the last term is non-zero for $\xi_{\mathbf{k}} < 0$. The first term thus corresponds to the scattering-out term in Eq. (15.44), while the second term corresponds to the scattering in term. If we furthermore approximate $\xi_{\mathbf{k}} - \xi_{\mathbf{k}+\mathbf{q}} \approx 0$ in ε^{RPA} we now see that the lifetime in Eq. (15.65) is equivalent to the Fermi's golden rule expression Eq. (15.44). We have thus verified that the imaginary part of the retarded Green's function indeed goes to zero. At least when employing the RPA approximation for the self-energy, but the RPA approximation in Chapter 12 was shown to be exact in the high-density limit. An explicit calculation of Eq. (15.65) was done by Quinn and Ferrell[55] who got

$$\frac{1}{\tau_{\mathbf{k}}} = \frac{\sqrt{3}\pi^2}{128}\omega_p\left(\frac{\xi_k}{\varepsilon_F}\right)^2.$$ (15.68)

Going beyond RPA, it can in fact be shown that the imaginary part vanishes to all orders in the interaction. This was done by Luttinger[56] who proved that the imaginary part of any diagram for the self-energy goes to zero as ξ^2 or faster. The derivation is rather lengthy and we do not give it here. It is however not hard to imagine that more complicated scattering events than the simple one depicted in Fig. 15.1 will have even more constraints on the energies. Hence after integration, they will result in higher powers of $\xi_{\mathbf{k}}$. This concludes our analysis of the single-particle Green's function. The analysis indeed confirmed the physical picture put forward by Landau in his Fermi liquid theory.

15.4.3 *Mass renormalization?*

In the previous section we saw how the assumption of weakly interacting quasiparticles was justified by the long lifetime of the single-particle Green's function. We also found that the effective mass of the quasiparticle was renormalized due to the interactions. This seems to contradict the postulate of the Fermi liquid theory that the current of the quasiparticles is independent of interactions, i.e., it is given by $\mathbf{k}/m$ and not $\mathbf{k}/m^*$. The bare velocity of the quasiparticles was important for obtaining the Drude formula for the conductivity, $\sigma = ne^2\tau^{\mathrm{tr}}/m$. How come the renormalized mass m^* appears in the Green's function whereas the physically observable conductivity contains the bare mass m? The answer to this question is found by studying how the conductivity is calculated diagrammatically. As we remember from the Kubo formula, the conductivity is related to the current-current correlation function. The calculation has to be done in a consistent way, such that the diagrams included in the irreducible self-energy, are also included in the diagrams for the two-particle correlation function. When the same type of diagrams are included both in the self-energy and in the lines that cross the two-particle "bubble" then the mass renormalization exactly cancels. In Section 16.4 we shall see an explicit example of this, by calculating diagrammatically the finite resistance due to impurity scattering starting from the fully microscopic theory.

15.5 Summary and outlook

We have developed the semi-classical Fermi liquid theory of interacting particles. The theory is valid whenever perturbation theory is valid, i.e., when the interaction does not induce a phase transition. Almost miraculously, the interacting system of particles can be described by a gas of non-interacting particles. We call these particles quasiparticles and they can be labelled by the same quantum numbers as those of the

[55] J. J. Quinn and R. A. Ferrell, Phys. Rev **112**, 812 (1958).

[56] J.M. Luttinger, Phys. Rev. **121**, 942 (1961).

non-interacting system, provided that the corresponding operators also commute with the full Hamiltonian. For a translation-invariant system the quantum numbers are $\mathbf{k}$ and σ.

On long length and timescales, we can use a semi-classical approach to study various properties. This approach is based on the Boltzmann equation

$$\partial_t(n_{\mathbf{k}}) + \dot{\mathbf{k}} \cdot \nabla_{\mathbf{k}} n_{\mathbf{k}} + \mathbf{v}_{\mathbf{k}} \cdot \nabla_{\mathbf{r}} n_{\mathbf{k}} = \left(\frac{\partial n_{\mathbf{k}}}{\partial t}\right)_{\text{collisions}}. \tag{15.69}$$

This equation is extremely useful, since in many situations it gives a sufficiently accurate description of the physics. It has been widely used to explain numerous transport phenomena in gases and solids. One can include both electric and magnetic fields driving the system out of equilibrium. The driving fields enter through the Lorentz force as $\dot{\mathbf{p}} = \hbar \dot{\mathbf{k}} = (-e)(\mathbf{E} + \mathbf{v} \times \mathbf{B})$. On the right-hand side of Eq. (15.69) we have included collisions due to impurities and particle-particle collisions. One can also include, e.g., particle-phonon scattering in solids and thus explain the temperature dependence of the different transport coefficients.

Landau's phenomenological theory was shown to be justified by a rigorous microscopic calculation, using the random phase approximation result for the self-energy. The result of this analysis was that even in the presence of interactions does the Fermi surface persist, and near the Fermi surface the imaginary part of the single-particle Green's function rapidly vanish as

$$\text{Im}\, \Sigma^R(k_F, \varepsilon) \propto \max(\varepsilon^2, T^2). \tag{15.70}$$

This explains why the Fermi liquid theory works: when the imaginary part goes to zero the single-particle Green's function is identical to that of a free particle.

16

IMPURITY SCATTERING AND CONDUCTIVITY

We now return to the problem of calculating the resistance of a metallic conductor due to scattering against impurities. The basic physics of impurity scattering was discussed in Chapter 12, where we saw how the single-particle Green's function acquired a finite lifetime after averaging over the positions of the impurities. In Chapter 15 the conductivity was calculated within the Boltzmann equation approach. We now rederive the Boltzmann equation result starting from a microscopic quantum approach. The advantage of this microscopic approach, besides giving a first principle justification of the Boltzmann equation, is that it can be extended to include correlation and coherence effects that cannot be described in the semi-classical Boltzmann approach. In order to get familiar with the techniques, we therefore start by deriving the semi-classical result. Then we go on to include the quantum mechanical effect known as weak localization, which is due to interference between time reversed paths. Weak localization involves coherent scattering on many impurities, and it can therefore not be explained semi-classically.

In 1979 the weak localization correction to resistivity was observed experimentally in large 2D samples at low temperatures. It was explained theoretically later the same year, and an extended research was initiated on the role of quantum coherence in transport properties. A few years later another low-temperature interference effect, the so-called universal conductance fluctuations, was discovered in small ($\sim \mu$m) phase-coherent structures. This discovery started the modern field of mesoscopic physics. To understand these smaller systems one must take into account the finite size of the conductors, which was the topic in Chapter 10. In this chapter we deal with extended systems and discuss the most important disorder-induced quantum corrections. The leading quantum correction is precisely the weak localization effect in 2D.[57] We also discuss quantum corrections to the average conductance of mesoscopic samples.

Based on the physical picture that emerged from the Fermi liquid description in Chapter 15, we assume in the first part of this chapter that we can describe the electrons as non-interacting. In the second part of the chapter we include electron-electron interactions together with impurity scattering and explicitly demonstrate that the non-interacting approximation is valid. This means that we shall see how the mass renormalization discussed in Sec. 16.4 is canceled out. Furthermore, we shall see that in order to obtain meaningful results, it is absolutely imperative to include vertex corrections to the current-current correlation bubble diagrams. These corrections cannot be treated by evaluating only single-particle Green's functions.

---

[57]In 1D, things are more complicated because there all states are localized. In 3D, the situation is again different in that at some critical amount of impurity scattering there exists a metal-insulator transition known as the Anderson localization.

They are thus genuine two-particle correlation effects, which can be described by diagrams where interaction lines "cross" the bubble diagrams.

16.1   Vertex corrections and dressed Green's functions

Let us start by the Kubo formula for the electrical conductivity tensor $\sigma_{\alpha\beta}$ given in Eq. (6.24) in terms of the retarded current-current correlation function (6.25). Here, we shall only look at the dissipative part of the conductivity, and therefore we take the real part of Eq. (6.24),

$$\operatorname{Re}\sigma_{\alpha\beta}(\mathbf{r},\mathbf{r}';\omega) = -\frac{e^2}{\omega}\operatorname{Im}\Pi_{\alpha\beta}^R(\mathbf{r},\mathbf{r}',\omega). \qquad (16.1)$$

Note that the last, so-called "diamagnetic," term of σ in Eq. (6.24) drops out of the real part. In the following we therefore only include the first, so-called "paramagnetic", term in Eq. (6.24), denoted σ^{∇}. For a translation-invariant system we consider as usual the Fourier transform

$$\sigma_{\alpha\beta}^{\nabla}(\mathbf{q};\omega) = \frac{ie^2}{\omega}\Pi_{\alpha\beta}^R(\mathbf{q},\omega). \qquad (16.2)$$

The DC-conductivity is then found by letting $q \to 0$ and then $\omega \to 0$.[58] The DC-response at long wavelengths is thus obtained as

$$\operatorname{Re}\sigma_{\alpha\beta} = -e^2 \lim_{\omega\to 0}\lim_{q\to 0}\frac{1}{\omega}\operatorname{Im}\Pi_{\alpha\beta}^R(\mathbf{q},\omega). \qquad (16.3)$$

In this chapter we consider only homogeneous translation-invariant systems, i.e. the conductivity tensor is isotropic and therefore diagonal,

$$\sigma_{\alpha\beta} = \sigma\,\delta_{\alpha\beta}. \qquad (16.4)$$

In particular, we have no magnetic field and take $\mathbf{A} = 0$. In the computation we can choose α to be the x direction. Note that the system is translation invariant even in the presence of impurities after performing the position average described in Chapter 12.

As usual, we calculate the retarded function starting from the corresponding Matsubara function. The Matsubara current-current correlation function is

$$\Pi_{xx}(\mathbf{q},\tau-\tau') = -\frac{1}{\mathcal{V}}\Big\langle T_\tau J_x(\mathbf{q},\tau)J_x(-\mathbf{q},\tau')\Big\rangle. \qquad (16.5)$$

In the frequency domain it is

$$\Pi_{xx}(\mathbf{q},iq_n) = -\frac{1}{\mathcal{V}}\int_0^\beta d(\tau-\tau')e^{iq_n(\tau-\tau')}\Big\langle J_x(\mathbf{q},\tau)J_x(-\mathbf{q},\tau')\Big\rangle, \qquad (16.6)$$

[58]If in doubt always perform the limit $q \to 0$ first, because having an electric field $E(\mathbf{q},\omega)$ where $\omega = 0$ and q finite is unphysical, since it would give rise to an infinite charge built up.

where the time-ordering operator T_τ is omitted, because $\tau > \tau'$. We can now express $J_x(\mathbf{q}, \tau)$ in terms of $J_x(\mathbf{q}, iq_n)$ and obtain

$$
\Pi_{xx}(\mathbf{q}, iq_n) = -\frac{1}{\mathcal{V}} \int_0^\beta d(\tau - \tau') \, e^{iq_n(\tau - \tau')}
$$
$$
\times \frac{1}{\beta} \sum_{iq_l} \frac{1}{\beta} \sum_{iq_m} \left\langle J_x(\mathbf{q}, iq_l) J_x(-\mathbf{q}, iq_m) \right\rangle e^{-iq_l \tau} e^{-iq_m \tau'}. \tag{16.7}
$$

The integration with respect to τ leads to $iq_n = iq_l$. Finally, since the result cannot depend on τ', we must have $iq_n = -iq_m$, and whence

$$
\Pi_{xx}(\mathbf{q}, iq_n) = -\frac{1}{\mathcal{V}\beta} \left\langle J_x(\mathbf{q}, iq_n) J_x(-\mathbf{q}, -iq_n) \right\rangle. \tag{16.8}
$$

This we conveniently rewrite using the four-vector notation $\tilde{q} = (iq_n, \mathbf{q})$

$$
\Pi_{xx}(\tilde{q}) = -\frac{1}{\mathcal{V}\beta} \left\langle J_x(\tilde{q}) J_x(-\tilde{q}) \right\rangle. \tag{16.9}
$$

In order to begin the diagrammatical analysis, we write the current density $J_x(\tilde{q})$ in four-vector notation

$$
J_x(\tilde{q}) = \int_0^\beta d\tau e^{iq_n \tau} \frac{1}{2m} \frac{1}{\mathcal{V}} \sum_{k\sigma} (2\mathbf{k} + \mathbf{q})_x c_{\mathbf{k}\sigma}^\dagger(\tau) c_{\mathbf{k}+\mathbf{q}\sigma}(\tau)
$$
$$
= \frac{1}{2m} \frac{1}{\beta} \sum_{ik_n} \frac{1}{\mathcal{V}} \sum_{k\sigma} (2\mathbf{k} + \mathbf{q})_x c_{\mathbf{k}\sigma}^\dagger(ik_n) c_{\mathbf{k}+\mathbf{q}\sigma}(ik_n + iq_n)
$$
$$
\equiv \frac{1}{2m} \frac{1}{\beta} \frac{1}{\mathcal{V}} \sum_{\tilde{k}} \sum_{\sigma} (2k_x + q_x) c_\sigma^\dagger(\tilde{k}) c_\sigma(\tilde{k} + \tilde{q}), \tag{16.10}
$$

which we represent diagrammatically as a vertex

$$
J_x(\tilde{q}) = \qquad\qquad . \tag{16.11}
$$

The vertex conserves four-momentum, and thus has the momentum $\tilde{q} = (iq_n, \mathbf{q})$ flowing out from it to the left.

We can now draw diagrams for the current-current correlation function using the Feynman rules. The procedure is analogous to that for the charge-charge correlation function in Chapter 13, however, here we include both the impurity lines ●– –✿ from Chapter 12 and the Coulomb interaction lines ●〜〜● from Chapter 13. We obtain

$$\Pi_{xx}(\tilde{q}) = $$

$$\equiv$$

$$(16.12)$$

We can perform a partial summation of diagrams to all orders by replacing each Green's function $\mathcal{G}_0$ by the full Green's function $\mathcal{G}$. In doing so, we have in one step resummed Eq. (16.12) and are left with bubble diagrams where the only interaction and impurity lines to be drawn are those connecting the lower and upper electron Green's functions. Eq. (16.12) then becomes

$$\Pi_{xx}(\tilde{q}) = $$

$$(16.13)$$

Here, the double lines represent full Green's functions expressed by Dyson's equation as in Eq. (13.22)

$$\mathcal{G}(\tilde{k}) = \qquad$$

$$= \qquad + \qquad$$

$$= \mathcal{G}_0(\tilde{k}) + \mathcal{G}_0(\tilde{k})\Sigma^{\mathrm{irr}}(\tilde{k})\mathcal{G}(\tilde{k}), \tag{16.14}$$

where $\Sigma^{\mathrm{irr}} = \;$ is the irreducible self-energy. For example in the case where we include impurity scattering within the first Born approximation and electron-electron interaction in the RPA approximation, the irreducible self-energy is simply

$$\text{1BA + RPA:} \quad \Sigma^{\mathrm{irr}}(\tilde{k}) = \qquad \approx \qquad + \qquad , \tag{16.15}$$

where RPA means the following screening of all impurity and interaction lines

$$\qquad = \qquad + \qquad , \tag{16.16}$$

$$\qquad = \qquad + \qquad . \tag{16.17}$$

The next step is to organize the diagrams according to the lines crossing the bubbles from the upper to the lower fermion line in a systematic way. These diagrams are denoted vertex corrections. To obtain a Dyson equation for Π_{xx}, we first introduce the irreducible line-crossing diagram Λ^{irr} consisting of the sum of all possible diagrams connecting the upper and lower fermion line, which cannot be cut into two pieces by cutting both the upper and the lower line just once,[59]

$$\Lambda^{\mathrm{irr}} \equiv \qquad \equiv \qquad + \qquad + \qquad + \qquad + \qquad + \qquad + \cdots$$

$$\tag{16.18}$$

Using Λ^{irr} we see that we can resum all diagrams in Π_{xx} in the following way

---

[59]We do not include diagrams like the terms 9, 10 , and 11 in Eq. (16.13). Diagrams of this type are proportional to q^{-2} and thus they vanish in the limit $q \to 0$.

$$\equiv -\int d\tilde{k}' \, \Gamma_{0,x}(\tilde{k}', \tilde{k}' + \tilde{q}) \mathcal{G}(\tilde{k}') \mathcal{G}(\tilde{k}' + \tilde{q}) \Gamma_x(\tilde{k}' + \tilde{q}, \tilde{k}'), \qquad (16.19)$$

where the unperturbed vertex is

$$\Gamma_{0,x}(\tilde{k}, \tilde{k} + \tilde{q}) = \frac{1}{2m}(2k_x + q_x), \qquad (16.20)$$

and the "dressed" vertex function is given by an integral equation, which can be read off from Eq. (16.19)

$$\equiv \Gamma_{0,x}(\tilde{k}+\tilde{q}, \tilde{k}) + \int d\tilde{q}' \Lambda^{\mathrm{irr}}(\tilde{k}, \tilde{q}, \tilde{q}') \mathcal{G}(\tilde{k}+\tilde{q}') \mathcal{G}(\tilde{k}+\tilde{q}'+\tilde{q}) \Gamma_x(\tilde{k}+\tilde{q}'+\tilde{q}, \tilde{k}+\tilde{q}').$$
$$(16.21\mathrm{b})$$

The notation for the arguments of the Γ functions is $\Gamma = \Gamma(\text{"out going", "in going"})$.

The question is now which diagrams to include in Λ^{irr}. We have seen examples of how to choose the physically most important self-energies, both for the impurity scattering problem in Chapter 12 and for the case of interacting particles in Chapter 13. In the present case, once the approximation for Σ^{irr} is chosen, the answer is simply that there is no freedom left in the choice for the vertex function Γ. If we include certain diagrams for the self-energy, we must include the corresponding diagrams in the vertex function. This follows from a general relation, called the Ward identity, between the self-energy and the vertex function. The Ward identity is derived from the continuity equation, so not fulfilling it is equivalent to not conserving the number of particles. Therefore, a physically sensible approximation must obey Ward's identity, and one uses the term "conserving approximation" for the correct choice for the vertex function. We shall not derive the Ward identity here,[60] but simply use the following rule dictated by the Ward identity: if an irreducible diagram is included in Σ^{irr}, the corresponding diagrams should also be included in Λ^{irr}.

If we consider the first Born approximation and RPA for Σ^{irr} as depicted in Eq. (16.15), we obtain for Λ^{irr}

$$
\Lambda^{\mathrm{irr}} = \boxed{} \approx \;\mathrel{\ast}\; + \;\{\; \equiv \tilde{W}, \tag{16.22}
$$

and in this case the integral function for Γ becomes

$$
\Gamma_x(\tilde{k}+\tilde{q},\tilde{k}) = \Gamma_{0,x}(\tilde{k}+\tilde{q},\tilde{k}) + \int d\tilde{q}'\, \tilde{W}(\tilde{q}')\mathcal{G}(\tilde{k}+\tilde{q}')\mathcal{G}(\tilde{k}+\tilde{q}'+\tilde{q})\Gamma_x(\tilde{k}+\tilde{q}'+\tilde{q},\tilde{k}+\tilde{q}'),
$$
$$
\tag{16.23}
$$

where

$$
\tilde{W}(\tilde{q}) = W^{\mathrm{RPA}}(\tilde{q}) + n_{\mathrm{imp}}\frac{u(\mathbf{q})}{\varepsilon^{\mathrm{RPA}}(q,0)}\frac{u(-\mathbf{q})}{\varepsilon^{\mathrm{RPA}}(-q,0)}\delta_{q_n,0}. \tag{16.24}
$$

This particular approximation is also known as the ladder sum, a name which perhaps becomes clear graphically if Eq. (16.22) for Λ^{irr} is inserted into the first line of Eq. (16.19) for Π_{xx}, and for clarity we consider only the impurity scattering lines:

$$
\Pi_{xx}(\tilde{q}) = \bigcirc + \bigcirc + \bigcirc + \bigcirc + \cdots
$$
$$
\tag{16.25}
$$

The vertex correction in this set of ladder diagrams is known as the diffuson.

16.2 The conductivity in terms of a general vertex function

Having the expressions for both the single-particle Green's function $\mathcal{G}$ and the vertex function Γ, we can obtain from Eq. (16.19) a general formula for the conductivity. This definition involves a summation over the internal Matsubara frequency. If we drop the

---

[60]The Ward identity reads

$$
iq_0\Gamma_0(\tilde{k}+\tilde{q},\tilde{k}) - i\mathbf{q}\cdot\mathbf{\Gamma}(\tilde{k}+\tilde{q},\tilde{k}) = -\mathcal{G}^{-1}(\tilde{k}+\tilde{q}) + \mathcal{G}^{-1}(\tilde{k}),
$$

where the function Γ_0 is the charge vertex function, and $\mathbf{\Gamma}$ is the current vertex function. For more details see, e.g., Schrieffer (1983).

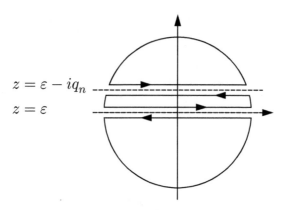

$z = \varepsilon - iq_n$

$z = \varepsilon$

FIG. 16.1. The tripartite contour used in the frequency summation in Eq. (16.27).

four-vector notation in favor of the standard notation, and furthermore treat the case $\mathbf{q} = 0$, the current-current correlation function is

$$\Pi_{xx}(0, iq_n) = -\frac{1}{\beta}\sum_{ik_n}\frac{1}{V}\sum_{\mathbf{k}}\Gamma_{0,x}(\mathbf{k},\mathbf{k})\mathcal{G}(\mathbf{k}, ik_n)\mathcal{G}(\mathbf{k}, ik_n + iq_n)\Gamma_x(\mathbf{k},\mathbf{k};ik_n + iq_n, ik_n).$$

(16.26)

The Matsubara sum over ik_n is performed in the usual way by a contour integration over $z = ik_n$. The presence of two $\mathcal{G}$'s in the summand leads to two branch cuts; one along $z = \varepsilon$ and one along $z = -iq_n + \varepsilon$, with ε being real. Therefore, we first study a summation of the form

$$S_{2F}(iq_n) = \frac{1}{\beta}\sum_{ik_n} f(ik_n, ik_n + iq_n)$$

$$= -\int_C \frac{dz}{2\pi i} n_F(z) f(z, z + iq_n),$$

(16.27)

where the integration contour C is shown in Fig. 16.1. It consists of three parts, leading to four integrals over ε,

$$S_{2F}(iq_n) = -\int_{-\infty}^{\infty}\frac{d\varepsilon}{2\pi i}n_F(\varepsilon)\big[f(\varepsilon + i\eta, \varepsilon + iq_n) - f(\varepsilon - i\eta, \varepsilon + iq_n)\big]$$

$$-\int_{-\infty}^{\infty}\frac{d\varepsilon}{2\pi i}n_F(\varepsilon - iq_n)\big[f(\varepsilon - iq_n, \varepsilon + i\eta) - f(\varepsilon - iq_n, \varepsilon - i\eta)\big].$$

(16.28)

At the end of the calculation we continue iq_n analytically to $\omega + i\eta$, and find

$$S_{2F}^R(\omega) = -\int_{-\infty}^{\infty}\frac{d\varepsilon}{2\pi i}n_F(\varepsilon)\big[f^{RR}(\varepsilon, \varepsilon + \omega) - f^{AR}(\varepsilon, \varepsilon + \omega)$$

$$+ f^{AR}(\varepsilon - \omega, \varepsilon) - f^{AA}(\varepsilon - \omega, \varepsilon)\big],$$

(16.29)

with the convention that $f^{AR}(\varepsilon, \varepsilon')$ means that the first argument is advanced, $\varepsilon - i\eta$, and the second argument is retarded, i.e. $\varepsilon + i\eta$, and so on. If we shift the integration variable $\varepsilon \to \varepsilon + \omega$ in the two last terms, we obtain

$$S_{2F}^R(\omega) = \int_{-\infty}^{\infty} \frac{d\varepsilon}{2\pi i} \left[n_F(\varepsilon) - n_F(\varepsilon + \omega) \right] f^{AR}(\varepsilon, \varepsilon + \omega)$$

$$- \int_{-\infty}^{\infty} \frac{d\varepsilon}{2\pi i} \left[n_F(\varepsilon) f^{RR}(\varepsilon, \varepsilon + \omega) - n_F(\varepsilon + \omega) f^{AA}(\varepsilon, \varepsilon + \omega) \right]. \qquad (16.30)$$

Since we are interested in the low-frequency limit, we expand to first order in ω. Furthermore, we also take the imaginary part as in Eq. (16.3). Since $\left(f^{AA} \right)^* = f^{RR}$, we find

$$\operatorname{Im} S_{2F}^R(\omega) = \omega \operatorname{Im} \int_{-\infty}^{\infty} \frac{d\varepsilon}{2\pi i} \left(-\frac{\partial n_F(\varepsilon)}{\partial \varepsilon} \right) \left[f^{AR}(\varepsilon, \varepsilon) - f^{RR}(\varepsilon, \varepsilon) \right]. \qquad (16.31)$$

At low temperatures, we can approximate the derivative of the Fermi–Dirac function by a delta function

$$\left(-\frac{\partial n_F(\varepsilon)}{\partial \varepsilon} \right) \approx \delta(\varepsilon), \qquad (16.32)$$

and hence

$$\operatorname{Im} S_{2F}^R(\omega) = \frac{\omega}{2\pi} \operatorname{Re} \left[f^{AA}(0,0) - f^{AR}(0,0) \right]. \qquad (16.33)$$

By applying this to Eq. (16.26) and then inserting into Eq. (16.3), one obtains

$$\operatorname{Re} \sigma_{xx} = 2 \operatorname{Re} \frac{e^2}{2\pi} \frac{1}{V} \sum_{\mathbf{k}} \Gamma_{0,x}(\mathbf{k}, \mathbf{k}) \Big[G^A(\mathbf{k}, 0) G^R(\mathbf{k}, 0) \Gamma_x^{RA}(\mathbf{k}, \mathbf{k}; 0, 0)$$

$$- G^A(\mathbf{k}, 0) G^A(\mathbf{k}, 0) \Gamma_x^{AA}(\mathbf{k}, \mathbf{k}; 0, 0) \Big], \qquad (16.34)$$

where we have included a factor of 2 due to spin degeneracy. This is how far one can go on general principles. To proceed further, one must look at the specific physical cases and then solve for the vertex function satisfying Eq. (16.21b) and insert the result into (16.34). In the following we consider various cases, but only those where the disorder is weak. In the next section it is shown that for this case, the product $G^R G^A$ exceeds $G^A G^A$ by a factor of order $1/(\tau E_F)$, where τ is the scattering time for impurity scattering. Hence in the weak-disorder limit, we may replace the general formula in Eq. (16.34) by the first term only.

16.3 The conductivity in the first Born approximation

The conductivity was calculated in Sec. 15.3 using a semi-classical approximation for the scattering against the impurities. The semi-classical approximation is similar to the first Born approximation in that it only includes scattering against a single impurity and neglects interference effects. Therefore, we expect to reproduce the semi-classical result, if we only include the first Born approximation in our diagrammatical calculation. The starting point in this section is non-interacting electrons scattering on impurities. The RPA part of the self-energy in Eq. (16.15) is *not* included in this section. Later we discuss what happens if interactions are included.

The vertex function is now solved using the first Born approximation, i.e. the first diagram in Eq. (16.22). In this case, again taking $\mathbf{q} = 0$, the integral equation (16.23) becomes

$$\Gamma_x(\mathbf{k}, \mathbf{k}; ik_n + iq_n, ik_n) = \Gamma_{0,x}(\mathbf{k}, \mathbf{k}) + \frac{1}{\mathcal{V}} \sum_{\mathbf{q}'} n_{\text{imp}} \left| u^{\text{RPA}}(\mathbf{q}') \right|^2 \mathcal{G}(\mathbf{k} + \mathbf{q}', ik_n) \quad (16.35)$$

$$\times \, \mathcal{G}(\mathbf{k} + \mathbf{q}', ik_n + iq_n) \Gamma_x(\mathbf{k} + \mathbf{q}', \mathbf{k} + \mathbf{q}'; ik_n + iq_n, ik_n),$$

where the second term in Eq. (16.24) has been inserted and where $u^{\text{RPA}} = u/\varepsilon^{\text{RPA}}$. The Green's functions $\mathcal{G}$ are, as we learned from the Ward identity, also those obtained in the first Born approximation. Note that there is no internal Matsubara sum, because the impurity scattering conserves energy. Since we do not expect the dynamical screening to be important for the elastic scattering, we set the frequency in $\varepsilon^{\text{RPA}}(\mathbf{q}, 0)$ to zero. Remembering that Γ_x is a component of a vector function $\mathbf{\Gamma}$, and that the unperturbed vertex is $\mathbf{\Gamma}_0(\mathbf{k}, \mathbf{k}) = \mathbf{k}/m$, we define for convenience a scalar function $\gamma(\mathbf{k}, \varepsilon)$ as $\mathbf{\Gamma}(\mathbf{k}, \varepsilon) = \mathbf{k}\gamma(\mathbf{k}, \varepsilon)/m$. In doing so, we in fact use that the system is isotropic which means that only the vector $\mathbf{k}$ can give the direction. When inserting this into Eq. (16.35), multiplying by $(1/k^2)\mathbf{k}\cdot$, and shifting the variable $\mathbf{q}'$ to $\mathbf{q}' = \mathbf{k}' - \mathbf{k}$, we obtain

$$\gamma(\mathbf{k}, \mathbf{k}; ik_n + iq_n, ik_n) = 1 + \frac{1}{\mathcal{V}} \sum_{\mathbf{k}'} n_{\text{imp}} \left| u^{\text{RPA}}(\mathbf{k}' - \mathbf{k}) \right|^2 \mathcal{G}(\mathbf{k}', ik_n)$$

$$\times \, \mathcal{G}(\mathbf{k}', ik_n + iq_n) \frac{\mathbf{k} \cdot \mathbf{k}'}{k^2} \gamma^{\text{1BA}}(\mathbf{k}', \mathbf{k}'; ik_n + iq_n, ik_n), \quad (16.36)$$

In the formula Eq. (16.34) for the conductivity, both Γ_x^{RA} and Γ_x^{RR} appear (or rather $\Gamma_x^{RR} = (\Gamma_x^{AA})^*$). They satisfy two different integral equations, which we obtain from Eq. (16.36) by letting $iq_n + ik_n \to \omega + \varepsilon + i\eta$ and $ik_n \to \varepsilon \pm i\eta$, and subsequently taking the DC-limit $\omega \to 0$. We arrive at

$$\gamma^{RX}(\mathbf{k}, \varepsilon) = 1 + \frac{1}{\mathcal{V}} \sum_{\mathbf{k}'} n_{\text{imp}} \left| u^{\text{RPA}}(\mathbf{k}' - \mathbf{k}) \right|^2 G^X(\mathbf{k}', \varepsilon) G^R(\mathbf{k}', \varepsilon) \frac{\mathbf{k} \cdot \mathbf{k}'}{k^2} \gamma^{RX}(\mathbf{k}', \varepsilon),$$

$$(16.37)$$

where $X = A$ or R. One immediately sees that the small factor n_{imp} tends to kill the sum, and in the weak scattering limit one should expect the solution of this equation to be simply $\gamma^{RX}(\mathbf{k}, \varepsilon) \approx 1$. It is immediately seen that this is a consistent solution for the imaginary part of both γ^{RA} and γ^{RR}, but it turns out that for the real part of γ^{RA} a factor $1/n_{\text{imp}}$ is contained in the Green's function. The lesson we learn here is that we have to be rather careful with products of Green's function carrying the same arguments, because in the limit of small n_{imp}, $\text{Im}\, G^X$ tends to a delta function, and the product of two delta functions has to be defined with care. Let us look more carefully into the products $G^A G^R$ and $G^R G^R$, which also appear in Eq. (16.34). The first combination is

$$G^A(\mathbf{k}, \varepsilon) G^R(\mathbf{k}, \varepsilon) = \left|G^R(\mathbf{k}, \varepsilon)\right|^2 \equiv \left|\frac{1}{\varepsilon - \xi_{\mathbf{k}} - \Sigma^R(\mathbf{k}, \varepsilon)}\right|^2$$

$$= \frac{1}{\operatorname{Im} \Sigma^R(\mathbf{k}, \varepsilon)} \operatorname{Im} \frac{1}{\varepsilon - \xi_{\mathbf{k}} - \Sigma^R(\mathbf{k}, \varepsilon)}$$

$$\equiv \frac{1}{-2 \operatorname{Im} \Sigma^R(\mathbf{k}, \varepsilon)} A(\mathbf{k}, \varepsilon) \equiv \tau A(\mathbf{k}, \varepsilon), \tag{16.38}$$

where $A = -2 \operatorname{Im} G^R$ is the spectral function, and where, as before, the lifetime τ is defined by $\tau^{-1} = -2 \operatorname{Im} \Sigma^R(\mathbf{k}, \varepsilon)$. For the case of weak impurity scattering the scattering rate τ^{-1} is so small that the spectral function can be approximated by a delta function. In the case of small n_{imp}, we therefore obtain

$$G^A(\mathbf{k}, \varepsilon) G^R(\mathbf{k}, \varepsilon) \approx \tau 2\pi \delta(\varepsilon - \xi_{\mathbf{k}}). \tag{16.39}$$

Because $\tau \propto n_{\text{imp}}^{-1}$, the product $n_{\text{imp}} G^A G^R$ in Eq. (16.37) is finite in the limit $n_{\text{imp}} \to 0$. On the other hand, the combination $G^R G^R$ is not divergent, and in fact $n_{\text{imp}} G^R G^R \to 0$ as $n_{\text{imp}} \to 0$. That $G^R G^R$ is finite, is seen as follows:

$$G^R(\mathbf{k}, \varepsilon) G^R(\mathbf{k}, \varepsilon) = \left(\frac{\varepsilon - \xi_{\mathbf{k}} - \frac{i}{2\tau}}{(\varepsilon - \xi_{\mathbf{k}})^2 + \left(\frac{1}{2\tau}\right)^2}\right)^2$$

$$= \frac{(\varepsilon - \xi_{\mathbf{k}})^2 - \left(\frac{1}{2\tau}\right)^2}{\left((\varepsilon - \xi_{\mathbf{k}})^2 + \left(\frac{1}{2\tau}\right)^2\right)^2} + i(\varepsilon - \xi_{\mathbf{k}}) A(\mathbf{k}, \varepsilon). \tag{16.40}$$

The last term clearly goes to zero when τ is large, and A can be approximated by a delta function. The first term is a peaked function at $\varepsilon - \xi_{\mathbf{k}} = 0$, but the integrated weight is in fact zero as can be checked by performing an integration over $\xi_{\mathbf{k}}$. From these arguments it follows that the terms with $G^R G^R$ can be omitted and only terms with $G^R G^A$ are kept. As explained above, we use the first Born approximation for the self-energy. In the following we therefore approximate τ with the first Born approximation lifetime τ_0,

$$\tau^{-1} \approx \tau_0^{-1} \equiv 2\pi n_{\text{imp}} \sum_{\mathbf{k}'} |u(\mathbf{k} - \mathbf{k}')|^2 \delta(\xi_{\mathbf{k}} - \xi_{\mathbf{k}'}). \tag{16.41}$$

Because all energies are at the Fermi energy, this lifetime is independent of $\mathbf{k}$. The conductivity Eq. (16.34) then becomes

$$\operatorname{Re} \sigma_{xx} = 2e^2 \operatorname{Re} \frac{1}{\mathcal{V}} \sum_{\mathbf{k}} \Gamma_{0,x}(\mathbf{k}, \mathbf{k}) \tau_0 \delta(\xi_{\mathbf{k}}) \Gamma_x^{RA}(\mathbf{k}, \mathbf{k}; 0, 0) \tag{16.42}$$

$$= 2e^2 \tau_0 \operatorname{Re} \frac{1}{\mathcal{V}} \sum_{\mathbf{k}} \frac{k_x}{m} \delta(\xi_{\mathbf{k}}) \frac{k_x}{m} \gamma^{RA}(\mathbf{k}, \mathbf{k}; 0, 0) = \frac{e^2 n}{m} \tau_0 \gamma^{RA}(k_F, k_F; 0, 0).$$

Here, we used the following identity: $\frac{2}{\mathcal{V}} \sum_{\mathbf{k}} (k_x^2/m^2) \delta(\xi_{\mathbf{k}}) = \frac{2}{3\mathcal{V}} \sum_{\mathbf{k}} (k^2/m^2) \delta(\xi_{\mathbf{k}}) = \frac{2m}{\pi^2} \int_0^{\infty} dk \, k^4 \delta(k^2 - k_F^2) = n/m$ (see Eq. (2.26)). The remaining problem is to find $\gamma^{RA}(\mathbf{k}, \mathbf{k}; 0, 0)$ for $|\mathbf{k}| = k_F$. The solution follows from the integral equation Eq. (16.36)

$$\gamma^{RA}(\mathbf{k}) = 1 + \frac{2\pi}{\mathcal{V}} \sum_{\mathbf{k}'} n_{\text{imp}} \left| u^{\text{RPA}}(\mathbf{k}' - \mathbf{k}) \right|^2 \tau_0 \delta(\xi_{\mathbf{k}'}) \frac{\mathbf{k} \cdot \mathbf{k}'}{k^2} \gamma^{RA}(\mathbf{k}'). \qquad (16.43)$$

Since this equation has no dependence on the direction of $\mathbf{k}$, and since the lengths of both $\mathbf{k}$ and $\mathbf{k}'$ are given by k_F, γ^{RA} depends only on k_F. But k_F is constant, so

$$\gamma^{RA} = 1 + \left[\frac{2\pi}{\mathcal{V}} \sum_{\mathbf{k}'} n_{\text{imp}} \left| u^{\text{RPA}}(\mathbf{k}' - \mathbf{k}) \right|^2 \delta(\xi_{\mathbf{k}'}) \frac{\mathbf{k} \cdot \mathbf{k}'}{k^2} \right] \tau_0 \gamma^{RA}, \qquad (16.44)$$

with the simple solution

$$\gamma^{RA} = \frac{1}{1 - \lambda \tau_0}, \qquad (16.45)$$

where

$$\lambda = \frac{2\pi}{\mathcal{V}} \sum_{\mathbf{k}'} n_{\text{imp}} \left| u^{\text{RPA}}(\mathbf{k}' - \mathbf{k}) \right|^2 \delta(\xi_{\mathbf{k}'}) \frac{\mathbf{k} \cdot \mathbf{k}'}{k^2} = (\tau_0)^{-1} - (\tau^{\text{tr}})^{-1}. \qquad (16.46)$$

Here, the transport time τ^{tr} is defined as

$$\left(\tau^{\text{tr}} \right)^{-1} \equiv \frac{2\pi}{\mathcal{V}} \sum_{|\mathbf{k}'|=k_F} n_{\text{imp}} \left| u^{\text{RPA}}(\mathbf{k}' - \mathbf{k}) \right|^2 \left(1 - \frac{\mathbf{k} \cdot \mathbf{k}'}{k^2} \right). \qquad (16.47)$$

This expression is precisely the transport time derived in the Boltzmann-equation approach leading to Eq. (15.38). When inserted back into Eq. (16.45) γ^{RA} becomes

$$\gamma^{RA} = \frac{\tau^{\text{tr}}}{\tau_0}. \qquad (16.48)$$

Finally, the conductivity formula (16.42) at zero temperature is

$$\sigma = \frac{e^2 \tau^{\text{tr}}}{m^2} \frac{1}{\mathcal{V}} \sum_{\mathbf{k}} \delta(\xi_{\mathbf{k}}) k_x^2 = \frac{e^2 n \tau^{\text{tr}}}{m}. \qquad (16.49)$$

As expected, this is in full agreement with the semi-classical result obtained in the previous chapter. Thus, having gained confidence in the mathematical structure of the theory, we can go on to calculate various quantum corrections to the Drude formula; corrections not obtainable in the Boltzmann approach.

16.4 Conductivity from Born scattering with interactions

In Chapter 15 and in the introduction to this chapter, it was discussed how the apparent contradiction that the mass, according to the microscopic derivation of the quasiparticle pole in Section 15.4.1, is *renormalized* by interactions and yet, general arguments based on momentum conservation show that it is the *bare mass* that should appear in the Drude formula.

This problem can in fact be resolved if we combine both impurity scattering and interactions in the diagrammatical analysis. The impurity scattering is again considered to be weak, such that the first Born approximation suffices, and the interactions are included in the RPA approximation for the self-energy. This treatment follows the spirit of Section 15.4.1. Furthermore, to be consistent, we must include the dressed electron lines inside the self-energy, such that the electron line in the first Born approximation diagram is the RPA-dressed electron line.

The vertex corrections, which are consistent with this set of self-energies, are precisely the ladder diagrams shown in Eqs. (16.22) and (16.23). To simplify the arguments will make to two assumptions about the wave vector dependence of the interactions. First, we take the impurities to be be short-range scatterer so that in $\mathbf{q}$-space we replace $u(\mathbf{q})$ by a constant potential u. Also, the Coulomb interaction is considered to be short-ranged due to screening, and again we replace $W^{\mathrm{RPA}}(\mathbf{q}, iq_n)$ by its small $\mathbf{q}$ value, $W^{\mathrm{RPA}}(0, iq_n)$. These approximations, which are not essential for the validity of our arguments, imply that $\tilde{W}(\tilde{q})$ in Eq. (16.24) does not depend on $\mathbf{q}$. The reason for using this simplification is that the vertex corrections in Eq. (16.23) cancels. This is seen by verifying that $\Gamma_x = \mathbf{k}/m$ is a solution to Eq. (16.23)

$$\Gamma_x(\mathbf{k}, \mathbf{k}; ik_n + iq_n, ik_n) = \Gamma_{0,x}(\mathbf{k}, \mathbf{k})$$
$$+ \frac{1}{\mathcal{V}} \sum_{\mathbf{q}'} \frac{1}{\beta} \sum_{iq_n'} \left[n_{\mathrm{imp}} \left| u^{\mathrm{RPA}}(0) \right|^2 \delta_{q_n',0} + W^{\mathrm{RPA}}(0, iq_n') \right]$$
$$\times \mathcal{G}(\mathbf{k} + \mathbf{q}', ik_n + iq_n' + iq_n) \mathcal{G}(\mathbf{k} + \mathbf{q}', ik_n + iq_n' + iq_n)$$
$$\times \Gamma_x(\mathbf{k} + \mathbf{q}', \mathbf{k} + \mathbf{q}'; ik_n + iq_n, ik_n),\qquad(16.50)$$

because the terms with $\mathcal{G}(\mathbf{k} + \mathbf{q}')$ only depends on the length of $\mathbf{k} + \mathbf{q}'$, and so the second term cancels upon the summation over $\mathbf{q}'$. See also Exercise 16.1.

The conductivity then follows from the general formula in Eq. (16.34)

$$\operatorname{Re}\sigma_{xx} = \operatorname{Re} \frac{e^2}{\pi} \frac{1}{\mathcal{V}} \sum_{\mathbf{k}} \frac{k^2}{m^2} \left[G^A(\mathbf{k}, 0) G^R(\mathbf{k}, 0) - G^A(\mathbf{k}, 0) G^A(\mathbf{k}, 0) \right], \qquad (16.51)$$

Since we also here consider the weak impurity-scattering limit, the imaginary part of the self-energy stemming from the impurity scattering is small. Moreover, at low temperatures it is valid to assume that the imaginary part of the self-energy due to electron-electron interactions is small as well. This was indeed the prime argument for the consistency of Fermi liquid theory in the preceding chapter. We take $\operatorname{Im}\Sigma^{\mathrm{RPA}} \ll \operatorname{Im}\Sigma^{\mathrm{1BA}}$, and thus our previous arguments holds: only the first term in Eq. (16.51) is important. Using Eq. (16.38) in this term results in

$$\sigma_{xx} = \frac{e^2}{\pi} \frac{1}{\mathcal{V}} \sum_{\mathbf{k}} \frac{k^2}{m^2} \tau_0^* A(\mathbf{k}, 0), \qquad (16.52)$$

where τ_0^* is the renormalized lifetime in the Born approximation, and $A(\mathbf{k}, 0)$ is the spectral function, which in the RPA is $A(\mathbf{k}, 0) = Z\delta(\tilde{\xi}_k)$, see Eq. (15.57). Performing the integral as in Eq. (16.42) then gives

$$\sigma_{xx} = \frac{Ze^2 nm^* \tau_0^*}{m^2}, \qquad (16.53)$$

where we used the renormalized result for $\tilde{\xi}$ while k_{F} is not renormalized, because it is fixed by the density. Thus apparently, the renormalized mass is still in the Drude formula. However, we must the carry the arguments through and find the renormalized lifetime

$$[\tau_0^*]^{-1} = n_{\text{imp}} \sum_{\mathbf{k}'} |u|^2 A(\mathbf{k}, 0)$$

$$= 2\pi Z i n_{\text{imp}} \sum_{\mathbf{k}'} |u|^2 \delta(\tilde{\xi}_k)$$

$$= 2\pi \frac{Z m^*}{m} i n_{\text{imp}} \sum_{\mathbf{k}'} |u|^2 \delta(\xi_k)$$

$$= \frac{Z m^*}{m} [\tau_0]^{-1}. \tag{16.54}$$

When this inserted back into Eq. (16.53), we see the renormalized mass cancels out, and we are left with the usual Drude formula in terms of the bare mass. Hence the apparent contradiction has been removed.

16.5 The weak localization correction to the conductivity

The Born approximation includes only scattering on one impurity at a time. We saw in Chapter 10 that there was in practice only little difference between the first Born and the full Born approximation. The reason is that even the full Born approximation depicted in Eq. (12.54), which does take into account multiple scattering does so only for multiple scatterings on the same impurity. Quantum effects such as interference between scattering on different impurities can therefore not be incorporated within the Born approximation scheme. In Sec. 12.5.4 it was hinted that such interference processes are represented by crossing diagrams as in Fig. 12.6. In this section we shall study in detail why that is so.

As the temperature is lowered, we expect quantum mechanical coherence to become more important because the phase coherence length ℓ_ϕ increases with decreasing temperature. When the coherence length ℓ_ϕ exceeds the mean free path ℓ_{imp} for impurity scattering, scattering on different impurities can interfere. Here, the coherence length means the scale on which the electrons preserve their quantum mechanical phase, i.e. the scale on which the wavefunction evolves according to the one-particle Schrödinger equation. If an electron interacts with another electron or with a phonon through an inelastic scattering event, its energy changes, and hence the evolution of its phase. Due to these processes, the phase of the electron wave acquires some randomization or "dephasing," and its coherence length becomes finite. At low temperatures the dominant dephasing mechanism is electron-electron scattering, and as we know from Chapter 13, the scattering rate for these processes is proportional to T^2. Hence $\ell_\phi \propto T^{-2}$ can become very large at sufficiently low temperatures. At liquid helium temperaturez, 4.2 K and below, typical coherence lengths are of the order 1–10 μm, equivalent to 10^4–10^5 atomic lattice spacings.

If the coherence length ℓ_ϕ is longer than the mean free path ℓ_0, but still smaller than the sample size $\mathcal{L}$, most of the interference effects disappear. This is because the limit $\ell_\phi \ll \mathcal{L}$ effectively corresponds to averaging over many small independent segments, the so-called self-averaging illustrated in Fig. 12.2. However, around 1980

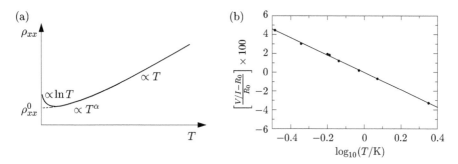

FIG. 16.2.   (a) A sketch of the electrical resistivity $\rho_{xx}(T)$ of a disordered metal as a function of temperature. As in Fig. 12.1, the linear behavior at high temperatures is due to electron-phonon scattering, but now at low temperatures we have added the small but significant increase due to the quantum interference known as weak localization. (b) An outline of the experimental data from measurements on a thin PdAu film showing that the low-temperature weak-localization correction to the resistivity increases logarithmically as the temperature decreases. The actual data can be found in Dolan and Osheroff, *Phys. Rev. Lett.* **43**, 721 (1979).

it was found through the observation of the so-called weak localization, shown in Fig. 16.2, that even in the case of large samples, $\ell_0 \ll \ell_\phi \ll \mathcal{L}$, one very important class of interference processes survive the self-averaging. Naturally, as discovered around 1985, much more dramatic quantum effects appear in small samples in the so-called mesoscopic regime (see also Chapter 7) given by $\mathcal{L} \simeq \ell_\phi$. In this regime all kinds of quantum interference processes become important, and most notably cause the appearance of the universal conductance fluctuations shown in Fig. (12.2).

In the following we study only the weak localization phenomenon appearing in large samples and not the universal conductance fluctuations appearing in small samples. To picture how averaging over impurity configurations influences the interference effects, we follow an electron after it has been scattered to a state with momentum $\mathbf{k}$ by an impurity positioned at $\mathbf{R}_1$. When the electron hits the next impurity at position $\mathbf{R}_2$ it has acquired a phase factor $e^{i\phi} = e^{i\mathbf{k}\cdot(\mathbf{R}_1 - \mathbf{R}_2)}$. Terms describing interference between the two scattering events will thus contain the factor $e^{i\mathbf{k}\cdot(\mathbf{R}_1 - \mathbf{R}_2)}$, and it is therefore intuitively clear that these terms vanish when one averages over $\mathbf{R}_1$ and $\mathbf{R}_2$. *Only the interference processes which are independent of the impurity positions survive self-averaging.*

Interference generally means that the amplitude for two paths t_1 and t_2 are added as $t_1 + t_2$, so that when taking the absolute square $|t_1 + t_2|^2 = |t_1|^2 + |t_2|^2 + 2|t_1 t_2| \cos(\phi_1 - \phi_2)$, the cross-term expresses the interference. The relative phase $\phi_1 - \phi_2$ determines whether the contributions from the two paths interfere constructively or destructively. If we can find two paths where the relative phase is independent of the position of the impurities, the cross-term would thus survive the impurity average. This is indeed possible, and two such paths are shown in Fig. 16.3. The key observation is that for each path that ends in the starting point after a specific sequence of scattering events, there is a corresponding reverse path which scatters on

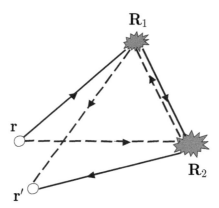

FIG. 16.3.   Illustration of the two interfering time-reversed paths discussed in the
text. When $\mathbf{r} = \mathbf{r}'$ the interference does not vanish upon impurity averaging.

the same impurities but in the opposite order. Most remarkably, these two paths pick
up exactly the same phase factor, and thus their relative phase $\phi_1 - \phi_2$ is always zero,
independent of the actual positions of the impurities. Thus for two such time-reversed
paths there is always *constructive* interference. As a consequence there is an enhanced
probability for returning to the same point, and the electrons therefore tend to be
localized in space, hence the name "weak localization".[61]

Having realized that the interference between time-reversed paths survive impurity
averaging, we now want to calculate the resulting correction to the conductivity.
In order to do so we need to identify the corresponding diagrams. First, we recall
the Dyson equation for the single-particle Green's functions in an external potential,
which was derived in Chapter 10. Here, the external potential is given by the impurity
potential, U_{imp}. Writing it in the frequency domain and making analytic continuation,
$ik_n \to \epsilon + i\eta$, we have for the retarded Green's function

$$G^R(\mathbf{r}, \mathbf{r}', \epsilon) = G_0^R(\mathbf{r}, \mathbf{r}', \epsilon) + \int d\mathbf{r}'' G_0^R(\mathbf{r}, \mathbf{r}'', \epsilon) U_{\mathrm{imp}}(\mathbf{r}'') G^R(\mathbf{r}'', \mathbf{r}', \epsilon). \qquad (16.55)$$

If we for simplicity assume $U_{\mathrm{imp}}(\mathbf{r}) \approx \sum_i U_0 \delta(\mathbf{r} - \mathbf{R}_i)$, i.e. short-range impurities
located at the positions $\{\mathbf{R}_i\}$, we have

$$G^R(\mathbf{r}, \mathbf{r}', \epsilon) = G_0^R(\mathbf{r}, \mathbf{r}', \epsilon) + \sum_i G_0^R(\mathbf{r}, \mathbf{R}_i, \epsilon) U_0 G^R(\mathbf{R}_i, \mathbf{r}', \epsilon). \qquad (16.56)$$

Let us look at a specific process where an electron scatters at, say, two impurities
located at $\mathbf{R}_1$ and $\mathbf{R}_2$. To study interference effects between scattering at these two
impurities we must expand to second order in the impurity potential. The interesting
second-order terms (there are also less-interesting ones where the electron scatters on
the same impurities twice) are

---

[61]The term "strong localization" is used for the so-called Anderson localization, where a metal-
insulator transition is induced in 3D at a critical strength of the disorder potential.

$$G^{R(2)}(\mathbf{r}, \mathbf{r}', \epsilon) = G_0^R(\mathbf{r}, \mathbf{R}_1, \epsilon) U_0 G_0^R(\mathbf{R}_1, \mathbf{R}_2, \epsilon) U_0 G_0^R(\mathbf{R}_2, \mathbf{r}', \epsilon)$$
$$+ G_0^R(\mathbf{r}, \mathbf{R}_2, \epsilon) U_0 G_0^R(\mathbf{R}_2, \mathbf{R}_1, \epsilon) U_0 G_0^R(\mathbf{R}_1, \mathbf{r}', \epsilon). \tag{16.57}$$

These two terms correspond to the transmission amplitudes t_1 and t_2 discussed above and illustrated in Fig. 16.3. The probability for the process is obtained from the absolute square of the Green's function, and because we want to find the correction $\delta|r|^2$ to the reflection coefficient, we set $\mathbf{r} = \mathbf{r}'$ at the end of the calculation. First the quantum correction due to interference to the transmission from $\mathbf{r}$ to $\mathbf{r}'$ is

$$\delta|t(\mathbf{r}, \mathbf{r}')|^2 \propto \mathrm{Re}\left[G_0^R(\mathbf{r}, \mathbf{R}_1, \epsilon) U_0 G_0^R(\mathbf{R}_1, \mathbf{R}_2, \epsilon) U_0 G_0^R(\mathbf{R}_2, \mathbf{r}', \epsilon) \right.$$
$$\left. \times \left(G_0^R(\mathbf{r}, \mathbf{R}_2, \epsilon) U_0 G_0^R(\mathbf{R}_2, \mathbf{R}_1, \epsilon) U_0 G_0^R(\mathbf{R}_1, \mathbf{r}', \epsilon) \right)^* \right]. \tag{16.58}$$

Now reflection is described by setting $\mathbf{r} = \mathbf{r}'$. Doing this and averaging over impurity positions $\mathbf{R}_1$ and $\mathbf{R}_2$ we find the quantum correction $\delta|r|^2$ to the reflection. In $\mathbf{k}$-space one gets

$$\langle \delta|r|^2 \rangle_{\mathrm{imp}} \equiv \langle \delta|t(\mathbf{r} = \mathbf{r}')|^2 \rangle_{\mathrm{imp}}$$
$$\propto \mathrm{Re}\, \frac{1}{\mathcal{V}^4} \sum_{\mathbf{p}_1 \mathbf{p}_2 \mathbf{p}_3 \mathbf{Q}} G_0^R(\mathbf{Q} - \mathbf{p}_1, \epsilon) U_0 G_0^R(\mathbf{Q} - \mathbf{p}_2, \epsilon) U_0 G_0^R(\mathbf{Q} - \mathbf{p}_3, \epsilon)$$
$$\times G_0^A(\mathbf{p}_1, \epsilon) U_0 G_0^A(\mathbf{p}_2, \epsilon) U_0 G_0^A(\mathbf{p}_3, \epsilon). \tag{16.59}$$

This formula can be represented by a diagram similar to the last one in Eq. (16.25) with the upper lines being retarded and the lower lines being advanced Green's functions. Notice however that, contrary to a usual diagram for conductance, the Green's functions in the lower and upper branch run in same direction. However, if we twist the lower branch such that the Green's functions run in opposite directions (thus allowing the impurity lines to cross) and furthermore join the retarded and advanced Green's function at the ends, then the diagram looks like a normal conductivity diagram,

$$\langle \delta|r|^2 \rangle_{\mathrm{imp}} = \quad\text{}\quad . \tag{16.60}$$

This hints that the interference term coming from time-reversed paths can be summed by taking diagrams of this form into account. These crossed diagrams were not included in the Born approximation, which we used to derive the Boltzmann equation result, and in fact they were shown in Sec. 12.5.4 to be smaller than the Born approximation by a factor $1/k_F\ell$. Nevertheless, at low temperatures they do play a role as the leading quantum correction. If we continue this line of reasoning we should include also diagrams where paths scattering on more than two impurities interfere with their time-reversed counter parts. It is straightforward to see that the corresponding diagrams are of the same type as (16.60) but with more crossing lines. This class of diagrams is called *the maximally crossed diagrams*. We have now identified which diagrams we need to sum in order to get the leading quantum correction to the conductivity. Most importantly, this is a contribution which does not disappear upon self-averaging.

Let us return to the Kubo formula for conductance, and let us sum the maximally crossed diagrams. We write the current-current correlation function as $\Pi = \Pi^B + \Pi^{\mathrm{WL}}$ where Π^B is the Boltzmann result derived in the previous section, and where

$$\Pi_{xx}^{\mathrm{WL}}(\tilde{q}) = \quad \text{(16.61)}$$

$$+ \quad \cdots \quad \text{(16.62)}$$

The full electron Green's functions in these diagrams are as before the full Green's function with an appropriately chosen self-energy. Since we include crossed diagrams in the vertex function we should in principle also include these in the self-energy. However, they can safely be ignored, since they only give a small contribution, down by a factor $1/k_{\mathrm{F}} v_{\mathrm{F}} \tau_0$ (see the discussion in Fig. 12.6). The crossed diagrams we are about to evaluate are also small by the same factor, but as we shall see they nevertheless yield a divergent contribution. This divergence stems from summing the interference of many time-reversed paths. This sum is different from the ladder diagrams that we summed in the Born approximation. There is however a trick which allows for a summation just like a ladder diagram. Let us twist the diagram in Eq. (16.62) with for example three impurity lines so as to make the impurity lines parallel,

$$\Pi_{xx}^{\mathrm{WL}(3)}(\tilde{q}) = \quad . \quad \text{(16.63)}$$

Then we see that the full series in Eq. (16.62) can be written as

$$\Pi_{xx}^{\mathrm{WL}}(\tilde{q}) =$$

$$= -\frac{1}{(2m)^2} \frac{1}{\mathcal{V}^2} \int d\tilde{k} \int d\tilde{k}' \, (2k_x + q_x) \, \mathcal{G}(\tilde{k}) \, \mathcal{G}(\tilde{k} + \tilde{q})$$
$$\mathcal{C}(\tilde{k}, \tilde{k}', \tilde{q}) \, \mathcal{G}(\tilde{k}') \mathcal{G}(\tilde{k}' + \tilde{q}) \, (2k_x' + q_x'), \quad \text{(16.64)}$$

where the box $\mathcal{C}$ is a sum of parallel impurity lines, i.e. analogous to the normal diffuson ladder sum of Eq. (16.25), but now with the fermion lines running in the same direction. This reversed ladder sum $\mathcal{C}$ is called a cooperon. The cooperon couples two electron lines or two hole lines, while a diffuson couples one electron line and one hole line. The solution for the cooperon $\mathcal{C}$ is found from the following Dyson-like equation:

$$(16.65)$$

In order to simplify our calculation, we only study the case $\mathbf{q} = 0$, and furthermore we restrict the analysis to the case of short-range impurities so that we can approximate $W(\mathbf{q})$ by a constant, $W_0 = n_{\mathrm{imp}}|u_0|^2$. With these approximations, and denoting $\mathbf{k} + \mathbf{k}' \equiv \mathbf{Q}$, the cooperon becomes

$$(16.66)$$

Because the impurity scattering conserves Matsubara frequencies the upper fermion lines all carry the frequency $ik_n + iq_n$, while the lower ones carry the frequency ik_n. It is now straightforward to solve Dyson equation for the cooperon ladder,

$$\mathcal{C}(\mathbf{Q}; ik_n + iq_n, ik_n) = \frac{\frac{1}{\mathcal{V}} \sum_{\mathbf{p}} W_0 \mathcal{G}(\mathbf{Q} - \mathbf{p}, ik_n + iq_n) \mathcal{G}(\mathbf{p}, ik_n) W_0}{1 - \frac{1}{\mathcal{V}} \sum_{\mathbf{p}} W_0 \mathcal{G}(\mathbf{Q} - \mathbf{p}, ik_n + iq_n) \mathcal{G}(\mathbf{p}, ik_n)}. \tag{16.67}$$

This can then be inserted into the expression for the current-current correlation function Π_{xx}^{WL} in Eq. (16.64)

$$\Pi_{xx}^{\mathrm{WL}}(0, iq_n) = -\frac{1}{(2m)^2} \frac{1}{\mathcal{V}^2} \frac{1}{\beta} \sum_{ik_n} \sum_{\mathbf{k}\mathbf{k}'} (2k_x) \mathcal{G}(\mathbf{k}, ik_n) \mathcal{G}(\mathbf{k}, ik_n + iq_n)$$

$$\times \mathcal{C}(\mathbf{k} + \mathbf{k}'; ik_n + iq_n, ik_n) \mathcal{G}(\mathbf{k}', ik_n) \mathcal{G}(\mathbf{k}', ik_n + iq_n)(2k_x'). \tag{16.68}$$

The Green's function $\mathcal{G}$ is here the Born approximation Green's function which after analytic continuation is

$$G^R(\mathbf{k}, \varepsilon) = \mathcal{G}(\mathbf{k}, ik_n \to \varepsilon + i\eta) = \frac{1}{\varepsilon - \xi_{\mathbf{k}} + i/2\tau_0}, \tag{16.69}$$

where $[\tau_0]^{-1} = 2\pi W_0 d(\varepsilon_F)$. It is now simple to find the solution for the cooperon $\mathcal{C}$.

In the previous section we learned that only the $G^A G^R$ term in Eq. (16.34) contributed in the limit of weak scattering and therefore we should replace $ik_n + iq_n$ by a retarded frequency and ik_n by an advanced frequency. Likewise, we obtain the weak localization correction from Eq. (16.68) by the replacements $ik_n + iq_n \to \varepsilon + \omega + i\eta$ and $ik_n \to \varepsilon - i\eta$, followed by insertion of the result into Eq. (16.34). Taking the DC-limit $\omega \to 0$ and the low temperature limit $T \to 0$, we have

$$\operatorname{Re} \delta\sigma_{xx}^{\text{WL}} = 2 \times \frac{e^2}{2\pi} \left(\frac{1}{m}\right)^2 \frac{1}{\mathcal{V}^2} \sum_{\mathbf{kk'}} (k_x k_x') G^R(\mathbf{k}, 0) G^A(\mathbf{k}, 0)$$
$$\times C^{AR}(\mathbf{k} + \mathbf{k'}; 0, 0) G^R(\mathbf{k'}, 0) G^A(\mathbf{k'}, 0). \tag{16.70}$$

As in the previous section we have factors of $G^A G^R$ appearing. However, here we cannot replace them by delta functions, because $\mathbf{k}$ and $\mathbf{k'}$ are connected through $C^{RA}(\mathbf{k} + \mathbf{k'})$. Instead we evaluate the cooperon as follows. After analytical continuation the cooperon in Eq. (16.67) becomes

$$C^{RA}(\mathbf{Q}) = \frac{W_0 \zeta(\mathbf{Q})}{1 - \zeta(\mathbf{Q})}, \tag{16.71a}$$

$$\zeta(\mathbf{Q}) \equiv \frac{n_{\text{imp}}}{\mathcal{V}} \sum_{\mathbf{p}} |u_0|^2 G^R(\mathbf{Q} - \mathbf{p}, 0) G^A(\mathbf{p}, 0), \tag{16.71b}$$

where we have introduced the auxiliary function $\zeta(\mathbf{Q})$. Using Eq. (16.69) $\zeta(\mathbf{Q})$ becomes

$$\zeta(\mathbf{Q}) = n_{\text{imp}} |u_0|^2 \frac{1}{\mathcal{V}} \sum_{\mathbf{p}} \frac{1}{-\xi_{\mathbf{Q}-\mathbf{p}} + i/2\tau_0} \frac{1}{-\xi_{\mathbf{p}} - i/2\tau_0}. \tag{16.72}$$

To proceed further we must now evaluate the $\mathbf{p}$-sum in $\zeta(\mathbf{Q})$. We begin by studying $\mathbf{Q} = 0$, in which case we have

$$\zeta(0) = n_{\text{imp}} |u_0|^2 d(\varepsilon_F) \int_{-\infty}^{\infty} d\xi \frac{1}{-\xi + i/2\tau_0} \frac{1}{-\xi - i/2\tau_0}$$
$$= n_{\text{imp}} |u_0|^2 d(\varepsilon_F) \int_{-\infty}^{\infty} d\xi \frac{1}{\xi^2 + (1/2\tau_0)^2} = n_{\text{imp}} |u_0|^2 d(\varepsilon_F) 2\pi\tau_0 = 1, \tag{16.73}$$

where we have used the definition of the lifetime τ_0 in the Born approximation. Combining Eqs. (16.71a) and (16.73), it follows that C^{RA} diverges in the limit of small Q and small frequency. The dc conductivity is therefore dominated by the contribution from values of Q near zero. Consequently, we study this contribution by expanding Eq. (16.72) for small Q. Here small means small compared the width τ_0^{-1} of the spectral function, i.e. we study the limit $Qv_F\tau_0 \ll 1$ or $Q \ll \ell_0^{-1} = 1/v_F\tau_0$. Furthermore, by symmetry arguments the term linear in Q vanish, so we need to go to second order in Q

$$\zeta(\mathbf{Q}) \approx 1 + n_{\text{imp}}|u_0|^2 \frac{1}{\mathcal{V}} \sum_{\mathbf{p}} \left(\frac{1}{-\xi_{\mathbf{p}} + i/2\tau_0}\right)^2 \frac{1}{-\xi_{\mathbf{p}} - i/2\tau_0} \left(-\mathbf{v_p} \cdot \mathbf{Q} + \frac{Q^2}{2m}\right)$$

$$+ n_{\text{imp}}|u_0|^2 \frac{1}{\mathcal{V}} \sum_{\mathbf{p}} \left(\frac{1}{-\xi_{\mathbf{p}} + i/2\tau_0}\right)^3 \frac{1}{-\xi_{\mathbf{p}} - i/2\tau_0} (\mathbf{v_p} \cdot \mathbf{Q})^2, \tag{16.74}$$

where it is indeed seen that the term linear in $\mathbf{Q}$ is zero because $\mathbf{v_p}$ is an odd function of $\mathbf{p}$. Now transforming the sum into integrations over ξ and performing the angular integrations, we find

$$\zeta(\mathbf{Q}) \approx 1 + \frac{1}{2\pi\tau_0} \int_{-\infty}^{\infty} d\xi \left(\frac{1}{-\xi + i/2\tau_0}\right)^2 \frac{1}{-\xi - i/2\tau_0} \left(\frac{Q^2}{2m}\right)$$

$$+ \frac{1}{2\pi\tau_0} \int_{-\infty}^{\infty} d\xi \left(\frac{1}{-\xi + i/2\tau_0}\right)^3 \frac{1}{-\xi - i/2\tau_0} \frac{Q^2 v_{\text{F}}^2}{N_{\text{dim}}}, \tag{16.75}$$

where N_{dim} is the number of dimensions. Closing the contour in the lower part of the complex ξ plan, we find that

$$\zeta(\mathbf{Q}) \approx 1 + \frac{2\pi i}{2\pi\tau_0} \left[\left(\frac{1}{i/\tau_0}\right)^2 \frac{Q^2}{2m} + \left(\frac{1}{i/\tau_0}\right)^3 \frac{Q^2 v_{\text{F}}^2}{N_{\text{dim}}}\right]. \tag{16.76}$$

To leading order in τ_0^{-1}, τ_0^3 dominates over τ_0^2, and we end up with

$$\zeta(\mathbf{Q}) \approx 1 - \frac{1}{N_{\text{dim}}} Q^2 \ell_0^2 \equiv 1 - D\tau_0 Q^2, \tag{16.77}$$

where

$$\ell_0 = v_{\text{F}}\tau_0, \qquad D = \frac{v_{\text{F}}^2\tau_0}{N_{\text{dim}}}, \tag{16.78}$$

D being the diffusion constant. We emphasize that Eq. (16.77) is only valid for $Q \ll \ell_0^{-1}$. With this result for $\zeta(\mathbf{Q})$ inserted into (16.71a) we obtain the final result for the cooperon

$$C^{RA}(\mathbf{Q}; 0, 0) = \frac{W_0(1 - D\tau_0 Q^2)}{D\tau_0 Q^2} \approx \frac{W_0}{\tau_0} \frac{1}{DQ^2}. \tag{16.79}$$

Because the important contribution comes from $Q \approx 0$, $\delta\sigma^{WL}$ in Eq. (16.70) becomes

$$\delta\sigma^{WL} = 2 \times \frac{e^2}{\pi} \left(\frac{1}{m}\right)^2 \frac{W_0}{\tau_0} \frac{1}{\mathcal{V}^2} \sum_{\mathbf{k}, Q < \ell_0^{-1}} (-k_x^2) G^R(\mathbf{k}, 0) G^A(\mathbf{k}, 0)$$

$$\times \frac{1}{DQ^2} G^R(\mathbf{Q} - \mathbf{k}, 0) G^A(\mathbf{Q} - \mathbf{k}, 0). \tag{16.80}$$

First we perform the sum over $\mathbf{k}$. Since $Q < \ell_0^{-1}$, and hence smaller than the width of the spectral function, we can approximate $\mathbf{Q} - \mathbf{k}$ by just $-\mathbf{k}$ and obtain

$$\frac{1}{V}\sum_{\mathbf{k}} k_x^2 G^R(\mathbf{k},0)G^A(\mathbf{k},0)G^R(-\mathbf{k},0)G^A(-\mathbf{k},0)$$

$$= d(\varepsilon_F)\frac{k_F^2}{N_{\mathrm{dim}}}\int_{-\infty}^{\infty} d\xi \left(\frac{1}{\xi^2 + (1/2\tau_0)^2}\right)^2 = \frac{4\pi k_F^2}{N_{\mathrm{dim}}} d(\varepsilon_F)\tau_0^3. \qquad (16.81)$$

From this follows

$$\delta\sigma^{\mathrm{WL}} = -\frac{e^2}{\pi}\left(\frac{k_F}{m}\right)^2 \frac{2\tau_0}{N_{\mathrm{dim}}} \frac{1}{V}\sum_{Q<\ell^{-1}} \frac{1}{DQ^2}. \qquad (16.82)$$

We are then left with the $\mathbf{Q}$-integration, which amounts to

$$\frac{1}{V}\sum_{Q<\ell_0^{-1}} \frac{1}{DQ^2} = \int_{Q<\ell_0^{-1}} \frac{d\mathbf{Q}}{(2\pi)^{N_{\mathrm{dim}}}} \frac{1}{DQ^2} \propto \int_{Q<\ell_0^{-1}} dQ \frac{Q^{N_{\mathrm{dim}}-1}}{DQ^2}. \qquad (16.83)$$

It is evident that this integral is divergent in the small Q limit in both 1D and 2D. Physically this is because we have allowed interference between path of infinite length, which does not occur in reality. In a real system the electron cannot maintain coherence over arbitrarily long distances due to scattering processes that cause decoherence. We must therefore find a method to cut off these unphysical paths. To properly describe the breaking of phase coherence between the time-reversed paths one should include coupling to other degrees of freedom such as coupling to phonons or electron-electron scatterings. Here, we choose to do this in a phenomenological fashion instead. Let us suppose that each path in the sum over paths in Eq. (16.66) has a probability of being destroyed by a scattering event and that this probability is proportional to the length of the path, or equivalently to the number of impurity scattering events involved in the path. This can be modeled by including a factor $e^{-\gamma}$ in the impurity potential so that instead of W_0 we write $W_0 e^{-\gamma}$. Clearly a path with n scatterings will then carry a factor $e^{-n\gamma}$. The parameter γ is then interpreted as the amount of decoherence experienced within a mean free path, i.e. $\gamma = \ell_0/\ell_\phi$. With this modification, the function $\zeta(\mathbf{Q})$ is changed into

$$\zeta(\mathbf{Q},\omega) \approx e^{-\gamma}\left(1 - D\tau_0 Q^2\right), \qquad (16.84)$$

and hence the cooperon gets modified as

$$C^{RA}(\mathbf{Q};0,0) = \frac{W_0 e^{-\gamma}}{1 - e^{-\gamma} + e^{-\gamma}DQ^2\tau_0}. \qquad (16.85)$$

In the limit of large ℓ_ϕ or small γ, we therefore have

$$C^{RA}(\mathbf{Q};0,0) \simeq \frac{W_0}{\tau_0}\frac{1}{1/\tau_\phi + DQ^2}. \qquad (16.86)$$

where $\tau_\phi = \ell_\phi/v_{\mathrm{F}}$. This is a physical sensible result. It says that the paths corresponding to a diffusion time longer than the phase breaking time cannot contribute to the interference effect. If the phase coherence length becomes larger than the sample, the sample size $\mathcal{L}$ must of course replace ℓ_ϕ as a cut-off length, because paths longer than the sample should not be included. We can now return to (16.83) and evaluate the integral in 1D, 2D and 3D, respectively

$$\int \frac{d\mathbf{Q}}{(2\pi)^{N_{\text{dim}}}} \frac{1}{1/\tau_\phi + DQ^2} = \int_0^{1/\ell_0} dQ \frac{1}{1/\tau_\phi + DQ^2} \begin{pmatrix} \frac{1}{\pi} \\ \frac{1}{2\pi} Q \\ \frac{1}{2\pi^2} Q^2 \end{pmatrix}$$

$$= \begin{cases} \frac{1}{\pi}\sqrt{\frac{\tau_\phi}{D}} \tan^{-1}\sqrt{\frac{D\tau_\phi}{\ell_0^2}}, & \text{1D} \\[2ex] \frac{1}{4D\pi} \ln\left(1 + \frac{D\tau_\phi}{\ell_0^2}\right), & \text{2D} \qquad (16.87) \\[2ex] \frac{1}{2\pi^2 D\ell_0} - \frac{1}{2\pi^2 D\sqrt{D\tau_\phi}} \tan^{-1}\sqrt{\frac{D\tau_\phi}{\ell_0^2}}, & \text{3D} \end{cases}$$

which in the limit of large τ_ϕ gives us information about the importance of the quantum corrections:

$$\delta\sigma^{WL} \propto \begin{cases} -(\tau_\phi)^{1/2}, & \text{1D} \\[1ex] -\ln\left(\frac{\tau_\phi}{\tau_0}\right), & \text{2D} \qquad (16.88) \\[1ex] (\tau_\phi)^{-1/2}, & \text{3D.} \end{cases}$$

This is an important result, which states that due to the localization correction the conductivity decreases with increasing phase coherence time. Furthermore, in the 1D case it tells us that in 1D the localization correction is enormously important and may exceed the Drude result. In fact it can be shown that a quantum particle in a 1D disordered potential is always localized. In 3D the situation is more subtle, there a metal-insulator transition occurs at a critical value of the disorder strength. 2D is in between these two cases, and it is in this case that the term "weak" localization makes sense, because here the correction is small. For the 2D case we have

$$\delta\sigma_{2D}^{WL} \approx -\frac{e^2}{2\pi^2} \ln\left(\frac{\tau_\phi}{\tau_0}\right). \qquad (16.89)$$

This result is "universal" since, apart from the logarithmic factor, it does not depend on the details of the material or the impurity concentration. That it is a small correction to the Drude conductivity can be seen from the ratio

$$\frac{\delta\sigma_{2D}^{WL}}{\sigma_0} = -\frac{1}{\pi k_F \ell_0} \ln\left(\frac{\tau_\phi}{\tau_0}\right). \qquad (16.90)$$

A way to measure this effect is to change the phase coherence time τ_ϕ and to look at the change of conductivity. The phase coherence can be changed in two ways. Foremost, one can apply a magnetic field which breaks the time-reversal symmetry giving rise to the fundamental interference between time-reversed paths. Second, decreasing the temperature increases the phase coherence time $\tau_\phi^{-1} \propto T^\alpha$, and a logarithmic increase of the conductivity is to be expected. Indeed $\delta\rho \propto -\delta\sigma \propto \ln\tau_\phi \propto -\ln T$ as is measured and shown in Fig. 16.2.

16.6 Disordered mesoscopic systems

So far we have studied averages over impurity configurations in bulk systems, and learned how to calculate the conductance due to scattering on random impurities. In this section, we briefly touch upon how to study the average properties of disordered mesoscopic systems.

In experiments on mesoscopic systems, such as quantum dots in semiconductor heterostructures, the geometry of the system is most likely not well-defined. One cannot precisely neither determine nor control the position of the walls and the impurities. Instead, one studies the statistical properties of the conductance for an ensemble of systems. The average and the variance of the conductance will turn out to exhibit interesting quantum phenomena, namely weak localization and universal conductance fluctuations, respectively. In order to understand these two phenomena, we must first learn about how to average over S-matrices.

In Chapter 7 we calculated the conductance for a given realization of a mesoscopic conductor by finding the transmission coefficients, or equivalently the scattering matrix, and then inserting these into the Landauer formula. The statistical properties are probed by varying either the geometry via the gate potentials, the Fermi level, or an external magnetic field. If the transmission coefficients are sufficiently sensitive to these external parameters, one often assumes that one effectively makes an ensemble average over all possible configurations.

Without any further knowledge or constraints, this implies an assumption about the ensemble of S-matrices, namely that all unitary matrixes are equally likely, or in other words that the distribution $P(\mathbf{S})$ of scattering matrices $\mathbf{S}$ is uniform in the group of unitary matrices of size $2N \times 2N$, denoted $\mathcal{U}(2N)$.

Such ensemble averaged have been studied for a long time, initially in atomic nuclei containing a large number of nucleons. The basic assumption being made there was that the Hamiltonians describing each of the systems of the ensemble are drawn randomly according to some probability distribution only constrained by the symmetry of the system. This statistical method is known as random matrix theory (RMT).

In this brief introduction, we do neither cover the large field of RMT nor its application to mesoscopic physics. The interested reader could consult the book by Mahta (1991) and the reviews by Stone *et al.* (1991), Beenakker (1997), and Alhassid (2000) for further studies.

16.6.1 *Statistics of quantum conductance, random matrix theory*

Here, we will not be concerned with the microscopic justification for the ensemble averaging, but simply say that since we have no information about the scattering matrix the most sensible thing to assume is that all scattering matrices in $\mathcal{U}(2N)$ will appear with equal probability only subject to symmetry constraints.

For the time-reversal symmetry case, we are therefore restricted to symmetric members of $\mathcal{U}(2N)$. The ensemble of S-matrixes for the time-reversal case can thus be realized by writing the S-matrix as a product of the two general unitary matrixes $\mathbf{S} = \mathbf{U}\mathbf{U}^T$, where $\mathbf{U} \in \mathcal{U}(2N)$. Clearly, this would generate all symmetric unitary matrixes.

For the statistical analysis we are going to carry out, we will need moments up to fourth order. One method to derive the ensemble average of some function $f(\mathbf{U})$ of a random matrix $\mathbf{U}$ is to utilize that if $\mathbf{U}$ is multiplied by some constant unitary $\mathbf{V}$, the average should be invariant, because we simply make a constant "shift" in $\mathcal{U}(2N)$, where all matrixes are equally likely. In other words, $\langle f(\mathbf{U}) \rangle = \langle f(\mathbf{VU}) \rangle = \langle f(\mathbf{UV}) \rangle$. The first moment of the unitary ensemble is the average of a single unitary matrix is $\langle U_{\alpha\beta} \rangle$, where $\langle \cdot \rangle$ here denotes ensemble average. According to the principle stated above, this average must for any matrix $V \in \mathcal{U}(2N)$ fulfill $\langle U_{\alpha\beta} \rangle = \sum_\gamma \langle U_{\alpha\gamma} \rangle V_{\gamma\beta} = \sum_\gamma V_{\alpha\gamma} \langle U_{\gamma\beta} \rangle$, the only solution being $\langle U_{\alpha\beta} \rangle = 0$.

The second moment is determined in a similar way

$$\langle U_{\alpha a}^* U_{\beta b} \rangle = \sum_{a'b'} \langle U_{\alpha a'}^* U_{\beta b'} \rangle V_{a'a}^* V_{b'b}. \tag{16.91}$$

A solution to this equation is $\langle U_{\alpha a}^* U_{\beta b} \rangle \propto \delta_{ab}$, which is seen by inspection. In the same spirit, we see that $\langle U_{\alpha a}^* U_{\beta b} \rangle \propto \delta_{\alpha\beta}$ by multiplying by $\mathbf{V}$ from the left. Now we know that $\langle U_{\alpha a}^* U_{\beta b} \rangle \propto C \delta_{ab} \delta_{\alpha\beta}$, where C is a normalization factor which we determine from the trace as

$$2N = \mathbf{U}\mathbf{U}^\dagger = \sum_{\alpha a} \langle U_{\alpha a}^* U_{\alpha a} \rangle = \sum_{\alpha a} C \times 1 = C(2N)^2, \tag{16.92}$$

and hence $C = 1/2N$.

The fourth moment can be derived using the same method and considerably more work. For a random unitary matrix of dimension $M = 2N$, we find the averages

$$\langle U_{\alpha\beta} \rangle = 0, \tag{16.93a}$$

$$\langle U_{\alpha a}^* U_{\beta b} \rangle = \frac{1}{M} \delta_{\alpha\beta} \delta_{ab}, \tag{16.93b}$$

$$\langle U_{\alpha a}^* U_{\alpha' a'}^* U_{\beta b} U_{\beta' b'} \rangle = \frac{1}{M^2 - 1} \left(\delta_{\alpha\beta} \delta_{ab} \delta_{\alpha'\beta'} \delta_{a'b'} + \delta_{\alpha\beta'} \delta_{ab'} \delta_{\alpha'\beta} \delta_{a'b} \right)$$
$$- \frac{1}{M(M^2 - 1)} \left(\delta_{\alpha\beta} \delta_{ab'} \delta_{\alpha'\beta'} \delta_{a'b} + \delta_{\alpha\beta'} \delta_{ab} \delta_{\alpha'\beta} \delta_{a'b'} \right). \tag{16.93c}$$

The first term in Eq. (16.93c) is equivalent to assuming the real and imaginary parts of $U_{\alpha a}$ to be independent, while the last term corrects for that because the unitarity condition gives some constraints on the elements of $\mathbf{U}$. These correlations however become less important in limit of large M.

16.6.2 Weak localization in mesoscopic systems

In the previous section we studied the weak localization in self-averaging macroscopic samples. The origin of this effect was found to be the constructive interference between time-reversed pairs of paths. The weak localization correction was the leading quantum correction that survived the impurity average. For the mesoscopic system, we also find quantum corrections that remain after the ensemble average. This is seen by using the random matrix theory for the S-matrix. According to the Landauer formula derived in Chapter 7, the average conductance is

$$\langle G \rangle = \frac{2e^2}{h} \left\langle \text{Tr} \left[\mathbf{t}^\dagger \mathbf{t} \right] \right\rangle = \frac{2e^2}{h} \sum_{n=1}^{N} \sum_{m=N+1}^{2N} \langle S_{mn}^* S_{mn} \rangle. \tag{16.94}$$

The result now depends on whether the system has time-reversal symmetry is present or not. As shown in Chapter 7 the time-reversal symmetry is broken by an applied **B**-field. First consider the case of broken time-reversal symmetry, $\mathbf{B} \neq 0$. In this case, there are no other constraints on **S** than that it is unitary, and we can therefore use Eq. (16.93b) directly,

$$\langle G \rangle_{\mathbf{B} \neq 0} = \frac{2e^2}{h} N^2 \frac{1}{2N} = \frac{2e^2}{h} \frac{N}{2}. \tag{16.95}$$

Next, we consider the case $\mathbf{B} = 0$ where, as shown in Section 7.1.4, the S-matrix is symmetric. This condition is fulfilled by representing S as $\mathbf{S} = \mathbf{U}\mathbf{U}^T$, and hence we have

$$\langle G \rangle_{\mathbf{B}=0} = \frac{2e^2}{h} \sum_{n=1}^{N} \sum_{m=N+1}^{2N} \sum_{i=1}^{2N} \sum_{j=1}^{2N} \langle U_{mi}^* U_{ni}^* U_{mj} U_{nj} \rangle. \tag{16.96}$$

Applying Eq. (16.93c), we find

$$\langle G \rangle_{\mathbf{B}=0} = \frac{2e^2}{h} \sum_{n=1}^{N} \sum_{m=N+1}^{2N} \sum_{i=1}^{2N} \sum_{j=1}^{2N} (\delta_{ij} + \delta_{mn}\delta_{ij}) \left(1 - \frac{1}{2N}\right) \frac{1}{4N^2-1} \tag{16.97}$$

$$= \frac{2e^2}{h} \frac{1}{4N^2-1} (2N^3) \left(1 - \frac{1}{2N}\right) = \frac{2e^2}{h} \frac{N^2}{2N+1}, \tag{16.98}$$

which is *smaller* than the $\mathbf{B} \neq 0$ result. It is natural to compare the conductance with the classical conductance, i.e., the conductance of two contacts each with $2N$ channels in series

$$\frac{\langle \delta G \rangle}{2e^2/h} = \frac{\langle G \rangle}{2e^2/h} - \frac{N}{2} = \begin{cases} -\frac{N}{2(2N+1)}, & \text{for } B = 0, \\ 0, & \text{for } B \neq 0. \end{cases} \tag{16.99}$$

This result clearly shows that quantum corrections, which come from the last term in Eq. (16.93c), give a reduced conductance and that the quantum coherence is destroyed by a magnetic field. Of course in reality the transition from the $\mathbf{B} = 0$ to the finite **B**-field case is a smooth transition. The transition happens when the flux enclosed by a typical trajectory is of order the flux quantum, which we see from the arguments leading to Eq. (7.64).

16.6.3 *Universal conductance fluctuations*

The fluctuations of the conductance contains some interesting information about the nature of the eigenstates of a chaotic system. Historically, the study of these fluctuations were the first in the field of mesoscopic transport. They were observed experimentally around 1980 and explained theoretically about five years later.

It is an experimental fact that the fluctuations turn out to be independent of the size of the conductance itself, which has given rise to the name universal conductance fluctuations (UCF). Naively, one would expect that if the average conductance is $\langle G \rangle = N_0(2e^2/h)$, corresponding to N_0 open channels, then the fluctuations in the number of open channels would be $\sqrt{N_0}$, so that $\langle \delta G \rangle = (2e^2/h)\sqrt{N_0}$. This is not seen experimentally, the reason being that the transmission probabilities are not independent. The number of conducting channels in a given energy window does therefore not follow a Poisson distribution.

For a completely random system without any symmetries, we do not expect degeneracies to occur. In fact one can show from RMT that the statistical measure vanishes when two eigenvalues coincide. Given an eigenvalue at $x = 0$, the probability for the next eigenvalue to be at x can be shown to be

$$P(x) = \frac{\pi}{2} x \exp\left(-\frac{\pi}{4} x^2\right), \tag{16.100}$$

for the case with time-reversal symmetry. This is called the Wigner surmise. The fluctuations of the number of eigenvalues in a given interval is therefore far from from the Poisson distribution, where $P(x) \propto \exp(-x)$. This "repulsion" between eigenvalues is the physical reason for the universal behavior.

In the following we calculate the fluctuations of G using the statistical RMT for the S-matrix as outlined above. The fluctuation of the conductance in the case without time-reversal symmetry is,

$$\langle G^2 \rangle_{\mathbf{B} \neq 0} = \left(\frac{2e^2}{h}\right)^2 \sum_{n=1}^{N} \sum_{m=N+1}^{2N} \sum_{n'=1}^{N} \sum_{m'=N+1}^{2N} \langle S_{mn}^* S_{mn} S_{m'n'}^* S_{m'n'} \rangle,$$

$$= \left(\frac{2e^2}{h}\right)^2 \sum_{n=1}^{N} \sum_{m=N+1}^{2N} \sum_{n'=1}^{N} \sum_{m'=N+1}^{2N} \frac{1}{4N^2 - 1}$$

$$\times \left(1 + \delta_{mm'} \delta_{nn'} - \frac{1}{2N} \left(\delta_{nn'} + \delta_{mm'}\right)\right),$$

$$= \left(\frac{2e^2}{h}\right)^2 \frac{N^4}{4N^2 - 1} \approx \left(\frac{2e^2}{h}\right)^2 \left(\frac{N}{2}\right)^2 \left(1 + \frac{1}{4N^2}\right), \quad \text{for } N \gg 1$$

$$\tag{16.101}$$

and the variance $\langle \delta G^2 \rangle = \langle G^2 \rangle - \langle G \rangle^2$ is

$$\frac{\langle \delta G^2 \rangle_{\mathbf{B} \neq 0}}{(2e^2/h)^2} \approx \frac{1}{16}, \quad \text{for } N \gg 1. \tag{16.102}$$

A similar calculation for the $\mathbf{B} = 0$ case gives

$$\frac{\langle \delta G^2 \rangle_{\mathbf{B} = 0}}{(2e^2/h)^2} \approx \frac{1}{8}, \quad \text{for } N \gg 1. \tag{16.103}$$

The variance is thus independent of the average value of G and furthermore, it is expected to decrease by a factor of 2, when a magnetic field is applied. Indeed, this is what is seen experimentally. For an experimental study of this effect see Chan et al. (1995).

16.7   Summary and outlook

In this chapter we have studied the formal theory of how to calculate conductivity for a self-averaging macroscopic sample and also the statistical properties conductance of disordered mesoscopic systems. We determined quantum effects of impurity scattering for both cases. We have presented a theory for ordinary conductivity as well as for the interesting quantum correction to this result known as weak localization.

The field of quantum effects from disorder or impurity scattering is huge. Many experiments have been performed, and much of the experimental data has been explained to a reasonable degree. Further development, which we have not discussed here, is the inclusion of interaction and the combined effect of electron-electron interactions and impurity scattering. For further reading and references on impurity scattering physics see, Bergmann (1984), Lee and Ramakrishnan (1985), and Rammer and Smith (1986), and for the physics of random mesoscopic systems, see Imry (1997), Beenakker (1997), and Alhassid (2000).

17

GREEN'S FUNCTIONS AND PHONONS

In this chapter, we develop and apply the Green's function technique for free phonons and for the electron-phonon interaction. The point of departure is the second quantization formulation of the phonon problem presented in Chapter 3, in particular the bosonic phonon creation and annihilation operators $b^{\dagger}_{-\mathbf{q},\lambda}$ and $b_{\mathbf{q},\lambda}$ introduced in Eqs. (3.10) and (3.22) and appearing in the jellium phonon Hamiltonian Eq. (3.4) and in the lattice phonon Hamiltonian Eq. (3.23).

We first define and study the Green's functions for free phonons in both the jellium model and the lattice model. Then we apply the Green's function technique to the electron-phonon interaction problem. We derive the one-electron Green's function in the presence of both the electron-electron and the electron-phonon interaction. We also show how the high-frequency Einstein phonons in the free-phonon jellium model become renormalized and become the usual low-frequency acoustic phonons once the electron-phonon interaction is taken into account. Finally, we prove the existence of the so-called Cooper instability of the electron gas, the phonon-induced instability which is the origin of superconductivity.

17.1 The Green's function for free phonons

It follows from all the Hamiltonians describing electron-phonon interactions, e.g., $H^{\text{INA}}_{\text{el}-\text{ph}}$ in Eq. (3.41) and $H^{\text{jel}}_{\text{el}-\text{ph}}$ in Eq. (3.43), that the relevant phonon operators to consider are not the individual phonon creation and annihilation operators, but rather the operators $A_{\mathbf{q}\lambda}$ and $A^{\dagger}_{\mathbf{q}\lambda}$ defined as

$$A_{\mathbf{q}\lambda} \equiv \left(b_{\mathbf{q}\lambda} + b^{\dagger}_{-\mathbf{q}\lambda} \right), \qquad A^{\dagger}_{\mathbf{q}\lambda} \equiv \left(b^{\dagger}_{\mathbf{q}\lambda} + b_{-\mathbf{q}\lambda} \right) = A_{-\mathbf{q}\lambda}. \qquad (17.1)$$

The phonon operator $A_{\mathbf{q}\lambda}$ can be interpreted as removing momentum $\mathbf{q}$ from the phonon system either by annihilating a phonon with momentum $\mathbf{q}$ or by creating one with momentum $-\mathbf{q}$. With these prerequisites, the non-interacting phonons are described by H_{ph} and the electron-phonon interaction by $H_{\text{el}-\text{ph}}$ as follows:

$$H_{\text{ph}} = \sum_{\mathbf{q}\lambda} \Omega_{\mathbf{q}\lambda} \left(b^{\dagger}_{\mathbf{q}\lambda} b_{\mathbf{q}\lambda} + \frac{1}{2} \right), \qquad (17.2a)$$

$$H_{\text{el}-\text{ph}} = \frac{1}{\mathcal{V}} \sum_{\mathbf{k}\sigma} \sum_{\mathbf{q}\lambda} g_{\mathbf{q}\lambda}\, c^{\dagger}_{\mathbf{k}+\mathbf{q},\sigma} c_{\mathbf{k}\sigma}\, A_{\mathbf{q}\lambda}. \qquad (17.2b)$$

Since H_{ph} does not depend on time, we can in accordance with Eq. (11.5) define the phonon operators $\hat{A}_{\mathbf{q}\lambda}(\tau)$ in the imaginary-time interaction picture,[62]

---

[62]This expression is also valid in the grand canonical ensemble governed by $H_{\text{ph}} - \mu N$. This is because the number of phonons can vary, and thus minimizing the free energy gives $\partial F/\partial N \equiv \mu = 0$.

$$\hat{A}_{\mathbf{q}\lambda}(\tau) \equiv e^{\tau H_{\mathrm{ph}}} A_{\mathbf{q}\lambda} \, e^{-\tau H_{\mathrm{ph}}}. \tag{17.3}$$

With this imaginary-time boson-like operator, we can follow Eq. (11.20) and introduce the bosonic Matsubara Green's function $\mathcal{D}_\lambda^0(\mathbf{q}, \tau)$ for free phonons,

$$\mathcal{D}_\lambda^0(\mathbf{q}, \tau) \equiv -\langle T_\tau \, \hat{A}_{\mathbf{q}\lambda}(\tau)\hat{A}_{\mathbf{q}\lambda}^\dagger(0)\rangle_0 = -\langle T_\tau \, \hat{A}_{\mathbf{q}\lambda}(\tau)\hat{A}_{-\mathbf{q}\lambda}(0)\rangle_0, \tag{17.4}$$

where T_τ is the bosonic time ordering operator defined in Eq. (11.21) with a plus sign. The frequency representation of the free-phonon Green's function follows by applying Eq. (11.28),

$$\mathcal{D}_\lambda^0(\mathbf{q}, iq_n) \equiv \int_0^\beta d\tau \, e^{iq_n\tau} \, \mathcal{D}_\lambda^0(\mathbf{q}, \tau), \quad q_n = \frac{2\pi}{\beta} n. \tag{17.5}$$

The specific forms for $\mathcal{D}_\lambda^0(\mathbf{q}, \tau)$ and $\mathcal{D}_\lambda^0(\mathbf{q}, iq_n)$ are found using the boson results of Section 11.3.1 with the substitutions $(\nu, \varepsilon_\nu, c_\nu) \to (\mathbf{q}\lambda, \Omega_{\mathbf{q}\lambda}, b_{\mathbf{q}\lambda})$. In the imaginary-time domain we find

$$\mathcal{D}_\lambda^0(\mathbf{q}, \tau) = \begin{cases} -\left[n_{\mathrm{B}}(\Omega_{\mathbf{q}\lambda}) + 1\right] e^{-\Omega_{\mathbf{q}\lambda}\tau} - n_{\mathrm{B}}(\Omega_{\mathbf{q}\lambda}) e^{\Omega_{\mathbf{q}\lambda}\tau}, & \text{for } \tau > 0, \\ -n_{\mathrm{B}}(\Omega_{\mathbf{q}\lambda}) e^{-\Omega_{\mathbf{q}\lambda}\tau} - \left[n_{\mathrm{B}}(\Omega_{\mathbf{q}\lambda}) + 1\right] e^{\Omega_{\mathbf{q}\lambda}\tau}, & \text{for } \tau < 0, \end{cases} \tag{17.6}$$

while in the frequency domain we obtain

$$\mathcal{D}_\lambda^0(\mathbf{q}, iq_n) = \frac{1}{iq_n - \Omega_{\mathbf{q}\lambda}} - \frac{1}{iq_n + \Omega_{\mathbf{q}\lambda}} = \frac{2\,\Omega_{\mathbf{q}\lambda}}{(iq_n)^2 - (\Omega_{\mathbf{q}\lambda})^2}, \tag{17.7}$$

where we have used that $n_{\mathrm{B}}(\Omega_{\mathbf{q}\lambda}) = 1/\left[\exp(\beta\Omega_{\mathbf{q}\lambda}) - 1\right]$.

17.2 Electron-phonon interaction and Feynman diagrams

Next, we treat the electron-phonon interaction perturbatively using the Feynman diagram technique. For clarity, we do not take the Coulomb interaction between the electrons into account in this section. The unperturbed Hamiltonian is the sum of the free-electron and free-phonon Hamiltonians, H_{el} and H_{ph},

$$H_0 = H_{\mathrm{el}} + H_{\mathrm{ph}} = \sum_{\mathbf{k}\sigma} \varepsilon_{\mathbf{k}} c_{\mathbf{k}\sigma}^\dagger c_{\mathbf{k}\sigma} + \sum_{\mathbf{q}\lambda} \Omega_{\mathbf{q}\lambda} \left(b_{\mathbf{q}\lambda}^\dagger b_{\mathbf{q}\lambda} + \frac{1}{2}\right). \tag{17.8}$$

When governed solely by H_0, the electronic and phononic degrees of freedom are completely decoupled, and as in Eq. (1.106) the basis states are given in terms of simple outer product states described by the electron occupation numbers $n_{\mathbf{k}\sigma}$ and the phonon occupation numbers $N_{\mathbf{q}\lambda}$,

$$|\Psi_{\mathrm{basis}}\rangle = |n_{\mathbf{k}_1\sigma_1}, n_{\mathbf{k}_2\sigma_2}, \ldots\rangle \, |N_{\mathbf{q}_1\lambda_1}, N_{\mathbf{q}_2\lambda_2}, \ldots\rangle. \tag{17.9}$$

What happens then as the electron-phonon interaction $H_{\mathrm{el-ph}}$ of Eq. (17.2) is turned on? We choose to answer this question by studying the single-electron Green's function $\mathcal{G}_\sigma(\mathbf{k}, \tau)$. In analogy with Eq. (13.8) we use the interaction picture representation, but now in momentum space, and substitute the two-particle interaction Hamiltonian $\hat{W}(\tau)$ with the electron-phonon interaction $\hat{P}(\tau)$,

$$
\mathcal{G}_\sigma(\mathbf{k},\tau) = -\frac{\sum_{m=0}^{\infty}\frac{(-1)^m}{m!}\int_0^\beta d\tau_1\ldots\int_0^\beta d\tau_m \left\langle T_\tau \hat{P}(\tau_1)\ldots\hat{P}(\tau_m)\hat{c}_{\mathbf{k}\sigma}(\tau)\,\hat{c}_{\mathbf{k}\sigma}^\dagger(0)\right\rangle_0}{\sum_{m=0}^{\infty}\frac{(-1)^m}{m!}\int_0^\beta d\tau_1\ldots\int_0^\beta d\tau_m \left\langle T_\tau \hat{P}(\tau_1)\ldots\hat{P}(\tau_m)\right\rangle_0},
$$

$$
\tag{17.10}
$$

where the $\hat{W}(\tau)$-integral of Eq. (13.9) is changed into a $\hat{P}(\tau)$-integral,

$$
\int_0^\beta d\tau_j\,\hat{P}(\tau_j) = \frac{1}{\mathcal{V}}\int d\tau_j\sum_{\mathbf{k}\sigma}\sum_{\mathbf{q}\lambda} g_{\mathbf{q}\lambda}\,\hat{c}_{\mathbf{k}+\mathbf{q},\sigma}^\dagger(\tau_j)\hat{c}_{\mathbf{k}\sigma}(\tau_j)\,\hat{A}_{\mathbf{q}\lambda}(\tau_j). \tag{17.11}
$$

At first sight, the two single-electron Green's functions in Eqs. (13.8) and (17.10) seems to be quite different since $\hat{W}(\tau)$ contains four electron operators and $\hat{P}(\tau)$ only two. However, we shall now show that the two expressions in fact are very similar. Since the electronic and phononic degrees of freedom decouple, the thermal average of the integrand in the m'th term of, say, the denominator in Eq. (17.10) can be written as a product of a phononic and an electronic thermal average,

$$
\left\langle T_\tau \hat{A}_{\mathbf{q}_1\lambda_1}(\tau_1)\ldots\hat{A}_{\mathbf{q}_m\lambda_m}(\tau_m)\hat{c}_{\mathbf{k}_1+\mathbf{q}_1\sigma_1}^\dagger(\tau_1)\hat{c}_{\mathbf{k}_1\sigma_1}(\tau_1)\ldots\hat{c}_{\mathbf{k}_m+\mathbf{q}_m\sigma_m}^\dagger(\tau_m)\hat{c}_{\mathbf{k}_m\sigma_m}(\tau_m)\right\rangle_0 =
$$

$$
\left\langle T_\tau \hat{A}_{\mathbf{q}_1\lambda_1}(\tau_1)\ldots\hat{A}_{\mathbf{q}_m\lambda_m}(\tau_m)\right\rangle_0\left\langle T_\tau \hat{c}_{\mathbf{k}_1+\mathbf{q}_1\sigma_1}^\dagger(\tau_1)\hat{c}_{\mathbf{k}_1\sigma_1}(\tau_1)\ldots\hat{c}_{\mathbf{k}_m+\mathbf{q}_m\sigma_m}^\dagger(\tau_m)\hat{c}_{\mathbf{k}_m\sigma_m}(\tau_m)\right\rangle_0.
$$

$$
\tag{17.12}
$$

From Eq. (17.1) follows that only an even number of phonon operators will lead to a non-zero contribution in the equilibrium thermal average, so we write $m = 2n$. Next, by Wick's theorem Eq. (11.82) for boson operators, the n-particle phonon Green's function becomes a product of n single-particle Green's functions of the form

$$
g_{\mathbf{q}_i\lambda_i}g_{\mathbf{q}_j\lambda_j}\left\langle T_\tau \hat{A}_{\mathbf{q}_i\lambda_i}(\tau_i)\hat{A}_{\mathbf{q}_j\lambda_j}(\tau_j)\right\rangle_0
$$

$$
= |g_{\mathbf{q}_i\lambda_i}|^2\left\langle T_\tau \hat{A}_{\mathbf{q}_i\lambda_i}(\tau_i)\hat{A}_{-\mathbf{q}_i\lambda_i}(\tau_j)\right\rangle_0 \delta_{\mathbf{q}_j-\mathbf{q}_i}\delta_{\lambda_i\lambda_j}
$$

$$
= -|g_{\mathbf{q}_i\lambda_i}|^2\,D_\lambda^0(\mathbf{q}_i,\tau_i-\tau_j)\delta_{\mathbf{q}_j-\mathbf{q}_i}\delta_{\lambda_i\lambda_j}. \tag{17.13}
$$

Note how the thermal average forces the paired momenta to add up to zero. In the final combinatorics, the prefactor $(-1)^m/m! = 1/(2n)!$ of Eq. (17.10) is modified as follows. A sign $(-1)^n$ appears from one minus sign in each of the n factors of the form Eq. (17.13). Then a factor $(2n)!/(n!\,n!)$ appears from choosing the n momenta $\mathbf{q}_j$ among the $2n$ to be the independent momenta. And finally, a factor $n!/2^n$ from all possible ways to combine the remaining n momenta to the chosen ones and symmetrizing the pairs, all choices leading to the same result. Hence, we end up with the prefactor $\frac{1}{n!}\left(-\frac{1}{2}\right)^n$. For each value of n the $2n$ operators, $\hat{P}(\tau_i)$ form n pairs, and we end with the following single-electron Green's function,

$$\mathcal{G}_\sigma(\mathbf{k},\tau) = -\frac{\displaystyle\sum_{n=0}^\infty \frac{(-1)^n}{n!} \int_0^\beta d\tau_1 \dots \int_0^\beta d\tau_n \left\langle T_\tau \hat{P}(\tau_1)\dots\hat{P}(\tau_n)\hat{c}_{\mathbf{k}\sigma}(\tau)\,\hat{c}_{\mathbf{k}\sigma}^\dagger(0)\right\rangle_0}{\displaystyle\sum_{n=0}^\infty \frac{(-1)^n}{n!} \int_0^\beta d\tau_1 \dots \int_0^\beta d\tau_n \left\langle T_\tau \hat{P}(\tau_1)\dots\hat{P}(\tau_n)\right\rangle_0},$$

(17.14)

where the $\hat{\mathcal{P}}(\tau)$-integral substituting the original $\hat{P}(\tau)$-integral of Eq. (17.10) is given by the effective two-particle interaction operator

$$\int_0^\beta d\tau_i\,\hat{\mathcal{P}}(\tau_i) = \int_0^\beta d\tau_i \int_0^\beta d\tau_j \sum_{\mathbf{k}_1\sigma_1}\sum_{\mathbf{k}_2\sigma_2}\sum_{\mathbf{q}\lambda} \frac{1}{2V^2}\,|g_{\mathbf{q}\lambda}|^2\,\mathcal{D}_\lambda^0(\mathbf{q},\tau_i-\tau_j)$$
$$\times\,\hat{c}_{\mathbf{k}_1+\mathbf{q},\sigma_1}^\dagger(\tau_j)\hat{c}_{\mathbf{k}_2-\mathbf{q},\sigma_2}^\dagger(\tau_i)\hat{c}_{\mathbf{k}_2\sigma_2}(\tau_i)\hat{c}_{\mathbf{k}_1\sigma_1}(\tau_j).$$

(17.15)

From this interaction operator we can identify a new type of electron-electron interaction $V_{\text{el}-\text{el}}^{\text{ph}}$ mediated by the phonons

$$V_{\text{el}-\text{el}}^{\text{ph}} = \frac{1}{2V}\sum_{\mathbf{k}_1\sigma_1}\sum_{\mathbf{k}_2\sigma_2}\sum_{\mathbf{q}\lambda} \frac{1}{V}|g_{\mathbf{q}\lambda}|^2\mathcal{D}_\lambda^0(\mathbf{q},\tau_i-\tau_j)\,\hat{c}_{\mathbf{k}_1+\mathbf{q},\sigma_1}^\dagger(\tau_j)\hat{c}_{\mathbf{k}_2-\mathbf{q},\sigma_2}^\dagger(\tau_i)\hat{c}_{\mathbf{k}_2\sigma_2}(\tau_i)\hat{c}_{\mathbf{k}_1\sigma_1}(\tau_j).$$

(17.16)

This interaction operator resembles the basic two-particle Coulomb interaction operator Eq. (2.34), but while the Coulomb interaction is instantaneous or local in time, the phonon-mediated interaction is retarded, i.e. non-local in time, regarding both the operators and the coupling strength $(1/V)\,|g_{\mathbf{q}\lambda}|^2\mathcal{D}_\lambda^0(\mathbf{q},\tau_i-\tau_j)$. However, the Feynman rules in Fourier space are derived as for the Coulomb interaction (13.27):

(1) Fermion lines with oriented four-momenta: $\bullet\!\!\longleftarrow\!\!\bullet \equiv \mathcal{G}_\sigma^0(\mathbf{k},ik_n)$
$\mathbf{k}\sigma, ik_n$

(2) Phonon lines with oriented four-momenta: $\bullet\!\!\text{∿∿∿}\!\!\bullet \equiv -\frac{1}{V}|g_{\mathbf{q}\lambda}|^2\,\mathcal{D}_\lambda^0(\mathbf{q},iq_n)$
$\mathbf{q}\lambda, iq_n$

(3) Conserve the spin and four-momentum at each vertex, i.e. incoming momenta must equal the outgoing, and no spin flipping.

(4) At order n, draw all topologically different connected diagrams containing n oriented phonon lines $-\frac{1}{V}|g_{\mathbf{q}\lambda}|^2\,\mathcal{D}_\lambda^0(\mathbf{q},iq_n)$, two external fermion lines $\mathcal{G}_\sigma^0(\mathbf{k},ik_n)$, and $2n-1$ internal fermion lines $\mathcal{G}_\sigma^0(\mathbf{p}_j,ip_j)$. All vertices must contain a phonon line and an incoming and an outgoing fermion line.

(5) Multiply each fermion loop by -1.

(6) Multiply by $\frac{1}{\beta V}$ for each internal four-momentum $\tilde{p}$; perform the sum $\sum_{\tilde{p}\sigma\lambda}$.

(17.17)

17.3 Combining Coulomb and electron-phonon interactions

We now discuss the effect of the long range Coulomb interactions between electrons and ions and between electrons themselves. For simplicity, we henceforth study only longitudinal phonons and hence drop all reference to the polarization index λ. In Fig. 3.1 we have already sketched the ion plasma oscillation that occurs, if we consider the interaction between the ions and the electron gas assuming the latter to be homogeneous and completely inert, i.e. disregarding all the dynamics of the electrons. A complete calculation is rather tedious, but in Section 3.1 we studied the ion plasma oscillations in the jellium model neglecting the electron dynamics. In the case of an ion density $\rho_{\text{ion}}^0 = N/\mathcal{V}$ we found the dispersionless jellium phonon modes in the long wavelength limit,

$$\Omega_{\mathbf{q}} = \Omega = \sqrt{\frac{Z^2 e^2 N}{\epsilon_0 M \mathcal{V}}}. \tag{17.18}$$

The coupling constant for the electron-electron interaction mediated by these jellium phonons is found by combining Eqs. (3.44) and (17.18),

$$\frac{1}{\mathcal{V}} |g_{\mathbf{q}}|^2 = \frac{1}{\mathcal{V}} \left(\frac{Ze^2}{q\epsilon_0}\right)^2 \frac{N\hbar}{2M\Omega} = \frac{e^2}{\epsilon_0 q^2} \frac{\hbar\Omega}{2} = \frac{1}{2} W(q)\,\Omega, \tag{17.19}$$

which, not surprisingly, is proportional to the Coulomb interaction $W(q)$. Note that we have dropped $\hbar$ in the last equality in accordance with the convention introduced in Section 5.1. The resulting, bare, phonon-mediated electron-electron interaction is

$$\frac{1}{\mathcal{V}} |g_{\mathbf{q}}|^2 \, \mathcal{D}^0(\mathbf{q}, iq_n) = W(q) \, \frac{\Omega^2}{(iq_n)^2 - \Omega^2}. \tag{17.20}$$

To discuss the role of the electron dynamics we now add the electron-electron Coulomb interaction $V_{\text{el-el}}$ of Eq. (2.34) and study the full Hamiltonian H for the electronic and phononic system,

$$H = H_{\text{el}} + V_{\text{el-el}} + H_{\text{ph}} + H_{\text{el-ph}}. \tag{17.21}$$

17.3.1 Migdal's theorem

When the electron-phonon coupling $H_{\text{el-ph}}$ is added, the question naturally arises of whether to study the influence of the electrons on the ions before that of the ions on the electrons, or vice versa. The answer is provided by Migdal's theorem. This theorem is the condensed matter physics analogue to the well-known Born–Oppenheimer approximation of molecular physics. The latter states that it is a good approximation to consider the coordinates $\mathbf{R}_i$ of the slowly moving, heavy ions as parameters in the Schrödinger equation for the fast-moving, light electrons, which is then solved. In the second stage the values of $\mathbf{R}_i$ are then changed adiabatically. Likewise, it can be proven by phase space arguments that renormalization of the electron-phonon vertex is suppressed at least by a factor $\sqrt{m/M} \sim 10^{-2}$, where m and M are the masses of the electron and ion, respectively. We will just outline the proof of Migdal's theorem here by studying the simplest phonon correction to the electron-phonon vertex,

$$\approx \sqrt{\frac{m}{M}} \times \qquad\qquad . \tag{17.22}$$

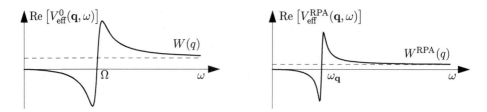

FIG. 17.1.   (a) The real part of the bare, effective electron-electron interaction
$V_{\text{eff}}^{0}(\mathbf{q}, \omega)$ as a function of the real frequency ω for a given momentum $\mathbf{q}$. Note
that the interaction is attractive for frequencies ω less than the jellium phonon
frequency Ω, and that $V_{\text{eff}}^{0}(\mathbf{q}, \omega) \to W(q)$ for $\omega \to \infty$. (b) The same for the RPA
renormalized effective electron-electron interaction $V_{\text{eff}}^{\text{RPA}}(\mathbf{q}, \omega)$, see Section 17.4.
Now, the interaction is attractive for frequencies ω less than the acoustic phonon
frequency $\omega_{\mathbf{q}}$, and $V_{\text{eff}}^{\text{RPA}}(\mathbf{q}, \omega) \to W^{\text{RPA}}(q)$ for $\omega \to \infty$.

The proof builds on a self-consistency assumption. We assume that the jellium phonons
with the high frequency Ω, get renormalized by electron screening processes to becomes the experimentally observed acoustic phonons with low frequency $\omega_{\mathbf{q}} = v_s\, q$.
If these phonons are used, we can prove Eq. (17.22) as done in the following section,
and given this, we can prove the assumed phonon renormalization.

   The important frequencies for acoustic phonons are smaller than the Debye fre-
quency ω_{D}, thus we concentrate on phonon frequencies $\omega_{\mathbf{q}} < \omega_{\text{D}}$. The diagram on
the left-hand side in Eq. (17.22) contains one phonon interaction line and two elec-
tron propagators more than the diagram on the right-hand side. Now according
to Eq. (17.37), the typical (acoustic) phonon interaction line for low frequencies,
$|iq_n| \ll \omega_{\text{D}}$, is given by $-W(q)/\varepsilon^{\text{RPA}}$. Furthermore, due to four-momentum conservation, the two internal electron propagators are confined within ω_{D} to the Fermi surface.
Consequently, a phase space factor of the order $\omega_{\text{D}}/\varepsilon_{\text{F}}$ must appear in front of the
usual unrestricted contribution from two such lines, the pair-bubble of Eq. (14.22),
$\chi_0 = -d(\varepsilon_{\text{F}})$. The ratio between the values of the two diagrams is therefore roughly
given by

$$\frac{W(q)}{\varepsilon^{\text{RPA}}} \times \frac{\hbar\omega_{\text{D}}}{\varepsilon_{\text{F}}} \times d(\varepsilon_{\text{F}}) = \frac{\hbar\omega_{\text{D}}}{\varepsilon_{\text{F}}} = \frac{v_s k_{\text{D}}}{\frac{1}{2}v_{\text{F}} k_{\text{F}}} = 2\sqrt{\frac{Z}{3}}\,\sqrt{\frac{m}{M}}\frac{k_{\text{D}}}{k_{\text{F}}} \approx \sqrt{\frac{m}{M}}, \qquad (17.23)$$

where we have used Eqs. (14.23) and (3.5) at the first and third equality sign, respectively. In the following we assume that we can neglect the phonon-induced renormalization of the electron-phonon vertex. We therefore study only the influence of the
electronic degrees of freedom on the bare phonon degrees of freedom. The result of
the analysis is that the assumption for Migdal's theorem indeed is fulfilled.

17.3.2   *Jellium phonons and the effective electron-electron interaction*

In more realistic calculations involving interacting electrons, we need to consider the
sum of the pure electronic Coulomb interaction and the phonon-mediated interaction.
This combined interaction will be the basis for our analysis of the interacting electron

gas henceforth. Combining the Feynman rules for these two interactions, Eqs. (13.27) and (17.17), yields the following bare, effective electron-electron interaction line,

$$-V_{\text{eff}}^0(\mathbf{q}, iq_n) \qquad\qquad -W(\mathbf{q}) \qquad\qquad -\frac{1}{\mathcal{V}}|g_{\mathbf{q}}|^2 \mathcal{D}^0(\mathbf{q}, iq_n)$$

$$\equiv \qquad\qquad + \qquad\qquad . \tag{17.24}$$

The specific form of V_{eff}^0 is obtained by inserting Eq. (17.20) into Eq. (17.24),

$$V_{\text{eff}}^0(\mathbf{q}, iq_n) = W(q) + W(q)\frac{\Omega^2}{(iq_n)^2 - \Omega^2} = W(q)\frac{(iq_n)^2}{(iq_n)^2 - \Omega^2}, \tag{17.25}$$

or going to real frequencies, $iq_n \to \omega + i\eta$,

$$V_{\text{eff}}^0(\mathbf{q}, \omega) = W(q)\frac{\omega^2}{\omega^2 - \Omega^2 + i\tilde{\eta}}. \tag{17.26}$$

The real part of $V_{\text{eff}}^0(\mathbf{q}, \omega)$ is shown in Fig. 17.1(a). It is seen that the bare, effective electron-electron interaction becomes negative for $\omega < \Omega$, i.e., at low frequencies the electron-phonon interaction combined with the originally fully repulsive Coulomb interaction results in an attractive effective electron-electron interaction. At high frequencies the normal Coulomb interaction is recovered.

17.4 Phonon renormalization by electron screening in RPA

The electronic Coulomb interaction renormalizes the bare, effective electron-electron interaction. Migdal's theorem leads us to disregard renormalization due to phonon processes and only to consider the most important electron processes. Since $V_{\text{eff}}^0(\mathbf{q})$ is proportional to the bare Coulomb interaction, our main result of Chapter 14 tells us that in the limit of high electron densities, these processes are given by RPA. Before we consider how the phonon propagator is renormalized by the electronic RPA, let us remind ourselves of the following expressions, Eqs. (14.62)–(14.67) from Chapter 14, between the dielectric function ε^{RPA}, the density-density correlation function $-\chi^{\text{RPA}} = $, and the simple pair-bubble $-\chi_0 = $,

$$\varepsilon^{\text{RPA}}(\mathbf{q}, iq_n) = 1 - W(q)\chi_0(\mathbf{q}, iq_n), \tag{17.27a}$$

$$\chi^{\text{RPA}}(\mathbf{q}, iq_n) = \frac{\chi_0(\mathbf{q}, iq_n)}{1 - W(q)\chi_0(\mathbf{q}, iq_n)} = \frac{\chi_0(\mathbf{q}, iq_n)}{\varepsilon^{\text{RPA}}(\mathbf{q}, iq_n)}, \tag{17.27b}$$

$$1 + W(q)\chi^{\text{RPA}}(\mathbf{q}, iq_n) = 1 + \frac{W\chi_0}{1 - W\chi_0} = \frac{1}{1 - W\chi_0} = \frac{1}{\varepsilon^{\text{RPA}}(\mathbf{q}, iq_n)}. \tag{17.27c}$$

Returning to the electron-phonon problem, we now extend the RPA-result Eq. (14.69) for W^{RPA} and obtain

$$-V_{\text{eff}}^{\text{RPA}}(\mathbf{q}, iq_n) \quad = \quad$$ $$\quad = \quad$$ $$+$$ $$. \tag{17.28}$$

The solution for $V_{\text{eff}}^{\text{RPA}}(\mathbf{q}, iq_n)$ has the standard form

$$-V_{\text{eff}}^{\text{RPA}}(\mathbf{q}, iq_n) = \text{〰〰} = \frac{\text{〰〰}}{1 - \text{〰}} = \frac{-V_{\text{eff}}^0(\mathbf{q})}{1 - V_{\text{eff}}^0(\mathbf{q})\,\chi_0(\mathbf{q}, iq_n)}. \qquad (17.29)$$

While this expression is correct, a physically more transparent form of $V_{\text{eff}}^{\text{RPA}}$ is obtained by expanding the infinite series Eq. (17.28), and then collecting all the diagrams containing only Coulomb interaction lines into one sum (this simply yields the RPA screened Coulomb interaction W^{RPA}), while collecting the remaining diagrams containing a mixture of Coulomb and phonon interaction lines into another sum,

$$
\begin{array}{ccc}
-V_{\text{eff}}^{\text{RPA}}(\mathbf{q}, iq_n) & -W^{\text{RPA}}(\mathbf{q}) & -\frac{1}{\mathcal{V}}|g_{\mathbf{q}}^{\text{RPA}}|^2 \mathcal{D}^{\text{RPA}}(\mathbf{q}, iq_n) \\
\text{〰〰〰〰} & = \quad \text{〜〜〜} & + \quad \text{〰〰〰}
\end{array}
$$
$$(17.30)$$

Here, the renormalized coupling $g_{\mathbf{q}}^{\text{RPA}}$ $[\{g_{\mathbf{q}}^{\text{RPA}}\}^*]$, given by

$$g_{\mathbf{q}}^{\text{RPA}} \equiv \text{〰} = \text{〰} + \text{〰} \qquad , \qquad (17.31)$$

is the sum of all diagrams between the outgoing left [incoming right] vertex and the first [last] phonon line, while the renormalized phonon line $\mathcal{D}^{\text{RPA}}(\mathbf{q}, iq_n)$, given by

$$-\mathcal{D}^{\text{RPA}}(\mathbf{q}, iq_n) = \text{〰〰} = \text{〰〰} + \text{〰〰}, \qquad (17.32)$$

is the sum of all diagrams between the first and the last phonon line, i.e. without contributions from the external coupling vertices. The solution for the RPA renormalized phonon line is

$$-\mathcal{D}^{\text{RPA}}(\mathbf{q}, iq_n) = \frac{\text{〰〰}}{1 - \text{〰〰}} = \frac{-\mathcal{D}^0(\mathbf{q}, iq_n)}{1 - \chi^{\text{RPA}}(\mathbf{q}, iq_n)\frac{1}{\mathcal{V}}|g_{\mathbf{q}}|^2 \mathcal{D}^0(\mathbf{q}, iq_n)}.$$
$$(17.33)$$

Using first Eqs. (17.7) and (17.20) and then Eq. (17.27c), leads to

$$\mathcal{D}^{\text{RPA}}(\mathbf{q}, iq_n) = \frac{2\,\Omega}{[(iq_n)^2 - \Omega^2] - \Omega^2 W(q)\chi^{\text{RPA}}(\mathbf{q}, iq_n)} = \frac{2\,\Omega}{(iq_n)^2 - \omega_{\mathbf{q}}^2}, \qquad (17.34)$$

where

$$\omega_{\mathbf{q}} \equiv \frac{\Omega}{\sqrt{\varepsilon^{\text{RPA}}(\mathbf{q}, iq_n)}} = \sqrt{\frac{Z^2 e^2 \rho_{\text{ion}}^0}{\varepsilon^{\text{RPA}} \epsilon_0 M}} = \sqrt{\frac{Z e^2 \rho_{\text{el}}^0}{\varepsilon^{\text{RPA}} \epsilon_0 M}}, \qquad (17.35)$$

is the renormalized phonon frequency due to electronic RPA-screening. In a moment we shall interpret this new frequency, but before doing so, we study how also the coupling constant $g_{\mathbf{q}}$ gets renormalized in RPA and acquire the value $g_{\mathbf{q}}^{\text{RPA}}$,

$$g_{\mathbf{q}}^{\mathrm{RPA}} \equiv \quad \text{[diagram]} \quad = \quad \text{[diagram]} \quad + \quad \text{[diagram]}$$

$$= (1 + W\chi^{\mathrm{RPA}})\, g_{\mathbf{q}} = \frac{g_{\mathbf{q}}}{\varepsilon^{\mathrm{RPA}}(\mathbf{q}, iq_n)}. \tag{17.36}$$

The final form of the RPA-screened, phonon-mediated electron-electron interaction is now obtained by combining Eqs. (17.34) and (17.36),

$$\frac{1}{\mathcal{V}}\,|g_{\mathbf{q}}^{\mathrm{RPA}}|^2\,\mathcal{D}^{\mathrm{RPA}}(\mathbf{q}, iq_n) = - \quad \text{[diagram]}$$

$$= \frac{|g_{\mathbf{q}}|^2/\mathcal{V}}{\left(\varepsilon^{\mathrm{RPA}}\right)^2}\,\frac{2\,\Omega}{(iq_n)^2 - \omega_{\mathbf{q}}^2} = \frac{W(q)}{\varepsilon^{\mathrm{RPA}}}\,\frac{\omega_{\mathbf{q}}^2}{(iq_n)^2 - \omega_{\mathbf{q}}^2}. \tag{17.37}$$

We now see that this renormalized propagator is identical to the free phonon propagator Eq. (17.20), where the unscreened phonon frequency Ω and the unscreened Coulomb interaction $W(q)$ have been replaced by their RPA-screened counterparts $\omega_{\mathbf{q}}$ and $W(q)/\varepsilon^{\mathrm{RPA}}$, respectively.

A further physical interpretation of this result is obtained by evaluating the expression Eq. (17.35) for $\omega_{\mathbf{q}}$ in the static, long wavelength limit. We note from Eqs. (14.67) and (14.24) that in this limit $\varepsilon^{\mathrm{RPA}}(\mathbf{q}, iq_n) \to k_s^2/q^2 = (4k_{\mathrm{F}}/\pi a_0)/q^2$. Inserting this into Eq. (17.35) and using $k_{\mathrm{F}}^3 = 3\pi^2\,\rho_{\mathrm{el}}^0$ yields the following explicit form of $\omega_{\mathbf{q}}$:

$$\omega_{\mathbf{q}}(q \to 0,\, 0) = \sqrt{\frac{Ze^2\rho_{\mathrm{el}}^0}{k_s^2\epsilon_0 M}}\, q = \sqrt{\frac{Zm}{3M}}\, v_{\mathrm{F}} q. \tag{17.38}$$

This we recognize as the Bohm–Staver expression Eq. (3.5) for the dispersion of acoustic phonons in the jellium model. The significance of this result is that starting from the microscopic Hamiltonian Eq. (17.21) for the coupled electron and phonon problem, we have used the Feynman diagram technique to show how the phonon spectrum gets renormalized by interacting with the electron gas. The long-range Coulomb forces of the non-interacting problem resulted in optical jellium phonons with the high frequency Ω. By introducing the electron-electron interaction, the Coulomb forces get screened, and as a result the phonon dispersion gets renormalized to the usual low-frequency acoustic dispersion $\omega_{\mathbf{q}} = v_s q$. In more elementary treatments, this spectrum is derived by postulating short-range forces following Hooke's law, but now we have proven it from first principles.

We end by stating the main result of this section, namely the explicit form of the effective electron-electron interaction due to the combination of the Coulomb and the electron-phonon interaction, see also Fig. 17.1(b),

$$\underset{\text{[diagram]}}{-V_{\mathrm{eff}}^{\mathrm{RPA}}(\mathbf{q}, iq_n)} = \underset{\text{[diagram]}}{-W^{\mathrm{RPA}}(\mathbf{q})} + \underset{\text{[diagram]}}{-\frac{1}{\mathcal{V}}|g_{\mathbf{q}}^{\mathrm{RPA}}|^2\mathcal{D}^{\mathrm{RPA}}(\mathbf{q}, iq_n)}$$

$$= -W^{\mathrm{RPA}}(q)\,\frac{(iq_n)^2}{(iq_n)^2 - \omega_{\mathbf{q}}^2}. \tag{17.39}$$

17.5 The Cooper instability and Feynman diagrams

In 1956 Cooper discovered that the electron gas in an ordinary metal would become unstable below a certain critical temperature T_c due to the phonon-induced attractive nature of the effective electron-electron interaction $V_{\text{eff}}^{\text{RPA}}(\mathbf{q}, \omega)$ at low frequencies. This discovery soon lead Bardeen, Cooper, and Schrieffer (BCS) to develop the microscopic theory explaining superconductivity, see Chapter 18.

In this section we will derive the Cooper instability using Feynman diagrams. The instability arises, because a certain class of electron-electron scattering processes, when added coherently, yields a divergent scattering amplitude. We will first derive this divergence, and then we will discuss its physical interpretation. The divergence is due to repeated scattering between electron pairs occupying time-reversed states of the form $|\mathbf{k}\uparrow\rangle$ and $|-\mathbf{k}\downarrow\rangle$. Using the four-momentum notation $\tilde{k} = (\mathbf{k}, ik_n)$ we consider the following pair scattering vertex $\Lambda(\tilde{k}, \tilde{p}) =$ given by the infinite ladder-diagram sum over scattering events between time-reversed electron pairs:

$$\qquad (17.40)$$

Suppressing the arguments and removing the external electron propagators, we can recast Eq. (17.40) in the form of a Dyson equation for the pair-scattering vertex Λ,

$$\qquad (17.41)$$

which is equivalent to the following integral equation

$$\Lambda(\tilde{k}, \tilde{p}) = -V_{\text{eff}}^{\text{RPA}}(\tilde{k}-\tilde{p}) + \frac{1}{\mathcal{V}\beta}\sum_{\tilde{q}}\left[-V_{\text{eff}}^{\text{RPA}}(\tilde{k}-\tilde{q})\right]\mathcal{G}_{\uparrow}^{0}(\tilde{q})\,\mathcal{G}_{\downarrow}^{0}(-\tilde{q})\,\Lambda(\tilde{q}, \tilde{p}). \quad (17.42)$$

To proceed, the functional form of $V_{\text{eff}}^{\text{RPA}}(\mathbf{q}, iq_n)$ is simplified by a physically motivated assumption. First we note that according to our analysis of the electron gas in Chapter 14, no instabilities arise due to the pure Coulomb interaction. Thus we are really only interested in the deviations of $V_{\text{eff}}^{\text{RPA}}(\mathbf{q}, iq_n)$ from $W^{\text{RPA}}(q)$. According to Eq. (17.39) and Fig. 17.1(b), $V_{\text{eff}}^{\text{RPA}}(\mathbf{q}, iq_n)$ rapidly approaches $W^{\text{RPA}}(q)$ for frequencies larger than the given acoustic phonon frequency $\omega_{\mathbf{q}}$, while it becomes attractive instead of repulsive for frequencies below $\omega_{\mathbf{q}}$. Further, according to the Debye model of acoustic phonons, Section 3.5, the density of phonon states, $D_{\text{ion}}(\varepsilon)$, is proportional to ε^2 or $\omega_{\mathbf{q}}^2$ for frequencies less than the Debye frequency $\omega_{\text{D}} = v_s k_{\text{D}}$ and zero otherwise, see Eq. (3.27). This means that most of the phonons encountered have a frequency of the order ω_D. It is therefore a reasonable approximation to set $\omega_\mathbf{q} = \omega_D$. Finally, as a last simplification, we set the interaction strength to be constant. Hence we arrive at a model equivalent to the one used by Cooper and by BCS,

$$V_{\text{eff}}^{\text{RPA}}(\mathbf{q}, iq_n) \approx \begin{cases} -V, & |iq_n| < \omega_D, \\ 0, & |iq_n| > \omega_D. \end{cases} \tag{17.43}$$

The integral equation for $\Lambda(\tilde{k}, \tilde{p})$ thus only involves frequencies less than ω_D. For those frequencies Λ is independent of frequency, $\Lambda(\tilde{k}, \tilde{p}) = \Lambda(\mathbf{k}, \mathbf{p})$, and the differential equation takes the form

$$\Lambda(\mathbf{k}, \mathbf{p}) = V + \frac{1}{\beta} \sum_{iq_n}^{\omega_D} \frac{1}{V} \sum_{\mathbf{q}} V \, \mathcal{G}_\uparrow^0(\mathbf{q}, iq_n) \, \mathcal{G}_\downarrow^0(-\mathbf{q}, -iq_n) \, \Lambda(\mathbf{q}, \mathbf{p}). \tag{17.44}$$

The summand on the right-hand side does not contain the external momentum $\mathbf{k}$, whence for the left-hand side we conclude that $\Lambda(\mathbf{k}, \mathbf{p}) = \Lambda(\mathbf{p})$, a result subsequently used in the summand. Now it is furthermore evident that the $\mathbf{p}$-dependence occurs only in the Λ-function, hence a consistent solution is obtained by taking $\Lambda(\mathbf{k}, \mathbf{p})$ to be a constant, which we naturally denote Λ. On the right-hand side of Eq. (17.44) we can take Λ outside the sum, and solve for it:

$$\Lambda = \frac{V}{1 - \dfrac{V}{\beta} \displaystyle\sum_{|iq_n| < \omega_D} \frac{1}{V} \sum_{\mathbf{q}} \mathcal{G}_\uparrow^0(\mathbf{q}, iq_n) \, \mathcal{G}_\downarrow^0(-\mathbf{q}, -iq_n)}. \tag{17.45}$$

We see that at high temperatures, i.e., $\beta \ll 1/\omega_D$, the resulting pair-interaction Λ equals the attractive pair-interaction strength V from Eq. (17.43). As T is lowered the denominator in Eq. (17.45) can approach zero from above, which leads to an arbitrarily strong or divergent pair-interaction strength Λ. In quantum mechanics an infinite scattering amplitude signals a resonance, i.e., in the present case the formation of a bound state between the time-reversed pair of electrons. But in our model, this would then happen simultaneously for all electron pairs within a shell of thickness ω_D of the Fermi surface, since the effective pair-interaction is attractive only for energy exchanges less than ω_D. The conclusion is clear: if the pair-interaction strength Λ diverges for a certain critical temperature T_c, the entire Fermi-surface becomes unstable at that temperature, and a new ground state is formed involving bound electron pairs in time-reversed states. This instability is called the Cooper instability, and the onset of it marks the transition from a normal metallic state to a superconducting state.

The critical temperature $T = T_c$, or equivalently $\beta = \beta_c$, for the onset of the Cooper instability, is obtained by setting the denominator in Eq. (17.45) to zero using $\mathcal{G}_\sigma^0(\mathbf{q}, iq_n) = 1/(iq_n - \varepsilon_\mathbf{q})$ and $q_n = \frac{2\pi}{\beta_c}(n + \frac{1}{2})$:

$$1 = \frac{V}{\beta_c} \sum_{|iq_n|<\omega_D} \frac{1}{V} \sum_{\mathbf{q}} \frac{1}{iq_n - \varepsilon_{\mathbf{q}}} \frac{1}{-iq_n - \varepsilon_{\mathbf{q}}} = \frac{V}{\beta_c} \sum_{|iq_n|<\omega_D} \frac{d(\varepsilon_F)}{2} \int_{-\infty}^{\infty} d\varepsilon \frac{1}{q_n^2 + \varepsilon^2}$$

$$= \frac{V d(\varepsilon_F)}{2\beta_c} \sum_{|iq_n|<\omega_D} \frac{\pi}{|q_n|} = \frac{1}{2} V d(\varepsilon_F) \left[\sum_{n=0}^{\frac{1}{2\pi}\beta_c\omega_D} \frac{1}{n + \frac{1}{2}} - 2 \right]$$

$$\approx \frac{V d(\varepsilon_F)}{2} \ln\left(4 \frac{\beta_c\omega_D}{2\pi}\right), \quad \beta_c\omega_D \gg 1, \tag{17.46}$$

where we use the density of states per spin, $d(\varepsilon_F)/2$. From this equation T_c is found to be

$$k_B T_c \approx \hbar\omega_D \exp\left[-\frac{2}{V d(\varepsilon_F)}\right]. \tag{17.47}$$

Two important comments can be made at this stage. The first is that although the characteristic phonon energy $\hbar\omega_D$ is of the order 100 K, see Fig. 3.6(b), the critical temperature T_c for the Cooper instability is lowered to about 1K by the exponential factor. The second comment is that T_c is a non-analytic function of the pair-interaction strength V, since $T_c(V) \propto \exp(-\mathrm{const}/V)$. Consequently, it is not possible to reach the new ground state resulting from the Cooper instability by perturbation theory in V of the normal metallic Fermi sea. These problems will be treated in some of the exercises of this chapter and in much greater detail in the next chapter concerning the BCS theory of superconductivity.

17.6   Summary and outlook

In this chapter we have presented the Green's function formalism and established the Feynman diagram rules for the free phonon system and for the electron-phonon coupling.

One important example concerned the renormalization of the phonon frequencies once the electron-phonon interaction was taking into account. The dispersion-less frequency

$$\Omega_{\mathbf{q}} = \Omega = \sqrt{\frac{Z^2 e^2 N}{\epsilon_0 M \mathcal{V}}},$$

of the non-interacting jellium phonons was changed into the linear phonon dispersion

$$\omega_{\mathbf{q}}(q \to 0, \, 0) = \sqrt{\frac{Z e^2 \rho_{\mathrm{el}}^0}{k_s^2 \epsilon_0 M}} \, q$$

for acoustic phonons in the long-wave limit.

A main result was to demonstrate the existence of the Cooper instability in the electron gas due to electron-phonon interaction. This instability forms the starting point of the BCS theory of ordinary superconductivity, which is the topic of the following chapter.

18

SUPERCONDUCTIVITY

The Bardeen–Cooper–Schrieffer (BCS) theory of superconductivity is a corner stone in theoretical physics. Since it appeared in 1957, its influence has reached far beyond its original scope, which was to give a coherent explanation at the microscopic level of a wide range of intricate and fascinating phenomena in metals at low temperature, known as, and related to, superconductivity. Besides metallic superconductivity BCS-like theories has been used to explain superfluid ^{3}He, the motion of nucleons in nuclei, and the dynamics of fundamental matter fields in high energy physics.

In this chapter, we will limit our discussion to conventional superconductivity in metallic systems. Superconductivity was discovered in 1911 by Kamerlingh–Onnes as the disappearance of the DC electrical resistance of mercury on cooling it below a critical temperature T_c (here $T_c = 4.2$ K, see Fig. 18.1(a)). In 1933 Meissner and Ochsenfeld observed that metals in the superconducting state are perfect diamagnets, i.e., they expel magnetic fields completely. The following year Gorter and Casimir postulated that superconductivity is due to electrons in a kind of superfluid state. In 1935 London and London extended this model to electromagnetic phenomena in their famous phenomenological theory. In 1950 Fröhlich in his electron-phonon theory emphasized the importance of lattice vibrations for superconductivity, a view supported by the independent observation by Reynolds *et al.* and Maxwell of the so-called isotope effect, i.e., the dependence of T_c on the ion mass. The BCS theory from 1957 was the first successful microscopic theory of superconductivity. In the following we give a brief introduction of the theory. For deeper studies we refer the reader to a number of classic text books on the subject such as Schrieffer (1983), de Gennes (1999) or Tinkham (1996).

18.1 The Cooper instability

The Cooper instability has already been addressed in Section 17.5 from the point of view of Feynman diagrams. Here we turn to the simpler wave function method to gain further insight as to why, the electron gas in an ordinary metal becomes unstable below a certain critical temperature T_c.

Cooper's analysis from 1956 of the instability of the Fermi sea due to pair-formation was a precursor of the BCS theory. The basic ingredient is the combination of a single electron state with its time-reversed counterpart into an electron-pair state of zero momentum and spin. Consider the following superposition of such zero momentum spin-singlets given by

$$\psi_{CP}(\mathbf{r}_1, \mathbf{r}_2) = \sum_{|\mathbf{k}| > k_F} a_{\mathbf{k}}\, e^{i\mathbf{k}\cdot\mathbf{r}_1} e^{-i\mathbf{k}\cdot\mathbf{r}_2}\, c^\dagger_{\mathbf{k}\uparrow} c^\dagger_{-\mathbf{k}\downarrow} |FS\rangle, \qquad (18.1)$$

or, in short, introducing the Cooper-pair ket $|\mathbf{k}\rangle$

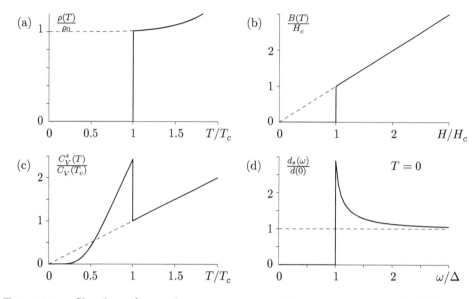

FIG. 18.1. Sketches of some key experiments with superconductors. (a) The transition to zero resistivity ρ below the critical temperature T_c. (b) The Meissner effect: the exclusion of magnetic induction B from a superconductor below the critical external magnetic field H_c. (c) The specific heat of electrons C_V^s in a superconductor compared to that of a normal metal. (d) The energy gap Δ in the density of states $d_s(\omega)$ of a superconductor.

$$|\psi_{CP}\rangle = \sum_{|\mathbf{k}|>k_F} a_{\mathbf{k}}|\mathbf{k}\rangle. \qquad (18.2)$$

Assume the pair state is added on top of a filled Fermi sea, which is felt only through Pauli's exclusion principle, thus disregarding Coulomb interactions. Let the Hamiltonian of the pair be

$$H_{CP} = H_0 + V_{\text{eff}}, \qquad (18.3)$$

where H_0 is the kinetic energy of the pair, $H_0|\mathbf{k}\rangle = 2(\varepsilon_{\mathbf{k}} - \mu)|\mathbf{k}\rangle = 2\xi_{\mathbf{k}}|\mathbf{k}\rangle$, and V_{eff} is the effective phonon-mediated interaction of Eq. (17.43) given by

$$\langle \mathbf{k}'|V_{\text{eff}}|\mathbf{k}\rangle = -Vw_{\mathbf{k}'}^* w_{\mathbf{k}}, \quad w_{\mathbf{k}} = \begin{cases} 1, & \text{for } |\xi_{\mathbf{k}}| < \omega_D \\ 0, & \text{otherwise} \end{cases}, \qquad (18.4)$$

where ω_D is the Debye frequency of the phonons. As shown in Exercise 18.1, the solution of the Schrödinger equation $H_{CP}|\psi_{CP}\rangle = E_{CP}|\psi_{CP}\rangle$ yields the eigenenergy

$$E_{CP} = -2\omega_D \left(\exp\left[\frac{2}{Vd(\varepsilon_F)}\right] - 1 \right)^{-1}, \qquad (18.5)$$

where $d(\varepsilon_F)$ is the electronic density of states at the Fermi level.

A remarkable conclusion of this little model calculation is that however weak the attractive interaction, a pair of electrons near the Fermi surface can gain energy by forming a Cooper pair. An intuitive explanation of the physics driving the Cooper-pair formation is the following: when an electron propagates through the crystal it attracts the positive ions and thus effectively creates a positive trace behind it. This trace is felt by the other electrons as an attractive interaction. It turns out that this effective interaction is most important for electrons occupying time reversed states and forming Cooper pairs, because in this case they can mutually take advantage of the positive trail of each other.

In the so-called weak coupling limit, $Vd(\varepsilon_F) \ll 1$, the energy gain Δ_{CP} of this bound pair state is

$$\Delta_{CP} = |E_{CP}| \approx 2\omega_D \exp\left[-\frac{2}{Vd(\varepsilon_F)}\right]. \tag{18.6}$$

Another result of the Cooper model, which agrees with Eq. (17.47), is that the characteristic energy scale seems not to be ω_D (~ 30 meV or ~ 400 K) as one might naïvely expect, but rather ω_D suppressed by the exponential factor $\exp[-2/Vd(\varepsilon_F)]$. This ultimately explains why the critical temperatures of conventional superconductors normally are below 10 K and not of the order of several hundred kelvin.

Finally, using the Heisenberg uncertainty relation and $\Delta E \approx (\partial E/\partial p)\,\Delta p$, we estimate in Exercise 18.2 the relatively large average distance ξ between the electrons forming an individual Cooper pair,

$$\xi \approx \frac{v_F}{\Delta_{CP}} \approx \frac{v_F}{k_B T_c}, \tag{18.7}$$

where we have anticipated the BCS result that the energy gain of a single Cooper pair sets the scale for the transition temperature T_c of the superconductor, $\Delta_{CP} \approx k_B T_c$.

The Cooper-pair size ξ plays an important role in the theory of superconductivity. It is commonly referred to as the coherence length of the superconductor.

18.2 The BCS groundstate

Two physical facts are crucial input to microscopic BCS theory of superconductivity: (i) The effective phonon-mediated electron-electron interaction Eq. (17.43) is attractive for energies $\xi_{\mathbf{k}}$ near the Fermi surface ($|\xi_{\mathbf{k}}| < \omega_D \ll \varepsilon_F$), and (ii) such an effective interaction renders the Fermi surface unstable under the formation of Cooper pairs, Sections 17.5 and 18.1. This led BCS to suggest the following variational wave function for the superconducting groundstate,

$$|\psi_{BCS}\rangle = \prod_{\mathbf{k}} \left(u_{\mathbf{k}} + v_{\mathbf{k}}\, c_{\mathbf{k}\uparrow}^{\dagger} c_{-\mathbf{k}\downarrow}^{\dagger}\right)|0\rangle, \tag{18.8}$$

where $u_{\mathbf{k}}$ and $v_{\mathbf{k}}$ are complex expansion coefficients. This wavefunction is a superposition of states each containing an integer number of Cooper pairs. The number of Cooper pairs in $|\psi_{BCS}\rangle$ is undetermined, and one should think of the superconductor being in contact with an electron reservoir allowing for fluctuations in the particle number. The concept of the Cooper pairs and the BCS groundstate is illustrated in Fig. 18.2.

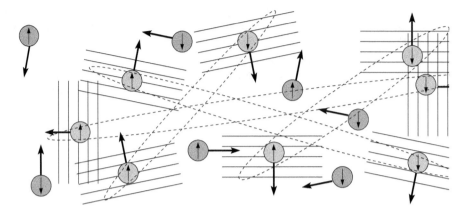

FIG. 18.2. A sketch in real space of the zero momentum spin-singlet Cooper pairs $|\mathbf{k}\rangle = |\mathbf{k}\uparrow, -\mathbf{k}\downarrow\rangle - |\mathbf{k}\downarrow, -\mathbf{k}\uparrow\rangle$ forming the BCS pair condensate. Although each electron is marked as a disk with a spin vector inside and a $\mathbf{k}$ vector just outside, they are in fact all moving in homogeneous plane-wave states spreading all over space. Each wave is indicated by five parallel lines. Pairs of counter-propagating electrons bound in Cooper pairs are marked with dark gray disks that are connected by a dashed ellipse. These ellipses are only serving as guides to the eye. The average size of each Cooper pair, the correlation length ξ, is much longer than the average distance $\bar{d}$ between the electrons. Some single electrons (light gray disks, without plane wave lines) that have been excited out of the condensate are also shown.

The variational calculation to determine $u_{\mathbf{k}}$ and $v_{\mathbf{k}}$ is not along the main line of this book, and we will therefore not pursue it further here. However, for completeness this analysis is carried out in Exercise 18.3 to Exercise 18.6. Instead, we will continue with the BCS mean-field theory, but before doing that it is important to discuss the physical reasoning leading to the BCS groundstate.

How can one guess such a strange-looking wave function as $|\psi_{\mathrm{BCS}}\rangle$? Experimental data like those of Fig. 18.1 are clear signatures of a phase transition in the electronic system. Integration of the electronic heat capacity curve Fig. 18.1(c) leads to the surprising result that the condensation energy gained by each electron, by undergoing this phase transition, is a meager 10^{-8} eV (see Exercise 18.7). This value should be compared to the binding energy of the electrons in the Fermi liquid of about 1 eV, Eq. (2.44). The largest energy that comes into play, is in fact the Debye energy $\omega_{\mathrm{D}} \approx 30$ meV $\ll \varepsilon_{\mathrm{F}}$, but the smallness of this energy allows us to use Fermi liquid theory.

Consider a homogeneous non-interacting electron gas. According to Chapter 15, when turning on the electron-electron interaction *without* the phonon-mediated pair-interaction, a low-energy electronic excitation $k^2/2m - \mu$ with momentum $\mathbf{k}$ develops into a well-defined quasiparticle excitation of energy $\xi_{\mathbf{k}}$, while keeping the momentum $\mathbf{k}$. One then consider the influence of the phonon-mediated pair interaction between these quasiparticles, which couples only Cooper-paired quasiparticles. This so-called BCS Hamiltonian is

$$H_{\text{BCS}} = \sum_{\mathbf{k}\sigma} \xi_{\mathbf{k}} c_{\mathbf{k}\sigma}^\dagger c_{\mathbf{k}\sigma} + \sum_{\mathbf{k}\mathbf{k}'} V_{\mathbf{k}\mathbf{k}'} c_{\mathbf{k}\uparrow}^\dagger c_{-\mathbf{k}\downarrow}^\dagger c_{-\mathbf{k}'\downarrow} c_{\mathbf{k}'\uparrow}, \tag{18.9}$$

where the coupling strength $V_{\mathbf{k}\mathbf{k}'} = -V < 0$ for states of energy $|\xi_{\mathbf{k}}|, |\xi_{\mathbf{k}'}| < \omega_{\text{D}}$ and zero otherwise as in Eq. (18.4). This effective Hamiltonian describing the interaction between the Cooper pairs is responsible for the formation of a new groundstate that is qualitatively different from a Fermi liquid. The effective pair interaction destabilizes the Fermi liquid. Cooper pairs are formed near the Fermi level, and these boson-like quasiparticles, each with zero momentum and zero spin, form the Bose—Einstein-like condensate described by the BCS groundstate $|\psi_{\text{BCS}}\rangle$.

18.3 Microscopic BCS theory

The mean-field assumption made by BCS is that due to the presence of many different Cooper pairs in the groundstate $|\psi_{\text{BCS}}\rangle$, the pair operator $c_{\mathbf{k}\uparrow}^\dagger c_{-\mathbf{k}\downarrow}^\dagger$ has a finite ground-state expectation value $\langle c_{\mathbf{k}\uparrow}^\dagger c_{-\mathbf{k}\downarrow}^\dagger \rangle \neq 0$, and its fluctuations around this average value are small. This picture is generalized to finite temperature with the understanding that as the temperature is raised more and more Cooper pairs are split and leave the Cooper-pair condensate as ordinary electrons. The critical temperature is the temperature where the condensate has "evaporated" fully. Below the critical temperature the thermodynamic average of $c_{\mathbf{k}\uparrow}^\dagger c_{-\mathbf{k}\downarrow}^\dagger$ is thus assumed to be non-zero.

Based on this assumption, the BCS mean-field Hamiltonian is derived from H_{BCS} in Eq. (18.9) in full analogy with Hartree–Fock mean-field theory described in Chapter 4

$$H_{\text{BCS}}^{\text{MF}} = \sum_{\mathbf{k}\sigma} \xi_{\mathbf{k}} c_{\mathbf{k}\sigma}^\dagger c_{\mathbf{k}\sigma} - \sum_{\mathbf{k}} \Delta_{\mathbf{k}} c_{\mathbf{k}\uparrow}^\dagger c_{-\mathbf{k}\downarrow}^\dagger - \sum_{\mathbf{k}} \Delta_{\mathbf{k}}^* c_{-\mathbf{k}\downarrow} c_{\mathbf{k}\uparrow}, \tag{18.10}$$

where

$$\Delta_{\mathbf{k}} = -\sum_{\mathbf{k}\mathbf{k}'} V_{\mathbf{k}'} \langle c_{-\mathbf{k}'\downarrow} c_{\mathbf{k}'\uparrow} \rangle, \tag{18.11}$$

and where the constant term $\sum_{\mathbf{k}\mathbf{k}'} V_{\mathbf{k}\mathbf{k}'} \langle c_{\mathbf{k}\uparrow}^\dagger c_{-\mathbf{k}\downarrow}^\dagger \rangle \langle c_{-\mathbf{k}\downarrow} c_{\mathbf{k}\uparrow} \rangle$ has been absorbed into the chemical potential. This mean-field Hamiltonian is quadratic in the electron operators and should be readily solvable. It is, however, somewhat unusual in that terms like $c^\dagger c^\dagger$ and cc appear. The way to solve it is by the so-called Bogoliubov transformation defined by the following unitary transformation of the c and $c^\dagger$ operators,

$$\begin{pmatrix} \gamma_{\mathbf{k}\uparrow} \\ \gamma_{-\mathbf{k}\downarrow}^\dagger \end{pmatrix} = \begin{pmatrix} u_{\mathbf{k}}^* & v_{\mathbf{k}} \\ -v_{\mathbf{k}}^* & u_{\mathbf{k}} \end{pmatrix} \begin{pmatrix} c_{\mathbf{k}\uparrow} \\ c_{-\mathbf{k}\downarrow}^\dagger \end{pmatrix} \tag{18.12a}$$

and the corresponding inverse transformation

$$\begin{pmatrix} c_{\mathbf{k}\uparrow} \\ c_{-\mathbf{k}\downarrow}^\dagger \end{pmatrix} = \begin{pmatrix} u_{\mathbf{k}} & -v_{\mathbf{k}} \\ v_{\mathbf{k}}^* & u_{\mathbf{k}}^* \end{pmatrix} \begin{pmatrix} \gamma_{\mathbf{k}\uparrow} \\ \gamma_{-\mathbf{k}\downarrow}^\dagger \end{pmatrix}. \tag{18.12b}$$

At zero temperature the values of $u_{\mathbf{k}}$ and $v_{\mathbf{k}}$ are the same in Eqs. (18.8) and (18.12a). As found in Exercise 18.8, $H_{\text{BCS}}^{\text{MF}}$ is diagonalized by the following values of $u_{\mathbf{k}}$ and $v_{\mathbf{k}}$,

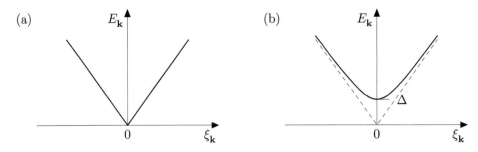

FIG. 18.3.   (a) The electronic quasiparticle excitation spectrum $\xi_{\mathbf{k}}$ near k_{F} for a normal metal. (b) The quasiparticle excitation spectrum $E_{\mathbf{k}}$ for a BCS superconductor near k_{F}. Note that $E_{\mathbf{k}} \to \xi_{\mathbf{k}}$ for $\Delta_{\mathbf{k}} \to 0$.

$$|u_{\mathbf{k}}|^2 = \frac{1}{2}\left(1 + \frac{\xi_{\mathbf{k}}}{E_{\mathbf{k}}}\right) \quad \text{and} \quad |v_{\mathbf{k}}|^2 = \frac{1}{2}\left(1 - \frac{\xi_{\mathbf{k}}}{E_{\mathbf{k}}}\right), \qquad (18.13)$$

where

$$E_{\mathbf{k}} = \sqrt{\xi_{\mathbf{k}}^2 + |\Delta_{\mathbf{k}}|^2}. \qquad (18.14)$$

With these values we obtain

$$H_{\mathrm{BCS}}^{\mathrm{MF}} = \sum_{\mathbf{k}} E_{\mathbf{k}}\left(\gamma_{\mathbf{k}\uparrow}^{\dagger}\gamma_{\mathbf{k}\uparrow} + \gamma_{\mathbf{k}\downarrow}^{\dagger}\gamma_{\mathbf{k}\downarrow}\right) + \text{constant}. \qquad (18.15)$$

The new quasiparticles described by the number operator $\gamma_{\mathbf{k}\uparrow}^{\dagger}\gamma_{\mathbf{k}\uparrow}$ are called bogoliubons after the Bogoliubov transformation spawning them. As is evident from the new Hamiltonian (18.15) and the solution (18.14), there are no fermion excitations possible with energy less than $|\Delta_{\mathbf{k}}|$. The mean-field parameter $\Delta_{\mathbf{k}}$ thus provides an energy gap denoted the superconducting gap. The existence of this gap has a number of important consequences.

The self-consistent solution for $\Delta_{\mathbf{k}}$ is found from Eq. (18.11), the so-called gap equation, by calculating the expectation value of the right-hand side using the diagonalized Hamiltonian.[63] Using Eqs. (18.13) and (18.14), we find

$$
\begin{aligned}
\Delta_{\mathbf{k}} &= -\sum_{\mathbf{k}'} V_{\mathbf{k}\mathbf{k}'}\langle c_{-\mathbf{k}'\downarrow}c_{\mathbf{k}'\uparrow}\rangle \\
&= -\sum_{\mathbf{k}'} V_{\mathbf{k}\mathbf{k}'}\left\langle\left(u_{\mathbf{k}'}^{*}\gamma_{-\mathbf{k}'\downarrow} - v_{\mathbf{k}'}\gamma_{\mathbf{k}'\uparrow}^{\dagger}\right)\left(u_{\mathbf{k}'}^{*}\gamma_{\mathbf{k}'\uparrow} + v_{\mathbf{k}'}\gamma_{-\mathbf{k}'\downarrow}^{\dagger}\right)\right\rangle \\
&= -\sum_{\mathbf{k}'} V_{\mathbf{k}\mathbf{k}'}\left(u_{\mathbf{k}'}^{*}v_{\mathbf{k}'}\langle\gamma_{-\mathbf{k}'\downarrow}\gamma_{-\mathbf{k}'\downarrow}^{\dagger}\rangle - v_{\mathbf{k}'}u_{\mathbf{k}'}^{*}\langle\gamma_{\mathbf{k}'\downarrow}^{\dagger}\gamma_{\mathbf{k}'\downarrow}\rangle\right) \\
&= -\sum_{\mathbf{k}'} V_{\mathbf{k}\mathbf{k}'}u_{\mathbf{k}'}^{*}v_{\mathbf{k}'}\left[1 - 2n_{\mathrm{F}}(E_{\mathbf{k}'})\right], \qquad (18.16)
\end{aligned}
$$

where in the last step it was used that according to the Hamiltonian (18.15), the bogoliubons are free fermions, and therefore their distribution function is the usual

---

[63] According to the general discussion in Section 4.1, this procedure is equivalent to minimizing the free energy with respect to the mean-field parameter.

Fermi–Dirac distribution, e.g., $\langle \gamma^\dagger_{\mathbf{k'}\downarrow}\gamma_{\mathbf{k'}\downarrow}\rangle = n_F(E_{\mathbf{k'}})$. We postpone the calculation of $\Delta_{\mathbf{k}}$ until the next section, where we rederive the BCS-self-consistency equation (18.16) by use of the Matsubara Green's function technique.

18.4 BCS theory with Matsubara Green's functions

In this section, we formulate the BCS mean-field theory in terms of Matsubara Green's functions. Being careful with the spin indices, we define the normal Green's function $\mathcal{G}_{\uparrow\uparrow}(\mathbf{k}, \tau)$ as

$$\mathcal{G}_{\uparrow\uparrow}(\mathbf{k}, \tau) = -\Big\langle T_\tau c_{\mathbf{k}\uparrow}(\tau) c^\dagger_{\mathbf{k}\uparrow}(0)\Big\rangle, \tag{18.17}$$

and the anomalous Green function $\mathcal{F}_{\downarrow\uparrow}(\mathbf{k}, \tau)$ as

$$\mathcal{F}_{\downarrow\uparrow}(\mathbf{k}, \tau) = -\Big\langle T_\tau c^\dagger_{-\mathbf{k}\downarrow}(\tau) c^\dagger_{\mathbf{k}\uparrow}(0)\Big\rangle. \tag{18.18}$$

As in the previous section, the BCS assumption about the existence of a Cooper-pair condensate ensures the possibility for non-zero values of the anomalous Green's function $\mathcal{F}$. Once this assumption has been made, the Matsubara Green's function technique can be applied: Fourier transformation of the equation of motion in τ-space to ik_n-space followed by Matsubara frequency summation.

From the BCS mean-field Hamiltonian $H^{\mathrm{MF}}_{\mathrm{BCS}}$ Eq. (18.10), it is straightforward to obtain the following equations of motion for $\mathcal{G}$ and $\mathcal{F}$ in τ-space:

$$\partial_\tau \mathcal{G}_{\uparrow\uparrow}(\mathbf{k}, \tau) = -\delta(\tau) - \xi_{\mathbf{k}}\mathcal{G}_{\uparrow\uparrow}(\mathbf{k}, \tau) + \Delta_{\mathbf{k}}\,\mathcal{F}_{\downarrow\uparrow}(\mathbf{k}, \tau), \tag{18.19}$$

$$\partial_\tau \mathcal{F}_{\downarrow\uparrow}(\mathbf{k}, \tau) = \qquad \xi_{\mathbf{k}}\mathcal{F}_{\downarrow\uparrow}(\mathbf{k}, \tau) + \Delta^*_{\mathbf{k}}\,\mathcal{G}_{\uparrow\uparrow}(\mathbf{k}, \tau). \tag{18.20}$$

Note, that there is no $\delta(\tau)$ term for $\partial_\tau \mathcal{F}$; why this is so, we leave as a little exercise for the reader. The subsequent Fourier transform to Matsubara frequency-space ik_n is also straightforward,

$$(-ik_n + \xi_{\mathbf{k}})\mathcal{G}_{\uparrow\uparrow}(\mathbf{k}, ik_n) = -1 + \Delta_{\mathbf{k}}\,\mathcal{F}_{\downarrow\uparrow}(\mathbf{k}, ik_n), \tag{18.21}$$

$$(-ik_n - \xi_{\mathbf{k}})\mathcal{F}_{\downarrow\uparrow}(\mathbf{k}, ik_n) = \qquad \Delta^*_{\mathbf{k}}\,\mathcal{G}_{\uparrow\uparrow}(\mathbf{k}, ik_n). \tag{18.22}$$

These algebraic equations are readily solved,

$$\mathcal{F}_{\downarrow\uparrow}(\mathbf{k}, ik_n) = \frac{-\Delta^*_{\mathbf{k}}}{(ik_n)^2 - (\xi^2_{\mathbf{k}} + |\Delta_{\mathbf{k}}|^2)}, \tag{18.23}$$

$$\mathcal{G}_{\uparrow\uparrow}(\mathbf{k}, ik_n) = \frac{ik_n + \xi_{\mathbf{k}}}{(ik_n)^2 - (\xi^2_{\mathbf{k}} + |\Delta_{\mathbf{k}}|^2)}. \tag{18.24}$$

The poles of $\mathcal{G}_{\uparrow\uparrow}(\mathbf{k}, ik_n)$ are seen to be

$$ik_n = \pm E_{\mathbf{k}} = \pm\sqrt{\xi^2 + |\Delta_{\mathbf{k}}|^2}, \tag{18.25}$$

which agrees with Eq. (18.14).

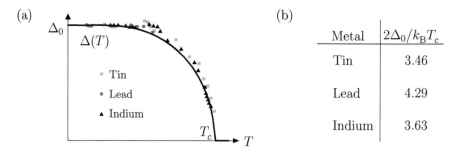

(a) Δ_0, $\Delta(T)$, Tin, Lead, Indium, T_c, T

(b)

Metal	$2\Delta_0/k_\mathrm{B}T_c$
Tin	3.46
Lead	4.29
Indium	3.63

FIG. 18.4. (a) A sketch showing measured values of the gap parameter for three
different metals compared to the BCS predictions. To the left the temperature
dependence is shown as it follows from the BCS gap equation in Eq. (18.28)
together with experimental values. (b) A table of the measured value of the ratio
between twice the gap at zero temperature and critical temperature, determined
from tunneling measurements. The theoretical BCS value is 3.53 as stated in
Eq. (18.33).

18.4.1 *Self-consistent determination of the BCS order parameter $\Delta_\mathbf{k}$*

The BCS interaction model Eq. (18.4) is now used in the definition (18.11) of the
BCS order parameter $\Delta_\mathbf{k}$. This gives

$$\Delta_\mathbf{k} = V \sum_\mathbf{k}^{|\xi_\mathbf{k}|<\omega_D} \langle c_{-\mathbf{k}\downarrow}c_{\mathbf{k}\uparrow}\rangle = V \sum_\mathbf{k}^{|\xi_\mathbf{k}|<\omega_D} \mathcal{F}^*_{\downarrow\uparrow}(\mathbf{k},\tau=0^+),\qquad(18.26)$$

where in the second equality we have used the complex conjugate of Eq. (18.18)
defining $\mathcal{F}(\mathbf{k},\tau)$. We then write $\mathcal{F}^*_{\downarrow\uparrow}(\mathbf{k},\tau=0^+)$ as a Fourier transform with Matsubara
frequency ik_n and insert the explicit expression (18.23) for $\mathcal{F}(\mathbf{k},ik_n)$. The resulting
Matsubara summation is carried out, and we find

$$\Delta_\mathbf{k} = V \sum_\mathbf{k}^{|\xi_\mathbf{k}|<\omega_D} \frac{1}{\beta}\sum_{ik_n} \frac{-\Delta_\mathbf{k}\,e^{-ik_n 0^+}}{(ik_n)^2 - (\xi_\mathbf{k}^2 + |\Delta_\mathbf{k}|^2)}$$

$$= -\Delta_\mathbf{k} V \sum_\mathbf{k}^{|\xi_\mathbf{k}|<\omega_D} \left(\frac{n_\mathrm{F}(E_\mathbf{k})}{2E_\mathbf{k}} + \frac{n_\mathrm{F}(-E_\mathbf{k})}{-2E_\mathbf{k}}\right).\qquad(18.27)$$

Dividing this equation with $\Delta_\mathbf{k}$ and converting the $\mathbf{k}$-sum into a $\xi_\mathbf{k}$-integral yields

$$1 = V \sum_\mathbf{k}^{|\xi_\mathbf{k}|<\omega_D} \frac{1-2n_\mathrm{F}(E_\mathbf{k})}{2E_\mathbf{k}} = Vd(\varepsilon_\mathrm{F}) \int_{-\omega_D}^{\omega_D} d\xi_\mathbf{k} \frac{\tanh\left(\frac{\beta}{2}\sqrt{\xi_\mathbf{k}^2+|\Delta_\mathbf{k}|^2}\right)}{2\sqrt{\xi_\mathbf{k}^2+|\Delta_\mathbf{k}|^2}}.\qquad(18.28)$$

The BCS energy gap $\Delta_\mathbf{k}(T)$ can be determined numerically from this equation. We
note that within the BCS interaction model, Eq. (18.4), the energy gap is the same
for all values of $\mathbf{k}$, so we can omit the subscript $\mathbf{k}$. The resulting BCS energy gap
$\Delta(T)$ is depicted in Fig. 18.4 together with some experimental values.

From Eq. (18.28) we can actually find analytical expressions for $\Delta(T)$ in the important limit $T = 0$. Since the square root factor is positive, the tanh-term approaches unity as $T \to 0$ or $\beta \to \infty$. This leads to

$$1 = Vd(\varepsilon_{\mathrm{F}}) \int_0^{\omega_{\mathrm{D}}} \frac{d\xi}{\sqrt{\xi^2 + |\Delta|^2}} = Vd(\varepsilon_{\mathrm{F}}) \sinh^{-1}\left(\frac{\omega_{\mathrm{D}}}{|\Delta|}\right), \qquad (18.29)$$

which in the weak coupling limit $Vd(\varepsilon_{\mathrm{F}}) \ll 1$ results in

$$|\Delta(0)| = 2\,\omega_{\mathrm{D}}\,\exp\left[\frac{-1}{Vd(\varepsilon_{\mathrm{F}})}\right]. \qquad (18.30)$$

18.4.2 Determination of the critical temperature T_c

From Eq. (18.28) it is also possible to derive an analytical expression for the critical temperature T_c, where the superconducting phase transition occurs. We note that as T rises from $T = 0$, the tanh-term in the numerator decreases from 1. Now, since the left-hand side of Eq. (18.28) remains 1, it follows that the denominator must decrease as well, and this can only happen if $|\Delta|$ decreases. At some temperature $T = T_c$ the energy gap reaches zero, and the superconductivity disappears. Thus for $T = T_c^-$ we have $\Delta = 0^+$, and Eq. (18.28) becomes

$$1 = Vd(\varepsilon_{\mathrm{F}}) \int_0^{\omega_{\mathrm{D}}} d\xi\, \frac{1}{\xi}\, \tanh\left(\frac{\xi}{2k_{\mathrm{B}}T_c}\right) = Vd(\varepsilon_{\mathrm{F}}) \int_0^{\frac{\omega_{\mathrm{D}}}{2k_{\mathrm{B}}T_c}} dx\, \frac{1}{x}\, \tanh(x). \qquad (18.31)$$

In the weak coupling limit, the integral can be performed analytically,[64]

$$k_{\mathrm{B}}T_c = \frac{2}{\pi}\, e^\gamma\, \omega_{\mathrm{D}} \exp\left[\frac{-1}{Vd(\varepsilon_{\mathrm{F}})}\right] \approx 1.13\,\omega_{\mathrm{D}} \exp\left[\frac{-1}{Vd(\varepsilon_{\mathrm{F}})}\right]. \qquad (18.32)$$

The BCS expressions for $|\Delta(0)|$ and T_c both contain the coupling constant V, which is very difficult both to calculate and to measure. However, by forming the ratio between the two quantities we arrive at a very strong prediction without any free parameters,

$$\frac{2\Delta(0)}{k_{\mathrm{B}}T_c} = 3.53. \qquad (18.33)$$

This ratio is in good agreement with the three experimental values listed in Fig. 18.4(b).

Both the gap and the critical temperature are reduced by the exponential factor $\exp[-1/Vd(\varepsilon_{\mathrm{F}})]$ compared to the bare energy scale of the interaction, ω_{D}. This strong renormalization is what generates the new scale, $\omega_{\mathrm{D}}\,\exp[-1/Vd(\varepsilon_{\mathrm{F}})]$. Note that the interaction strength appears in a non-perturbative fashion in this expression; the function $\exp[-1/x]$ has no Taylor expansion at $x = 0$. This tells us that the result could never have been derived using perturbation theory in the interaction, no matter

---

[64]We find $\int_0^{\frac{\omega_{\mathrm{D}}}{2k_{\mathrm{B}}T_c}} dx\, \frac{1}{x}\, \tanh(x) = \ln[\omega_{\mathrm{D}}/(2k_{\mathrm{B}}T_c)] - \int_0^{\frac{\omega_{\mathrm{D}}}{2k_{\mathrm{B}}T_c}} dx\, \ln(x)\cosh^{-2}(x)$ by partial integration. In the weak coupling limit, where $k_{\mathrm{B}}T_c \ll \omega_{\mathrm{D}}$, the upper limit of the integral can be set to ∞, resulting in $\ln[\omega_{\mathrm{D}}/(2k_{\mathrm{B}}T_c)] - \ln\frac{\pi}{4} + \gamma$, where $\gamma \approx 0.577$ is Euler's constant.

how many orders were included. This is in fact a general feature of phase transitions. It is not possible by perturbation expansions to cross a phase transition line, because the two states have no analytic connection. Once again, we see that there is no automatic way to predict the phase diagram of a given physical system, and one must rely on a combination of technical skill and, most importantly, physical intuition.

18.4.3   Determination of the BCS quasiparticle density of states

Quasiparticle excitations from the BCS groundstate are single un-bound electrons no longer part of the condensate of Cooper-pairs. The density of states $d_s(\omega)$ of these quasiparticles in the superconductor is given by the spectral function Eqs. (8.50) and (8.62) as

$$d_s(\omega) = \frac{1}{2\pi\mathcal{V}} \sum_{\mathbf{k}\sigma} A(\mathbf{k}\sigma, \omega) = -\frac{1}{\pi\mathcal{V}} \sum_{\mathbf{k}\sigma} \mathrm{Im}\ G_{\sigma\sigma}^{\mathrm{R}}(\mathbf{k}, \omega) = -\frac{1}{\pi\mathcal{V}} \sum_{\mathbf{k}\sigma} \mathrm{Im}\ \mathcal{G}_{\sigma\sigma}(\mathbf{k}, \omega + i\eta).$$

(18.34)

Using the BCS result Eq. (18.24) for $\mathcal{G}_{\sigma\sigma}$ with $E(\xi) = \sqrt{\xi^2 + |\Delta|^2}$ we obtaint

$$d_s(\omega) = \frac{-1}{\pi\mathcal{V}} \sum_{\mathbf{k}\sigma} \mathrm{Im}\ \frac{\omega + i\eta + \xi_{\mathbf{k}}}{(\omega + i\eta)^2 - E_{\mathbf{k}}^2}$$

$$= \frac{-1}{\pi\mathcal{V}} \sum_{\mathbf{k}\sigma} \mathrm{Im}\ \left\{ \left[\frac{1}{\omega - E_{\mathbf{k}} + i\eta} - \frac{1}{\omega + E_{\mathbf{k}} + i\eta} \right] \frac{\omega + \xi_{\mathbf{k}}}{2E_{\mathbf{k}}} \right\}.$$

(18.35)

At this stage, we take the imaginary part followed by a change of variables from the discrete wavenumber $\mathbf{k}$ to the continuous energy ξ,

$$d_s(\omega) = \frac{1}{\mathcal{V}} \sum_{\mathbf{k}\sigma} \left[\delta(\omega - E_{\mathbf{k}}) - \delta(\omega + E_{\mathbf{k}}) \right] \frac{\omega + \xi_{\mathbf{k}}}{2E_{\mathbf{k}}}$$

$$= d(0) \int_{-\infty}^{\infty} d\xi\ \frac{1}{2} \left\{ \left[1 + \frac{\xi}{E} \right] \delta(\omega - E) + \left[1 - \frac{\xi}{E} \right] \delta(\omega + E) \right\}.$$

(18.36)

We now focus on positive excitation energies $\omega > 0$. In this case only the first of the two delta functions can differ from zero. Moreover, $[\xi/E(\xi)]\ \delta[\omega - E(\xi)]$ is antisymmetric and does therefore not contribute to the integral. The symmetric part $1\ \delta[\omega - E(\xi)]$ leads to

$$\frac{d_s(\omega)}{d(0)} = 2 \int_0^{\infty} d\xi\ \frac{1}{2} \delta(\omega - E) = \int_{|\Delta|}^{\infty} dE\ \frac{\partial \xi}{\partial E} \delta(\omega - E) = \int_{|\Delta|}^{\infty} dE\ \frac{E}{\sqrt{E^2 - |\Delta|^2}} \delta(\omega - E).$$

(18.37)

The BCS quasiparticle density of states is thus

$$\frac{d_s(\omega)}{d(0)} = \frac{\omega}{\sqrt{\omega^2 - |\Delta|^2}}\ \theta(\omega - |\Delta|),$$

(18.38)

which is plotted in Fig. 18.1(d). It is left as an exercise for the reader to show that $d_s(-\omega) = d_s(\omega)$. The main conclusion is that there is a gap of size $2|\Delta|$ around the

Fermi level. There are no quasiparticles with an excitation energy between $-|\Delta|$ and $|\Delta|$, and on each side of this interval there is spike in the density of states.

It is possible to measure the peculiar BCS density of states in electron tunneling experiments. Indeed, such experiments constitute one of the main methods to measure the superconducting gap Δ, see Fig. 18.4. The main result from Section 8.4.1 regarding tunneling currents is Eq. (8.67),

$$I_p = \int_{-\infty}^{\infty} \frac{d\omega}{2\pi} \sum_{\nu\mu} |T_{\nu\mu}|^2 A_1(\nu, \omega) A_2(\mu, \omega + eV) \left[n_F(\omega + eV) - n_F(\omega) \right]. \quad (18.39)$$

We now imagine performing tunneling spectroscopy where electrode 2 is always the same normal metal m with a simple spectral function $A_2(\mu, \omega)$ yielding a constant density of states as in Eq. (8.75). Electrode 1 is then either a metal m or a superconductor s with spectral functions $A_{1m}(\nu, \omega)$ and $A_{1s}(\nu, \omega)$, respectively. The differential conductance dI/dV Eq. (8.76) for metal-metal and metal-superconductor tunneling thus become

$$\frac{dI_{mm}}{dV} = C_1 \sum_{\nu} A_{1m}(\nu, -eV) = C_2 \, d(-eV), \quad (18.40)$$

$$\frac{dI_{ms}}{dV} = C_1 \sum_{\nu} A_{1s}(\nu, -eV) = C_2 \, d_s(-eV). \quad (18.41)$$

The ratio of these differential conductances becomes

$$\frac{\frac{dI_{mm}}{dV}}{\frac{dI_{ms}}{dV}} = \frac{d_s(-eV)}{d(-eV)} \approx \frac{d_s(-eV)}{d(0)} = \frac{|eV|}{\sqrt{(eV)^2 - |\Delta|^2}} \, \theta(|eV| - |\Delta|), \quad (18.42)$$

which gives a direct test of the BCS expression for the quasiparticle density of states.

18.5 The Nambu formalism of the BCS theory

The spin indices on the Matsubara Green's functions $\mathcal{G}_{\uparrow\uparrow}$ and $\mathcal{F}_{\downarrow\uparrow}$ in Eqs. (18.17) and (18.18) suggest that we could define a matrix Green's function. This is indeed possible, and by doing so we obtain the compact Nambu formalism. We shall use this formalism to calculate the linear-response current density in a BCS superconductor, in analogy with the Kubo-calculation of the conductivity in Chapter 15.

18.5.1 *Spinors and Green's functions in the Nambu formalism*

The starting point of the Nambu formalism is the introduction of the spinors $\alpha_{\mathbf{k}}(\tau)$ and $\alpha_{\mathbf{k}}^\dagger(\tau)$ with imaginary time argument τ,

$$\alpha_{\mathbf{k}}(\tau) = \begin{pmatrix} c_{\mathbf{k}\uparrow}(\tau) \\ c_{-\mathbf{k}\downarrow}^\dagger(\tau) \end{pmatrix} \quad \text{and} \quad \alpha_{\mathbf{k}}^\dagger = \begin{pmatrix} c_{\mathbf{k}\uparrow}^\dagger(\tau), & c_{-\mathbf{k}\downarrow}(\tau) \end{pmatrix}. \quad (18.43)$$

Using these spinors the Nambu Green's function $\bar{\mathcal{G}}(\mathbf{k}, \tau)$ is defined as a matrix generalization of the standard Matsubara Green's function,

$$\bar{\bar{\mathcal{G}}}(\mathbf{k}, \tau) = -\langle T_\tau \alpha_{\mathbf{k}}(\tau)\alpha_{\mathbf{k}}^\dagger(0)\rangle = \begin{pmatrix} \mathcal{G}_{\uparrow\uparrow}(\mathbf{k}, \tau) & \mathcal{F}_{\downarrow\uparrow}^*(\mathbf{k}, \tau) \\ \mathcal{F}_{\downarrow\uparrow}(\mathbf{k}, \tau) & \mathcal{G}_{\downarrow\downarrow}^*(-\mathbf{k}, \tau) \end{pmatrix}. \tag{18.44}$$

From the BCS mean-field Hamiltonian $H_{\mathrm{BCS}}^{\mathrm{MF}}$ Eq. (18.10), or more directly Eqs. (18.19) and (18.20), it is straightforward to obtain the equation of motion for $\bar{\bar{\mathcal{G}}}(\mathbf{k}, \tau)$,

$$\partial_\tau \bar{\bar{\mathcal{G}}}(\mathbf{k}, \tau) = -\delta(\tau)\begin{pmatrix} 1 & 0 \\ 0 & 1 \end{pmatrix} - \begin{pmatrix} \xi_{\mathbf{k}} & -\Delta_{\mathbf{k}} \\ -\Delta_{\mathbf{k}}^* & -\xi_{\mathbf{k}} \end{pmatrix}\bar{\bar{\mathcal{G}}}(\mathbf{k}, \tau). \tag{18.45}$$

By Fourier transformation and collecting terms containing $\bar{\bar{\mathcal{G}}}$ we obatin

$$\begin{pmatrix} i\omega_k - \xi_{\mathbf{k}} & \Delta_{\mathbf{k}} \\ \Delta_{\mathbf{k}}^* & i\omega_k + \xi_{\mathbf{k}} \end{pmatrix}\bar{\bar{\mathcal{G}}}(\mathbf{k}, i\omega_k) = \begin{pmatrix} 1 & 0 \\ 0 & 1 \end{pmatrix}. \tag{18.46}$$

Matrix inversion leads to the final form for $\bar{\bar{\mathcal{G}}}$,

$$\bar{\bar{\mathcal{G}}}(\mathbf{k}, i\omega_k) = \frac{1}{(i\omega_k)^2 - E_{\mathbf{k}}^2}\begin{pmatrix} i\omega_k + \xi_{\mathbf{k}} & -\Delta_{\mathbf{k}} \\ -\Delta_{\mathbf{k}}^* & i\omega_k - \xi_{\mathbf{k}} \end{pmatrix}, \tag{18.47}$$

where as before $E_{\mathbf{k}}^2 = \xi_{\mathbf{k}}^2 + |\Delta_{\mathbf{k}}|^2$.

18.5.2 The Meissner effect and the London equation

Before employing the Nambu formalism in a calculation of the paramagnetic current response of a superconductor, it is necessary to give a brief description of the Meissner effect. In 1933 Meissner and Ochsenfeld observed that metals in the superconducting state are perfect diamagnets, i.e., they expel magnetic fields completely as long as the temperature T and the external magnetic field H are not too large. For a given T the superconducting state initially remains intact as H is increased from zero. However, it breaks down and becomes a normal metallic state when H exceeds a critical value $H_c(T)$. This critical field is given by the approximate expression

$$H_c(T) = H_c(0)\left[1 - \frac{T^2}{T_c}\right], \tag{18.48}$$

where T_c is the usual critical temperature in zero magnetic field, and where $H_c(0)$ is the critical magnetic field at zero temperature. A superconductor with this magnetic response is denoted a type 1 superconductor. A typical response curve is shown in Fig. 18.1(b).

A major step towards a theoretical explanation of the Meissner effect was given by London and London in 1935. They considered the electric current density $\mathbf{J}_e$ (the particle current density of Eq. (6.20) times the electron charge) being a sum of the paramagnetic and the diamagnetic contribution,

$$\mathbf{J}_e(\mathbf{r}, \mathbf{A}) = \mathbf{J}_e^\nabla(\mathbf{r}, \mathbf{A}) - \frac{e^2\rho(\mathbf{r})}{m}\mathbf{A}, \tag{18.49}$$

where $\rho(\mathbf{r})$ is the electron density, and where $\mathbf{A}$ is the vector potential in some specific gauge. For normal metals the paramagnetic and diamagnetic terms nearly cancel out leaving only a very weak diamagnetic response. London and London speculated that if for some reason the superconducting wave function were rigid upon turning on the vector potential, it would not generate a paramagnetic response different from its zero field value

$$\mathbf{J}_e^{\nabla}(\mathbf{r}, \mathbf{A}) = \mathbf{J}_e^{\nabla}(\mathbf{r}, 0) = 0. \tag{18.50}$$

The current density would then be governed by the diamagnetic term

$$\mathbf{J}_e(\mathbf{r}) = -\frac{e^2 \rho(\mathbf{r})}{m} \mathbf{A}. \tag{18.51}$$

If the density ρ_s of superconducting electrons is smaller the ρ, the former is used, as we shall derive in the next subsection, i.e.,

$$\mathbf{J}_e(\mathbf{r}) = -\frac{e^2 \rho_s(\mathbf{r})}{m} \mathbf{A}. \tag{18.52}$$

This is the famous London equation. The Maxwell equation $\mu_0 \mathbf{J}_e = \nabla \times \mathbf{B} = \nabla \times \nabla \times \mathbf{A}$ valid for static fields is inserted on the left-hand side, and choosing the transverse gauge

$$\nabla \cdot \mathbf{A} = 0, \tag{18.53}$$

the following equation for $\mathbf{A}$ is easily derived, given that ρ_s is constant,

$$\nabla^2 \mathbf{A} = \frac{\mu_0 e^2}{m} \rho_s \mathbf{A} = \frac{1}{\lambda_L^2} \mathbf{A}, \tag{18.54}$$

where we have introduced the charactetistic London length λ_L,

$$\lambda_L = \sqrt{\frac{m}{\mu_0 e^2 \rho_s}}. \tag{18.55}$$

In Exercise 18.9, we consider the specific case of a superconductor taking up the positive half-space $x > 0$ with vacuum in the negative half-space $x < 0$. It is shown that a magnetic induction in the z direction, $\mathbf{B} = B(x)\hat{\mathbf{e}}_z$, which has the constant value B outside the superconductor, will decay exponentially on the length scale λ_L inside it leaving the interior (almost) completely field-free,

$$B(x) = \begin{cases} B, & \text{for } x < 0, \\ B\, e^{-x/\lambda_L}, & \text{for } x > 0. \end{cases} \tag{18.56}$$

Thus the Meissner effect can be explained once the rigidity of the wave function Eq. (18.50) has been proven. This we will do using the Nambu formalism in the next subsection.

18.5.3 *The vanishing paramagnetic current response in BCS theory*

Building on the linear response results of Sections 6.2 and 16.1, it is relatively simple to set up the calculation of the paramagnetic current response in the Nambu formalism. It is just a question of demonstrating that in the basic formulae for the current operator and the Green's functions, the ordinary electron operators $c^\dagger$ can be substituted by the Nambu spinors $\alpha^\dagger$. From Eq. (1.99) we find

$$
\begin{aligned}
\mathbf{J}(\mathbf{q}) &= \frac{1}{2m}\sum_{\mathbf{k}\sigma}(2\mathbf{k}+\mathbf{q})\,c^\dagger_{\mathbf{k}\sigma}c_{\mathbf{k}+\mathbf{q}\sigma} \\
&= \frac{1}{2m}\sum_{\mathbf{k}}(2\mathbf{k}+\mathbf{q})\left[c^\dagger_{\mathbf{k}\uparrow}c_{\mathbf{k}+\mathbf{q}\uparrow} - c^\dagger_{-\mathbf{k}-\mathbf{q}\downarrow}c_{-\mathbf{k}\downarrow}\right] \\
&= \frac{1}{2m}\sum_{\mathbf{k}}(2\mathbf{k}+\mathbf{q})\left[\left(c^\dagger_{\mathbf{k}\uparrow},\; c_{-\mathbf{k}\downarrow}\right)\begin{pmatrix} c_{\mathbf{k}+\mathbf{q}\uparrow} \\ c^\dagger_{-\mathbf{k}-\mathbf{q}\downarrow}\end{pmatrix} + \delta_{\mathbf{q}0}\right] \\
&= \frac{1}{2m}\sum_{\mathbf{k}}(2\mathbf{k}+\mathbf{q})\,\alpha^\dagger_{\mathbf{k}}\alpha_{\mathbf{k}+\mathbf{q}},
\end{aligned}
\tag{18.57}
$$

where in the third line $c^\dagger_{-\mathbf{k}-\mathbf{q}\downarrow}$ and $c_{-\mathbf{k}\downarrow}$ have been anti-commuted, and where in the fourth line the delta function term vanishes upon summation due to its antisymmetric prefactor. Moreover, the definition of the Nambu Green's function Eq. (18.44) gives immediately

$$
\bar{\bar{\mathcal{G}}}(\mathbf{k},\tau) = -\left\langle T_\tau \alpha_{\mathbf{k}}(\tau)\alpha^\dagger_{\mathbf{k}}(0)\right\rangle.
\tag{18.58}
$$

We thus see that the ordinary electron operators $c^\dagger$ have been substituted by the Nambu spinors $\alpha^\dagger$, and we can take over the entire linear response formalism from Sections 6.2 and 16.1 directly,

$$
J^\alpha_e = \sum_\beta \left[(-e^2)\Pi_{\alpha\beta} - \frac{e^2\rho}{m}\delta_{\alpha\beta}\right]A^\beta.
\tag{18.59}
$$

Neglecting all other interactions than the Cooper-pair interaction, the current-current correlation function $\Pi_{\alpha\beta}(\mathbf{q}, iq_n)$ follows directly from the first Feynman diagram in Eq. (16.13),

$$
= \frac{1}{V\beta}\sum_{\mathbf{k},ik_n}\left[\frac{1}{2m}(2\mathbf{k}+\mathbf{q})_\alpha\right]\left[\frac{1}{2m}(2\mathbf{k}+\mathbf{q})_\beta\right]\mathrm{Tr}\left[\bar{\bar{\mathcal{G}}}(\mathbf{k}, ik_n)\bar{\bar{\mathcal{G}}}(\mathbf{k}+\mathbf{q}, ik_n+iq_n)\right],
\tag{18.60}
$$

where summation over all internal variables has been performed, in particular the Nambu indices now represented by the trace.[65]

---

[65]Due to the particular nature of the pair correlations the simple polarization bubble is an excellent approximation. The proof of this statement is rather involved and beyond the scope of this book.

To perform the calculation, we use the spectral form Eq. (8.52a) of the Nambu Green's function,

$$\bar{\bar{\mathcal{G}}}(\mathbf{q}, iq_n) = \int_{-\infty}^{\infty} \frac{d\omega}{(-\pi)} \frac{\operatorname{Im}\bar{\bar{\mathcal{G}}}(\mathbf{q}, \omega + i\eta)}{iq_n - \omega}. \tag{18.61}$$

Inserting this into Eq. (18.60) and performing the Matsubara frequency summation, we arrive at

$$\Pi_{\alpha\beta}(\mathbf{q}, iq_n) = \frac{1}{4Vm^2} \sum_{\mathbf{k}} (2\mathbf{k}+\mathbf{q})_\alpha (2\mathbf{k}+\mathbf{q})_\beta \tag{18.62}$$

$$\times \int_{-\infty}^{\infty} \frac{d\omega}{\pi} \int_{-\infty}^{\infty} \frac{d\tilde{\omega}}{\pi} \operatorname{Tr}\left[\operatorname{Im}\bar{\bar{\mathcal{G}}}(\mathbf{k}, \omega + i\eta)\operatorname{Im}\bar{\bar{\mathcal{G}}}(\mathbf{k}+\mathbf{q}, \tilde{\omega} + i\eta)\right]\left(\frac{n_{\mathrm{F}}(\omega) - n_{\mathrm{F}}(\tilde{\omega})}{iq_n + \omega - \tilde{\omega}}\right).$$

For simplicity, we assume in the following that the gap parameter Δ is real and obtain

$$\operatorname{Im}\bar{\bar{\mathcal{G}}}(\mathbf{k}, \omega + i\eta) = \frac{-\pi}{2E_{\mathbf{k}}}\begin{pmatrix} \omega + \xi_{\mathbf{k}} & -\Delta_{\mathbf{k}} \\ -\Delta_{\mathbf{k}} & \omega - \xi_{\mathbf{k}} \end{pmatrix}\left[\delta(\omega - E_{\mathbf{k}}) - \delta(\omega + E_{\mathbf{k}})\right], \tag{18.63}$$

and further

$$\operatorname{Tr}\left[\operatorname{Im}\bar{\bar{\mathcal{G}}}(\mathbf{k}, \omega + i\eta)\operatorname{Im}\bar{\bar{\mathcal{G}}}(\mathbf{k}+\mathbf{q}, \tilde{\omega} + i\eta)\right] \tag{18.64}$$

$$= \pi^2\left[\delta(\omega - E) - \delta(\omega + E)\right]\left[\delta(\tilde{\omega} - \tilde{E}) - \delta(\tilde{\omega} + \tilde{E})\right]\frac{\omega\tilde{\omega} + \xi\tilde{\xi} + \Delta^2}{2E\tilde{E}},$$

where

$$\xi = \xi_{\mathbf{k}}, \quad \tilde{\xi} = \xi_{\mathbf{k}+\mathbf{q}}, \quad E = E_{\mathbf{k}} \quad \text{and} \quad \tilde{E} = E_{\mathbf{k}+\mathbf{q}}. \tag{18.65}$$

For a static $\mathbf{A}$-field $iq_n = 0$. With this simplification and carefully evaluating the four delta function terms, the current-current correlation function becomes

$$\Pi_{\alpha\beta}(\mathbf{q}, 0) = \frac{1}{4Vm^2}\sum_{\mathbf{k}}(2\mathbf{k}+\mathbf{q})_\alpha(2\mathbf{k}+\mathbf{q})_\beta \tag{18.66}$$

$$\times\left[\left(1 + \frac{\xi\tilde{\xi} + \Delta^2}{E\tilde{E}}\right)\frac{n_{\mathrm{F}}(E) - n_{\mathrm{F}}(\tilde{E})}{E - \tilde{E}} + \left(1 - \frac{\xi\tilde{\xi} + \Delta^2}{E\tilde{E}}\right)\frac{n_{\mathrm{F}}(E) - n_{\mathrm{F}}(-\tilde{E})}{E + \tilde{E}}\right].$$

In the long wavelength limit $q \to 0$ further simplifications occur, $\tilde{\xi} \to \xi$, $\tilde{E} \to E$ and $\xi\tilde{\xi} + \Delta^2 \to E^2$, and we obtain the simple result,

$$\Pi_{\alpha\beta}(0, 0) = \frac{1}{4Vm^2}\sum_{\mathbf{k}}4\mathbf{k}_\alpha\mathbf{k}_\beta\left[(1 + 1)\frac{\partial n_{\mathrm{F}}(E)}{\partial E} + 0\right] = -\frac{2}{3m}\frac{1}{V}\sum_{\mathbf{k}\sigma}\xi_{\mathbf{k}}\left(-\frac{\partial n_{\mathrm{F}}}{\partial E}\right)\delta_{\alpha\beta}, \tag{18.67}$$

where the off-diagonal elements in $\mathbf{k}_\alpha\mathbf{k}_\beta$ add up to zero by the summation, and where a factor of 2 has been absorbed in the reappearing spin summation index σ.

Combining this result for the paramagnetic current-current correlation function with the diamagnetic contribution, we obtain the final answer for the electric current density in the static, long wavelength limit:

$$\mathbf{J}_e = -\frac{e^2}{m}(\rho - \delta\rho)\mathbf{A}, \tag{18.68}$$

where $\delta\rho$, defined by

$$\delta\rho = \frac{2}{3}\frac{1}{V}\sum_{\mathbf{k}\sigma}\xi_{\mathbf{k}}\left(-\frac{\partial n_F}{\partial E_{\mathbf{k}}}\right) = \frac{2}{3}\int_{-\infty}^{\infty}d\xi\,\xi d(\xi)\left[-n_F'\left(\sqrt{\xi^2 + |\Delta|^2}\right)\right], \tag{18.69}$$

can be interpreted as the density of the few quasiparticle excitations that have left the otherwise completely rigid Cooper-pair condensate and thereby are able to exhibit a paramagnetic response. The form of Eq. (18.68) makes it natural to define a density of superconducting electrons ρ_s as

$$\rho_s = \rho - \delta\rho, \tag{18.70}$$

whereby we have derived the London equation (18.52).

We can calculate the density of superconducting electrons in two important limits. Firstly, at zero temperature in a superconductor $|\Delta|^2 > 0$ yielding a strictly zero Fermi function derivative for all values of ξ. In this case

$$\rho_s(T = 0) = \rho - \frac{2}{3}\int_{-\infty}^{\infty}d\xi\,\xi\,d(\xi)\,\delta\left(\sqrt{\xi^2 + |\Delta|^2}\right) = \rho. \tag{18.71}$$

So in an ideal superconductor, all electrons are superconducting in the sense that none of them yield any paramagnetic response. Secondly, for normal metals at any temperature, e.g., a superconductor at $T > T_c$, the energy gap Δ vanishes, and the Fermi function derivative is well-approximated by the usual delta function,

$$\rho_s(\Delta = 0) = \rho - \frac{2}{3}\int_{-\infty}^{\infty}d\xi\,\xi\,d(\xi)\,\delta(\xi - \varepsilon_F) = 0. \tag{18.72}$$

So in a normal metal, all electrons yield a fully compensating paramagnetic response. In general, according to the BCS theory, as temperature rises two things happen: the superconducting gap Δ is lowered and the Fermi function derivative broadens. Therefore the integral begins to pick up some weight, and $\delta\rho$ is no longer strictly zero and consequently the normal paramagnetic response begins to set in; however, note that as long as it is smaller than the diamagnetic one, the wave function is rigid enough to exhibit the Meissner effect.

If we return to the example given in Eq. (18.56) and Exercise 18.9, the solutions inside the superconductor ($x > 0$) for the magnetic induction, the vector potential, and the current density are

$$\mathbf{B} = Be^{-x/\lambda_L}\hat{\mathbf{e}}_z, \quad \mathbf{A} = -B\lambda_L e^{-x/\lambda_L}\hat{\mathbf{e}}_y, \quad \mathbf{J}_e = \frac{B}{\mu_0\lambda_L}e^{-x/\lambda_L}\hat{\mathbf{e}}_y. \tag{18.73}$$

It is thus clear that the Meissner effect is the result of magnetic-field screening due to currents induced in the superconductor in a surface layer of thickness λ_L. Note, that this current is flowing in the groundstate at thermodynamic equilibrium, i.e., at minimal free energy. This current must therefore be dissipationless, and it is one example of the remarkable property of superconductors: the lack of electrical resistivity. This we will discuss further in the following sections.

18.6 Gauge symmetry breaking and zero resistivity

In Chapter 4, the relation between phase transitions and broken symmetries was discussed. In this section, we apply these concepts for the transition from a normal to a superconducting state. It turns out that the relevant broken symmetry for superconductors is the global gauge symmetry. This is apparent already from the London equation 18.52, which is clearly not gauge invariant. The way to understand this is that the superconducting order parameter is linked to the gauge choice, which we discuss next.

18.6.1 *Gauge transformations*

Consider first non-interacting electrons in a magnetic field given by the vector postential $\mathbf{A}$. The Hamiltonian is then given by Eq. (1.90),

$$H_{\mathbf{A}} = \frac{1}{2m} \sum_{\sigma} \int d\mathbf{r} \, \Psi_{\sigma}^{\dagger}(\mathbf{r}) \left(\frac{1}{i} \nabla_{\mathbf{r}} + e\mathbf{A} \right)^2 \Psi_{\sigma}(\mathbf{r}). \tag{18.74}$$

It is seen by direct substitution that upon the transformation

$$\mathbf{A} \quad \rightarrow \quad \tilde{\mathbf{A}} = \mathbf{A} + \nabla \chi, \tag{18.75a}$$

$$\Psi_{\sigma}(\mathbf{r}) \quad \rightarrow \quad \tilde{\Psi}_{\sigma}(\mathbf{r}) = \Psi_{\sigma}(\mathbf{r}) \, e^{-ie\chi}, \tag{18.75b}$$

both the Hamiltonian $H_{\mathbf{A}}$ as well as the magnetic field $\mathbf{B}$ remain invariant. Such a transformation is denoted a gauge transformation, or more specifically a U(1) gauge transformation since the phase factor $\exp(-ie\chi)$ represents a unitary scalar transformation. Here, the phase factor has been multiplied on $\Psi_{\sigma}(\mathbf{r})$, but it would have been the same in the $\mathbf{k}$-space representation for $c_{\mathbf{k}\sigma}$.

In the case of the Meissner effect, the London equation (18.52) is not gauge invariant. If we perform the transformation Eq. (18.75a) the physical current $\mathbf{J}_e$ changes. But we were careful to state that the calculation was explicitly for the transverse part of $\mathbf{A}$ fulfilling $\nabla \cdot \mathbf{A} = 0$. However, the longitudinal part $\nabla \chi$ appearing in $\mathbf{A}$ by the gauge transformation affects the order parameter Δ,

$$c_{\mathbf{k}\sigma} \rightarrow c_{\mathbf{k}\sigma} e^{-ie\chi} \Rightarrow \Delta_{\mathbf{k}} = -\sum_{\mathbf{k}'} V_{\mathbf{k}\mathbf{k}'} \langle c_{-\mathbf{k}'\downarrow} c_{\mathbf{k}'\uparrow} \rangle \rightarrow \tilde{\Delta}_{\mathbf{k}} = \Delta_{\mathbf{k}} \, e^{-i2e\chi}. \tag{18.76}$$

This result indicates that the Meissner-effect calculation should have been done self-consistently with the longitudinal component of the vector potential appearing in the order parameter. Such a calculation corresponds to a generalization of the random phase approximation with pair correlation taken into account. We will not carry this analysis through in this book, but instead refer the reader to the book by Schrieffer

	Superconductivity	Ferromagnetism
Type of Transformation	Global U(1) gauge transformation	Global SO(3) spatial rotation
Global symmetry of the Hamiltonian	$c_\nu \to c_\nu e^{i\varphi} \Rightarrow H \to H$	$\mathbf{S} \to \mathbf{US} \Rightarrow H \to H$
Broken symmetry by the order parameter	$0 \neq \langle c_\nu c_{\nu'} \rangle \to \langle c_\nu c_{\nu'} \rangle e^{2i\varphi}$	$0 \neq \langle \mathbf{S} \rangle \to \mathbf{U} \langle \mathbf{S} \rangle$

Table 18.1 *The analogy between the global transformations and broken symmetries of a superconductor and a ferromagnet.*

(1983). Fortunately, however, by performing this calculation the expression presented above is not significantly altered.

We have seen that gauge transformations will alter the superconducting order parameter Δ. Even in the simplest case of zero magnetic field and a constant χ (yielding $\nabla \chi = 0$) the order parameter is modified, because if $-e\chi = \varphi = \text{const.}$, then

$$\mathbf{A} = 0 \ \to \ \tilde{\mathbf{A}} = 0, \tag{18.77a}$$

$$\Delta_{\mathbf{k}} \ \to \ \tilde{\Delta}_{\mathbf{k}} = \Delta_{\mathbf{k}} \, e^{i2\varphi}. \tag{18.77b}$$

The appearance of the pair-field Δ in the BCS Hamiltonian (18.10) seems to break the global gauge symmetry of the original Hamiltonian, thus indicating that this is the relevant symmetry to consider in connection with the superconducting phase transition.

18.6.2 *Broken gauge symmetry and dissipationless current*

Above the critical temperature the Hamiltonian is invariant under a global gauge transformation. Below T_c the symmetry is broken, *cf.* the definition in Section 4.4, and the order parameter together with its phase acquire a finite value, which means that a specific "gauge angle" φ has been picked. This global phase coherence is a combination of the intricate superposition and correlations in Eqs. (18.8) and (18.11).

An analogous situation is realized in a ferromagnetic system. For temperatures above the Curie temperature all spins can be rotated by the same constant angle without changing the Hamiltonian, which therefore has a global SO(3) symmetry. When cooling below the Curie temperature the SO(3) symmetry is broken, because the expectation value of the spin $\langle \mathbf{S} \rangle$ is *not* invariant under rotation. Similarly, phase rotations change the superconducting order parameter, which is of the form $\langle c_\nu c_{\nu'} \rangle$. Schematically the analogies between superconductors and ferromagnetic systems are shown in Table 18.1.

While it is clear why the finite expectation $\langle \mathbf{S} \rangle$ gives a magnetization in the case of a ferromagnetic system, it is not so clear why broken symmetry in the superconducting case leads to a phase without resistance. We have argued that the superconducting state is sensitive to a change of global phase, but it is also clear that a constant phase cannot have any measurable effect, because all expectation values are given by the absolute square of the wave function. However, phase gradients can have an effect.

For example, let us assume that the phase $\varphi(\mathbf{r})$ of the superconducting order parameter depends on position, but that it varies so slowly that $\varphi(\mathbf{r})$ changes substantially only over macroscopic distances. For any non-superconducting system it would not make sense to talk about a quantum mechanical phase difference over macroscopic distances, simply because quantum coherence is destroyed over large distances. For the superconducting case it is different, because the phase is a thermodynamic variable, i.e., it is given by the value that minimizes the free energy. In fact, it turns out that if the phase variable is slowly varying, the system chooses to carry a current in order to minimize the free energy. Alternatively, one can say that if a current is forced through the superconductor it responds by tilting the phase of the order parameter, thus maintaining the superconducting order. It turns out that the current and the phase variable are related as

$$\langle \mathbf{J} \rangle = \frac{\rho_s}{m_e} \nabla \varphi, \tag{18.78}$$

where ρ_s is the density of electrons in participating in the condensed state.

Since the superconducting system can minimize its energy by carrying a current even in thermodynamical equilibrium, this current must be dissipationless, because maintaining thermodynamical equilibrium means maintaining constant entropy, and constant entropy means no heat exchange. Of course, the condensation energy is decreased compared to the state with zero current, but as long as this cost is smaller than the alternative, which is to go out of the superconducting state, the current-carrying state is chosen. The critical current is reached when the energies are equal, and for higher currents the superconductor goes into the normal state.

Furthermore, given this framework it should also be clear that the appearance of the excitation gap is not the reason for the superconductivity itself. The superconductivity is, as we have argued, due to the lack of gauge invariance, and in fact, gapless superconductors do exist.

Naturally, the above arguments are only qualitative and in order to derive the existence of a supercurrent, i.e., a current without voltage, more rigorously we discuss in the following section the so-called Josephson effect. The Josephson supercurrent can be viewed as a discrete and extreme version of Eq. (18.78), because it occurs between two superconductors connected by an insulator.

18.7 The Josephson effect

When two superconductors are separated by a thin insulating barrier, a so-called tunnel junction being either vacuum or some oxide, it is still possible to observe a supercurrent flowing between them. This is known as the Josephson effect. We will use a method based on a derivative of the free energy to emphasize that the supercurrent is an equilibrium quantity, and not, as for normal conductors, a non-equilibrium dissipative current.

The electron operators on each side of the barrier are denoted $c_{\mathbf{k}\sigma}$ and $f_{\mathbf{k}\sigma}$, respectively. Both the c and the f system are superconducting BCS systems. The tunneling operator describing the electron transfer between the two systems is given by Eq. (8.65),

$$H_t = \sum_{\mathbf{kp}\sigma} \left(t_{\mathbf{kp}} c_{\mathbf{k}\sigma}^\dagger f_{\mathbf{p}\sigma} + t_{\mathbf{kp}}^* f_{\mathbf{p}\sigma}^\dagger c_{\mathbf{k}\sigma} \right). \tag{18.79}$$

In the following, we assume, for simplicity, that in the energy range of interest, the tunnel matrix element depends weakly on the states $\mathbf{k}$ and $\mathbf{p}$, so we have

$$t_{\mathbf{kp}} \approx t, \quad \text{a constant.} \tag{18.80}$$

The Hamiltonian for the two sides are the usual BCS Hamiltonians with constant order parameters,

$$H_c = \sum_{\mathbf{k}\sigma} \xi_k c_{\mathbf{k}\sigma}^\dagger c_{\mathbf{k}\sigma} - \Delta e^{-i\phi_c} \sum_{\mathbf{k}} c_{\mathbf{k}\uparrow}^\dagger c_{-\mathbf{k}\downarrow}^\dagger - \Delta e^{i\phi_c} \sum_{\mathbf{k}} c_{-\mathbf{k}\downarrow} c_{\mathbf{k}\uparrow}, \tag{18.81a}$$

$$H_f = \sum_{\mathbf{k}\sigma} \xi_k f_{\mathbf{k}\sigma}^\dagger f_{\mathbf{k}\sigma} - \Delta e^{-i\phi_f} \sum_{\mathbf{k}} f_{\mathbf{k}\uparrow}^\dagger f_{-\mathbf{k}\downarrow}^\dagger - \Delta e^{i\phi_f} \sum_{\mathbf{k}} f_{-\mathbf{k}\downarrow} f_{\mathbf{k}\uparrow}, \tag{18.81b}$$

where the two superconductors are assumed to be identical. The order parameters on each side of the junction have different phases, ϕ_c and ϕ_f, but the same (real) modulus Δ. By the gauge transformations

$$c \to e^{i\frac{1}{2}\phi_c} c \quad \text{and} \quad f \to e^{i\frac{1}{2}\phi_f} f, \tag{18.82}$$

the phase difference between the two sides can be absorbed as a phase shift of the tunnel matrix element

$$t \quad \to \quad e^{-i\frac{1}{2}(\phi_c - \phi_f)} t = e^{-i\frac{1}{2}\phi} t, \tag{18.83}$$

where the phase difference $\phi = \phi_c - \phi_f$ has been introduced. After inserting the transformations (18.82) and (18.83) in the tunneling Hamiltionian (18.79), it is a straightforward exercise to show that in equilibrium, i.e., without any voltage bias applied, there is a current, the so-called Josephson current I_J, flowing across the junction,

$$I_J = \langle I \rangle = (-2e) \left\langle \frac{\partial}{\partial \phi} H_t \right\rangle = (-2e) \frac{\partial F}{\partial \phi}, \tag{18.84}$$

where I is the operator for the electrical current $I = (-e)\dot{N}_c$, and F is the free energy.

The Josephson current runs in thermodynamical equilibrium, and hence it is dissipationless in the sense that it runs without an applied bias; the chemical potential of the two sides is by definition identical in equilibrium.

In the following, we calculate the Josephson current Eq. (18.84) to second-order in the tunneling amplitude. We begin with the expectation value of the derivative of H_t. Since the perturbation is given by H_t itself we arrive at a second order expression in H_t by going to first order in the expansion (11.13) of the thermal weight operator $\exp(-\beta H) \approx \exp(-\beta H_0) T_\tau [1 - \int_0^\beta d\tau \, \hat{H}_t(\tau)]$:

$$\left\langle \frac{\partial}{\partial \phi} H_t \right\rangle \approx -\int_0^\beta d\tau \left\langle T_\tau \hat{H}_t(\tau) \frac{\partial}{\partial \phi} H_t \right\rangle_0 = -\frac{1}{2} \frac{\partial}{\partial \phi} \int_0^\beta d\tau \left\langle T_\tau \hat{H}_t(\tau) H_t \right\rangle_0, \qquad (18.85)$$

where the expectation value $\langle . \rangle_0$ is evaluated with respect to the unperturbed Hamiltonian Eqs. (18.81a) and (18.81b). By straightforward manipulations, we obtain

$$\frac{1}{2} \frac{\partial}{\partial \phi} \int_0^\beta d\tau \langle T_\tau \hat{H}_t(\tau) H_t \rangle_0 = \frac{\partial}{\partial \phi} \left(\int_0^\beta d\tau \sum_{kp} t^2 e^{i\phi} \bar{\bar{\mathcal{G}}}_{21}(\mathbf{k},\tau) \bar{\bar{\mathcal{G}}}_{12}(\mathbf{p},-\tau) + \text{c.c.} \right)$$

$$= \frac{\partial}{\partial \phi} \left(\frac{1}{\beta} \sum_{ik_n} \sum_{kp} t^2 e^{i\phi} \bar{\bar{\mathcal{G}}}_{21}(\mathbf{k}, ik_n) \bar{\bar{\mathcal{G}}}_{12}(\mathbf{p}, ik_n) + \text{c.c.} \right), \qquad (18.86)$$

where $\bar{\bar{\mathcal{G}}}_{12}$ and $\bar{\bar{\mathcal{G}}}_{21}$ are the off-diagonal Nambu Green's functions (18.44) and (18.47). These are evaluated as follows:

$$\sum_k \bar{\bar{\mathcal{G}}}_{21}(\mathbf{k}, ik_n) = \sum_p \bar{\bar{\mathcal{G}}}_{12}(\mathbf{p}, ik_n)$$

$$= \frac{1}{2} \Delta d(\varepsilon_F) \int_{-\infty}^\infty d\xi \, \frac{1}{(ik_n)^2 - E^2}$$

$$= -\frac{1}{2} \Delta d(\varepsilon_F) \frac{\pi}{\sqrt{k_n^2 + \Delta^2}}. \qquad (18.87)$$

It is then a simple matter to find I_J,

$$I_J = -\left(\frac{\pi \Delta d(\varepsilon_F) t}{2} \right)^2 \frac{1}{\beta} \sum_{ik_n} \frac{1}{k_n^2 + \Delta^2}$$

$$= \frac{1}{2} e [\pi d(\varepsilon_F) t]^2 \Delta \tanh \left(\frac{\Delta \beta}{2} \right) \sin \phi$$

$$= \frac{\pi}{2} \frac{\Delta}{eR_N} \tanh \left(\frac{\Delta \beta}{2} \right) \sin \phi, \qquad (18.88)$$

where the normal state tunnel resistance is given by

$$\frac{1}{R_N} = \pi d^2 t^2 \frac{e^2}{\hbar}. \qquad (18.89)$$

It is interesting to note that the expression (18.88 is a variant of the general result $\mathbf{J} \propto \nabla \varphi$ stated in Eq. (18.78). In the limit of a small phase difference, we find namely $I_J \propto \sin \varphi \approx \varphi = \varphi_c - \varphi_f \approx \ell \nabla \varphi$, where ℓ is the characteristic length separating the two superconductors.

The calculated Josephson current is a dc response to a constant difference in the superconducting phases surrounding the insulating barrier. There exists also an ac Josephson effect, where the ac current is the response to constant voltage bias applied across the junction. This effect is studied in Exercise 18.10.

18.8   Summary and outlook

Superconductivity is a result of condensation of pairs of electrons, and hence it is a macroscopic quantum state. This situation is similar to Bose–Einstein condensation, where a finite fraction of the total number of particles occupies the zero momentum state.

The Bose–Einstein condensate is often described in following way: the occupation of the zero momentum state is $N_0 = \langle a_0^\dagger a_0 \rangle$, but if this state is macroscopically occupied it is fair to say that removing a single particle from the condensate does not change the macroscopic state and hence $\langle a_0 \rangle \neq 0$, which is in fact the order parameter for Bose–Einstein condensation.[66] For the superconductor, the condensate consists of pairs of electrons and in analogy with the Bose–Einstein condensate, the phase transition occurs when the expectation value $\langle c_{-\mathbf{k}\downarrow} c_{\mathbf{k}\uparrow} \rangle$ is non-zero. In this chapter we have described the mean-field theory for this transition and, furthermore, discussed a number of consequences of the superconducting phase. We have seen that a gap appears in the density of states as result of the BCS theory, but in fact this is not always the case, because in some cases the gap vanishes at special points in the reciprocal space. We have also emphasized the importance of breaking of global gauge symmetry. The broken symmetry has two important consequences: the Meissner effect and the possibility for the superconductor to sustain a supercurrent. This is demonstrated in detail for the so-called Josephson effect, where a supercurrent runs between separated superconductors with zero voltage bias.

The physics of superconductors is covered in a number of classical textbooks such as Schrieffer (1983), de Gennes (1999), and Tinkham (1996). In recent years, also the physics of mesoscopic and nanoscale superconductors have been exploited and new and interesting phenomena have been identified.

---

[66]The size of $\langle a_0 \rangle$ is determined as $N_0 = \langle a_0^\dagger a_0 \rangle = \langle a_0^\dagger \rangle \langle a_0 \rangle$ and thus $\langle a_0 \rangle = \sqrt{N_0}$.

19

1D ELECTRON GASES AND LUTTINGER LIQUIDS

The 1D interacting electron gas is very different from its higher dimensional counterparts. The main difference being the breakdown of Fermi liquid theory in 1D. In Chapter 15 we saw that the main achievement of the Fermi liquid theory is that it enables us to picture the elementary excitations in terms of quasiparticles. As we shall see in this chapter, the quasiparticle excitations are in 1D electron systems replaced by the Luttinger–Tomonaga excitations, a sort of density waves, having a completely different nature. The corresponding electron liquid is denoted the Luttinger liquid.

It is not merely an academic exercise to study the interacting 1D electron gas. There are several systems, listed below, in condensed matter physics, where interacting Fermions confined to 1D can be investigated and thus motivating our present study of the difference between 1D and 3D physics.

In recent years, the concept of Luttinger liquids has been of much interest because of the possibility to observe the unusual properties of 1D electron gases in artificially fabricated nanostructures and in carbon nanotubes. This activity was also influenced by the study of how a single impurity can lead to drastic changes of the transport properties of an 1D system (Fisher and Glazman 1997), which is in contrast to the usual understanding of phase coherent transport in mesoscopic systems as we saw in Chapter 7.

The purpose of this chapter is to provide a short presentation of some of the key points, for more comprehensive overview of Luttinger-liquid theory and its many applications we refer the reader to the review papers listed at the end of the chapter.

19.1 What is a Luttinger liquid?

In 1D the so-called Luttinger liquid, which was conjectured by Haldane (Haldane 1981), replaces the Fermi liquid as a generic many-particle state. In the Luttinger liquid the elementary excitations are very different from those of free electrons, and an attempt to define quasiparticles as in the 3D case fails.

The general Luttinger liquid picture is in nature similar to the Fermi liquid picture. In both pictures, the low-energy excitations of a many-particle ground state characterize the systems, but for Luttinger liquids the low-energy excitations are not quasiparticles, but rather the density waves given by Luttinger–Tomonaga model (see Sections 19.3 and 19.5). It is often the case that the simple form of the effective low-energy Hamiltonian of this model (see Eq. (19.52c)) remains the same, even if we include more complicated interactions such as spin-flip backscattering (see Section 19.9), hence the generic name "Luttinger liquid". That such processes often are unimportant at low energies, can be shown by a careful renormalization-group treatment of the more general Hamiltonian. The discussion of this is, however, outside the scope of the present book.

19.2 Experimental realizations of Luttinger liquid physics

Naturally, there is no true 1D system in nature. However, there are a number of physical systems in condensed matter physics that are well described by 1D models. The crucial requirement is that the quantum confinement in the two perpendicular directions is strong enough, so that, at low energies, one can disregard all quantum levels except a single 1D mode. Below we list some systems that have been assigned properties characteristic of 1D metals. More references can be found in the review papers listed at the end of this chapter.

19.2.1 Example: Carbon Nanotubes

Experimentally, one of the most promising candidates for a detailed study of 1D physics is the carbon nanotube (see Section 2.3.3). In the Tomonaga–Luttinger model a linear dispersion relation is assumed, see Eq. (19.15). In metallic carbon nanotubes this an unusually good approximation, because the dispersion relation, which is inherited from the semi-metallic graphene, is linear over a large energy range of order several hundred meV (see, e.g., (Saito *et al.* 1998) for details). For example, tunneling data on nanotubes show power-law tunneling density of states similar to the one derived in Section 19.8 (Egger *et al.* 2001).

19.2.2 Example: semiconductor wires

Another example of quasi 1D system, where also signatures of the famous spin charge separation has been observed is thin wires made in semiconductor heterostructures. More precisely, the so-called cleaved edge overgrowth systems, where tunneling experiments between two wires have been argued to show the spin-charge separation characteristic of Luttinger liquids (Tserkovnyak *et al.* 2003).

19.2.3 Example: quasi 1D materials

A number of organic crystals and other crystals have very large anisotropies, where the conduction along one crystal direction is much larger than along the two others. One example is the so-called Bechgaard salts. It has been argued that the optical response of such systems can indeed be described by the power laws that are characteristic of Luttinger-liquids (Giamarchi 2003).

19.2.4 Example: Edge states in the fractional quantum Hall effect

The last example is the 1D edge states of a 2D electron gas in a perpendicular magnetic field in the fractional quantum Hall (FQH) state. It turns out that one can formulate a theory for the 1D edge states, which is similar to the Luttinger liquid model that we shall consider in Sections 19.4 and 19.5. The theory of Luttinger liquids in FQH states was suggested by Wen (1992, 2004). In Exercise 19.6, we go through a simplistic version of this theory.

19.3 A first look at the theory of interacting electrons in 1D

In the RPA treatment of plasmons in dimensions higher than one, we saw previously in Fig. 14.2 that the plasmon excitation was clearly separated from the electron-hole pair excitations at low wave numbers. This fact made the plasmon excitation stable, since it could not decay by emitting electron-hole pairs. In dimensions 2 and 3 the plasmon dispersion has the form $\omega_{\mathrm{pl}}(q) \propto q^{(3-d)/2}$, see Exercise 14.3. The dispersion relation for a classical 1D system of charges with short-ranged interaction is $\omega_{\mathrm{pl}} \propto q$. Since the plasmon dispersion is linear it is not as obvious as in 2D and 3D that we can make a clear distinction between collective excitations and single particle excitations.

In Eq. (8.79) and Section 14.5 we have seen that the ability of an electron to dissipate energy is given by the imaginary part of the polarizability. From Eq. (8.79) and the RPA result in Eq. (14.66) the conductivity σ becomes

$$\mathrm{Re}\,\sigma\,(q,\omega) = -\frac{\omega e^2}{q^2}\,\mathrm{Im}\left[\frac{\chi_0^R(q,\omega)}{1 - V(q)\chi_0^R(q,\omega)}\right], \tag{19.1}$$

where $V(q)$ is the Fourier transformed interaction, and where $\chi_0^R(q,\omega)$ is the usual Lindhard function, but now both expressions are given in 1D. The imaginary part in Eq. (19.1) can be rewritten as

$$\mathrm{Im}\left[\frac{\chi_0^R(q,\omega)}{1 - V(q)\chi_0^R(q,\omega)}\right] = \frac{1}{V(q)}\,\mathrm{Im}\left[\frac{1}{\varepsilon^{\mathrm{RPA}}}\right] \equiv -\frac{1}{V(q)}S(q,\omega), \tag{19.2}$$

where $\varepsilon^{\mathrm{RPA}} = 1 - V(q)\chi_0^R(q,\omega)$ and where we have defined the dynamical structure factor $S(q,\omega)$.[67] Excitations are thus possible whenever S is different from zero. Now there are two distinct ways that S can become non-zero: (i) if the real part of ε is zero and the imaginary part is infinitely small, or (ii) if the imaginary part of χ_0^R is finite. Excitations of type (i), the plasmons, are given by

$$S(q,\omega) \approx \pi\delta\left(\mathrm{Re}\left[\varepsilon(q,\omega_q)\right]\right). \tag{19.3}$$

We can find the dielectric function in RPA, and in the long wavelength limit we get

$$\varepsilon(q,\omega) = 1 - V(q)\frac{2}{\mathcal{L}}\sum_k \frac{n_F(\xi_{k+q}) - n_F(\xi_k)}{\xi_{k+q} - \xi_k - \omega}$$

$$\approx 1 - V(q)\int\frac{dk}{\pi}\left(-\frac{\partial n_F(\xi_k)}{\partial\xi_k}\right)\frac{kq}{m}\frac{1}{\omega - \frac{kq}{m}}$$

$$\approx 1 - V(q)\left(\frac{qv_F}{\omega}\right)^2\frac{2}{\pi v_F}. \tag{19.4}$$

Now the Fourier transform of the 1D interaction $V(q)$ depends on the details of the wavefunction, since for the strictly 1D case

$$V(q) = e_0^2\int dx\,\frac{1}{|x|}\,e^{-iqx}, \tag{19.5}$$

we encounter a divergence. However, this divergence will be cut-off by the finite extent of the wavefunction in the transverse directions, and in order not to worry about these

---

[67]Note that the structure factor is some times defined with a different prefactor $S = \frac{1}{d_0 V(\mathbf{q})}\,\mathrm{Im}\,\varepsilon^{-1}$.

details we simply say that the transverse direction provides a short-distance cut-off denoted a,

$$V(q) \approx 2e_0^2 \int_a^\infty dx \frac{1}{x} e^{-iqx} \approx 2e_0^2 \int_a^{\frac{1}{q}} \frac{1}{x} dx = 2e_0^2 \ln\left(\frac{1}{qa}\right). \tag{19.6}$$

Hence we arrive at the dispersion relation $\omega_{\mathrm{pl}}(q)$ for a 1D electron gas in the RPA approximation. It is linear and given by

$$\omega_{\mathrm{pl}}(q) = qv, \tag{19.7a}$$

where the velocity v of the wave is given by

$$v \approx v_{\mathrm{F}} \sqrt{\frac{4e_0^2}{\pi v_{\mathrm{F}}} \ln\left(\frac{1}{qa}\right)}, \tag{19.7b}$$

which apart from the logarithmic factor agrees with the dispersion relation discussed in the beginning of this section.

19.3.1 The "quasiparticles" in 1D

Next, we discuss the excitations of type (ii), the electron-hole excitations. For these we express $S(q,\omega)$ by the polarization function,

$$S(q,\omega) = -V(q)\,\mathrm{Im}\left[\frac{\chi_0^R(q,\omega)}{1 - V(q)\chi_0^R(q,\omega)}\right] = -\frac{V(q)}{|\varepsilon(q,\omega)|^2}\,\mathrm{Im}\left[\chi_0^R(q,\omega)\right]. \tag{19.8}$$

At small frequencies $\omega \ll \varepsilon_{\mathrm{F}}$ and temperatures $T \ll \varepsilon_{\mathrm{F}}$, the imaginary part of χ_0^R is

$$-\mathrm{Im}\left[\chi_0(q,\omega)\right] = -\frac{2\pi}{\mathcal{L}}\sum_k \left[n_F(\xi_{k+q}) - n_F(\xi_k)\right]\delta(\xi_{k+q} - \xi_k - \omega)$$

$$\approx 2\pi\omega \int \frac{dk}{2\pi} \delta(\xi_k)\,\delta\left(\frac{qk}{m} + \frac{q^2}{2m} - \omega\right),$$

$$= \frac{\omega}{v_{\mathrm{F}}}\left[\delta\left(\frac{q^2}{2m} - \omega - qv_{\mathrm{F}}\right) + \delta\left(\frac{q^2}{2m} - \omega + qv_{\mathrm{F}}\right)\right]. \tag{19.9}$$

First, we note that for small q, excitations can only occur on the straight line given by $|\omega| = v_F q$ in $q-\omega$ space. This is contrast to the 2D and 3D cases, where there was an area below the line $\omega = v_F q$, see Fig. 19.1. More importantly, the weight of the single-particle excitations in Eq. (19.9) is proportional to frequency. Thus low-energy behavior cannot be described by a single-particle picture. The standard arguments for having well defined quasiparticles separated from the collective modes is simply not correct. In fact, the low-energy excitations that are left are the plasmons, which precisely is also the finding of the exact solution based on bosonization to be presented below in Section 19.5. Therefore in this respect, RPA is exact when it comes to describing the low-energy excitations in 1D.

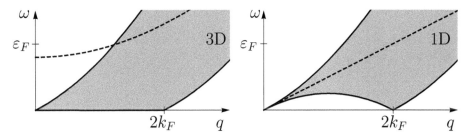

FIG. 19.1. The dispersion of excitation in 3D and 1D. The gray areas are the phase regions in $q - \omega$ space where single-particle excitations are possible. The dashed lines are plasmon dispersions. While in 3D these are clearly separated, this is not the case in 1D. Also, 1D is special in that at low frequencies, it only allows for a narrow band of single particle excitations near $q = 0$ (forward scattering) and $q = 2k_F$ (backward scattering).

19.3.2 The lifetime of the "quasiparticles" in 1D

If we had some hope of justifying Fermi liquid theory in 1D, this is finally destroyed by the following consideration. In Chapter 15 on Fermi liquids, we saw that the decay-rate of quasiparticles vanishes as energy squared. Now, let us try the same calculation in 1D starting from Eq. (15.65)

$$\frac{1}{\tau_k} = -\frac{4\pi}{\mathcal{L}} {\sum_{k',q}}' \left| \frac{V(q)}{\varepsilon(q)} \right|^2 \delta(\xi_{k'} - \xi_{k'-q} - \xi_{k+q} + \xi_k)$$

$$\times \left[n_B(\xi_{k+q} - \xi_k) + n_F(\xi_{k+q}) \right] \left[n_F(\xi_{k'}) - n_F(\xi_{k'-q}) \right]. \qquad (19.10)$$

We exclude $q = 0$ from the sum in accordance with the usual definition of the electron-electron interaction in a positive background. We can perform the k' integral by first using the delta function, which is

$$\delta\left[\frac{q}{m}(q + k - k') \right] = \frac{m}{|q|} \delta(q + k - k') + \frac{m}{|k - k'|} \delta(q). \qquad (19.11)$$

From this expression we see that the final state of the particle with momentum k is the state k', and consequently momentum and energy conserving scattering between two particles in 1D can only occur if the particles exchange momentum. The phase space for scattering in 1D is thus very limited, unlike in 3D where the integration over possible initial and final states leads to the relation $1/\tau_k \propto \xi_k^2$, see Eq. (15.68), which means that the lifetime in 3D diverges at the Fermi energy. in 1D, the lifetime turns out to be energy independent, which renders the usual Fermi liquid argument invalid.

The term corresponding to $q = 0$ does not contribute to Eq. (19.10). From the other term we obtain after using the relation Eq. (10.19a) that

$$\frac{1}{\tau_k} = \frac{4\pi}{\mathcal{L}} {\sum_{q}}' \frac{m}{|q|} \left| \frac{V(q)}{\varepsilon(q)} \right|^2 n_F(\xi_{k+q}) \left[1 - n_F(\xi_{k+q}) \right]. \qquad (19.12)$$

The factor $n_F(\xi_{k+q})\left[1 - n_F(\xi_{k+q})\right]$ requires that for k close to the Fermi points, also $k + q$ must be close to one of the Fermi points, meaning that q is either close to zero or close to $-2k_F$, corresponding to forward scattering and backward scattering, respectively. If we perform the integral and take the limit of small temperature,

$$n_F(\xi)\left[1 - n_F(\xi)\right] = -kT\frac{\partial n_F(\xi)}{\partial \xi} \quad \rightarrow \quad kT\delta(\xi - \xi_F), \tag{19.13}$$

we obtain $\tau^{-1} \propto T/\xi$, which for typical energies, $\xi \sim kT$, gives a constant lifetime. Therefore, in contrast to the case in higher dimensions, there is no energy window where $\xi_k \gg \tau_k^{-1}$.

From the above considerations on the life time and the excitation spectrum we are thus forced to conclude that Fermi liquid theory does not work in 1D. The very concept of a quasiparticle has to be discarded. In what follows, we will see that this interesting situation reveals a new type of quantum liquid, where we have to learn how to deal with interacting many-body systems in a different way.

19.4   The spinless Luttinger–Tomonaga model

Our starting point in this section will be the Hamiltonian for a gas of interacting spinless electrons. The spinful case is briefly dealt with in Section 19.9. We will map the 1D system to the so-called Tomonaga–Luttinger model by the interaction in terms of density operators for the left- and right-moving particles. This rearrangement is possible under the assumption that the physics is restricted to energies much lower than the Fermi energy, and therefore only small momenta exchanges are included. Next, we show that the density operators for left- and right-moving particles obey a boson like algebra, or to be specific a so-called Kac–Moody algebra. For the Kac–Moody algebra to be realized, we must assume that the occupation of the electrons with energies of order the Fermi energy is not affected by the interactions. The small parameter of the theory is thus the ratio of the interaction strength to the Fermi energy. However, it does not mean that we are making a perturbation theory in this small parameter, and the end result will in fact be non-perturbative in the interaction strengths.

The next key point is to realize that the usual kinetic energy of free electrons can be written as a bilinear form of density operators and as a consequence, the entire Hamiltonian is now quadratic in density operators. This last step is made possible by the additional assumption that for the energies of interests, it suffices to linearize the dispersion relation near the Fermi energy. Now, by a simple transformation into usual boson operators with canonical commutations, we will thus have "bosonized" the model and in fact solved it.

19.4.1   The Luttinger–Tomonaga model Hamiltonian

Consider a 1D system of spinless interacting electrons. The Hamiltonian for the system is

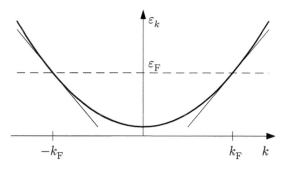

FIG. 19.2. Dispersion relation of the 1D electron gas showing the linear dispersion approximation as tangents to the parabolic free-electron dispersion relation at the two Fermi points $k = \pm k_F$.

$$H = H_0 + H_{\text{int}}, \tag{19.14a}$$

$$H_0 = \sum_k \xi_k c_k^\dagger c_k, \tag{19.14b}$$

$$H_{\text{int}} = \frac{1}{2\mathcal{L}} \sideset{}{'}\sum_{kk'q} V(q) c_k^\dagger c_{k'}^\dagger c_{k'-q} c_{k+q}. \tag{19.14c}$$

The interaction $V(q)$ depends on the physical situation, but we assume that it has well-defined values for forward scattering (q near zero) and backscattering ($q \approx \pm 2k_F$). Furthermore, as in Eq. (2.34) we use a prime in the q-summation sign to indicate that we have taken out the Hartree term corresponding to $q = 0$.

The basic idea of the so-called Tomonaga model is to linearize the dispersion relation ε_k. The linearized single-particle dispersion is

$$\xi_k = \varepsilon_k - \mu \approx (|k| - k_F)\, v_F, \tag{19.15}$$

which is illustrated in Fig. 19.2 for a parabolic dispersion ε_k. Thus, the Tomonaga model has a finite energy (or momentum) cut-off given by the range where the linear dispersion is applicable. Another model is the Luttinger model, which assumes an infinite bandwidth with linear dispersion for *all* real values of k, and that the left- and right-moving branches represent two different species of electrons. This rather artificial model is nevertheless useful, because it can be solved exactly. The solution is found by the boson representation to be introduced in Section 19.5, and this remarkable fact makes it therefore sometimes advantageous to work with the model. Here, we choose to include the cut-offs explicitly and refer to the resulting low-energy Hamiltonian as the Tomonaga–Luttinger model. However, when deriving the Tomonaga–Luttinger liquid from a "real" interacting electron gas, we must invoke a set of approximations and clever ways of parameterizing the interaction.

The next step is to separate the electrons in two species: the right-moving electrons with $k > 0$ and the left-moving ones with $k < 0$. The electron operator is therefore written as

$$c_k = c_{kR}\Theta(k) + c_{kL}\Theta(-k). \tag{19.16}$$

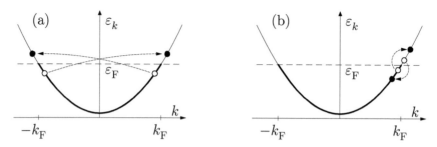

FIG. 19.3. The two possible types of scattering near the Fermi surface: (a) inter-branch scattering where the electrons exchange branch, and (b) intra-branch scattering where they stay on the same branch. The states below the Fermi energy are indicated by the thick part of the dispersion.

With this separation, the free-particle Hamiltonian becomes[68]

$$H_0 = v_F \sum_{k>0} (c_{kR}^\dagger k c_{kR} - c_{kL}^\dagger k c_{kL}) - (N_R + N_L) k_F v_F, \tag{19.17}$$

where N_L and N_R are the number operators for the two branches.

With the decomposition given in Eq. (19.16), the interaction part of the Hamiltonian splits up into 2^4 terms, but only six of them conserve the number of left- and right-moving particles. The "non-conserving terms" are not possible for scatterings around the Fermi points and can be ignored at small energies compared to the Fermi energy. The six conserving terms can be grouped in different ways, which will be done in the following three subsections.

19.4.2   Inter-branch interaction

The first grouping that we consider describes the inter-branch interaction, see Fig. 19.3,

$$H_{\text{int}}^{(1)} = \frac{1}{2\mathcal{L}} \sideset{}{'}\sum_{\substack{k>0,q \\ k'<0}} V(q) \left(c_{kR}^\dagger c_{k'L}^\dagger c_{k'-qL} c_{k+qR} + c_{kR}^\dagger c_{k'L}^\dagger c_{k'-qR} c_{k+qL} \right) + (R \leftrightarrow L),$$

$$= \frac{1}{2\mathcal{L}} \sideset{}{'}\sum_{\substack{k>0,q \\ k'<0}} V(q) \left(c_{kR}^\dagger c_{k+qR} c_{k'L}^\dagger c_{k'-qL} - c_{kR}^\dagger c_{k'-qR} c_{k'L}^\dagger c_{k+qL} \right) + (R \leftrightarrow L).$$

$$\tag{19.18}$$

In the last term we have $k \sim k_F$ and $k' \sim -k_F$ and hence q must be close to $-2k_F$. After a change of variables to $q \to -q - k + k'$, the result is

---

[68]It is interesting to note that this form is analogous to the 1D Dirac equation for relativistic *massless* fermions.

$$H_{\text{int}}^{(1)} = \frac{1}{2\mathcal{L}} \sum_{\substack{k>0,q \\ k'<0}}' \left[V(q)c_{kR}^\dagger c_{k+qR} c_{k'L}^\dagger c_{k'-qL} - V(-q+k'-k)c_{kR}^\dagger c_{k+qR} c_{k'L}^\dagger c_{k'-qL} \right]$$

$$+ (R \leftrightarrow L). \tag{19.19}$$

For $q \neq 0$, we introduce the left- and right-mover density operators as

$$\rho_R(q) = \sum_{k>0} c_k^\dagger c_{k+q} \approx \sum_{k>0} c_{kR}^\dagger c_{k+qR}, \tag{19.20a}$$

$$\rho_L(q) = \sum_{k<0} c_k^\dagger c_{k+q} \approx \sum_{k<0} c_{kL}^\dagger c_{k+qL}, \tag{19.20b}$$

and we can thus rewrite Eq. (19.19) as

$$H_{\text{int}}^{(1)} \approx \frac{1}{2\mathcal{L}} \sum_q' \left[V(0) - V(2k_F) \right] \rho_R(q)\rho_L(-q) + (R \leftrightarrow L). \tag{19.21}$$

For the last step it is assumed that the interaction is smooth on the energy scale of interest. Note that the exchange term tends to decrease the interaction and that for an "ideal" point-like interaction there is no effect of the interaction, which is in agreement with the Pauli exclusion principle that two particles are never at the same position.

It is worthwhile to stress that as indicated in Eq. (19.20) the left- and right-moving charge densities are defined *without* subscripts L, R on the fermion operators c and $c^\dagger$. The difference between the two definitions is a question of whether the right and left-moving electrons are regarded as two different species of particles or not. In practice the distinction matters only when $k + q \sim 0$, but since the important k's lie near the Fermi level, and we have assumed $q \ll k_F$, we can use either at our convenience.

19.4.3 Intra-branch interaction and charge conservation

The second possible grouping contains the intra-branch interaction, i.e., all operators defined on the same branch

$$H_{\text{int}}^{(2)} = \frac{1}{2\mathcal{L}} \sum_{\substack{k>0,q \\ k'>0}}' V(q) \, c_{kR}^\dagger c_{k'R}^\dagger c_{k'-qR} c_{k+qR} + (R \to L). \tag{19.22}$$

Above, we replaced the interaction function with two parameters $V(0)$ and $V(2k_F)$, and we want to make a similar approximation for the intra-branch interaction. We start by noting that the intra-branch interaction becomes ambiguous, if $V(q)$ is replaced by a constant V_0. To see that we collect the operators in Eq. (19.22) in two different ways. The first way is simply to identify the sums over k and k' with two density operators,

$$H_{\text{int}}^{(2)} = \frac{1}{2\mathcal{L}} \sum_q' V_0 \rho_R(q)\rho_R(-q) + (R \to L), \tag{19.23}$$

where we omit the constant shift of the chemical potential one gets when commuting c_{k+qR} with $c^\dagger_{k'R}$. The second way is to group c_{k+qR} with $c^\dagger_{k'R}$ instead, and then write the interaction as

$$H^{(2)}_{\text{int}} = -\frac{1}{2\mathcal{L}} {\sum_{\substack{k>0,q \\ k'>0}}}' V_0 c^\dagger_{kR} c_{k'-qR} c^\dagger_{k'R} c_{k+qR} + (R \to L). \qquad (19.24)$$

If we then shift q as $q \to -q - k + k'$, we arrive at Eq. (19.23) but with the opposite sign. The ambiguity we have encountered is a consequence of the artificial linearization of the dispersion and the separation into left and right movers. We therefore need a physical principle to fix the value of the parameter V_0. Above, we saw that in accordance with the Pauli principle, the inter-branch interaction vanishes for short-range interaction. Since the intra-branch interaction must also have this property, we can expect that V_0 should be proportional to $V(0) - V(2k_F)$ (see also Starykh et al. (1999)). This is indeed seen to be correct when we use the requirement that the model must conserve charge or, in other words, that the continuity equation $\partial_t \rho + \partial_x J = 0$ does not change form. This is ensured if

$$\left[H_{\text{int}}, \rho(q)\right] = 0, \qquad (19.25)$$

where $\rho(q)$ is the density operator. The density operator for long wavelength excitations is the sum of excitations in the left or the right movers (at $T = 0$),

$$\rho(q) = \rho_L(q) + \rho_R(q), \quad \text{for } q \approx 0. \qquad (19.26)$$

The condition (19.25) fixes V_0, because we see that if $V_0 = V(0) - V(2k_F)$, the combined intra- and inter-branch interaction terms become

$$H_{\text{int}} = \frac{1}{2\mathcal{L}} \sum_{q \neq 0} \left[V(0) - V(2k_F)\right] \left[\rho_L(q) + \rho_R(q)\right] \left[\rho_L(-q) + \rho_R(-q)\right], \qquad (19.27)$$

and this does indeed commute with the density operator in Eq. (19.26). To verify this we must, however, first find the commutation relation between $\rho_L(q)$ and $\rho_L(-q)$ and similarly for right movers. This is done in Section 19.5, and using these results the commutator in Eq. (19.25) is indeed seen to be zero. The interaction in Eq. (19.27) is the final form of the interaction that we are going to use in the following.

19.4.4   Umklapp processes in the half-filled band case

We note in passing that there is one additional term, which is important for electrons in a half-filled Bloch band, namely that corresponding to the so-called umklapp process.[69] This process comes about because the state at k_F is equivalent to the state at $k_F + G$, where G a reciprocal lattice vector, see Eq. (2.8). For the half-filled band case

---

[69]"Umklapp" is the German word for "abrupt turn around"

$k_F = \pi/2a$, and therefore the scattering processes $k_F \to k_F + 2k_F$ and $k_F \to -k_F$ are equivalent, and there is then one more way to group the interaction, namely as

$$H^{(3)}_{\text{int}} = \frac{1}{2\mathcal{L}} \sum_{kk'q}{}' V(q) c^\dagger_{kR} c^\dagger_{k'R} c_{k'-qL} c_{k+qL} + (R \leftrightarrow L). \qquad (19.28)$$

In order to have k' being right going and $k' - q$ being left going, q must be near $2k_F$. Since also k is near $+k_F$ the last operator has $k + q$ close to $3k_F$. For a nearly half filled periodic band $3k_F$ is equivalent to $-k_F$ and if we shift $q \to 2k_F + q$ and insert a reciprocal lattice vector G we arrive at

$$H^{(3)}_{\text{int}} \approx \frac{1}{2\mathcal{L}} \sum_{kk'q}{}' V(2k_F) c^\dagger_{kR} c^\dagger_{k'R} c_{k'-2k_F-qL} c_{k+2k_F+q-GL} + (R \leftrightarrow L), \qquad (19.29)$$

where q is now small and G should be close to $-3k_F$. This term only exists if the band is close to half-filled, and we will not consider it. However when present, it is indeed important and can change the ground state of the system.

19.5 Bosonization of the Tomonaga model Hamiltonian

A most remarkable aspect of the 1D Hamiltonian derived above is that it can be expressed as a bilinear form of bosonic operators. Once this is done it is trivially solved and we have the astonishing outcome: an "exact" solution of an interacting many-particle problem.

19.5.1 Derivation of the bosonized Hamiltonian

First, let us study the commutation relation of the left- and right-mover densities. Consider, e.g.,

$$\begin{aligned}
\left[\rho_R(q), \rho_R(-q)\right] &= \sum_{kk'>0} \left[c^\dagger_k c_{k+q}, c^\dagger_{k'} c_{k'-q}\right] \\
&= \sum_{kk'>0} \left(c^\dagger_k c_{k'-q} \delta_{k+q,k'} - c^\dagger_{k'} c_{k+q} \delta_{k,k'-q}\right) \\
&= \sum_{k>0} \Theta(k+q) \left(c^\dagger_k c_k - c^\dagger_{k+q} c_{k+q}\right) \\
&= \Theta(q) \sum_{q>k>0} n_k - \Theta(-q) \sum_{-q>k>0} n_k \\
&\approx \frac{\mathcal{L}}{2\pi} \int_0^q dk = \frac{\mathcal{L}q}{2\pi}. \qquad (19.30)
\end{aligned}$$

The last crucial approximation was to assume that for small k (much smaller than k_F), we can replace the operator n_k by its expectation value for the non-interacting problem, i.e., $n_k = 1$ for $k \ll k_F$. The basic assumption is here that the interactions only modify the properties near the Fermi surface. Similarly we obtain

$$\left[\rho_L(q), \rho_L(-q)\right] \approx -\frac{\mathcal{L}q}{2\pi}. \qquad (19.31)$$

For the non-diagonal parts, $q \neq q'$, we find

$$
\begin{aligned}
\left[\rho_R(q), \rho_R(-q')\right] &= \sum_{k>0} \left(c_k^\dagger c_{k+q-q'} \Theta(k+q) - c_{k+q'}^\dagger c_{k+q} \Theta(k+q') \right) \\
&= \sum_{k>0} c_k^\dagger c_{k+q-q'} \Theta(k+q) - \sum_{k>0} c_k^\dagger c_{k+q-q'} \\
&= -\Theta(-q) \sum_{|q|>k>0} c_k^\dagger c_{k+q-q'} \\
&\approx -\Theta(-q) \sum_{|q|>k>0} \left\langle c_k^\dagger c_{k+q-q'} \right\rangle_0 \\
&\approx 0,
\end{aligned}
\tag{19.32}
$$

where we use the same approximation as before to replace the operator deep in the Fermi sea with its expectation value for the free-particle case. Finally, we need to find

$$
\begin{aligned}
\left[\rho_R(q), \rho_L(-q')\right] &= \sum_{k>0, k'<0} \left[c_k^\dagger c_{k+q}, c_{k'}^\dagger c_{k'-q'} \right] \\
&= \sum_{k>0} \left(c_k^\dagger c_{k+q-q'} \Theta(-k-q) - c_{k+q'}^\dagger c_{k+q} \Theta(-k-q') \right) \\
&= \Theta(-q) \sum_{0<k<-q} c_k^\dagger c_{k+q-q'} - \Theta(-q') \sum_{0<k<-q'} c_{k+q'}^\dagger c_{k+q} \\
&\approx \delta_{qq'} \Theta(-q) \sum_{0<k<-q} \left(\langle n_k \rangle_0 - \langle n_{k+q} \rangle_0 \right) \\
&\approx 0.
\end{aligned}
\tag{19.33}
$$

In summary, the commutation relations between the ρ operators are

$$
\left[\rho_R(q), \rho_R(-q')\right] = +\delta_{qq'} \frac{q\mathcal{L}}{2\pi},
\tag{19.34a}
$$

$$
\left[\rho_L(q), \rho_L(-q')\right] = -\delta_{qq'} \frac{q\mathcal{L}}{2\pi},
\tag{19.34b}
$$

$$
\left[\rho_R(q), \rho_L(-q')\right] = 0.
\tag{19.34c}
$$

It is now clear that the interaction part of the Hamiltonian can be expressed in terms of the density operators ρ_R and ρ_L. In the hope that we can express the kinetic part in terms of density operators, we begin by examining the commutators,

$$[H_0, \rho_R(q)] = v_F \sum_{kk'>0} k\left[c_k^\dagger c_k, c_{k'}^\dagger c_{k'+q}\right]$$

$$= v_F \sum_{kk'>0} k(c_k^\dagger c_{k'+q}\delta_{k',k} - c_{k'}^\dagger c_k \delta_{k,k'+q})$$

$$= v_F \sum_{k>0} k\left[c_k^\dagger c_{k+q} - c_{k-q}^\dagger c_k \Theta(k-q)\right],$$

$$= v_F\Theta(q) \sum_{k>0} \left[kc_k^\dagger c_{k+q} - (k+q)c_k^\dagger c_{k+q}\right]$$

$$+ v_F\Theta(-q)\left[\sum_{k>0} kc_k^\dagger c_{k+q} - \sum_{k>-q}(k+q)c_k^\dagger c_{k+q}\right]$$

$$= -v_F q\rho_R(q) + v_F\Theta(-q)\sum_{0<k<-q}(k+q)c_k^\dagger c_{k+q} \approx -qv_F\rho_R(q), \quad (19.35)$$

where we used the crucial approximation that far below the Fermi energy we have $c_k^\dagger c_{k+q} \approx \langle c_k^\dagger c_k\rangle\delta_{q,0}$. Similarly, we obtain for the left movers,

$$[H_0, \rho_L(q)] \approx qv_F\rho_L(q), \quad (19.36)$$

which shows that the eigenmodes of the unperturbed Hamiltonian are described by

$$\langle\rho(q,t)\rangle = e^{-iqv_F(t-t_0)}\langle\rho(q,t_0)\rangle. \quad (19.37)$$

This together with Eqs. (19.34a) and (19.34c) and the fact that the density operators commute with the number operator, shows that the free-electron part can be written as

$$H_0 = \frac{2\pi v_F}{\mathcal{L}} \sum_{q>0} [\rho_R(-q)\rho_R(q) + \rho_L(q)\rho_L(-q)] + C(N), \quad (19.38)$$

where $C(N)$ is a constant that only depends on the total number of particles.

In the derivations, we have used the filled Fermi sea $|N,0\rangle$ with N electrons as the reference state. It can be shown that any many-particle state with fixed number of electrons can be reached by acting on the reference state with linear combinations of density operators ρ_L and ρ_R. In other words, states of the form $f(\rho_R, \rho_L)|N\rangle$ constitute a complete set within the subspace with a given number of particles. The proof is to show that the partition function is the same in the two representations (Haldane 1981; von Delft and Schoeller 1998) since if they were not equal, states would be missing. The Hamiltonian does not bring us out of the subspace with constant N and therefore the Hamiltonian in Eq. (19.38) correctly describes excitation energies within Hilbert spaces of fixed number of particles.

All we have left to do now is to find the constant $C(N)$. It is simply given by the energy of a filled non-interacting Fermi sea, and it is easily found to be

$$C(N_L, N_R) = \frac{\pi v_F}{\mathcal{L}}(N_L^2 + N_R^2) = \frac{\pi v_F}{2\mathcal{L}}\left(N^2 + J^2\right), \quad (19.39)$$

where we have allowed for different number of electrons in the two branches and defined N as the total number of electrons and J as the current, $J = N_R - N_L$. Both N and J are of course conserved quantities.

The final result for bosonized Hamiltonian $H = H_0 + H_{int}$ is

$$H_0 = \frac{2\pi v_F}{\mathcal{L}} \sum_{q>0} [\rho_R(-q)\rho_R(q) + \rho_L(q)\rho_L(-q)] + C(N_L, N_R), \tag{19.40a}$$

$$H_{int} = \frac{1}{2\mathcal{L}} \sum_{q\neq 0} V_1 [\rho_R(q) + \rho_L(q)] (\rho_R(-q) + \rho_L(-q)), \tag{19.40b}$$

$$V_1 = V(0) - V(2k_F). \tag{19.40c}$$

19.5.2 Diagonalization of the bosonized Hamiltonian

The following boson operators are now defined for $q > 0$,

$$a_q = \sqrt{\frac{2\pi}{q\mathcal{L}}} \rho_R(q), \qquad a_q^\dagger = \sqrt{\frac{2\pi}{q\mathcal{L}}} \rho_R(-q), \tag{19.41a}$$

$$b_q = \sqrt{\frac{2\pi}{q\mathcal{L}}} \rho_L(-q), \qquad b_q^\dagger = \sqrt{\frac{2\pi}{q\mathcal{L}}} \rho_L(q), \tag{19.41b}$$

where a and b obey usual canonical boson commutation relations,

$$[a_q, a_{q'}^\dagger] = [b_q, b_{q'}^\dagger] = \delta_{qq'}, \tag{19.42a}$$

$$[a_q, a_{q'}] = [b_q, b_{q'}] = [a_q, b_{q'}] = [a_q, b_{q'}^\dagger] = 0. \tag{19.42b}$$

In terms of these operators, the Hamiltonian becomes (up to a constant),

$$H_0 = v_F \sum_{q>0} q \left(a_q^\dagger a_q + b_q^\dagger b_q \right), \tag{19.43a}$$

$$H_{int} = \frac{1}{2\pi} \sum_{q>0} q V_1 \left(a_q^\dagger a_q + b_q^\dagger b_q + a_q b_q + b_q^\dagger a_q^\dagger \right), \tag{19.43b}$$

which is easily diagonalized using a bosonic Bogoliubov transformation (see Exercise 1.8) analogous to the fermionic one studied in Eq. (18.12a). The transformed boson operators γ_q and $\gamma_q^\dagger$ defines the Luttinger Liquid ground state $|LLG\rangle$ and the excitations $|q\rangle$,

$$\gamma |LLG\rangle = 0, \qquad |q\rangle = \gamma_q^\dagger |LLG\rangle. \tag{19.44}$$

The excitation eigenenergies E_q of the interacting Luttinger–Tomonaga model are given by

$$(H_0 + H_{int})|q\rangle = E_q|q\rangle = q\tilde{v}|q\rangle, \quad \text{with } \tilde{v} = v_F \sqrt{1 + \frac{V_1}{\pi v_F}}. \tag{19.45}$$

19.5.3 Real space representation

Often a real space formulation of the bosonized Tomonaga–Luttinger model is easier to work with because of its more appealing form. The real space representation is obtained by the following Fourier transforms $\rho_{R,L}(x)$ of the density operators $\rho_{R,L}(q)$,

$$\rho_{R,L}(x) = \frac{1}{\mathcal{L}} \sum_{q}' e^{iqx} \rho_{R,L}(q) + \rho_0, \tag{19.46}$$

again removing the $q = 0$ term originating from the constant background. Strictly speaking, this Fourier transform is not well-defined in the Tomonaga model, because for large values of $|q|$ the bottom of the band is reached. Therefore, Eq. (19.46) should be viewed as valid at long wave lengths only, i.e., that contributions from large $|q|$ must not be important. Alternatively, one can simply define the left- and right-mover dispersion relations to be linear for all values of q, i.e., also below the bottom of the bands. This is then equivalent to the Luttinger model. For now we simply work with Eq. (19.46) as the definition of the Fourier transform of left- and right-mover densities, but as we go on we must remember that the theory is to be thought of as an effective theory for the long wavelength excitations only.

The commutation relations for the real space quantities $\rho_{R,L}(x)$ are given by

$$\left[\rho_R(x), \rho_R(x')\right] = \frac{1}{\mathcal{L}^2} \sum_{qq'}' e^{iqx - iq'x'} \left[\rho_R(q), \rho_R(-q)\right]$$

$$= \frac{1}{2\pi\mathcal{L}} \sum_{q}' q e^{iq(x-x')} = \frac{1}{2\pi\mathcal{L}} \frac{\partial}{i\partial x} \sum_{q}' e^{iq(x-x')}$$

$$= \frac{1}{2\pi i} \frac{\partial}{\partial x} \delta(x - x'), \tag{19.47a}$$

$$\left[\rho_L(x), \rho_L(x')\right] = -\frac{1}{2\pi i} \frac{\partial}{\partial x} \delta(x - x'). \tag{19.47b}$$

These relations are sometimes referred to as a Kac–Moody algebra, while the non-zero commutators between the densities are called anomalous commutators. They are, as we have seen, consequences of the separation of electrons into two different species with a momentum cut-off (or infinite bands for the Luttinger model).

In terms of the density operators $\rho_L(x)$ and $\rho_R(x)$, the Hamiltonians in Eqs. (19.40a) and (19.40b) become

$$H_0 = \pi v_{\rm F} \int_{-\mathcal{L}/2}^{\mathcal{L}/2} dx \left[\rho_R(x)\rho_R(x) + \rho_L(x)\rho_L(x)\right] + C(N_L, N_R), \tag{19.48a}$$

$$H_{\rm int} = \frac{V_1}{2} \int_{-\mathcal{L}/2}^{\mathcal{L}/2} dx \left[\rho_R(x) + \rho_L(x)\right] \left[\rho_R(x) + \rho_L(x)\right], \tag{19.48b}$$

This real space Hamiltonian can be cast into a familiar form by introducing two new fields $\phi(x)$ and $P(x)$,

$$\frac{1}{\sqrt{\pi}} \partial_x \phi(x) \equiv \rho_R(x) + \rho_L(x) - \rho_{R0} - \rho_{L0}, \tag{19.49}$$

$$-\frac{1}{\sqrt{\pi}} P(x) \equiv \rho_R(x) - \rho_L(x) - (\rho_{R0} - \rho_{L0}), \tag{19.50}$$

where $\rho_{(R,L)0}$ are the background charges for the left and right movers, respectively. By integration of the commutator

$$[\partial_x \phi, P] = i\partial_x \, \delta(x - x'), \tag{19.51a}$$

the fields $\phi(x)$ and $P(x)$ can be defined such that they become conjugate variables

$$[\phi(x), P(x')] = i\delta(x - x'), \tag{19.51b}$$

$$[\phi(x), \phi(x')] = [P(x), P(x')] = 0. \tag{19.51c}$$

With the new field operators $\phi(x)$ and $P(x)$ the Hamiltonian is (up to a constant)

$$H_0 = \frac{v_F}{2} \int_{-L/2}^{L/2} dx \left[P^2(x) + (\partial_x \phi(x))^2 \right], \tag{19.52a}$$

$$H_{\text{int}} = \frac{V_1}{2\pi} \int_{-L/2}^{L/2} dx \, (\partial_x \phi(x))^2, \tag{19.52b}$$

which can be collected as follows

$$H = \frac{\tilde{v}}{2} \int_{-L/2}^{L/2} dx \left[g P^2(x) + \frac{1}{g} (\partial_x \phi(x))^2 \right], \tag{19.52c}$$

$$g^{-1} = \sqrt{1 + \frac{V_1}{\pi v_F}}, \qquad \tilde{v} = \frac{1}{g} \, v_F \tag{19.52d}$$

The unperturbed Hamiltonian H_0 is equivalent to that of a string, where P plays the role of the momentum operator and ϕ describes the local displacement field. Notice the close analogy with the transmission line model discussed in Exercise 19.2. The corresponding equations of motion for the unperturbed Hamiltonian are

$$\dot{P}(x) = -\frac{\delta H_0}{\delta \phi(x)} = v_F \partial_x^2 \phi(x), \tag{19.53a}$$

$$\dot{\phi}(x) = +\frac{\delta H_0}{\delta P(x)} = v_F P(x). \tag{19.53b}$$

By combining Eqs. (19.53) and (19.49) we obtain the continuity equation, $\partial_x J + \dot{\rho} = 0$, where the current has been identified as $J = v_F(\rho_R - \rho_L)$ and the total charge density as $\rho = \rho_R + \rho_L$. The Hamilton equations Eq. (19.53) leads to the wave equation $\ddot{\phi} - v_F^2 \partial_x^2 \phi = 0$, with dispersion relation $\omega = v_F q$.

The model in Eq. (19.52c) includes interactions, and it gives $(g\tilde{v}/v_F)\partial_x J + \rho = 0$, which means that the physical current is $(g\tilde{v}/v_F)J$. Going back to the definition, we see that $g\tilde{v}/v_F = 1$, and the interacting model thus conserves charge as it should. The velocity for the plasmon waves has however been renormalized by the interactions, which is seen from the corresponding wave equation: $\ddot{\phi} - \tilde{v}^2 \partial^2 \phi = 0$, and which is in accordance with the dispersion relation in Eq. (19.45). The interaction parameter is directly related to the compressibility κ of the system, which is defined as

$$\kappa \equiv \left(\frac{\partial^2 E}{\partial \rho^2} \right)^{-1}. \tag{19.54}$$

According to the Hamiltonian Eq. (19.52c), a density change gives rise to an energy change $\delta E = \pi \tilde{v}(\delta \rho)^2/2g$, and hence $\kappa = g/\pi \tilde{v}$. The ratio $g v_F / \tilde{v}$ is $1/(1 + V_1/[\pi v_F])$, which is less than unity for repulsive interactions, and in this case the compressibility is thus *decreased* by interactions, as we expect physically.

Finally, we can diagonalize the Hamiltonian Eq. (19.52c) in analogy with the phonon problem of Section 3.3. First, we introduce the Fourier transforms ϕ_k and P_k of $\phi(x)$ and $P(x)$, respectively, which gives

$$H = \frac{\tilde{v}}{2\mathcal{L}} \sum_k \left[g P_{-k} P_k + \frac{k^2}{g} \phi_{-k} \phi_k \right], \tag{19.55}$$

together with

$$\left[\phi_k, P_{-k'} \right] = i \mathcal{L} \delta_{kk'}. \tag{19.56}$$

Next, we define the boson operators which diagonalize Eq. (19.55),

$$\gamma_k = \frac{1}{\sqrt{2\mathcal{L}}} \left(\sqrt{\frac{|k|}{g}} \phi_k + i \sqrt{\frac{g}{|k|}} P_k \right), \quad \gamma_k^\dagger = \frac{1}{\sqrt{2\mathcal{L}}} \left(\sqrt{\frac{|k|}{g}} \phi_{-k} - i \sqrt{\frac{g}{|k|}} P_{-k} \right), \tag{19.57}$$

with the usual relations $[\gamma_k, \gamma_{k'}^\dagger] = \delta_{kk'}$. The inverse transformation is easily found to be

$$\phi_k = \sqrt{\frac{\mathcal{L}g}{2|k|}} \left(\gamma_k + \gamma_{-k}^\dagger \right), \quad P_k = i \sqrt{\frac{\mathcal{L}|k|}{2g}} \left(\gamma_{-k}^\dagger - \gamma_k \right). \tag{19.58}$$

The Hamiltonian has now finally been diagonalized,

$$H = \sum_k \tilde{v} |k| \, \gamma_k^\dagger \gamma_k, \tag{19.59}$$

where the eigenmodes, of course, are in agreement with the previous result Eq. (19.45) obtained by a Bogoliubov transformation.

19.6   Electron operators in bosonized form

What have we achieved so far, is to identify the charge-density eigenmodes of the interacting model and, therefore, the collective excitations. What we have not yet found is the one-particle properties, e.g., the local density of states as being probed by tunneling experiments and other transport experiments. To this end, we need an explicit form of the single electron operator in terms of boson operators, which would allow for the computation of the single-electron Green's functions and higher order correlation functions. This problem was solved by Luther and Peschel, by Mattis, and later Haldane (1981). Here, we shall construct such an operator by using the requirements that it must obey anti-commutation relations, and that it creates a displacement of the charge density by one electron charge (see Fig. 19.4).

The bosonized form of the Hamiltonian Eq. (19.40a) refers to a representation of the Fock space, where a given state is represented by a filled Fermi sea with a given number of left- and right-moving electrons, $|N_R, N_L\rangle$, plus some linear combination

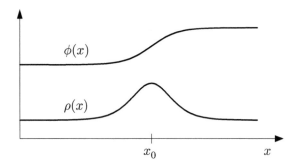

FIG. 19.4. Illustration of a 1D electron gas with a single charge excitation created by adding an electron at position x_0. The corresponding peak in the density $\rho(x)$ around x_0 is associated with a kink at x_0 in the phase field $\phi(x)$.

of bosons operators, ρ_L and ρ_R, which rearranges the filled Fermi sea. In fact, any state can be written as

$$\mathcal{O}\Big(\{\rho_L(x)\}, \{\rho_R(x)\}\Big) |N_R, N_L\rangle, \tag{19.60}$$

where $\mathcal{O}$ is some function. This representation of the Fock space should be used to construct a Fermion operator. First we split the electron field operator into left- and right-moving components

$$\Psi(x) = \frac{1}{\sqrt{\mathcal{L}}}\sum_k c_k e^{ikx} = \frac{1}{\sqrt{\mathcal{L}}}\sum_{k>0} c_{Lk} e^{ikx} + \frac{1}{\sqrt{\mathcal{L}}}\sum_{k<0} c_{Rk} e^{ikx} = \Psi_R(x) + \Psi_L(x). \tag{19.61}$$

So a field operator $\Psi_R(x)$ operating on some state, does two things: it decreases the number of electrons by one and it displaces the boson configuration for that particular state. Let us introduce the "ladder operators" that change the number of electrons by one:

$$U_R|N_R, N_L\rangle = |N_R - 1, N_L\rangle, \quad U_R^\dagger|N_R, N_L\rangle = |N_R + 1, N_L\rangle, \tag{19.62}$$

and similarly for the set $(U_L, U_L^\dagger)$.

The other operation of the electron field operator is to displace the boson modes corresponding to the sudden disappearance of one electron charge. How this works is most easily seen by looking at its commutator with the density operator

$$\big[\Psi_\eta(x), \rho_\eta(x')\big] = \delta(x - x')\Psi_\eta(x), \tag{19.63}$$

where η indicates the electron species. In our case η labels left- or right-moving particles, $\eta = L, R$.

We note that since the density operator has been expressed as a boson operator, we can try to express the single electron operator in the form of a displacement operator e^B, where B is some linear function of the boson operator. We then know from Exercise 19.1(a) that the commutation with the operator for the density has a form similar to Eq. (19.63):

$$[e^B, \rho_\eta(x')] = [B, \rho_\eta(x')] \, e^B, \tag{19.64}$$

The form of the ρ-operators at Eqs. (19.47a) and (19.47b) then suggests that

$$\Psi_\eta(x) = A e^{-i\theta_\eta(x) \mp i k_F x} \, U_\eta, \quad \eta = R, L, \tag{19.65a}$$

where the x-dependent phase-operator $\theta_\eta(x)$ is given by

$$\theta_\eta(x) = \mp 2\pi \int_{-\infty}^{x} dx' [\rho_\eta(x') - \rho_{\eta 0}], \quad \eta = R, L. \tag{19.65b}$$

Here the upper sign goes with the $\eta = R$ and the lower with $\eta = L$. We have taken special care of the constant part of the density which gives the factor $\mp i k_F x$ in the definition of $\Psi_\eta(x)$.

Let us check that this, in fact, works:

$$[\Psi_\eta(x), \rho_\eta(x')] = A [e^{-i\theta_\eta(x)}, \rho_\eta(x')] e^{\mp i k_F x} U_\eta, \tag{19.66}$$

where A is a normalization constant yet to be found and

$$[e^{-i\theta_\eta(x)}, \rho_\eta(x')] = \pm i e^{-i\theta_\eta(x)} (2\pi) \int_{-\infty}^{x} dy [\rho_\eta(y), \rho_\eta(x')]$$

$$= \pm 2\pi i e^{-i\theta_\eta(x)} \int_{-\infty}^{x} dy \left[\pm \frac{1}{2\pi i} \frac{\partial}{\partial y} \delta(y - x') \right]$$

$$= \delta(x - x') e^{-i\theta_\eta(x)}. \tag{19.67}$$

So we conclude that the commutator, in fact, evaluates to the expected result. Furthermore, we see that the anti-commutation relation also comes out correctly

$$\Psi_\eta(x)\Psi_\eta(x') = \Psi_\eta(x')\Psi_\eta(x) \exp\left(- [\theta(x), \theta(x')] \right)$$

$$= -\Psi_\eta(x')\Psi_\eta(x), \tag{19.68}$$

because by using the Kac–Moody algebra relations, Eqs. (19.47a) and (19.47b), we obtain for the commutator appearing in the exponential,[70]

$$[\theta_\eta(x), \theta_\eta(x')] = \pm i\pi \, \text{sign}(x' - x). \tag{19.69}$$

The same argument applies to $\{\Psi_\eta^\dagger(x), \Psi_\eta(x')\}$ at different positions, $x \neq x'$. At equal positions $x = x'$, the anticommutator is not well-defined unless we put in a finite momentum cut-off in the Fourier transform of the fields θ_η:

$$\theta_\eta(x) = \pm \frac{2\pi i}{L} \sum_{k \neq 0} \frac{1}{k} e^{ikx - \alpha|k|/2} \rho_\eta(k). \tag{19.70}$$

In order to determine the relationship between the cut-off α and the proportionality constant in Eq. (19.65a), we compute the anti-commutation relation for $x \to x'$. In

---

[70] In order to derive this we must symmetrize the derivative of the delta function as $\partial_x \delta(x - x') = \frac{1}{2} \partial_x \delta(x - x') - \frac{1}{2} \partial_{x'} \delta(x - x')$ or alternatively be more careful and work with the explicit momentum cut-off introduced below.

order to do this carefully, we separate the annihilation and creation operators in the product $\Psi_\eta(x)\Psi_\eta^\dagger(x')$, so that the annihilation operators stand to the right of the creation operators. Why this is sensible is explained below. Here, we write out the right-mover electron operators in terms of the bosons defined in Eq. (19.41a),

$$\Psi_R(x) = A \; e^{-ik_F x} \exp\left(\frac{2\pi}{\mathcal{L}} \sum_{k>0} \sqrt{\frac{kL}{2\pi}} \frac{e^{-\alpha k/2}}{k} \left[e^{ikx} a_k - e^{-ikx} a_k^\dagger \right] \right), \qquad (19.71a)$$

$$\Psi_R^\dagger(x') = A^* e^{ik_F x'} \exp\left(\frac{2\pi}{\mathcal{L}} \sum_{k'>0} \sqrt{\frac{k'L}{2\pi}} \frac{e^{-\alpha k'/2}}{k'} \left[e^{-ik'x'} a_{k'} - e^{ik'x'} a_{k'}^\dagger \right] \right), \quad (19.71b)$$

and then collect the exponents as

$$\{\Psi_R(x), \Psi_R^\dagger(x')\} \equiv |A|^2 e^{-ik_F(x-x')} \left(e^B e^C + e^C e^B \right)$$
$$= |A|^2 e^{-ik_F(x-x')} e^{B+C} \left(e^{[B,C]/2} + e^{-[B,C]/2} \right), \qquad (19.72)$$

where

$$\frac{1}{2}[B,C] = \frac{1}{2} \frac{2\pi}{\mathcal{L}} \sum_{k>0} \frac{e^{-k\alpha}}{k} \left(e^{ik(x-x')} - e^{-ik(x-x')} \right)$$
$$= i \int_0^\infty \frac{dk}{k} e^{-k\alpha} \sin(k(x-x')) = i \tan^{-1}\left(\frac{x-x'}{\alpha} \right). \qquad (19.73)$$

The commutator in Eq. (19.72) is thus so far given by

$$\{\Psi_\eta(x), \Psi_\eta^\dagger(x')\} = |A|^2 e^{-ik_F(x-x')} e^{B+C} \frac{2\alpha}{\sqrt{(x-x')^2 + \alpha^2}}. \qquad (19.74)$$

Because the term e^{B+C} still contains the cut-off parameter α we cannot take the limit $\alpha \to 0$ yet. In order to do this in a controlled way, we must separate annihilation and creation in the expression

$$e^{B+C} \equiv \exp\left(\sum_{k>0} \left[a_k^\dagger f_k - a_k f_k^* \right] \right), \qquad (19.75)$$

where f_k is

$$f_k = -\frac{2\pi}{k\mathcal{L}} \sqrt{\frac{kL}{2\pi}} \left(e^{-ikx} - e^{-ikx'} \right) e^{-\frac{1}{2}\alpha k}. \qquad (19.76)$$

This is done as follows,

$$e^{B+C} = \exp\left(\sum_{k>0} a_k^\dagger f_k \right) \exp\left(-\sum_{k>0} a_k f_k^* \right) \exp\left(-\frac{1}{2} \sum_{k>0} |f_k|^2 \right). \qquad (19.77)$$

For the last term in this expression, we have

$$\frac{1}{2}\sum_k |f_k|^2 = -\frac{1}{2}\frac{2\pi}{\mathcal{L}}\sum_{k>0}\frac{e^{-k\alpha}}{k}\left(e^{-ikx} - e^{-ikx'}\right)\left(e^{ikx} - e^{ikx'}\right)$$

$$= -\int_0^\infty \frac{dk}{k}e^{-k\alpha}\left\{1 - \cos\left[k(x - x')\right]\right\} = \ln\sqrt{\frac{\alpha^2}{(x - x')^2 + \alpha^2}}, \quad (19.78)$$

whereas for the first two factors in Eq. (19.77) we can now safely take the limit $\alpha \to 0$, because at energies much lower than the cut-off energy, $\propto 1/\alpha$, the k-sums in those two terms are cut-off by the annihilation operators. In other words, for a typical matrix element

$$\langle n|\exp\left(\sum_k a_k^\dagger f_k\right)\exp\left(-\sum_k a_k f_k^*\right)|m\rangle, \quad (19.79)$$

the occupied k-states of $|n\rangle$ and $|m\rangle$ are well below $1/\alpha$ and we can neglect α in the matrix element. We are thus left with

$$\{\Psi_R(x), \Psi_R^\dagger(x')\} = |A|^2 e^{-ik_F(x-x')}\exp\left(\sum_k a_k^\dagger f_k\right)\exp\left(-\sum_k a_k f_k^*\right)\frac{2\alpha^2}{(x - x')^2 + \alpha^2}.$$
$$(19.80)$$

Taking the limit $\alpha \to 0$, which for the last factor gives a delta function $2\pi\alpha\delta(x - x')$, and assuming that $f_k(x = x') = 0$, we finally arrive at

$$\{\Psi_\eta(x), \Psi_\eta^\dagger(x')\} \to 2\pi\alpha|A|^2\delta(x - x'), \quad \text{for } \alpha \to 0. \quad (19.81)$$

The normalization constant $|A|^2$ thus follows as

$$|A|^2 = \frac{1}{2\pi\alpha}. \quad (19.82)$$

Since we wish to express the single particle operator in terms of the fields ϕ and P, we find the correspondence between the pair $\{\theta_L, \theta_R\}$ and the pair $\{\phi, P\}$

$$\left.\begin{array}{l}\phi = -\dfrac{1}{2\sqrt{\pi}}(\theta_R - \theta_L),\\[2mm] P = \dfrac{1}{2\sqrt{\pi}}(\partial_x\theta_R + \partial_x\theta_L),\end{array}\right\} \Leftrightarrow \left\{\begin{array}{l}\theta_R = \sqrt{\pi}\left[-\phi + \displaystyle\int^x dx' P(x')\right],\\[2mm] \theta_L = \sqrt{\pi}\left[\phi + \displaystyle\int^x dx' P(x')\right],\end{array}\right. \quad (19.83)$$

The final form of the normalized electron operator now reads

$$\Psi_{R,L}(x) = \frac{1}{\sqrt{2\pi\alpha}}U_{R,L}e^{\pm ik_F x}e^{-i\sqrt{\pi}[\mp\phi(x)+\Phi(x)]} = \frac{1}{\sqrt{2\pi\alpha}}U_{R,L}e^{\pm ik_F x}e^{-i\theta_{R,L}(x)},$$
$$(19.84)$$

where

$$\Phi(x) = \int^x dx' P(x'). \quad (19.85)$$

It is important to keep in mind that the operator identity Eq. (19.84) is to be understood as an identity in the limit $\alpha \to 0$.

What Eq. (19.84) expresses is that creation or annihilation of a localized charge involves physics on two length scales. The usual de Broglie wavelength of a particle moving with the Fermi momentum gives the $e^{\pm i k_F x}$ factor, while the exponent with the boson fields causes a shake-up of a lot of low-energy collective modes of the electron gas. The length scale for the latter process depends on the energy involved, and is given by $q^{-1} \approx v_F/\varepsilon$, where ε is the available energy. The decomposition of the single electron operator into an infinite number of low lying states is what gives rise to many of the peculiar properties of 1D metals.

19.7 Green's functions

Above we saw how the Hamiltonian of interacting electrons was diagonalized by a transformation into a bilinear form of boson operators. However, we had to pay the price that the single electron operators became rather complicated objects in terms of these bosons. Nevertheless, calculations with these single particle operators are possible, because they are nothing but coherent states of bosons. We will now compute the Green's function analytically using this form for the single-particle operators. This result is rather remarkable: it is rare that we are able to compute a Green's function for a complicated interacting many-body problem. The non-trivial form of the Green's function is shown to have interesting consequences for, e.g., the tunneling current which is considered in the next section.

We now consider the retarded Green's function $G^R(x,t)$. According to Eq. (8.32), we can write

$$G^R(x,t) = \Theta(t)\big[G^>(x,t) - G^<(x,t)\big], \qquad (19.86)$$

where

$$
\begin{aligned}
G^>(x,t) &= -i\Big\langle \Psi(x,t)\Psi^\dagger(0,0)\Big\rangle \\
&\approx -i\Big\langle \Psi_R(x,t)\Psi_R^\dagger(0,0)\Big\rangle - i\Big\langle \Psi_L(x,t)\Psi_L^\dagger(0,0)\Big\rangle \\
&= G_R^>(x,t) + G_L^>(x,t),
\end{aligned}
\qquad (19.87)
$$

and similarly

$$G^<(x,t) = i\Big\langle \Psi^\dagger(0,0)\Psi(x,t)\Big\rangle \approx G_R^<(x,t) + G_L^<(x,t). \qquad (19.88)$$

Now we insert the electron operator expressed in terms of the bosons from Section 19.6

$$G_R^>(x,t) = \qquad\qquad\qquad\qquad\qquad\qquad\qquad\qquad\qquad (19.89)$$
$$\frac{-i}{2\pi\alpha}\, e^{ik_F x}\Big\langle \exp\Big(-i\sqrt{\pi}\big[-\phi(x,t)+\Phi(x,t)\big]\Big)\exp\Big(i\sqrt{\pi}\big[-\phi(0,0)+\Phi(0,0)\big]\Big)\Big\rangle.$$

Note that the ladder operators U do not play a role, because they only change the number of particles. This has no effect, since the expectation value in Eq. (19.89) is independent of the number of particles in the reference state.

Next, we utilize a very nice property of free bosons, namely that the second-order cumulant expansion is exact, i.e.,

$$\langle e^{iA}e^{-iB}\rangle = \exp\left(i\langle A\rangle - i\langle B\rangle - \frac{1}{2}\langle A^2\rangle - \frac{1}{2}\langle B^2\rangle + \langle AB\rangle\right), \tag{19.90}$$

if A and B are linear in the boson operators. Because the linear terms vanish, $\langle A\rangle = \langle B\rangle = 0$, and the equal time products are reduced as $\langle[A(t)]^2\rangle = \langle[A(0)]^2\rangle$, this nice formula gives for our Green's function

$$G_R^>(x,t) = \frac{-i}{2\pi\alpha}\, \exp(ik_F x)\, \exp\left(\pi[D(x,t) - D(0,0)]\right), \tag{19.91}$$

$$\tag{19.92}$$

where $D(x)$ is

$$D(x,t) = \left\langle\left[-\phi(x,t) + \Phi(x,t)\right]\left[-\phi(0,0) + \Phi(0,0)\right]\right\rangle$$

$$= \frac{1}{L^2}\sum_{kk'\neq 0} e^{ikx}e^{-\alpha(|k|+|k'|)/2}$$

$$\times \left\langle\left(-\phi_k(t) + \frac{1}{ik}P_k(t)\right)\left(-\phi_{-k'} - \frac{1}{ik'}P_{-k'}\right)\right\rangle. \tag{19.93}$$

Using Eq. (19.58) to express ϕ_k and P_k in terms of the Bogoliubov operators γ_k, we obtain after some algebra

$$D(x,t) = \frac{1}{L}\sum_{k>0}\frac{\cos kx}{k}e^{-\alpha k}\left(g + \frac{1}{g}\right)\left[e^{-i\tilde{v}kt} + 2n_B(\tilde{v}k)\cos(\tilde{v}kt)\right] \tag{19.94}$$

$$+ \frac{2i}{L}\sum_{k>0}\frac{\sin kx}{k}e^{-\alpha k}\left[e^{-i\tilde{v}kt} - 2in_B(\tilde{v}k)\sin(\tilde{v}kt)\right]. \tag{19.95}$$

For $T = 0$, the exponent in Eq. (19.91) becomes

$$\pi[D(x,t) - D(0,0)] = \frac{1}{2}\left(g + \frac{1}{g}\right)\int_0^\infty dk\, e^{-\alpha k}\frac{e^{-i\tilde{v}kt}\cos kx - 1}{k}$$

$$+ i\int_0^\infty dk\, e^{-i\tilde{v}kt - \alpha k}\frac{\sin kx}{k}. \tag{19.96}$$

Below, this result is utilized to calculate the density of states.

19.8 Measuring local density of states by tunneling

Some of the most convincing observations of Luttinger liquid behavior are the measurements of power laws in the tunneling density of states. In many of these experiments the tunneling setup is as illustrated in Fig. 19.5. Tunneling between two

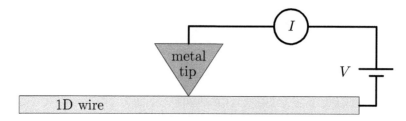

FIG. 19.5. Tunneling into a 1D wire from a metal tip. The contact can be made by, e.g., a scanning tunneling microscope or by attachment of a lead to the wire.

conducting systems was discussed in Section 8.4.1, where also the tunneling Hamiltonian was introduced. For tunneling into 1D wires we will use a real-space formulation of the tunneling Hamiltonian,

$$H_T = \int_2 d\mathbf{r} \int_1 dx \left[T(x, \mathbf{r}) \Psi_1^\dagger(x) \Psi_2(\mathbf{r}) + \Psi_2^\dagger(\mathbf{r}) \Psi_1(x) \right], \tag{19.97}$$

where the x-integral runs over the 1D metal, and the $\mathbf{r}$-integral is defined by the tip. The tunnel matrix element $T(x, \mathbf{r})$ describes the amplitude for tunneling between positions x and $\mathbf{r}$. Now we follow the same line of arguments as in Section 8.4.1, where the tunneling matrix element is assumed to be small and therefore one can use linear response theory in H_T, and we find

$$I = 2\,\mathrm{Re} \int_{-\infty}^{\infty} dt'\theta(t-t') \int_2 d\mathbf{r}\,d\mathbf{r}' \int_1 dx\,dx'\, [T(x,\mathbf{r})]^* \, T(x',\mathbf{r}')$$
$$\times \left(\left\langle \hat{\Psi}_1(x,t)\hat{\Psi}_1^\dagger(x',t') \right\rangle \left\langle \hat{\Psi}_2^\dagger(\mathbf{r},t)\hat{\Psi}_2(\mathbf{r}',t') \right\rangle \right.$$
$$\left. - \left\langle \hat{\Psi}_1^\dagger(x',t')\hat{\Psi}_1(x,t) \right\rangle \left\langle \hat{\Psi}_2(\mathbf{r}',t')\hat{\Psi}_2^\dagger(\mathbf{r},t) \right\rangle \right). \tag{19.98}$$

The voltage dependence is pulled out similarly to Eq. (8.71) and furthermore, using the fact that the tip can be described as a metal we use a diagonal basis for the tip field operators: $\Psi_2(\mathbf{r}) = \sum_\nu c_\nu \phi_\nu(\mathbf{r})$. We then find an expression similar to Eq. (8.72),

$$I = 2\,\mathrm{Re} \int_{-\infty}^{\infty} dt'\theta(t-t') \sum_\nu \int_1 dx\,dx'\, [T(x,\nu)]^* \, T(x',\nu) e^{-ie(V_1-V_2)(t-t')}$$
$$\times \left[G_1^>(xt, x't') G_2^<(\nu, t'-t) - G_1^<(xt, x't') G_2^>(\nu, t'-t) \right], \tag{19.99}$$

where

$$T(x,\nu) = \int d\mathbf{r}\, T(x,\mathbf{r}) \phi_\nu(\mathbf{r}). \tag{19.100}$$

This current formula becomes more transparent after transformation to Fourier transforms (using $\mathrm{Re} \int_0^\infty dt e^{i\omega t} = \pi\delta(\omega)$):

$$I = \int_{-\infty}^{\infty} \frac{d\omega}{2\pi} \sum_\nu \int_1 dx\,dx'\, [T(x,\nu)]^* \, T(x',\nu)$$
$$\times \left[G_1^>(x, x', \omega) G_2^<(\nu, \omega + V) - G_1^<(x, x', \omega) G_2^>(\nu, \omega + V)) \right], \tag{19.101}$$

with $V = V_1 - V_2$.

Finally, we specialize to the case where the spatial extent of the tip and hence the spatial variations of the tunneling matrix element is small compared with typical length scales, over which the 1D Green's function changes, i.e., smaller than the wavelength of charge fluctuations: $\max\{eV, kT\}/\tilde{v}$. In this case, we set $x = x'$, and furthermore, we choose $x = 0$. The lesser and greater Green's functions can now be expressed in terms of the spectral functions using Eq. (8.51). The same steps that lead to Eq. (8.74) now give us

$$I = \frac{1}{e} \int d\omega\, G(\omega) \left[n_F(\omega + V) - n_F(\omega) \right] \nu(\omega), \tag{19.102a}$$

$$G(\omega) = e^2 A(0,0) \sum_\nu \int_1 dx dx' \left[T(x,\nu) \right]^* T(x',\nu) \delta(\omega + V - \xi_n), \tag{19.102b}$$

$$\nu(\omega) = \frac{A(0,\omega)}{A(0,0)}, \tag{19.102c}$$

where we have defined an energy-dependent conductance $G(\omega)$, which characterizes the tip, and a normalized density of states for the 1D system $\nu(\omega)$. If the tip is a simple metal, and we are at low voltages and low temperatures, it is a good approximation to neglect the energy dependence of $G(\omega)$, and we see that tunneling spectrum directly measures the density of state of the 1D metal. The spectral function $A(0,\omega)$ is given by

$$A(0,\omega) = -2\mathrm{Im}\, G^R(0,\omega) = 2\mathrm{Im} \left[i \int_0^\infty dt \left\langle \{ \Psi(0,t), \Psi^\dagger(0,0) \} \right\rangle \right]. \tag{19.103}$$

The density of states is thus related to the Green's function in Section 19.7. We found there that

$$G^>(0,t) = G_R^>(0,t) + G_L^>(0,t), \tag{19.104}$$

$$G_R^>(0,t) = G_L^>(0,t) = -i \left\langle \Psi_R(0,t) \Psi_R^\dagger(0) \right\rangle = \frac{-i}{2\pi\alpha} \exp[F(t)], \tag{19.105}$$

where $F(t)$ is

$$F(t) = \frac{1}{2} \left(g + \frac{1}{g} \right) \int_0^\infty \frac{dk}{k} e^{-\alpha k} \left[\left(\cos(k\tilde{v}t) - 1 \right) \coth\left(\frac{k\tilde{v}}{2kT} \right) - i \sin(k\tilde{v}t) \right]. \tag{19.106}$$

For simplicity we now take $T = 0$, which leads to

$$\begin{aligned} F(t) &= \frac{1}{2} \left(g + \frac{1}{g} \right) \int_0^\infty \frac{dk}{k} e^{-\alpha k} \left[\exp\left(-ik\tilde{v}t \right) - 1 \right] \\ &= -\frac{1}{2} \left(g + \frac{1}{g} \right) \ln \left(\frac{\alpha + i\tilde{v}t}{\alpha} \right). \end{aligned} \tag{19.107}$$

In the same way,

$$G_R^<(0,t) = G_L^<(0,t) = i \left\langle \Psi_R^\dagger(0)\Psi_R(0,t) \right\rangle = \frac{i}{2\pi\alpha} \exp[F(-t)]. \qquad (19.108)$$

Using the fact that the lesser and the greater function obey $[G^{0,\lessgtr}(t)]^* = -G^{\lessgtr}(-t)$, as can be seen in Eqs. (19.105) and (19.108), we find for the spectral function

$$A(0,\omega) = 2i \int_{-\infty}^{\infty} dt \, [G_R^>(0,t) - G_R^<(0,t)] \, e^{i\omega t}. \qquad (19.109)$$

This integral is performed as follows

$$
\begin{aligned}
A(0,\omega) &= \frac{1}{\pi\alpha} \int_{-\infty}^{\infty} dt \, e^{i\omega t} \left(\exp\left[F(t)\right] + \exp\left[F(-t)\right] \right) \\
&= \frac{1}{\pi\alpha} \int_{-\infty}^{\infty} dt \, \cos(\omega t) \left(\frac{\alpha + i\tilde{v}t}{\alpha} \right)^{-\frac{1}{2}\left(g+\frac{1}{g}\right)} \\
&= \frac{1}{\tilde{v}\,\Gamma\left(\frac{g}{2} + \frac{1}{2g}\right)} e^{-|\omega|\alpha/\tilde{v}} \left(\frac{|\omega|\alpha}{\tilde{v}} \right)^{\frac{1}{2}\left(g+\frac{1}{g}-2\right)}, \qquad (19.110)
\end{aligned}
$$

which gives for the total tunneling density of states $\nu(\omega)$ in Eq. (19.102c),

$$\nu(\omega) = e^{-|\omega|\alpha/\tilde{v}} \left(\frac{|\omega|\alpha}{\tilde{v}} \right)^{\frac{1}{2}\left(g+\frac{1}{g}-2\right)}. \qquad (19.111)$$

At the end of the calculation we are supposed to take the limit $\alpha \to 0$. It is, however, not clear how to do that in the expression above. Therefore we have to take one step back and consider how the cut-off was introduced. The cut-off α, appearing in Eqs. (19.106) and (19.107), was a mathematical cut-off, which was included in order to make the single-particle operators behave correctly. But we should also include a physical cut-off, saying that the interactions are not important for large energies, which we do by giving g a momentum dependence,

$$g + \frac{1}{g} \quad \rightarrow \quad \left(g + \frac{1}{g}\right) e^{-k/k_c} + 2(1 - e^{-k/k_c}), \qquad (19.112)$$

such that for large k-values the non-interacting case is recovered. Instead of doing the integrals, we simply argue that for small energies (small values of k/k_c), we can let α play the role of the momentum cut-off and replace α with k_c^{-1} in Eq. (19.111), i.e., the tunneling density of states becomes

$$\nu(\omega) = \left(\frac{|\omega|}{\varepsilon_c} \right)^{\frac{1}{2}\left(g+\frac{1}{g}-2\right)}, \qquad (19.113)$$

where ε_c is a high-energy cut-off, $\varepsilon_c = \tilde{v}k_c$.

So for tunneling between two identical Luttinger liquids (LL) or tunneling from a metal (MT) to a Luttinger liquid, we thus get the I–V characteristics

$$I \sim \int_0^V d\omega \left[\nu_{LL}(\omega)\right]^2 \sim \int_0^V d\omega\, \omega^{g+\frac{1}{g}-2} \quad \sim V^{g+\frac{1}{g}-1}, \quad \text{(LL to LL)}, \quad (19.114)$$

$$I \sim \int_0^V d\omega\, \nu_{LL}(\omega) \quad \sim \int_0^V d\omega\, \omega^{\frac{1}{2}\left(g+\frac{1}{g}-2\right)} \sim V^{\frac{1}{2}\left(g+\frac{1}{g}\right)}, \quad \text{(LL to MT)}, \quad (19.115)$$

where constants that only depend on the tunneling matrix element, the high-energy cut-off, and the renormalized velocity have been omitted. The different power laws for the different tunneling setups indeed seem to be supported by experimental observations on carbon nanotubes.

In Exercise 19.4, a similar characteristic is derived for a somewhat different situation, namely a transmission line model. The exercise thus shows an analogy with a 1D metal and 1D electromagnetic system.

19.9   Luttinger liquid with spin

So far we have only discussed the spinless case. However the spin-full in also very interesting but of course a bit more complicated. Let us here briefly discuss what happens in the case with spin. First of all, it is clear that the unperturbed Hamiltonian can be written in the same but with an additional sum over spin index, σ,

$$H_0 = \frac{\pi v_F}{\mathcal{L}} \sum_{\sigma q}{}' \left[\rho_{R\sigma}(q)\rho_{R\sigma}(-q) + \rho_{L\sigma}(q)\rho_{L\sigma}(-q)\right]. \quad (19.116)$$

The interaction can now be written as products of density operators along the same lines as for the spinless case. However, there is one thing we must ensure, namely that the SU(2) rotational invariance of the spin sector is fulfilled by the bosonized interaction. The SU(2) symmetry implies that the Hamiltonian is invariant under the rotation

$$c_\sigma \to \sum_\sigma (\mathbf{U})_{\sigma\sigma'} c_{\sigma'}, \quad (19.117)$$

where $\mathbf{U}$ is a unitary matrix. As the most simple example of a "bosonized" interaction that satisfies this symmetry and at the same type has the same structure as the interaction we derived in the spinless case, we take

$$H_{\text{int}} = \frac{V}{\mathcal{L}} \sum_{q\neq 0} \sum_{\sigma\sigma'} \left(\left[\rho_{R\sigma}(q) + \rho_{L\sigma}(q)\right] \left[\rho_{R\sigma'}(-q) + \rho_{L\sigma'}(-q)\right] \right). \quad (19.118)$$

There is however one type of scattering which cannot be cast into the form of products of density operators. This is backscattering terms between electrons with opposite spin, i.e., a term like

$$\sum c^\dagger_{R\uparrow} c^\dagger_{L\downarrow} c_{R\downarrow} c_{L\uparrow}. \quad (19.119)$$

One can argue with renormalization group methods that this term is not important for repulsive interactions. This is however beyond the scope of this book, and for now we will simply neglect it.

Now the Hamiltonian separates into charge and spin degrees of freedom, because if one introduces charge and spin densities,

$$\rho = \rho_\downarrow + \rho_\uparrow, \qquad \sigma = \rho_\downarrow - \rho_\uparrow, \qquad (19.120)$$

the Hamiltonian reduces to

$$H_0 = \frac{\pi v_F}{2L} \sum_{q\neq 0} \left[\rho_R(-q)\rho_R(q) + \rho_L(q)\rho_L(-q) + \sigma_R(-q)\sigma_R(q) + \sigma_L(q)\sigma_L(-q)\right],$$

$$(19.121a)$$

$$H_{\text{int}} = \frac{V}{L} \sum_{q\neq 0} \left(\left[\rho_R(q) + \rho_L(q)\right] \left[\rho_R(-q) + \rho_L(-q)\right]\right). \qquad (19.121b)$$

The Hamiltonian has indeed two decoupled parts describing spin and charge, respectively. Many of the calculations are now similar to the spinless case: one can find the commutation relations between ρ and σ operators to be similar to the spinless counter parts, and hence introduce boson operators in the same way. Also single electron operators can be constructed straight forwardly but now being functions of both spin and charge bosons. As a consequence the tunneling density will have similar form, as we saw in the previous section, i.e., a power law dependence, the only difference being different powers.

The important lesson we learn here is that spin and charge degrees of freedom live independent lives, which is known as *charge and spin separation*. This unusual feature has no simple counterpart in higher dimensions.

19.10   Summary and outlook

We have discussed the behavior of 1D metals. In 1D the Fermi liquid theory that was argued to be realized in 3D systems, does not apply. This fact became clear when we looked at the lifetime in 1D, and we saw that it does not diverge sufficiently rapidly at the Fermi surface. Furthermore, the collective excitations were very different from the 3D counterparts. The dispersion relation was, in fact, in RPA shown to be linear in q, rather than having a gap. The RPA result indeed agrees with the Tomonaga–Luttinger model, where we found that the elementary excitations all had the form of plasmon modes, leaving no room for electron-hole pair excitations.

The Tomonaga–Luttinger model was solvable, because the interactions and the kinetic energy both could be written as a quadratic forms of the left- and right-mover densities, and that these densities obeyed boson-like commutation relations. In fact, the Hamiltonian for the 1D electron gas could be written in form of a Hamiltonian of an elastic medium,

$$H = \frac{\tilde{v}}{2} \int_{-\mathcal{L}/2}^{\mathcal{L}/2} dx \left[gP^2(x) + \frac{1}{g}\left(\partial_x\phi(x)\right)^2\right], \qquad (19.122)$$

where g is an effective interaction parameter, which smaller is than unity for repulsive interactions and larger than one for attractive interactions. The derivative of the field

ϕ gives the density of electrons, while its conjugated momentum π describes the current in electron gas.

The bosonic formulation of the Hamiltonian allows us to solve for the eigenmodes and eigenenergies $\omega(q) = \tilde{v}q$ of the charge density modes. However, in order to describe single-particle Green's functions it was necessary to express the single-electron operator in terms of the bosons, i.e., to the fermion operators as a function of boson operator. This was indeed possible and the result of this analysis was that the left- and right-mover field operators became

$$\Psi_{R,L}(x) = \frac{1}{\sqrt{2\pi\alpha}} U_{R,L} e^{\pm i k_F x} e^{-i\sqrt{\pi}[\mp\phi(x)+\Phi(x)]}, \quad \Phi(x) = \int^x dx' P(x'). \quad (19.123)$$

This form of the electron operator was used to find the single-particle Green's functions and the tunneling density of states. The result of this analysis was that the tunneling current exhibits a power law dependence on the voltage. Such power laws are indeed a very common feature of Luttinger liquids and similar field theories.

This concludes our treatment of 1D physics, which is a interesting field of mathematical physics in itself, and the present chapter is in many respect only an appetizer to this vast field. For the interested we refer to, e.g., the books by Gogolin et al. (1998) and Giamarchi (2003), or the reviews by Solyum (1970), von Delft and Schoeller (1998), Fisher and Glazman (1997), Voit (1995), Egger et al. (2001), and references therein.

APPENDIX A

FOURIER TRANSFORMATIONS

Fourier transformation is useful to employ in the case of homogeneous systems or to change linear differential equations into linear algebraic equations. The idea is to resolve the quantity $f(\mathbf{r}, t)$ under study on plane wave components,

$$f_{\mathbf{k},\omega}\, e^{i(\mathbf{k}\cdot\mathbf{r}-\omega t)}, \tag{A.1}$$

traveling at the speed $v = \omega/|\mathbf{k}|$.

A.1 Continuous functions in a finite region

Consider a rectangular box in 3D with side lengths L_x, L_y, L_z and a volume $\mathcal{V} = L_x L_y L_z$. The central theorem in Fourier analysis states that any well-behaved function fulfilling the periodic boundary conditions,

$$f(\mathbf{r} + L_x \mathbf{e}_x) = f(\mathbf{r} + L_y \mathbf{e}_y) = f(\mathbf{r} + L_z \mathbf{e}_z) = f(\mathbf{r}), \tag{A.2}$$

can be written as a Fourier series

$$f(\mathbf{r}) = \frac{1}{\mathcal{V}} \sum_{\mathbf{k}} f_{\mathbf{k}}\, e^{i\mathbf{k}\cdot\mathbf{r}}, \quad \begin{cases} k_x = \frac{2\pi n_x}{L_x},\ n_x = 0, \pm 1, \pm 2, \dots, \\ \text{likewise for } y \text{ and } z, \end{cases} \tag{A.3}$$

where

$$f_{\mathbf{k}} = \int_{\mathcal{V}} d\mathbf{r}\, f(\mathbf{r})\, e^{-i\mathbf{k}\cdot\mathbf{r}}. \tag{A.4}$$

Note the prefactor $1/\mathcal{V}$ in Eq. (A.3). It is our choice to put it there. Another choice would be to put it in Eq. (A.4), or to put $1/\sqrt{\mathcal{V}}$ in front of both equations. In all cases the product of the normalization constants should be $1/\mathcal{V}$.

An extremely important and very useful theorem states

$$\int d\mathbf{r}\, e^{-i\mathbf{k}\cdot\mathbf{r}} = \mathcal{V}\, \delta_{\mathbf{k}0}, \qquad \frac{1}{\mathcal{V}} \sum_{\mathbf{k}} e^{i\mathbf{k}\cdot\mathbf{r}} = \delta(\mathbf{r}). \tag{A.5}$$

Note the dimensions in these two expressions so that you do not forget where to put the factors of $\mathcal{V}$ and $1/\mathcal{V}$. Note also that by using Eq. (A.5) you can prove that Fourier transforming from $\mathbf{r}$ to $\mathbf{k}$ and then back brings you back to the starting point: insert $f_{\mathbf{k}}$ from Eq. (A.4) into the expression for $f(\mathbf{r})$ in Eq. (A.3) an reduce by use of Eq. (A.5).

A.2 Continuous functions in an infinite region

If we let $\mathcal{V}$ tend to infinity the $\mathbf{k}$-vectors become quasicontinuous variables, and the $\mathbf{k}$-sum in Eq. (A.3) is converted into an integral,

$$f(\mathbf{r}) = \frac{1}{\mathcal{V}} \sum_{\mathbf{k}} f_{\mathbf{k}}\, e^{i\mathbf{k}\cdot\mathbf{r}} \quad \xrightarrow[\mathcal{V}\to\infty]{} \quad \frac{1}{\mathcal{V}} \frac{\mathcal{V}}{(2\pi)^3} \int d\mathbf{k}\, f_{\mathbf{k}}\, e^{i\mathbf{k}\cdot\mathbf{r}} = \int \frac{d\mathbf{k}}{(2\pi)^3}\, f_{\mathbf{k}}\, e^{i\mathbf{k}\cdot\mathbf{r}}. \tag{A.6}$$

Now you see why we choose to put $1/\mathcal{V}$ in front of $\sum_{\mathbf{k}}$. We have

$$f(\mathbf{r}) = \int \frac{d\mathbf{k}}{(2\pi)^3}\, f_{\mathbf{k}}\, e^{i\mathbf{k}\cdot\mathbf{r}}, \qquad f_{\mathbf{k}} = \int d\mathbf{r}\, f(\mathbf{r}) e^{-i\mathbf{k}\cdot\mathbf{r}}, \tag{A.7}$$

and also

$$\int \frac{d\mathbf{k}}{(2\pi)^3}\, e^{i\mathbf{k}\cdot\mathbf{r}} = \delta(\mathbf{r}), \qquad \int d\mathbf{r}\, e^{-i\mathbf{k}\cdot\mathbf{r}} = (2\pi)^3\, \delta(\mathbf{k}). \tag{A.8}$$

Note that the dimensions are alright. Again it is easy to use these expression to verify that Fourier transforming twice brings you back to the starting point.

A.3 Time and frequency Fourier transforms

The time t and frequency ω transforms can be thought of as an extension of functions periodic with the finite period $\mathcal{T}$, to the case where this period tends to infinity. Thus t plays the role of $\mathbf{r}$ and ω that of $\mathbf{k}$, and in complete analogy with Eq. (A.7) — but with the opposite sign of i due to Eq. (A.1) — we have

$$f(t) = \int_{-\infty}^{\infty} \frac{d\omega}{2\pi}\, f_\omega\, e^{-i\omega t}, \qquad f_\omega = \int_{-\infty}^{\infty} dt\, f(t) e^{i\omega t}, \tag{A.9}$$

and also

$$\int_{-\infty}^{\infty} \frac{d\omega}{2\pi}\, e^{-i\omega t} = \delta(t), \qquad \int_{-\infty}^{\infty} dt\, e^{i\omega t} = 2\pi\, \delta(\omega). \tag{A.10}$$

Note again that the dimensions are okay.

A.4 Some useful rules

We can think of Eqs. (A.5), (A.8), and (A.10) as the Fourier transform of the constant function $f = 1$ to delta functions (and back):

$$1_{\mathbf{r}} \longleftrightarrow \mathcal{V}\,\delta_{\mathbf{k}0}, \qquad\qquad 1_{\mathbf{k}} \longleftrightarrow \delta(\mathbf{r}), \qquad\quad \text{discrete } \mathbf{k}, \tag{A.11a}$$

$$1_{\mathbf{r}} \longleftrightarrow (2\pi)^3\,\delta(\mathbf{k}), \qquad 1_{\mathbf{k}} \longleftrightarrow \delta(\mathbf{r}), \qquad\quad \text{continuous } \mathbf{k}, \tag{A.11b}$$

$$1_t \longleftrightarrow 2\pi\,\delta(\omega), \qquad\qquad 1_\omega \longleftrightarrow \delta(t), \qquad\quad \text{continuous } \omega. \tag{A.11c}$$

Another useful rule is the rule for Fourier transforming convolution integrals. By direct application of the definitions and Eq. (A.8) we find

$$f(\mathbf{r}) = \int d\mathbf{s}\, h(\mathbf{r}-\mathbf{s})\, g(\mathbf{s}) = \int d\mathbf{s}\, \frac{1}{\mathcal{V}^2} \sum_{\mathbf{k},\mathbf{k}'} h_{\mathbf{k}} e^{i\mathbf{k}\cdot(\mathbf{r}-\mathbf{s})} g_{\mathbf{k}'} e^{i\mathbf{k}'\cdot\mathbf{s}} = \frac{1}{\mathcal{V}} \sum_{\mathbf{k}} h_{\mathbf{k}} g_{\mathbf{k}}\, e^{i\mathbf{k}\cdot\mathbf{r}}, \tag{A.12}$$

or in words: a convolution integral in $\mathbf{r}$-space becomes a product in $\mathbf{k}$-space.

$$\int d\mathbf{s}\, h(\mathbf{r}-\mathbf{s})\, g(\mathbf{s}) \quad \longleftrightarrow \quad h_\mathbf{k}\, g_\mathbf{k}. \qquad (A.13)$$

A related rule, the invariance of inner products going from $\mathbf{r}$ to $\mathbf{k}$, is derived in a similar way (and here given in three different versions):

$$\int d\mathbf{r}\, h(\mathbf{r})\, g^*(\mathbf{r}) = \int \frac{d\mathbf{k}}{(2\pi)^3}\, h_\mathbf{k} g_\mathbf{k}^*, \qquad (A.14)$$

$$\int d\mathbf{r}\, h(\mathbf{r})\, g(\mathbf{r}) = \int \frac{d\mathbf{k}}{(2\pi)^3}\, h_\mathbf{k} g_{-\mathbf{k}}, \qquad (A.15)$$

$$\int d\mathbf{r}\, h(\mathbf{r})\, g(-\mathbf{r}) = \int \frac{d\mathbf{k}}{(2\pi)^3}\, h_\mathbf{k} g_\mathbf{k}. \qquad (A.16)$$

Finally we mention the Fourier transformation of differential operators. For the gradient operator we have

$$\nabla_\mathbf{r} f(\mathbf{r}) = \nabla_\mathbf{r} \frac{1}{\mathcal{V}} \sum_\mathbf{k} f_\mathbf{k}\, e^{i\mathbf{k}\cdot\mathbf{r}} = \frac{1}{\mathcal{V}} \sum_\mathbf{k} f_\mathbf{k}\, \nabla_\mathbf{r} e^{i\mathbf{k}\cdot\mathbf{r}} = \frac{1}{\mathcal{V}} \sum_\mathbf{k} i\mathbf{k} f_\mathbf{k}\, e^{i\mathbf{k}\cdot\mathbf{r}}. \qquad (A.17)$$

Similarly for ∇^2, $\nabla\times$ and ∂_t (remember the sign change of i in the latter):

$$\nabla_\mathbf{r} \longleftrightarrow i\mathbf{k}, \qquad\qquad \partial_t \longleftrightarrow -i\omega, \qquad (A.18)$$

$$\nabla^2 \longleftrightarrow -\mathbf{k}^2, \qquad\qquad \nabla\times \longleftrightarrow i\mathbf{k}\times. \qquad (A.19)$$

A.5   Translation-invariant systems

We study a translation invariant system. Any physical observable $f(\mathbf{r},\mathbf{r}')$ of two spatial variables $\mathbf{r}$ and $\mathbf{r}'$ can only depend on the difference between the coordinates and not on the absolute position of any of them,

$$f(\mathbf{r},\mathbf{r}') = f(\mathbf{r}-\mathbf{r}'). \qquad (A.20)$$

The consequences in $\mathbf{k}$-space from this constraint are:

$$f(\mathbf{r},\mathbf{r}') = \int \frac{d\mathbf{k}}{(2\pi)^3} \int \frac{d\mathbf{k}'}{(2\pi)^3} f_{\mathbf{k},\mathbf{k}'} e^{i\mathbf{k}\cdot\mathbf{r}} e^{i\mathbf{k}'\cdot\mathbf{r}'} = \int \frac{d\mathbf{k}}{(2\pi)^3} \int \frac{d\mathbf{k}'}{(2\pi)^3} f_{\mathbf{k},\mathbf{k}'} e^{i\mathbf{k}\cdot(\mathbf{r}-\mathbf{r}')} e^{i(\mathbf{k}'+\mathbf{k})\cdot\mathbf{r}'}.$$
$$(A.21)$$

Since this has to be a function of $\mathbf{r}-\mathbf{r}'$, it is obvious from the factor $e^{i(\mathbf{k}'+\mathbf{k})\cdot\mathbf{r}'}$ that any reference to the absolute value of $\mathbf{r}'$ only can vanish if $\mathbf{k}' = -\mathbf{k}$, and thus $f_{\mathbf{k},\mathbf{k}'} \propto \delta(\mathbf{k}+\mathbf{k}')$. To find the proportionality constant, we can also find the Fourier transform of f by explicitly using that f only depends on the difference $\mathbf{r}-\mathbf{r}'$

$$f(\mathbf{r},\mathbf{r}') = \tilde{f}(\mathbf{r}-\mathbf{r}') = \int \frac{d\mathbf{k}}{(2\pi)^3} \tilde{f}_\mathbf{k}\, e^{i\mathbf{k}\cdot(\mathbf{r}-\mathbf{r}')}, \qquad (A.22)$$

and by comparing the two expressions (A.21) and (A.22) we read off that

$$f_{\mathbf{k},\mathbf{k}'} = (2\pi)^3 \delta(\mathbf{k} + \mathbf{k}')\tilde{f}_{\mathbf{k}}, \tag{A.23}$$

or in short

$$f(\mathbf{r},\mathbf{r}') \longleftrightarrow f_{\mathbf{k},-\mathbf{k}}, \qquad \text{translation-invariant systems.} \tag{A.24}$$

For the discrete case, we can go through the same arguments or use the formulae from above to get

$$f_{\mathbf{k},\mathbf{k}'} = \mathcal{V}\delta_{\mathbf{k},-\mathbf{k}'}\tilde{f}_{\mathbf{k}}. \tag{A.25}$$

This result is used several times in the main text when we consider correlation functions of the form

$$g(\mathbf{r},\mathbf{r}') = \langle \mathcal{A}(\mathbf{r})\mathcal{B}(\mathbf{r}')\rangle, \tag{A.26}$$

where $\mathcal{A}$ and $\mathcal{B}$ are some operators. For a translation-invariant system we know that $g(\mathbf{r},\mathbf{r}') = g(\mathbf{r} - \mathbf{r}')$, and by using the result in Eq. (A.25) for $\mathbf{k} = -\mathbf{k}'$ we get that

$$g(\mathbf{k}) = \frac{1}{\mathcal{V}}\langle \mathcal{A}(\mathbf{k})\mathcal{B}(-\mathbf{k})\rangle. \tag{A.27}$$

EXERCISES

Exercises for Chapter 1

Exercise 1.1

Single-particle operators in second quantization. Prove Eq. (1.61) for fermions: $T_{\text{tot}} = \sum_{\nu_i,\nu_j} T_{\nu_i\nu_j} c^\dagger_{\nu_i} c_{\nu_j}$. Hints: write Eq. (1.58) with fermion operators $c^\dagger_\nu$. Argue why in this case one has $c^\dagger_{\nu_b} = c^\dagger_{\nu_b} c_{\nu_{n_j}} c^\dagger_{\nu_{n_j}}$. Obtain the fermion analogue of Eq. (1.60) by moving the pair $c^\dagger_{\nu_b} c_{\nu_{n_j}}$ to the left. What about the fermion anti-commutator sign?

Exercise 1.2

Find the current density operator **J** in terms of the arbitrary single particle basis states ψ_ν and the corresponding creation and annihilation operators $a^\dagger_\nu$ and a_ν. Hint: use the basis transformations Eq. (1.65) in the real space representation Eq. (1.98a).

Exercise 1.3

We often refer to a non-interacting problem or a single-particle problem as solvable. This is equivalent to a Hamiltonian, which is equivalent in creation and annihilation operators. This exercise deals with Hamiltonian

$$H = \sum_{ij} a^\dagger_i h_{ij} a_j \tag{E.1.1}$$

Now since **h** is a hermitian matrix there exists a transformation, $\alpha_i = \sum_j U_{ij} a_j$, with **U** being a unitary matrix, such that **h** is diagonal. Find an equation that U must fulfil, such H becomes

$$H = \sum_i \alpha^\dagger_i \epsilon_i \alpha_i, \tag{E.1.2}$$

where $\{\epsilon_i\}$ are the eigenvalues of **h**.

Exercise 1.4

The tight-binding Hamiltonian in second quantization in 1D and 2D
In some crystals the valence electrons can be tightly bound to their host ions. A good starting point for analyzing such systems is to describe the kinetic energy by hopping processes, where with the probability amplitude t one valence electron can hop from an ion j to one of the nearest neighbor ions $j + \delta$ (as usual $\{c^\dagger_j, c_{j'}\} = \delta_{jj'}$):

$$H = -t \sum_{j\delta} c^\dagger_{j+\delta} c_j, \tag{E.1.3}$$

This Hamiltonian is known as the tight-binding Hamiltonian.
    (a) Consider a 1D lattice with N sites, periodic boundary conditions, and a lattice constant a. Here $j = 1, 2, \ldots, N$ and $\delta = \pm 1$. Use the discrete Fourier transformation

$c_j = (1/\sqrt{N}) \sum_k e^{ikja} c_k$ to diagonalize H in k-space and plot the eigenvalues ε_k as a function of k.

(b) In high-temperature superconductors the conduction electrons are confined to parallel CuO-planes, where the ions form a 2D square lattice. In this case the 2D tight-binding model is applicable. Generalize the 1D model to a 2D square lattice also with the lattice constant a and plot contours of constant energy $\varepsilon_{k_x k_y}$ in the $k_x k_y$ plane.

Exercise 1.5

The displaced harmonic oscillator
Consider a bosonic particle moving in 1D with the Hamiltonian

$$H = \hbar\omega \left(a^\dagger a + \frac{1}{2} \right) + \hbar\omega_0 \left(a^\dagger + a \right), \tag{E.1.4}$$

where $[a, a^\dagger] = 1$, while ω and ω_0 are positive constants. Diagonalize H by introducing the operator $\alpha \equiv a + \omega_0/\omega$ and its Hermitian conjugate $\alpha^\dagger$, and determine the eigenenergies. What might be the physical origin of the second term in H (see Section 1.4.1)? Compare the result to a classical and a first quantized treatment of the problem.

Exercise 1.6

The Yukawa potential is defined as $V^{k_s}(\mathbf{r}) = \frac{e_0^2}{r} e^{-k_s r}$, with k_s being some real positive constant with the dimensions of a wavevector. Prove that the Fourier transform is $V_{\mathbf{q}}^{k_s} = \frac{4\pi e_0^2}{q^2 + k_s^2}$. Relate the result to the Coulomb potential. Hints: work in polar coordinates $\mathbf{r} = (r, \theta, \phi)$, and perform the $\int_0^{2\pi} d\phi$ and $\int_{-1}^{+1} d(\cos\theta)$ integrals first. The remaining $\int_0^\infty r^2\, dr$ integral is a simple integral of the sum of two exponential functions.

Exercise 1.7

The general Bogoliubov transformation for fermions
We study the Bogoliubov transformation in second quantization by considering a bilinear Hamiltonian of the general form

$$H = E_0(a^\dagger a + b^\dagger b) + E_1(a^\dagger b^\dagger + ba), \tag{E.1.5}$$

where a, b ($a^\dagger, b^\dagger$) denote fermionic annihilation (creation) operators. We introduce a new set of Fermi operators denoted $\hat{\alpha}$ and $\hat{\beta}$, and seek a linear transformation of the form

$$a = u\hat{\alpha} - v\hat{\beta}^\dagger, \quad b = u\hat{\beta} + v\hat{\alpha}^\dagger \tag{E.1.6}$$

that diagonalizes the Hamiltonian (here u and v are both real). Show that by inserting these expressions in the anticommutation relations for $a, a^\dagger, b$, and $b^\dagger$ one obtains

$$u^2 + v^2 = 1, \tag{E.1.7}$$

and further that by inserting them in the Hamiltonian and demanding

$$E_1(u^2 - v^2) - 2uvE_0 = 0, \tag{E.1.8}$$

the unwanted cross terms $\hat{\alpha}\hat{\beta}$ and $\hat{\alpha}^\dagger\hat{\beta}^\dagger$ vanish. Finally, show by parametrizing $u = \cos t$ and $v = \sin t$ that the eigenenergy λ is given by

$$\lambda = \sqrt{E_0^2 + E_1^2}, \tag{E.1.9}$$

so that the diagonalized Hamiltonian takes the form

$$H = \lambda(\hat{\alpha}^\dagger\hat{\alpha} + \hat{\beta}^\dagger\hat{\beta}) + \text{const.} \tag{E.1.10}$$

The Bogoliubov transformation for bosons is treated in Exercise 1.8.

Exercise 1.8

The general Bogoliubov transformation for bosons
We study the Bogoliubov transformation in second quantization by considering a bilinear Hamiltonian of the general form

$$H = E_0(a^\dagger a + b^\dagger b) + E_1(a^\dagger b^\dagger + ba), \tag{E.1.11}$$

where a, b $(a^\dagger, b^\dagger)$ denote bosonic annihilation (creation) operators. We introduce a new set of boson operators denoted $\hat{\alpha}$ and $\hat{\beta}$, and seek a linear transformation of the form

$$a = u\hat{\alpha} - v\hat{\beta}^\dagger, \quad b = u\hat{\beta} - v\hat{\alpha}^\dagger \tag{E.1.12}$$

that diagonalizes the Hamiltonian. Show that by inserting these expressions in the anticommutation relations for $a, a^\dagger, b$, and $b^\dagger$ one obtains

$$u^2 - v^2 = 1, \tag{E.1.13}$$

and further that by inserting them in the Hamiltonian and demanding

$$E_1(u^2 + v^2) - 2uvE_0 = 0, \tag{E.1.14}$$

the unwanted cross terms $\hat{\alpha}\hat{\beta}$ and $\hat{\alpha}^\dagger\hat{\beta}^\dagger$ vanish. Finally, show by parametrizing $u = \cosh t$ and $v = \sinh t$ that the eigenenergy λ is given by

$$\lambda = \sqrt{E_0^2 - E_1^2}, \tag{E.1.15}$$

so that the diagonalized Hamiltonian takes the form

$$H = \lambda(\hat{\alpha}^\dagger\hat{\alpha} + \hat{\beta}^\dagger\hat{\beta}) + \text{const.} \tag{E.1.16}$$

The Bogoliubov transformation for fermions is treated in Exercise 1.7

Exercises for Chapter 2

Exercise 2.1

Iron (Fe) in its metallic state has valence II, and X-ray measurements have revealed that it forms a body-centered-cubic (BCC) crystal with side length $a = 0.287$ nm. Calculate the density n of the resulting gas of valence electrons, and use this value to determine the microscopic parameters k_F, ε_F, v_F, and λ_F.

Exercise 2.2

Use the variational principle to argue that although the expression Eq. (2.43) is not exact near the energy minimum density $r_s = r_s^* = 4.83$, the result $E^*/N = -1.29$ eV nevertheless ensures the stability of the electron gas.

Exercise 2.3

Starting from Eqs. (2.34) and (2.45) derive the expression Eq. (2.47) for the contributions from the direct Coulomb interaction processes to the interaction energy in second order perturbation theory.

Exercise 2.4

In Section 2.3.2 we saw an example of the existence of 2D electron gases in GaAs–Ga$_{1-x}$Al$_x$As heterostructures. Derive, in analogy with the 3D case, the relation between the 2D Fermi wave vector k_F and the 2D electron density: $k_F{}^2 = 2\pi n$. Use the result to derive the 2D density of states per area, $d(\varepsilon)$.

Exercise 2.5

In Section 2.3.3 we saw an example of the existence of 1D electron gases in carbon nanotubes. Derive, in analogy with the 3D case, the relation between the 1D Fermi wave vector k_F and the 1D density of states per length, $d(\varepsilon)$. Use the result to derive the 1D electron density: $k_F = n\pi/2$.

Exercises for Chapter 3

Exercise 3.1

Electron-phonon scattering I
We want to study the influence of electron-phonon scattering on a given electron state $|k\sigma\rangle$ using the simple Hamiltonian H_{el-ph}^{INA} of Eq. (3.41). For simplicity we restrict our study to processes that scatter electrons out of $|k\sigma\rangle$.

(a) Argue that in this case we need only consider the simple phonon absorption and emission processes given by

$$H_{el-ph} = H_{el-ph}^{abs} + H_{el-ph}^{emi} = \sum_q g_q\, c_{k+q,\sigma}^\dagger c_{k\sigma} b_q + \sum_q g_q\, c_{k+q,\sigma}^\dagger c_{k\sigma} b_{-q}^\dagger. \quad (E.3.1)$$

(b) The scattering rate corresponding to the emission processes is denoted τ_k^{emi}. It can be estimated using Fermi's Golden Rule (suppressing the unimportant spin index):

$$\frac{1}{\tau_k^{emi}} = \frac{2\pi}{\hbar} \sum_f \left|\langle f|H_{el-ph}^{emi}|i\rangle\right|^2 \delta(E_f - E_i), \quad (E.3.2)$$

involving a sum over all possible final states with energy $E_f = E_i$, and an initial state $|i\rangle$ having the energy E_i and being specified by the occupation numbers $n_{k\sigma}^i$ and N_q^i for electron states $|k\sigma\rangle$ and phonon states $|q\rangle$ (see Eq. (1.108)). Assume that $|i\rangle$ is a simple but unspecified product state, i.e. $|i\rangle = \left(\prod_{\{k\sigma\}_i} c_{k\sigma}^\dagger\right)\left(\prod_{\{q\}_i} \frac{1}{\sqrt{N_q^i!}} [b_q^\dagger]^{N_q^i}\right)|0\rangle,$

and show that for a given $\mathbf{q} \neq 0$ in $H_{\text{el}-\text{ph}}^{\text{emi}}$ the only possible normalized final states is

$$\frac{1}{\sqrt{N_{\mathbf{q}}+1}}\, c_{\mathbf{k}+\mathbf{q}}^{\dagger} c_{\mathbf{k}} b_{\mathbf{q}}^{\dagger} |i\rangle.$$

(c) Show for the state $|i\rangle$ that

$$\frac{1}{\tau_{\mathbf{k}}^{\text{emi}}} = \frac{2\pi}{\hbar} \sum_{\mathbf{q}} |g_{\mathbf{q}}|^2\, (N_{\mathbf{q}}^i + 1)\, (1 - n_{\mathbf{k}+\mathbf{q}}^i)\, n_{\mathbf{k}}^i\, \delta(\varepsilon_{\mathbf{k}+\mathbf{q}} - \varepsilon_{\mathbf{k}} + \hbar\omega_{\mathbf{q}}). \qquad \text{(E.3.3)}$$

Derive the analogous expression for the scattering rate $1/\tau_{\mathbf{k}}^{\text{abs}}$ due to absorption.

(d) Keeping $n_{\mathbf{k}} = 1$ fixed for our chosen state, argue why thermal averaging leads to

$$\frac{1}{\tau_{\mathbf{k}}^{\text{emi}}} = \frac{2\pi}{\hbar} \sum_{\mathbf{q}} |g_{\mathbf{q}}|^2\, [n_{\text{B}}(\omega_{\mathbf{q}}) + 1]\, [1 - n_{\text{F}}(\varepsilon_{\mathbf{k}+\mathbf{q}})]\, \delta(\varepsilon_{\mathbf{k}+\mathbf{q}} - \varepsilon_{\mathbf{k}} + \hbar\omega_{\mathbf{q}}). \qquad \text{(E.3.4)}$$

Exercise 3.2

Electron-phonon scattering II

We now determine the temperature dependence of the scattering rate $\tau_{\mathbf{k}}^{\text{emi}}$ in the high and low temperature limits. This immediately gives us the behavior of the total scattering rate $1/\tau_{\mathbf{k}} = 1/\tau_{\mathbf{k}}^{\text{emi}} + 1/\tau_{\mathbf{k}}^{\text{abs}}$, since at low T, due to the lack of phonons, $1/\tau_{\mathbf{k}}^{\text{abs}} \approx 0$, while at high T we have $1/\tau_{\mathbf{k}}^{\text{emi}} \approx 1/\tau_{\mathbf{k}}^{\text{abs}}$.

(a) To obtain realistic results we need to use the screened Coulomb or Yukawa potential for the ionic potential $V_{\mathbf{q}}$ (see Eq. (3.42) and Exercise 1.6). The electrons redistribute in an attempt to neutralize the ionic potential. As we shall see in Chapter 14 they succeed to do so for distances further away than $1/k_s$ from the ion. Show by dimensional analysis involving the Fourier component $e^2/(\epsilon_0 k_s^2)$, the Fermi energy ε_{F}, and the electron density n that $k_s^2 \approx k_{\text{F}}/a_0$.

(b) Show how Eq. (3.42) together with $\mathbf{k}' = \mathbf{k} + \mathbf{q}$ change $\frac{1}{\tau_{\mathbf{k}}^{\text{emi}}}$ from Exercise 3.1d to

$$\frac{1}{\tau_{\mathbf{k}}^{\text{emi}}} \propto \frac{2\pi}{\hbar} \sum_{\mathbf{k}'} \omega_{\mathbf{q}}\, [n_{\text{B}}(\omega_{\mathbf{q}}) + 1]\, [1 - n_{\text{F}}(\varepsilon_{\mathbf{k}'})]\delta(\varepsilon_{\mathbf{k}'} - \varepsilon_{\mathbf{k}} + \hbar\omega_{\mathbf{q}}), \qquad \text{(E.3.5)}$$

where we here and in the following do not care about the numerical prefactors.

(c) As usual, we are mainly interested in electrons moving relatively close to the Fermi surface (why?), i.e. $k', k \approx k_{\text{F}}$. Furthermore, we employ the Debye model of the phonon spectrum (see Section 3.5): $\omega_{\mathbf{q}} = v_{\text{D}} q$. We note that since $\mathbf{k}'$ and $\mathbf{k}$ are tied to the Fermi surface the largest q is $2k_{\text{F}}$, and the corresponding largest phonon energy is denoted $\hbar\omega_{\text{max}} \equiv 2v_{\text{D}} k_{\text{F}}$. Now use polar coordinates to obtain $\sum_{\mathbf{k}'} \propto \int d\varepsilon_{\mathbf{k}'} \int_{-1}^{1} d(\cos\theta)$, and show using $q^2 = |\mathbf{k}' - \mathbf{k}|^2$ that $d(\cos\theta) \propto q\, dq$. With this prove that

$$\int_{-1}^{1} d(\cos\theta)\delta(\varepsilon_{\mathbf{k}'} - \varepsilon_{\mathbf{k}} + \hbar\omega_{\mathbf{q}}) \propto \int_{0}^{2k_{\text{F}}} q\, dq\, \delta(\varepsilon_{\mathbf{k}'} - \varepsilon_{\mathbf{k}} + \hbar\omega_{\mathbf{q}})$$

$$\propto \begin{cases} \omega_{\mathbf{q}}, & \varepsilon_{\mathbf{k}} - \varepsilon_{\mathbf{k}'} < \hbar\omega_{\text{max}} \\ 0, & \varepsilon_{\mathbf{k}} - \varepsilon_{\mathbf{k}'} > \hbar\omega_{\text{max}} \end{cases} \qquad \text{(E.3.6)}$$

Since $d\varepsilon_{\mathbf{k'}} = \hbar\, d\omega_{\mathbf{q}}$ show in the limit $\hbar\omega_{\max} \ll \varepsilon_{\mathbf{k}} - \varepsilon_F \ll \varepsilon_F$ how to obtain

$$\frac{1}{\tau_{\mathbf{k}}^{\text{emi}}} \propto \int_0^{\omega_{\max}} d\omega_{\mathbf{q}}\, \omega_{\mathbf{q}}^2 \left[n_B(\omega_{\mathbf{q}}) + 1 \right] \left[1 - n_F(\varepsilon_{\mathbf{k'}}) \right] \approx \int_0^{\omega_{\max}} d\omega_{\mathbf{q}}\, \omega_{\mathbf{q}}^2 \left[n_B(\omega_{\mathbf{q}}) + 1 \right]. \quad \text{(E.3.7)}$$

(d) Show that the result in (c) leads to the following temperature dependencies:

$$\frac{1}{\tau_{\mathbf{k}}^{\text{emi}}} \propto \begin{cases} T, & \text{for } T \gg \hbar\omega_{\max}/k_B \\ T^3 + \text{const.}, & \text{for } T \ll \hbar\omega_{\max}/k_B. \end{cases} \quad \text{(E.3.8)}$$

Exercise 3.3

Phonons in the bi-ionic 1D chain
In analogy with the homogeneous 1D chain of Section 3.3 we now want to find the eigenmodes of the linear 1D chain with lattice constant a mentioned in Fig. 3.3(c). The ionic lattice has a unit cell with two different ions • and ∘, respectively. All spring constants are the same, namely K. The masses, the momenta, and the displacements of the • ions are denoted m, p_j and u_j, while for the ∘ ions they are denoted M, P_j and U_j. The sites are numbered by j as $\ldots, u_{j-1}, U_{j-1}, u_j, U_j, u_{j+1}, U_{j+1}, \ldots$.
(a) Verify that the Hamiltonian of the two-atoms-per-unit-cell chain is

$$H = \sum_j \left[\frac{1}{2m}p_j^2 + \frac{1}{2M}P_j^2 + \frac{1}{2}K(u_j - U_{j-1})^2 + \frac{1}{2}K(U_j - u_j)^2 \right] \quad \text{(E.3.9)}$$

(b) Use Hamilton's equations $\dot{u}_j = \frac{\partial H}{\partial p_j}$ and $\dot{p}_j = -\frac{\partial H}{\partial u_j}$ (similar for $\dot{U}_j$ and $\dot{P}_j$), to obtain the equations for $\ddot{u}_j$ and $\ddot{U}_j$.
(c) Assume the harmonic solutions $u_j \equiv u_k e^{i(kja - \omega t)}$ and $U_j \equiv U_k e^{i(kja - \omega t)}$ to derive a 2×2 matrix eigenvalue equation for (u_k, U_k). Verify the dispersion curve $\omega_{\mathbf{k}}$ displayed in Fig. 3.3(c) and the eigenmode displayed in Fig. 3.4.
(d) Check that in the limit $M = m$ the dispersion ω_k in Eq. (3.9) of the one-atom-per-unit-cell is recovered.

Exercise 3.4

The Bohm-Staver expression
The task is to prove the Bohm-Staver expression Eq. (3.5). We study the situation described in Section 3.2, where the light and mobile electrons always follow the motion of the slow and heavy ions to maintain local charge neutrality. The ions are treated as the jellium of Section 3.1.
(a) Multiply the continuity equation by the ion mass M to obtain

$$M\partial_t \rho_{\text{ion}} + \nabla \cdot \boldsymbol{\pi} = 0, \quad \text{(E.3.10)}$$

where $\boldsymbol{\pi}$ is the momentum density.
(b) Take the time derivative and note that $\dot{\boldsymbol{\pi}}$ is the force density $\mathbf{f}$, which on the other hand is equal to the pressure gradient $-\nabla P$ due to the compression of the electron gas following the ionic motion:

$$\dot{\pi} = \mathbf{f} = -\nabla P = \nabla \left(\frac{\partial E^{(0)}}{\partial \mathcal{V}} \Big|_N \right), \tag{E.3.11}$$

where the electron gas ground state energy $E^{(0)}$ is given in Eq. (2.28).

(c) Combine the equations and derive the wave equation for ρ_{ion}, from which the (square of the) sound velocity v_s is read off:

$$M \partial_t^2 \rho_{\text{ion}} - \frac{2Z}{3} \varepsilon_{\text{F}} \nabla^2 \rho_{\text{ion}} = 0. \tag{E.3.12}$$

Exercise 3.5

Electrons and phonons in the jellium model

In this exercise we quantize the jellium model of the ion system in a solid and derive the electron phonon interaction in a way which is somewhat different from the method used in the main text.

We take the case of a monovalent metal, i.e. each ion has charge $+e$. Because the system is charge neutral as a whole, we need only take into account interactions between deviations from equilibrium. We define

$$\rho_{\text{ion}}(\mathbf{r}) = \rho_{\text{ion}}^0 + \delta\rho_{\text{ion}}(\mathbf{r}) \equiv \rho_{\text{ion}}^0 + \rho_{\text{ion}}^0 \nabla \cdot \mathbf{u}(\mathbf{r}), \tag{E.3.13}$$
$$\rho(\mathbf{r}) = \rho_{\text{ion}}^0 + \delta\rho(\mathbf{r}), \tag{E.3.14}$$

where $\mathbf{u}$ is a displacement field describing the deviation of the ion density from equilibrium, and $\rho(\mathbf{r})$ is the electron density. The potential energy contributions which involves the ionic system are the ion-ion interaction and the electron-ion interaction

$$E_{\text{pot}}^{\text{ion-ion}} = \frac{1}{2} \int d\mathbf{r} d\mathbf{r}' \, \delta\rho_{\text{ion}}(\mathbf{r}) V(\mathbf{r} - \mathbf{r}') \delta\rho_{\text{ion}}(\mathbf{r}'), \tag{E.3.15}$$

$$E_{\text{pot}}^{\text{ion-el}} = -\int d\mathbf{r} d\mathbf{r}' \, \delta\rho(\mathbf{r}) V(\mathbf{r} - \mathbf{r}') \delta\rho_{\text{ion}}(\mathbf{r}'), \tag{E.3.16}$$

with the usual definition

$$E_{\text{pot}}^{\text{el-el}} = \frac{1}{2} \int d\mathbf{r} d\mathbf{r}' \, \delta\rho(\mathbf{r}) V(\mathbf{r} - \mathbf{r}') \delta\rho(\mathbf{r}'). \tag{E.3.17}$$

We have not explicitly included the electron-electron interaction here. When included it gives rise to the term in Eq. (2.34).

(a) First we quantize the ion system and we start by looking at the isolated ion system. The classical Lagrangian for this system is

$$L_{\text{ion}}^0 = T_{\text{ion}} - V_{\text{ion}} \tag{E.3.18}$$
$$T_{\text{ion}} = \frac{1}{2} \int d\mathbf{r} \, M \rho_{\text{ion}}(\mathbf{r}) v^2(\mathbf{r}) \tag{E.3.19}$$
$$V_{\text{ion}} = E_{\text{pot}}^{\text{ion-ion}}, \tag{E.3.20}$$

where $\mathbf{v}$ is the velocity field of the ions. Because we are interested in the low energy excitations we linearize the kinetic such that

$$T = \frac{1}{2} \int d\mathbf{r} \, M \rho_{\text{ion}}^0 v^2(\mathbf{r}).$$ (E.3.21)

Using particle conservation show that

$$\mathbf{v}(\mathbf{r}) = -\dot{\mathbf{u}}(\mathbf{r}).$$ (E.3.22)

(b) Using Eq. (E.3.22), derive the Lagrangian as a functional of $\mathbf{u}$

$$L_{\text{ion}}^0[\mathbf{u}] = \frac{1}{2} \int d\mathbf{r} \, M \rho_{\text{ion}}^0 |\dot{\mathbf{u}}(\mathbf{r})|^2 - \frac{(\rho_{\text{ion}}^0)^2}{2} \int d\mathbf{r} d\mathbf{r}' \, [\nabla \cdot \mathbf{u}(\mathbf{r})] \, V(\mathbf{r} - \mathbf{r}') \, [\nabla \cdot \mathbf{u}(\mathbf{r}')] .$$ (E.3.23)

(c) Because the Lagrangian is quadratic in $\mathbf{u}$ it is equivalent to a set of harmonic oscillators that describe sound waves of the ion system. What dispersion relation do you expect?

(d) Next, you should quantize the ion system and find the Hamiltonian. First show that the canonical momentum corresponding to the field $\mathbf{u}$ is

$$\mathbf{p}(\mathbf{r}) = M \rho_{\text{ion}}^0 \dot{\mathbf{u}}(\mathbf{r}),$$ (E.3.24)

and show, using the well known relation $H = \int \mathbf{p} \cdot \dot{\mathbf{u}} - L$, that

$$H_{\text{ion}}^0 = \frac{1}{2} \int d\mathbf{r} \, \frac{p^2(\mathbf{r})}{M \rho_{\text{ion}}^0} + \frac{(\rho_{\text{ion}}^0)^2}{2} \int d\mathbf{r} d\mathbf{r}' \, [\nabla \cdot \mathbf{u}(\mathbf{r})] \, V(\mathbf{r} - \mathbf{r}') \, [\nabla \cdot \mathbf{u}(\mathbf{r}')]$$

$$= \frac{1}{2V} \sum_{\mathbf{q}} \frac{\mathbf{p}(\mathbf{q}) \cdot \mathbf{p}(-\mathbf{q})}{M \rho_{\text{ion}}^0} + \frac{(\rho_{\text{ion}}^0)^2}{2V} \sum_{\mathbf{q}} V(\mathbf{q}) \, [\mathbf{q} \cdot \mathbf{u}(\mathbf{q})] \, [\mathbf{q} \cdot \mathbf{u}(-\mathbf{q})] .$$ (E.3.25)

(e) The system is quantized by the condition (canonical quantization)

$$[u_i(\mathbf{r}), p_j(\mathbf{r}')] = i \delta_{ij} \delta(\mathbf{r} - \mathbf{r}').$$ (E.3.26)

Show that in $\mathbf{k}$-space this becomes

$$[u_i(\mathbf{q}), p_j(-\mathbf{q}')] = i \delta_{ij} V \delta_{\mathbf{q}\mathbf{q}'} .$$ (E.3.27)

(f) Now follow the standard scheme for diagonalization of the phonons modes. Define

$$\mathbf{u}_{\mathbf{q}} = \boldsymbol{\epsilon}_{\mathbf{q}} \sqrt{\frac{V}{2M \rho_{\text{ion}}^0 \Omega}} \left(b_{\mathbf{q}} + b_{-\mathbf{q}}^\dagger \right), \quad \mathbf{p}_{-\mathbf{q}} = i \boldsymbol{\epsilon}_{\mathbf{q}} \sqrt{\frac{V M \rho_{\text{ion}}^0 \Omega}{2}} \left(b_{\mathbf{q}}^\dagger - b_{-\mathbf{q}} \right),$$ (E.3.28)

where the polarization vector has the property that $\boldsymbol{\epsilon}_{\mathbf{q}} = \boldsymbol{\epsilon}_{-\mathbf{q}}$, i.e. one must choose a positive polarization direction. The polarization is here parallel to $\mathbf{q}$ i.e. only longitudinal modes exist in the jellium model[71]. Furthermore, Ω is chosen such that H is diagonal. Verify that

[71] In the general non-jellium case the polarization vectors are eigenvectors of the dynamical matrix $\mathbf{D}(\mathbf{q})$ in Eq. (3.14). In the jellium case, one thus has $D_{\alpha\beta}(\mathbf{q}) = q_\alpha q_\beta \, V(\mathbf{q})$ which only has one eigenvector with non-zero eigenvalue, namely $\boldsymbol{\epsilon}_{\mathbf{q}}$, which you can easily check.

$$H = \sum_q \Omega \left(b_q^\dagger b_q + \frac{1}{2} \right); \qquad \Omega = \sqrt{\frac{4\pi e_0^2 \rho_{\text{ion}}^0}{M}}. \tag{E.3.29}$$

(g) Explain the physics of this result.

Exercise 3.6

Bare electron-phonon interaction in the jellium model

Using the quantization from the previous exercise verify that the Hamiltonian describing the electron-phonon interaction is given by

$$
\begin{aligned}
H_{\text{el-ph}} &= \int d\mathbf{r} d\mathbf{r}' \, \delta\rho(\mathbf{r}) V(\mathbf{r}-\mathbf{r}') \delta\rho_{\text{ion}}(\mathbf{r}') \\
&= \frac{\rho_{\text{ion}}^0}{\mathcal{V}} \sum_q \rho(-\mathbf{q}) V(\mathbf{q})(i\mathbf{q} \cdot \mathbf{u}(\mathbf{q})) \\
&= \frac{1}{\sqrt{\mathcal{V}}} \sum_q g(\mathbf{q}) \rho(-\mathbf{q}) A_\mathbf{q},
\end{aligned} \tag{E.3.30}
$$

where

$$g(\mathbf{q}) = iqV(\mathbf{q})\sqrt{\frac{\rho_{\text{ion}}^0}{2M\Omega_0}}, \tag{E.3.31}$$

and

$$A_\mathbf{q} = b_\mathbf{q} + b_{-\mathbf{q}}^\dagger. \tag{E.3.32}$$

Exercises for Chapter 4

Exercise 4.1

The Hartree–Fock approximation

Consider the Hartree–Fock solution of the homogeneous electron gas in a positive background. After the mean-field approximation the Hamiltonian can be written as

$$H^{\text{HF}} = \sum_k \varepsilon_k^{\text{HF}} c_{k\sigma}^\dagger c_{k\sigma} + \text{constant}. \tag{E.4.1}$$

(a) Argue why it is that in this case the Hartree–Fock energy follows from Eq. (4.24b) and is given by

$$\varepsilon_k^{\text{HF}} = \varepsilon_k + V_{\text{HF}}(k), \quad V_{\text{HF}}(k) = -\sum_{k'} V(k-k') n_{k'\sigma}, \quad V(q) = \frac{4\pi e_0^2}{q^2} \tag{E.4.2}$$

The occupation numbers should of course be solved self-consistently. What is the self-consistency condition?

(b) Consider the zero temperature limit, and assume that $n_{k'\sigma} = \theta(k_F - k')$, which then gives

$$V_{\text{HF}}(k) = -\frac{e_0^2 k_F}{\pi} \left(1 + \frac{k_F^2 - k^2}{2k_F k} \ln \left| \frac{k+k_F}{k-k_F} \right| \right). \tag{E.4.3}$$

$V_{\text{HF}}(k)$ is increasing monotonously with k (which you might check, e.g. graphically). Use this to argue that the guess $n_{k'\sigma} = \theta(k_F - k')$ is in fact the correct solution.

(c) Now find the energy of the electron gas in the Hartree–Fock approximation. Is it given by

$$E_{\mathrm{HF}} \stackrel{?}{=} \sum_{\mathbf{k}} \varepsilon_{\mathbf{k}}^{\mathrm{HF}} \, n_{\mathrm{F}}\left(\varepsilon_{\mathbf{k}}^{\mathrm{HF}}\right),\qquad\qquad (\text{E.4.4})$$

and why not? Hint: show that the correct energy reduces to $E^{(1)}$ given in Eq. (2.39).

(d) The conclusion is so far that Hartree–Fock and first order perturbation theory are in this case identical. Is that true as well for the mean field solution of the Stoner model?

Exercise 4.2

Unphysical aspects of the Hartree–Fock approximation

The Hartree–Fock energies derived in the previous exercise have however some unphysical features. Using the result from Exercise 4.1, show that the velocity $d\varepsilon_{\mathbf{k}}^{\mathrm{HF}}/d\mathbf{k}$ diverges at the Fermi level.

This conclusion contradicts both experiments and the Fermi liquid theory discussed in Chapter 15. It also warns us that the single-particle energies derived from a mean-field Hamiltonian are not necessary a good approximation of the excitation energies of the system, even if the mean-field approach gives a good estimate of the groundstate energy.

Exercise 4.3

Landau theory of symmetry breaking phase transitions

In 1937 Landau developed a general phenomenological theory of symmetry breaking phase transitions. The basic idea is to expand the free energy in powers of the order parameter. Consider a transition to a state with a finite order parameter, η. For second order phase transitions only even terms are present in the free energy expansion

$$F(T,\eta) = F_0\left(T\right) + A\left(T\right)\eta^2 + C\left(T\right)\eta^4. \qquad\qquad (\text{E.4.5})$$

(a) At the transition point η vanishes. Use this to argue that A also vanishes at the transition point, $T = T_C$, and that $A < 0$ for $T < T_C$, while $A > 0$ for $T > T_C$. Then write A and C as

$$A\left(T\right) = \left(T - T_C\right)\alpha, \quad C\left(T\right) = C. \qquad\qquad (\text{E.4.6})$$

(b) Use the principle of minimal free energy to show that

$$\eta = \sqrt{\frac{-A}{2C}} = \sqrt{\frac{\left(T_C - T\right)}{T_C}}\eta\left(0\right), \quad \eta\left(0\right) = \sqrt{\frac{T_C\alpha}{2C}}. \qquad\qquad (\text{E.4.7})$$

(c) Finally, make a sketch of the specific heat of the system and show that it is discontinuous at the transition point. Hint: recall that

$$C_V = -T\frac{\partial^2 F}{\partial T^2}. \qquad\qquad (\text{E.4.8})$$

Exercises for Chapter 5

Exercise 5.1

We return to the bosonic particle described by the Hamiltonian of Exercise 1.5. Write down the Heisenberg equations of motion for $a^\dagger$ and a. Solve these equations by introducing the operator $\alpha^\dagger \equiv a^\dagger + \omega_0/\omega$. Express H in terms of $\alpha^\dagger(t)$ and $\alpha(t)$. Interpret the change of the zero point energy.

Exercise 5.2

Show that the third-order term $\hat{U}_3(t, t_0)$ of $\hat{U}(t, t_0)$ in Eq. (5.18) indeed has the form

$$\hat{U}_3(t, t_0) = \frac{1}{3!}\left(\frac{1}{i}\right)^3 \int_{t_0}^t dt_1 \int_{t_0}^t dt_2 \int_{t_0}^t dt_3\, T_t\!\left(\hat{V}(t_1)\hat{V}(t_2)\hat{V}(t_3)\right). \tag{E.5.1}$$

Hint: study Eqs. (5.16) and (5.17) and the associated footnote.

Exercise 5.3

Use the Heisenberg picture to show that for the diagonal Hamiltonian H of Eq. (5.22) we have

$$H = \sum_{\nu'} \varepsilon_{\nu'} a_{\nu'}^\dagger a_{\nu'} \quad \Rightarrow \quad H(t) = \sum_{\nu'} \varepsilon_{\nu'} a_{\nu'}^\dagger(t) a_{\nu'}(t). \tag{E.5.2}$$

Exercise 5.4

Due to the equation of motion for operators Eq. (5.6) we will often need to calculate commutators of the form $[AB, C]$, for some operators A, B, and C. Show the very important relations

$$[AB, C] = A[B, C] + [A, C]B, \qquad \text{useful for boson operators,} \tag{E.5.3}$$
$$[AB, C] = A\{B, C\} - \{A, C\}B, \qquad \text{useful for fermion operators.} \tag{E.5.4}$$

Exercise 5.5

In the jellium model of metals the kinetic energy of the electrons is described by the Hamiltonian H_{jel} of Eq. (2.19), while the interaction energy is given by V_{el-el}' of Eq. (2.34). In the Heisenberg picture the time evolution of the electron creation and annihilation operators $c_{k\sigma}^\dagger$ and $c_{k\sigma}$ is governed by the total Hamiltonian $H = H_{jel} + V_{el-el}'$. In analogy with Eq. (5.31) derive the equation of motion for $c_{k\sigma}(t)$. Apply the Hartree–Fock approximation to the result.

Exercise 5.6

Assume bosonic operators a_ν in the Hamiltonian H of Eq. (5.30) and derive $\dot{a}_\nu(t)$ in analogy with Eq. (5.31).

Exercise 5.7

Cauchy's principle part and integrals of simple singularities
We study integrals of the form $\int_{-\infty}^\infty dx \, \frac{1}{x+i\eta} f(x)$, where $f(x)$ is any function with a well behaved Taylor expansion around $x = 0$, and $\eta = 0^+$ is a positive infinitesimal.

Show that in this context $\frac{1}{x+i\eta}$ can be decomposed as the following real and imaginary parts

$$\frac{1}{x+i\eta} = \mathcal{P}\frac{1}{x} - i\pi\,\delta(x). \tag{E.5.5}$$

Here $\mathcal{P}$ means Cauchy principle part:

$$\mathcal{P}\int_{-\infty}^{\infty} dx\,\frac{1}{x}f(x) \equiv \int_{-\infty}^{-\eta} dx\,\frac{1}{x}f(x) + \int_{\eta}^{\infty} dx\,\frac{1}{x}f(x). \tag{E.5.6}$$

Exercise 5.8

The Fourier transformation of retarded functions. As an example of how to Fourier transform retarded functions that do not decay at large times, we study the following function

$$f^R(t) = -i\theta(t)\exp(-i\varepsilon t). \tag{E.5.7}$$

(a) Show that the Fourier transform $f^R(\omega)$ of $f^R(t)$ as defined in Eq. (5.46) is

$$f^R(\omega) = \frac{1}{\omega - \varepsilon + i\eta}. \tag{E.5.8}$$

Here you see that the positive infinitesimal is crucial because it moves the location of the pole away from the real axis.

(b) Next, perform the inverse Fourier transform Eq. (5.47), and show that you recover $f^R(t)$ after letting $\eta \to 0^+$. Hint: the frequency integration can be done using contour integration in the complex ω plane.

Exercise 5.9

Fermi's Golden Rule for transition rates.
Verify the steps that lead to Eq. (5.34).

(a) Show that to first order in V and for $t_0 \to -\infty$ that Eq. (5.33) follows from Eq. (5.32).

(b) Calculate the probability $P_f(t) = |\langle f|V|i\rangle|^2$ to find the system in state $|f\rangle$ at time t.

(c) Show that the rate at which P_f changes, i.e. $dP_f(t)/dt$, gives (after using that $\eta t \ll 1$) the well-known Fermi's golden rule in Eq. (5.34)

(d) Above, we have used the convention $\hbar = 1$. Is there in fact an $\hbar$ in the final result? And if so, where?

Exercises for Chapter 6

Exercise 6.1

As in Exercise 5.1 we consider a harmonic oscillator influenced by an external force $f(t)$, but now we treat this force as a time-dependent perturbation

$$H' = f(t)\,x. \tag{E.6.1}$$

Express x in terms of a and $a^\dagger$ and calculate the linear response result for the expectation value $\langle x(t)\rangle$. Argue that this result is in fact exact, for example by considering the equation of motion for $\langle x(t)\rangle$.

Exercise 6.2

Spin susceptibility in linear response theory
The spin susceptibility measures the response to a magnetic field. Suppose that a piece of some material is perturbed by external magnetic moments. These moments could for example be in the form of a neutron beam in a neutron scattering experiment. The perturbation is in this case given by

$$H' = -g\mu_B \int d\mathbf{r}\, \mathbf{B}_{\text{ext}}(\mathbf{r}, t) \cdot \mathbf{S}(\mathbf{r}), \tag{E.6.2}$$

where $\mathbf{S}$ is the spin density operator $\mathbf{S}(\mathbf{r}) = \Psi^\dagger(\mathbf{r})\mathbf{s}\Psi(\mathbf{r})$, see Section 1.4.3. Find a formal expression for the response to linear order in $\mathbf{B}$ for the induced spin density in the material, $\langle \mathbf{S}(\mathbf{r}, t)\rangle$. Express your result in both real space and momentum space.

Neutron scattering experiments are the main source for obtaining experimental information about the distribution of spins in condensed matter systems.

Exercise 6.3

The conductivity of isotropic media (I)
In this exercise we consider the conductivity of a translation- and rotational-invariant system. This means that the conductivity $\sigma(\mathbf{r}, \mathbf{r}')$ is a function of $\mathbf{r} - \mathbf{r}'$ only and that the conductivity tensor is diagonal with identical diagonal components. Show that in the Fourier domain

$$\mathbf{J}_e(\mathbf{q}, \omega) = \sigma(\mathbf{q}, \omega)\mathbf{E}(\mathbf{q}, \omega). \tag{E.6.3}$$

Find the relation between the conductivity, i.e. $\sigma(\mathbf{q}, \omega)$ and the correlation function

$$C_{JJ}(t) = \left\langle \left[J^\alpha(\mathbf{q}, t), J^\alpha(-\mathbf{q}, 0)\right]\right\rangle, \tag{E.6.4}$$

where $\mathbf{J}(\mathbf{q})$ is the particle current operator in momentum space.

Exercise 6.4

The conductivity of isotropic media (II)
Consider again the conductivity of a translation-invariant and rotational-invariant system.

(a) First consider the conductivity of a non-interacting electron gas at long wave lengths, $\mathbf{q} \to 0$. Derive the expression for the particle current operator in this limit,

$$\mathbf{J}(0, t) = \frac{1}{m}\sum_{\mathbf{k}\sigma} \mathbf{k}\, c^\dagger_{\mathbf{k}\sigma}(t)\, c_{\mathbf{k}\sigma}(t). \tag{E.6.5}$$

(b) Show that it is time-independent in the Heisenberg picture. From this you can derive obtain the long wavelength conductivity

$$\sigma^{\alpha\beta}(0, \omega) = i\delta_{\alpha\beta}\frac{ne^2}{\omega m}. \tag{E.6.6}$$

(c) How does this fit with the Drude result (13.42) in the clean limit, where the impurity induced scattering time τ tends to infinity (i.e. $\omega\tau \to \infty$)?

(d) How does the conclusions change for an *interacting* translation-invariant system?

Exercise 6.5

Symmetry of the retarded current-current correlation function. In this exercise, we show that the following property for the frequency dependent conductance holds

$$G(\xi, \xi', \omega) = G(\xi', \xi, \omega). \tag{E.6.7}$$

Here the conductance function G is defined as

$$G(\xi, \xi', \omega) = \text{Re}\left[\frac{ie^2}{\omega} C^R_{I(\xi)I(\xi')}(\omega)\right]. \tag{E.6.8}$$

(a) Start by showing the following identity

$$\text{Im}C^R_{AB} = \frac{1}{2} \int_{-\infty}^{\infty} dt \left\langle [\hat{A}(t), \hat{B}(0)] \right\rangle. \tag{E.6.9}$$

Hint: insert a complete set of states (and eigenstates of H_0) between $\hat{A}(t)$ and $\hat{B}(0)$ and the perform explicitly the time integration in the definition of the Fourier transform of the retarded function, and then compare with the result you when the same procedure in Eq. (E.6.9).

(b) Use the result from **(a)** to show Eq. (E.6.7), which is the dc-limit $\omega = 0$ of the property assumed in the main text.

Exercises for Chapter 7

Exercise 7.1

Specific shapes of perfect leads
Consider the rectangular cross section given by $\Omega = [0, w_y] \times [0, w_z]$ in cartesian coordinates $\mathbf{r}_\perp = (y, z)$.

(a) Show that the corresponding transverse wavefunction χ_{nm} given by

$$\chi_{nm}(\mathbf{r}_\perp) \propto \sin\left(n\pi\frac{y}{w_y}\right) \sin\left(m\pi\frac{z}{w_z}\right), \quad (y, z) \in \Omega = [0, w_y] \times [0, w_z], \tag{E.7.1}$$

solves Eqs. (7.2c) and (7.2d), and find an expression for the transverse eigenenergy ε_{nm}.

(b) Next, consider the circular cross section $\Omega = [0, a] \times [0, 2\pi]$ in cylindrical coordinates $\mathbf{r}_\perp = (r, \phi)$. Show that the corresponding transverse wavefunction χ_{ln} given by

$$\chi_{ln}(\mathbf{r}_\perp) \propto J_l\left(\gamma_{ln}\frac{r}{a}\right) e^{il\phi}, \quad (r, \phi) \in \Omega = [0, a] \times [0, 2\pi], \tag{E.7.2}$$

likewise solves Eqs. (7.2c) and (7.2d), and find an expression for the transverse eigenenergy ε_{ln}. Here, J_l is the l'th Bessel function of the first kind and γ_{ln} its n'th root.

Exercise 7.2

Conductance of a delta function barrier

As a model system of a mesoscopic 1D channel take the following Hamiltonian

$$H = -\frac{1}{2m}\frac{\partial^2}{\partial x^2} + V_0\,\delta(x). \tag{E.7.3}$$

Consider a scattering state ψ_k^+ (and likewise for ψ_k^-)

$$\psi_k^+(x) = \begin{cases} \frac{1}{\sqrt{\mathcal{L}}}\left(e^{ikx} + re^{-ikx}\right), & x < 0, \\ \frac{1}{\sqrt{\mathcal{L}}}te^{ikx}, & x > 0, \end{cases} \tag{E.7.4}$$

(a) Show that

$$\psi'(0^+) - \psi'(0^-) = 2mV_0\,\psi(0) \tag{E.7.5}$$

and use it to find r and t.

(b) Suppose now that the two ends of the wire have different chemical potential, so that the distribution function for electrons in the states ψ_k^+ is given by $n_{\rm F}(\varepsilon_k - \mu_L)$, while the distribution function for electrons in states ψ_k^- is $n_{\rm F}(\varepsilon_k - \mu_R)$. Show that the current at low temperature becomes

$$I = 2e\sum_{k>0}\left(v_k^+ n_{\rm F}(\varepsilon_k - \mu_L) - v_k^- n_{\rm F}(\varepsilon_k - \mu_R)\right) = \frac{2e^2}{h}\frac{1}{1+(V_0/v_F)^2}V, \tag{E.7.6}$$

where $eV = \mu_R - \mu_L \ll \varepsilon_{\rm F}$.

Hints: $v_k^+ = \frac{1}{2mi}\left[\left(\psi_k^+\right)^* \partial_x\psi_k^+ - {\rm c.c.}\right] = v_k|t|^2/\mathcal{L} = v_k(1-|r|^2)/\mathcal{L}$ with $t = (1+iz)^{-1}$, $z = mV_0/k$.

Exercise 7.3

The Landauer formula at finite temperatures

We study a general mesoscopic system.

(a) Show that the linear conductance at finite temperature is generalized to

$$G = \frac{2e^2}{h}\sum_n\int_0^\infty dE\left(-\frac{\partial n_{\rm F}(E - \mu)}{\partial E}\right)T_n(E), \tag{E.7.7}$$

where $T_n(E)$ is the transmission probability of a given mode $T_n(E) = \sum_{n'} t_{nn'}^* t_{n'n}$.

(b) Using a model electron waveguide, where the potential in the transverse direction is a parabolic confinement, the transmission coefficient is supposed to be

$$T_n(E) = \Theta(E - E_n) = \Theta\left(E - (n+\frac{1}{2})\omega_T\right), \tag{E.7.8}$$

where ω_T is the frequency of transverse oscillation. Find an expression for the conductance G and plot the G as a function of μ. Plot for example $G(h/2e^2)$ versus μ/ω_T for two different temperatures, $k_{\rm B}T/\omega_T = 0.05$ and 0.15.

(c) How would the result look if the transverse confinement were a square well?

Exercises for Chapter 8

Exercise 8.1

Verify that the self-consistent equations in Eqs. (8.16) and (8.17) both are solution to the Schrödinger equation in Eq. (8.13).

Exercise 8.2

Prove that the propagator in Eq. (8.22) is, in fact, is identical to the Green's function by showing that it obeys the same differential equation, namely Eq. (8.14b). Hint: differentiate (8.22) with respect to time using that the derivative of the theta function is a delta function, and that

$$\langle \mathbf{r}|H|\phi \rangle = H(\mathbf{r})\langle \mathbf{r}|\phi \rangle \tag{E.8.1}$$

Which you can see for example by inserting a complete set of eigenstates of H.

Exercise 8.3

Find the greater propagator, $G^>(\mathbf{r},\mathbf{r}';\omega)$ similar to Eq. (8.44), but now in one- and two dimensions. Can you suggest an experiment (at least in principle) that measures this propagator.

Exercise 8.4

Eqs. (8.51) are valid for fermions. Show that the corresponding results for bosons are

$$iG^>(\nu,\omega) = A(\nu,\omega)\left[1 + n_B(\omega)\right], \tag{E.8.2}$$
$$iG^<(\nu,\omega) = A(\nu,\omega)\, n_B(\omega). \tag{E.8.3}$$

Exercise 8.5

The dc conductance of a perfect 1D wire

Consider a one-dimensional electron gas with $H = (1/L)\sum_k (k^2/2m - \mu)c_k^\dagger c_k$.

From Sec. 6.3 we have that conductance G is given by

$$G = \frac{ie^2}{\omega}\Pi^R(\omega), \qquad \Pi^R(x-x';t-t') = -i\theta(t-t')\langle[I_p(xt), I_p(x't')]\rangle \tag{E.8.4}$$

where I_p is the operator for the particle current through the wire.

(a) Show that he one-dimensional version of the particle current operator is

$$I_p(x) = \frac{\hbar}{mL}\sum_{kq\sigma}\left(k+\frac{q}{2}\right)c_{k\sigma}^\dagger c_{k+q\sigma}\, e^{iqx}. \tag{E.8.5}$$

(b) Use the method in Sec. 8.5 to find that

$$\Pi^R(x-x';t-t') = -i\theta(t-t')\left(\frac{\hbar}{mL}\right)^2 \sum_{kq\sigma}(n_F(\varepsilon_k) - n_F(\varepsilon_{k+q}))$$
$$\times \left(k+\frac{q}{2}\right)^2 e^{i(\varepsilon_k - \varepsilon_{k+q})(t-t')} e^{iq(x-x')}. \tag{E.8.6}$$

(c) Now, set $x = x'$ find $\Pi^R(0, \omega)$ and study it in the low frequency limit. Show that

$$\lim_{\omega \to 0} \mathrm{Im}\, \Pi^R(\omega) = -\hbar\omega\pi \left(\frac{\hbar}{mL}\right)^2 \sum_{kq\sigma} \left[-\frac{\partial n_\mathrm{F}(\varepsilon_k)}{\partial \varepsilon_k}\right] \delta(\varepsilon_k - \varepsilon_{k+q}) \left(k + \frac{q}{2}\right)^2. \quad (\mathrm{E}.8.7)$$

(d) In the above expression do the q-integral first and find

$$\lim_{\omega \to 0} \mathrm{Im}\, \Pi^R(\omega) = \hbar\omega\pi \left(\frac{\hbar}{m}\right)^2 \left(\frac{m}{\hbar^2}\right) \frac{1}{2\pi L} \sum_{k\sigma} \left(-\frac{\partial}{\partial \varepsilon_k} n_\mathrm{F}(\varepsilon_k)\right) \frac{k^2}{|k|}$$

$$= \frac{\omega}{\hbar\pi} \frac{1}{e^{-\beta\mu} + 1}. \quad (\mathrm{E}.8.8)$$

(e) Show that in the limit $\mu \gg kT$, the famous result for the conductance G of a perfect 1D channel is recovered:

$$G = \frac{2e^2}{h}. \quad (\mathrm{E}.8.9)$$

Exercise 8.6

2D electron gases and Friedel oscillations Consider a 2D electron gas in the xy plane confined to the strip $0 < x < L$. What is the electron density as a function of the distance x from the left edge? Take for simplicity $T = 0$. What will change at larger temperatures? The oscillations that you will find are called Friedel oscillations.

Hints: Use standing waves in the x-direction fulfilling the proper boundary conditions, and assume quasi-continuous states with periodic boundary conditions in the y direction. Find the x-dependent density as $n(x) = \int dy \sum_\nu \langle c_\nu^\dagger c_\nu \rangle |\langle xy|\nu\rangle|^2$, where the ν-sum runs over the appropriately normalized states $|\nu\rangle$. You may need to know the integral $\int ds \sqrt{1 - s^2} \sin^2(xs) = \frac{\pi}{8x}[x - J_1(2x)]$.

Exercise 8.7

A relation between Green's function and the transmission coefficient
In this exercise, we show the property of the Green's function stated in Eq. (8.27).

(a) First insert the eigenstates for the one-dimensional scattering problem mentioned in Section 8.2.1 into Eq. (8.26) and show for $x' < 0$ and $x > W$ that

$$G^R(x, x', E) = \int_0^\infty \frac{dk}{2\pi} \frac{1}{E - k^2/2m + i\eta} \left(t_k e^{ik(x-x')} + [t_k']^* e^{-ik(x-x')}\right). \quad (\mathrm{E}.8.10)$$

In deriving this you should use the property of the S-matrix stated in Eq. (7.13). Remember to insert the positive infinitesimal η.

(b) Calculate the integral in Eq. (E.8.10) in the limit $|x - x'| \to \infty$ using a contour integral. Hints: for the first term use the quarter circle contour $0 \to \infty \to +i\infty \to 0$ in the complex k plane. Argue that the integral along the circle arc and the imaginary axis tends to zero for large and positive $x - x'$. For the second term use this quarter circle contour: $0 \to \infty \to -i\infty \to 0$.

(c) Assuming t_k to be an analytic function of k (clearly it cannot have poles because $|t|^2 < 1$) show by using the two contours in (b) that

$$G^R(x, x', \omega) \to -i \frac{t_{k_\omega}}{k_\omega} e^{ik_\omega(x - x')}. \tag{E.8.11}$$

Hints: the poles of the integrand are $k = \pm(\sqrt{2m\omega} + i0^+)$ for positive E. For negative E the poles are at $k = \pm i\sqrt{2m|\omega|} \pm 0^+$.

Exercise 8.8

Relation between retarded and advanced Green's functions. Show Eq. (8.54) starting from the Fourier transform to frequency domain of the definition of the retarded Green's function in Eq. (8.34) and the corresponding one for the advanced Green's function. Then by complex conjugation of the latter and change of variable $t \to -t$, you can show Eq. (8.54).

Exercises for Chapter 9

Exercise 9.1

A fermionic two-orbital system. Consider a physical system consisting of fermions allowed to occupy two orbitals. The Hamiltonian is given by

$$H = E_1 c_1^\dagger c_1 + E_2 c_2^\dagger c_2 + t c_1^\dagger c_2 + t^* c_2^\dagger c_1. \tag{E.9.1}$$

Find the Green's function $G^R(ij, \omega)$, where i and j can be both be either 1 or 2 and where $G^R(ij, t - t') = -i\theta(t - t')\langle\{c_i(t), c_j^\dagger(t')\}\rangle$. Use the equation of motion method. Do not forget to interpret the result.

Exercise 9.2

Derive Eqs. (9.30) and (9.31) by differentiating the Green's functions in (9.28) and (9.29).

Exercise 9.3

An atom on a metal surface. The electronic states of an atom on a metal surface will hybridize with the conduction electrons in the metal. If we assume that only a single orbital couples to the metal states, then the atom and the metal can be described by the Anderson model Hamiltonian Eq. (9.26).

When a scanning tunneling microscope (STM) is placed near the atom current will flow from the STM tip through the atom to the metal. Since the atom is strongly coupled to the metal surface the bottleneck for the current is the tunneling from STM to atom, which we can describe by a tunneling Hamiltonian as in Eq. (8.65), and not the tunneling between atom and metal, described by Eq. (9.25). It is therefore a good approximation to assume that the atom is in equilibrium with the metal, and to use tunneling theory for the current between tip and atom.

Sketch the resulting dI/dV using the expression derived in Chapter 8 for the tunnel current and the mean field expression for the d electron Green's function, derived in Sec. 9.3.

Exercises for Chapter 10

Exercise 10.1

Show that Eqs. (10.28) and (10.29) follow from Eq. (10.27).

Exercise 10.2

A quantum dot in the sequential tunneling regime

Consider a dot described by the Anderson Hamiltonian Eq. (10.8).

(a) Set up the master equations for the probabilities to find the system in one of the four possible states, i.e. $P(0), P(\uparrow), P(\downarrow)$, and $P(d)$, where the four state are empty, singly occupied spin up or down, and doubly occupied, respectively. Define the different transition rates as $\Gamma^L_{\uparrow 0}, \Gamma^R_{\uparrow 0}, \Gamma^L_{d\downarrow}, \Gamma^R_{d\downarrow}$ et cetera.

(b) Now suppose that U is very large so that $P(d) = 0$ and let $\xi_{d\uparrow} = \xi_{d\downarrow}$. Show that

$$P(0) = \frac{\Gamma_{01}}{\Gamma_{01} + 2\Gamma_{10}}, \quad P(\uparrow) = P(\downarrow) = \frac{\Gamma_{10}}{\Gamma_{01} + 2\Gamma_{10}}. \quad \text{(E.10.1)}$$

where $\Gamma_{\alpha\beta} = \Gamma^L_{\alpha\beta} + \Gamma^R_{\alpha\beta}$.

(c) Suppose now that a magnetic field is applied to the system so that the energy of the two singly occupied states split and suppose that only $P(0)$ and $P(\uparrow)$ are non-zero. Show that in this case

$$P(0) = \frac{\Gamma_{0\uparrow}}{\Gamma_{0\uparrow} + \Gamma_{\uparrow 0}}, \quad P(\uparrow) = \frac{\Gamma_{\uparrow 0}}{\Gamma_{0\uparrow} + \Gamma_{\uparrow 0}}. \quad \text{(E.10.2)}$$

(d) We now want to find the current. Using Eq. (10.26), show that the current is given by

$$I = -e\frac{\Gamma^R_{0\uparrow}\Gamma^L_{\uparrow 0} - \Gamma^L_{0\uparrow}\Gamma^R_{\uparrow 0}}{\Gamma_{0\uparrow} + \Gamma_{\uparrow 0}}. \quad \text{(E.10.3)}$$

Then use Eq. (10.29) to show

$$\Gamma^x_{\uparrow 0} = \Gamma^x n_F(\varepsilon - \mu_x), \quad \Gamma^x_{0\uparrow} = \Gamma^x[1 - n_F(\varepsilon - \mu_x)], \quad \text{(E.10.4)}$$

where $\varepsilon_\uparrow = \varepsilon$ and $x = L$ or R, and show also that

$$I = e\left[n_F(\varepsilon - \mu_R) - n_F(\varepsilon - \mu_L)\right]\frac{\Gamma^R\Gamma^L}{\Gamma^L + \Gamma^R}. \quad \text{(E.10.5)}$$

(e) Finally, consider linear response, i.e. lowest order in the voltage difference. If $\mu_L = \mu - eV/2$ and $\mu_R = \mu + eV/2$. Show that the conductance $G = I/V$ becomes

$$G = \frac{e^2}{\hbar}\frac{\Gamma^R\Gamma^L}{\Gamma^L + \Gamma^R}\left(-\frac{\partial n_F(\varepsilon - \mu)}{\partial \varepsilon}\right) = \frac{e^2}{\hbar}\frac{\Gamma^R\Gamma^L}{\Gamma^L + \Gamma^R}\frac{1}{4kT\cosh^2((\varepsilon - \mu)/2kT)}, \quad \text{(E.10.6)}$$

where $\hbar$ has been reinserted. Discuss how the conductance peak behaves as a function of the dot potential ε and of the temperature T.

Exercise 10.3

Show that Eq. (10.35) follows from Eq. (10.33) when using the same set of arguments and simplifications as in Eqs. (10.16) and (10.17).

Exercise 10.4

Take the limit of small bias and $T = 0$ in Eq. (10.35), which allows one to neglect the ε's in the denominator, perform the integrals and derive Eq. (10.36).

Exercise 10.5

Derive Eq. (10.68) from Eqs. (10.65) and (10.67).

Exercise 10.6

Starting from Eqs. (10.56) and (10.69) show that master equation result in Exercise 10.2 follows in the limit $\Gamma \ll k_{\mathrm{B}} T$ for large U.

Exercise 10.7

The self-consistent solution of the simple Anderson model
Show that the self-consistent solution of Eq. (10.70) for $\varepsilon_{d\uparrow} = \varepsilon_{d\downarrow} = \varepsilon$ is

$$\langle n_\uparrow \rangle = \langle n_\downarrow \rangle = \frac{n_0}{1 - n_U + n_0} , \tag{E.10.7}$$

with

$$n_0 = \int \frac{d\omega}{2\pi} \frac{\Gamma \, n_{\mathrm{F}}(\omega)}{(\omega - \varepsilon + \mu)^2 + (\Gamma/2)^2} , \qquad n_U = \int \frac{d\omega}{2\pi} \frac{\Gamma \, n_{\mathrm{F}}(\omega)}{(\omega - \varepsilon + \mu - U)^2 + (\Gamma/2)^2} , \tag{E.10.8}$$

which at $T = 0$ becomes

$$n_0 = \frac{1}{2} - \frac{1}{\pi} \tan^{-1}\left(\frac{2(\varepsilon - \mu)}{\Gamma} \right) , \qquad n_U = \frac{1}{2} - \frac{1}{\pi} \tan^{-1}\left(\frac{2(\varepsilon - \mu + U)}{\Gamma} \right) . \tag{E.10.9}$$

Exercise 10.8

The Kondo effect in bulk metals vs. quantum dots
By using the equation of motion technique in Chapter 9 we want to shed some light on the connection between the Kondo effect in bulk metals and in quantum dots.

    (a) Show that the retarded Green's function of the lead electrons,

$$G^R(\nu\alpha\sigma, \nu'\alpha'\sigma, t - t') = -i\theta(t - t')\langle[c_{\nu\sigma\alpha}(t), c_{\nu'\sigma'\alpha'}^\dagger(t')]\rangle, \tag{E.10.10}$$

in the frequency domain is given by

$$\begin{aligned} G^R(\nu\alpha\sigma, \nu'\alpha'\sigma, \omega) &= G_0^R(\nu\alpha\sigma, \omega)\delta_{\nu\nu'}\delta_{\alpha\alpha'} \\ &\quad + G_0^R(\nu\alpha\sigma, \omega)t_\alpha G^R(d\sigma, \omega)t_{\alpha'}^* G_0^R(\nu'\alpha'\sigma, \omega). \end{aligned} \tag{E.10.11}$$

Hint: in the second equation that you generate on the way, you should differentiate with respect to t' and not t.

    (b) Argue from Eq. (E.10.11) that the dot Green's function is directly related to the the self-energy $\Sigma^R(\nu\alpha\sigma, \nu'\alpha'\sigma, \omega)$ of the lead electrons, and hence

$$G^R(d\sigma, \omega) = \frac{1}{t_\alpha t_{\alpha'}} \Sigma^R(\nu\alpha\sigma, \nu'\alpha'\sigma, \omega). \tag{E.10.12}$$

From this formula we can now understand the connection between the Kondo effect in metal and in quantum dots. In metals the resistivity contribution is related to the self energy of the lead electrons (see Chapter 16) and for the quantum dot the same quantity is related to the conductance.

Exercise 10.9

The current operator and the Schrieffer-Wolff transformation
In this exercise, we derive the current operator in the Schrieffer-Wolff canonical transformation. When the performing the canonical transformation that lead as from the Anderson model to the Kondo model, we should of course also perform the transformation on the operators of the observable that we wish to calculate. Therefore the current operator transform as

$$I_S = e^{iS} I e^{-iS}, \tag{E.10.13}$$

where the current operators for the current through one of the two junctions are given in Eq. (10.46). Looking at the current through the contact β, where β is left or right, we have

$$I_{S\beta} = e^{iS} I_\beta e^{-iS} = i e^{iS} (H_{T\beta}^- - H_{T\beta}^+) e^{-iS} \approx [S^-, H_{T\beta}^+] + \text{h.c.}, \tag{E.10.14}$$

where the approximate sign indicates that we neglected the terms $S^+ H_T^+$ that does not correspond to the single occupied subspace. The calculation now follows from Eq. (10.85) and onward, the only difference being that the index α' is not to be summed over but instead replaced by β.

(a) Thus show that

$$I_{S\beta} = -\frac{i}{2} \left(\sum_{\nu\nu',\alpha} J_{\alpha\beta} \, \mathbf{S}_d \cdot \mathbf{S}_{\nu\alpha,\nu'\beta} + \sum_{\nu\nu'\sigma,\alpha\beta} W_{\alpha\beta} \, c_{\nu\alpha\sigma}^\dagger c_{\nu'\beta\sigma} - \text{h.c.} \right). \tag{E.10.15}$$

Here we note that the term with $\alpha = \beta$ is cancelled by the Hermitian conjugate term, which of course agrees with the expectation that an electron must scatter from one side to other to other in order to contribute to the current. This also means that the operators for the currents through the left or the right junctions become identical (apart for a sign difference, of course), i.e. $I_{SL} = -I_{SR}$. If we count the current as the electrical current from left to right, $I = -eI_L$, we therefore have the final expression for the transformed current operator

$$I_{Se} = \frac{ie}{2} \left(\sum_{\nu\nu'} J_{RL} \, \mathbf{S}_d \cdot \mathbf{S}_{\nu R,\nu' L} + \sum_{\nu\nu'\sigma} W_{RL} \, c_{\nu R\sigma}^\dagger c_{\nu' L\sigma} - \text{h.c.} \right). \tag{E.10.16}$$

Thus the current has two contributions: a scattering from left to right either via an interaction with the dot spin and a direct scattering term which is independent of the spin state.

Exercises for Chapter 11

Exercise 11.1

Find the Fermi-Dirac distribution by starting from the Matsubara Green's function and setting $\tau = 0^-$. Then show that

$$\langle c_\nu^\dagger c_\nu \rangle = \mathcal{G}_\sigma^0(\nu, \tau = 0^-) = \frac{1}{\beta} \sum_{ik_n} e^{-ik_n 0^-} \mathcal{G}_\sigma^0(\nu, ik_n) = n_F(\varepsilon_\nu) \qquad \text{(E.11.1)}$$

How would you calculate $\langle c_\nu c_\nu^\dagger \rangle$?

Exercise 11.2

Repeat Exercise 8.5 but this time using the imaginary time formalism. Use the procedure going from Eq. (11.83) to Eq. (11.88).

Exercise 11.3

According to Eq. (11.66) the equation of motion for the Matsubara Green's function of a free particle is

$$\left(-\partial_\tau - \frac{\mathbf{p}^2}{2m} + \mu\right) \mathcal{G}_\sigma^0(\mathbf{r} - \mathbf{r}', \tau - \tau') = \delta(\mathbf{r} - \mathbf{r}')\,\delta(\tau - \tau'). \qquad \text{(E.11.2)}$$

Show (11.42) by Fourier transforming Eq. (E.11.2).

Exercises for Chapter 12

Exercise 12.1

Single-impurity scattering.
The Dyson equation for otherwise free electrons scattering against an external potential is written in Eqs. (12.5) and (12.9). Suppose now that the electrons are confined to move in one dimension and that the external potential can be represented by a delta-function impurity potential, $U(x) = U_0 \delta(x)$.

(a) Show that in this case the solution of the Dyson equation becomes

$$\mathcal{G}_\sigma(xx', ik_n) = \mathcal{G}_\sigma^0(xx', ik_n) + \mathcal{G}_\sigma^0(x0, ik_n) \frac{U_0}{1 - \mathcal{G}_\sigma^0(00, ik_n)\,U_0} \mathcal{G}_\sigma^0(0x', ik_n). \quad \text{(E.12.1)}$$

Hint: solve for $\mathcal{G}(0x', ik_n)$ first and insert that in the Dyson equation for $\mathcal{G}(xx', ik_n)$.

To find the retarded Green's we thus need the unperturbed Green's function, which is

$$G_\sigma^{0R}(xx', \omega) = \mathcal{G}_\sigma^0(xx', \omega + i\eta) = \frac{1}{L} \sum_k \frac{e^{ik(x-x')}}{\omega - \varepsilon_k + \mu + i\eta} = \frac{1}{iv_\omega} e^{ik_\omega |x-x'|}, \quad \text{(E.12.2)}$$

(do you agree?) where $k_\omega = \sqrt{2m(\omega + \mu)}$ and $v_\omega = \partial\varepsilon_k/\partial k|_{k=k_\omega}$. Since the retarded Green's function tells us about the amplitude for propagation from point x' to point x, we can in fact extract the transmission and reflection amplitudes t and r.

(b) Show that for $x' < 0$ $G_\sigma^R(xx', \omega)$ can be written as

$$G_\sigma^R(xx', \omega) = t\, G_\sigma^{0R}(xx', \omega)\,\theta(x) + \left[1 + re^{i\phi(x,x')}\right] G_\sigma^{0R}(xx', \omega)\,\theta(-x), \quad \text{(E.12.3)}$$

where $e^{i\phi(x,x')}$ is a phase factor, which is determined by the calculation.

(c) Find r and t and discuss the phase shifts that the electrons acquire when they are scattered.

Exercise 12.2

Resonant tunneling.
In for example semiconductor heterostructures one can make quantum-well systems
which to a good approximation can be described by a one-dimensional model of
free electrons with two tunneling barriers. Here we simplify it somewhat further by
representing the tunneling barriers by delta-function potentials situated at a_1 and a_2.
The Hamiltonian is then given by

$$H = H_0 + \int_{-\infty}^{\infty} dx\, \rho(x)\, U_0\big[\delta(x - a_1) + \delta(x - a_2)\big], \qquad \text{(E.12.4)}$$

where H_0 is the Hamiltonian for free electrons in one dimension.

(a) Find a formal expression for the Matsubara Green's function using Dyson's
equation.

(b) From the Dyson equation show that the retarded Green's function for $x' <
a_1 < a_2 < x$ is

$$G_\sigma^R(xx', \omega) = \frac{e^{ik(x-x')}}{iv_\omega} \left[1 + \alpha \left(e^{-ika_1},\, e^{-ika_2} \right) \begin{pmatrix} 1-\alpha & -\alpha e^{i\theta} \\ -\alpha e^{i\theta} & 1-\alpha \end{pmatrix}^{-1} \begin{pmatrix} e^{ika_1} \\ e^{ika_2} \end{pmatrix} \right],$$
$$\text{(E.12.5)}$$

where $k = k_\omega$ (see previous exercise) and where $\alpha = U_0/iv$ and $\theta = k(a_2 - a_1)$.

(c) Use this to show that the transmission is unity for the particular values of θ
satisfying

$$\alpha = i \cot \theta. \qquad \text{(E.12.6)}$$

(d) Derive the same result using the following simple argument involving two
paths for an electron to go from x' to x: (1) $x' \to a_1 \to a_2 \to x$, and (2) $x' \to
a_1 \to a_2 \to a_1 \to a_2 \to x$. The transmission is unity when these two paths interfere
constructively – as does paths with any number of trips back and forth in the "cavity".

Exercises for Chapter 13

Exercise 13.1

Matsubara frequency summation.
Use the rule Eq. (11.57) for summing over functions with simple poles to perform the
Matsubara frequency summation appearing in the following diagrams of Eqs. (13.33)
and (13.37):

$$\Sigma_\sigma^F(\mathbf{k}, ik_n) \equiv \quad\quad\quad\quad \Pi^0(\mathbf{q}, iq_n) \equiv \quad\quad\quad \text{(E.13.1)}$$

Exercise 13.2

The cancellation of disconnected diagrams in $\mathcal{G}(b, a)$
We study the one-particle Green's function, which in the interaction picture in the
presence of the particle-particle interaction $W(\mathbf{r} - \mathbf{r}')$ becomes:

$$\mathcal{G}(b,a) = -\frac{\left\langle T_\tau \left[\hat{U}(\beta,0)\, \hat{\Psi}(b)\, \hat{\Psi}^\dagger(a) \right] \right\rangle_0}{\left\langle \hat{U}(\beta,0) \right\rangle_0}, \quad \text{with } \hat{U}(\beta,0) = T_\tau \exp\left(-\int_0^\beta d\tau\, \hat{W}(\tau) \right).$$

(E.13.2)

As in Eq. (13.17) use the Feynman rules to expand the denominator and the numerator, but now to second order in W, and show explicitly the cancellation of the disconnected diagrams. Hints. (1) Start with the simpler denominator (how many terms?). (2) Draw topologically identical diagrams only once and multiply with the number of them. (3) Get most of the diagrams in the numerator by cutting open and stretching out a Fermion line in the diagrams from the denominator (how many terms?).

Exercise 13.3

Feynman diagrams and Dyson's equation for the Anderson model

We return to Anderson's model for localized magnetic moments in metals, see Section 9.3. We wish to derive the Dyson equation Eq. (9.37) using Feynman diagrams. The unperturbed Hamiltonian is given by $H_0 = \sum_\sigma (\varepsilon_d - \mu) d_\sigma^\dagger d_\sigma + \sum_{\mathbf{k}\sigma} (\varepsilon_\mathbf{k} - \mu) c_{\mathbf{k}\sigma}^\dagger c_{\mathbf{k}\sigma}$, while the interaction part is given by $H_\text{int} = H_\text{hyb} + H_U^\text{MF}$, the sum of the hybridization Eq. (9.25) and on-site repulsion Eq. (9.35). We employ the mean-field approximation given by Eq. (9.35) where the σ spins only interact with the average density $\langle n_{d\bar\sigma} \rangle$ of the opposite $\bar\sigma$ spins.

We introduce the following rather obvious diagrammatic notation for the Matsubara Green's functions and interactions:

$$\begin{aligned} &\longleftarrow && \equiv && \mathcal{G}^0(d\sigma) & & \overset{t_\mathbf{k}}{\bullet} && \equiv && \sum_\mathbf{k} t_\mathbf{k} \\ &\Longleftarrow && \equiv && \mathcal{G}(d\sigma) \\ &\text{-}\!\!\blacktriangleleft\text{-}\text{-} && \equiv && \mathcal{G}^0(k\sigma) & & \overset{U}{\wwww} && \equiv && U \end{aligned}$$

(E.13.3)

We write the diagrammatic expansion (here shown up to second order in $|t_\mathbf{k}|^2$ and U) for the full d-orbital spin up Green function $\mathcal{G}(d\sigma)$ as:

$$\text{(E.13.4)}$$

(a) Express the self-energy as a sum of diagrams using a definition analogous to Eq. (13.21), and derive in analogy with Eq. (13.22) Dyson's equation graphically.

(b) Use the obtained Dyson equation to verify the solution Eqs. (9.38a) and (9.38b). The tedious work with the equation of motion has been reduced to simple manipulations with diagrams.

Exercise 13.4

The third order selfenergy $\Sigma^{(3)}$ in the Kondo problem

In this exercise we study the integration of the internal time variable in the third order selfenergy $\Sigma^{(3)}$ of the Kondo problem.

(a) Starting from Eq. (13.58) perform the integral over the internal time variable τ_3. Write the unperturbed Green's function $\mathcal{G}^0$, given in Eq. (11.40), as

$$\mathcal{G}^0(\mathbf{k}\sigma, \tau) = -\left[\theta(\tau) - n_{\mathrm{F}}(\xi_{\mathbf{k}})\right] e^{-\xi_{\mathbf{k}}\tau}, \qquad \text{(E.13.5)}$$

and verify that

$$\int_0^\beta d\tau_3\, \mathcal{G}^0(\mathbf{k}_1\sigma, \tau_1 - \tau_3)\mathcal{G}^0(\mathbf{k}_2\sigma_2, \tau_3)\mathrm{sign}(\tau - \tau_3)$$

$$= \frac{(1 - n_1)(1 - n_2)}{\xi_1 - \xi_2}\left(e^{-\xi_2\tau} - e^{-\xi_1\tau}\right) + \frac{n_1(1 - n_2)}{\xi_1 - \xi_2}\left(e^{\xi_1(\beta-\tau)}e^{-\xi_2\beta} - e^{-\xi_2\tau}\right)$$

$$= \frac{1}{\xi_1 - \xi_2}\left[-(1 - 2n_1)\mathcal{G}_2(\tau) + (1 - 2n_2)\mathcal{G}_1(\tau)\right]. \qquad \text{(E.13.6)}$$

For brevity use the notation $\xi_i = \xi_{\mathbf{k}_i}$, $n_i = n_{\mathrm{F}}(\xi_i)$, and $\mathcal{G}_i(\tau) = -(1 - n_i)\exp(-\xi_i\tau)$, where $i = 1, 2$.

(b) Calculate the Fourier transform of Eq. (E.13.6) and obtain Eq. (13.59).

Exercises for Chapter 14

Exercise 14.1

A classical treatment of the plasma oscillation
The electronic plasma frequency $\omega_p \equiv \sqrt{ne^2/m\epsilon_0}$ introduced in Eq. (14.77) does not contain Planck's constant and is therefore not a quantum object. Derive ω_p from the following purely classical argument.

Consider an electron gas of density n confined in a rectangular box of length L_x in the x direction and having a large surface area $L_y L_z$ in the yz plane ($L_x \ll L_y, L_z$). Treat the ions as an inert, charge compensating jellium background. Imagine now the electron gas being translated a tiny distance ξ in the x direction ($\xi \ll L_x$), leaving the ion jellium fixed. The resulting system resembles a plate capacitor. The electron gas is then released.

(a) Find the equation of motion for the coordinate ξ using Newtons law and classical electrostatics.

(b) Give a physical interpretation of the resulting motion of the electron gas.

Exercise 14.2

Interactions in two dimensions
Consider a translation-invariant electron gas in two dimensions fabricated in a GaAs heterostructure (see Section 2.3.2). The electron mass for this material is $m^* = 0.067\, m$, the relative permittivity is $\varepsilon_r = 13$, while the electron density ranges from $n^{2D} = 1 \times 10^{15}$ m^{-2} to 5×10^{15} m^{-2}.

The electron wave function for the two dimensional electron gas is restricted to be

$$\psi_{\mathbf{k}}(\mathbf{r}, z) = \frac{1}{\sqrt{L_x L_y}} e^{i\mathbf{k}\cdot\mathbf{r}} \zeta_0(z), \tag{E.14.1}$$

where $\mathbf{k} = (k_x, k_y)$ and $\mathbf{r} = (x, y)$, while $\zeta_0(z)$ is the lowest eigenstate in the z direction, i.e. $n = 0$ in Eq. (2.50). Write down the interaction part of the Hamiltonian and show that it is of the form

$$H^{2D} = \frac{1}{2A} \sum_{\mathbf{k}\mathbf{k}'\mathbf{q}} \sum_{\sigma\sigma'} W^{2D}(\mathbf{q})\, c^\dagger_{\mathbf{k}+\mathbf{q},\sigma} c^\dagger_{\mathbf{k}'-\mathbf{q},\sigma'} c_{\mathbf{k}',\sigma'} c_{\mathbf{k},\sigma} \tag{E.14.2}$$

where $\mathbf{q} = (q_x, q_y)$. For a strictly 2D system, i.e. $|\zeta_0(z)|^2 = \delta(z)$, show that

$$W^{2D}(\mathbf{q}) = \frac{e^2}{2\varepsilon_r\epsilon_0 q}. \tag{E.14.3}$$

Hint: use $\int_0^\pi d\theta \cos(\alpha\cos\theta) = \pi J_0(\alpha)$, where J_0 is the Bessel function of the first kind of order zero.

Exercise 14.3

Plasmons in two dimensions
Consider a translation-invariant electron gas in two dimensions fabricated in a GaAs

heterostructure. The electron mass for this material is $m^* = 0.067\,m$, the relative permittivity is $\varepsilon_r = 13$, while the electron density ranges from $n^{2D} = 1 \times 10^{15}\,\mathrm{m}^{-2}$ to $5 \times 10^{15}\,\mathrm{m}^{-2}$.

For such a system the RPA dielectric function is given by

$$\varepsilon_{\mathrm{RPA}}^{2D}(\mathbf{q}, iq_n) = 1 - W^{2D}(\mathbf{q})\,\chi_0^{2D}(\mathbf{q}, iq_n), \tag{E.14.4}$$

with $\mathbf{q} = (q_x, q_y)$ and where $\chi_0^{2D}(\mathbf{q}, iq_n)$ is the 2D version of the 3D pair bubble χ^0 given in Eq. (14.21).

(a) Show that at low temperatures, $k_{\mathrm{B}}T \ll \varepsilon_{\mathrm{F}}$, and long wave lengths, $q \ll \omega/v_F$, the plasmon dispersion relation is $\omega = v_F \sqrt{k_s^{2D}\,q/2}$, where k_s^{2D} is the Thomas-Fermi screening wavenumber in 2D.

(b) Find the relation between k_{F} and the electron density, n^{2D}.

(c) Express k_s^{2D} in terms of the parameters of the electron gas. Is it larger or smaller than k_F for $n^{2D} = 2 \times 10^{15}\,\mathrm{m}^{-2}$?

Exercise 14.4

Static screening in two dimensions

Show that in 2D the static RPA screened interaction at small wavevectors, $q \ll k_{\mathrm{F}}$, and low temperatures, $k_{\mathrm{B}}T \ll \varepsilon_{\mathrm{F}}$, is given by

$$W_{\mathrm{RPA}}^{2D}(\mathbf{q}, 0) \equiv \frac{W^{2D}(\mathbf{q})}{\varepsilon_{\mathrm{RPA}}^{2D}(\mathbf{q}, 0)} = \frac{e^2}{2\varepsilon_r \epsilon_0\,(q + k_s^{2D})}. \tag{E.14.5}$$

Exercise 14.5

Damping of two dimensional plasmons

The electron-hole pair continuum is the region in $q - \omega$ space where $\mathrm{Im}\,\chi_0^{2D} \neq 0$. Find the condition for the plasmons not to be damped by single-particle excitations for $q < k_{\mathrm{F}}$. In the estimate you can use the small-q expressions for the plasmon frequency and the polarization, that you found above. Are the plasmons damped in the region $q < k_{\mathrm{F}}$ in GaAs with the parameters given above?

Exercises for Chapter 15

Exercise 15.1

Semi-classical motion

We study Eqs. (15.16), (15.17), and (15.18). If the quasiparticles behaves like non-interacting particles why is then the number of quasiparticles conserved on the semi-classical level?

To answer this question we introduce the concept of a wave packet, i.e. a wave function fairly localized in both space and momentum space:

$$\psi(\mathbf{r}, t) = \int d\mathbf{k}\, f(\mathbf{k} - \mathbf{k}_0)\, e^{i[\mathbf{k}\cdot\mathbf{r} - \omega(\mathbf{k})t]}, \qquad \text{e.g. with } f(\mathbf{k} - \mathbf{k}_0) = \exp\left(-\frac{(\mathbf{k} - \mathbf{k}_0)^2}{2\,(\Delta\mathbf{k})^2}\right). \tag{E.15.1}$$

(a) Taylor expand $\omega(\mathbf{k})$ to first order and show that the wave packet can be written as

$$\psi(\mathbf{r},t) \approx e^{i[\mathbf{k}_0\cdot\mathbf{r}-\omega(\mathbf{k}_0)t]}\, F(\mathbf{r}-\partial_{\mathbf{k}_0}\omega(\mathbf{k}_0)\,t), \qquad\qquad \text{(E.15.2)}$$

where F is some envelope function.

(b) Based on Eq. (E.15.2) argue that the physical interpretation of $\omega(\mathbf{k}_0)$ and $\partial_{\mathbf{k}_0}\omega(\mathbf{k}_0)$ is that the wave packet has the energy $\varepsilon_{\mathbf{k}}$ and the velocity $\mathbf{v}_{\mathbf{k}}$ given by

$$\varepsilon_{\mathbf{k}} = \hbar\,\omega_{\mathbf{k}}, \qquad\qquad \mathbf{v}_{\mathbf{k}} = \partial_{\mathbf{k}}\omega_{\mathbf{k}} = \frac{1}{\hbar}\,\partial_{\mathbf{k}}\varepsilon_{\mathbf{k}}. \qquad\qquad \text{(E.15.3)}$$

(c) For external forces $\mathbf{F}(\mathbf{r},t) = -\nabla V(\mathbf{r},t)$ varying slowly in space and time, we can through the power $P_{\mathbf{k}}$ absorbed by the wave packet centered around $\mathbf{k}$ deduce the time evolution of $\mathbf{k}$ as follows. Combine the two classical expressions for the power, $P_{\mathbf{k}} = \mathbf{F}\cdot\mathbf{v}_{\mathbf{k}}$ and $P_{\mathbf{k}} = \dot{\varepsilon}_{\mathbf{k}}$, to show

$$\dot{\mathbf{k}} = \frac{1}{\hbar}\,\mathbf{F}. \qquad\qquad \text{(E.15.4)}$$

Exercise 15.2

Measuring the discontinuity of the distribution function

For an interacting electron gas we discuss the spectral function $A(\mathbf{k},\omega)$ in Eq. (15.57) and use it to calculate the distribution function $\langle n_{\mathbf{k}} \rangle$. We demonstrate the existence of a Fermi surface characterized by the renormalization parameter Z. The value of Z can be inferred from X-ray Compton scattering on the electron gas, see Fig. (a).

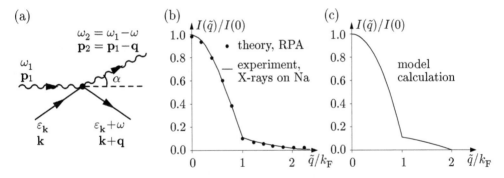

In the so-called impulse approximation for Compton scattering, the intensity $I(\omega_1,\omega,\mathbf{q})$ of incoming photons of energy ω_1 being scattered with the energy and momentum loss ω and $\mathbf{q}$, respectively, is proportional to the number of scattering events on all electrons fulfilling the simple kinematic constraint: conservation of energy and momentum,

$$I(\omega_1,\omega,\mathbf{q}) = \mathcal{N}(\omega_1,\omega)\int d\mathbf{k}\,\langle n_{\mathbf{k}}\rangle\,\delta(\omega+\varepsilon_{\mathbf{k}}-\varepsilon_{\mathbf{k}+\mathbf{q}}) \propto \int d\mathbf{k}\,\langle n_{\mathbf{k}}\rangle\,\delta\!\left(\omega-\frac{1}{2m}q^2-\frac{1}{m}\mathbf{q}\cdot\mathbf{k}\right).$$
$$\text{(E.15.5)}$$

We omit the explicit reference to the fixed ω_1 and work with $I(q,\tilde{q}) \equiv I(\omega_1,\omega,\mathbf{q})$.

(a) Show that

$$I(q, \tilde{q}) \propto \frac{1}{q} \int_{\frac{\tilde{q}^2}{2m}}^{\infty} d\varepsilon_{\mathbf{k}} \, \langle n_{\mathbf{k}} \rangle = \frac{1}{q} \int_{\frac{\tilde{q}^2}{2m}}^{\infty} d\varepsilon_{\mathbf{k}} \int_{-\infty}^{\infty} \frac{d\omega}{2\pi} A(\mathbf{k}, \omega) \, n_F(\omega). \qquad (\text{E.15.6})$$

where $A(\mathbf{k}, \omega)$ is the spectral function and $\tilde{q} \equiv m\omega/q - q/2$. Fig. (b) contains an experimental determination of $I(q, \tilde{q})$ from X-ray scattering on sodium. The experimental result is compared to theory based on RPA calculations of $A(\mathbf{k}, \omega)$.

(b) Instead of using RPA, discuss the following simple model for $A(\mathbf{k}, \omega)$ containing the essential features. At low energies, $\varepsilon_{\mathbf{k}} < 4\varepsilon_F$, a renormalized quasiparticle pole of weight Z coexists with a broad background of weight $1 - Z$, while at higher energies, $\varepsilon_{\mathbf{k}} > 4\varepsilon_F$, no renormalization occurs, and the quasiparticle is in fact the bare electron:

$$A(\mathbf{k}, \omega) = Z_{\mathbf{k}} 2\pi \delta(\omega - \xi_{\mathbf{k}}) + (1 - Z_{\mathbf{k}}) \frac{\pi}{W} \theta(W - |\omega|), \qquad Z_{\mathbf{k}} = \begin{cases} Z, & \text{for } k < 2k_F \\ 1, & \text{for } k > 2k_F. \end{cases} \qquad (\text{E.15.7})$$

Here W is the large but unspecified band width of the conduction band. Explain Fig. (c).

Exercise 15.3

Semiclassical equation of motion and detailed balance

The scattering life time in Eq. (15.44) expresses the time between scatterings assuming some unknown distribution function $n(\mathbf{k})$. The Boltzmann equation with inclusion of e-e scattering therefore reads

$$\partial_t(n_{\mathbf{k}}) + \dot{\mathbf{k}} \cdot \nabla_{\mathbf{k}} n_{\mathbf{k}} + \mathbf{v}_{\mathbf{k}} \cdot \nabla_{\mathbf{r}} n_{\mathbf{k}} = -\left(\frac{1}{\tau_{\mathbf{k}}}\right)_{\text{collisions}}. \qquad (\text{E.15.8})$$

In the homogenous and static case, i.e. absence of external forces, the left hand side is expected to be zero. Show that the usual Fermi-Dirac equation solves the Boltzmann equation in this case, i.e. that the right hand side is also zero if we use $n = n_F$. Hint: show and use that $n_F(\varepsilon)[1 - n_F(\varepsilon')] \exp(\beta(\varepsilon - \varepsilon')) = [1 - n_F(\varepsilon)] n_F(\varepsilon')$.

Exercise 15.4

Why are metals shiny?

According to Eq. (14.76) we have in the semiclassical high frequency, long wave limit that $\varepsilon(0, \omega) = 1 - \omega_p^2/\omega^2$. Consider a monochromatic electromagnetic wave with $\mathbf{E} = E(x) e^{-i\omega t} \hat{\mathbf{e}}_z$ incident on a metal occupying the half-space $x > 0$.

(a) Use the high-frequency limit of Maxwell's equations in matter. Set $\mathbf{D} = \epsilon_0 \varepsilon(0, \omega) \mathbf{E}$ and prove that $\nabla^2 E(x) = \frac{\omega_p^2 - \omega^2}{c^2} E(x)$. Hint: you may need $\nabla \times \nabla \times \mathbf{E} = -\nabla^2 \mathbf{E}$.

(b) For which frequencies does the wave propagate through the metal, and for which is it reflected?

(c) From X-ray diffraction we know that the unit-cell of Na is body-centered cubic (i.e. one atom in each corner and one in the center of the cube) with a side-length of 4.23 Å. It is observed that Na is transparent for UV-light with a wavelength shorter than 206 nm. Explain this, and explain why (polished) metals appear shiny. Hint: Each Na atom donates one electron to the conduction band.

Exercises for Chapter 16

Exercise 16.1

The integral equation for the vertex function in the Born approximation.
This exercise deals with the Kubo formula method applied to impurity scattering in
metals. The conductivity is in the weak scattering limit given by

$$\sigma_{xx} = \frac{e^2}{\pi} \frac{1}{V} \sum_{\mathbf{k}} \left(\frac{k_x}{m}\right) G^R(\mathbf{k},0) G^A(\mathbf{k},0) \Gamma_x^{RA}(\mathbf{k},\mathbf{k};0,0). \tag{E.16.1}$$

In the Born approximation the Dyson equation for the vertex function is

$$\Gamma_x^{RA}(\mathbf{k},\mathbf{k};0,0) = \frac{k_x}{m} + \frac{1}{V} \sum_{\mathbf{k'}} n_{\text{imp}} |u(\mathbf{k}-\mathbf{k'})|^2 G^R(\mathbf{k'},0) G^A(\mathbf{k'},0) \Gamma_x^{RA}(\mathbf{k'},\mathbf{k'};0,0). \tag{E.16.2}$$

(a) If the impurity potential is short ranged argue that we can approximate it by
a constant $|u(\mathbf{k}-\mathbf{k'})|^2 \approx |u_0|^2$. Prove that in this case

$$\Gamma_x^{RA} = \frac{k_x}{m}, \tag{E.16.3}$$

and use this to find that

$$\sigma = \frac{e^2 n \tau_0}{m}, \tag{E.16.4}$$

where τ_0 is the Born approximation life time

$$\tau_0^{-1} = 2\pi d(\varepsilon_F) n_{\text{imp}} |u_0|^2. \tag{E.16.5}$$

(b) Now relax the assumption of short range scatterers but assume instead that
$u(\mathbf{k}-\mathbf{k'})$ is slowly varying on the scale given by the width of the spectral function,
i.e. τ_0^{-1}. In this more realistic case, you will find for $|\mathbf{k}| = k_F$ that

$$\vec{\Gamma}^{RA}(\mathbf{k}) = \frac{\mathbf{k}}{m} + \frac{d(\varepsilon_F)}{2} \int d\Omega' \, |u(\mathbf{k}-\mathbf{k'})|^2 \vec{\Gamma}^{RA}(\mathbf{k'}), \tag{E.16.6}$$

with $|\mathbf{k'}| = k_F$ and $\int d\Omega' = \int d\phi' d\theta' \sin\theta'$ is an integration over the angle of $\mathbf{k'}$. Now
show that

$$\sigma = \frac{e^2 n \tau^{\text{tr}}}{m}, \tag{E.16.7}$$

where

$$(\tau^{\text{tr}})^{-1} = 2\pi d(\varepsilon_F) n_{\text{imp}} \int \frac{d\Omega'}{4\pi} |u(\mathbf{k}-\mathbf{k'})|^2 \left(1 - \frac{\mathbf{k} \cdot \mathbf{k'}}{k_F^2}\right). \tag{E.16.8}$$

Hint: use the ansatz $\vec{\Gamma}^{RA}(\mathbf{k}) = (\mathbf{k}/m)\gamma$.
(c) Explain the physical meaning of the last term and why it does not appear in
the result for point-like impurities.

Exercise 16.2

Life time of the Green's function in the Born approximation
Show that the retarded impurity averaged Green's function in the Born approximation
decays exponentially in time and given a physical interpretation of this result.

Exercise 16.3

Weak localization at finite frequency
Consider the weak localization correction at finite frequency. The only change in the
formula for the weak localization is through the Cooperon which becomes instead

$$C^{RA}(\mathbf{Q}, \omega) = \frac{W_0 \zeta(\mathbf{Q}, \omega)}{1 - \zeta(\mathbf{Q}, \omega)}, \tag{E.16.9}$$

$$\zeta(\mathbf{Q}, \omega) = \frac{1}{\mathcal{V}} \sum_{\mathbf{p}} |u_0|^2 G^R(\mathbf{p} - \mathbf{Q}, \omega) G^A(\mathbf{p}, 0). \tag{E.16.10}$$

(a) Show that in this case, the low frequency $\omega\tau_0 \ll 1$ and long wavelength
$Q v_F \tau_0 \ll 1$ limit of $\zeta(\mathbf{Q}, \omega)$ is

$$\zeta(\mathbf{Q}, \omega) \approx 1 + i\omega\tau_0 - D\tau_0 Q^2. \tag{E.16.11}$$

(b) Show that the frequency provide a small Q cut-off in the conductivity correction and try to explain why.

Exercises for Chapter 17

Exercise 17.1

Phonon Green's function. Prove Eq. (17.7).

Exercise 17.2

Cooper's instability
In this exercise we shall see that an attractive electron-electron interaction leads to
an instability of the Fermi surface.

In Chapter 17 it is shown that the electron-phonon interaction leads to an effective
electron-electron interaction for small energies. The scale of this energy is given by
the Debye energy which lead Cooper to study the following model Hamiltonian

$$H = H_0 + H' \tag{E.17.1}$$

$$H_0 = \sum_{k\sigma} (\varepsilon_{\mathbf{k}} - \mu) c_{\mathbf{k}\sigma}^\dagger c_{\mathbf{k}\sigma}, \tag{E.17.2}$$

$$H' = -\frac{V_0}{2} \sum_{}' c_{\mathbf{k}+\mathbf{q}\sigma}^\dagger c_{\mathbf{k}'-\mathbf{q}\sigma'}^\dagger c_{\mathbf{k}'\sigma'} c_{\mathbf{k}\sigma}. \tag{E.17.3}$$

Here the sum is restricted such that all initial and final states lie in an interval given by
$[\mu - \omega_D, \mu + \omega_D]$. Anticipating the physical idea that the electrons will form pairs with
zero total momentum and spin, we look specifically at the interaction between such

pairs. The pairs are thus supposed to consist of electrons with opposite momentum, which means that we choose $\mathbf{k}' = -\mathbf{k}$ and $\sigma = -\sigma'$. After relabelling we have

$$H' = -V_0 \sum {}' c^\dagger_{\mathbf{k}'\uparrow} c^\dagger_{-\mathbf{k}'\downarrow} c_{-\mathbf{k}\downarrow} c_{\mathbf{k}\uparrow}. \tag{E.17.4}$$

We wish to find the energy of a pair of electrons added to a filled Fermi sea state, and with interactions according to (2). In order to do so the $\sum'$ sum is further restricted to involve only states outside the Fermi sea. Thus Eq. (E.17.4) becomes

$$H' = -V_0 \sum_{k,k'>k_F} {}' c^\dagger_{\mathbf{k}'\uparrow} c^\dagger_{-\mathbf{k}'\downarrow} c_{-\mathbf{k}\downarrow} c_{\mathbf{k}\uparrow}. \tag{E.17.5}$$

We look at the difference between the two situations: 1) The electron pair is added to the Fermi surface, i.e. with $|\mathbf{k}|, |\mathbf{k}'| = k_F$ and energy equal to zero. 2) The electron pair forms a coherent superposition of pairs not necessary at the Fermi surface.

For situation 2 we start by an Ansatz wavefunction, which is a superposition of so-called Cooper pairs

$$|\psi\rangle = \sum_{\mathbf{k}} \alpha_{\mathbf{k}} c^\dagger_{\mathbf{k}\uparrow} c^\dagger_{-\mathbf{k}\downarrow} |FS\rangle. \tag{E.17.6}$$

'(a) Show that if $\alpha_{\mathbf{k}}$ satisfies the following equation, then $|\psi\rangle$ is an eigenstate of H.

$$2\alpha_{\mathbf{k}}(\varepsilon_{\mathbf{k}} - \mu) - V_0 \sum_{k'>k_F} {}' \alpha_{\mathbf{k}'} = E\alpha_{\mathbf{k}}, \tag{E.17.7}$$

where E is the energy measured relatively to the energy of the filled Fermi sea. This now leads to a condition for E given by

$$1 = V_0 \sum_{k'>k_F} {}' \frac{1}{2(\varepsilon_{\mathbf{k}} - \mu) - E}. \tag{E.17.8}$$

(b) In order to find the energy E you should make use of the following hierarchy of energy scales

$$E \ll \omega_D \ll E_F, \tag{E.17.9}$$

where the validity of the first one of course must be checked at the end of the calculation. Find after reinserting $\hbar$ that

$$E = -2\hbar\omega_D \exp\left[-2/V_0 d(E_F)\right]. \tag{E.17.10}$$

(c) Discuss the following two important issues: Why does this result indicate an instability of the Fermi surface when compared to situation 1)? Could this result have been reached by perturbation theory in V_0?

Exercises for Chapter 18

Exercise 18.1

A single Cooper pair in the Fermi sea
Based on Eqs. (18.1) to (18.4) calculate the eigenenergy E_{CP} of a single Cooper pair in a Fermi sea and verify the results Eqs. (18.5) and (18.6).

Hints: Multiply $H_{\mathrm{CP}}|\psi_{\mathrm{CP}}\rangle = E_{\mathrm{CP}}|\psi_{\mathrm{CP}}\rangle$ with $\langle \mathbf{k}'|$. Then after some rearrangement multiply with $w'_{\mathbf{k}}$ and sum over $\mathbf{k}'$. Note how the number $\sum_{\mathbf{k}} w_{\mathbf{k}} a_{\mathbf{k}}$ can be divided out. How does the answer imply an instability of the Fermi sea?

Exercise 18.2

The size ξ of a Cooper pair

The charateristic energy spread δE of the two electrons forming a Cooper pair is given by the binding energy $|\Delta E_{\mathrm{CP}}|$.

(a) Make a Taylor expansion in momentum p around the Fermi energy $\varepsilon_F = p_F^2/2m$ to find the corresponding characteristic momentum spread δp of the Cooper pair. Use Heisenbergs uncertainty principle, $\delta p\, \delta x \sim \hbar$ to express the spatial uncertainty (and hence size) $\xi = \delta x$ of a Cooper pair.

(b) Insert numbers for a typical metal and compare the size of the Cooper pair with the typical distance between electrons in the Fermi sea. Is the Cooper pair small or large on that length scale?

Exercise 18.3

The BCS ground state: the mean value $\langle N \rangle$ of the particle number

The celebrated variational BCS ground state manifestly contains pair correlations:

$$|\psi_{\mathrm{BCS}}\rangle = \prod_{\mathbf{k}}(u_{\mathbf{k}} + v_{\mathbf{k}} c^{\dagger}_{\mathbf{k}\uparrow} c^{\dagger}_{-\mathbf{k}\downarrow})|0\rangle, \qquad (\text{E.18.1})$$

where $|0\rangle$ is the vacuum state, and where $u_{\mathbf{k}}$ and $v_{\mathbf{k}}$ are the amplitudes for the pair state $\mathbf{k}$ to be unoccupied and occupied, respectively (see the figure). The wavefunction is normalized by $|u_{\mathbf{k}}|^2 + |v_{\mathbf{k}}|^2 = 1$. The pair state is occupied (unoccupied) for $\varepsilon_{\mathbf{k}} \ll \varepsilon_F$ (for $\varepsilon_{\mathbf{k}} \gg \varepsilon_F$), while for $\varepsilon_{\mathbf{k}} \sim \varepsilon_F \pm 2k_B T_c$ a superposition of the pair state being both occupied and unoccupied results.

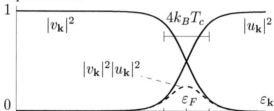

Show that the average particle number in $|\psi_{\mathrm{BCS}}\rangle$ is

$$\langle N \rangle = \langle \psi_{\mathrm{BCS}}|\hat{N}|\psi_{\mathrm{BCS}}\rangle = 2\sum_{\mathbf{q}}|v_{\mathbf{q}}|^2, \quad \text{with } \hat{N} = \sum_{\mathbf{q},\sigma} c^{\dagger}_{\mathbf{q},\sigma} c_{\mathbf{q},\sigma}. \qquad (\text{E.18.2})$$

Hints: Use that the symmetric appearance of spin up and down in $|\psi_{\mathrm{BCS}}\rangle$ allows for $\hat{N}$ to be written in the simpler form $\hat{N} = 2\sum_{\mathbf{q}} c^{\dagger}_{\mathbf{q}\uparrow} c_{\mathbf{q}\uparrow}$. Arrange $\prod_{\mathbf{k}}$ in $|\psi_{\mathrm{BCS}}\rangle$ in factors $\mathbf{k} \neq \mathbf{q}$ and $\mathbf{k} = \mathbf{q}$, and show that the terms with $\mathbf{k} \neq \mathbf{q}$ yield 1 while $\mathbf{k} = \mathbf{q}$ yields $|v_{\mathbf{q}}|^2$.

Exercise 18.4

The BCS ground state: the variance $\delta N^2 = \langle N^2 \rangle - \langle N \rangle^2$

Show that the variance of the particle number in $|\psi_{\mathrm{BCS}}\rangle$ is

$$\delta N^2 = \langle \psi_{\text{BCS}} | \hat{N}^2 | \psi_{\text{BCS}} \rangle - \langle N \rangle^2 = 4 \sum_{\mathbf{q}} |u_{\mathbf{q}}|^2 |v_{\mathbf{q}}|^2, \quad \text{with } \hat{N}^2 = 4 \sum_{\mathbf{q},\mathbf{p}} c^\dagger_{\mathbf{p}\uparrow} c_{\mathbf{p}\uparrow} c^\dagger_{\mathbf{q}\uparrow} c_{\mathbf{q}\uparrow},$$

where we again have used the spin symmetry to simplify $\hat{N}^2$. Hints: Use that $\sum_{\mathbf{p},\mathbf{q}}$ (E.18.3) $\sum_{\mathbf{p}=\mathbf{q}} + \sum_{\mathbf{p}\neq\mathbf{q}}$ and arrange as above $\prod_{\mathbf{k}}$ in $|\psi_{\text{BCS}}\rangle$ in factors $\mathbf{k} \neq \mathbf{q}, \mathbf{p}$ as well as $\mathbf{k} = \mathbf{p}$ and $\mathbf{k} = \mathbf{q}$. Use at some point the normalization condition $1 - |v_{\mathbf{k}}|^2 = |u_{\mathbf{k}}|^2$. Argue, based on the figure, that $\delta N^2 = 4 \sum_{\mathbf{q}} |u_{\mathbf{q}}|^2 |v_{\mathbf{q}}|^2 \approx (k_B T_c / \varepsilon_F) \langle N \rangle$. Calculate the relative standard deviation $\delta N / \langle N \rangle$ for a typical superconducting metal and discuss the result.

Exercise 18.5

The BCS ground state: Projection into a state with fixed particle number
We now specify the relative phase φ of the amplitudes $u_{\mathbf{k}}$ and $v_{\mathbf{k}}$, which becomes the phase of the BCS superconductor. We also define a new state $|\psi_N\rangle$:

$$|\psi^\varphi_{\text{BCS}}\rangle = \prod_{\mathbf{k}} (|u_{\mathbf{k}}| + |v_{\mathbf{k}}| e^{i\varphi} c^\dagger_{\mathbf{k}\uparrow} c^\dagger_{-\mathbf{k}\downarrow}) |0\rangle, \qquad |\psi_N\rangle = \int_0^{2\pi} d\varphi \, e^{-i\frac{1}{2}N\varphi} |\psi^\varphi_{\text{BCS}}\rangle. \tag{E.18.4}$$

Argue why $|\psi_N\rangle$ is the part of $|\psi^\varphi_{\text{BCS}}\rangle$ which contains exactly N particles (or $N/2$ pairs).

Exercise 18.6

The BCS ground state: determination of $u_{\mathbf{k}}$ and $v_{\mathbf{k}}$
We study the BCS groundstate $|\psi_{\text{BCS}}\rangle$ Eq. (18.8) and the BCS mean-field Hamiltonian $H^{\text{MF}}_{\text{BCS}}$ Eq. (18.10). For simplicity the parameters $u_{\mathbf{k}}$, $v_{\mathbf{k}}$, and Δ are all taken to be real:

$$|\psi_{\text{BCS}}\rangle = \prod_{\mathbf{k}} (u_{\mathbf{k}} + v_{\mathbf{k}} c^\dagger_{\mathbf{k}\uparrow} c^\dagger_{-\mathbf{k}\downarrow}) |0\rangle, \tag{E.18.5}$$

$$H^{\text{MF}}_{\text{BCS}} = \sum_{\mathbf{k}} \left[\xi_{\mathbf{k}} (c^\dagger_{\mathbf{k}\uparrow} c_{\mathbf{k}\uparrow} + c^\dagger_{\mathbf{k}\downarrow} c_{\mathbf{k}\downarrow}) - \Delta (c^\dagger_{\mathbf{k}\uparrow} c^\dagger_{-\mathbf{k}\downarrow} + c_{-\mathbf{k}\downarrow} c_{\mathbf{k}\uparrow}) \right]. \tag{E.18.6}$$

Use the techniques of Exercise 18.3 and Exercise 18.4 to find the free energy F:

$$F \equiv \langle \psi_{\text{BCS}} | H^{\text{MF}}_{\text{BCS}} | \psi_{\text{BCS}} \rangle = \sum_{\mathbf{k}} \left[\varepsilon_{\mathbf{k}} \, 2v_{\mathbf{k}}^2 - \Delta \, 2u_{\mathbf{k}} v_{\mathbf{k}} \right]$$

$$= \sum_{\mathbf{k}} \left[\varepsilon_{\mathbf{k}} \{ \cos(2\theta_{\mathbf{k}}) + 1 \} - \Delta \sin(2\theta_{\mathbf{k}}) \right], \tag{E.18.7}$$

where the parameter $\theta_{\mathbf{k}}$ is introduced to write $u_{\mathbf{k}} = \sin(\theta_{\mathbf{k}})$ and $v_{\mathbf{k}} = \cos(\theta_{\mathbf{k}})$ and explicitly ensure the normalization constraint $u_{\mathbf{k}}^2 + v_{\mathbf{k}}^2 = 1$. Regard $|\psi_{\text{BCS}}\rangle$ as a variational wave function and vary $\theta_{\mathbf{k}}$ to minimize F. Define $E_{\mathbf{k}} = \sqrt{\varepsilon_{\mathbf{k}}^2 + \Delta^2}$ and show:

$$\frac{\partial F}{\partial \theta_{\mathbf{k}}} = 0 \quad \Rightarrow \quad \tan(2\theta_{\mathbf{k}}) = -\frac{\Delta}{\varepsilon_{\mathbf{k}}} \quad \Rightarrow \quad \begin{cases} 2u_{\mathbf{k}} v_{\mathbf{k}} = \frac{\Delta}{E_{\mathbf{k}}}, \\ v_{\mathbf{k}}^2 - u_{\mathbf{k}}^2 = -\frac{\varepsilon_{\mathbf{k}}}{E_{\mathbf{k}}}. \end{cases} \tag{E.18.8}$$

Combine the very last equation with $v_{\mathbf{k}}^2 + u_{\mathbf{k}}^2 = 1$ and obtain (and comment on) the result:

$$v_{\mathbf{k}}^2 = \frac{1}{2}\left(1 - \frac{\varepsilon_{\mathbf{k}}}{E_{\mathbf{k}}}\right), \qquad u_{\mathbf{k}}^2 = \frac{1}{2}\left(1 + \frac{\varepsilon_{\mathbf{k}}}{E_{\mathbf{k}}}\right). \tag{E.18.9}$$

Exercise 18.7

The condensation energy per electron in a superconductor
At $T = 0$ the superconductor energy gap Δ is maximal. Roughly speaking, the N_Δ electrons, which occupy the energy shell $\varepsilon_F - \Delta < \varepsilon_{\mathbf{k}} < \varepsilon_F + \Delta$, each gains the Cooper binding energy Δ when the system undergoes the transition from the normal metal state to the superconducting state. Calculate in terms of Δ and ε_F the fraction N_Δ/N, where N is the total number of electrons in the system. Show that the total energy gain, also denoted the condensation energy, per electron is the following and comment on the result:

$$\frac{N_\Delta}{N}\Delta = 10^{-8} \text{ eV}. \tag{E.18.10}$$

Exercise 18.8

Diagonalizing the BCS Hamilton by a Bogoliubov transformation
To diagonalize the BCS Hamiltonian it is convenient to write it in matrix form

$$H_{BCS}^{MF} = \sum_{\mathbf{k}} \left(c_{\mathbf{k}\uparrow}^\dagger \ c_{-\mathbf{k}\downarrow} \right) \begin{pmatrix} \xi_{\mathbf{k}} & \Delta_{\mathbf{k}} \\ \Delta_{\mathbf{k}}^* & -\xi_{\mathbf{k}} \end{pmatrix} \begin{pmatrix} c_{\mathbf{k}\uparrow} \\ c_{-\mathbf{k}\downarrow}^\dagger \end{pmatrix}$$

$$+ \sum_{\mathbf{k}} \xi_{\mathbf{k}} + \sum_{\mathbf{kk'}} V_{\mathbf{kk'}} \langle c_{\mathbf{k}\uparrow}^\dagger c_{-\mathbf{k}\downarrow}^\dagger \rangle \langle c_{\mathbf{k'}\downarrow} c_{-\mathbf{k'}\uparrow} \rangle, \tag{E.18.11}$$

$$= \sum_{\mathbf{k}} \mathbf{A}_{\mathbf{k}}^\dagger \mathbf{H}_{\mathbf{k}} \mathbf{A}_{\mathbf{k}} + \text{constant}, \tag{E.18.12}$$

where

$$\mathbf{A}_{\mathbf{k}} = \begin{pmatrix} c_{\mathbf{k}\uparrow} \\ c_{-\mathbf{k}\downarrow}^\dagger \end{pmatrix}, \qquad \mathbf{H}_{\mathbf{k}} = \begin{pmatrix} \xi_{\mathbf{k}} & \Delta_{\mathbf{k}} \\ \Delta_{\mathbf{k}}^* & -\xi_{\mathbf{k}} \end{pmatrix}. \tag{E.18.13}$$

To bring the Hamiltonian into a diagonal form, we introduce the unitary transformation

$$\mathbf{B}_{\mathbf{k}} = \begin{pmatrix} \gamma_{\mathbf{k}\uparrow} \\ \gamma_{-\mathbf{k}\downarrow}^\dagger \end{pmatrix} = \mathbf{U}_{\mathbf{k}}^{-1} \mathbf{A}_{\mathbf{k}}, \qquad \mathbf{U}_{\mathbf{k}} = \begin{pmatrix} u_{\mathbf{k}} & -v_{\mathbf{k}} \\ v_{\mathbf{k}}^* & u_{\mathbf{k}}^* \end{pmatrix}, \tag{E.18.14}$$

which diagonalizes the problem if

$$\mathbf{U}_{\mathbf{k}}^\dagger \mathbf{H}_{\mathbf{k}} \mathbf{U}_{\mathbf{k}} = \begin{pmatrix} E_{\mathbf{k}} & 0 \\ 0 & \tilde{E}_{\mathbf{k}} \end{pmatrix}. \tag{E.18.15}$$

Use this expression to prove Eqs. (18.13) to (18.15).

Exercise 18.9

The London equation and the Meissner effect

Derive the differential equation Eq. (18.54) for the vector potential $\mathbf{A}$ starting from the London equation Eq. (18.52) for a static magnetic field.

Hint: you may need to rewrite $\nabla \times \nabla \times \mathbf{A}$. Then consider the following simple geometry in xyz space: the half-space $x < 0$ is vacuum, while a superconducting metal fills up the other half-space $x > 0$. A magnetic field $\mathbf{B}(x, y, z) = B(x)\hat{\mathbf{e}}_z$, with $B(x) = B_0$ for $x < 0$, is present. Calculate its form in the superconductor $(x > 0)$ and discuss what happens in the boundary layer of thickness $\lambda_L = \sqrt{m/(\mu_0 \rho_s e^2)}$. The parameter λ_L is denoted the London length. How large is λ_L for a typical superconducting metal at $T = 0$, where all the electrons have condensed into the superconducting condensate?

Exercise 18.10

RSJ model of a Josephson junction: the ac Josephson effect

Consider a Josephson junction as in Section 18.7, however now with a finite bias voltage applied across the junction. One can still have a dc Josephson current I_J running given by Eq. (18.88)

$$I_J = I_C \sin \phi. \tag{E.18.16}$$

This relation is also known as the first Josephson relation.

The finite voltage changes the energy and hence the phase of the electrons in the two superconductors of the junction. We can simply include this phase change in the time dependence of by the following substitution

$$c(t) \quad \rightarrow \quad c(t)e^{iVt/2}, \qquad f(t) \quad \rightarrow \quad f(t)e^{-iVt/2}, \tag{E.18.17}$$

which corresponds to

$$\phi \quad \rightarrow \quad \phi + 2eVt, \tag{E.18.18}$$

or, after reintroducing $\hbar$,

$$\dot{\phi} = \frac{2e}{\hbar}V, \tag{E.18.19}$$

which is called the second Josephson relation.

The second Josephson relation adds interesting dynamics to the Josephson junction because of the intrinsic frequency $2eV/\hbar$. One can measure this frequency by applying external RF radiation to the junction. The Josephson junction thus acts as a voltage to frequency converter, which has many applications.

Now we look at the current-voltage characteristic of a Josephson junction in the resistively shunted unction (RSJ) model. The current is carried by two kinds of electrons: those that are paired and those that are not. The pair current is described by the Josephson relations while the normal current is supposed to be given by Ohm's law.

Consider a current biased setup, i.e. a junction with a fixed current I. This current is made up by the sum of the supercurrent and the normal current. Thus

$$I = I_N + I_J = \frac{1}{R}V + I_C \sin \phi = \frac{\hbar}{2eR}\dot{\phi} + I_C \sin \phi. \tag{E.18.20}$$

Write this equation in the dimensionless form

$$\eta = \frac{I}{I_C} = \frac{d\phi}{d\tau} + \sin\phi, \qquad \tau = \frac{2eI_cR}{\hbar}t. \qquad \text{(E.18.21)}$$

The voltage is time dependent, but in a dc measurement one measures the average voltage. Integrate the equation for η and show that the average voltage becomes

$$\langle V \rangle = \begin{cases} 0, & I < I_C, \\ RI_c\sqrt{(I/I_C)^2 - 1}, & I > I_C. \end{cases} \qquad \text{(E.18.22)}$$

Hint: first find solutions for $\dot{\phi} = 0$ and then a "running" solution where $\dot{\phi} \neq 0$. For the last situation the average voltage is $\langle V \rangle = \frac{1}{T}\int_0^T dt\frac{d\phi}{dt} = \frac{2\pi}{T}$. Here T is the period of the voltage or the time it takes to increase ϕ by 2π.

Exercises for Chapter 19

Exercise 19.1

Some useful expressions for boson operators
The following three expressions are useful for further studies of the 1D transmission line model of the Luttinger liquid.

(a) Let O and B be two operators which commutator $[O, B]$ is a c-number. Prove the commutator for displacement operator e^B

$$[O, e^B] = [O, B]\, e^B. \qquad \text{(E.19.1)}$$

Hint: consider the function $f(\lambda) = e^{-\lambda B}Oe^{\lambda B}$ and differentiate with respect to λ.

(b) Prove the Baker-Hausdorff formula:

$$e^{A+B} = e^A e^B e^{-\frac{1}{2}[A,B]} \qquad \text{(E.19.2)}$$

Hint: use the analogy with the time-evolution operator in the interaction picture, e.g. by writing $U(\tau) = e^{-\tau A}e^{\tau(A+B)}$ and differentiate with respect to τ and then just keep going.

(c) Consider coherent states of the form $\exp(g\, a_k^\dagger)|0\rangle$, where g is some constant. Show that the overlap between two such states is given by

$$\left\langle 0\left|e^{fa}e^{ga_k^\dagger}\right|0\right\rangle = \exp(gf). \qquad \text{(E.19.3)}$$

Hint: you can do it by expanding the exponentials and by showing that $a^n\left(a^\dagger\right)^n|0\rangle = na^{n-1}\left(a^\dagger\right)^{n-1}|0\rangle$ and hence $a^n\left(a^\dagger\right)^n|0\rangle = n!|0\rangle$.

Exercise 19.2

A transmission line model for an 1D electron system
To get a better physical understanding of the physics of Luttinger liquids, we look at a simpler and maybe more familiar case: a line of inductors and capacitors. We will quantize this transmission line and see that it has a lot in common with the

Luttinger liquid. The quantum fluctuations of transmission lines was an important piece of physics that went into the understanding of Coulomb blockade in single tunnel junction, now known as the environment theory of Coulomb blockade.[72]

As shown in the following figure, a transmission line consists of a series of inductors and capacitors. If we go to a continuum description, we have a wire with a distributed capacitance (capacitance per unit length) c and an inductance per unit length, l.

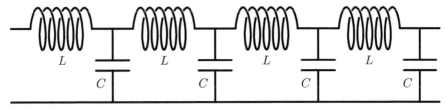

The energy for such a system is

$$E = \int dx \left(\frac{l}{2} [j(x)]^2 + \frac{1}{2c} [\rho(x)]^2 \right), \qquad (E.19.4)$$

where $j(x)$ is the local current and $\rho(x)$ is the local charge density at position x. The first part corresponds to the kinetic energy while the later is the potential energy. The current and the charge densities are linked together through the continuity equation:

$$\dot{\rho} + \partial_x j = 0, \qquad (E.19.5)$$

and if we define a displacement density, u, given by $\rho = \partial_x u$, then from the continuity equation the current is $j = -\dot{u}$. Let us now write down a Lagrangian for our transmission line and identifying the inductive terms as the kinetic energy

$$L = E_{\text{kin}} - E_{\text{pot}} = \int dx \left(\frac{l}{2} [\dot{u}(x,t)]^2 - \frac{1}{2c} [\partial_x u(x,t)]^2 \right). \qquad (E.19.6)$$

(a) Derive from L the equation of motion,

$$\ddot{u} - v^2 \partial_x^2 u = 0, \quad v = \sqrt{1/lc}, \qquad (E.19.7)$$

using the principle of least action: let $u = u_0 + \delta u$ and find the extremum condition for u_0. (You can also derive the same equation from the Euler-Lagrange equation.)

(b) Defining the canonical momentum field $\pi(x)$ of the field variable u as $\pi = \frac{\delta L}{\delta \dot{u}} = l\dot{u}$, show from the usual connection between the Lagrangian L and the Hamiltonian H that

$$H = \int dx \left(\frac{[\pi(x)]^2}{2l} + \frac{1}{2c} [\partial_x u(x)]^2 \right), \qquad (E.19.8)$$

which in fact is the Hamiltonian of string.

---

[72] see e.g. G.L. Ingold and Y.V. Nazarov in Single Charge Tunneling, p. 21-107, NATO ASI Series B, Vol. 294, ed. by H. Grabert and M.H. Devoret (Plenum, 1992)

Next let us quantize the transmission mode and find the electromagnetic eigenmodes. The commutators between a variable and its canonical conjugate in real space and k-space is

$$[u(x), \pi(x')] = i\delta(x - x'), \qquad [u(k), \pi(-k')] = i\mathcal{L}\delta_{kk'}, \qquad \text{(E.19.9)}$$

where we defined the Fourier transforms as

$$(u(k), \pi(k)) = \int_0^{\mathcal{L}} dx \, e^{ikx}(u(x), \pi(x)) \qquad \text{(E.19.10)}$$

(c) Write down the Fourier transform of H and employ the usual trick of defining creation and annihilation operators to bring H on the diagonal form

$$H = \sum v|k| \left(a_k^\dagger a_k + \frac{1}{2} \right), \qquad \text{(E.19.11)}$$

where the operators a and $a^\dagger$ obey $[a_k, a_{k'}^\dagger] = \delta_{kk'}$ and where

$$u(k) = \left(a_k + a_{-k}^\dagger \right) \sqrt{\frac{\mathcal{L}}{2|k|lv}}, \qquad \pi(-k) = i \left(a_k^\dagger - a_{-k} \right) \sqrt{\frac{\mathcal{L}lv|k|}{2}}. \qquad \text{(E.19.12)}$$

The modes that we have found are the collective excitations of charged system and we can identify them as the plasmon excitations. Unlike in three dimensions plasmons where the plasmons are gapful, we here in 1D have a gapless modes with linear dispersion relation. In 2D the situation is intermediate since $w_{pl}^{(2D)} \propto \sqrt{q}$, see Exercise 14.3.

Exercise 19.3

Charge displacement operator for sudden addition of charges
Consider the physical situation where a charge q suddenly appears at position x_0. Define the displacement operator, $D_q(x) = e^{-iqA(x)}$, where A is a Hermitian operator, such that $[\rho(x), D_q(x_0)] = qD_q(x)\delta(x - x_0)$.
    (a) Show that the expectation value of $\rho(x)$ in the state $|\psi_q(x_0)\rangle = D_q(x_0)|0\rangle$ is

$$\langle \psi_q(x_0)|\rho(x)|\psi_q(x_0)\rangle = q\delta(x - x_0). \qquad \text{(E.19.13)}$$

    (b) Show that a possible choice of A is the operator

$$A(x) = -\int_{-\infty}^{x} dx' \pi(x'). \qquad \text{(E.19.14)}$$

Exercise 19.4

Correlation function of charge displacement operator
Next we study the correlation function of the charge excitation corresponding to the displacement operator $D_q(x)$. Suppose a charge was created at position $x = 0$ at time $t = 0$, then the amplitude for finding the charge at some later time, t, at the

same position is given by the overlap: $\langle\psi_q(0,t)|\psi_q(0,0)\rangle$. This overlap defines an equal position charge propagator or Green's function:

$$G(t) = \langle 0|D_q^\dagger(0)e^{-iHt}D_q(0)|0\rangle. \tag{E.19.15}$$

(a) Use result from Exercise 19.1(b) to decompose the displacement operator so that the G is of the form $G = \prod_k c_k e^{-|f_k|^2}$, where $c_k = \langle 0|e^{f_k a_k(t)}e^{f_k^* a_k^\dagger}|0\rangle$.

(b) Use result from Exercise 19.1(c) to write c_k as $c_k = \exp(|f_k|^2 e^{-i\tilde{v}|k|t})$ and collect terms as an integral in the exponent.

(c) Now you will realize that the integral does not converge for large k, therefore include a cut-off as $f_k \rightarrow f_k e^{-|k|/k_0}$ and, finally, get a power-law at long times, $G \propto t^{-g}$, where $g = Ze^2/h$. Here $Z = \sqrt{l/c}$ is the impedance of the transmission line. Hint: use the integral

$$\int_0^\infty dx \frac{\cos(xt) - 1}{x} e^{-x} = -\frac{1}{2}\ln(1 + t^2).$$

(d) Evaluate G from Eq. (E.19.15)

Exercise 19.5

Analogy between the transmission line model and the Luttinger liquid
Go back to the Hamiltonian for the transmission line and try to write it in a form similar to the Tomonaga-Luttinger model

$$H = \frac{\tilde{v}}{2} \int dx \left(gP^2(x) + \frac{1}{g}(\partial_x\phi(x))^2 \right). \tag{E.19.16}$$

How are the g-parameters of the two models related physically?

Exercise 19.6

Luttinger liquid theory of fractional quantum Hall edge states
In this exercise we work with a simplistic version of the Luttinger liquid theory of fractional quantum Hall edge states (Wen 1992, 2004).

As briefly mentioned in Eq. (1.5) and Fig. 1.1 the eigenstates ψ_{nk} of the free electrons in the xy plane subjected to a strong perpendicular magnetic field B are given by

$$\psi_{nk}(x, y) = e^{ikx}\phi_n(y - k\ell^2), \tag{E.19.17}$$

where n is the integer Landau level index, $\ell = \sqrt{\hbar/eB}$ is the magnetic length, and k is the wavenumber which also gives the center $k\ell^2$ of the harmonic oscillator functions ϕ_n. Each Landau level is degenerate, with density of states per area given by $1/2\pi\ell^2$. The filling, ν, per available state is called the filling fraction

$$\nu = \rho_0 \, 2\pi\ell^2, \tag{E.19.18}$$

where ρ_0 is the electron density per area. In terms of ν the classical Hall resistance is given by

$$R_H = \frac{1}{\nu}\frac{h}{e^2}. \qquad (E.19.19)$$

Near special filling fractions, such as $\nu = 1/3$, the Hall resistance shows a plateau as a function of v due to properties of the many-body state, but this is not the issue here. We focus entirely on the edges. Only states close to the edges are influenced by the confining potential, and there they acquire a energy dispersion

$$\varepsilon_{nk} \approx \omega_c(n+\frac{1}{2}) + V_{\text{conf}}(k\ell^2) + \frac{1}{2}mv_D^2. \qquad (E.19.20)$$

$V_{\text{conf}}(y)$ is the confining potential that is flat in the bulk of the sample but bends upwards at the edge, and $v_D = V'_{\text{conf}}(y)/eB$ is the drift velocity, which is non-zero only for states near the edge.

Following the analysis of Section 19.5 our analysis of the fractional Hall state edges begins by studying the density operator for electrons near the edge. Consider the density near the, say, right edge

$$\rho_R(q) = \sum_{k>0} c_k^\dagger c_{k+q}. \qquad (E.19.21)$$

(a) Show that the commutator between density operators is

$$[\rho_R(q), \rho_R(-q')] \approx \frac{L}{2\pi}\text{sign}(q) \sum_{0<k<|q|} \langle n_k \rangle \approx \frac{Lq}{2\pi}\nu. \qquad (E.19.22)$$

The special nature of the FQH groundstate only enters in the last approximation: it would not make sense to replace the occupation $\langle n_k \rangle$ by unity, but it should rather be the occupation of the FQH bulk state, i.e., $\langle n_k \rangle = \nu$. From this observation now follows some interesting consequences.

(b) In analogy with the treatment in Section 19.5.3 introduce the fields $\phi(x)$ and $P(x)$ through Fourier transformations of the density operators,

$$\frac{1}{\sqrt{\pi}}\partial_x\phi = \rho_L(x) + \rho_R(x), \qquad \frac{1}{\sqrt{\pi}}P(x) = [\rho_L(x) - \rho_R(x)]\frac{1}{\nu}, \qquad (E.19.23)$$

and show that in order to preserve the canonical commutation relations, an additional factor of $1/\nu$ must be included in the definition of P.

(c) The second basic assumption is that the energy of edge excitations is quadratic in charge density. In other words, we assume that there is no gap for charge fluctuations near the edge, and then we expand the energy in density fluctuations. Show in analogy with Section 19.5 that this this results in a bosonic Hamiltonian with a structure similar to that of the Luttinger liquid Hamiltonian:

$$H = \frac{\pi v_D'}{L}\sum_q \rho(q)\rho(-q) = \frac{v_D'\nu}{2}\int dx \left(\nu P(x)^2 + \frac{1}{\nu}[\partial_x\phi]^2\right), \qquad (E.19.24)$$

where v_D' is some velocity of the edge states, related to the drift velocity v_D.

(d) In order to probe this interesting structure of the edge states, tunneling experiments were suggested by Wen and has been performed recently. To calculate the tunneling density of states we need a single electron operator $\Psi(x)$, and as in Section 19.6 we must demand that $[\rho(x), \Psi(x')] = \Psi(x)\delta(x - x')$. Show that because of the unusual commutation relation Eq. (E.19.22) $\Psi(x)$ takes the form

$$\Psi(x) \sim e^{-2\pi i \int^x dx' \rho(x')/\nu},\qquad\text{(E.19.25)}$$

with an additional ν appearing.

(e) Show that the anti-commutator of $\Psi(x)$ and $\Psi(x')$ is

$$\Psi(x)\Psi(x') = \Psi(x')\Psi(x)\exp(i\pi/\nu).\qquad\text{(E.19.26)}$$

(f) Determine which values of the filling fraction ν in the former question that lead to proper Fermi statistics.

SELECTED BIBLIOGRAPHY

Abrikosov, A.A., Gorkov, L.P., and Dzyaloshinski, I.E. (1975), *Methods of quantum field theory in statistical physics*, Dover Publications (New York).

Alhassid, Y., (2000), *The statistical theory of quantum dots*, Review of Modern Physics **72**, 895.

Aleiner, I.L., Brouwer P.W., and Glazman, L.I., (2002), *Quantum effects in Coulomb blockade*, Physics Reports **358**, p. 309.

Altshuler, B.L., Lee, P.A., and Webb, R.A. (1991), *Mesoscopic phenomena in solids*. North-Holland (Amsterdam).

Anderson, P.W., (1984), *Basic notions of condensed matter physics*, The Benjamin/Cummings Publishing Company (London).

Ashcroft, N.W., and Mermin, N.D. (1981), *Solid state physics*, Holt-Saunders International Editions (Tokyo).

Averin, D.V., Likharev, K.K. (1990), in *Mesoscopic Phenomena in Solids*, edited by Althuler, B.L, Lee, P.A., and Webb, R.A. (Elsevier, Amsterdam).

Averin, D.V., and Nazarov, Yu.V., (1990), *Virtual electron diffusion during quantum tunneling of the electric charge*, Physical Review Letters **65**, 2446.

Beenakker, C.W.J., and van Houten, H. (1991), *Quantum transport in semiconductor nanostructures*, Solid State Physics **44**, 1, eds. H. Ehrenreich and D. Turnbull, (Academic Press, Boston).

Beenakker, C.W.J, (1991), *Theory of Coulomb oscillations in the conductance of a quantum dot*, Physical Review B **44**, 1646.

Beenakker, C.W.J. (1997), *Random-matrix theory of quantum transport*, Review of Modern Physics **69**, 731.

Breuer, H.-P, and Petruccione, F., (2002). *The theory of open quantum systems*, Oxford University Press (Oxford).

Büttiker, M (1990), *Quantized transmission of a saddle-point constriction*, Physical Review B **41**, 7906.

Datta, S. (1997), *Electronic transport in mesoscopic systems*, Cambridge University Press (Cambridge).

de Gennes, P.-G. (1999), *Superconductivity of Metals and Alloys*, Perseus Books Group (New York).

Dirac, P.A.M. (1989) *The principles of quantum mechanics* Oxford University Press (Oxford, fourth revised edition).

Doniach, S., and Sondheimer, E.H. (1974), *Green's functions for solid state physicists*, The Benjamin/Cummings Publishing Company (London).

Egger, R., Bachtold, A., Fuhrer, M., Bockrath, M., Cobden, D. and McEuen, P. (2001). *Luttinger liquid behavior in metallic carbon nanotubes*, in *Interacting Electrons in Nanostructures*, eds. R. Haug and H. Schoeller. (Springer, New York).

Ferry, D.K., and Goodnick, S.M. (1999), *Transport in nanostructures*, Cambridge University Press (Cambridge).

Fetter, A.L., and Walecka, J.D. (1971), *Quantum theory of many-particle systems*, McGraw-Hill (New York).

Feynman, R.P. (1972), *Statistical Mechanics*, Addison-Wesley (Readding MA).

Fisher, M.P.A, and Glazman, L.I., (1997), *Transport in a one-dimensional Luttinger liquid*, in *Mesoscopic Electron Transport*, eds. Sohn, L.L, Kouwenhoven, L.P., and Schön, G. NATO ASI Series E, Appl. Sci., No. 345. (Kluwer Academic Publ.).

Giamarchi, T. (2003), *Quantum Physics in One Dimension*, Oxford University Press (Oxford).

Gogolin, A., Nersesyan A., and Tsvelik A. (1998), *Bosonization and Strongly Correlated Systems*, Cambridge University Press (Camdridge).

Grabert, H., and Devoret, M.H., (1992), *Single Charge Tunneling: Coulomb Blockade Phenomena in Nanostructures*, Proc. ASI, Les Houches (France), 1991 (Plenum publishing corporation).

Haldane, F. D. M. (1981), *'Luttinger liquid theory' of one-dimensional quantum fluids: I. Properties of the Luttinger model and their extension to the general 1D spinless Fermi sea*, Journal of Physics C: Solid State Physics **14**, 2585.

Halperin, B.I. (1982), *Quantized Hall conductance, current-carrying edge states, and the existence of extended states in a two-dimensional disordered potential* , Physical Review B **25**, 2185.

Haug, H., and Jauho, A.-P. (1996), *Quantum kinetics in transport and optics of semiconductors*, Springer-Verlag (Berlin).

Hewson, A.C. (1993), *The Kondo problem to heavy fermions*, Cambridge University Press (Cambridge).

Imry, Y. (1997), *Introduction to Mesoscopic Physics* Oxford University Press (New York).

Jensen, J., and Mackintosh, A.R. (1991), *Rare earth magnetism*, Oxford University Press (Oxford).

Kittel, C. (1995), *Introduction to solid state physics*, Wiley Text Books (New York).

Kittel, C., and Kroemer, H., (2000), *Thermal physics*, W. H. Freeman and Company (New York).

Landau, L.D., and Lifshitz, E.M. (1977), *Quantum mechanics*, Pergamon Press (Oxford).

Landau, L.D., and Lifshitz, E.M. (1982), *Statistical physics, part 1*, 3rd ed., Pergamon Press (Oxford).

Mahan, G.D. (1990), *Many-particle physics*, 2nd ed., Plenum Press (New York).

Mattuck, R.D. (1976), *A guide to Feynman diagrams in the many-body problem*, Dover Publications (New York).

Mehta, M.L. (1991), *Random matrices and the statistical theory of energy levels*, 2nd ed., Academic (New York).

Meir, Y., and Wingreen, N.S., (1992), *Landauer formula for current through an interacting electron region*, Physical Review Letters **68**, 2512.

Merzbacher, E. (1970), *Quantum Mechanics*, John Wiley & Sons (New York).

Nozières, P. (1997), *The many-body problem*, Addison-Wesley (Reading MA).

Pines, D. (1997), *Theory of interacting Fermi systems*, Addison-Wesley (Reading MA).

Pustilnik, M., and Glazman, L.I. (2004), *Kondo effect in quantum dots*,
  Journal of Physics: Condensed Matter **16**, R513.

Rammer, J. (2004). *Quantum transport theory*, Perseus Books (New York).

Rickayzen, G. (1991), *Green's functions and condensed matter*,
  Academic Press (London).

van Ruitenbeek, J.M. (1999), *Conductance quantization in metallic point contacts*, in:
  Metal Clusters on Surfaces: Structure, Quantum Properties, Physical Chemistry,
  K.-H. Meiwes-Broer, ed., Springer-Verlag (Berlin).

Schrieffer, J.R. (1983), *Theory of superconductivity*, 3rd revised printing,
  Addison-Wesley (Readding MA).

Schrieffer, J.R., and Wolff, P.A. (1966), *Relation between the Anderson
  and Kondo Hamiltonians*, Physical Review **149**, 491.

Sohn, L.L, Kouwenhoven, L.P., Schön, G. (1997), *Mesoscopic Electron Transport*
  NATO ASI Series E, Applied Sciences, No. 345., Kluwer Academic Publishers.

Starykh, O. A., Maslov, D. L., Häusler, W. and Glazman, L. I. (2000). *Gapped Phases
  of Quantum Wires*, in: *Low-dimensional systems interactions and transport properties*, ed. Brandes, Lecture Notes in Physics **544**, Springer-Verlag (New York).

Stone, A.D., Mello, P.A., Muttalib, K.A., and Pichard, J.-L. (1991),
  in: *Mesoscopic phenomena in solids*,
    eds. Altshuler, B.L., Lee, P.A., and Webb, R.A., North Holland (Amsterdam).

Sólyom, J. (1970), *The Fermi gas model of one-dimensional conductors*,
  Advances in Physics **28**, 201.

Tinkham, M. (1996), *Introduction to superconductivity*, McGraw-Hill (New York).

Tserkovnyak, Y., Halperin, B.I., Auslaender, O.M., and Yacoby, A. (2003),
  Interference and zero-bias anomaly in tunneling between Luttinger-liquid wires,
  Physical Review B **68**, 125312.

Tsvelick, A.M., and Wiegmann, P.B. (1983),
  Exact results in the theory of magnetic alloys, Advances in Physics **32**, p. 453.

Voit, J. (1995), *One-Dimensional Fermi liquids*,
  Reports on Progress in Physics **58**, 977.

von Delft, J. and Schoeller, H.(1998), *Bosonization for
  Beginners – Refermionization for Experts*, Annalen der Physik **7**, 225.

Weiss, U. (1999). *Quantum dissipative systems*, World Scienticfic (Singapore).

Wen, X.G. (1992), *Theory of the edge states in fractional quantum Hall effects*,
  International Journal of Modern Physics **6**, 1711.

Wen, X.G. (2004), *Quantum Field Theory of Many-Body Systems - From the Origin
  of Sound to an Origin of Light and Electrons*, Oxford University Press (Oxford).

Wingreen N.S. and Meir Y. (1992), *Landauer formula for the current through
  an interacting region*, Physical Review Letters **68**, 2512.

Yosida, K. (1996), *Theory of magnetism*, Springer-Verlag (Berlin).

INDEX

acoustic phonons
  Cooper instability, 322
  Debye phonons, 53
  graphical representation, 52, 57
  Green's functions, 313
  in second quantization, 56
  jellium model, 321
  Migdal's theorem, 318
adiabatic continuity, 266
advanced function, 191
  Fourier transformation, 88
  Green's function, 124
Aharonov-Bohm effect, 117
analytic continuation, 189
analytic function, 189
Anderson model, 153
  for magnetic impurities, 142
  general current formula, 161
  relation to Kondo model, 168
annihilation operators
  bosons, 10
  fermions, 13
  phonons, 55
  time dependence, 84
  time-derivative, 139
anti-commutator
  1D, 365
  definition, 13
anti-symmetrization operator, 7
antiferromagnetism, 74
anyons, 5
atom
  Bohr radius a_0, 41
  electron orbitals, 2
  ground state energy E_0, 41
  in metal, 32
attractive pair-interaction, 319

Baker-Hausdorff formula, 416
band structure diagram
  extended zone scheme, 36
  metal, semiconductor, insulator, 46
basis states
  change in second quantization, 16
  complete basis set, 3
  Green's function, 123
  many-particle boson systems, 13
  many-particle fermion systems, 14
  orthonormal basis set, 2
  scattering states, 106
  systems with different particles, 25

BCS theory
  critical temperature, 333
  interaction potential model, 323
  mean-field Hamiltonian, 329
  Nambu formalism, 335
  quasiparticle density of states, 334
  self-consistent gap equation, 330, 332
  tunneling spectroscopy, 134
Bloch
  band structure, 36
  Bloch theory of lattice electrons, 33
  Bloch's equation, density matrix, 185
  Bloch's theorem, 35
Bogoliubov transformation
  BCS Hamiltonian, 414
  bosons, 360, 382
  fermions, 381
Bohm–Staver sound velocity
  from RPA-screened phonons, 321
  semi-classical, 54
Bohr radius a_0, 41
Boltzmann distribution, 27, 153
Boltzmann equation
  collision free, 271
  introduction, 266
  with impurity scattering, 274
Born approximation
  first Born approximation, 217, 291
  full Born approximation, 220
  in conductivity, 293
  self-consistent Born approximation, 222
  spectral function, first order, 219
Born–Oppenheimer approximation, 317
Bose–Einstein distribution, 29, 52
Bose–Einsten condensation, 346
boson
  creation/annihilation operators, 10
  defining commutators, 12
  definition, 5
  frequency, 189, 193
  many-particle basis, 13
bosonization, 357, 364
bra state, 2
Brillouin zone
  band structure diagram, 36
  definition, 35
  for 1D phonons, 54
broadening of the spectral function, 130
broken symmetry, 72, 341

canonical

ensemble, 28
momentum, 21
partition function, 27
carbon nanotubes
1D electron gas, 49
Luttinger liquids, 348
charge-charge correlation function, 99, 256
charging energy, 151
chemical potential
definition, 28
temperature dependence, 40
coherence length
in mesoscopic systems, 102
in superconductivity, 327
in weak localization, 298
coherent state, 416
collapse of wavefunction, 1
commutator
$[AB, C] = A[B, C] + [A, C]B$, 85
$[AB, C] = A\{B, C\} - \{A, C\}B$, 85
1D left/right movers, 357
defining bosons, 12
defining fermions, 13
general definition, 11
complete
basis states, 3
set of quantum numbers ν, 3
compressibility, 1D, 362
conductance
Anderson model, 161
conductance fluctuations, 212
Kubo formalism, 97
mesoscopic system, 108
quantization, 113
quantum dot device, 162
universal fluctuations, 308
conductivity
cooperons, 303
diffusons, 291
introduction, 285
Kubo formalism, 95
relation to dielectric function, 100
semi-classical approach, 272
connected Feynman diagrams, 231, 239
conservation of four-momentum, 235
conserving approximation, 291
continuity equation
for electric current, 100
for ions in the jellium model, 53
for quasiparticles, 271
contour integral, 194
convergence factor
retarded function, 88
convergence of Matsubara functions, 188
Cooper
Cooper pairs, 325
instability, Feynman diagrams, 322

instability, wave function, 325
cooperons in conductivity, 303
core electron, 32
correlation function
charge-charge correlation, 99
current-current correlation, 96, 286
general Kubo formalism, 94
correlation hole around electrons, 66
correlation, in transport, 151, 165
cotunneling
definition, 158
elastic, 160
inelastic, 159
Coulomb blockade
and Kondo effect, 178
in the Anderson model, 165
metallic dot, 154
Coulomb interaction
combined with phonons, 316
direct process, 44
divergence, 45, 246
exchange process, 45
in conductivity, 287
RPA renormalization, 250, 260
screened impurity scattering, 208
second quantization, 23
Yukawa potential, RPA-screening, 251
coupling constant
electron interaction strength e_0^2, 23
electron-phonon, general, 63
electron-phonon, jellium model, 65
electron-phonon, lattice model, 64
electron-phonon, RPA-renormalized, 320
integration over, 253
creation operators
bosons, 10
fermions, 13
phonons, 55
time dependence, 84
critical temperature
BCS theory, 333
Cooper instability, 323
ferromagnetism, 76
crossed diagram
definition, 301
maximally crossed, 302
suppressed in the Born approx., 223
current density operator
dia- and paramagnetic terms, 96
second quantization, 22
current-current correlation function, Π
BCS theory, 337
definition, 96
diagrammatics, 286
cut-off
in Anderson model, 146
in Kondo model, 176

momentum in 1D, 365
Tomonaga model, 361

d-shell, 143
Debye
    acoustical Debye phonons, 53
    Debye energy or frequency ω_D, 60
    Debye model, 53, 59, 322
    Debye temperature T_D, 60
    Debye wave number k_D, 60
    density of states, Debye model, 60
    frequency cut-off, BCS, 323
delta function $\delta(\mathbf{r})$, 4
density in second quantization, 22
density matrix operator, 27
density of states
    BCS quasiparticles, 334
    measured by tunneling, 134, 369
    non-interacting electrons, 39
    phonons, Debye model, 60
    spectral function, 130
density waves, 73
density-density correlation function
    in dielectric function, 99
    the pair-bubble $\chi_0 \equiv -\Pi^0$, 249
    the RPA-bubble χ^{RPA}, 260
    the RPA-bubble and phonons, 319
dephasing, 102, 298, 306
determinant
    first quantization, 7
    in Wick's theorem, 200
    Slater, 7
diagonal Hamiltonian, 126
diagonalization of H
    1D phonons, 54
    3D phonons, 57
    Bogoliubov, bosons, 382
    Bogoliubov, fermions, 381
    bosonization in 1D, 360
    harmonic oscillator, 18
    photons, 19
    quadratic Hamiltonian, 380
diamagnetic term in current density, 96
dielectric function ε
    1D, 349
    equation of motion derivation, 150
    irreducible polarization function χ^{irr}, 259
    Kubo formalism, 98
    relation to polarization function χ, 257
    relation to conductivity, 100
differential conductance, 134
    Coulomb blockade, 156
    Kondo effect, 179
differential equation
    classical Green's function, 120
    many-body Green's function, 124
    single-particle Green's function, 140

diffusons in conductivity, 291
Dirac
    bra(c)ket notation for quantum states, 2
    delta function $\delta(\mathbf{r})$, 4
    relativistic equation, 354
disconnected Feynman diagrams, 231, 239
disorder, mesoscopic systems, 308
dissipation
    due to electron-hole pairs, 137, 263
    of electron gas, 137
distribution function
    Boltzmann, 27
    Boltzmann, Gibbs, 27
    Bose–Einstein, 29
    electron reservoir, 109
    Fermi–Dirac, 29
    Maxwell–Boltzmann, 46
    non-interacting bosons, 30
    non-interacting fermions, 29
donor atoms, 47
Drude formula, 272, 283, 296
Dulong–Petit value for specific heat, 60
dynamical matrix $\mathbf{D}(\mathbf{k})$, 57
dynamical structure factor, 349
Dyson equation
    Feynman diag., external potential, 206
    first Born approximation, 217
    for Π_{xx}, 289
    for cooperon, 303
    full Born approximation, 220
    impurity and interaction, 289
    impurity-averaged electrons, 216
    pair interactions in Fourier space, 236
    pair interactions in real space, 233
    pair-scattering vertex Λ, 322
    polarization function χ, 259
    self-consistent Born approximation, 222
    single-particle in external potential, 205

edge states, 348
effective electron-electron interaction
    Coulomb and phonons, jellium, 318
    Coulomb and phonons, RPA, 321
    phonon mediated, RPA, 321
effective mass
    approximation, 36
    renormalization, 279, 283, 296
eigenmodes
    electromagnetic field, 20
    lattice vibrations, 58
eigenstate
    definition, 1
    superposition, 1
eigenvalue, definition of, 1
Einstein model of specific heat, 61
Einstein phonons
    in the jellium model, 52

optical phonons, 53
elastic scattering
    general formalism, 204
    Matsubara Green's function, 206
    random impurities, 211
electric potential
    classical theory, 120
    external and induced, 270
electron
    core electrons, 32
    density of states, 39
    operator 1D, 363
    phase coherence, 211
    valence electrons, 32
electron gas, in general
    0D: quantum dots, 50
    1D: carbon nanotubes, 49
    2D: GaAs heterostructures, 47
    3D: metals and semiconductors, 45
    introduction, 32
electron gas, interacting
    1D, 347
    attractive interaction, 319
    dielectric properties and screening, 256
    first-order perturbation, 42, 44
    full self-energy diagram, 246
    full theory, 246
    general considerations, 40
    groundstate energy, 253, 256
    Hartree Fock mean-field Hamiltonian, 71
    infinite perturbation series, 246, 255
    Landau damping, 263
    plasma oscillations, 262
    second-order perturbation, 44
    thermodynamic potential Ω, 254
electron gas, non-interacting
    Bloch theory, 33
    density of states, 39
    Feynman diagrams, 204
    finite temperature, 39
    ground state energy, 39
    jellium model, 36
    motion in external potentials, 204
    static ion lattice, 33
electron interaction strength e_0^2, 23
electron wave guides, 113
electron-electron interaction, 151
electron-electron scattering
    attractive interaction, 319
    Cooper instability, 322
    dephasing, 298, 306
    lifetime, 276
electron-hole pairs
    1D, 348
    excitations, 137, 150
    interpretation of RPA, 253
    Landau damping, 263

electron-phonon interaction
    adiabatic electron motion, 53
    basis states, 314
    combined with Coulomb interaction, 316
    Feynman diagrams, 314
    general introduction, 52
    graphical representation, 64
    in nanostructures, 151
    the jellium model, 64, 313
    the lattice model, 61, 313
    the sound velocity, 53
    umklapp process, 63
electronic plasma oscillations
    graphical representation, 52
equation of motion
    Anderson's model, 144
    derivation of RPA, 148
    for ions, 58
    frequency domain, 140
    Heisenberg operators, 81
    in proof of Wick's theorem, 199
    introduction, 139
    Matsubara Green's function, 197
    non-interacting particles, 141
    single-particle Green's function, 139
ergodicity assumption, 26
extended zone scheme, 36

Fermi
    Fermi energy ε_F, 37
    Fermi sea diagrams, 38
    Fermi sea with interactions, 44
    Fermi sea, Cooper instability, 324
    Fermi sea, definition, 37
    Fermi sea, excitations, 138
    Fermi velocity v_F, 37
    Fermi wave length λ_F, 37
    Fermi wavenumber k_F, 37
    Fermi's golden rule, 86, 273, 276, 282
    Thomas–Fermi screening, 251, 252
Fermi liquid theory
    1D, 351
    introduction, 266
    microscopic basis, 278
Fermi–Dirac distribution, 29, 269
    in Coulomb blockade problem, 154
fermion
    creation/annihilation operators, 13
    defining commutators, 13
    definition, 5
    fermion loop, 229
    frequency, 189, 193
    many-particle basis, 14
ferromagnetism
    critical temperature, 76
    introduction, 74
    order parameter, 73

Stoner model, 76
Feynman diagrams
    cancel disconnected diagrams, 231
    cancellation of disconnected diagrams, 239
    Cooper instability, 322
    electron-impurity scattering, 209
    electron-phonon interaction, 314
    external potential scattering, 205
    first Born approximation, 217
    full Born approximation, 220
    impurity-averaged single-particle, 215
    interaction line in Fourier space, 235
    interaction line in real space, 231
    irreducible diagrams, imp. scattering, 216
    irreducible diagrams, pair interaction, 233
    Kondo model, 241
    pair interactions, 226
    polarization function χ, 258
    self-consistent Born approximation, 222
    single-particle, external potential, 204
    topologically different diagrams, 231, 240
Feynman rules
    electron-impurity scattering, 211
    external potential scattering, 205
    impurity-averaged Green's function, 215
    pair interactions in Fourier space, 235
    pair interactions in real space, 231
    pair interactions, $\mathcal{G}$ denominator, 229
    pair interactions, $\mathcal{G}$ numerator, 230
    phonon mediated pair interaction, 316
first quantization
    many-particle systems, 4
    name, 1
    single-particle systems, 2
fluctuation-dissipation theorem, 128
Fock
    approximation for interactions, 71
    Fock self-energy for pair interactions, 237
    Fock space, 10, 28
    Hartree–Fock approximation, 70
four-vector/four-momentum, 234
four-vector/four-momentum notation, 287
Fourier transformation
    basic theory, 376
    complex frequency, 88
    Coulomb interaction, Matsubara, 234
    equation of motion, 140
    ion vibrations, 54
    Matsubara functions, 188
    retarded and advanced functions, 88
Frölich, 325
fractional quantum Hall effect, 348, 420
free energy
    definiton, 28
    in mean-field theory, 68
Friedel oscillations, 396

GaAs/GaAlAs heterostructures, 47
gauge
    breaking of gauge symmetry, 341
    Landau gauge, 2
    radiation field, 19
    transversality condition, 19
Gauss box, 48
Gibbs distribution, 27
grand canonical
    density matrix, 28
    ensemble, 28
    partition function, 28
gravitation, 1
Greek letters, 185
Green's function
    S-matrix, 123
    n-particle, 198
    1D electron gas, 368
    bosonization, 368
    classical, 120
    dressed, 286
    free electrons, 125
    free phonons, 313
    greater and lesser, 124
    imaginary time, 187
    introduction, 120
    Lehmann representation, 127
    Nambu formalism, 335
    Poisson's equation, 120
    renormalization, 278
    retarded, equation of motion, 139
    retarded, many-body system, 124
    retarded, one-body system, 122
    RPA-screened phonons, 320
    Schrödinger equation, 120
    single-particle, many-body system, 124
    translation-invariant system, 125
    two-particle, 135

Hamiltonian
    bosonization, 360
    diagonal, 126
    non-interacting particles, 129
    quadratic, 129, 140, 145, 197
harmonic oscillator
    length, 18
    second quantization, 18
Hartree
    approximation for interactions, 71
    Hartree self-energy, pair interactions, 236
    Hartree–Fock approximation, 70
Hartree–Fock approximation
    introduction, 70
    mean-field Hamiltonian, 71
    the interacting electron gas, 71
heat capacity
    for electrons, 40

for ions, 53
  superconductor, 328
Heaviside's step function $\theta(x)$, 4
Heisenberg
  Heisenberg picture, 81
  model of ferromagnetism, 74
helium, Hamiltonian, 9
heterostructures, GaAs/GaAlAs, 47, 308
Hilbert space, 1
hopping, 144
Hubbard model, 77
hybridization, 143
hydrogen atom
  Bohr radius a_0, 41
  electron orbitals, 2
  ground state energy E_0, 41

imaginary time
  discussion, 185
  Greek letters, 185
  Green's function, 187
impurities, magnetic, 142
impurity scattering, conductivity, 285
impurity self-average, 211
impurity-scattering line
  Feynman rules, 215
  in conductivity, 287
  renormalization by RPA-screening, 260
inelastic light scattering, 137
infinite perturbation series
  breakdown at phase transitions, 334
  electron gas groundstate energy, 255
  Matsubara Green's function, 227
  self-energy for interacting electrons, 246
  single-particle Green's function, 205
  time-evolution operator $\hat{U}(t, t_0)$, 83
infinitesimal shift η, 88, 189
integration over the coupling constant, 253
interaction line
  general pair interaction in real space, 231
  pair interaction in Fourier space, 235
  RPA screened Coulomb line, 250, 260
  RPA screened impurity line, 260
interaction picture
  imaginary time, 185
  introduction, 81
  real space Matsubara Green's fct., 227
interference, 298, 299
ions
  forming a static lattice, 33
  Heisenberg model, ionic ferromagnets, 74
  in a metal, 32
  ionic plasma oscillations, 52
irreducible Feynman diagrams
  impurity scattering, 216
  pair interaction, 233
  polarization function χ^{irr}, 258

iterative solution, integral eqs., 82, 121

jellium model
  effective electron-electron interaction, 318
  Einstein phonons, 52
  electron-phonon interaction, 64
  full electronic self-energy, 246
  oscillating background, 52
  static case, 36
Josephson effect, 343, 415

Kac–Moody algebra, 352, 361
Kamerlingh–Onnes, 325
ket state, 2
kinetic energy operator
  including a vector potential, 21
  second quantization, 21
kinetic equation, 154
kinetic momentum, 21
Kondo effect
  beyond perturbation theory, 181
  bulk metals versus quantum dots, 173
  conductance, second order $H_S^{(2)}$, 173
  conductance, third order $H_S^{(2)}$, 174
  in quantum dots, 168
  relation to Anderson model, 168
  self-energy, 241
Kondo model, 241
Kronecker's delta function δ_{ij}, 4
Kubo formalism
  conductance, 97
  conductivity, 95, 286
  correlation function, 94
  dielectric function, 98
  general introduction, 92
  Landauer formula, 111
  RPA-screening in the electron gas, 256
  time evolution, 93
  tunnel current, 133
Kubo formula
  Fourier transformation, 88

ladder diagram
  Cooper instability, 322
  direct (diffuson), 291
  reversed (cooperon), 303
ladder operator, 364
Landau
  and Fermi liquid theory, 266
  damping and plasma oscillations, 263
  eigenstates, 3
  gauge, 2
  phase transitions, 389
Landauer formula
  heuristic derivation, 109
  linear response derivation, 111

multiprobe, 112
Landauer–Büttiker formalism
    introduction, 102
    multiprobe, 112
    two-probe, 103
lattice model
    basis in real space, 34
    basis in reciprocal space, 34
    Hamiltonian, 33
lattice vibrations
    electron-phonon interaction, 61
    phonon Hamiltonian, 54
left movers, 355
Lehmann representation
    definition, 127
    for $G^>$, $G^<$, and G^R, 127
    Matsubara function, 189
Levi–Civita symbol ϵ_{ijk}, 4, 177
lifetime, 142, 268, 276, 295, 351
Lindhard function, 136, 150
    in 1D systems, 349
linear response theory
    introduction, 92
    Landauer formula, 111
    mesoscopic system, 108
    time evolution, 84
    tunnel current, 133
London equation, 336
Luttinger liquid
    experimental realizations, 348
    general definition, 347
    general introduction, 347
    tunneling density of states, 369
    with spin, 373
Luttinger–Tomonaga model, 352
    real space representation, 360

magnetic impurities, 142
magnetic length, 3
magnetic moment, 75, 142, 144
magnetization, 73, 144
many-body system
    first quantization, 2
    second quantization, 10
    single-particle Green's function, 124
mass renormalization, 283, 285, 296
Matsubara
    convergence of, 188
    Fourier transformation, 188
    frequency, 189
    function, equation of motion, 197
    Green's function, 187
    relation to retarded function, 189
    sums, evaluation of, 193
    sums, simple poles, 194
    sums, with branch cuts, 196
Matsubara Green's function

elastic scattering, 206
electron-impurity scattering, 209
    first Born approximation, 217
    free phonons, 313
    full Born approximation, 220
    impurity-averaged single-particle, 215
    interacting elec. in Fourier space, 236
    interacting electrons in Fourier space, 234
    interacting electrons in real space, 226
    RPA-screened phonons, 320
    self-consistent Born approximation, 222
    superconductivity, 331
    two-particle polarization function χ, 257
maximally crossed diagrams, 302
MBE, molecular beam epitaxy, 47
mean free path, 102
mean-field theory
    Anderson's model, 145
    BCS mean-field Hamiltonian, 329
    broken symmetry, phase transistions, 72
    discussion, 165
    general Hamiltonian H_{MF}, 67
    Hartree–Fock mean-field Hamiltonian, 71
    introduction, 66
    mean-field approximation, 67
    partition function Z_{MF}, 68
    the art of mean-field theory, 69
measuring the spectral function, 131
Meissner effect, 336
mesoscopic
    disordered systems, 308
    interacting system, 151
    physics, 285
    regime, 299
    systems, introduction, 102
    transport, 151
metal
    disordering and random impurities, 208
    electrical resistivity, 208
    general description, 32
    Hamiltonian, 32
    observation of plasmons, 263
    Thomas–Fermi screening in metals, 252
Migdal's theorem, 317
molecular beam epitaxy, MBE, 47
momentum
    canonical, 21
    kinetic, 21
    relaxation, 272, 276
momentum cut-off in 1D, 365
MOSFET, 47
multiprobe Landauer formula, 112

Nambu formalism
    introduction, 335
    paramagnetic current response, 337
    spinors and Green's functions, 335

nanostructure, 151
nanotechnology, 212
Newton's second law
   for ions in the jellium model, 53
non-equilibrium
   master equation, 153
non-interacting particles
   distribution functions, 29
   equation of motion, 141
   Green's functions, 125
   Hamiltonian, 129
   in conductivity, 293
   Matsubara Green's function, 192
   quasiparticles, 266
   retarded Green's function $G^R(\mathbf{k}\sigma, \omega)$, 129
   spectral function $A_0(\mathbf{k}\sigma, \omega)$, 129
normalization
   bosonized electron operator, 367
   quantum states, 3
   scattering state, 104
nucleons, superconductivity, 325
nucleus, 32

occupation number operator
   bosons, 13
   fermions, 14
   introduction, 10
occupation number representation, 10
operator
   adjoint, 2
   boson creation/annihilation, 10
   electromagnetic field, 19
   electron bosonized, 363
   expansion of e^{-iHt}, 80
   fermion creation/annihilation, 13
   first quantization, 8
   Heisenberg equation of motion, 81
   Hermitian, 1
   real time ordering T_t, 83
   second quantization, 14
   time evolution operator $\hat{U}(t, t_0)$, 82
   trace Tr, 28
optical phonons
   Einstein phonons, 53
   graphical representation, 57
optical theorem, scattering theory, 221
order parameter
   definition, 73
   list of order parameters, 73, 342
overlap of wavefunctions
   localized/extended states, 143
   particle propagation, 126
   tunneling, 132

pair condensate, 73
pair interactions

Dyson equation in Fourier space, 236
Dyson equation in real space, 233
Feynman diagrams, 226
Feynman rules in Fourier space, 235
Feynman rules in real space, 231
self-energy in Fourier space, 236
self-energy in real space, 233
pair-bubble
   calculation of the pair-bubble, 251
   Feynman diagram $\Pi^0(\mathbf{q}, iq_n)$, 239
   in the RPA self-energy, 249
   self-energy diagram, 238
   the correlation function $\chi_0 \equiv -\Pi^0$, 249
paramagnetic term in current density, 96
particle-particle scattering
   in the collision term, 284
   lifetime, 276
partition function
   canonical ensemble, 27
   grand canonical ensemble, 28
   in mean-field theory, 68
Pauli
   exclusion principle, 5, 41, 71, 355, 356
   matrix product rule, 177
   spin matrices, 22
perfect diamagnetism, 325
periodic boundary conditions
   electrons, 36
   phonons, 54
   photons, 20
permanent
   for bosons, 7
   in first quantization, 7
   in Wick's theorem, 200
permutation, 198
permutation group S_N, 7, 83
perturbation theory
   first-order, electron gas, 42
   infinite order, Green's function, 205
   infinite order, groundstate energy, 255
   infinite order, interacting electrons, 246
   linear response, Kubo formula, 92
   second-order, electron gas, 44
   single-particle wavefunction, 121
   third-order Kondo model, 173
   time-evolution operator $\hat{U}(t, t_0)$, 83
phase coherence, 211, 298
phase coherence length l_φ, 212
phase space, 277
phase transition
   breakdown of perturbation theory, 334
   broken symmetry, 72
   order parameters, 73
phonons
   annihilation/creation operators, 55
   Debye model, 53
   density of states, Debye model, 60

dephasing, 298, 306
eigenmodes in 3D, 58
Einstein model of specific heat, 61
free Green's function, 313
general introduction, 52
Hamiltonian for jellium phonons, 53
lattice vibrations, 54
phonon branches, 56
relevant operator $A_{q\lambda}$, 313
RPA renormalization, 319
RPA-renormalized Green's function, 320
second quantization, 56, 58
plasma frequency
  1D, 350
  for electron gases in a metals, 262
  ionic plasma frequency, 53
plasma oscillations
  electronic plasma oscillations, 52
  interacting electron gas in RPA, 262
  ionic plasma oscillations, 52
  Landau damping, 263
  plasmons, 262
plasmons
  1D, 350
  dynamical screening, 271
  experimental observation in metals, 263
  plasma oscillations, 262
  semi-classical treatment, 269
Poisson's equation
  GaAs heterostructures, 48
  Green's function, 120
polarization function χ
  1D, 350
  Dyson equation, 259
  Feynman diagrams, 258
  free electrons, 136, 201
  irreducible Feynman diagrams, 258
  Kubo formalism, 99
  momentum space, 136
  relation to dielectric function ε, 257
  two-particle Matsubara Green's fct., 257
polarization vectors
  phonons, 58
  photons, 20
probability current conservation, 106
probability distribution, 129
propagator
  Green's function, 122
  single-particle in external potential, 205

quadratic Hamiltonian, 67, 129, 140, 145,
  197, 380
quantization of conductance, 113
quantum coherence, 211, 343
quantum correction, 285, 296, 307
quantum dot
  introduction, 50

transport, 151
tunneling spectroscopy, 134
quantum effects, 102
quantum field operator
  definition, 17
  Fourier transform, 17
quantum fluctuations
  in conductance, 212, 285
quantum number ν
  Feynman rules, Dyson equation, 208
  general introduction, 3
  sum over, 4
quantum point contact, 113
quantum state
  bra and ket state, 2
  free particle, 2
  hydrogen, 2
  Landau states, 3
  orthogonal, 2
  time evolution, 1
quasiparticle
  1D, 350
  BCS density of states, 334
  definition, 269
  discussion, 268
  introduction, 266
  lifetime, 276
quasiparticle-quasiparticle scattering, 276

radiation field, 19
Raman scattering, 137
random impurities, 208
random matrix theory, 308
random phase approximation (see RPA), 246
rational function, 190
reciprocal lattice basis, 34
reciprocal space, 34
reduced zone scheme, 36
reflection amplitude, 106
reflectionless contact, 103, 109
relaxation time approximation, 275
renormalization
  constant Z, 279
  effective mass, 279, 283, 296
  Green's function, 278
  of phonons by RPA-screening, 319
reservoir, 26, 103
resistivity (see conductivity), 272
resummation of diagrams
  current-current correlation, 288
  impurity scattering, 216
  the RPA self-energy, 248
retarded function
  asymptotics, 89
  convergence factor, 141
  Fourier transformation, 88
  Green's function, 124, 125

relation to Matsubara function, 189
right movers, 355
rigidity of wave function, 337
Roman letters, 185
RPA for the electron gas
  1D, 348
  Coulomb and impurity lines, 289
  deriving the equation of motion, 148
  electron-hole pair interpretation, 253
  Fermi liquid theory, 270, 278
  plasmons and Landay damping, 260
  renormalized Coulomb interaction, 250
  resummation of the self-energy, 248
  the dielectric function ε^{RPA}, 260
  the polarization function χ^{RPA}, 260
  vertex corrections, 291
Rydberg, unit of energy (Ry), 41

S-matrix, 103
scattering length, 220
scattering matrix, S, 103
  Green's function, 123
scattering state, 106
scattering theory
  optical theorem, 221
  Schrödinger equation, 121
  T-matrix, 87
  transition matrix, 221
Schrödinger equation
  Green's function, 120
  quantum point contact, 114
  scattering theory, 121
  time reversal symmetry, 107
  time-dependent, 1
Schrödinger picture, 80
Schrieffer-Wolff transformation, 168
screening
  dieelectric properties of the elec. gas, 256
  RPA-screened Coulomb interaction, 251
  semiclassical, dynamical, 271
  semiclassical, static, 270
  Thomas–Fermi screening, 251
second quantization
  basic concepts, 10
  basis for different particles, 25
  change of basis, 16
  Coulomb interaction, 23
  electromagnetic field, 19
  electron-phonon interaction, 61
  free phonons in 1D, 56
  free phonons in 3D, 58
  harmonic oscillator, 18
  kinetic energy, 21
  name, 1
  operators, 14
  particle current density, 22
  particle density, 22

  spin, 22
  statistical mechanics, 26
  thermal average, 27
self-average for impurity scattering
  basic concepts, 211
  weak localization, 299
self-consistent equation
  Anderson's model, 146
  BCS gap equation, 332
  general mean-field theory, 68
self-energy
  due to hybridization, 142
  first Born approximation, 217
  Fock diagram for pair interactions, 237
  full Born approximation, 220
  Hartree diagram for pair interactions, 236
  impurity-averaged electrons, 216
  interacting electrons, jellium model, 246
  irreducible, 289
  Kondo model, 241
  pair interactions in Fourier space, 236
  pair interactions in real space, 233
  pair-bubble diag., pair interactions, 238
  RPA self-energy, interacting electrons, 249
  self-consistent Born approximation, 222
semi-classical
  approximation, 293
  screening, 269
  transport equation, 272
sequential tunneling, 153
single-particle states
  as N-particle basis, 6
  free particle state, 2
  hydrogen orbital, 2
  Landau state, 3
Slater determinant, fermions, 7
Sommerfeld expansion, 40
sound velocity
  Bohm–Staver formula, RPA, 321
  Bohm–Staver formula, semi-classical, 54
  Debye model, 53
sounds waves, 52
space-time, points and integrals, 204
spectral function
  1D, 371
  Anderson's model, 146
  broadening, 130
  definition, 128
  first Born approximation, 219
  in sums with branch cuts, 197
  measurement, 131
  non-interacting particles, 129
  physical interpretation, 129
  sum rule, 129
spectroscopy, tunneling, 132
spin
  Heisenberg model, 74

Kondo model, 172
Pauli matrices, 22
second quantization, 22
Stoner model, 76
spin flip, 168
Spinors, Nambu formalism, 335
spontaneous symmetry breaking
breaking of gauge symmetry, 341
introduction, 73
statistical mechanics
second quantization, 26
step function $\theta(x)$, 4
STM, 132
Stoner model of metallic ferromagnetism, 76
structure factor, 349
sum rule, spectral function, 129
superconductivity
BCS groundstate, 327
coherence length, 327
critical temperature, 327, 333
density of states, 334
introduction, 325
London equation, 336
Matsubara Green's functions, 331
Meissner effect, 336
microscopic BCS theory, 329
order parameter, 73
self-consistent gap equation, 332
tunneling, 335
symmetrization operator, 7

T-matrix
cotunneling, 158
definition, 88
derivation, 87
thermal average, 27
thermodynamic potential Ω
definition, 28
for the interacting electron gas, 254
Thomas–Fermi screening, 251, 252, 256
time dependent Hamiltonian, 92
time evolution
creation/annihilation operators, 84
Heisenberg picture, 81
in linear response, 84
interaction picture, 81
linear response, Kubo, 93
operator, imaginary time, 185
Schrödinger picture, 80
unitary operator $\hat{U}(t, t_0)$, 82
time-ordering operator
imaginary time T_τ, 187
real time T_t, 83
time-reversal symmetry, 107, 307
time-reversed paths, 300, 302, 306
Tomonaga model, 352
topologically different diagrams, 231, 240

trace of operators, 28
transition matrix, scattering theory, 221
translation-invariant system
conductivity, 286
Green's function, 125
transmission amplitude, 106, 123, 301
transmission coefficients, 108
transmission line, 362
transport equation, 266
transport time, 275
transversality condition, 19
triangular potential well, 48
truncation
derivation of RPA, 149
discussion, 139
Equation of motion theory, 139
tunneling
BCS superconductor, 134, 335
current, 132
Hamiltonian, 132, 151
scanning microscope, 132
sequential, 153
spectroscopy, 132

UCF, conductance fluctuations, 310
umklapp process
1D, 356
electron-phonon scattering, 63
unit cell, 56
unitary
S-matrix, 106
transformation, 161
universal conductance fluctuations, 308, 310

valence electrons, 32
vector potential
electromagnetic field, 19
kinetic energy, 22
Kubo formalism, 95
vertex
current vertex, 287
dressed vertex function, 290
electron-phonon vertex, 317
pair-scattering vertex Λ, 322
vertex correction, 286, 289
vertex function, 302

Ward identity, 291, 294
wavefunction
collapse of, 1
Cooper pair, 325
rigidity, 337
weak localization
and conductivity, 298
introduction, 285
mesoscopic systems, 308, 309

Wick's theorem
    derivation, 198
    in mean-field theory, 70
    interacting electrons, 228
    phonon Green's function, 315
    spin, absence of, 241
WKB approximation, 116

Yukawa potential
    definition, 24, 246
    Fourier transform, 381
    RPA-screened Coulomb interaction, 251